Innovations in Materials Chemistry, Physics, and Engineering Research

Eugene de Silva
Virginia Research Institute, USA

Pramudi Abeydeera
Polymeric Ltd., USA

IGI Global
PUBLISHER of TIMELY KNOWLEDGE

A volume in the Advances in Chemical and Materials Engineering (ACME) Book Series

Published in the United States of America by
IGI Global
Engineering Science Reference (an imprint of IGI Global)
701 E. Chocolate Avenue
Hershey PA, USA 17033
Tel: 717-533-8845
Fax: 717-533-8661
E-mail: cust@igi-global.com
Web site: http://www.igi-global.com

Library of Congress Cataloging-in-Publication Data

Names: De Silva, Eugene, editor. | Abeydeera, Pramudi, 1989- editor.
Title: Innovations in materials chemistry, physics, and engineering research / edited by Eugene de Silva, Pramudi Abeydeera.
Description: Hershey, PA : Engineering Science Reference, [2023] | Includes bibliographical references and index. | Summary: "This book will cover the latest theoretical and experimental research work in the field of material chemistry, physics, engineering, and allied fields. In addition to the current ongoing research, this book will also include critical reviews of past and current research work in the field. The chapters will include basic mechanisms of reactions, applications, synthesis, properties, and innovations in materials chemistry"-- Provided by publisher.
Identifiers: LCCN 2023019958 (print) | LCCN 2023019959 (ebook) | ISBN 9781668468302 (h/c) | ISBN 9781668468319 (s/c) | ISBN 9781668468326 (eISBN)
Subjects: LCSH: Materials--Technological innovations. | Chemical engineering--Technological innovations. | Physical metallurgy--Technological innovations.
Classification: LCC TA403.8 .I47 2023 (print) | LCC TA403.8 (ebook) | DDC 660--dc23/eng/20230918
LC record available at https://lccn.loc.gov/2023019958
LC ebook record available at https://lccn.loc.gov/2023019959

This book is published in the IGI Global book series Advances in Chemical and Materials Engineering (ACME) (ISSN: 2327-5448; eISSN: 2327-5456)

British Cataloguing in Publication Data
A Cataloguing in Publication record for this book is available from the British Library.

For electronic access to this publication, please contact: eresources@igi-global.com.

Advances in Chemical and Materials Engineering (ACME) Book Series

J. Paulo Davim
University of Aveiro, Portugal

ISSN:2327-5448
EISSN:2327-5456

Mission

The cross disciplinary approach of chemical and materials engineering is rapidly growing as it applies to the study of educational, scientific and industrial research activities by solving complex chemical problems using computational techniques and statistical methods.

The **Advances in Chemical and Materials Engineering (ACME) Book Series** provides research on the recent advances throughout computational and statistical methods of analysis and modeling. This series brings together collaboration between chemists, engineers, statisticians, and computer scientists and offers a wealth of knowledge and useful tools to academics, practitioners, and professionals through high quality publications.

Coverage

- Multifuncional and Smart Materials
- Metallic Alloys
- Numerical Techniques
- Ceramics and glasses
- Polymers Engineering
- Electrochemical and Corrosion
- Thermo-Chemical Treatments
- Coatings and surface treatments
- Nanomaterials
- Chemical Engineering

IGI Global is currently accepting manuscripts for publication within this series. To submit a proposal for a volume in this series, please contact our Acquisition Editors at Acquisitions@igi-global.com or visit: http://www.igi-global.com/publish/.

The Advances in Chemical and Materials Engineering (ACME) Book Series (ISSN 2327-5448) is published by IGI Global, 701 E. Chocolate Avenue, Hershey, PA 17033-1240, USA, www.igi-global.com. This series is composed of titles available for purchase individually; each title is edited to be contextually exclusive from any other title within the series. For pricing and ordering information please visit http://www.igi-global.com/book-series/advances-chemical-materials-engineering/73687. Postmaster: Send all address changes to above address.

Titles in this Series

For a list of additional titles in this series, please visit: www.igi-global.com/book-series

Handbook of Research on Advanced Functional Materials for Orthopedic Applications
R. Ranjith (SNS College of Technology, India) and J. Paulo Davim (University of Aveiro, Portugal)
Engineering Science Reference • copyright 2023 • 400pp • H/C (ISBN: 9781668474129) • US $325.00 (our price)

Production, Properties, and Applications of Engineered Cementitious Composites
S. Praveenkumar (PSG College of Technology, India) and J. Paulo Davim (University of Aveiro, Portugal)
Engineering Science Reference • copyright 2023 • 300pp • H/C (ISBN: 9781668481820) • US $250.00 (our price)

Innovative Multifunctional Nanomaterial for Photocatalysis, Sensing, and Imaging
Azad Kumar (Siddharth University, India)
Engineering Science Reference • copyright 2023 • 300pp • H/C (ISBN: 9781668487433) • US $270.00 (our price)

Sustainable Utilization of Nanoparticles and Nanofluids in Engineering Applications
Sampath Boopathi (Muthayammal Engineering College, India) and J. Paulo Davim (University of Aveiro, Portugal)
Engineering Science Reference • copyright 2023 • 300pp • H/C (ISBN: 9781668491355) • US $270.00 (our price)

Modeling, Characterization, and Processing of Smart Materials
Ajay . (Mechanical Engineering Department, School of Engineering and Technology, JECRC University, India) Parveen . (Mechanical Engineering Department, Rawal Institute of Engineering and Technology, India) Ashish Kumar Srivastava (G.L. Bajaj Institute of Technology & Management, India) and Vikas Goyat (Mechanical Engineering Department, SRM Institute of Science and Technology, India)
Engineering Science Reference • copyright 2023 • 320pp • H/C (ISBN: 9781668492246) • US $270.00 (our price)

Fundamental and Biomedical Aspects of Redox Processes
Gheorghe Duca (Institute of Chemistry, Moldova State University, Moldova) and Ashok Vaseashta (International Clean Water Institute, USA)
Engineering Science Reference • copyright 2023 • 495pp • H/C (ISBN: 9781668471982) • US $295.00 (our price)

Handbook of Research on Corrosion Sciences and Engineering
Younes El Kacimi (University Ibn Tofail, Morocco) and Lei Guo (Tongren University, China)
Engineering Science Reference • copyright 2023 • 673pp • H/C (ISBN: 9781668476895) • US $325.00 (our price)

IGI Global
PUBLISHER of TIMELY KNOWLEDGE

701 East Chocolate Avenue, Hershey, PA 17033, USA
Tel: 717-533-8845 x100 • Fax: 717-533-8661
E-Mail: cust@igi-global.com • www.igi-global.com

To our parents:

Angelo and Latha de Silva

Sunil and Priyanka Abeydeera

Table of Contents

Detailed Table of Contents

The chapter provides an overview of polymer science, covering synthesis, polymer characterization, rheology, mechanical properties, and applications. The chapter delves into the definition of polymers, their historical development, and various classification systems. It also explores the importance of polymers for human survival and discusses the fundamentals of polymer science. The chapter continues by discussing the properties and applications of polymers, including an examination of polymer solution properties, mechanical properties, and the wide range of applications in different industries. Also explored is synthesis and construction of polymers, which form the foundation of their physical nature. This leads to the crucial topic of structure-property relationships, examining how molecular interactions and arrangements impact polymer properties. Furthermore, the chapter explores different types of polymers and their applications, spanning from thermoplastics to thermoset polymers, natural polymers to biopolymers. The chapter concludes by discussing polymer processing techniques.

This chapter is devoted to experimental studies of changes in structural-phase blocks and tribological properties of combustion of layers of high-speed steels R6M5, R9, and R18 during electrolytic-plasma nitriding. Systematized experimental data on the effect of electrolytic-plasma nitriding on the structure, phase composition, and tribological properties of the surface layer of high-speed steels have been obtained. In addition, the patterns of formation of modified layers in high-speed steels during electrolytic-plasma nitriding, revealed in this work, can be used by researchers when choosing the modes of electrolytic-plasma treatment of alloyed steels, as well as when analyzing structural transformations of high-speed steels.

This chapter presents an analytical review of literature sources devoted to research on the rapid heating of steels using various heat sources; in particular, on plasma surface hardening of materials. The results of studying the effect of a hardened surface layer during plasma hardening in the cathodic mode on the mechanical and tribological characteristics of grade two wheel steel are presented. Based on the experimental data obtained, a method for electrolytic-plasma surface hardening of wheel steel was developed and the optimal mode was determined. A description of the phase composition and morphology of grade two steel in the initial state and after electrolytic-plasma surface hardening is given. These descriptions are accompanied by images obtained by transmission electron microscopy. The authors present bright-field images and microdiffraction patterns obtained from these areas, as well as their dark-field patterns.

This chapter is devoted to experimental research on the investigated of the structural-phase, morphology, elemental composition, physical-mechanical, and tribological properties (friction, wear, and adhesion) of hard and superhard micro- and nanostructured coatings of systems based on (Zr-Ti-Nb)N, (Zr-Ti-Cr-Nb)N, and (Zr-Ti-Cr-Nb-Si)N were fabricated by vacuum-arc deposition in the nitrogen atmosphere. The authors in the work used advanced proven experimental research methods (SEM-EDX, TEM-EDS, XRD, SIMS, GDMS, PIXE, XPS, AFM, hardness measurements, and adhesion testing), as well as theoretical methods for analysing the results. Based on the results obtained, according to the experimental data, an optimal mode was selected that provides an increase in mechanical and tribological characteristics, and a method for vacuum-arc surface hardening was also developed. Such coatings seem to have prospects as protective ones for couples of friction and cutting tools.

Transmission electron microscopy (TEM) investigations of the structure and phase composition of ferritic-pearlitic wheel steel mark two surface after electrolyte-plasma surface hardening are presented. Initially the morphology of the steel matrix consists of lamellar perlite and non-fragmented and fragmented ferrite. Electrolytic plasma quenching of the steel surface results in the martensite transformation, steel self-tempering, and the formation of cementite particles in all martensite crystals. This treatment also leads to the diffusion transformation of gamma to alpha phases, the release of residual austenite along the low-temperature martensite laths and lamellas and in all crystals of lamellar martensite, the formation of M23C6 special carbides and, finally, to the enhancement of all parameters of the steel fine structure.

Scanning electron microscopy (SEM), EBSD-analysis, and X-ray structure (XRD) analysis were used to investigate the microstructure, morphology, elemental composition, phase composition, and crystal structure of M2 high speed steel after standard thermal treatment. It has been shown that the microstructure of the M2 high speed steel after hardening and three-time tempering consists of tempered martensite and solid carbide M6C-type and MC spherical shape. The volume fraction of carbides and their distribution have been defined. The main carbides in the study of steel after heat treatment – M6C and MC carbides who have a complex FCC lattice and space group Fd3m, have been established by X-ray structural analysis. Carbides are homogeneous and single crystal. EBSD-analysis method with the support of X-ray structural analysis has established that the bright carbides spherical shape M6C correspond to the composition Fe3W3C, and grey carbides spherical shape MC correspond to the composition of the VC.

This study investigated the structural-phase, morphology, elemental composition, and tribological properties of coatings systems based on Al2O3, NiCr, and NiCr-Al2O3 fabricated by detonation-gas spraying method in a nitrogen/propane atmosphere. The coatings were characterised using SEM-EDX, XRD, Raman spectroscopy, and hardness measurements, and results analysed are presented. According to the experimental data, an optimal mode was selected that provides an increase in mechanical and tribological characteristics, and a method for detonation spraying surface hardening was also developed. Tribology tests have shown that the wear rates and friction coefficients of the coatings are highly dependent on the degree of filling of the gun.

Water scarcity has been exacerbated in recent years due to factors including the expansion of industrialization and agricultural activities. The contaminants released by industries into surrounding water bodies include toxic aromatic molecules, dyes, and heavy metals, which upon consumption pose a threat to human life. The development of water treatment technologies and catalysts targeting these pollutants has thus emerged as a critical area of research and development. This chapter therefore focuses on providing comprehensive information on the application of metal oxide semiconductor photocatalysis for the treatment of a range of industrial pollutants. With focus mainly being on titanium dioxide (TiO2), zinc oxide (ZnO), tin oxide (SnO2), cadmium oxide (CdO), and tungsten trioxide (WO3), the synthesis and modification methods for enhancement of the photocatalytic efficiencies of the semiconductor materials is discussed. Their physical, morphological, and optical properties are compared, and the mechanisms behind their activities are reported.

The issue of ensuring that people have access to water that is free from hazardous microbes and toxic chemicals presents a significant challenge that calls for a comprehensive investigation. Nanotechnology has enormous potential applications in water purification because it paves the way for the exploration and production of nanomaterials that have never-before-seen characteristics in the size range of one nanometer to one hundred nanometers. These nanomaterials make the process of water treatment more effective, which in turn reduces the amount of money, time, and effort that are required by conventional methods. This short review chapter focuses on the limitations of current methodologies for wastewater treatment, nanotechnology as an alternative for wastewater treatment, advancements made in nanomaterials for water treatment, and a brief look at potential challenges and future aspects of nanotechnology-based water treatment processes.

It has become urgently necessary to develop energy storage technologies and materials due to the rising need for advanced sustainable energies. Among the emerging advances in materials chemistry is the research and development of new electrode materials for rechargeable batteries and supercapacitors. To increase their energy density, service life, specific power, charge and/or discharge capacity, and environmental

The growing rate of cardiovascular disease all over the globe is so alarming that urgent measures need to be put in place. Approximately 18 million deaths are recorded every year globally due to cardiovascular disease. People living with high blood pressure, diabetes, high cholesterol level, or living poor are at high risk of developing cardiovascular disease. Other factors include age, gender, family history, and ethnic group. Understanding the risks of cardiovascular disease can help individuals, organizations, and governments promote a healthy lifestyle. Cleveland heart disease dataset was collected from online and open source for the purpose of this study. The dataset includes 297 medical records of patients between the ages of 29-77. The dataset was processed and analyzed in Microsoft Excel using frequency counts and percentages. The risk factors are important risk factors of cardiovascular disease for the participant of the study. Discussion on the use of biomaterials in fabricating tissue-engineered nanofibers for the treatment of cardiovascular diseases is also presented.

Foreword

I am delighted to present an enthralling book meticulously crafted by Prof. de Silva and Dr. Abeydeera together with international teams of esteemed experts in their respective fields. This compilation showcases applied research that holds immense relevance to various industries, and the content is presented in a lucid manner accessible to non-specialists.

Through the lens of multidisciplinary approaches, the book tackles a diverse array of challenges. It addresses issues such as wear and lifespan of components in the manufacturing industry through innovative surface engineering techniques. Furthermore, it explores the potential of coatings in enhancing battery materials, particularly for electric vehicles and other energy storage applications, aiming to extend their range and durability.

The first chapter serves as an enlightening overview of pivotal advancements in the polymer industry, setting the stage for subsequent exploration.

This book encompasses three chapters dedicated to the significant applications and fabrications of materials, with a specific focus on nanotechnology's role in wastewater treatment. The authors shed light on the remarkable challenges faced in this crucial area of global environmental concern, while also presenting recent advances. Furthermore, this contribution discusses the use of 2D materials for electrodes in supercapacitors and rechargeable batteries, offering insights into their potential for wastewater treatment.

Another notable chapter delves into the manufacture of leather and proposes strategies to reduce waste during the manufacturing process, aligning with the global net zero agenda. Employing mathematical modeling techniques, this chapter uncovers potential solutions and highlights areas for further investigation.

Two chapters of this book cast their spotlight on polymers and printing inks, as well as factors influencing cardiovascular disease (CVD). As CVD continues to impact societies worldwide, becoming a leading cause of death, this section examines its complex dynamics. It explores the role of stress, diet, exercise, and lifestyle as contributing factors while highlighting the profound impact of biocompatible materials and treatments on recovery and prevention.

This book caters to a wide readership in the fields of science and policymakers. Its engaging and informative content promises to captivate readers from various backgrounds and interests.

I hope this timely book encourages more research in the field of materials and polymer science.

Waqar Ahmed
College of Science, University of Lincoln, UK

Preface

We are pleased to present this remarkable book, a collaborative effort of esteemed experts from around the globe who have come together to share their groundbreaking research and insights. This book offers a comprehensive exploration of diverse topics that span multiple disciplines and industries.

The chapters in this compilation represent an extraordinary collection of applied research, each contributing valuable knowledge and innovative approaches to their respective fields. With a strong emphasis on relevance to industry, the authors have sought to address practical challenges and provide solutions that can drive positive change.

The book commences with an introductory chapter on polymers science and technology. The next chapter provides an in-depth examination of the influence of electrolytic-plasma treatment on the tribological properties of high-speed steel. This chapter sheds light on the impact of surface engineering techniques on high-speed steel and explores the high-temperature testing of nitrided layers. Their findings offer crucial insights into enhancing the performance and durability of materials.

Continuing the exploration of surface treatments, the third chapter focuses on changes in the structural-phase state of ferrite-perlite steel through electrolyte-plasma surface quenching. This chapter provides a comprehensive analysis of the effects of surface quenching on steel grade 2, opening new avenues for improving material properties.

The book further delves into the realm of coatings and their impact on materials. The fourth chapter explores the deposition of hard and superhard multicomponent nitride coatings using vacuum-arc evaporation. This research sheds light on the structure and properties of nanostructured coatings, providing valuable insights into materials fabrication.

The fifth and sixth chapters discuss surface hardening on wheel steel using electrolytic plasma to the electrolytic plasma treatment of Turkurite-phase in high-speed steels. These chapters delve into the transformative potential of advanced surface treatments and offer insights into fine structure formation and phase composition in various materials.

Another captivating chapter explores the detonation-gun spraying technology for multilayer coatings. The research described in this chapter seven unravels the potential of this method in obtaining multilayer coatings, offering new possibilities for advanced materials fabrication.

Additionally, the book addresses the pressing issue of industrial wastewater treatment through the eighth and ninth chapters on the photocatalytic treatment using semiconductor-based materials and nanotechnological approaches for water and wastewater treatment. These chapters highlight the potential of innovative technologies in mitigating environmental challenges.

The role of two-dimensional nanomaterials in electrodes for rechargeable batteries and supercapacitors is addressed in the tenth chapter. This work provides valuable insights into this emerging field, shedding light on the exciting possibilities of nanomaterials in energy storage applications.

Shifting gears, the book addresses the critical importance of the leather industry in achieving net-zero goals. This eleventh chapter highlights the environmental impact of the leather industry and proposes potential solutions to reduce waste during the manufacturing process. Their innovative approach, including mathematical modeling, paves the way for sustainable practices and aligns with the global net-zero agenda.

The book also covers the exciting field of epoxy-based composite materials for innovative biomedical applications. This twelfth chapter explores the use of epoxy-based composites in the biomedical field, opening doors to advancements in healthcare and medical technology.

Lastly, the book explores the use of material science in preventing cardiovascular diseases.

This book aims to captivate readers from diverse backgrounds, including engineers, scientists, policymakers, students, teachers, and decision-makers. Its accessible language and informative content make it an engaging resource for all those seeking to expand their knowledge and gain insights into the latest advancements across various fields.

We extend our sincere appreciation to the authors who have contributed their expertise and dedication to making this book a reality. Their invaluable contributions have enriched the content and ensured its high quality. We also express our gratitude to the esteemed reviewers who provided valuable feedback and guidance throughout the publication process.

We hope that this book sparks inspiration, fosters collaboration, and stimulates further research and innovation in the areas covered. As editors, it has been a privilege to work with such a talented group of authors and professionals who are at the forefront of their fields.

We extend our heartfelt thanks to the publishers for their support and commitment to disseminating knowledge and promoting advancements in science and technology.

We invite you, the readers, to embark on a captivating journey through the chapters of this book. May the insights and discoveries within these pages expand your horizons, inspire new ideas, and contribute to the advancement of knowledge and the betterment of society.

Eugene de Silva
Virginia Research Institute, USA

Pramudi Abeydeera
Polymeric Ltd., USA

Chapter 1
Polymer Science and Technology

Pramudi Abeydeera
Nazdar Ink Technologies, USA

Eugene de Silva
Virginia Research Institute, USA

ABSTRACT

The chapter provides an overview of polymer science, covering synthesis, polymer characterization, rheology, mechanical properties, and applications. The chapter delves into the definition of polymers, their historical development, and various classification systems. It also explores the importance of polymers for human survival and discusses the fundamentals of polymer science. The chapter continues by discussing the properties and applications of polymers, including an examination of polymer solution properties, mechanical properties, and the wide range of applications in different industries. Also explored is synthesis and construction of polymers, which form the foundation of their physical nature. This leads to the crucial topic of structure-property relationships, examining how molecular interactions and arrangements impact polymer properties. Furthermore, the chapter explores different types of polymers and their applications, spanning from thermoplastics to thermoset polymers, natural polymers to biopolymers. The chapter concludes by discussing polymer processing techniques.

INTRODUCTION

Polymers, which are large molecules composed of repeated smaller chemical units, have been fundamental to the development of life (Alberts et al., 2014). They can exist in linear chains or interconnected structures, forming 3D networks. The basic unit of a polymer is called a monomer. For example, vinyl chloride is the monomer of poly(vinyl chloride), which has a repeating unit of -CH2CHCl-. The length of a polymer chain is determined by the number of repeat units and is referred to as the degree of polymerization (DP) (Callister Jr. & Rethwisch, 2018).

DOI: 10.4018/978-1-6684-6830-2.ch001

Polymers can be classified into biological and non-biological materials. Biological polymers are essential components of life, including enzymes, nucleic acids, and proteins, while cellulose, natural rubber, wool, and starch are examples of biopolymers with simpler structures found in plants (Alberts et al., 2014; Callister Jr. & Rethwisch, 2018). Polymers play indispensable roles in various aspects of human life, from clothing and shelter to food. Synthetic polymers, such as fibers, plastics, and elastomers, are crucial for modern living (Alberts et al., 2014). Biopolymers can be further categorized as naturally occurring polymers found in nature, like cellulose and starch, or as biobased polymers artificially synthesized from natural resources. Natural polymers, such as proteins, cellulose, and natural rubber, are abundant in nature and can be extracted from renewable sources like plants and animals (Alberts et al., 2014). Fibers encompass a wide range of materials, including natural and synthetic, organic and inorganic, and are defined based on their shape (Callister Jr. & Rethwisch, 2018). For example, fibers are generally required to have a length at least 100 times greater than their diameter. Nylon, polyester, and polyolefins are examples of artificial fibers, while cotton, wool, and flax are natural fibers with lengths typically 1000-3000 times their diameter (Callister Jr. & Rethwisch, 2018).

In recent years, polymers have become prevalent in daily life, both in the form of natural and synthetic polymers. Polymers can be found in various forms such as solid plastics, fibers, elastomers, foams, films, coatings, and adhesives, with a wide range of properties from hard to soft. One useful classification of polymers is based on their response to thermal treatments. Thermoplastic polymers can melt when heated and resolidify upon cooling without undergoing significant structural or property changes. Examples include polyethylene, polystyrene, and nylon. In contrast, thermoset polymers do not melt but undergo irreversible chemical changes and form crosslinked networks when exposed to high temperatures (Callister Jr. & Rethwisch, 2018).

Polymer classification can also be based on the polymerization mechanism, distinguishing between condensation (step-growth) and addition (chain-growth) polymerization. Condensation polymerization involves the reaction between two functionalized monomers, leading to the formation of larger molecules and the elimination of a small molecule like water. This process is typically slower and requires longer reaction times to achieve high molecular weight polymers (Callister Jr. & Rethwisch, 2018).

Understanding the physical structure of polymers, referred to as the secondary structure, is crucial for understanding their properties. The size of a polymer molecule is described by its molecular weight, while the shape is influenced by the nature of the repeating unit and how the units are linked together, described by configuration and conformation (Callister Jr. & Rethwisch, 2018).

As we know small molecules have three states of matter namely, solid state, liquid state and gaseous state. Solid state particles are very tightly packed and have very little or no motion. Solid state particles are very tightly packed and have very little or no motion. In other hand particles in liquid state are more loosely packed and can move. Particles of gas are very loosely packed and move faster with great energy. Unlike small molecules, polymers behave differently due to their high molecular weight and involve a large number of atoms. The principal properties of polymers can be classified into mechanical, electrical, thermal, surface contact, and optical properties, which determine the flexibility and strength of the polymer (Callister Jr. & Rethwisch, 2018).

Materials can respond to external forces in three ways: by directly adding stress to the forces holding the constituent atoms or molecules together (as observed in rigid materials like metals, crystalline polymers and ceramics), by changing shape while absorbing energy (as seen in nanocrystalline polymers), or by deforming semi-permanently (as in viscoelastic materials) or permanently (as in plastics) under external forces. The viscoelastic behavior of polymeric materials depends on time and temperature, which must be considered when using polymers for structural applications. This behavior is characterized by creep, stress relaxation, and dynamic mechanical response, which can be measured and quantified using various experimental methods and mathematical models (Callister Jr. & Rethwisch, 2018).

Polymers do not undergo rust like metals, but they can undergo degradation through different physiochemical reactions caused by environmental effects. Polymer degradation processes are distinct from the electrochemical reaction of metallic corrosion. Polymer degradation can occur through physiochemical reaction which physical and chemical processes such as swelling, dissolving, thermal degradation, photo-oxidation, hydrolysis, ultrasonic treatment, or radiation exposure. Polymer degradation strongly influences the stability and durability of polymers, and while it represents a failure of polymer properties, it can also be useful for polymer recycling to reduce environmental pollution (Alberts et al., 2014; Callister Jr. & Rethwisch, 2018).

Many polymers, such as plastics, are not useful in their pure chemical form but can be transformed into highly usable products by incorporating different additives. Polymer additives are employed to enhance stability against aging processes, improve product quality and performance. Common additives include flame retardants, light stabilizers, antioxidants, and thermal stabilizers. Additionally, other additives, such as plasticizers, lubricants, fillers, modifiers, pigments, and dyes, can be incorporated into polymer blends to enhance their mechanical and physical properties and reduce costs. Polymer blends, as well as copolymerization, can be utilized to achieve desired performance characteristics (Callister Jr. & Rethwisch, 2018).

THERMOPLASTICS, ELASTOMERS, AND THERMOSET

Polymers can be classified into different categories based on their end use. One way to classify them is as plastics (thermoplastics or thermosets), elastomers (rubber), and fibers, coatings, adhesives, forms, and films. Some polymers can be used in multiple categories depending on their properties. Depending on its properties, a particular polymer can be used in two or more of these application categories. For example, a plastic, if crosslinked and utilized above its glass transition temperature, may make a satisfactory elastomer or a fiber material may be used as a plastic. Plastics are the largest category and are used in everyday applications. Examples of plastics include polyethylene, polypropylene, polystyrene, poly(vinyl chloride), polyesters, phenolics, fluorocarbons, and epoxides. Plastics can be rigid or flexible, and their molecular structures can vary, such as linear, branched, isotactic, or with different degrees of crystallinity. Plastics can be further classified into thermoplastics and thermoset polymers (Callister Jr. & Rethwisch, 2018).

Thermoplastics are composed of a large number of independent molecular chains that can be melt processed using various methods. When heated, the chains can slide and cause plastic flow. Some thermoplastics solidify in an orderly arrangement called semi-crystalline thermoplastics. Examples of thermoplastics include polyethylene, polypropylene, and polyamide. Crystalline thermoplastics are more susceptible to solvent crazing, while amorphous thermoplastics are transparent and can be used as glass replacements. Examples of amorphous thermoplastics include polycarbonate and polymethylmethacrylate (PMMA) (Callister Jr. & Rethwisch, 2018).

Thermoset plastics are used when higher rigidity is required. These materials are brittle and often used in reinforced form as load-bearing solids. In thermoset plastic formation, at least one of the monomers must be trifunctional or higher. Thermosets differ from thermoplastics in that their molecular chains are

Figure 1. Thermoplastic heating process

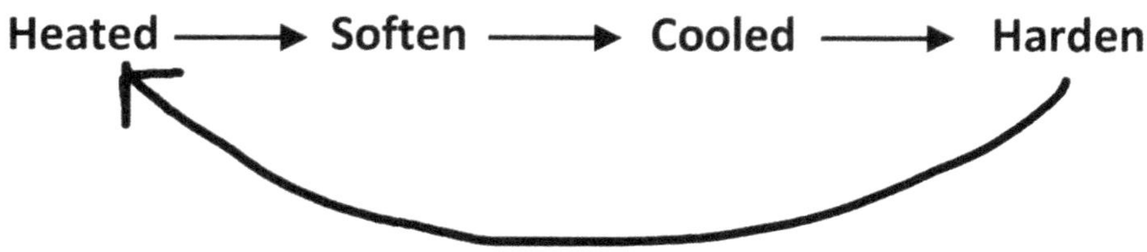

Figure 2. Thermoset heating process

crosslinked by primary covalent bonds and are fully amorphous. Examples of thermoset plastics include phenol formaldehyde, epoxy resins, and polyurethane. Unlike thermoplastics, thermosets can only be softened once during initial heating (Callister Jr. & Rethwisch, 2018).

Elastomers exhibit rubber-like elasticity, meaning they can be deformed to a large extent and then spring back to their original form. Elastomers have long, flexible chains with low intermolecular attraction, allowing localized segment mobility. To prevent permanent chain flow, the chains are crosslinked to form a 3D network. By controlling the crosslinking density, the properties of elastomers can be adjusted. Natural rubber, for example, can be made more rigid by increasing the crosslink density through vulcanization with sulfur. Elastomers generally have glass transition temperatures below -50°C (Callister Jr. & Rethwisch, 2018).

POLYMER SYNTHESIS

Polymerization is the process of creating polymers. Addition or chain growth polymerization occurs through a chain reaction involving unsaturated molecules (vinyl monomers). The chain carrier can be an ion (anion, cation) or a reactive substance with an unpaired electron called a free radical. Polymerization proceeds through initiation, propagation, and termination steps. Initiation involves the formation of free radicals from initiators by splitting up of initiators into free radicals in the presence of heat UV or electromagnetic irradiation, which then react with monomers. Propagation adds monomers to the growing polymer chain, generating new reactive centers at the end of the growing polymer chain resulting longer polymer chains. Termination occurs when free radicals lose their radical activity, either by combination or disproportionation. Combination involves the coupling of two growing polymer chain radicals. Additionally in chain polymerization. Chain transfer reactions can also occur, where the reactivity of free radicals is transferred to neutral molecules (Callister Jr. & Rethwisch, 2018).

Ionic polymerization differs from free radical polymerization in that anionic polymerization involves a negatively charged growing polymer chain end, while cationic polymerization involves a positively charged growing polymer chain (carbonium ion). The polymerization process requires reactants such as monomers, initiators/catalysts, and solvent medium to form the polymer chain which can be classified based on the reaction medium, such as bulk, solution, suspension, slurry, emulsion, and gas polymerization (Callister Jr. & Rethwisch, 2018).

Polymerization of pure liquid or gaseous monomer is called *bulk polymerization* which can be used to produce free-radical polymers and some condensation polymers. In bulk polymerization process only monomer, polymer and initiators are present therefore, through this technique the highest purity polymers can be obtained. This process produces highly transparent polymers such as PMMA. In *solution polymerization*, the monomer is added to an inert solvent with a boiling point that corresponds the polymerization temperature. During the polymerization process some solvent evaporates and helps to remove the heat of polymerization. Since the boiling point of the temperature is constant, this ensures a constant polymerization temperature. Therefore, compare to bulk polymerization, solution polymerization provides easier temperature control because of the added heat capacity of solvent and lower viscosity. Next is *suspension polymerization* which is essentially a bulk polymerization carried out in droplets in an aqueous solution in which the monomer is dispersed. This polymerization begins by the radical initiators in the monomer droplets which used in the manufacture of PVC. The *slurry polymerization* process is mainly used in the process of polyolefins. Then in *emulsion polymerization*, as in the suspension polymerization the monomer is also dispersed in water but in smaller droplets created with the use of emulsifiers, e.g., soaps. In the presence of emulsifier, monomer droplets and subsequently formed polymers are kept apart and dispersed rather than coalesce. Finally in *gas polymerization,* the monomer is in the gaseous state and a heterogeneous coordination catalyst such as a *Ziegler-Natta* catalyst is used. In gas polymerization and slurry polymerization processes, catalysts such as Ziegler-Natta and metallocene should be supported on a suitable substance as the polymer is formed on the active sites of the catalyst in to a gradually expanding catalyst-polymer unit. Gas polymerization process is also mainly used for the production of polyolefins.

Polymer properties are influenced by chain structure details, including monomer unit sequence, chemical composition, functional groups, stereochemistry, and chain geometry. Copolymers can have completely random or alternating arrangements of repeat units in the chain. During the polymerization process the repeating units in the copolymer chain can be arranged in completely random or alternating order. Types of copolymers include random copolymers, alternating copolymers, block copolymers, and graft copolymers. The length of the polymer chain is specified by the degree of polymerization (DP), and the molecular weight of the polymer is the product of the repeat unit's molecular weight and the degree of polymerization. For example, if we take poly (vinyl chloride) as an example, a polymer of DP 1000 has a molecular weight of 63 x 100 = 63000. The configuration of a polymer chain is fixed by chemical bonding, while the conformation is influenced by rotation around primary valence bonds (Callister Jr. & Rethwisch, 2018). The size of the polymer chain can be determined by the molecular weight while the shape or geometry of the polymer chain can be determined by the nature of the repeating units and how they are linked together. This can be divided in to two categories,

Configuration - The stereochemical arrangement of atoms along the polymer chain which is fixed by the chemical bonding (valence bonds) in the molecule, such as **cis** and *trans* isomers. The configuration of a polymer chain can be altered only by the breaking or reforming of chemical bonds.

Conformation - Arrangement established by rotation about primary valence bonds. Polymer chains in solutions are free to rotate around individual bonds, therefore large number of conformations/chain orientations can occur. Number of conformations depends on, steric hindrance, polymer is amorphous or crystalline and Polymer is in a solution or solid state.

Mechanical properties such as tensile strength, elongation at break, solubility, softening point, and viscosity depend on the molecular weight of the polymer. Polymers have a distribution of molecular weights due to the random nature of polymer formation. At very low molecular weights, these properties can be hardly seen. However, beyond this MW or DP, there is a steep rise in the performance until a certain level and beyond this point change of the properties is very minimum with the increase of molecular weight. During the polymerization not all polymer chains obtain the same length because the formation of polymer is a purely random occurrence and do not consist of identical molecules, which results a distribution of molecular weights. Therefore, the *average molecular weight* of the polymer (M') is related to the average degree of polymerization (DP)

$$M' = M \times DP$$

M = molecular weight of the repeat unit/monomer

The molecular weight of the polymer can be determined using various chemical and physical methods, such as end group analysis, colligative property measurements, light scattering, ultracentrifugation, dilute solution viscosity, and gel permeation chromatography (GPC) (Callister Jr. & Rethwisch, 2018).

The solubility parameter of substances in liquids depends on intermolecular forces which can be soluble in liquids if they overcome the forces holding the molecules together in the solid. Polymers have primary covalent bonds within their chains, but secondary forces between chains. These secondary forces are weaker and can be overcome by appropriate solvents, resulting in polymer solutions (Callister Jr. & Rethwisch, 2018).

POLYMER SOLID STATE PROPERTIES

When a polymer undergoes cooling from a melted state or concentration from a diluted solution, its molecules are attracted to each other, resulting in the formation of a solid mass. The solid state of polymers can be categorized into two main arrangements: amorphous and crystalline structures. The morphology of a polymer plays a crucial role in distinguishing between these two types of solids.

Amorphous polymers have atoms held together in a loosely structured manner, with the polymer chains being randomly coiled and entangled. These polymers exhibit a hard and glassy nature. On the other hand, crystalline polymers consist of individual chains that are folded and packed regularly in a three-dimensional order, resembling a diamond-like structure. The closely packed regions of long chain crystalline polymers are referred to as "lamellae." However, even in crystalline polymers, a small number of chains remain randomly coiled, creating amorphous regions. Therefore, crystalline polymers are never 100% crystalline and are often referred to as semi-crystalline, with a range of crystallinity from 0% to 100%. Most polymers fall somewhere between these extremes, making them generally known as semi-crystalline polymers (Callister Jr. & Rethwisch, 2018).

The degree of crystallinity is used to describe where a polymer stands on this spectrum. The tendency of a polymer to crystallize, the percentage of crystallinity, and the stability of crystallization depend on intermolecular forces within the polymer chain, such as hydrogen bonds, as well as the size and stereoregularity of side groups. The degree of crystallinity directly influences whether a polymer melts like

Figure 3. Two crystalline polymers

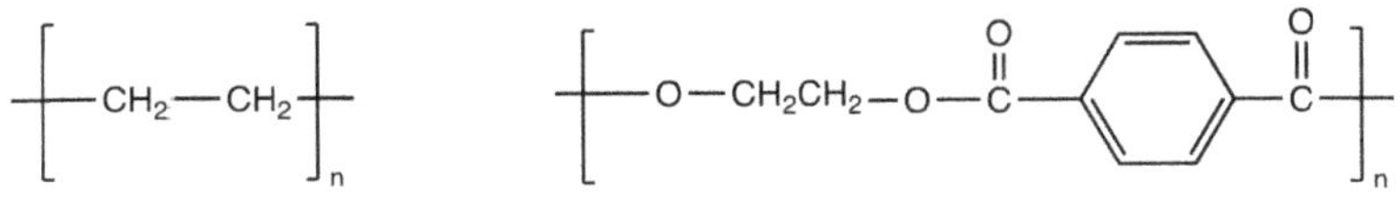

Polyethylene   Poly(ethylene terephthalate

a typical solid or undergoes a transition between a glassy and rubbery state (Callister Jr. & Rethwisch, 2018). Highly crystalline polymers exhibit a traditional melting point, where the orderly arrangement of long chain polymer structures converts to a random and disorganized arrangement upon heating. This transition temperature is known as the melting point (Tm). In contrast, amorphous polymers do not have a distinct melting point but instead exhibit a temperature range in which the material becomes less glassy and more rubbery, or vice versa. This temperature, which marks the transition from the amorphous solid state to the melt state, is called the glass transition temperature (Tg). Below Tg, the amorphous polymer chains are unable to rotate or move in space, resulting in a glassy state that is hard, rigid, and brittle. As the temperature rises above Tg, the entangled chains start moving, leading to a rubbery state where the amorphous polymer becomes soft and flexible. Examples of crystalline polymers include polyethylene and poly(ethylene terephthalate) (Callister Jr. & Rethwisch, 2018).

In contrast to crystalline polymers, amorphous polymers lack ordered arrangements of chains and are characterized in the solid state by short-range order of repeating units. Therefore, it can be concluded that the solid states of crystalline and amorphous polymers are characterized by long-range order of molecular chains and short-range order of repeating units, respectively.

When examining structure-property relationships, the molecular structure plays a significant role in determining the melting point (Tm) and glass transition temperature (Tg) of a polymer. For instance, linear polyethylene has a crystalline melting point of approximately 140°C. By replacing the methylene groups in polyethylene's chain backbone with other moieties, higher or lower Tm values can be observed. Two factors primarily govern the magnitude of Tm: chain flexibility and inter-chain forces.

Figure 4. Amorphous polymer

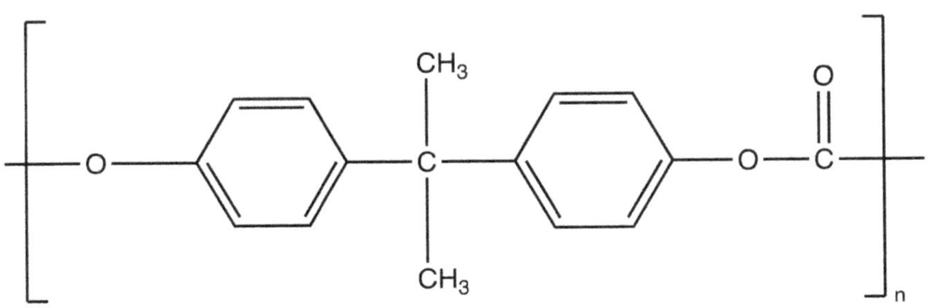

Polycarbonate

Flexible units, such as ether, ester, or sulfide groups, result in lower melting points, while rigid units, such as p-phenylene, result in higher melting points. Strong intermolecular forces, such as hydrogen bonds, also yield higher melting points. Generally, both Tm and Tg increase with decreasing flexibility of the polymer chain. Polymer chain flexibility can be reduced by adding bulky groups or incorporating aromatic groups into the chain (Callister Jr. & Rethwisch, 2018).

When a polymer is used as a structural material, its ability to withstand applied stress and strain over time becomes crucial. The mechanical properties of a polymer describe its response to deforming forces and external forces of various types, including tensile strength, flexural strength, compressive strength, and toughness. Polymers are viscoelastic materials, exhibiting properties of both solids and viscous liquids. These properties are dependent on time and temperature. Several factors influence the mechanical properties of polymers, such as molecular weight, composition, temperature, and distribution of crystallinity. Various methods are employed to determine the mechanical properties of polymers, including static testing, transient tests (such as stress relaxation and creep), impact testing, and fatigue testing (Callister Jr. & Rethwisch, 2018).

1. Static testing- this method is used to measure the force response when the sample is strained, compressed, or sheared at a constant rate which provide the mechanical properties of polymer in terms of elastic modulus, tensile strength, and elongation at break.
2. Transient test- this method measures the time response of the force or stress on a polymer when it is rapidly stretched to a given length (stress relaxation) or the time response of a strain when a stress is rapidly applied (creep)
3. Impact test- this method measures the energy required for a sample to fail under different loading conditions. Impact test provides useful information in the selection of polymer for a specific application such as determining the suitability of a given plastic as a substitute for glass bottles or replacement for window glass.
4. Fatigue test- number of cycles of applied stress required for failure.

VISCOELASTICITY AND RUBBER ELASTICITY

Viscoelastic materials possess characteristics of both elastic and viscous materials. All polymers exhibit viscoelastic properties due to the internal structure created by interlinked polymers. A viscoelastic material can return to its original shape after the removal of any deforming force, but this process takes some time, indicating an elastic response. At low temperatures, an amorphous polymer can behave like a solid glass, while at intermediate temperatures (above the glass transition temperature), it exhibits rubbery characteristics. At higher temperatures, it behaves as a viscous liquid. Viscoelasticity allows for mechanical characteristics that encompass both of these extremes.

According to Hooke's law, the stress applied to an elastic material is proportional to the strain produced, up to a certain limit. However, no material is entirely elastic. An ideal elastic material exhibits no time effects and responds instantaneously to applied stress. When the stress is removed, the material fully and instantaneously returns to its original dimensions. The induced strain (ε) is proportional to the applied stress (σ) and is independent of the strain rate or testing speed, as stress is a function of strain only.

$$\sigma = E \times \varepsilon$$

σ = stress (Pa),
E = Young's modulus of the material (Pa),
ε = Strain

When a viscoelastic material is subjected to stress, three deformation responses can be observed: an initial instant elastic response, time-dependent delayed elasticity (which is fully recoverable), and viscous non-recoverable flow. Viscoelasticity can be characterized experimentally through creep (creep compliance), stress relaxation (stress relaxation modulus), and dynamic mechanical response (storage and loss). Two common methods used to measure viscoelastic properties are transient tests and dynamic tests.

Transient tests involve deforming the material through elongation or shear and observing its response over time. There are two transient methods,

1. Stress-relaxation experiments – In this method material is rapidly deformed to a specific strain, and the force or stress required to maintain a constant deformation as the material's molecules move relative to each other. Deformation at a constant value is measured as, with time and molecule of the material move relative to one another. However, the stress required to hold the material at constant deformation dies with the time and said to be relaxed. Figure 5 shows how the stress varies with constant strain over wide range of time. Constant strain over wide range of time. These two transient experiments are possible to plot because there are three variables namely, *stress, strain, and time*.
2. In creep experiments - a sample is subjected to a constant stress/load, and the resulting time-dependent strain is observed for a relatively long period. The strain increases with time as segments of the material rotate and flow relative to one another at a rate determined by the material's viscosity, stress, temperature, and the duration of stress application. Many polymeric materials are susceptible to time-dependent deformation when to time-dependent deformation when the stress level is maintained constant. Such deformation is termed as *viscoelastic creep*. Figure 6 shows how the strain varies with constant stress over wide range of time.

Dynamic mechanical experiments, on the other hand, vary the stress or strain cyclically with time, usually in a sinusoidal manner, and measure the material's response at different frequencies of deformation. These tests are typically conducted at low strains to ensure that the deformation is not permanent. Dynamic mechanical tests cover a wide range of temperature and frequency and provide insights into the chemical and physical structure of polymeric materials. These tests can be used to study relaxation transitions, such as the glass transition temperature of polymers. Instruments for dynamic mechanical testing include torsional pendulums- is a free vibrational instrument and produces a decaying trace of amplitude of vibration with time, vibrating reed instruments- based on forced resonance vibration where if a strip of material is excited electrostatically at different frequencies, using an audio frequency oscillator, and non-resonance forced vibration instruments- often a flat specimen is clamped at both ends and is deformed in a sinusoidal manner.

Figure 5. Stress relaxation curve

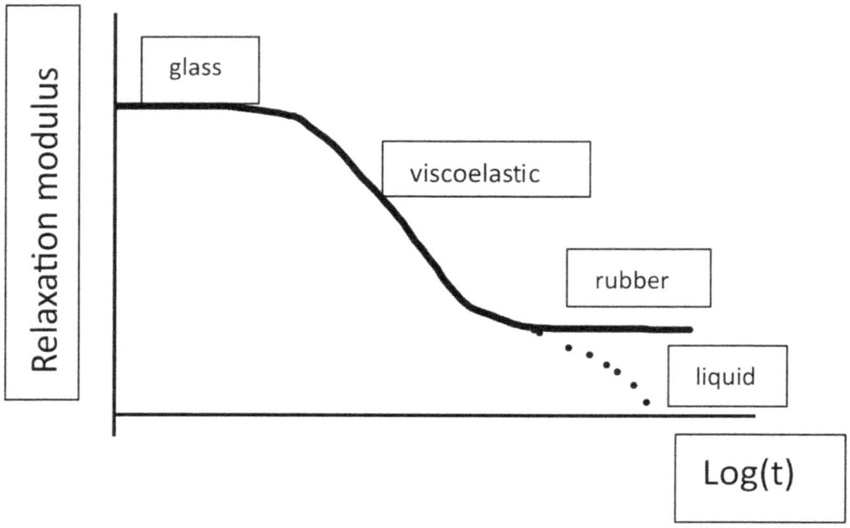

Figure 6. Creep experiment curve

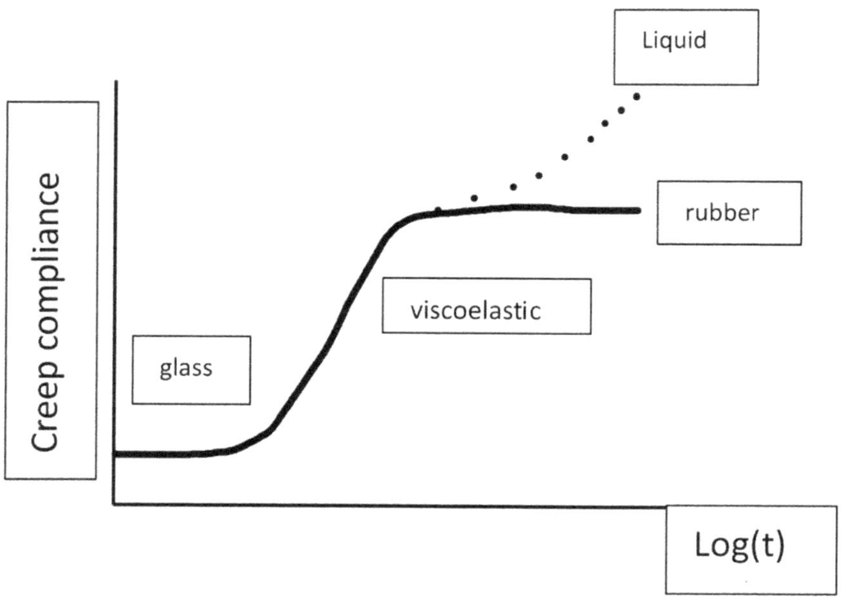

Viscoelastic behavior can be described using three approaches:

1. Boltzmannsuperposition principle (integral representation of linear viscoelasticity), differential representation (utilizing Hookean elastic elements and Newtonian viscous elements as models), and models based on assumptions about the molecules.
2. Based on assumptions about the molecules.
3. Differential representation involves the use of different types of arrangements to model viscoelastic materials. This includes an ideal spring to represent Hookean elastic elements and a dashpot to represent viscous elements.Various mathematical models are employed, with the most popular ones being:

Maxwell model: This is the simplest way to model viscoelastic behavior. It combines a Hookean spring and a Newtonian dashpot in series, where the stress relaxation behavior of the material is similar to its creep behavior. In stress relaxation measurements, a given fixed strain or stress is applied to the material, and the strain or stress required to maintain this deformation is measured over time.

Voigt (Kelvin) model: This model consists of a parallel combination of a spring and a dashpot. It describes the creep behavior of a real material. When externally loaded, the model is assumed to undergo a uniform strain, with the model and its components experiencing the same strain.

Figure 7. Maxwell and Voigt models

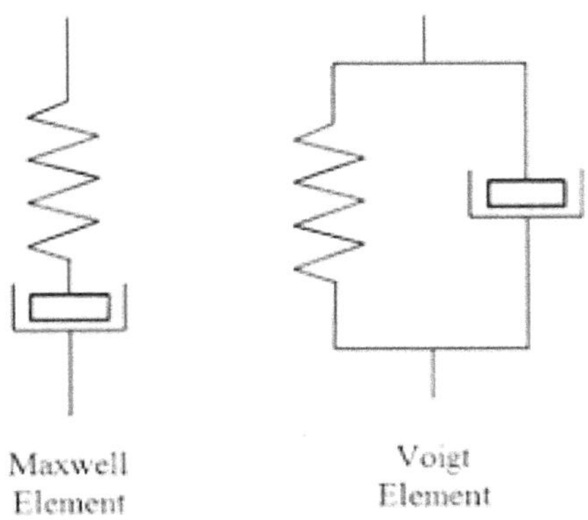

The characterization of viscoelastic behavior can also be achieved through Boltzmann superposition principle, integral representation, or by employing differential representation using Hookean elastic elements (springs) and Newtonian viscous elements (dashpots) as models. These models are based on assumptions about the behavior of polymer molecules.

Figure 8. Polymerization and depolymerization

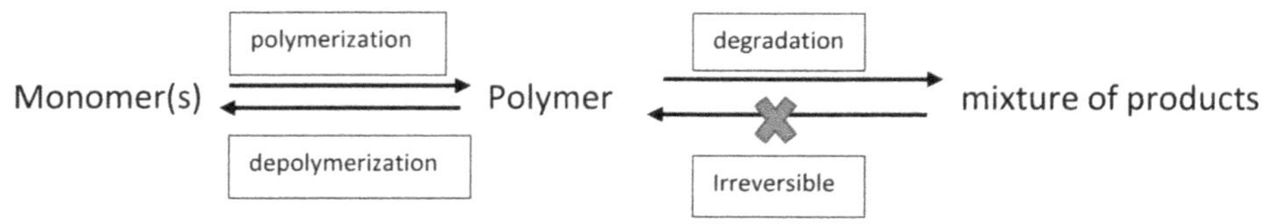

POLYMER DEGRADATION

Polymer degradation refers to the alteration of the chemical and physical structure of a polymer chain, resulting in changes to its properties such as color, shape, tensile strength, and molecular weight. This degradation occurs in the presence of various environmental factors, including heat, light, chemicals, radiation, and applied forces. The breakdown of macromolecules into smaller fragments of different structures and sizes is commonly referred to as polymer degradation. There are two primary mechanisms involved in polymer degradation: hydrolysis and oxidation. However, if the final products are monomers, the process is referred to as depolymerization.

Thermal Degradation

Thermal degradation occurs when elevated temperatures cause molecular deterioration in polymers. Increased temperatures amplify intramolecular vibrations and accelerate conformational changes in macromolecules. Thermal degradation typically involves three distinct reactions: depolymerization, chain scission, and side-group elimination. Depolymerization involves the breakdown of polymer chains into monomers, dimers, and oligomers at high temperatures. Chain scission occurs through random or chain end mechanisms, resulting in the fragmentation of the polymer and a rapid decrease in molecular weight and mechanical strength. Side-group elimination involves the breaking of bonds between side groups and the main chain, leading to the formation of volatile byproducts.

Depolymerization - The polymer chains are degraded to monomers, dimers, and/or oligomers at high temperature which generally starts at chain ends where the free radical is generated (initiation) and then the monomers are detached from the polymer main chain. (de-propagation). Depolymerization process will not stop until the polymer chain is completely depolymerized (termination). Thermal depolymerization is characteristic for some polymers obtained by chain polymerization, such as poly (methyl methacrylate)/PMMA and polystyrene.

Chain scission – The chain scission can occur by one of three mechanisms. Such as *random chain scission,* for random scission to occur, the polymer chain does not necessarily require carrying any active sites. In this multistep process, the free radicals are randomly created along the main polymer chains, which causes the fragmentation of polymer into smaller molecules, as a result rapid decrease in the molecular weight and mechanical strength of the polymer can be noticed. Except for random scission, the *chain end scission process*, monomers are effectively removed from the chain ends resulting volatile products. Furthermore, intermolecular chain transfer and disproportion termination reaction can occur. Random chain scission is a typical degradation process of polyolefins.

Side group elimination- In this reaction, the bonds between side groups and the main chain are broken and the detached side groups reacts with each other. Typical example of a common non-chain scission reaction is *dehydrohalogenation.* For example, is dehydrohalogenation of poly (vinyl chloride) PVC.

Oxidative Degradation

Oxidative degradation occurs when polymers are exposed to oxygen or other oxidizing agents. The presence of oxygen can lead to the formation of free radicals within the polymer chains. These free radicals react with oxygen, initiating a chain reaction that results in the breakdown of the polymer structure. The degradation process can be accelerated by factors such as temperature, light, and the presence of catalysts. Oxidative degradation can cause discoloration, embrittlement, and a decrease in mechanical properties.

Photochemical Degradation and Oxidative Degradation

When photochemical initiation is followed by oxidation of polymer, that process called as photooxidative degradation. Exposure to UV radiation may cause the significant degradation of many materials. UV radiation caused photooxidative degradation which results breaking of the polymer chains, produce free radicals and reduces the molecular weight resulting the failure of mechanical properties. The initiation of polymer degradation via UV light depends mainly on the presence of chromophores such as organic molecules containing double bonds, flame retardants and catalyst residues. These chromophores absorb UV light and release some of the UV energy they absorbed by braking bonds and releases free radicals in the polymer which starts the degradation process. In general, polyolefins lose its properties relatively fast under UV light. However, by introducing a bulky additive to act as a shield or blocking UV light penetrating into the polymer can prevent or interrupt the polymer degradation process initiated by UV light. Furthermore, oxidative degradation is the branch radical chain reaction that always participate in corrosion of polymers exposed to air. This can be either initiated intentionally or it can occur spontaneously by autooxidation process.

Other Degradation Mechanisms

In addition to thermal and oxidative degradation, polymers can undergo degradation through various other mechanisms, including hydrolysis, photodegradation, and environmental stress cracking, mechanodegradation, Biological degradation.

Hydrolysis

Hydrolysis involves the cleavage of chemical bonds in a polymer chain through the addition of water molecules. This process is particularly relevant for polymers containing ester or amide groups. Hydrolytic degradation can occur under conditions of high humidity, exposure to water, or in the presence of acidic or alkaline solutions. The hydrolysis of polymers can lead to a decrease in molecular weight and the loss of mechanical properties.

Photodegradation

Photodegradation occurs when polymers are exposed to ultraviolet (UV) radiation, typically from sunlight. UV radiation can cause the excitation of electrons within the polymer chains, leading to the formation of free radicals and other reactive species. These reactive species can initiate chain reactions that result in the degradation of the polymer. Photodegradation can cause discoloration, surface cracking, and a decrease in mechanical strength.

Environmental Stress Cracking

Environmental stress cracking is a phenomenon in which polymers undergo cracking and failure when exposed to certain chemicals in combination with mechanical stress. The presence of certain chemicals can cause the polymer chains to become more susceptible to mechanical stress, leading to crack initiation and propagation. Environmental stress cracking can occur in polymers such as polycarbonate, acrylics, and polyethylene when exposed to chemicals such as alcohols, detergents, or solvents.

Mechano-Degradation

Polymer degradation can also achieve by the application of stress such as, in the case of solids, stress induced degradation may result from grinding, milling, or crushing, machining, stretching, tearing or abrasion. For example, when breaks a rubber band, it actually breaks lot of chemical bonds of original network. Likewise, a linear macromolecule can be mechanical brake into radical fragments if sufficient mechanical force is supplied to it via intermolecular interactions under flow shear stress.

Biological Degradation

Biodegradation utilizes the function of microbial species to convert organic substrate in this case polymers, to small molecular weight fragments that can be further degraded to carbon dioxide and water. The physical and chemical properties of polymer is important to biodegradation process. The efficiency of this process is achieved by the microorganisms is directly related to the key properties such as molecular weights and crystallinity of the polymers. Oxygen is the required terminal electron acceptor for the aerobic degradation process. Aerobic conditions lead to the formation of CO_2 and H_2O in addition to the cellular biomass of microorganisms during the degradation of plastics.

Polymer degradation can have significant implications for the performance and lifespan of polymer-based materials. Understanding the degradation mechanisms and factors that influence degradation can help in the selection of appropriate materials and the implementation of strategies to mitigate degradation effects, such as the use of stabilizers and protective coatings. Additionally, polymer recycling and waste management techniques play an important role in minimizing the environmental impact of polymer degradation.

ADDITIVES, BLENDS, BLOCK COPOLYMERS, AND COMPOSITES: ENHANCING POLYMER PROPERTIES

Additives, blends, block copolymers, and composites play a crucial role in modifying and enhancing the properties of polymers, making them more versatile and suitable for various applications. These substances can transform many plastics into high-quality products by incorporating fibrous solids or reinforcing fillers. For example, materials like wood flour, clay, talc, mica, and glass beads significantly enhance dimensional stability, impact resistance, tensile strength, abrasion resistance, and thermal stability in polymers (Chanda & Roy, 2020). The use of glass fibers as a reinforcing filler for polyester resin is another well-known example of utilizing fillers to improve polymer performance (Agrawal & Agrawal, 2019). Additionally, soft thermoplastics like hydrocarbon resins are often blended with minerals such as quartz, limestone, or clay in large quantities to achieve excellent abrasion resistance and dimensional stability, albeit with relatively poor tensile strength (Srinivasan, 2018). Carbon black and silica are widely employed additives to enhance the strength and abrasion resistance of elastomers (Mallick, 2017).

Plasticizers are additives commonly used in polymers, particularly in plastics, to enhance flow and processability while reducing brittleness. By lowering the glass transition temperature below room temperature, plasticizers transform polymers from hard, brittle, glass-like solids into soft, flexible materials (Jin, Liu, Wu, & Xiao, 2018). For instance, plasticization of poly(vinyl chloride) and vinyl chloride acetate copolymers using appropriate plasticizers improves their properties. Compatibility and permeance are the basic requirements for using plasticizers in specific polymer materials. Plasticizers need to be miscible with the polymer, although nonpolar polymers like polyethylene can pose challenges due to intermolecular forces (Hussain & Iqbal, 2019). Permeance requirements demand low vapor pressure and diffusion rates within the polymer, which can be achieved by using high molecular weight plasticizers (Jin et al., 2018). The efficiency of the plasticizer is also crucial in determining the required amount for achieving desired properties in each application.

Stabilizers are additives used to prevent or reduce the degradation of polymers. Examples include vinyl resin and carbon black, which are employed to prevent photochemical degradation in polymers (Mallick, 2017). Flame retardants are another important type of additive that reduce flammability by eliminating volatile fuel through cooling, creating thermal barriers to reduce heat transfer, and quenching chain reactions in flames by adding suitable radical scavengers (Babrauskas & Peacock, 2016). Antimony trioxide is a widely used material for imparting flame retardancy to plastics (Xu, Xu, & Qian, 2019).

Copolymerization is a process that enables the modification of polymer chain structures by reacting more than one type of monomer. Copolymers can be classified as random copolymers, alternating copolymers, block copolymers, or graft copolymers based on the arrangement of repeat units in the polymer molecular chain. Copolymerization can have a significant influence on properties such as glass transition temperature (Tg), melting temperature (Tm), mechanical toughness, crystallinity, and various other polymer properties (Li, Yang, Wei, Zhang, & Cheng, 2019). Block copolymers, such as styrene-butadiene-styrene, exhibit regular long-range elasticity similar to rubber without requiring vulcanization. This class of polymers is known as thermoplastic elastomers, which behave like conventional cross-linked rubbers at ambient temperatures but exhibit reversible thermal behavior (Jain, Jain, & Narula, 2016).

Polymer blends refer to the physical mixture of two or more polymers. Blending finished thermoplastics can serve as an alternative to copolymerization. Examples of polymer blends include PE/PP, PC/PP, and PC/thermoplastic PU. Moreover, combining crystalline and amorphous polymers in a single material can achieve a balance of chemical resistance and easy flow from the crystalline polymer, along with low shrinkage and strength from the amorphous polymer (Hussain & Iqbal, 2019). Most polymer blends exhibit a two-phase system, and stabilizers can be added to control phase mixing and separation (Ray, 2017).

POLYMER PROCESSING AND RHEOLOGY

Polymer materials offer remarkable versatility, and a wide range of processing and forming techniques have been employed to harness their potential. Various methods, such as extrusion, injection molding, thermoforming/vacuum forming, rotational molding, and compression/transfer molding, are utilized to manufacture products like pipes, tubes, cables, multilayer packaging, and fibers.

Viscosity plays a crucial role in polymer processing, as it is influenced by different states of deformation. In injection molding, for instance, the polymer melt experiences significant shear and strain, leading to shear viscosity. The relationship between shear stress, shear viscosity, and the rate of strain is given by the equation:

shear stress = shear viscosity × rate of strain

While low molecular weight simple liquids adhere to Newton's law of viscosity, wherein viscosity remains independent of shear stress and strain magnitude, high molecular weight liquids such as polymer melts and solutions exhibit non-Newtonian flow. In such cases, viscosity is no longer a constant and is referred to as apparent viscosity. Most polymer melts exhibit pseudoplastic behavior, where viscosity decreases with shear rate. However, by incorporating pigments or fillers into the polymer resin solution, dilatant behavior, characterized by viscosity increasing with shear rate (shear thickening), can be observed.

Polymer melts are typically processed under different conditions, with the applied shear rate depending on the specific technique. Compression molding operates at shear rates of 1-10 s^(-1), calendaring at 10-100 s^(-1), extrusion at 100-1000 s^(-1), and injection molding at 1000-10000 s^(-1). Thermoplastics, unlike thermosetting plastics, do not crosslink upon heating, allowing them to maintain a softened state while being shaped under pressure. Several processing and forming methods are designed specifically for thermoplastics, including injection molding (e.g., gas-assisted injection molding and blow molding), extrusion (e.g., blow molding, calendering sheet and film, extrusion and thermoforming, fiber melt spinning, and mesh), thermoforming and vacuum forming, rotational molding, coating, and dispensing forms.

Engineering and specialty polymers, distinguished by their outstanding properties compared to regular polymers like polystyrene and polyolefins, are gaining prominence. These high-performance polymers exhibit attributes such as high thermal stability, excellent chemical resistance, high tensile strength, good weatherability, and superior toughness. Consequently, engineering plastics are replacing traditional materials in numerous engineering applications, offering cost reduction, weight reduction, enhanced durability, and improved performance. Examples of engineering plastics include polyamides, ABS resin, polycarbonate, and engineering polyester. Chemical modifications like crosslinking, blending, and alloy formation can further enhance the performance of engineering and specialty polymers.

Nylon, commonly used as a generic term for synthetic polyamides, finds extensive application, particularly nylon-6 and nylon-6,6, both of which are primarily produced using cyclohexane. Nylon plastics excel as engineering materials due to their high tensile strength, ability to withstand high temperatures, good abrasion resistance, and favorable lubrication properties. Specialty grades of nylons, including high molecular weight polymers and copolymers with olefin-based resins for increased impact resistance at lower temperatures, are also gaining traction. Additionally, certain polyamides like nylon-610, nylon-611, and nylon-612 exhibit lower stiffness and heat resistance than nylon-6 and 6,6 variants but offer improved chemical resistance and lower moisture absorption, making them ideal as specialty polymers.

ABS (acrylonitrile-butadiene-styrene) resins are produced through continuous polymerization, involving the grafting of styrene and acrylonitrile onto polybutadiene latex. ABS resin consists of two distinct phases: a continuous glassy matrix of styrene-acrylonitrile copolymer and a dispersed phase of butadiene rubber. Each component contributes unique characteristics to the final product. Acrylonitrile provides heat and chemical resistance, along with high strength, while butadiene acts as a reinforcing agent, offering impact strength and toughness at low temperatures. ABS resins are particularly suitable for high-abuse applications that demand prolonged use under severe conditions. Injection-molded ABS finds applications in housewares, small tools, pipe fittings, and telephones.

Polycarbonate is an amorphous polymer characterized by the carbonate (-o-coo-) interunit linkage. It boasts exceptional engineering properties such as high impact strength, even at low temperatures, low moisture absorption, good heat resistance, thermal stability, and resistance to oxidative degradation during melting. Typical applications of polycarbonate include telephone parts, machinery for residential and commercial use, and the automotive industry. Specialty polymers exhibit excellent mechanical properties over extended periods at high temperatures. Several methods can increase the melting point of polymers, including stiffening the polymer chain by introducing ring structures or other stiffening elements like inorganic compounds, crosslinking the chains, and inducing crystallization. Examples of specialty polymers include high-temperature polyamides, high-performance fibers, electrically conductive polymers, and inorganic polymers such as silicones.

CONCLUSION

In current industries, the selection of materials primarily consists of ceramics, metals, and polymers (Pappu, Roy, & Dey, 2009). Among these options, polymers have evolved from inexpensive alternatives to natural products to offering high-quality choices for a wide range of applications (Vasanthan & Wang, 2013). With diverse molecular structures and properties, polymers have the potential to replace traditional materials such as wood, metals, ceramics, and natural fibers (Lewis, 2004). Due to advanced research and knowledge, polymer-based materials have become the preferred choice for numerous applications, swiftly replacing other materials in sectors like construction, automobiles, aerospace, and medicine (Ibrahim, Ismail, & Mazumder, 2020). However, selecting the appropriate material for a specific production requires careful consideration of factors such as material availability, mechanical, thermal, electrical, optical, and chemical properties, process conditions, environmental impact, health and safety considerations, and production costs (Al-Othman, 2013). These considerations are crucial to ensure the optimal performance of the selected polymer material in the desired application (Balazs, Emrick, & Russell, 2006).

Polymeric materials offer several advantages, including high strength and stiffness-to-weight ratios, corrosion resistance, electrical insulation/conduction, safety, and ease of fabrication into complex shapes, all at a low production cost (Rowell, 2006). Derived from petroleum, polymers' cost-effectiveness stems from the abundance of their feedstock, making petroleum likely to remain the principal raw material (Ruzaidi, Maulud, & Azhari, 2013). Polymer production encompasses various forms such as film packaging, solid molded forms for automobile body parts, composites for aircraft components, refrigerator insulation, fibers for clothing and carpets, adhesives, rubber for tires and tubing, paints, and other coatings (Rusli & Chee, 2019).

Polymers have become integral to modern society, and their unique properties depend on factors such as molecular composition, chemical makeup, molecular size, branching, cross-linking, and processing techniques employed during production (Callister & Rethwisch, 2014). These factors significantly impact the properties of the final product (Brant & Keener, 2017).

Structural polymers, including plastics, fibers, rubbers, and adhesives, exhibit diverse arrangements of synthetic and natural polymers. These materials are termed structural polymers as their macroscopic mechanical behavior is essential to their function (Fried, 2014). They come in various chemical and physical forms, such as cross-linked and non-cross-linked, amorphous and crystalline, rubbery and glassy (Fakirov, Bhattacharyya, & Bhattacharya, 2017). Additives are often incorporated in small quantities to enhance their properties (Sisson, 2016). Structural polymers are combined with composites or blends to further enhance their performance in industries like aerospace, automotive, electrical, construction, sports, chemical, medicine, and engineering (Maazouz, Notta-Cuvier, & Krawczak, 2013). There are two main classes of structural polymers, 'thermoplastics' and 'thermosets. Thermoplastic polymers consist of discrete macromolecules capable of flowing above their glass transition temperature (Tg) or melting point (Tm) (Harper, 2003). They can soften and flow upon heating and regain hardness upon cooling (Callister & Rethwisch, 2014). Thermoset materials form rigid, three-dimensional networks that do not flow upon heating (Ibrahim, et al., 2020). Elastomers, also known as rubbers, are compliant polymers capable of experiencing reversible deformations (Fried, 2014). Moreover, engineering thermoplastics, which include the higher performance polymers such as polyacetals, polycarbonates, nylon, polyesters, polysulfones, polyetherimides, acrylonitrile butadiene styrene (ABS) materials, generally have greater heat resistance and better mechanical properties than the commodity thermoplastics. Moreover, thermoset materials defined as three dimensional, chemically resistance networks which can be referred to as gels, vulcanizates or cured materials. These are rigid network materials below their glass transition temperature. Thermosets can be formed when polyfunctional reactants form 3D networks through linear growth, branching, gelation reactions. In this phenomenon, the starting monomer must include reactive functionality > 2 in order to ensure the reaction proceeds and number of chain ends will increase. These chains will eventually interconnect to produce gelled networks. However, as the reaction proceeds, the glass transition temperature raises to meet the reaction temperature and material become glassy, which is the motion of the main chain stop. Thermosetting polymers can be produced through chain polymerization or step polymerization reactions. Thermoset polymer materials although are less commonly use than the thermoplastics, they are used in a number of very high-performance applications, such as matrix resins and structural adhesives in composite systems.

In the automotive industry, polymer-based materials are extensively used to reduce weight, promote fuel efficiency, and minimize environmental pollution while meeting mechanical requirements (Ali, 2020). In the aerospace industry, polymer-based materials offer weight reduction, cost reduction, and radiation protection, influencing factors such as fuel efficiency and speed (Kumar, Singh, & Singh, 2015).

the appropriate selection of polymer-based materials has revolutionized various industries, providing exceptional properties and performance. Choosing the right polymeric material involves considering material properties, appearance, production processes, and potential environmental impacts (Al-Othman, 2013). By appropriate selection of polymer materials, significant advancements have been achieved, such as improved strength-to-weight ratios, cost reduction, corrosion resistance, electrical insulation, and enhanced environmental sustainability (Vasanthan & Wang, 2013). Ongoing research focuses on developing biodegradable and biobased polymers to address environmental concerns (Echegoyen, 2020). Additionally, innovative polymers like ultrahigh molecular weight polyethylene, liquid crystal polymers, and thermoplastic elastomers offer unique combinations of properties for diverse applications (Harper, 2003). The continuous evolution of polymer science and technology holds great promise for the future, leading to further advancements and sustainable solutions in material selection and development (Brant & Keener, 2017).

REFERENCES

Agrawal, P., & Agrawal, R. (2019). Glass Fiber Reinforced Polymer Composites: Manufacturing Methods, Microstructural Aspects, and Mechanical Behavior. In Reference Module in Materials Science and Materials Engineering (pp. 1-11). Elsevier. doi:10.1016/B978-0-12-803581-8.11443-4

Al-Othman, A. (2013). Polymer matrix materials. In *Materials for Transportation Technology* (pp. 309–328). Elsevier.

Alberts, B., Johnson, A., Lewis, J., Raff, M., Roberts, K., & Walter, P. (2014). *Molecular Biology of the Cell* (6th ed.). Garland Science.

Ali, A. M. (2020). Application of Polymers in Automotive Industry. *Polymer Science: Advanced Technologies*, *1*(1), 13–24.

Babrauskas, V., & Peacock, R. D. (2016). Heat Release Rate: The Single Most Important Variable in Fire Hazard. *Fire Technology*, *52*(4), 851–872. doi:10.100710694-015-0512-1

Balazs, A. C., Emrick, T., & Russell, T. P. (2006). Polymer-based materials for environmental and energy applications. *Science*, *314*(5802), 1107–1110. doi:10.1126cience.1130557 PMID:17110567

Brant, P., & Keener, T. (2017). *Polymer Processing Principles and Design*. Carl Hanser Verlag GmbH & Co. KG.

Callister, W. D. Jr, & Rethwisch, D. G. (2014). *Materials science and engineering: an introduction*. John Wiley & Sons.

Callister, W. D. Jr, & Rethwisch, D. G. (2018). *Materials Science and Engineering: An Introduction* (10th ed.). John Wiley & Sons.

Chanda, M., & Roy, S. K. (2020). Enhancing Polymer Properties through Fillers and Reinforcements. In *Plastics Technology Handbook* (pp. 107–141). CRC Press.

Echegoyen, Y. (2020). Biodegradable Polymers. In Handbook of Bioplastics and Biocomposites Engineering Applications (pp. 51-76). CRC Press.

Fakirov, S., Bhattacharyya, D., & Bhattacharya, S. N. (2017). *Polymer composites: from nano- to macro-scale*. CRC Press.

Fried, J. R. (2014). *Polymer Science and Technology*. Pearson.

Harper, C. A. (2003). *Handbook of plastics, elastomers, and composites*. McGraw-Hill Professional.

Hussain, M. M., & Iqbal, M. (2019). Polymer Additives and Blends. In Advanced Polymers in Medicine (pp. 141-162). Springer.

Ibrahim, S. D., Ismail, R., & Mazumder, M. A. J. (2020). Polymers in Aerospace Applications. In Handbook of Polymers for Aerospace Applications (pp. 3-24). Elsevier.

Jain, A., Jain, A., & Narula, S. (2016). *Introduction to Polymer Chemistry and Applications*. CRC Press.

Jin, L., Liu, T., Wu, C., & Xiao, R. (2018). Plasticizers: Types, Properties, and Applications. In Functional Polymeric Composites (pp. 255-278). Elsevier.

Kumar, V., Singh, R. K., & Singh, D. (2015). Polymer-based nanocomposites for aerospace applications. In *Handbook of Polymer Nanocomposites for Industrial Applications* (pp. 491–508). Elsevier.

Lewis, P. (2004). *Polymer Science: A Comprehensive Reference*. Elsevier.

Li, Y., Yang, X., Wei, G., Zhang, Z., & Cheng, S. (2019). Advanced Functional Block Copolymers. In Functional Polymers in Food Science: From Technology to Biology, (pp. 41-75). Springer.

Maazouz, A., Notta-Cuvier, D., & Krawczak, P. (2013). *Polypropylene-based polymer blends and composites: processing, morphology and properties*. CRC Press.

Mallick, P. K. (2017). Introduction to Polymer Composites Manufacturing. In Fiber-Reinforced Composites (pp. 3-14). CRC Press.

Pappu, A., Roy, A., & Dey, S. (2009). Applications of polymers in drug delivery. *International Journal of Green Pharmacy*, *3*(1), 9–22.

Ray, S. S. (2017). Polymer Blends and Composites. In Polymer Science: A Comprehensive Reference (pp. 317-357). Elsevier.

Rowell, R. M. (2006). *Handbook of Wood Chemistry and Wood Composites*. CRC Press.

Rusli, N. I., & Chee, W. S. (2019). *Handbook of Polymer Applications in Medicine and Medical Devices*. Elsevier.

Ruzaidi, C. M., Maulud, A. S., & Azhari, C. H. (2013). Feasibility of using biomass-derived polymer in composite materials: A review. *International Journal of Automotive and Mechanical Engineering*, *8*(1), 1354–1366.

Sisson, R. D. (2016). *The Plastics Compendium*. Carl Hanser Verlag GmbH & Co KG.

Srinivasan, M. (2018). Filled Polymer Composites. In Polymeric Composite Materials (pp. 319-377). Wiley.

Vasanthan, N., & Wang, Y. (2013). Polymer nanocomposites for energy storage. *Energy Storage Materials*, *1*(2), 82–102.

Xu, Y., Xu, Y., & Qian, J. (2019). Flame Retardants: Mechanisms, Current Trends, and Future Directions. In Advances in Flame Retardant Polymeric Materials (pp. 3-30). Elsevier.

ADDITIONAL READINGS

Agarwal, S., & Wong, C. (2013). Environmental degradation of polymer composites. In *Polymer Science: A Comprehensive Reference* (Vol. 7, pp. 429–463). Elsevier.

Bunsell, A. R. (2005). *Handbook of Properties of Textile and Technical Fibres*. Woodhead Publishing.

Carraher, C. E. (2008). *Introduction to Polymer Chemistry* (3rd ed.). CRC Press.

Carraher, C. E. Jr. (2013). *Seymour/Carraher's Polymer Chemistry* (7th ed.). CRC Press.

Cowie, J. M. G., & Arrighi, V. (2008). *Polymers: Chemistry and Physics of Modern Materials* (3rd ed.). CRC Press.

Crawford, R. J. (2012). *Plastics Engineering* (3rd ed.). Elsevier.

Ghosh, P., & Roy, M. (Eds.). (2017). *Handbook of Polymer Blends and Composites* (2nd ed.). CRC Press.

Gibson, R. F. (2013). *Principles of Composite Material Mechanics* (4th ed.). CRC Press.

Huang, S. J., Yu, T., & Li, L. (Eds.). (2015). *Handbook of Thermoplastic Elastomers* (2nd ed.). William Andrew.

Manas-Zloczower, I., Menczel, J., Wilkie, C. A., & Penczek, S. (Eds.). (2015). *Fire Retardancy of Polymeric Materials* (2nd ed.). CRC Press.

Mark, J. E., & Erman, B. (2013). *Science and Technology of Rubber* (4th ed.). Academic Press.

Odian, G. (2004). *Principles of Polymerization* (4th ed.). Wiley. doi:10.1002/047147875X

Papadopoulou, M., & Bikiaris, D. (2016). Natural polymers: The past future of plastics. In *Polymeric Materials with Antimicrobial Activity* (pp. 3–33). Elsevier.

Paul, D. R., & Newman, S. (Eds.). (2012). *Polymer Blends and Composites: Chemistry and Technology* (2nd ed.). John Wiley & Sons.

Paul, D. R., & Robeson, L. M. (2008). Polymer nanotechnology: Nanocomposites. *Polymer*, *49*(15), 3187–3204. doi:10.1016/j.polymer.2008.04.017

Roy, R. (1999). Structural and physical properties of polymers. In *Introduction to Polymers and Composites* (pp. 89–107). Cambridge University Press.

Stevens, M. P. (Ed.). (2015). *Polymer Chemistry: An Introduction* (3rd ed.). Oxford University Press.

Chapter 2
Study of the Influence of Electrolytic-Plasma Treatment on the Tribological Properties of High-Speed Steel:
High Temperature Testing of Nitrided Layers High-Speed Steel

Zarina Aringozhina
https://orcid.org/0009-0001-8428-4033
D. Serikbayev East Kazakhstan Technical University, Kazakhstan

Yerkezhan Tabiyeva
https://orcid.org/0000-0002-9726-7187
D. Serikbayev East Kazakhstan Technical University, Kazakhstan

Bauyrzhan Rakhadilov
PlasmaScience LLP, Kazakhstan

Waqar Ahmed
University of Lincoln, UK

Didar Yeskermessov
https://orcid.org/0000-0002-2206-8132
D. Serikbayev East Kazakhstan Technical University, Kazakhstan

ABSTRACT

This chapter is devoted to experimental studies of changes in structural-phase blocks and tribological properties of combustion of layers of high-speed steels R6M5, R9, and R18 during electrolytic-plasma nitriding. Systematized experimental data on the effect of electrolytic-plasma nitriding on the structure, phase composition, and tribological properties of the surface layer of high-speed steels have been obtained. In addition, the patterns of formation of modified layers in high-speed steels during electrolytic-plasma nitriding, revealed in this work, can be used by researchers when choosing the modes of electrolytic-plasma treatment of alloyed steels, as well as when analyzing structural transformations of high-speed steels.

DOI: 10.4018/978-1-6684-6830-2.ch002

INTRODUCTION

The development of engineering technologies largely depends on the technical level of tool production. Thus, one of the problems of environmental protection in mechanical engineering is the high wear resistance of metalworking tools under various conditions of the loading process when machining parts by cutting. The durability of the tool depends not only on the properties of the material, implemented by the technology of formation and volumetric hardening, but also in the manifestation of the degree of surface properties. Its role in ensuring the performance properties of products is constantly increasing, which, along with the widespread use of traditional methods of chemical-thermal treatment, has contributed to the emergence and development of a new direction-surface engineering by methods of energy and physical-chemical effects. The implementation of this concept when choosing a material will improve the performance properties of the tool, and in some cases reduce the consumption of expensive materials. So, recently, due to the use of protective coatings and surface hardening, high-speed steels are increasingly being used and produced, which has made it possible to reduce the cost of expensive hard alloys. At the same time, an important role in the application of protective coatings and surface hardening is played by the use of resource-saving technologies that help reduce resource and energy costs and increase labor productivity.

The most promising method of surface hardening is the method of chemical-thermal treatment in an electrolyte plasma, which makes it possible to intensify the saturation process. Reducing the duration of the saturation process is one of the important directions for the further development of this process. As is known, nitriding is the most common process of chemical-thermal treatment, which makes it possible to increase the tool life of high-speed steels. At present, a large number of technological processes of nitriding based on liquid nitriding in cyanide baths, gas nitriding and ion-plasma nitriding have been developed. However, despite the many advantages of these processes, they have significant drawbacks - a long process time, high cost and environmental hazard. Since nitriding consumes a large amount of energy and technological materials, and in many cases ammonia is used as a saturating medium, which is harmful to the environment. Therefore, it is advisable to use the electrolytic-plasma method for nitriding tools made of high-speed steels, which makes it possible to intensify the process of surface saturation, reduce processing time, reduce energy costs, improve the environmental performance of the process and reduce the consumption of the saturating medium.

Recently, quite intensive research has been carried out on the effect of electrolytic-plasma treatment on the structure and properties of steels and alloys, work is underway to develop and implement an electrolytic-plasma heating process for surface treatment, in particular, surface chemical-thermal treatment of various products. At the same time, research in this direction is limited to surface carburizing, nitrocarburizing, boriding, sulfiding, and nitriding of low-carbon and low-alloyed structural steels, carbon tool steels, and stainless steels. Literature analysis showed that there is no information about the chemical-thermal treatment of high-speed steels in an electrolyte plasma. This may be due to the laboriousness of electrolytic plasma treatment at relatively low temperatures, which does not allow the use of this method as a finishing treatment, which is an important condition for some alloyed steels, in particular, for high-speed steels. In addition, there is very little information about the structural-phase states and tribological characteristics of steels nitrided in electrolytic plasma, as well as about the physical processes that determine the formation of the composition, structure, and properties of the nitrided layer during electrolytic-plasma nitriding. Moreover, the experimental data are not sufficiently systematized in the literature and are often very contradictory. It should also be noted the insufficient level of development of the metallurgical approach to the tribological properties of nitrided steels, structural criteria responsible for the tribological characteristics of nitrided steels, which hinders the possibility of a reasonable choice of processing technology and the full implementation of the existing hidden reserves of this effective method of chemical-thermal treatment.

1. ANALYSIS OF THE CURRENT STATE OF THE PROBLEM

This section presents a literature analysis of the main methods for modifying the surface layers of high-speed steels based on chemical-thermal treatment, which have been implemented over the past 10–20 years, as well as recent data on the formation of structural-phase states and properties of modified layers of structural and tool steels in as a result of electrolytic-plasma nitriding. Attention was paid to the features of hardening and changes in the structural-phase states of high-speed steels during nitriding.

1.1 Modification of Surface Layers of High-Speed Steels by Chemical-Thermal Treatment Methods

At present, high-speed steels, which combine high strength, hardness, wear resistance and sufficiently high ductility, are a universal material for the manufacture of metal-cutting tools. High-speed steel and the method of its hardening have been known for more than 100 years, and it should be noted that during this time no material has been found that can replace it. Such tool materials as hard alloys, superhard materials and ceramic-metal composites have limited applications and cannot compete with high-speed steels in the mass production of tools for metal processing. The production of cutting tools from high-speed steels goes through a large number of technological operations. At the same time, one of the most important is various types of heat treatments. This includes pre-heat treatment and final heat treatment. Preliminary heat treatment involves reducing the hardness of workpieces, improving their machinability and preparation for subsequent heat treatment. The final heat treatment in the classic version, depending on the type of steel used, includes stepped heating for hardening to reduce internal stresses and prevent deformations. After the final heating, the tool is subjected to hardening and tempering according to individual modes, taking into account a specific type of steel (Kremnev, 1985). Over the past 20 years, special attention has been paid to the development and research of methods for additional surface treatment, which are the final stage in the technological chain for the manufacture of high-speed tools and are designed to additionally increase hardness, create useful compressive stresses, increase tensile strength, reduce frictional interaction between the processed and tool materials. during the cutting process. All methods of hardening surface treatment of high-speed tools can be divided into two large groups, which are fundamentally different from each other. The first group of methods is aimed at modifying the surface properties of a high-speed tool. The second group of methods involves applying additional coatings with improved physical and mechanical properties to the surface of the tool. The first group of methods is more promising.

Methods of surface modification of a high-speed tool can be conditionally divided into 3 groups (Grigoriev et al., 2011): mechanical, physical, and chemical-thermal. Each of these methods has its advantages and disadvantages, however, for economic reasons and due to technological simplicity and efficiency, chemical-thermal methods for modifying the surface of tools made of high-speed steels have found the widest application. Cutting tools, first of all, must be wear-resistant, hard, durable, have high heat resistance and corrosion resistance. The most acceptable and providing these requirements are the processes of nitriding, carbonitriding (cyanidation), boriding and sulfiding. Diffusion saturation of the metal surface with boron compounds in the form of iron borides Fe_2B and FeB is used to increase the wear resistance and heat resistance of products, including those operating at elevated or low temperatures, alternating and shock loads, or in aggressive and abrasive media. Boriding is mainly used to harden metal surfaces that work on abrasion: cutting tools made of high-speed steels, stamping tools, molds, crushing machine parts, screen troughs, coke pusher shoes and parts operating at 500-850 ° C. Boriding is carried out in a mixture of boron-containing powders, pastes, gases or in molten salts. As a result of boriding, a change in the dimensions of the part is possible. Therefore, in the manufacture of precise and high-precision parts, it is advisable to process them in the following sequence: preliminary machining (peeling); improvement (hardening and low tempering according to modes corresponding to steel grades); machining with a finishing allowance, boriding, heat treatment to ensure the required set of properties of the base metal; final fine-tuning of the dimensions of hardened parts by mechanical processing methods (grinding, polishing). As a result of boriding, the durability of a die tool made of carbon steels reaches up to 10 times, of alloyed tool steels - 3-4 times (Guriev et al., 2009; Guryev et al., 2001; Guryev et al., 2007). General disadvantages of the surface boriding process: chipping at operating temperatures above 800°C, punching and chipping of high-hardness boride layers based on a softer base, difficulty in restoring a worn surface due to the high hardness of boride layer residues. Sulfiding is applied to technological equipment and tools made of high-speed steels of all grades and steels of the 105WCr6 and 90CrSi types in order to improve the anti-friction properties of surfaces. As the analysis showed, the sulfiding of cutting tools gives conflicting results: in some cases (Zubtsov & Korsakov, 1971), an increase in wear resistance by 1.5–2 times is achieved, in others (Loladze, 1982) it is not detected. Therefore, in the factory, to increase the durability of cutting tools, sulfiding has so far found very little use. Most often, in the practice of tool production, to harden tools made of high-speed steels, methods are used to jointly saturate their surface layers with nitrogen and carbon (cyanidation, carbonitration, nitrocarburizing).

Cyaniding is a process of surface saturation of steel with nitrogen and carbon from melts of cyanide salts. Cyanides and cyanates of alkali and alkaline earth metals, their mixtures, urea, cyanamides, and other compounds are used as components of saturating media (Chatterjee-Fisher, & Supova,, 1990; Lakhtin, et al., 1973). Cyanidation of high-speed steels is constantly being improved, mainly in the direction of intensifying the process, for which baths with accelerating additives to cyanide salts are offered, molten salts are blown with ammonia (Novikov, 1986), bath compositions with urea have been developed (Minkevich, & Suchevyanu,, 1968), carbonitration of steels at elevated temperatures has been developed - so called "tenifer process". This process also has some disadvantages. Maintaining the activity of saturating media during cyanidation requires periodic introduction of highly toxic compounds (KCN, etc.) into their composition (Neustroyev et al., 1972; Solntseva, 1988). In the case of using non-toxic starting components of the bath, based on carbamide, cyanide compounds are formed in the melt as a result of chemical reactions, the amount of which is about 20% of the total volume of the melt (Lakhtin, & Kogan, 1982). In addition, as a result of the interaction of the melt with the tool material and the crucible, their dissolution is observed. This shortcoming forces the use of crucibles made of special alloys, which leads to an increase in the cost of technology. Currently, cyanidation is gradually being phased out of production, as it is considered unpromising due to the extremely high toxicity of cyanide salts and their decay products and the ensuing environmental problems.

Along with cyanidation in tool production, the saturation of high-speed steels in gaseous media is widely used – low-temperature nitrocarburizing. Low-temperature nitrocarburizing is more convenient in mass production, and its results are close in hardening effect to those obtained in cyanide baths. Tool processing during gas nitrocarburizing is carried out in an atmosphere consisting of a mixture of ammonia with carbon-containing gases such as methane, pyrolysis products of liquid hydrocarbons (synthine, kerosene, etc.). The process is carried out at a temperature of 570-580°C for 2-5 hours in sealed shaft chamber or continuous furnaces. Intense forced circulation of the gas atmosphere is used to uniformly saturate all surfaces of the instruments. As a result of nitrocarburizing, carbonitride zones 7–25 mm thick are formed on the surface of cutting tools (Chatterjee-Fisher, & Supova,, 1990). Nitrocarburizing is very effective for hardening cutting tools made of high speed steels. It is shown that nitrocarburizing of steels R9, R18, R9K5, R6M5, etc. at temperatures of 550-580°C for 1-1.5 hours, allows you to increase their wear resistance by 1.5-3 times.

There are quite a few examples of the positive experience of using nitrocarburizing to improve the physical and mechanical properties of tool steels in the literature, but it is necessary to mention the disadvantages of this process. Firstly, nitrocarburizing treatment noticeably embrittles tool steels, which are already not very tough. The mechanism of such embrittlement has not yet been sufficiently studied, although it is believed that the reason for the reduced impact strength of nitrocarburized steels is the presence of hard and brittle carbonitride crusts on their surfaces (Przhenosil, 1969). There is another explanation for this phenomenon associated with the presence of hydrogen in diffusion layers (Astaf'yev, 1997). The second disadvantage is associated with environmental problems, since during nitrocarburizing, the exhaust gas atmosphere contains such harmful substances as ammonia, carbon monoxide, carbonic and hydrocyanic (hydrogen cyanide) acids. In order to protect the environment during carbonitriding, intensive ventilation (extraction of gases) and neutralization or disposal of harmful substances before they are released into the atmosphere are necessary. Solving environmental problems associated with nitrocarburizing significantly (several times) increases the cost of its implementation. Nitriding of high-speed tools, i.e. saturation of its surface with pure nitrogen can have a significant effect of increasing tool life without changing the structure of the tool core. The low temperature of the process allows nitriding the finished tool, which has undergone preliminary mechanical and thermal processing. The nitrided steel surface has a reduced coefficient of friction and improved anti-friction properties, which ensures easy chip removal, reduces tool heating and reduces tool wear. All this makes it possible to increase the feed and cutting speed, and, consequently, the speed of metalworking. As reported in (Rakhshtadt et al., 2005), as a result of nitriding, the durability of cutters, drills, and other tools increases from 2 to 8 times. In its traditional form, nitriding is rarely used to increase the tool life of high-speed and other tool steels. This is due to the fact that when nitriding by traditional methods such as gas nitriding, the reactions of nitrogen with iron and alloying elements present in steel occur at a low temperature quite slowly, therefore, in order to obtain a sufficient depth of a nitrided layer on the surface of the part, a long exposure of about 15 -20 hours or more. The need for long exposure during nitriding is the main disadvantage of this process. In order to intensify the process of saturation of the treated surface, in recent years, high-temperature ion nitriding has been increasingly used to improve the performance properties of tool steels, especially for cutting (R6M5, R18, R9, etc.) and stamping (Kh 2, Kh12F1, Kh6VF, etc.) tool (Arzamasov, et al., 1999). The process of high-temperature nitriding is carried out in a nitrogen-containing environment, in the temperature range from 600 to 1200°C, i.e. above the eutectoid transformation temperature of Al in the Fe-N system, when the γ-phase is formed in the diffusion layer (Arzamasov, et al., 1999; Lakhtin, & Arzamasov, 1985; Lakhtin, et al., 1991). As a result of high-temperature nitriding, a structure is formed in structural steels and alloys, consisting of dispersed nitride particles distributed in a solid solution. Thus obtained finely dispersed stable nitride particles in steels and alloys provide high performance (Petrova, 2001; Petrova, 2004). The authors of (Budilov, Agzamov, & Ramazanov, 2004; Budilov, Agzamov, & Ramazanov, 2004) proposed to carry out high-temperature nitriding of tool steels with a glow discharge plasma with a hollow cathode at temperatures of 700–720°C, when the diffusion mobility of the elements increases sharply, and the nitriding time is reduced to 1–2 hours, as well as combined treatment, consisting of high-temperature nitriding and subsequent bright quenching under vacuum conditions in one technological cycle. Under such conditions of nitriding, a significant increase in microhardness occurs on the surface of specimens of R6M5 high-speed steel (Ramazanov et al., 2010). In world practice, the most widely used cutting tool nitriding is treatment in glow discharge plasma (ion nitriding) (Budilov et al., 2007). The energy of a glow discharge in a rarefied gas is used to transport nitrogen ions to the surface of the workpiece, which serves as the cathode. In this case, the moving ions provide the cathode with sufficient energy to heat it up to the nitriding temperature. Ionic nitriding provides a high rate of saturation, obtaining diffusion layers of a given composition and minor deformations during processing. Ion nitriding at 560°C for 10–30 hours makes it possible to noticeably increase the tool life of high speed steel tools in s-6–5–2 (R6M5 type) (Lakhtin, 1993). Nitriding in the plasma of a two-stage vacuum arc discharge has become a new stage in the development of ion nitriding of high-speed steels (Sablev et al., 2000). Such a discharge can exist in various gaseous media, i.e. it is possible to carry out nitriding in various saturating gases and their mixtures. The nitriding temperature is controlled by the value of the potential on the tool or the value of the arc discharge current. The main advantage is that the rate of nitriding at the initial stage is 3 times higher than in glow discharge nitriding.

Thus, it can be established that nitriding is the most effective method of chemical-thermal treatment, which provides a high range of operational properties of machined cutting tools made of high-speed steels: hardness, wear resistance, extreme pressure properties, heat resistance, corrosion resistance, fatigue strength. Recently, various methods and methods of plasma nitriding have been increasingly developed and used, which allow not only to get rid of the shortcomings of traditional nitriding methods. Plasma nitriding methods ensure the formation of a nitrided layer with a given structure on the surface of workpieces. This increases the wear resistance of the tool and its heat resistance. The nitrided surface of the tool, which has a reduced coefficient of friction and improved antifriction properties, provides easier chip removal, and also prevents chip sticking to the cutting edges and the formation of wear holes, which makes it possible to increase the feed and cutting speed (Sal'nikova & Rudman, 1987).

1.2 Analysis of Existing Methods of Nitriding High-Speed Steels

Nitriding is widely used to harden various steels and alloys, machine parts and tools operating under various operating conditions. At present, many technological variants of the process have been developed. Nitriding processes are classified according to (Lakhtin, & Arzamasov,, 1985; Prokoshkin, 1984):

- the composition of the medium - nitriding in gases, liquid and solid media (in pastes or powders);
- process temperature t - low-temperature, medium-temperature and high-temperature;
- diffusing element - nitriding, carbonitriding, hydroxynitriding, oxycarbonitriding, sulfonitriding;
- pressure in the reactor - nitriding at high, low and pulsating pressure;

- method of energy supply - heating due to convection, radiation, low-temperature plasma, induction electronic and laser heating;
- type of furnace device - nitriding in chamber, shaft, crucible and other furnaces;
- the structure of the resulting layers on iron and steel - nitriding to obtain a composite layer (nitride zone ε - and γ'-phases and a diffusion sublayer consisting of α-phase, excess γ'-phase of special nitrides and carbides (carbonitrides) - zone of internal nitriding ; internal nitriding (only diffusion sublayer - $\alpha + \gamma' + MN(M_2N)$ phases + carbides (carbonitrides); anticorrosion nitriding (non-porous dense surface nitride ε-phase or oxycarbonitride zone (Fe, M) (N, C, O);
- combinations of technological processes - nitriding + hardening (nitriding, carbonitriding); nitriding + hardening when heated Current high frequency, laser emitter; nitriding + mechanical hardening surface plastic deformation; nitriding + oxidation; carburizing + nitriding.

Liquid nitriding in cyanide baths has the highest productivity of the saturation process, however, it is expensive and environmentally harmful, therefore, at present, they are trying to replace it with gas nitriding. The prospects for the development of gas nitriding are also due to the fact that it is a more economical and more environmentally friendly process than nitriding in salt baths, and also allows you to control the process, i.e. makes it possible to obtain hardened layers with a given thickness and composition (Lakhtin, & Arzamasov, 1985). However, gas nitriding has the following disadvantages:

- long duration of the technological cycle. The main reason for the slow flow of the process is the low efficiency of the use of the saturating medium;
- formation on the surface of hardened tools (parts) of a developed nitride zone, which requires mandatory subsequent grinding, which reduces the strength properties of the surface. The search for ways to eliminate the above disadvantages led to the development of a fundamentally new method - nitriding using an electric discharge in a gas as an intensifier, both glow and arc (ion or ion-plasma nitriding). The use of these processes made it possible to significantly reduce the duration of nitriding, reduce the consumption of electricity and gas, and obtain a high-quality nitrided layer even on parts of complex configuration. The process is easily automated and provides the possibility of nitriding steels without special surface preparation. Nitriding in a glow discharge is realized in a gaseous medium (ammonia or a mixture of nitrogen and hydrogen) of reduced pressure when an electric field between the anode and cathode is applied to the gas gap. In general, the workpiece can be either a cathode (predominantly occurring) or an anode, or alternating this function. Active particles generated by a glow discharge in the working gas form various compounds or solid solutions with the metal components of the surface. Application of the glow discharge nitriding process in practice has revealed the following advantages. Firstly, the consumption of energy, working gas and processing time are reduced by one or two orders of magnitude. However, the process of nitriding in a glow discharge has its own significant drawback. This is due to the use of hydrogen-containing gas media (ammonia or a mixture of nitrogen and hydrogen) in the technology. The presence of hydrogen in a glow discharge stimulates hydrogen embrittlement. For cutting tools, gear wheels, as well as friction pairs operating at significant stresses in the surface layer, not only high hardness values, but also plasticity characteristics are essential. A fragile tool will not work at the slightest dynamic loads, and they are inevitable in real cutting processes. The lack of plastic properties at the root of the tooth leads to breakage of the gears. The chipping off of particles of high hardness from a nitrided surface subjected to significant contact stresses is essen-

tially the process of generating an abrasive that destroys the surface of metal parts (Pastukh,, 2006; Pastukh & Zdybel', 2008). In order to eliminate hydrogen embrittlement during nitriding, a new glow discharge nitriding technology has been developed using working gaseous media that do not contain hydrogen. This process is called "hydrogen-free glow-discharge nitriding". Recently, methods of ion-plasma (ion) nitriding have been increasingly developed and used, which make it possible to get rid of the shortcomings of traditional nitriding methods. At present, a number of technological processes of ion-plasma nitriding have been developed, such as hydrogen-free nitriding in a glow discharge and nitriding of a vacuum-arc discharge, etc. However, ion-plasma (ion) nitriding, along with advantages over conventional furnace nitriding, also has some disadvantages, namely: high cost of equipment, difficulty in processing parts of complex geometric shapes, the need for individual connection of each large-sized part with a high-voltage electric generator, and also the fact that the loading of parts into the nitriding container is carried out taking into account mutual shielding. Thus, the literature analysis showed that the existing nitriding processes used to harden cutting tools, despite a lot of advantages, have significant drawbacks - a long process time, high cost, labor intensity and environmental hazard of the process. Since nitriding consumes a large amount of energy and technological materials, and in many cases ammonia is used as a saturating medium, which is harmful to the environment. Therefore, at present, an urgent task is to develop resource-saving methods for nitriding high-speed steels, which make it possible to intensify the process of surface saturation, reduce processing time, reduce energy consumption, improve the environmental performance of the process, and reduce the consumption of saturating gas. One of the promising methods of nitriding, which practically does not have the above disadvantages, is the method of nitriding in electrolyte plasma. However, this method of nitriding is still not well understood. In addition, the literature does not provide information on the nitriding of high-speed steels in electrolytic plasma. In this regard, research and development of resource-saving methods of electrolytic-plasma nitriding for hardening high-speed steels seems to be an urgent and timely task.

1.3 Electrolyte-Plasma Nitriding as a Promising Direction of the Nitriding Process

One of the promising methods of nitriding is nitriding by electrolytic-plasma heating in an electrolyte. This method is based on the features of the flow of an electric current of high density (several amperes per square centimeter) at the metal-electrolyte interface. Over the past 15 years, an environmentally friendly method of electrolytic-plasma processing of metal products has been intensively introduced into production. New ways of implementing this method are being improved and developed. Electrolyte-plasma hardening is one of the methods of high-speed heating, in which the workpiece is a cathode or anode relative to an aqueous electrolyte (Gupta, Tenhundfeld, Daigle, & Ryabkov, 2007). Depending on the heating mode, electrolyte composition, design parameters of the equipment, it is possible to perform hardening, chemical-thermal and thermocyclic processing of materials. Chemical-thermal treatment by electrolytic-plasma heating is carried out by heating and holding in a saturating medium due to a change in the electric potential in the plasma layer created between the electrolyte and the sample surface. The closure of the electrical circuit between the electrodes occurs through the electrolyte (aqueous salt solution). The conversion of electrical energy into thermal energy takes place mainly in the layer adjacent to the product. As a result of heating, this layer passes into a vapor-gas state, and microarcs are excited in it under the influence of an applied voltage. The power density reaches up to $3{\cdot}10^3$W/cm^2. To increase productivity, the processing is performed by several electrolyte heaters. The technology makes it possible to vary the rate of heating and cooling and the thickness of the hardened layer within a wide range. By controlling the temperature-speed modes of plasma surface heating and cooling, as well as by using various electrolytes, it is possible to change the structural-phase state of the surface, creating an optimal set of microstructures that provide high mechanical properties (Gupta, Tenhundfeld, Daigle, & Ryabkov, 2007; Yerokhin et al., 1999).

The specific features of heating metals in an electrolyte plasma are:

- the possibility of surface and local hardening, while you can freely adjust the degree and depth of hardening;
- the possibility of high-speed heating, reducing the time of heat treatment;
- exclusion of oxidation of the heating surface of the part and improvement of the quality of its surface;
- the possibility of using an aqueous solution in which heating is carried out as a quenching medium;
- the presence of electroerosive phenomena that contribute to an increase in the efficiency of heating metals in an electrolyte plasma;
- the possibility of local hardening of products with such shapes that do not allow for high-frequency hardening.

The heating of metals in an electrolyte plasma during the anode process as a result of a favorable combination of high temperature of the active electrode, the flow of electrical discharges in the vapor-gas shell between the metal and electrolyte electrodes allows a number of processes of local accelerated thermal and chemical-thermal treatment of steel parts to be carried out. These processes include: heat treatment followed by cooling in the same electrolyte (i.e., to carry out the hardening process), nitriding, carburizing, carburizing, tungstening, molybdenation, and other types of complex diffusion saturation (Suminov et al., 2011). Many parts of machines and mechanisms subjected to bulk heat and chemical-thermal treatment require local surface hardening: hardening or chemical-thermal treatment. The traditional technology of local chemical-thermal treatment with heating in furnaces followed by volumetric hardening is characterized not only by the long duration of the diffusion saturation process (tens of hours during nitriding or hours during carburization and nitrocarburizing), but also the need to perform many additional operations (applying protective coatings, intermediate mechanical processing, etc.), which causes an increase in inter-shop transportation, a significant consumption of electricity, additional labor and other unproductive costs (Arzamasov, et al., 1999). One of the most important advantages of thermal and chemical-thermal treatment with heating in an electrolyte plasma during the anode process is the possibility of local treatment without applying protective coatings to the workpiece, which can significantly simplify technological processes and improve the quality of products (Bakovets, Polyakov, & Dolgovesova, 1991). Among the many methods of chemical-thermal treatment for strengthening various steels and alloys, nitriding has been widely used for more than 60 years. The main advantage of nitriding is the ability to control the composition and structure of the diffusion layer by controlling the saturating atmosphere; this makes it possible to strengthen a whole range of machine parts and tools operating in various operating conditions (Petrova & Zyuzin, 2005). In industry, both furnace and ion nitriding (nitriding in a glow discharge) are actively used. These processes, with all their advantages, have a common drawback - a long saturation time (tens of hours), which requires solving the problem of intensifying the technological process (Lakhtin, 1995; Petrova et al., 2003). The approach to solving the problem of intensifying the processes of chemical-thermal treatment (carburizing, nitriding, carbonitriding, etc.) is to activate elementary processes that occur during the formation of a diffusion layer: dissociation and formation of an active element in the gas phase, adsorption of atoms of a diffusing element on a metal surface, diffusion element to metal. One of the promising methods for accelerating nitrogen saturation during nitriding is the use of a glow-spark discharge, which is both a powerful source of energy and a good surface activator. The use of a glow-spark discharge makes it possible to form active nitrogen in the gas phase with an elevated temperature.

Electrolyte-plasma nitriding consists in processing products in a glow-spark discharge (low-temperature plasma) created between the electrolyte and the surface of the product. To create and maintain a low-temperature plasma, an electrolyte is used, which is a source of ions for the nitriding process: an aqueous solution of salts containing nitrogen ions, an aqueous solution of ammonia and other nitrogen-containing media. During the nitriding process, the workpiece is immersed in an open reactor filled with electrolyte. A rectified voltage is applied to the part (cathode) and the solid electrode (anode). Near the surface of the part, a gas-vapor jacket approximately 50–120 μm thick (Suminov et al., 2011) is created, consisting of electrolyte vapors in which water and salts interact to form ions. The gas-vapor shell enveloping the workpiece contains electrolyte ions, and, accordingly, nitrogen ions. It is a low-temperature plasma and is an active saturating atmosphere for nitriding. This saturating medium is dense; the pressure in the reactor corresponds to atmospheric pressure (Slovetskiy, 1991). In low-temperature plasma, microdischarges occur on the surface of the product, and a possibility is created for the flow of a stationary electric current. For efficient plasma combustion, the area of the cathode must be five times smaller than the area of the solid anode (Bakovets, Polyakov, & Dolgovesova, 1991). It was known from the literature analysis that electrolytic-plasma nitriding is possible in both anodic and cathodic processes.

Anode processes have recently been used much more frequently (Belkin & Pasinkovskiy, 1989; Plotnikov et al., 2016). Anode heating differs from cathode heating in the absence of erosion and the impossibility of melting the part. However, anodic thermal or chemical-thermal treatment processes are accompanied by the inevitable formation of an oxide layer. This layer, which significantly affects the processes of dissolution and diffusion saturation, does not allow increasing the hardness and wear resistance of the material at a sufficient level (Belkin & Ganchar, 1988). In addition, during anodic processes, anodic dissolution of the surface takes place, which leads to a change in the size of the processed material. Traditional cathodic processes are characterized by erosion of the surface of the workpiece, temperature control in the range of 400–700°C is difficult, there is a danger of edge melting (Suminov et al., 2011; Yasnorodskiy, 1971), and low reproducibility of results (Slovetskiy et al., 1986) prevents the widespread use of this process even when machining simple-shaped parts. low accuracy. However, the cathodic process favorably differs from the anodic one in the absence of an oxide layer, as well as in the high wear resistance and hardness of the processed products (Duradzhi & Bryantsev, 1979; Lazarenko, et al., 1979). It is assumed that a deeper study and development of a more efficient cathodic heating process allows the widespread application of the cathodic chemical-thermal treatment process. Nitriding in electrolyte plasma accelerates the process so much that it exceeds liquid cyanidation in baths in speed, and at the same efficiency ensures high environmental cleanliness. In this regard, it is advisable to use the method of electrolytic-plasma nitriding to harden the working surface of a tool made of high-speed steels. On the other hand, it should be noted that there is no information in the literature about the electrolytic-plasma nitriding of high-speed steels, so this issue requires a special study. Perhaps this is due to the fact that the known methods of electrolytic-plasma modification do not allow the processing process to be carried out at a relatively low temperature, which does not allow it to be used as a finishing treatment for heat-treated high-speed steels. In connection with the foregoing, one of the main tasks in the application of electrolytic-plasma technology for surface hardening of high-speed steels is to improve the cathodic heating process and develop a highly efficient method of electrolytic-plasma nitriding, which does not have the above disadvantages. An important direction in improving the cathodic process of electrolytic-plasma nitriding is the development of a new environmentally friendly electrolyte and an energy-saving heating method that allows you to control the temperature in the range of 400–700°C and does not lead to edge melting. This must be taken into account when developing resource-saving technologies, the main task of which is to obtain modified layers with desired properties.

2. MATERIALS, EQUIPMENT, AND METHODS OF EXPERIMENTAL STUDIES

2.1 Study Materials

In accordance with the tasks set, R6M5, R9 and R18 tool high-speed steels were chosen as the research material. The use of high-speed steels for cutting tools makes it possible to increase the cutting speed by several times, and tool life by dozens of times (Poznyak, et al., 1977). The main distinguishing feature of high speed steels is their high heat resistance or red hardness (600-700°C) in the presence of high hardness (63-70 HRC) and tool wear resistance. The unique properties of high-speed steels are achieved through special alloying and complex heat treatment, providing a certain phase composition (Gol'dshteyn et al., 1985). Table 1 shows the chemical composition of high speed steels R6M5, R9 and R18.

The choice of research materials is also justified by the fact that high-speed steels R6M5, R9, R18 are the most common in metalworking, typical high-speed steels of moderate heat resistance (Gol'dshteyn et al., 1985). Table 2 shows data on the use of selected research materials. At present, in order to reduce the consumption of expensive and scarce elements, especially tungsten, economically alloyed steels are mainly used. Of these, R6M5 steel has the widest application (Anel'nik et al., 1981), so a more detailed study was carried out on high-speed tool steel R6M5. The main alloying elements of high-speed steels R6M5, R9, R18, providing high red hardness, are tungsten, molybdenum, vanadium. In addition, these steels are alloyed with chromium. An important component is carbon (Gol'dshteyn et al., 1985).

Sample blanks for research in the form of parallelepipeds with dimensions of 10x20x20 mm^3 were cut from cutting tools (disc cutters) from steels R6M5, R9 and R18 of steels R6M5, R12 and R18 with a

Table 1. Chemical composition of high-speed steel R6M5, R9 and R18 (State standard 19265-73)

Steel Grade	C	Mn	Si	Cr	W	V	Co	Mo	Ni	Cu	S	P
R9	0.85 - 0.95	up to 0.50	up to 0.50	3.80 - 4.40	8.50 - 9.50	2.30-2.70	up to 0.50	up to 1.00	up to 0.40	-	up to 0.03	up to 0.03
R6M5	0.82-0.9	0,20 - 0.50	0,20 - 0.50	3.80 - 4.40	5.50 - 6.50	1.70-2.10	up to 0.50	4.80-5.30	up to 0.60	up to 0.25	up to 0.025	up to 0.03
R18	0.73-0.83	0,20 - 0.50	0,20 - 0.50	3.80 - 4.40	17.00 - 18.50	1.00 - 1.40	up to 0.50	up to 1.00	up to 0.60	up to 0.25	up to 0.03	up to 0.03

Table 2. Data on the use of selected research materials

Steel Grade	Application
R9	For simple shaped tools that do not require a large amount of grinding operations; it is applied to processing of usual constructional materials; has increased plasticity and can be used for the manufacture of tools by plastic deformation methods; reduced grindability.
R6M5	It can be used for all types of cutting tools, especially those subjected to increased grinding, when processing common structural materials. Has high technology.
R18	Steel for all kinds of cutting tools. Used for tools working with shock loads; increased tendency to decarburization.

diamond disk (thickness 1 mm), which was previously immersed in a coolant. At low cutting speeds n = 350 rpm and low load m = 250 g, the sample does not experience deformation and thermal effects. High service properties of high-speed steel tools are achieved by heat treatment. Therefore, sample blanks for electrolytic plasma treatment were cut from tools made of R6M5, R9, and R18 steels subjected to heat treatment common for these steels (Arshinger, 1982) (Table 3).

A prerequisite for all metallographic examinations is the preparation of sections that can be used for microscopic examination using light and electron microscopes, for determining microhardness, as well as for quantitative measurement of structural components. The correct preparation of thin sections is extremely important, since the correct interpretation of microstructures depends on this. The main point in the manufacture of metallographic sections is to prevent damage to the surface of the section, which consists in changing the microstructure of the surface layer of the material as a result of deformation or heating. Grinding of metallographic samples by mechanical means is carried out on special grinding machines. The grinding machine is one or more metal wheels driven by an electric motor. On metal circles impose or stick twists cut out of sanding paper. Sometimes carborundum wheels are used instead of coarse sanding paper. During mechanical grinding, as well as during manual grinding, grinding skins are changed, successively moving from coarse-grained to fine-grained. During mechanical grinding, samples often heat up; in this case, they are periodically cooled in water. Grinding of metallographic samples was carried out on a LaboPol-5 grinding and polishing machine with a device for automatic grinding and polishing of samples LaboForce-3 manufactured by Struers (Denmark). Electrolytic-plasma nitriding of high-speed steel specimens was carried out on a setup at LLP PlasmaScience.

Table 3. Modes of preliminary heat treatment of high-speed steels

Steel Grade	Types of Heat Treatment	
	Hardening	Vacation
R9	from 1240°C into oil	560°C (three times: duration of each vacation 1 hour, cooling in air)
R6M5	from 1230°C into oil	560°C (three times: duration of each vacation 1 hour, cooling in air)
R18	from 1270°C to oil	560°C (three times: duration of each vacation 1 hour, cooling in air)

2.2 Methods for Studying the Nitrided Layer

2.2.1 Metallographic Analysis

Optical metallography was used for the structure of natural materials. Metallographic researches were conducted on optical microscope "ALTAMI-MET-1M". Sections for research were prepared according to the standard technique, including mechanical grinding and mechanical polishing (Skakov & Rakhadilov, 2013). Grinding is carried out on sandpaper (sandpaper) of different grain sizes, successively moving from paper with large abrasive grains to papers of ever smaller grain sizes. Grinding is done manually or on a mechanical machine with rotating discs covered with sandpaper, to which the surface of the sample is gently pressed. When switching from coarse-grained sandpaper to fine-grained sandpaper, the sample is cleaned of emery dust, rotated 90° relative to the original grinding direction, and continue to grind until the scratches (scratches) from the previous processing completely disappear. Polishing was carried out on a grinding disk covered with felt, pre-washed and soaked in water for 1-2 hours. Further, in order to reveal the structure, sections were subjected to etching. A 4% alcohol solution of nitric acid was used for etching. Qualitative and quantitative metallographic analysis was carried out according to the recommendations given in (Baranova & Demina, 1986; Veynberga, 1973; Yelyutina, 1979).

2.2.2 X-Ray Diffraction Phase Analysis

X-ray diffraction studies of steel samples were performed using known methods of X-ray diffraction analysis on X'PertPRO and D8 ADVANCE diffractometers. Diffraction patterns were taken using CuK_α radiation (λ=2.2897 A°) at a voltage of 35 kV. In the D8 ADVANCE diffractometer, reflection from a flat pyrographite monochromator was used in front of the detector to isolate a narrow portion of the spectrum (monochromatization), as a result of which Kα radiation with certain wavelengths was selected into the detection unit. The method of X-ray diffraction analysis used in this work made it possible to analyze the following important parameters of the structure of the samples under study: determination of the crystal structure (unit cell type); accurate determination of unit cell parameters; determination of the phase composition of the material under study; determination of fine crystal structure parameters (sizes of coherent scattering regions). The interpretation of the diffraction patterns was carried out manually using standard methods and the PDF-4 database, and the quantitative analysis was performed using the Powder Cell program. Qualitative phase analysis was carried out by the X-ray diffraction method based on the comparison of the relative integrated intensity of diffraction lines and experimental values of interplanar spacings with reference interplanar spacings (Chernyavskiy, 1977; Gorelik, Skakov, and Rastorguyev, 2002; Mirkin, 1961). Quantitative phase analysis is based on the fact that the intensity of the lines of a given phase is proportional, apart from all intensity factors, to the volume fraction of the given phase in the mixture. The sizes of crystallites and microdistortions of the crystal lattice were determined from the broadening of the profiles of the Bragg maxima (Chernyavskiy, 1977).

The fact of detection of one or another phase was recorded if the two most intense lines of this phase were found on the diffraction patterns. X-ray diffraction is a common technique that determine a sample's composition or crystalline structure. For larger crystals such as macromolecules and inorganic compounds, it can be used to determine the structure of atoms within the sample. If the crystal size is too small, it can determine sample composition, crystallinity, and phase purity. This technique sends x-ray beams through it. X-ray beams are chosen because their wavelength is similar to the spacing between atoms in the sample, so the angle of diffraction will be affected by the spacing of the atoms in the molecule, as opposed to using much larger wavelengths, which would be unaltered by the spacing between atoms. The x-rays then pass through the sample, "bouncing" off of the atoms in the structure, and changing the direction of the beam at some different angle, theta, from the original beam. This is the angle of diffraction. Some of these diffracted beams cancel each other out, but if the beams have similar wavelengths, then constructive interference occurs. Constructive interference is when the x-ray beams that are whole number integers of the same wavelength add together to create a new beam with a higher amplitude. The greater amplitude of the wave translates into a greater signal for this specific angle of diffraction. The angle of diffraction can then be used to determine the difference between atomic planes using Bragg's law $\sin\theta = n\lambda/2d$, where lambda is the wavelength added, theta is the angle of diffraction, and d is the distance between atomic planes. The distance between atomic plates can then be used to determine composition or crystalline structure.

An X-ray incident upon a sample will either be transmitted, in which case it will continue along its original direction, or it will be scattered by the electrons of the atoms in the material. All the atoms in the path of the X-ray beam scatter X-rays. We are primarily interested in the peaks formed when scattered X-rays constructively interfere. Constructive interference occurs when two X-ray waves with phases separated by an integer number of wavelengths add to make a new wave with a larger amplitude.When two parallel X-rays from a coherent source scatter from two adjacent planes their path difference must be an integer number of wavelengths for constructive interference to occur.

Path difference = n λ

Therefore:

$$2d_{hkl}\sin\theta = n\lambda$$

(Popilov & Zaytseva, 1963) or dh_{kl} - interplanar spacing, θ - diffraction beam reflection angle, n- is the diffraction order, λ- wavelength added.

2.2.3 Electron Microscopic Research Methods

The morphology and elemental composition of the sample treated in electrolyte plasma was studied using a JSM-6390LV scanning electron microscope (JEOL, Japan) with an INCAEnergy energy dispersive microanalysis attachment from "OXFORD Instruments". The phase composition of the carbide phases and their sizes were studied by electron backscatter diffraction analysis on a system with electron and focused ion beams Quanta 200 3D. The principle of operation of electron backscatter diffraction analysis is as follows: the sample, tilted at an angle of 70°, is placed in the microscope column (system), and the surface under study is subjected to automatic step-by-step scanning "from point to point". The diffracted electrons backscattered from each scan point form a diffraction (Kikuchi) pattern on the fluorescent screen of a digital video camera mounted inside the microscope. These diffraction patterns are digitized and automatically displayed. Before the study, the samples were ground and polished. To identify the boundaries of grains and particles of carbide phases, chemical etching of thin sections in a 4% alcohol solution of nitric acid was used (etching time 5-7 s.). An EM-125 transmission electron microscope was used to study the fine structure of the nitrided layer in order to reveal the morphology, phase composition, and sizes of the second phases. The studies were carried out by diffraction electron microscopy of thin foils at an accelerating voltage of 125 kV. The working magnification in the microscope column was chosen from 10,000 to 80,000 times. To prepare a foil for studying the treated sample surface by the electrospark method, plates ~ 0.2–0.3 mm thick were cut out. The resulting plates were electrolytically drowned in an electrolyte of 90% phosphoric acid (H_3PO_4) and 10% chromic anhydride (CrO_3) at a voltage of 15–20 V and a current density of 3–4 A/cm^2. Subsequent electropolishing was carried out in a solution of 22% perchloric acid ($HClO_4$) + 78% acetic acid ($C_2H_4O_22$). To preserve the surface under study, polishing was performed only on the non-nitrided side. Microdiffraction analysis was used to identify the phases present in the material. The phases were identified according to the procedures described in (Endryus et al., 1971; Kheyker & Zevin, 1963). To do this, we used schemes of microdiffraction patterns calculated from the tabular values of the parameters of the corresponding crystal lattices. For each specific place under study on the sample, a large area of the foil was examined and up to 20–30 micrographs and microdiffraction patterns for them were taken.

2.3 Methods for Conducting Physical and Mechanical Tests

2.3.1 Method for Determining Microhardness

Microhardness, along with fatigue strength and wear resistance, is one of the most widely used mechanical characteristics of metals and alloys (Grigorevich, 1976; Utevskiy, 1973). At the same time, the microhardness test differs from other methods for determining the mechanical properties of materials by its high sensitivity and short testing time. It is carried out without destruction of samples, requires a minimum of costs in their preparation. Measurement of the microhardness of steel samples, carried out on the device PMT-3M in accordance with GOST 9450–76, with a load on the indenter P = 1 N and a holding time at this load of 10 sec. A regular tetrahedral diamond pyramid with an apex angle between opposite faces of 136° was used as an indenter for measuring microhardness, similarly to the determination of Vickers hardness. The number of measurements per sample was at least 20. The microhardness Hμ was determined in accordance with GOST 9450-76 using the formula:

$$H_{\mu} = \frac{P}{S} = \frac{1854P}{d^2}\left(kgf\ /\ mm^2\right) = \frac{18,2 \times 10^6 P\left(kg\right)}{d^2\left(\mu m\right)}$$

or d – is the arithmetic mean of the pyramid indentation diagonals, (mm); P – is the value of the load acting on the pyramid, N (kgf); $S=(s^2/1854)$ – the area of the lateral surface of the resulting pyramidal imprint.

When testing in accordance with the recommendations of GOST 9450-76, the following rules were observed:

- the distance from the center of the imprint to the edge of the object must be at least twice the size of the imprint;
- the distance between the centers of adjacent prints should exceed the size of the print by more than 3 times;

- the minimum thickness of the object or layer must exceed the depth of the imprint by at least 10 times.

2.3.2 Tribological Testing Methods

Tribological tests for sliding friction were carried out on a high-temperature tribometer THT-S-BE-0000 using the standard «ball-disk» technique (Figure 1 a) (international standards ASTM G 133-95 and ASTM G 99). As a counterbody, a ball with a diameter of 6.0 mm was used, from a certified material – Al_2O_3. The tests were carried out at a load of 1 N and a linear velocity of 2 cm/sec, a wear curvature radius of 5 mm, and a friction path of 31.4 m. Tribological tests were carried out at room temperature, as well as at high temperatures at 500, 550, and 600°C. The tribological characteristics of the modified layer were characterized by wear intensity and friction coefficient (Boyarskaya et al., 1986). Tribological tests of samples of high-speed steel were carried out under conditions of dry sliding friction on a tribometer (friction machine) CETR UMT-3 using the "scratch-testing" method at the Institute of Materials Science and Connecting Technologies of the University. Otto-von-Guerike (Magdeburg, Germany). The CETR UMT-3 tribometer is designed to determine the coefficients of friction of flat and cylindrical specimens when varying the level of the contact load, the trajectory of the relative motion of the specimens, the temperature, and the medium. Тестирование соответствует международному стандарту ASTM G 99. The tribometer makes it possible to test specimens for friction in the range of contact loads from 1 to 100 N at a rate of longitudinal displacement of specimens from 0.001 to 1 mm/s at temperatures up to 350°C. Tests of flat samples of high-speed steel were carried out with their translational movement along the X axis at a speed of 0.2 mm / s with a uniform increase in the contact load from 0 to 32 N (Figure 1b). The length of the friction path along the X axis was 8 mm. In order to study the friction coefficient at a constant load, tribological tests were carried out using the "Pin-disk" method on the "Pin-on-DiscTribotester" tribometer (Figure 1c). A pin with a contact size of 100 μm was used as a counterbody, made of a certified material – Al_2O_3. The tests were carried out in air at room temperature with a load of 10 N and a linear speed of 20 mm/sec, a wear curvature radius of 4 mm, a friction path of 60 m. Testing complies with the international standard ASTM G 99-95a and ASTM G 133-95.

Figure 1. Tribological tests according to the "ball-disk" scheme (a), according to the "scratch-testing" method (b), according to the "pin-disk" scheme (c)

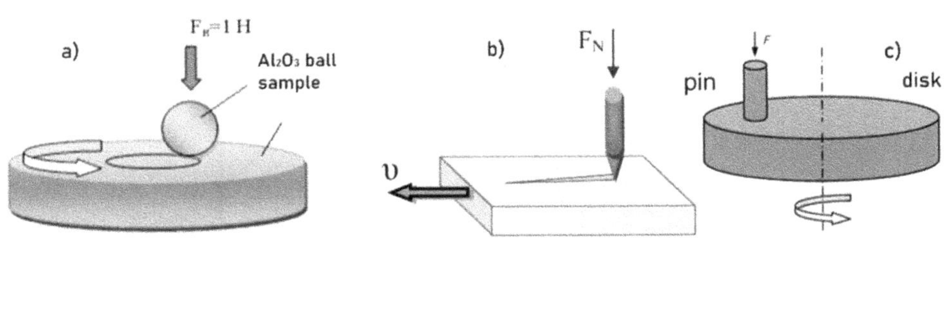

2.3.3 Test Method for Abrasion

Tests of samples for abrasive wear were carried out on an experimental setup for testing abrasive wear during friction against non-rigidly fixed abrasive particles according to the "rotating roller - flat surface" scheme in accordance with GOST 23.208-79, which coincides with the American standard ASTM C 6568. For testing abrasion on a rubber wheel, the surfaces of the samples were ground and polished, they were also cleaned with acetone and dried (Fillipov et al., 2009). A cylindrical rubber roller, pressed with a radial surface to a flat surface of the test sample with a force of 22 N, rotated at a frequency of 1 s^{-1}. The appearance and scheme of the device is shown in Figure 2. The rate of entry of abrasive particles between the rubber wheel and the sample, that is, into the test area, was 41-42 g/min. Electrocorundum with a grain size of 200–250 μm was used as abrasive particles. The wear resistance of the tested nitrided sample was evaluated by comparing its wear with the wear of the reference sample (not nitrided sample). Wear was measured by the gravimetric method on an ADV-200 analytical balance with an accuracy of 0.0001 g. The samples were weighed every minute and tested for three minutes, the total wear length was 28.8 m. Before weighing, the samples were burned with compressed air to remove the remaining sand particles on the samples. The wear resistance of the tested material was evaluated by comparing its wear with the wear of the reference sample (steel 45) by the weight loss of the samples during the test according to GOST-23.208-79. The relative wear resistance was calculated by the formula

$$K_{wr} = \frac{g_r \rho_{wr} N_{wr}}{g_{wr} \rho_r N_r}$$

or g_r, g_{wr} – values of mass loss during testing of the reference sample and samples of the test material; ρ_r, ρ_{wr} – density of reference and test material, g/cm^2; N_r, N_{wr} – the number of revolutions of the roller when testing the reference and test materials.

Figure 2. Scheme for testing samples for abrasive wear

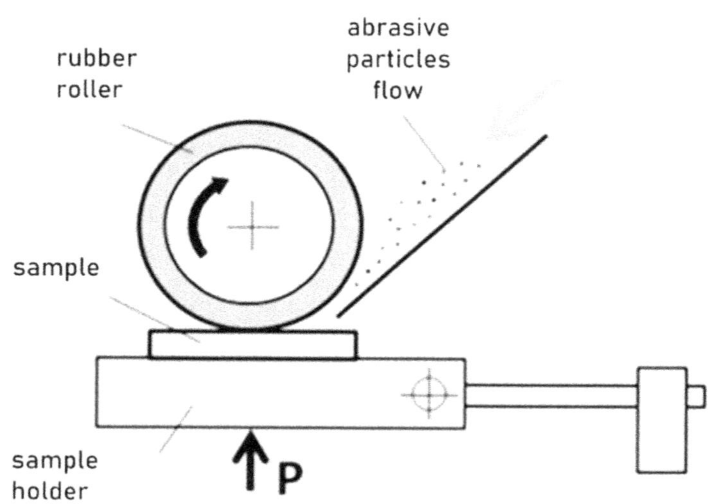

3. EFFECT OF ELECTROLYTIC-PLASMA NITRIDING ON THE TRIBOLOGICAL PROPERTIES OF THE SURFACE OF HIGH-SPEED STEELS

Ensuring maximum wear resistance of a metalworking tool under various loading conditions when machining parts by cutting is one of the most important problems of modern mechanical engineering. Since the working surfaces of the cutting tool wear out both from the mechanical impact of the material being processed on it, and as a result of molecular-thermal processes occurring in the cutting zone on the contact surfaces of the tool with the material being processed. In addition, the wear resistance of metallic materials is a structurally sensitive characteristic. For the technological support of the process of forming a wear-resistant structure during nitriding, a necessary condition is the knowledge of the laws that describe the relationship between the microstructure of the surface layer and the tribotechnical characteristics of the interface. Therefore, a relevant and promising direction in the development and improvement of the technology of electrolytic-plasma nitriding is the tribological approach to assessing the quality of the nitrided layer and developing structural criteria for the wear resistance of surface layers to improve the performance of nitrided high-speed steels under friction and wear conditions (Khrushchov & Babichev, 1970; Poznyak, 1996).

3.1 Influence of Electrolyte-Plasma Nitriding Modes on Tribological Characteristics and Microhardness of High-Speed Steels

The treatment process was carried out in an electrolyte from an aqueous solution containing 20% carbamide, 10% sodium carbonate in the following mode: the temperature of the nitriding of the samples is 450-550°C, the applied voltage between the anode and the sample when heated to the nitriding temperature of -320 V, 450-550°C-180-200 V, nitriding time varied from 3 to 12 minutes. As you know, the wear resistance of high-speed steel is one of its most important characteristics, since it determines the tool life and associated features of metalworking technology. Increased tool life reduces the time required to reinstall and regrind the tool, thereby increasing productivity and reducing the consumption of expensive cutting tools, which is most beneficial for the economy. It should be noted that wear resistance is a very complex property. It depends not only on the structure and properties of tool steel, but also on the properties of the material being processed, as well as on the coefficient of friction and the external conditions under which wear occurs: temperature in the friction zone and mechanical influences. When changing some of these conditions, the wear resistance of tool steel changes in turn (Baranchikov et al., 1990).

In this work, the tribological characteristics of high-speed steels before and after nitriding in an electrolyte plasma are studied. Figure 3 shows the wear rate (mm^3/Hm) of samples of R6M5 high-speed steel after testing according to the "ball-disk" scheme. It can be seen from the figure that the wear intensity of nitrided samples varies significantly depending on the nitriding time. All nitrided samples show a significant reduction in wear rate compared to the original state. It was determined that the wear intensity of samples nitrided for 7 min is 5 orders of magnitude less than that of the original. Figure 4 shows changes in the wear intensity of samples of high-speed steels R6M5, R9 and R18 depending on the nitriding temperature at a constant process duration of 7 minutes. All nitrided samples show a significant decrease in wear intensity compared to the original. It is known that wear resistance is affected not only by microstructure and hardness, but also by surface roughness, which depends on the nitriding temperature.Thus, we can conclude that the wear intensity of high-speed steels depends on the nitriding

Figure 3. Wear intensity of samples of high-speed steels before and after nitriding at a temperature of 550° C

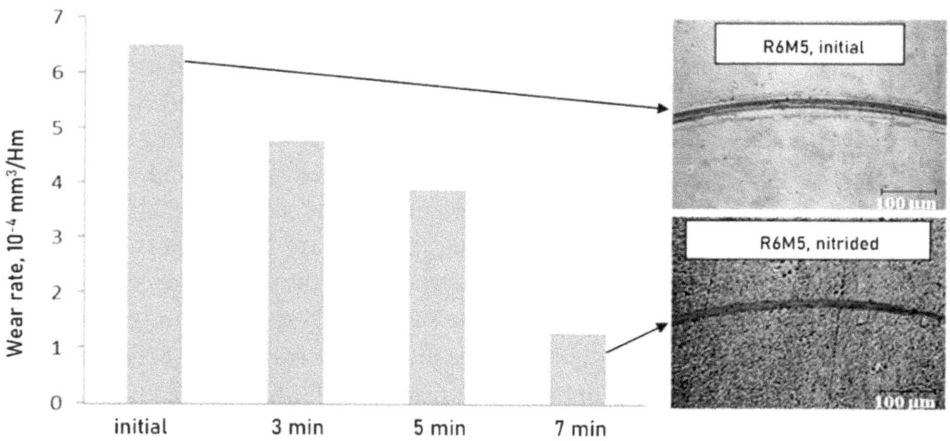

Figure 4. Dependence of wear intensity of high-speed steel specimens on nitriding temperature

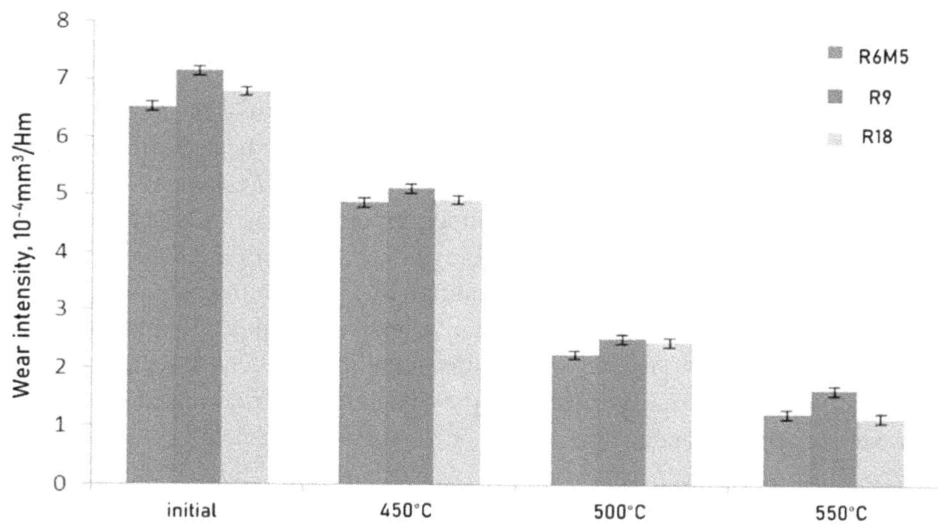

temperature. At a nitriding temperature of 550°C, the wear intensity of nitrided specimens of R9, R6M5, and R18 high-speed steels decreases by 77%, 81%, and 83%, respectively, which indicates a significant increase in the wear resistance of high-speed steels after surface nitriding (Geller, 1983; Rakhadilov and Skakov, 2012; Skakov et al., 2014). This may be due to with the formation of finely dispersed nitrides. Since, it is known (Badisch & Mitterer, 2003; Rakhadilov & Skakov, 2014; Ustinovshchikov, 1988) that dispersed precipitates, such as finely dispersed nitrides of alloying elements, protect the grain volume of a relatively soft matrix from abrasion.

As is known, in practice, the acting loads leading to wear are in most cases very complex and diverse, it is impossible to give a universal recipe for choosing steel, for the most optimal state of the structure and nitriding parameters. In this regard, in the work the tribological characteristics of samples of high-speed steels are also a certain method of scratch testing (method of controlled scratching), which is one of the modern methods of testing materials for wear resistance. The method is based on controlled scratching with a diamond indenter on a selected area of the sample (product).

Friction tests of R6M5 steel samples were carried out according to the procedure described in Section 3, with the translational movement of the sample along the X axis at a speed of 0.2 mm/s and a stroke length of 8 mm. In this case, the indenter tip moved over the sample surface with increasing load. Typical graphs of the friction coefficient, wear depth and contact load versus time (friction) of the translational motion of R6M5 steel specimens in the initial state and after nitriding are shown in Figure 5. It can be seen from the figure that the maximum depth of the scar on the surface of the R6M5 steel sample formed during friction is 9.4 μm before nitriding, and 4.4 μm after nitriding, which shows a significant increase in the hardness and wear resistance of the steel surface after nitriding. The graph shown in Figure 5 shows a decrease in the coefficient of friction after nitriding. The decrease in the friction coefficient is possibly associated with an increase in wear resistance (Chaus, 2008). Since, an increase in the wear resistance of the surface layer of steel during nitriding is manifested in a decrease in the coefficient of friction.

Thus, tests carried out by the scratch testing method showed that after nitriding, the friction coefficient decreases. However, this test method does not accurately determine the coefficient of friction, as by constantly increasing the load, the coefficient of friction is greater. In this regard, to study the coefficient of friction of nitrided samples, tests were carried out according to the "pin-disk" scheme on a tribometer TRIBO tester. Figure 6 shows the change in the coefficient of friction of the samples over time before and after nitriding at different temperatures. The figure shows that the coefficient of

Figure 5. Graphs of friction coefficient f, scar depth h and load P versus friction time τ:
a – before, b – after nitriding of R6M5 steel

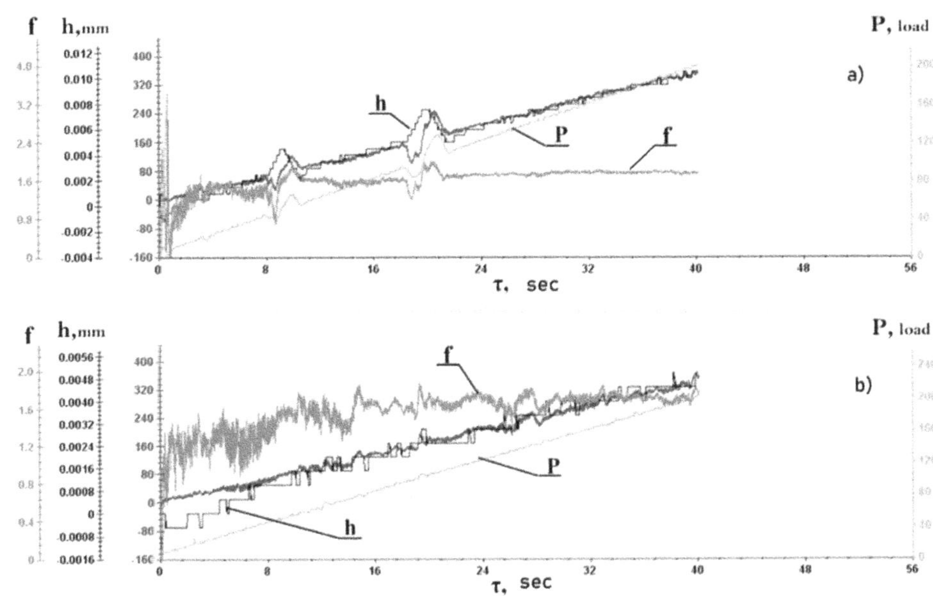

friction after nitriding decreases depending on the nitriding temperature. The friction coefficient of the original sample was 6.8±0.4. After electrolytic-plasma nitriding at 450 and 500°C, the friction coefficient decreases to 5.5±0.5 and 5.3±0.2, respectively. And after electrolytic-plasma nitriding at 550°C, there is a significant decrease in the coefficient of friction of the surface of the steel sample to 4.1 ± 0.2.

The decrease in the friction coefficient is explained by an increase in the wear resistance of the sample surface with the formation of particles of iron nitrides and alloying elements.As is known (Braun, Bushe, & Buyanovsky, 2001), with a decrease in the coefficient of friction during sliding (in the form of friction, which is most typical for the operation of most tools), their durability increases due to the less intense heat release during cutting and deformation, a decrease in the intensity of adhesion and sticking of the material being processed, especially soft, on the work surface. In addition, it is known (Belyy et al., 1991) that an increase in the amount and hardness of carbides and nitrides is accompanied by a more significant decrease in the friction coefficient. It is known (Belyy et al., 1991) that an important criterion for the wear resistance of nitrided steels is the high hardness of the surface layer. It is believed that with a higher hardness of the layer, its wear resistance increases. This principle underlies the choice of the chemical composition of steels and nitriding regimes.

At the same time, the analysis of works (Belyy et al., 1991; Braun, Bushe, & Buyanovsky, 2001; Chaus, 2008) devoted to the issues of friction and wear of various materials shows that it is not always necessary to strive for obtaining high hardness to ensure maximum wear resistance, since the structure of the material corresponding to maximum hardness and maximum wear resistance may be different. Considering that one of the most important properties of the nitrided layer, which largely affects wear resistance, is hardness, in this section, changes in the microhardness of the surface layer of high-speed steels during electrolytic-plasma nitriding are studied. Figure 7 shows the dependence of the microhardness of the surface layer of R6M5 steel on the nitriding temperature. Measurements of the microhardness of alloy samples were carried out on the PMT-3M device in accordance with GOST 9450-76, with loads on the indenter P = 1H (100g) and the holding time at this load of 10 seconds. One can see a significant increase in the microhardness of the surface of the steel samples after nitriding. In this case, the microhardness increases depending on the nitriding temperature. This is apparently due to the structural-phase state of the modified surface layer.

Figure 8 shows the distribution of microhardness over the depth of the nitrided layer of high-speed steels R6M5, R9 and R18 after nitriding at 550°C for 7 minutes. One can see a significant increase in microhardness near the surface of the treated steel samples. The nature of the transition zone has a smooth transition from the hardened layer to the base, while the microhardness of the base does not change significantly. After nitriding, the microhardness of the surface of steels R6M5, R9 and R18 increases up to 1.6 times.

According to the nature of the distribution of microhardness, it was revealed that electrolytic-plasma nitriding has an important advantage over the conventional gas nitriding process. Thus, during saturation in an electrolyte plasma, there is no sharp decrease in hardness, which is characteristic of ordinary saturation in ammonia. Since, during nitriding in ammonia, a sharp decrease in hardness is observed upon transition from the nitrided layer to the base (Lakhtin, & Silina, 1977). Figure 8 shows that the microhardness of the nitrided layer of steels R9, R6M5 and R18 differs from each other. This is due to the fact that alloying elements are present in these steels in various concentrations. The high hardness of the nitrided layer of steel R18 compared with steels R6M5 and R9 is due to the fact that alloying elements are present in greater amounts in steel R18. Since, the more alloying elements in the steel, the greater will be the hardness of the nitrided layer (Belyy et al., 1991). As a result of electrolytic-plasma nitriding

Figure 6. Changes in the coefficient of friction f of samples of R6M5 steel over time, before (a) and after nitriding at 450°C (b), 500°C (c) and 550°C (d) for 7 min

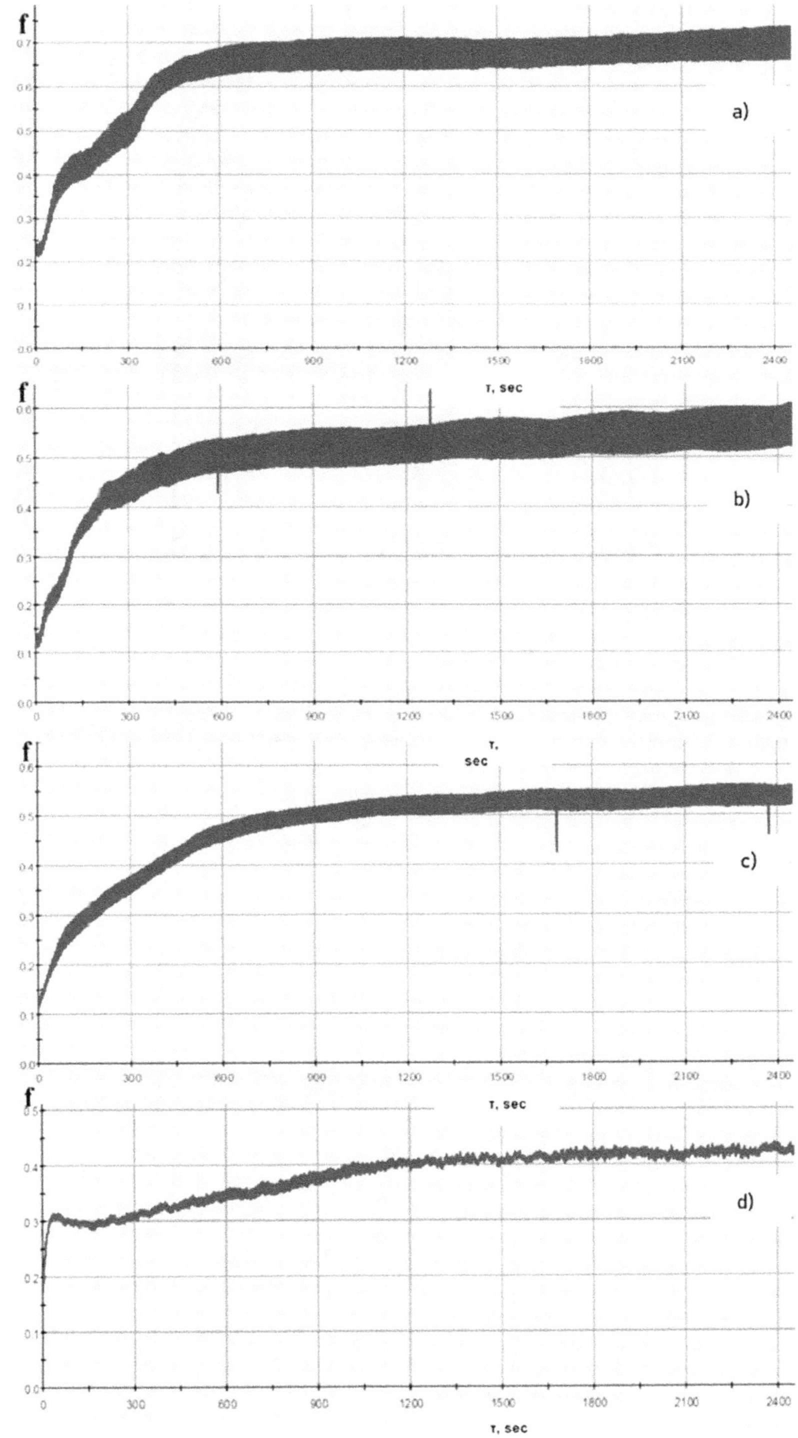

Figure 7. Dependence of the microhardness of the surface layer of R6M5 steel on the nitriding temperature

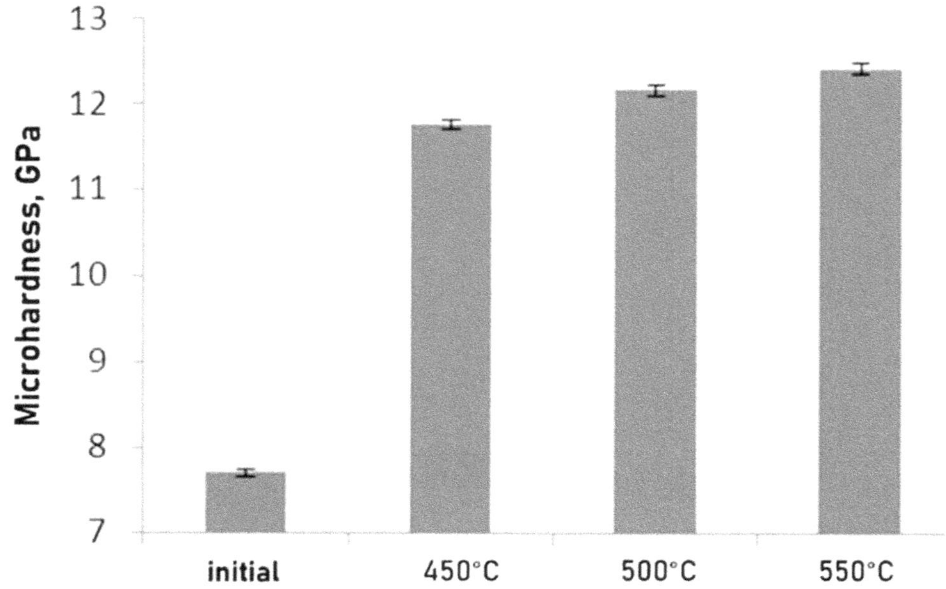

Figure 8. Distribution of microhardness over the depth of the nitrided layer of high-speed steels

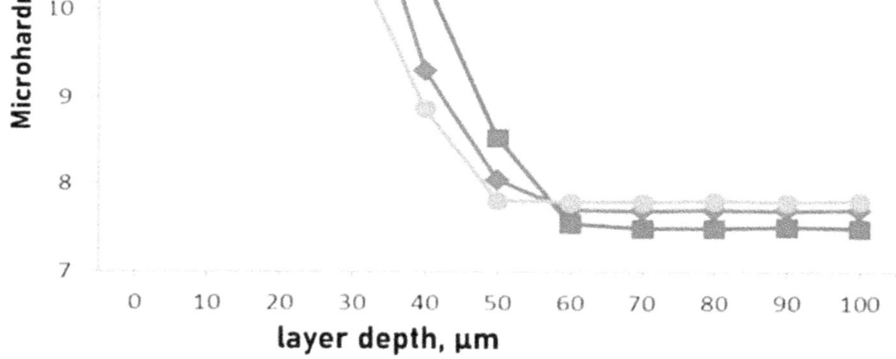

at 550°C, samples from high-speed steels, there was an increase in surface microhardness by 1.6 times. It has been experimentally established that the main increase in the microhardness of the diffusion layer occurs after nitriding at 550°C, which may be the result of the precipitation of finely dispersed, fully or partially coherent with the matrix, nitrides in the supersaturated α phase, which distort the crystal lattice of the matrix. Since it is known from (Belotskiy & Permyakov, 1972; Du et al., 2002; Kardonina et al., 2010; Lakhtin,, 1995) that the high hardness of the nitrided layer on steels alloyed with nitride-forming elements (Al, Cr, Mo, B, V) is the result of the total hardening effect:

a) the formation of nitrides of alloying elements in the process of nitriding;
b) separation and pre-isolation of dispersed nitrides of alloying elements in the process of intermediate γ®α-transformation;
c) isolation or pre-isolation of nitrides of alloying elements due to a decrease in solubility with decreasing temperature.

3.2 High-Temperature Testing of Nitrided Samples of High-Speed Steel for Wear Resistance

Modern mechanical engineering is characterized by difficult operating conditions for cutting tools associated with a high level of operating stresses, a wide temperature range, aggressive media, etc. Therefore, it is necessary to comply with special requirements for materials, in particular, high wear resistance at high temperatures and durability of the cutting tools from which they are made. In this regard, in this section, the wear resistance of nitrided samples of high-speed steel at high test temperatures is studied.

Figure 9. Dependence of the wear intensity of high-speed steel specimens on the test temperature

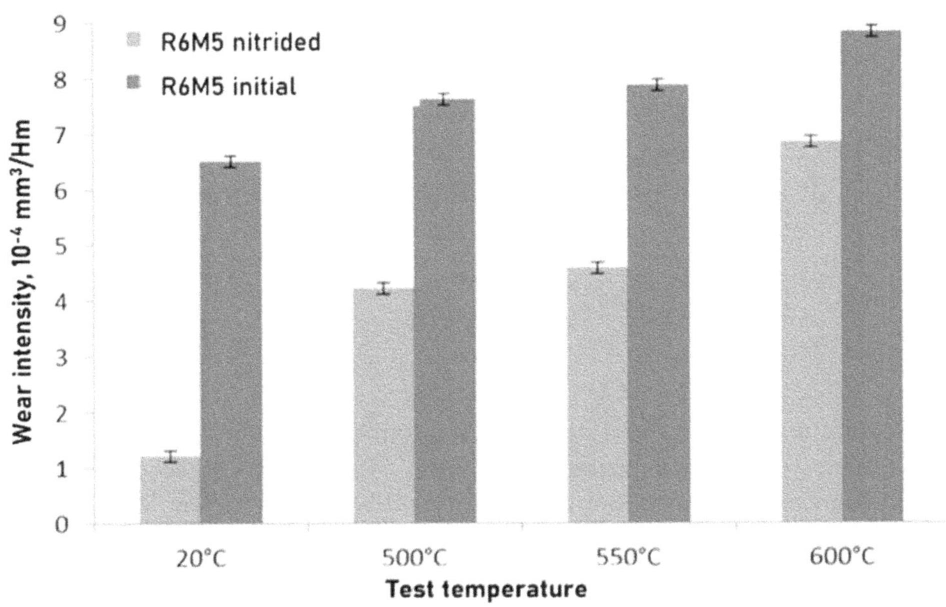

Figure 10. Graphs of friction coefficient f, counterbody penetration depth h, sample temperatures t1 and furnace t2 versus friction time τ during tests at temperatures of 20°C (a), 500°C (b), 550°C (c) of nitrided R6M5 steel specimens

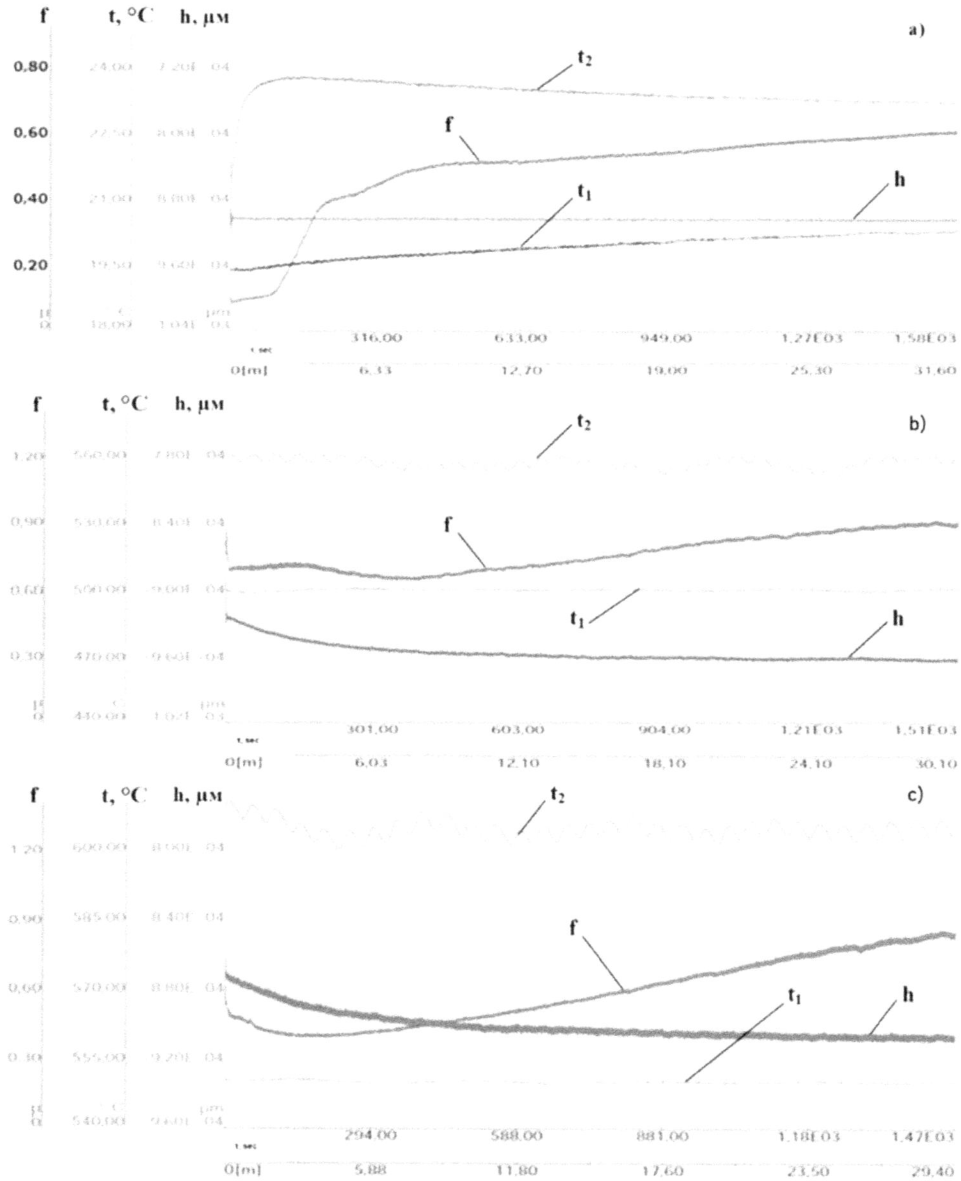

High-temperature tests were carried out according to the procedure described in section 2 of this work. The tests were carried out at temperatures of 500°C, 550°C and 600°C according to the "ball-disk" scheme. Wear resistance was evaluated by wear intensity and friction coefficient. The wear intensity values in each position are averaged over the results of testing three samples. Figure 9 shows the dependence of the wear intensity of the initial and nitrided samples of R6M5 steel on the test temperature. When tested at room temperature, the wear intensity of nitrided samples of R6M5 steel turned out to be significantly lower than that of non-nitrided samples. The wear intensity of nitrided specimens when tested at 500^{0}C and 550^{0}C averages $3.1*10^{-4}$ and $3.5*10^{-4}$ mm^3/Hm, and when tested at 600^{0}C - $6.7*10^{-4}$ mm^3/ Hm. At test temperatures of 500^{0}C and 550^{0}C, there is a noticeable increase in the wear intensity value for nitrided and non-nitrided samples. And at the test temperature at 600^{0}C, a more significant increase in the wear intensity value is observed. Perhaps this is due to the coagulation of carbides, which reduces hardness. Since it is known (Younesi, Bahrololoom, and Fooladfar, 2010; Grigor'yev, 2004; Badisch and Mitterer, 2003) that dispersed precipitates, such as finely dispersed nitrides of alloying elements, protect the grain volume of a relatively soft matrix from abrasion.

The results of tribological tests at high temperatures of R6M5 steel specimens that have undergone nitriding at a temperature of 550°C are shown in Figures 10 and 11. Figure 10 shows the curves of the change in the coefficient of friction, the depth of penetration of the counterbodies depending on the friction time during tests at temperatures of 20°C, 500°C and 550°C of nitrided samples of R6M5 steel. At high temperatures, an increase in the coefficient of friction is observed. However, changing the test temperature in the range from 500 to 550°C has practically no effect on the coefficient of friction. Figure 11 shows the curves of the coefficient of friction, the depth of penetration of the counterbodies, depending

Figure 11. Graphs of friction coefficient t, counterbody penetration depth h, sample temperatures t1 and furnace t2 versus friction time τ during testing at a temperature of 600°C of R6M5 steel specimens before (a) and after (b) nitriding

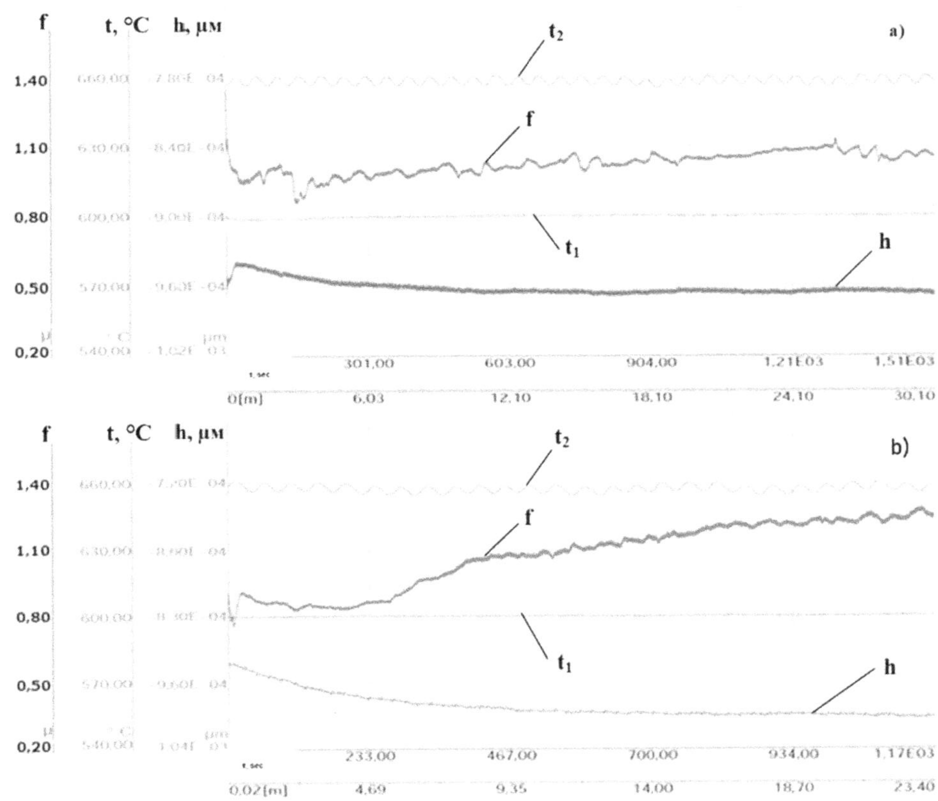

on the time of friction during tests at a temperature of 600°C of samples of steel R6M5 before and after nitriding. Increasing the test temperature to 600°C leads to a significant increase in the coefficient of friction of the original and nitrided samples compared to the samples tested at room temperature and at 500,550°C. At the same time, the coefficient of friction is of great importance during the entire testing process, remaining insufficiently high in magnitude. A decrease in wear resistance, as well as an increase in the friction coefficient at a test temperature of 600°C, may be associated with processes similar to the processes of coagulation or diffusion growth of coherent particles of the nitride phase. Thus, it is known from (Gol'dshmidt, 1971) that at temperatures above 550°C, diffusion processes develop on the surface layer of high-speed steel, as a result of which special carbides coagulate, which reduce hardness, as well as surface oxidation and decarburization.

Comparison of the wear intensity and wear coefficient of nitrided layers for different test temperatures (500°C, 550°C and 600°C) gives grounds to conclude that the wear resistance of nitrided steel varies depending on the ambient temperature, i.e. operating temperature, and the nitrided layer is more stable up to a temperature of 550°C.

3.3 Resistance of High-Speed Steels to Abrasive Wear Before and After Electrolytic-Plasma Nitriding

The mechanism of tool wear when cutting metals is complex and includes abrasive, adhesive and diffusion wear. The specific influence of each of them depends on the properties of the material, tool, workpiece, and processing conditions (Belotskiy et al., 1973). However, the material of the cutting tool during operation is more characterized by abrasive wear. In this regard, nitrided high-speed steels were tested for resistance to abrasive wear according to the "flat surface – rotating disk" scheme. Samples from high-speed steels were tested by abrasion of their surface with a rubber roller. The results of testing the samples for abrasive wear were characterized by the weight loss of the samples after the test. Figure 12 shows the dependences of the weight loss value of R6M5 steel samples before and after nitriding on the abrasion time. In order to investigate the wear resistance throughout the depth of the modified layer, the wear was evaluated every minute. In this case, after each minute, the sample was removed and weighed. It can be seen that the wear intensity of the nitrided samples is less than that of the original sample in the first and second minutes. And in the third minute, the wear intensity of nitrided and non-nitrided samples is equal. Further, the wear of all samples is almost the same, since the modified layer wears out after the 3rd minute, as experience has shown.

Thus, it is determined that in the surface modified layer, abrasive wear resistance is greater than in the base. It has been established that the surface modified layer of the nitrided sample has high wear resistance, and it decreases with the depth of the sample. Based on the studies of wear resistance in depth (every minute), it was determined that for nitrided steel samples, the optimal time for testing for abrasive wear is 3 minutes. Figure 13 shows the dependence of the mass loss value of R6M5 steel samples on the nitriding temperature. The tests were carried out for 3 minutes. It can be seen that the weight loss of the samples after nitriding at 500°C and 550°C is low compared to the samples nitrided at 450°C. Since after nitriding at 450°C, a modified layer with a small thickness is formed on the surface of R6M5 steel and Fe_4N nitrides are not formed, which provide high wear resistance. And the mass loss of steel samples nitrided at 500°C and 550°C is close, lies within the error. In this case, it can be seen that fine nitrides of alloying elements, which are formed only after nitriding at 550°C, do not play a special role in this

Figure 12. Dependence of the value of weight loss on the test time for samples of R6M5 steel nitrided at 550°C for 3, 5, and 7 min

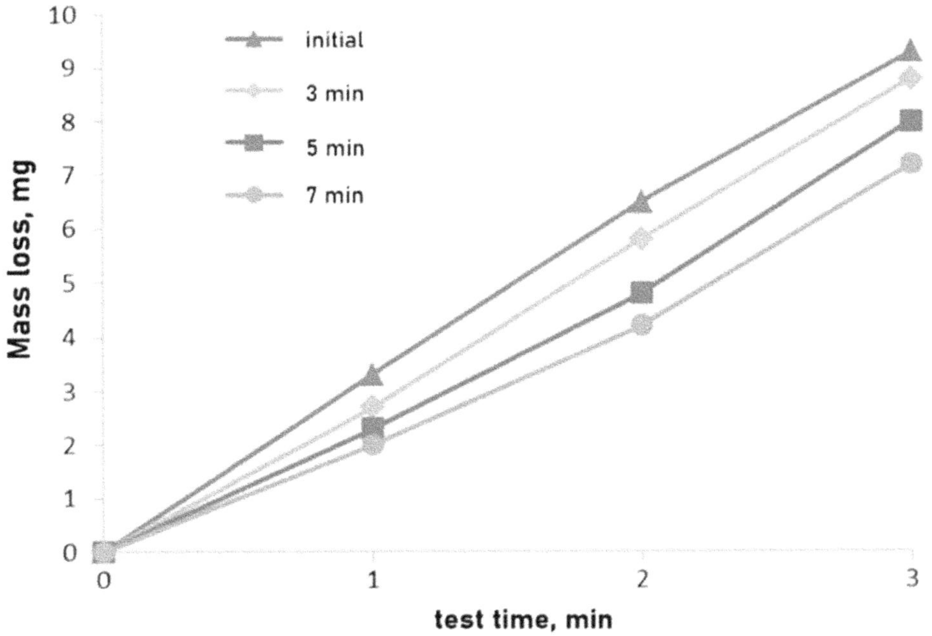

Figure 13. Dependence of the mass loss value of R6M5 steel samples on the nitriding temperature (nitriding was carried out for 7 min)

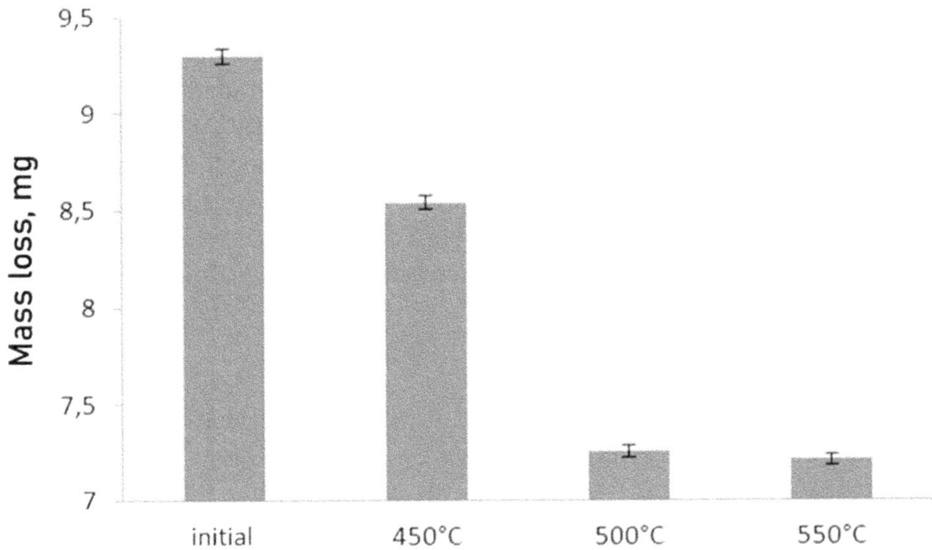

Figure 14. Resistance of samples of high-speed steels to abrasive wear

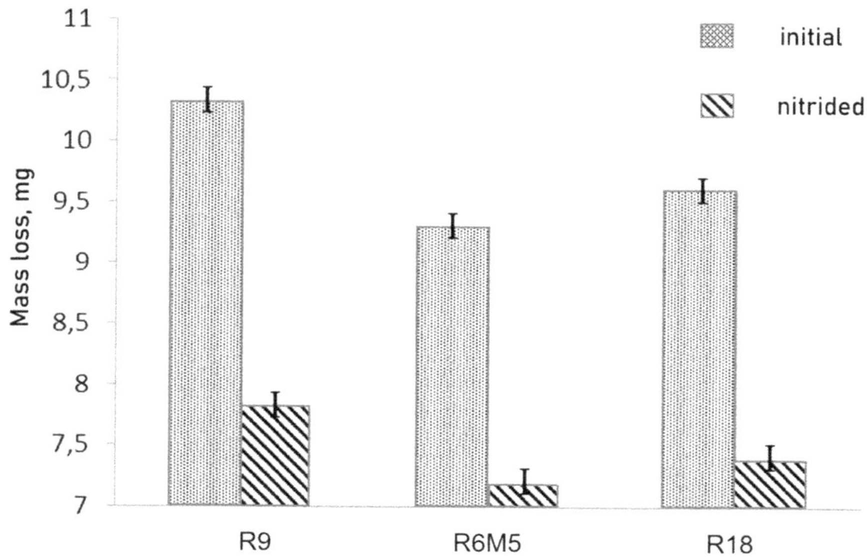

case. Perhaps this is due to the fact that the sizes of these nitrides are very small compared to abrasive particles, also due to their small formation depth.

Figure 14 shows the values of weight loss for samples of steels R9, R6M5 and R18 before and after nitriding at a temperature of 550°C, lasting 7 minutes. It can be seen that the mass loss of nitrided samples is less than that of non-nitrided samples, which indicates an increase in the abrasive wear resistance of high-speed steels after nitriding. According to the weight loss, the relative wear resistance of samples of high-speed steels was determined (Table 4).

Table 4 also shows that the abrasive wear resistance of nitrided high-speed steels turned out to be 1.3 times higher than the wear resistance of the same steels in the initial state. At the same time, the relative wear resistance of R6M5 and R18 steels is higher compared to R9 steel before and after nitriding. This is due to the fact that R9 steel is characterized by a smaller amount of hard carbides. It is known (Grigor'yev, 2004; Badisch and Mitterer, 2003; Gnyusov et al., 2008) that the nature, amount, distribution pattern, and size of carbides have a significant effect on the wear resistance of high-speed steels. Thus, it was found that the surface layer of the nitrided sample has a high resistance to abrasive wear.

Conclusions for Section 3

Thus, based on the analysis of the results of mechanical and tribological tests of high-speed steel samples that have undergone electrolytic-plasma nitriding, we can draw the following conclusions:

1. As a result of electrolytic-plasma nitriding at 550°C, samples of high-speed steels, there was an increase in surface microhardness by 1.6 times.

Table 4. Abrasion test results

N°	Sample	Relative Wear Resistance K_{wr}
1	R9, initial	1,84
2	R9, nitrided	2,43
3	R6M5, initial	2,04
4	R6M5, nitrided	2,64
5	R18, initial	1,98
6	R18, nitrided	2,57

2. It was determined that the wear intensity of nitrided samples of high speed steels R9, R6M5 and R18 is reduced by 77%, 81% and 83%, respectively, which indicates a significant increase in the wear resistance of high speed steels after electrolytic-plasma nitriding;
3. Scratch testing showed that after nitriding, the friction coefficient decreases, and the depth of the scar of the surface of the high-speed steel specimen formed during friction decreases, which shows a significant increase in the hardness and wear resistance of the steel surface after nitriding. Based on the pin-disk test, it was found that the coefficient of friction after nitriding decreases depending on the nitriding temperature. At the same time, there is a significant decrease in the coefficient of friction of samples nitrided at 550°C.
4. 4. It has been established that the wear resistance of nitrided steel varies depending on the ambient temperature, i.e. operating temperature, and the nitrided layer is more resistant to wear up to temperatures of 550°C, and an increase in temperature to 600°C leads to a decrease in wear resistance, as well as an increase in the coefficient of friction.
5. It has been established that after electrolytic-plasma nitriding, the resistance of high-speed steels to abrasive wear increases. The relative wear resistance of samples of high-speed steels after nitriding increases up to 1.3 times compared to the original samples

4. CHANGES IN THE STRUCTURAL-PHASE STATE OF HIGH-SPEED STEELS AS A RESULT OF ELECTROLYTE-PLASMA NITRIDING

This section is devoted to the study of changes in the structure, phase composition of the surface layers of high-speed steels R6M5, R9 and R18 as a result of electrolytic-plasma nitriding. As is known (Bakovets, Polyakov, & Dolgovesova, 1991; Parfenov et al., 2007; Petukhov, 2001), in the process of treatment with an electrolytic-plasma effect, significant changes occur in the structural-phase states of the material in thin surface layers due to the physical effect of low-temperature plasma ions and electric discharge. The processes of structural rearrangement developing in this case, structural phase transformations occur under conditions far from thermodynamically equilibrium states, and they make it possible to obtain modified surface layers with a unique set of physical and mechanical properties (Gupta, Tenhundfeld, Daigle, & Ryabkov, 2007). In this regard, the study of changes in the structural-phase state of high-speed steels as a result of electrolytic-plasma nitriding is of scientific and practical interest.

4.1 Structural-Phase States of High-Speed Steels in the Initial State

High cutting properties of high-speed steels (high heat resistance in the presence of high hardness and wear resistance) are achieved through special alloying and complex heat treatment, providing a certain phase composition (Suminov et al., 2005). The main alloying elements of high speed steels are carbon, tungsten, molybdenum, vanadium and chromium. These elements under certain temperature-time conditions form particles of the carbide phase in steel, which are the strengthening phase of the material. As a rule, in high-speed steels, the main carbide phase is a complexly alloyed carbide of the composition $(Fe,M)_6C$, where "M" are the elements that form the metal base of the carbide (W, Cr, Mo, V). According to the temperature range of formation, carbide particles are divided into a) primary, formed during the crystallization of steel from the melt; b) secondary, released during crystallization from austenite and c) tertiary, formed during the decomposition of martensite. The volume fraction of carbides is quite large and can reach up to 30% in high-speed steels, according to (Geller,, 1983). In addition, as the cutting speed increases, the requirements for the heat resistance of steel increase. The heat resistance of high-speed steels is due to alloying them with carbide-forming elements: tungsten, vanadium, molybdenum and chromium. These elements under certain temperature and time conditions form particles of the carbide phase in steel, which are the strengthening phase of the material (Moiseyev & Grigor'yev, n.d.). A high-temperature tool made of high-speed steels acquires high heat resistance after hardening and repeated tempering. Tempering after quenching within the temperature limits established for cutting tools leads to a decrease in the carbon content in martensite and the formation of ultramicroscopic carbides (Schastlivtsev et al., 1994). These carbides play an important role in the mechanical properties of steel, including hardness, wear resistance, and heat resistance (Liu, et al., n.d.).

In this regard, in this work, as the initial state of high-speed steels, we chose the state that the studied steels acquire after standard heat treatment, i.e. after quenching and three times tempering. As is known, the analysis of the microstructure and phase composition of steel after quenching followed by tempering makes it possible to evaluate its properties under operating conditions. At present, a large number of concepts have been developed, a number of patterns have been identified that relate the microstructure of a material to its physical and mechanical properties. However, the features of carbide phases, which play an important role in the mechanical properties of high speed steels, have been little studied. In this regard, in order to establish the main regularities of structural-phase transformations in high-speed steels during electrolytic-plasma nitriding, the structural-phase states of samples of high-speed steels R6M5, R9 and R18 in the initial state were previously studied. Figure 15 shows the microstructures of steels R6M5, R9 and R18 in the initial state, i.e. after standard heat treatment. The heat treatment mode is described in detail in the second section. It can be seen from the figure that the microstructures of steels R6M5, R9 and R18 are very similar to each other and consist of tempered martensite and special carbides (Figure 15). Previously, in (Maksakova et al.,2016; Skakov Rakhadilov, and Karipbaeva, 2013; Skakov and Rkhadilove, 2012), we found that martensite and special carbides are present in the steel structure after standard heat treatment. No residual austenite was observed in the matrix. The preliminary estimate was confirmed by X-ray diffraction phase analysis.

Figure 16 shows SEM images of the surface of steels R6M5, R9 and R18. The structure of steels consists of martensite and carbides. Carbide particles are evenly distributed in the matrix and are close to regular spherical shape. In the structure of high-speed steels R6M5 and R9, two types of carbides are observed: light and gray carbides (Figure 16 a, b, d). And in the structure of steel R18 there are only light carbides (Figure16 c). Figure 16d shows the microstructure of R6M5 steel obtained using a semi-

Figure 15. Microstructure of steels R6M5 (a), R9 (b), and R18 (c) in the initial state after standard heat treatment obtained using an optical microscope

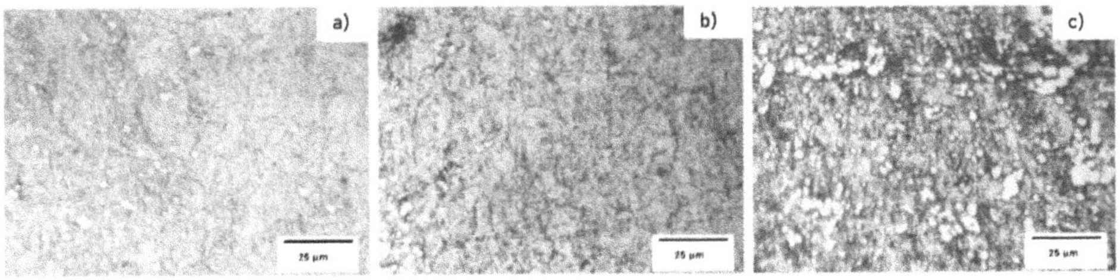

Figure 16. Surface microstructure of high-speed steels R6M5(a, d), R9 (b), R18 (c)

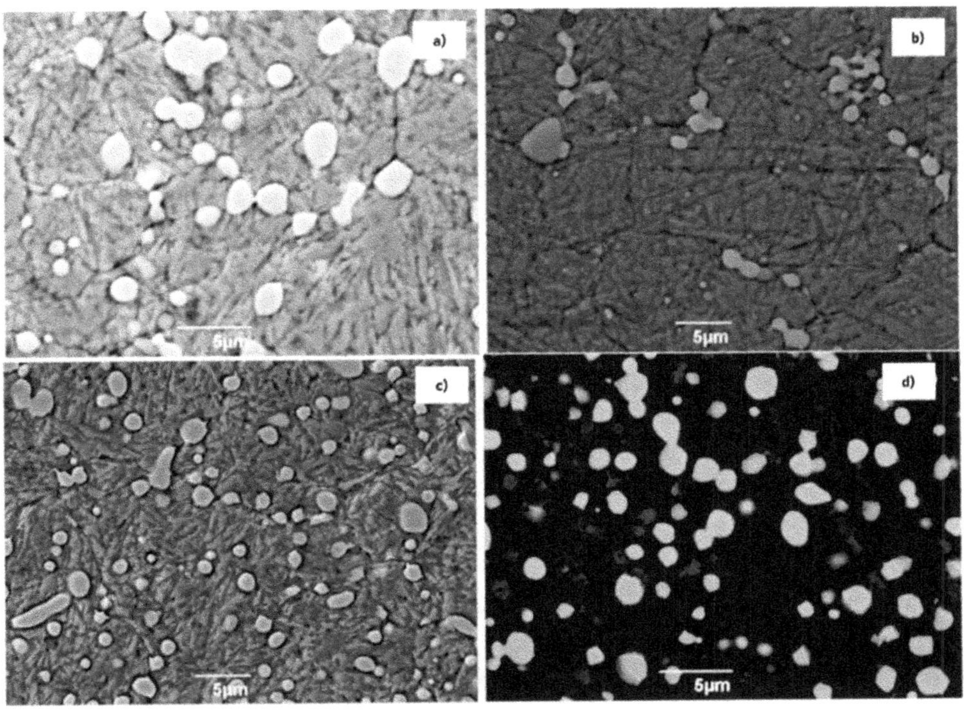

Table 5. Quantitative parameters of the structure of steels R6M5, R9, and R18

N°	Steel	Carbides	Volume Fraction	Average Particle Size
1	R6M5	Light colored carbides	10.4±0.6%	2,1 μm
		Dark carbides	2,3±0.4%	0,8 μm
2	R9	Light colored carbides	3.1±0.6%	1,6 μm
		Dark carbides	1,7±0.4%	1,8 μm
3	R18	Light colored carbides	13.4±0.6%	1,9 μm

conductor backscattered electron detector, giving surface patterns with chemical contrast display with high spatial resolution. Gray carbides are clearly visible from this image. Thus, it can be established that after standard heat treatment, two types of carbides are present in the structure of R6M5 and R9 steels: very bright colors of carbides (light carbides) containing elements above the atomic number and gray carbides containing elements below the atomic number, and in the steel structure R18 contains only light carbides. The volume fraction of each fraction was estimated. The sizes of carbide particles in the studied steels are also determined. The results of the quantitative parameters of the steel structure are shown in Table 5.

To reveal the composition of carbides and their distribution, a map of the distribution of alloying elements in the steel structure was obtained. The general nature of the distribution of alloying elements in the structure of steels R6M5, R9 and R18 is shown in Figures 17, 18 and 19, respectively. It can be seen from the figures that light spherical carbides are enriched in tungsten and molybdenum, and gray ones are enriched in vanadium. The obtained maps of the distribution of alloying elements confirmed that in the structure of steels R6M5 and R9 there are two types of carbides - light and dark, and in the structure of steel R18 only light. The absence of gray carbides, which are enriched with vanadium in the structure of steel R18, is possibly due to the fact that the content of vanadium in the composition of steel R18 is low compared to steels R6M5 and R9. In addition, Figure 19 shows that the bulk of vanadium is in light carbides in the structure of R18 steel. It should be kept in mind that V, W, Mo and Cr are carbide-forming elements. In other words, the carbides of these metals have a high binding energy and stability (Skakov & Rakhadilov, 2012; Skakov, Rakhadilov, Karipbayeva et al, 2014). That is why most of the alloying elements are in carbides, and not in solid solution.

To determine the elemental composition of particles of precipitated carbides and matrix (martensite), microprobe analysis was carried out (Figure 20). Table 6 shows the content of alloying elements in carbides and matrix for steels R6M5, R9 and R18.Tungsten, molybdenum, vanadium and chromium form special carbides in steel: M_6C based on tungsten and molybdenum, MC based on vanadium and $M_{23}C_6$ based on chromium. The results of mapping and microprobe analysis show that the structure of R6M5 steel after

Figure 17. Surface microstructure and distribution map of alloying elements of R6M5 steel in initial states

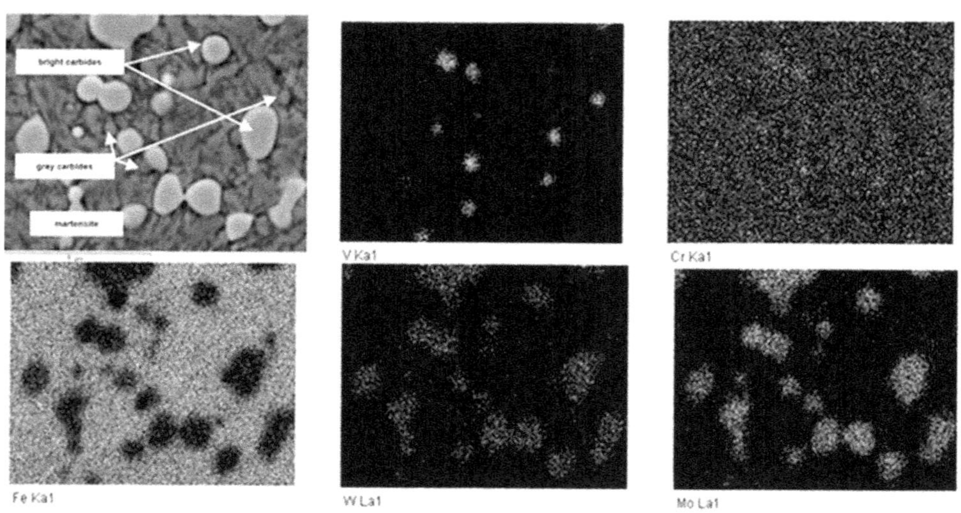

Figure 18. Surface microstructure and distribution map of alloying elements of R9 steel in initial states

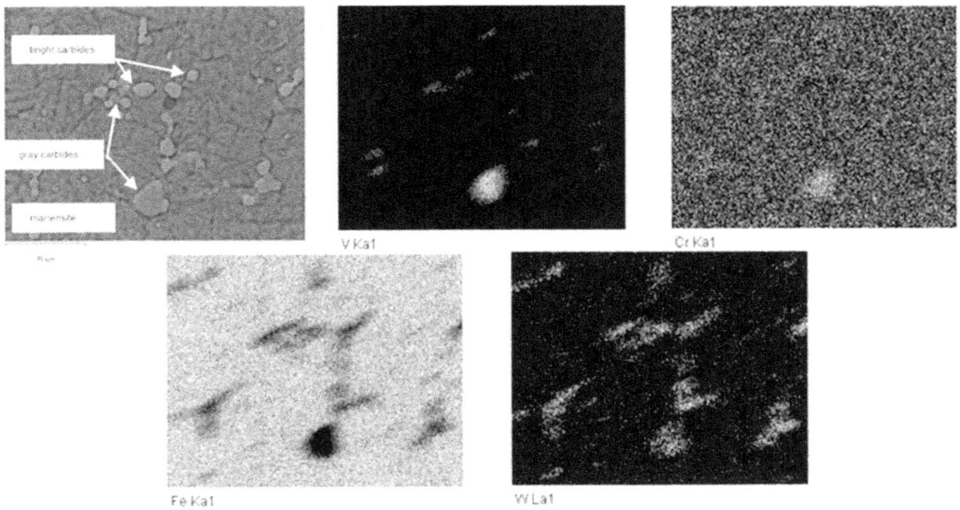

standard heat treatment contains M_6C and MC carbides and no $M_{23}C_6$ carbides, which is in good agreement with the literature data (Kremnev, 2008; Skakov, Rakhadilov, Karipbayeva et al, 2014). However, in some works (Guenzel et al., 2000; Vorob'yeva & Skladnova, 2003) it is stated that after standard heat treatment in the structure of R6M5 steel there are only particles of M_6C type carbide. Apparently, this is due to the small volume fraction of particles of MC type carbides and the similarity of these particles with the matrix, which makes it impossible to detect them. In addition, the methods used in these works have limitations in the detection of carbide particles with a low concentration. Therefore, in this work, along with X-ray diffraction phase analysis, special methods of scanning electron microscopy are used.

Figure 19. Surface microstructure and distribution map of alloying elements of steel R18 in initial states

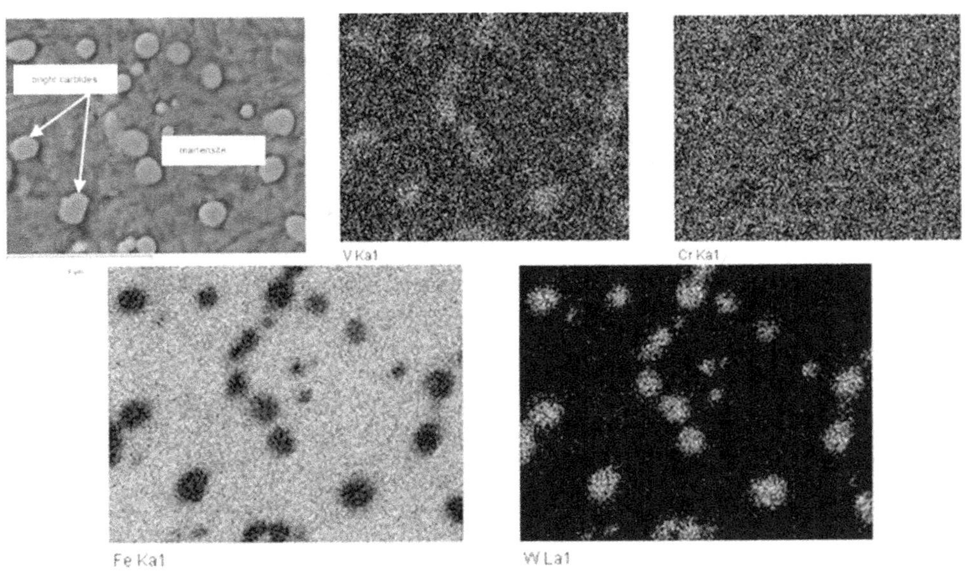

Table 6. The content of alloying elements in the structural components of high speed steels

Structural Components	Content of Elements, % (wt.)				
	V	Cr	Fe	Mo	W
R6M5					
Light colored carbides	3.42	3.31	30.85	26.05	36.37
Dark carbides	26.47	4.45	30.82	16.62	21.64
Martensite	1.33	4.62	84.50	4.25	5.30
R9					
Light colored carbides	3.15	3.52	28.83	3.10	61.4
Dark carbides	28.64	4.55	40.99	-	25.82
Martensite	3.11	4.95	80.78	-	11.16
R18					
Light colored carbides	3.50	3.80	34.77	2.70	55.23
Martensite	1.07	4.64	84.39	-	9.90

The proposed configuration of M_6C carbide is between the formulas $Fe_3(W,Mo)_3C$ - $Fe_3(W,Mo)_2C$ (Ivanov et al., 2002). In other words, along with tungsten and molybdenum atoms, M_6C carbide can contain up to 2/3 iron atoms of the total number of metal atoms. In addition, chromium and vanadium atoms can be dissolved, which replace iron atoms. According to the results of microprobe analysis, it can be assumed that gray carbide particles are vanadium-based MC carbides.

Figure 20 shows the diffraction patterns of steels R6M5, R9 and R18. X-ray diffraction analysis showed that in the initial state, i.e. after heat treatment, in the structure of steels R6M5 and R9 there is an α-phase and carbides M_6C, MC, and in the structure of steel R18 only M_6C carbides.

Thus, X-ray diffraction analysis confirmed that the main carbides in the studied steel are M_6C and MC carbides. It has been determined that M_6C type carbides, which have a complex FCC crystal lattice and Fd_3m space group, correspond to the Fe_3W_3C composition, and MC type carbides, which have a cubic crystal lattice and Fm_3m space group, correspond to the VC composition. In this case, it should be borne in mind that the M_6C type carbide can have the form of both Fe_3W_3C and Fe_3Mo_3C. As is known, one of the advantages of X-ray diffraction analysis (XRD) is that the peak position and lattice parameters can be determined quite accurately. Nevertheless, in this case, when studying individual carbides, it is advisable to apply the EBSD analysis. Therefore, to confirm the obtained results of X-ray diffraction analysis, the crystal structure of M_6C and MC carbides was studied by EBSD analysis using a reflected electron detector on a scanning microscope (system) with electron and focused ion beams. Figure 21 shows the results of an EBSD analysis of the surface of R6M5 steel. The EBSD analysis showed that tungsten-rich M_6C carbides are most optimally combined with the Fe_3W_3C cubic phase, while MC-type carbides correspond to the VC phase. However, it is worth noting that in this case Fe_3W_3C may also mean that other carbide-forming elements are present in the form of M_6C carbides.

Thus, we have characterized the structures and carbide phases of R6M5, R9, and R18 high-speed steels in the initial state, i.e., after standard heat treatment. The study of structural-phase states before a certain treatment is necessary in terms of identifying patterns of structure changes and its effect on

Figure 20. X-ray diffraction patterns of high-speed steels in the initial state (after heat treatment)

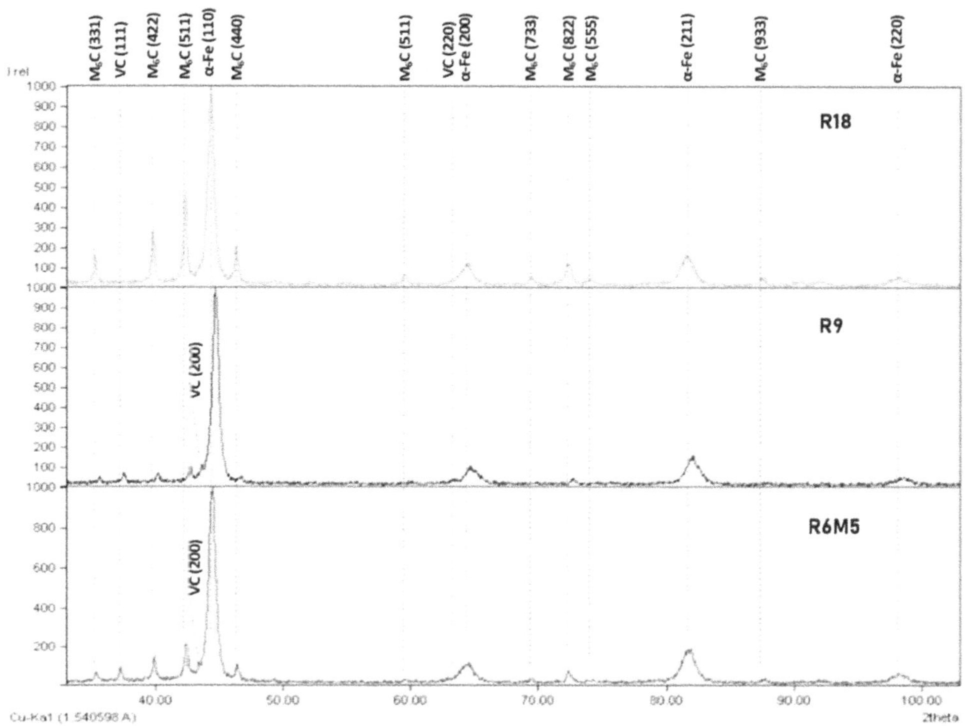

Figure 21. Results of EBSD analysis of steel R6M5

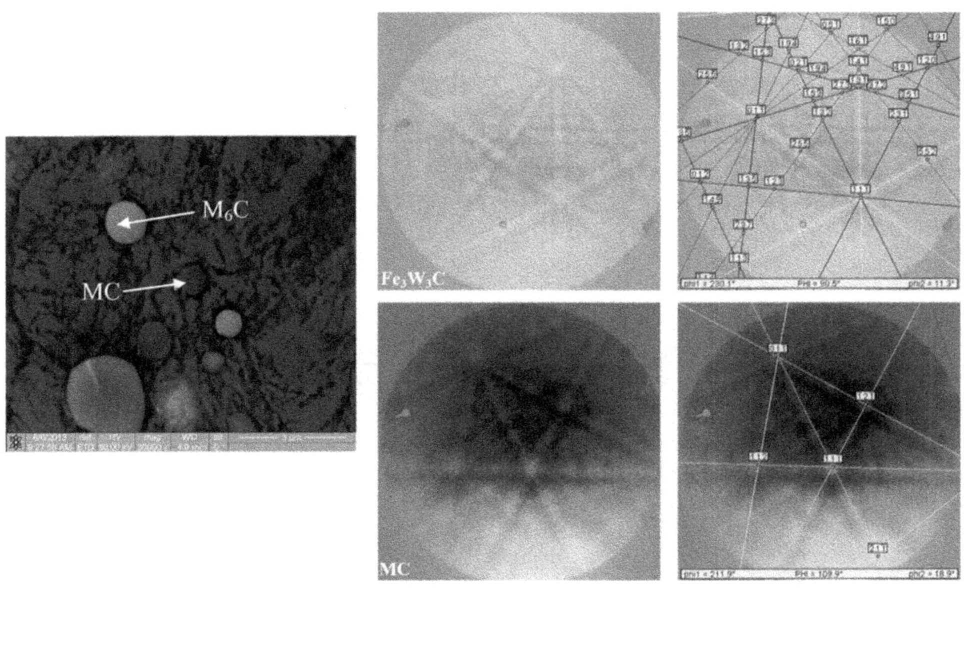

properties. Since the physical and mechanical properties of high-speed steel are largely determined by the structure and state of the carbide phases and their shape, size, distribution in volume.

4.2 Study of the Influence of Electrolytic-Plasma Nitriding on the Microstructure of High-Speed Steels

Figure 22 shows the microstructures of the modified surface layer of samples of high-speed steels R6M5, R9 and R18, nitrided for 7 minutes. It can be seen that after nitriding, a dark-etched nitrided layer is observed on the surface, which is a zone of internal nitriding. At the same time, the dark-etched zone smoothly transitions into the base. After nitriding at temperatures of 450°C, 500°C and 550°C, the thickness of the nitrided layer of R6M5 steel is 25–40 µm and increases with increasing nitriding temperature. And the thickness of the nitrided layer of steels R9 and R18 is on average 45 µm, 35 µm, respectively, after nitriding at 550°C. The thickness of the diffusion layer depends on: the composition of the treated metal (alloy), saturation temperature, duration of the process, the concentration of the diffusing element on the surface, and the nature of the formed solid solution. The higher the concentration of the diffusing element on the surface, the greater the layer thickness at a given temperature and the duration of the saturation process.

Thus, by metallographic analysis it has been established that during nitriding by electrolytic-plasma action, the process of diffusion saturation of steel with nitrogen proceeds with accelerated formation of a modified layer, consisting only of a well-developed zone of internal nitriding, i.e. diffusion layer. In this case, the diffusion layer is nitrogenous martensite, since the initial structure of the steel consisted of tempered martensite. The formation of a diffusion layer without a continuous nitride layer on high-speed steels makes it possible, on the one hand, to increase the hardness and wear resistance of the tool, and, on the other hand, to avoid high stresses leading to increased brittleness of the surface layer. Since it is known (Gol'dshmidt, 1971) that the formation of a nitride zone and an internal nitriding zone on the surface of nitrided steel, which have a larger specific volume than the core, leads to the appearance of large residual compressive stresses in the diffusion layer, and tensile stresses in the core. The dependence

Figure 22. Microstructure of the modified layer of high-speed steels after nitriding a) R6M5, at 450°C; b) R6M5, at 500°C; c) R6M5, at 550°C; d) R9, at 550°C; e) R18, at 550°C

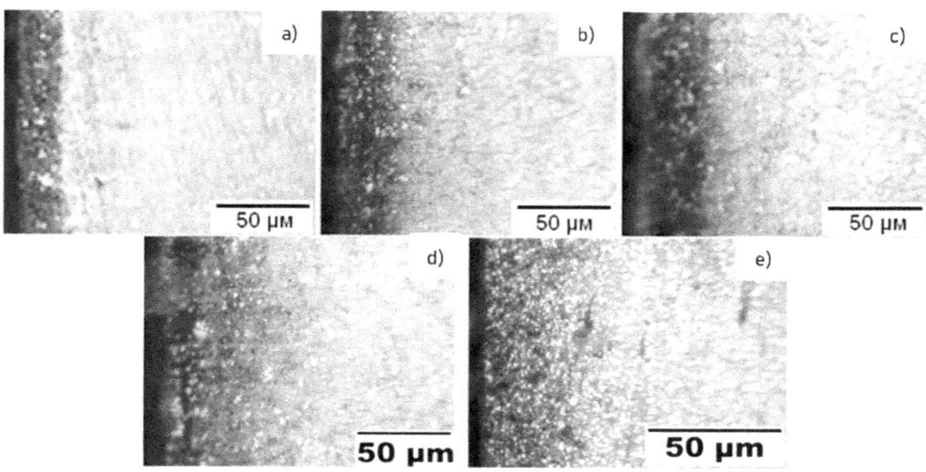

Figure 23. Growth kinetics of nitrided layers of R6M5 steel at different temperatures depending on the holding time, 1 -450°C; 2 -500°C; 3 -550°C

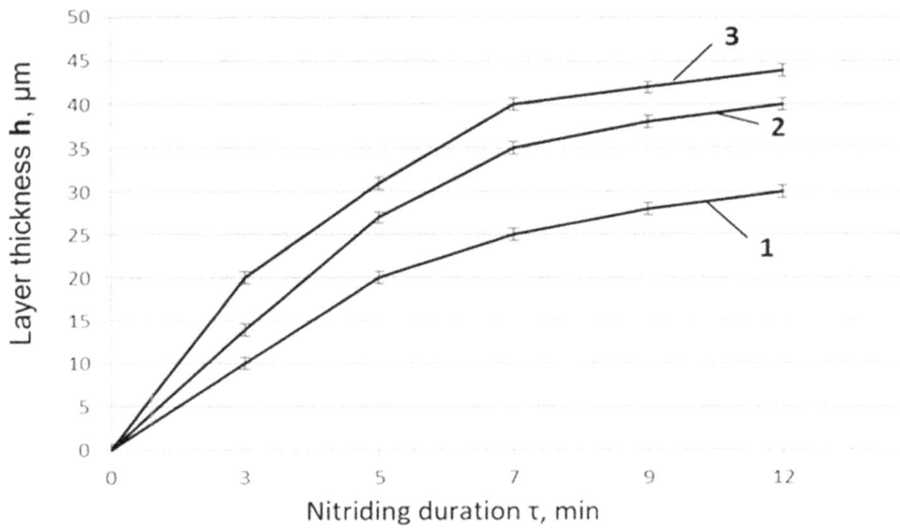

Figure 24. Microstructure of R6M5 high-speed steel before (a) and after electrolytic-plasma nitriding at temperatures of 450°C (b), 500°C (c), 550°C (d)

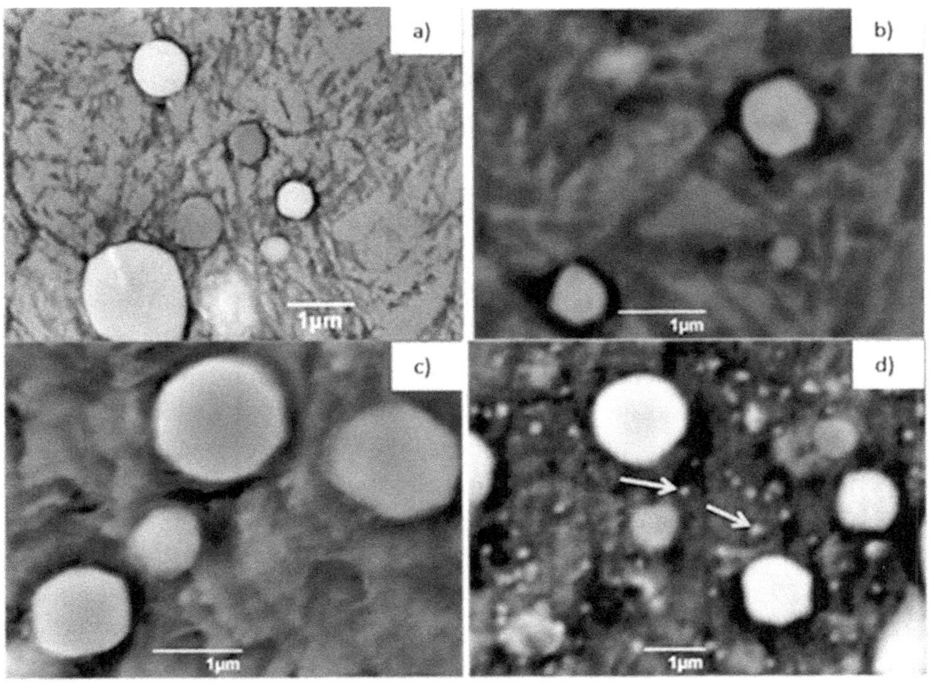

Figure 25. Microstructure of high-speed steel R9 before (a) and after (b) electrolytic-plasma nitriding at temperatures of 550°C

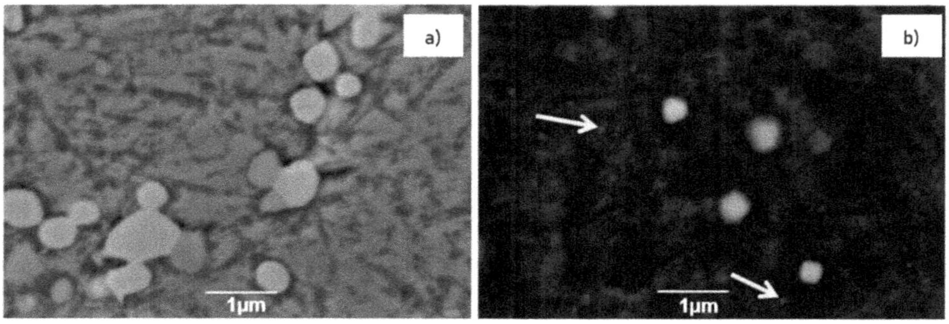

of the thickness of the diffusion layer h of the R6M5 steel on the time of its exposure at various nitriding temperatures was obtained (Figure 23). Figure 23 shows that the thickness of the modified layer increases with increasing nitriding temperature and holding time, which makes it possible to obtain a modified layer of a given thickness by varying the temperature and nitriding time. According to literature sources (Goldshmidt, 1971; Artinger, 1982; Lakhtin, 2009), for a cutting tool, the value of h should not exceed 50–60 µm. Therefore, the considered modes of nitriding are quite suitable for hardening cutting tools. It has been determined that the optimal time for electrolytic-plasma nitriding for high-speed steel is 7 minutes. In this regard, only samples nitrided with a duration of 7 minutes were studied further.

Figures 24, 25, 26 show SEM images of the surface of steels R6M5, R9 and R18 before and after nitriding with a duration of 7 minutes. The results of scanning electron microscopy showed that after electrolytic-plasma nitriding at 550°C, fine particles are formed on the surface of R6M5 steel (Figure 24 d, fine particles are marked by arrows). And when nitriding at 450°C and 500°C, they are not observed (Figure 24 b, c).

In the structure of steels R9 and R18 after nitriding at 550°C, fine particles are also observed (Figures 25 b, 26 b, fine particles are indicated by arrows). It is assumed that these particles are finely dispersed nitrides of alloying elements, in particular, chromium. Since alloying elements, such as tungsten, molybdenum and vanadium, which are part of high-speed steels, are 95% bound into carbides, it can be assumed that the main share of the formed nitrides of alloying elements falls on chromium nitrides. Since in all grades of high-speed steels considered in this work, the chromium content is the same and amounts to ~ 4%. The formation of fine nitrides of alloying elements at a temperature of 550°C is possible due to the fact that this temperature corresponds to the tempering temperature of this steel. Since, during the tempering of R6M5 steel at a temperature of 550–560°C, precipitation hardening occurs as a result of the partial decomposition of martensite and the precipitation of finely dispersed inclusions of strengthening phases (Arshinger,, 1982).

Figures 27, 28 and 29 show the results of surface mapping of R6M5 steel after nitriding at 450°C, 500°C and 550°C. The mapping results showed that after nitriding, the surface of R6M5 steel is saturated with nitrogen. In this case, after nitriding at 500°C and 550°C, areas are observed in the surface where the nitrogen content is higher. These areas are most consistent with the matrix. However, mapping does not allow one to accurately determine which structural component is more saturated with nitrogen, i.e. where the nitrides are located. Therefore, to identify the nitrogen content in the structural components and

Figure 26. Microstructure of high-speed steel R18 before (a) and after (b) electrolytic-plasma nitriding at temperatures of 550°C

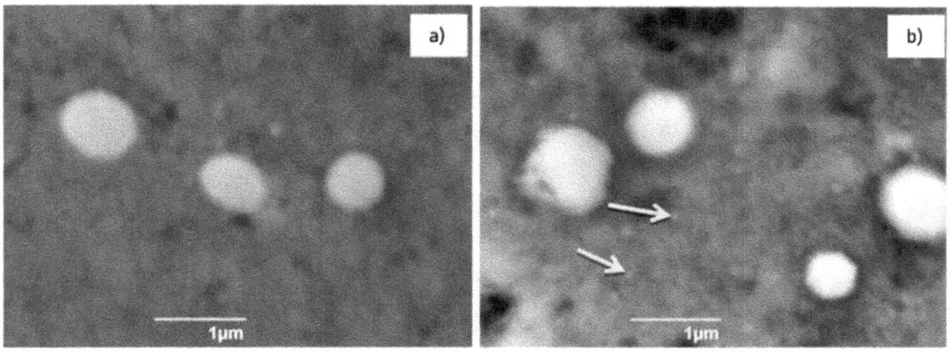

Figure 27. Surface microstructure and distribution map of alloying elements of R6M5 steel after nitriding at 450°C

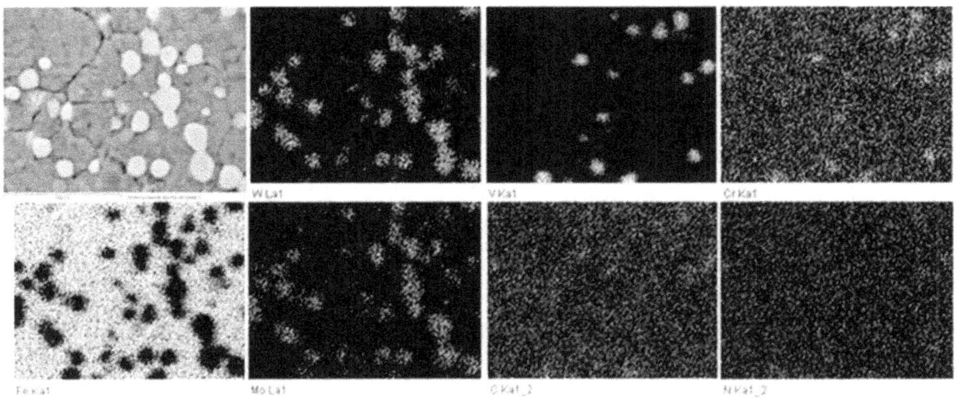

Figure 28. Surface microstructure and distribution map of alloying elements of R6M5 steel after nitriding at 500°C

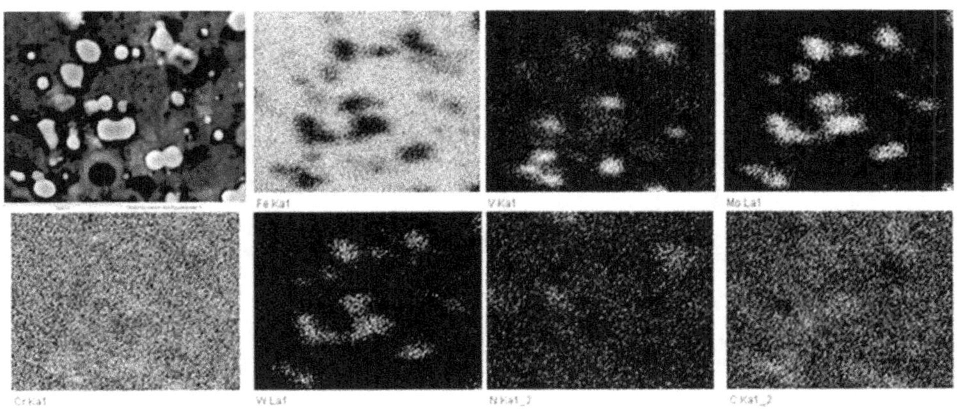

Figure 29. Surface microstructure and distribution map of alloying elements of R6M5 steel after nitriding at 550°C

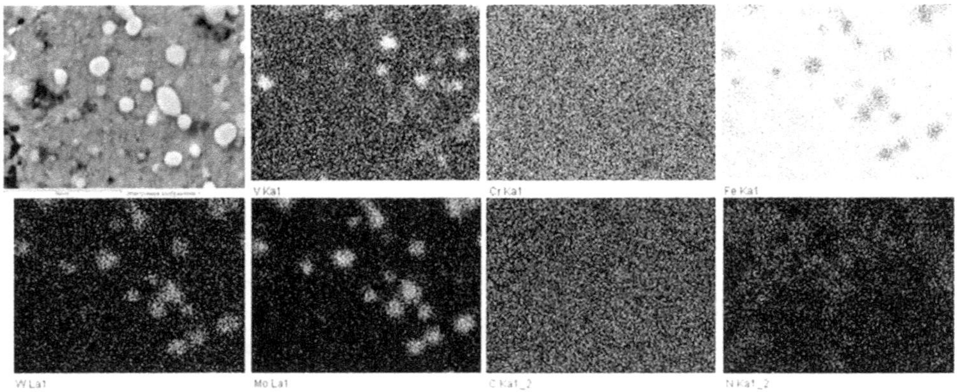

changes in the chemical composition of the surface as a whole, it is necessary to conduct a quantitative energy dispersive analysis of the surface. In order to identify changes in the elemental composition of the surface after nitriding, an energy dispersive analysis of the surface of R6M5 steel samples nitrided at 450°C, 500°C, and 550°C was carried out. Figures 30, 31 and 32 show the results of the energy dispersive analysis.

The analysis showed that nitrogen is observed only in the matrix, as expected from the mapping results. The nitrogen content in the surface of R6M5 steel reaches up to 7% (in weight%). The results of mapping and energy dispersive analysis showed a more uniform distribution of nitrogen over the matrix of the nitrided layer. The observed phenomenon of uniform adsorption and uniform growth of the nitrided layer should be explained by the appearance of a special defect substructure in the grain volume (dislocation-disclination structure of grains in the near-surface zone). The resulting defectiveness, apparently, approaches the defectiveness of the grain boundaries. Since, electrolytic-plasma nitriding is carried out under conditions of excessive excitation of the metal surface and subsurface layers. An

Figure 30. SEM image of R6M5 steel after nitriding at 450°C (a), characteristic radiation spectra (b), and microanalysis results, at% (c)

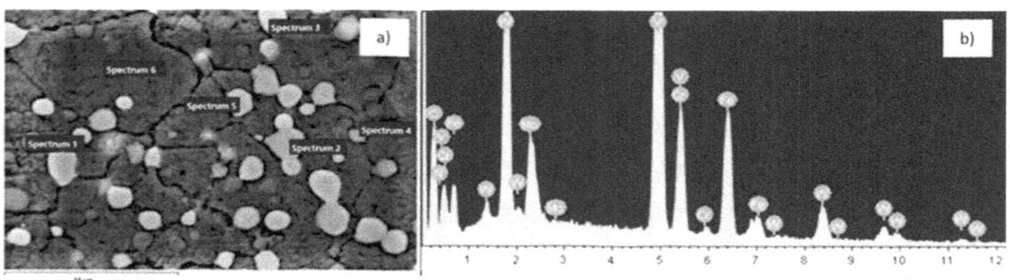

Spectrum	N	V	Cr	Fe	Mo	W	Total
Spectrum 1	0.00	39.66	5.78	17.06	9.87	27.63	100.00
Spectrum 2	0.00	2.31	3.40	26.73	13.74	53.82	100.00
Spectrum 3	0.00	2.42	3.25	26.60	13.55	54.18	100.00
Spectrum 4	0.06	4.06	4.69	78.71	3.33	9.15	100.00
Spectrum 5	0.12	12.33	5.05	65.71	3.91	12.88	100.00
Spectrum 6	0.40	1.03	4.44	85.09	1.97	7.07	100.00

c)

Figure 31. SEM image of R6M5 steel after nitriding at 500°C (a), characteristic radiation spectra (b), and microanalysis results at% (c)

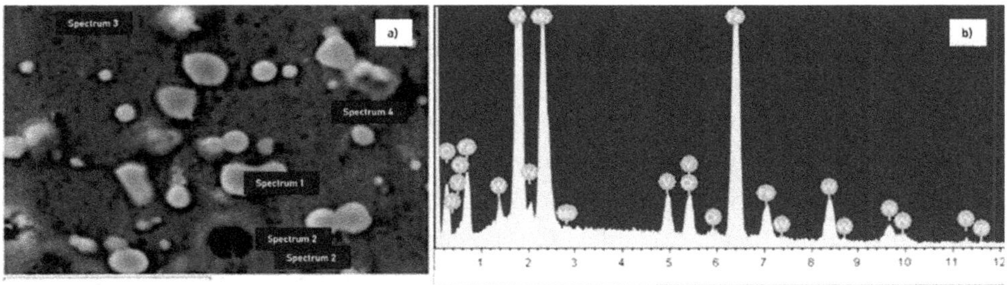

Spectrum	N	V	Cr	Fe	Mo	W	Total
Spectrum 1	0.00	3.25	3.51	32.04	25.91	35.29	100.00
Spectrum 2	6.36	9.86	6.14	62.59	7.12	7.93	100.00
Spectrum 3	5.87	1.45	4.70	80.41	3.60	3.97	100.00
Spectrum 4	6.00	1.56	4.91	78.45	3.97	5.11	100.00

c)

Figure 32. SEM image of R6M5 steel after nitriding at 550°C (a), characteristic radiation spectra (b), and microanalysis results (c) at% (c)

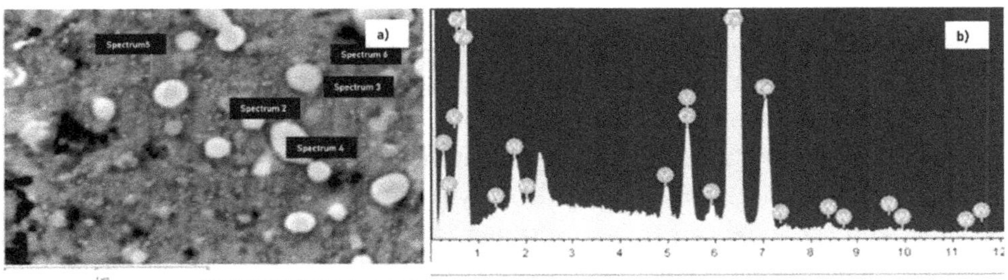

Spectrum	N	V	Cr	Fe	Mo	W	Total
Spectrum 2	0.14	10.46	5.19	72.12	5.29	6.80	100.00
Spectrum 3	7.08	1.82	5.12	77.05	4.47	4.46	100.00
Spectrum 4	0.00	1.47	4.53	33.93	24.26	35.81	100.00
Spectrum 5	5.91	2.47	5.12	78.02	3.94	4.54	100.00
Spectrum 6	3.58	2.59	4.78	81.23	3.21	4.61	100.00

c)

important factor influencing the growth rate of the nitrided layer and its structure during this process is the exclusion of the predominant role of boundary diffusion. Plasma, accelerating the directed mass transfer of ions to the surface of the sample, creates conditions for the uniform adsorption of nitrogen atoms over the entire surface of the metal, and not selectively along the grain boundaries, as is observed during conventional nitriding by traditional methods. Thus, the conducted studies have shown that after electrolytic-plasma nitriding, a modified layer is formed on the surface layer of high-speed steels, consisting of a diffusion layer, which is nitrogenous martensite with finely dispersed nitrides of alloying elements. It is determined that the surface is saturated with nitrogen up to 7% (in weight%). Nevertheless, in order to verify the results obtained and to establish the regularities of the formation of the modified layer, it is necessary to study the phase transformations in the surface layers of high-speed steel during nitriding by the X-ray diffraction analysis.

4.3 Phase Transformations in the Surface Layers of High-Speed Steels During Electrolytic-Plasma Nitriding

One of the most effective ways to improve the physical and mechanical properties of the surface layer of high-speed steels is the development of optimal nitriding modes. It is practically impossible to choose the necessary nitriding regimes without a detailed study of phase transformations in the material during nitriding. Figure 33 shows X-ray diffraction patterns of R6M5 steel before and after nitriding. X-ray diffraction analysis showed that in the initial state, that is, after standard heat treatment, the R6M5 steel structure contains martensite (α-phase) and M_6C, MC carbides. After nitriding, broadening, a decrease in intensity, and a shift towards smaller Bragg angles of the interference line (110) of the α-phase are observed, which indicates the formation of a solid solution of nitrogen in iron, i.e. internal nitriding zone. On the diffraction patterns of R6M5 steel samples nitrided at 500°C and 550°C, interference lines of the Fe_4N phase were found.

Figures 34 and 35 show X-ray diffraction patterns of R9 and R18 high-speed steels. X-ray diffraction analysis showed that in the initial state, that is, after standard heat treatment, the structure of R9 steel contains martensite (α-phase) and carbides M_6C, MC. And in the structure of steel R18 there are - martensite (α-phase) and M_6C carbides. After nitriding, interference lines of the Fe_4N phase were found on the diffraction patterns (Figure 34 b, 35 b). Thus, it can be established that the diffusion layer of high-speed steels R6M5, R9 and R18 consists of an α-phase with excess nitrides - Fe_4N, formed during cooling in the surface layer. Since metallographic analysis showed that after nitriding, a modified layer is formed on the surface of high-speed steels, consisting only of a diffusion layer. The results of X-ray diffraction analysis are shown in Table 7. The analysis showed that the main carbides in the studied steels are M_6C and MC carbides. It has been determined that M_6C type carbides are most optimally combined with the Fe_3W_3C phase, and MC type carbides correspond to the VC phase. Quantitative analysis showed that after nitriding, there is no noticeable change in the content of carbide phases. At the same time, it was determined that the content of the γ'-phase (Fe_4N) in steels R9, R6M5 and R18 is 5.5%, 3.9% and 3.7%, respectively.

Using the method of scanning electron microscopy, it was found that finely dispersed nitrides of alloying elements are formed in the surface layer of steels after nitriding. However, X-ray diffraction studies did not reveal nitride phases of alloying elements, possibly due to their low concentration and small size. Perhaps this is also due to the formation of finely dispersed nitrides, the distribution of which over the layer does not ensure their detection with the existing sensitivity of X-ray diffraction phase analysis. In this case, to study these particles, it is necessary to apply the method of transmission electron microscopy. Thus, the main advantage of nitriding by electrolytic-plasma action is the possibility of obtaining a diffusion hardened layer from the α-phase with excess Fe_4N nitrides (γ'-phase), in contrast to gas nitriding in ammonia, where a nitride layer is formed, consisting of two γ'- and ε-phases, which is a source of internal stresses at the phase boundary and causes brittleness and flaking of the hardened layer of cutting tools during operation (Arzamasova, & Mukhina,, 2001; Kuksenova et al., 2004; Skakov, Rakhadilov, & Scheffler, 2013; Skakov, Rakhadilov, & Rakhadilov, 2014). The formation of a diffusion layer of nitrogenous martensite in the surface layers will have a positive effect on the performance properties of a cutting tool made of R6M5 high-speed steel. For example, iron nitrides have a higher heat capacity than iron (Skakov, Rakhadilov, Batyrbekov, & Manapbayeva, 2014). This creates favorable conditions for preventing temperature flashes on the surface of the cutting tool.

Figure 33. XRD patterns of samples of R6M5 high-speed steel before (a) and after nitriding at temperatures of 450°C (b), 500°C (c), and 550°C (d)

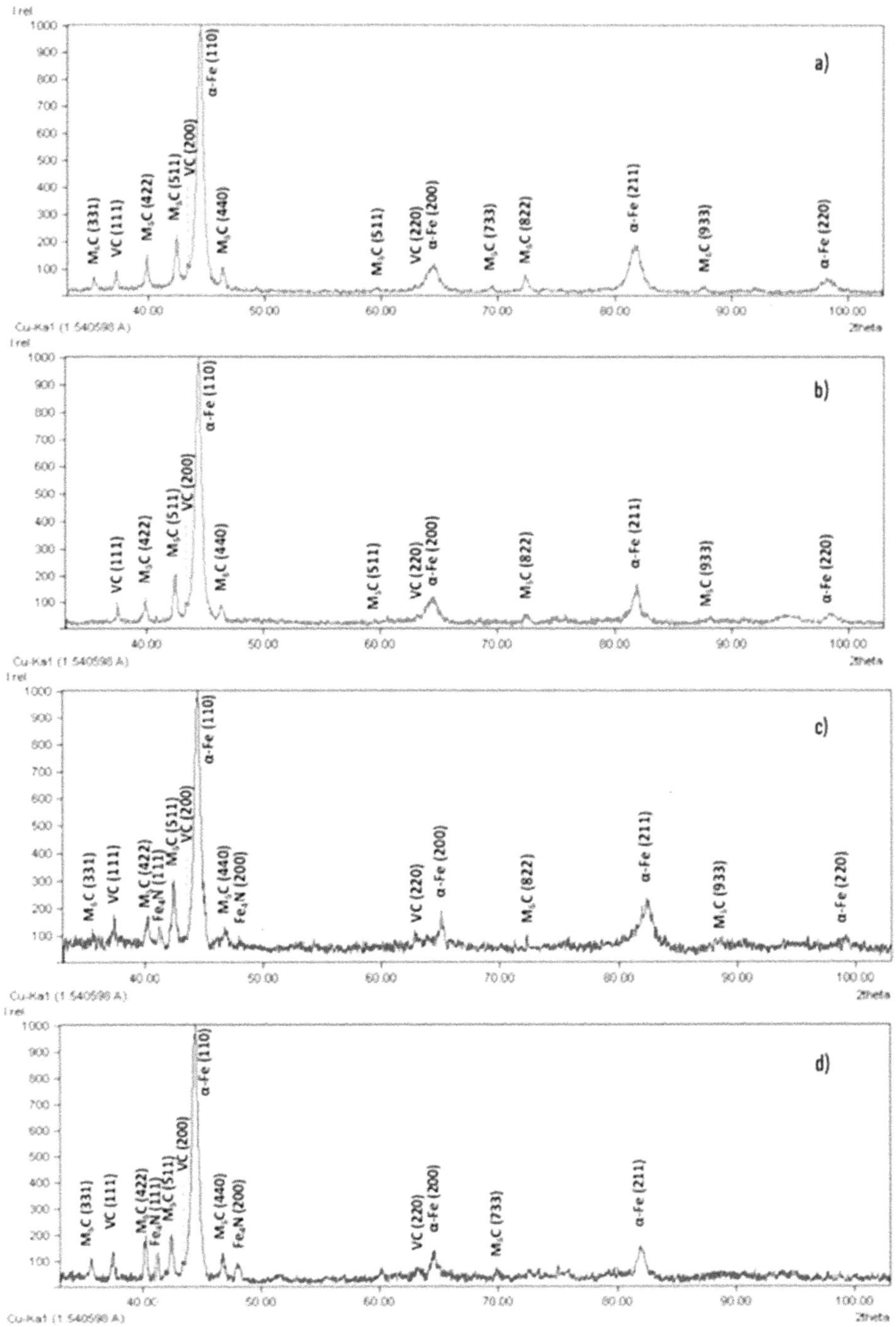

Figure 34. X-ray diffraction patterns of high-speed steel R9 before (a) and after (b) nitriding at a temperature of 550°C

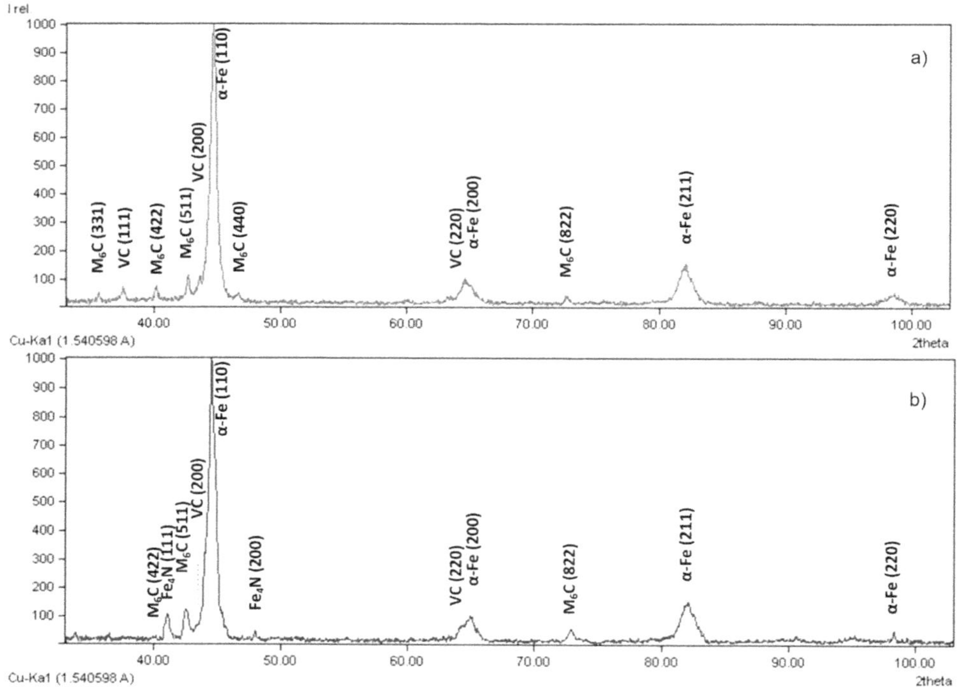

Figure 35. X-ray diffraction patterns of high-speed steel R18 before (a) and after (b) nitriding at a temperature of 550°C

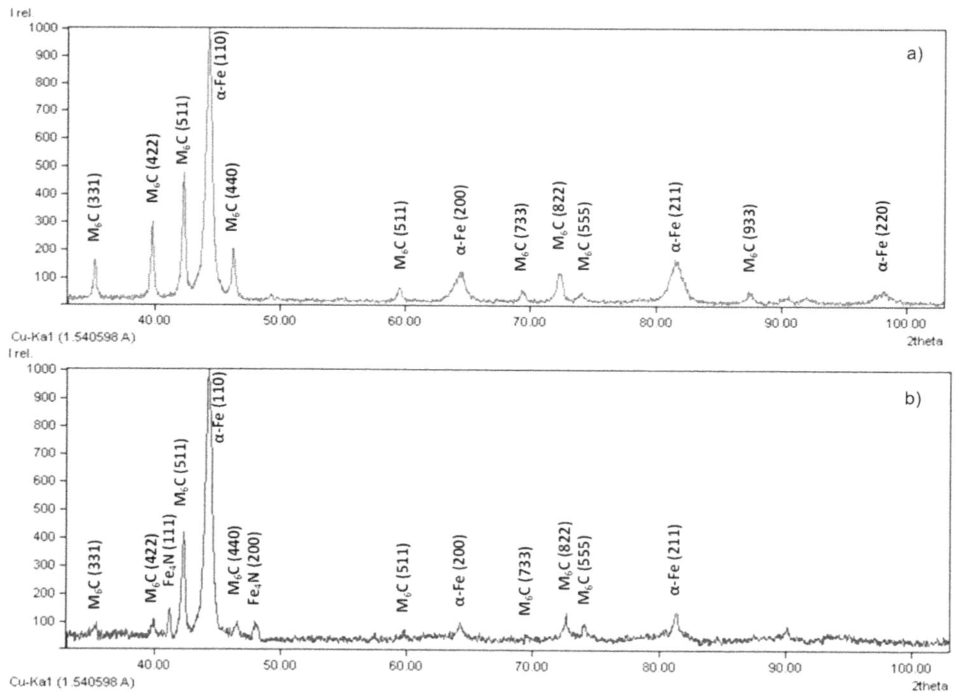

4.4 Studies of the Fine Structure and Second Phases of High-Speed Steel After Electrolytic-Plasma Nitriding

Intensive progress in the physical materials science of steels created the foundations of the science of their strength, which continue to be intensively developed at the present time. At the same time, a number of important issues in the physical materials science of steels have not been adequately developed. In this regard, it is necessary to note the clearly insufficient attention to the fine structure of steels and its evolution during chemical-thermal treatment. This is especially true for high speed steels. Of particular interest is the formation of secondary phases in the surface layers of high-speed steels during plasma nitriding. As already noted, the structure of high-speed steel in the initial (heat-treated) state is tempered martensite with hard carbides. X-ray diffraction analysis established that Fe_4N nitride is formed in the surface layer after nitriding. Also, by scanning electron microscopy, it was determined that after nitriding at 550°C, fine nitrides of alloying elements are formed in the surface layer of high-speed steel. However, X-ray diffraction analysis did not show the presence of nitrides of alloying elements, possibly due to the low concentration and shallow depth of their formation (Skakov, & Rakhadilov,, 2013; Usmanov & Yakunin, 1984). In this regard, in order to reveal the crystal structures of the second phases formed during nitriding, this section examines the fine structure of R6M5 steel samples before and after nitriding at 550°C for 7 min using transmission electron microscopy.

Previously, X-ray diffraction studies, the results of which are discussed in (Maksakova et al., 2016; Skakov, Rakhadilov, and Karipbaeva, 2013; Skakov and Rakhadilov, 2012), showed that the structure of R6M5 steel in the initial state, i.e. after standard heat treatment, it consists of an α'-phase based on iron with special carbides M_6C and MC. The results of TEM studies confirm these data. The fine structure of R6M5 steel, observed by electron microscopy, is shown in Figure 36 It is clearly seen that the main phase is the α'-phase (martensite) and that carbide particles of globular morphology occupy a significant part of the volume. Morphologically, the carbide phases are more clearly faceted. It should be noted that the α'-phase has a body-centered cubic lattice, a crystal lattice, and can be solid solutions based on iron of interstitial (C, N, B, S, P, etc.) and substitution atoms (Si, Mn, Ni, Cr, Mo, V, W, etc.) at the same time. In the case of the steel under study, the α'-phase is possibly a substitutional solid solution of Cr, W and Mo and an interstitial solid solution of carbon. After nitriding, the martensite structure is more fragmented and is characterized by a defective substructure. The resulting imperfection, apparently, approaches the defectiveness of the grain boundaries. The possibility of forming such a substructure of grains during electrolytic-plasma nitriding is provided by the increased energy state of the surface and subsurface layers, which are bombarded with ions and neutral atoms of low-temperature plasma throughout the treatment.

Figure 37 shows electron microscopic images of the fine structure of R6M5 steel before nitriding and their microdiffraction patterns obtained from the matrix and M_6C and MC carbides and their indexing schemes. Microdiffraction studies have confirmed that the main carbides in steel are M_6C carbide, which has a complex FCC crystal lattice and Fd_3m space group, and MC carbide, which has an FCC NaCl-type lattice and Fm_3m space group.

Figure 38 shows electron microscopic images of the fine structure of R6M5 steel after nitriding at 550^0C. It was found that after nitriding, particles are formed in the surface layer in the form of thin layers (Figure 4.24 a, indicated by arrows) and finely dispersed precipitates of irregular shape (Figure 4.24 b, indicated by arrows), randomly located in the martensitic matrix. Thus, the formation of fine precipitates of secondary phases in the surface layer after nitriding was confirmed by the electron microscopic method.

Table 7. Results of X-ray diffraction analysis of steel samples before and after electrolytic-plasma nitriding

Sample	Detected Phases	Volume Fraction of Phases, %	Lattice Parameters, nm	Size Coherent Scattering Region, nm
R9, initial	α-Fe	81	0,28680	23,30 (all ref)
	M_6C	11,8	1,10543	26,14 (100)
	MC	7,2	0,41566	25,01 (100)
R9, nitrided	α-Fe	76,6	0,28712	34,90 (all ref)
	M_6C	11.4	1,10804	32.79 (100)
	MC	6,5	0,41625	17.43 (100)
	γ'-Fe_4N	5.5	0.37945	27.43 (100)
R6M5, initial	α-Fe	75,2	0,28781	19,20 (all ref)
	M_6C	13,7	1,10867	38,77 (all ref)
	MC	11,	0,41663	18,72 (100)
R6M5, nitrided	α-Fe	73,2	0,28784	26,53 (all ref)
	M_6C	13,1	1,10818	40,98 (100)
	MC	9,8	0,41576	16,61 (100)
	γ'-Fe_4N	3,9	0.37923	40,22 (100)
R18, initial	α-Fe	72,2	0,28720	50,36 (all ref)
	M_6C	27,8	1,10429	39,43 (100)
R18, nitrided	α-Fe	70,5	0,28822	41,68 (all ref)
	Fe_3W_3C	25,8	1,10472	106,98 (100)
	γ'-Fe_4N	3,7	0.37756	97,44 (100)

Figure 36. Electron microscopic image of the fine structure of R6M5 steel before (a) and after (b) nitriding at 550^{0}C

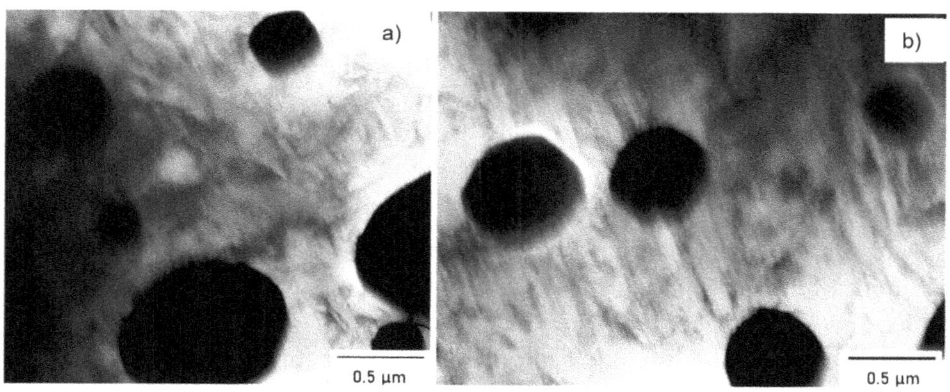

Figure 37. Electron microscopic images of the fine structure of R6M5 steel before nitriding and their microdiffraction patterns obtained from the matrix (a) and M_6C (b) and MC (c) carbides and their indexing scheme

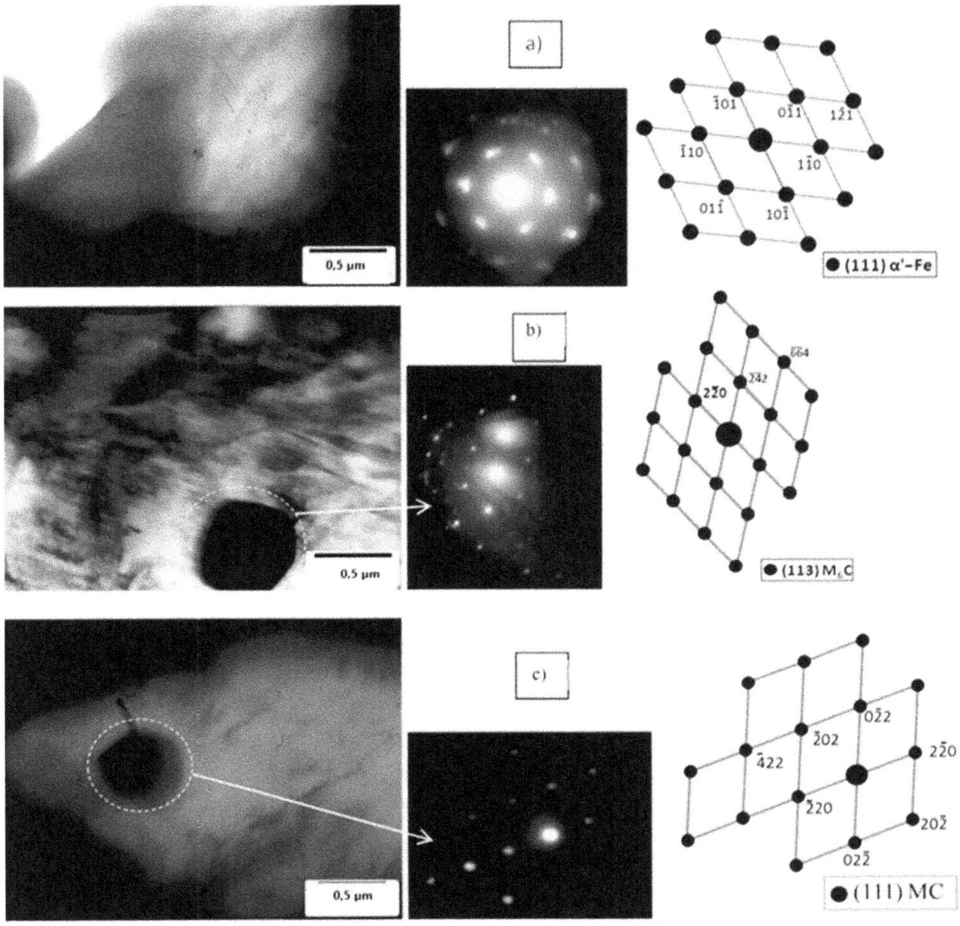

Figure 38. Electron microscopic images of the fine structure of R6M5 steel after nitriding at 550⁰C (arrows mark particles of the γ/-phase (a) and fine precipitates of secondary phases (b))

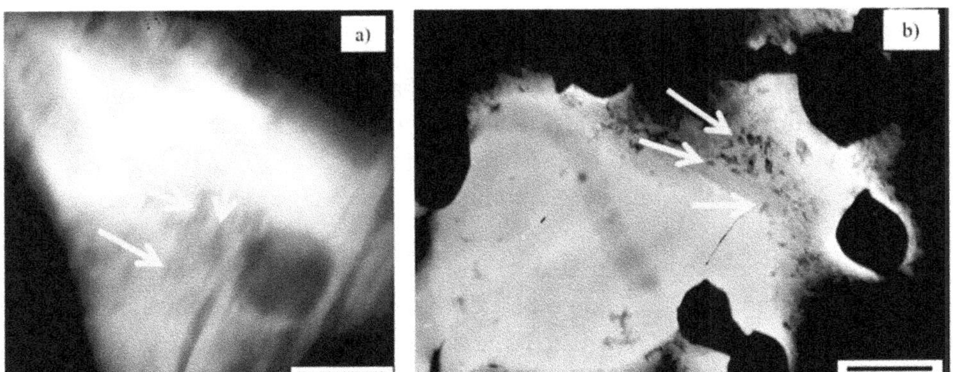

Electron microscopic images of the fine structure of R6M5 steel after nitriding and their microdiffraction patterns obtained from carbide, matrix and fine nitrides and their indexing schemes are shown in Figure 39. X-ray diffraction analysis determined that after nitriding at 500 and 550^{0}C, γ'-nitride is formed in the structure of high-speed steels. From the electron microscopic image obtained from the surface of R6M5 steel nitrided at 550°C, one can see particles of γ'-nitrides in the form of thin layers. The indexing of the microelectron diffraction pattern shown in Figure 39 showed that the formed thin layers (plate-shaped particles) are iron nitrides of the composition Fe_4N (γ'-phase). Iron nitrides of composition Fe_4N (γ'-phase) are formed upon cooling. Thus, in the process of cooling to room temperature, due to the low solubility of nitrogen in the α-phase at low temperatures, a strongly supersaturated solid solution appears. And this leads to the precipitation, preferably on dislocations, of α''-nitride particles of a lamellar shape with a certain orientation, which are subsequently transformed into γ'-nitrides (Pastukh,, 2006). The mechanism corresponds to the release of carbon and the formation of carbides during the tempering of hardened steel. Electron microscopic studies have shown that after nitriding, fine precipitates with a size of 50–100 nm are formed on the surface of R6M5 steel, which are in a nanocrystalline state.

Microdiffraction analysis also showed that the crystal structure of these finely dispersed precipitates corresponds to chromium nitride of the co imposition CrN, which has a FCC lattice of the NaCl type and space group Fm3m. The formation of such CrN nitrides in the surface layers of R6M5 and X210Cr12 tool steels after nitriding was observed earlier in (Rakhadilov et al., 2011; Shestopalova, 2011). The formation of CrN nitride in high-speed steels is possible in the presence of fluctuations with a sufficiently high concentration (Ramazanov & Vafin, 2011). The presence of concentration fluctuations in solutions in general and in solid solutions in particular is beyond doubt and has been confirmed experimentally in many systems. Under conditions when fluctuations of chromium occur in a medium supersaturated with carbon, they are stabilized at the initial stage due to interaction with carbon (Permyakov et al., 1972). This is all the more likely that chromium is a carbide-forming element. Stabilization of fluctuations by carbon creates prerequisites during nitriding for the transformation of the chromocarbon cluster into the nitride phase. The latter is natural in connection with the higher affinity of chromium for nitrogen compared with the affinity for carbon. Thus, we have characterized all secondary phases formed after nitriding in the surface layers of R6M5 high-speed steel. These studies are practically significant, because the mechanical properties of high-speed steel are largely determined by the structure and composition of the secondary phases (Ivanov et al., 2002). It is assumed that the formation of the γ'-phase (Fe_4N) and finely dispersed CrN nitride leads to an increase in the hardness and wear resistance of R6M5 steel.

4.5 Mechanisms That Provide High Wear Resistance of the Nitrided Layer of High-Speed Steels

In the previous subsections of this section, it was shown that high-speed steels R6M5, R9 and R18 after electrolytic-plasma nitriding have high wear resistance and increased microhardness. As is known (Barabash et al., 1974), the main criterion for the wear resistance of nitrided steels is their high hardness. However, the surface layer structure corresponding to maximum wear resistance and the structure corresponding to maximum hardness are not always the same. Wear resistance is a structurally sensitive characteristic. Therefore, it is not always necessary to strive to achieve the maximum hardness of steel during nitriding. In the process of friction, the structure and properties of the material of the surface plastic deformation zone change, and a friction structure is created that controls the degree of wear. In turn, the structure of the deformation zone during friction depends on the initial structure of the nitrided layer. Therefore,

Figure 39. Electron microscopic images of the fine structure of the surface layer of R6M5 steel formed after nitriding and their microdiffraction patterns obtained from carbide (a), matrix (b), and fine nitrides (c) and schemes for their indexing

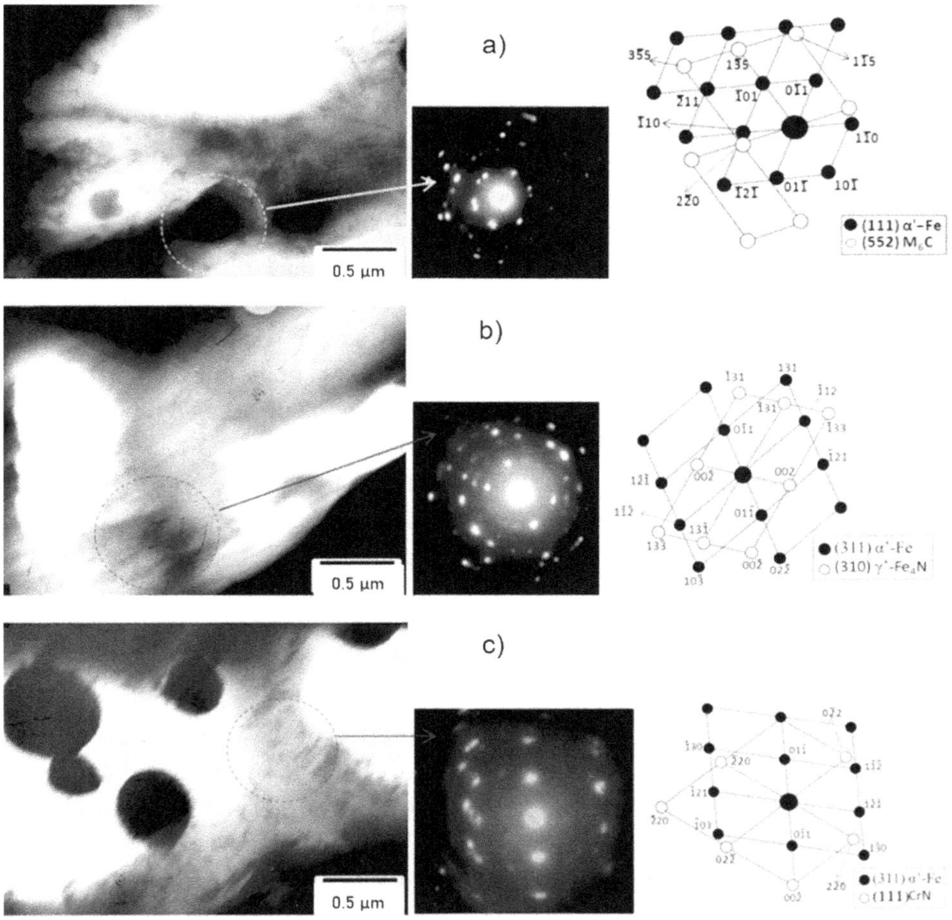

Table 8. Experimental data on the structure and tribological characteristics of the nitrided layer of R6M5 steel

t_{azot}, °C	Characteristics					
	h, µm	**Phase Composition**	**Hµ, MPa**	**f**	**j, 10^4 mm³/Hm**	**K_{wr}**
initial	-	αc', M6C, MC	7,7	6,8±0,4	6,52	2,04
450°C, 7 min	25	αN', M6C, MC	11,75	5,5±0,5	4,85	2,24
500°C 7 min	35	αN', γ', M6C, MC	12,15	5,3±0,2	2,23	2,58
550°C 7 min	40	αN', γ', NC, M6C, MC	12,4	4,1±0,2	1,21	2,64

there is a connection between the conditions of pre-treatment of steel, which forms a certain structural state of the surface layer, the conditions of nitriding, which create the structural state of the nitrided layer characteristic of them, and the tribotechnical characteristics of the structural state of the deformation zone during friction. Determining this relationship is the key to solving the problem of optimizing the technological process of nitriding for processing materials for tribotechnical products. In addition, there is no consensus in the literature on the structure and causes of a sharp change in the properties of the nitrided layer on high-alloy steels of various compositions. A correct understanding of the mechanisms of formation and hardening of the nitrided layer makes it possible to assume the structure of the layer and to foresee the changes that can occur with it depending on the nature of the alloying. In this regard, it is necessary to find out the main mechanisms that ensure high wear resistance of the nitrided layer of high-speed steels. The properties are determined by the nature of the structure formed during nitriding and by the transformations that take place in the layer upon cooling. We have carried out a systematic study of the change in tribological characteristics (wear intensity and friction coefficient) of the nitrided layer of high-speed steels depending on the nitriding temperature and holding time during nitriding, as well as on the test temperature. The paper shows that by varying the nitriding temperature, it is possible to obtain different structural states and different phase composition of the surface modified layer. Table 8 shows data on the structure and tribological characteristics of the nitrided layer obtained at nitriding temperatures of 450, 500 and 550°C. The experimental data clearly illustrate the correlation between the structural and tribological characteristics of the nitrided layer.

Note - h is the thickness of the nitrided layer; Hμ - microhardness; f - is the coefficient of friction; j - is the wear rate; K_{wr} - relative wear resistance to abrasive wear. From the generalized data given in the table, the dependence of the wear intensity of the modified layers on the structural-phase states is clearly traced. The wear of the modified layer consisting of the nitrided α'-phase (nitrogenous martensite) obtained after nitriding at 450°C is significantly lower than that of the layer consisting of the nitrided α'-phase with excess γ'-phase particles obtained after nitriding at 500°C. Particularly noteworthy are the results on the wear resistance of the modified layer obtained after nitriding at 550°C. This layer, consisting of a nitrided α'-phase with γ'-phase particles and fine particles of secondary phases, showed increased wear resistance when tested both at room temperature and at high test temperatures. It is known from (Barabash et al., 1974; Du et al., 2002) that the effects of an increase in hardness and wear resistance are directly related to the size and number of dispersed inclusions that are formed during the saturation of the steel surface with nitrogen. In this case, a special role is given to finely dispersed particles of secondary phases of CrN composition, which are in the nanocrystalline (ultrafinely crystalline) state. Under such states, the CrN particles turn out to be coherently bound to the matrix. Similar results were obtained in (Permyakov et al., 1972; Ramazanov & Vafin, 2011), where a significant amount of dispersed CrN particles coherently bonded to the matrix was found on chromium-alloyed steels.

Thus, the totality of experimental results on the evaluation of the parameters of the structural state and wear resistance suggests that in ensuring high wear resistance of nitrided high-speed steel under sliding friction, an important role belongs under the surface layer with dispersed nitrides and the change in its substructure, in general, has three main components: fineness particles, the level of microdeformation of the crystal lattice, a component of microdiffusion processes. Therefore, on the basis of experimental data, it can be said that a special role in the processes of surface hardening is assigned to finely dispersed particles of secondary phases of the CrN composition. Since, according to modern concepts of the dislocation nature of plastic deformation (Chaus & Hudakova, 2009), dispersed precipitates of secondary phases are effective barriers to moving dislocations and contribute to metal strengthening. The high wear resistance of the nitrided layer was also confirmed by fractographic analysis, which showed that in the initial state, crumbling particles of special M6C carbide are observed on the friction surface, which have an abrasive effect on the surface of the mating sample, the wear of which is significant. However, such particles are not observed on nitrided friction surfaces, which may be due to the formation of a nitrogenous α'-phase, as well as excess particles of the γ'-phase and finely dispersed particles of secondary phases. Thus, summarizing the data available in the literature and the results obtained in this study, we can formulate the following conclusions:

- a modified layer consisting of a nitrided α'-phase with a fragmented substructure, M_6C and MC carbides, excess particles of the γ'-phase (Fe_4N) and finely dispersed chromium nitride CrN, obtained after nitriding at 550°C, is wear-resistant, hard, antifriction and has the effect of extremely high wear resistance at high temperatures;
- the main mechanisms that ensure high wear resistance of the nitrided layer of high-speed steels obtained by electrolytic-plasma nitriding are, firstly, the formation of a modified layer consisting only of a diffusion zone (i.e., an internal nitriding zone), and secondly, precipitation during cooling in diffusion layer, preferably on dislocations of γ'-nitrides of lamellar shape; thirdly, the formation of fine particles of chromium nitride composition CrN in the diffusion layer due to fluctuations of chromium.

Conclusions for Section 4

Thus, based on the analysis of the research results obtained in the section, the following conclusions can be drawn:

1. It has been established that the structure of high-speed steels R6M5, R9 and R18 in the initial state, i.e. after standard heat treatment, consists of α'-phase and special carbides. At the same time, carbides of the M_6C and MC types are present in the structure of steels R6M5 and R9, and only carbides of the M_6C type are present in the structure of steel R18. Using X-ray diffraction analysis and EBSD analysis, it was found that carbides of the M_6C type, which have a complex FCC crystal lattice and space group Fd_3m, correspond to the Fe_3W_3C composition, and MC-type carbides, which have a cubic crystal lattice and space group Fm3m, correspond to the VC composition.
2. Metallographic analysis has established that during electrolytic-plasma nitriding, the process of diffusion saturation of steel with nitrogen occurs with accelerated formation of a modified layer, consisting only of a well-developed zone of internal nitriding, i.e. diffusion layer.
3. X-ray diffraction analysis has established that on the surface of steels R9, R6M5 and R18 after electrolytic-plasma nitriding at a temperature of 450°C, a diffusion layer is formed, consisting of the α'-phase ($Fe_{\alpha(N)}$), and at temperatures of 500°C and 550 °C along with the α'-phase, the γ'-phase (Fe_4N) is formed.
4. It was found that after electrolytic-plasma nitriding at a temperature of 550°C for 7 minutes, fine precipitates form on the surface of high-speed steels R 9, R6M5 and R18. By microdiffraction analysis, it was found that these fine precipitates are particles of chromium nitride of composition CrN with an FCC lattice NaCl type.

5. It has been established that after electrolytic-plasma nitriding at a temperature of 550°C for 7 minutes, the structure of R6M5 steel consists of nitrogenous martensite with a fragmented substructure, carbides (M_6C and MC), Fe4N nitride, and finely dispersed CrN nitride.
6. It has been established that the main structural factors affecting the increase in the tribological properties of high-speed steels during electrolytic-plasma nitriding are, firstly, the formation of a modified layer consisting only of a diffusion zone (i.e., an internal nitriding zone), and secondly, the formation of diffusion layer of excess particles of the γ'-phase, thirdly, the formation in the diffusion layer of fine particles of chromium nitride composition CrN;
7. It was determined that the modified layer, consisting of a nitrided α'-phase with a fragmented substructure, M_6C and MC carbides, excess particles of the γ'-phase (Fe_4N) and finely dispersed CrN chromium nitride, obtained after nitriding at 550°C, are wear-resistant, hard, antifriction and have the effect of extremely high wear resistance at high temperatures. The main mechanisms that ensure high wear resistance of the nitrided layer of high-speed steels obtained by electrolytic-plasma nitriding are, firstly, the formation of a modified layer consisting only of a diffusion zone (i.e., an internal nitriding zone), and secondly, precipitation during cooling in a diffusion layer, preferably on dislocations of plate-like γ'-nitrides.

CONCLUSION

Based on the results of this study, the following conclusions can be drawn:

1. It has been established that during electrolytic-plasma nitriding of high-speed steels, the process of diffusion saturation of steel with nitrogen occurs with accelerated formation of a modified layer, consisting only of a well-developed zone of internal nitriding, i.e. diffusion layer. It was found that after electrolytic-plasma nitriding at a temperature of 450°C, a modified layer is formed on the surface of high-speed steel, consisting of α'-phase ($Fe_{\alpha(N))}$ and carbides (M_6C and MC), with an increase in the nitriding temperature from 450^{0}C to 500^{0}C in the modified layer particles of the γ′-phase (Fe_4N) are formed, and at a nitriding temperature of 550^{0}C, finely dispersed particles of chromium nitride of composition CrN with an FCC lattice of the NaCl type are formed.
2. As a result of electrolytic-plasma nitriding at 550°C, samples of high-speed steels, there was an increase in surface microhardness by 1.6 times. It has been experimentally established that the main increase in the hardness of the diffusion layer occurs after nitriding at 550°C, which is the result of the precipitation of finely dispersed nitrides in the supersaturated α-phase, which distort the crystal α-lattice. It has been determined that the main factors influencing the increase in the microhardness of R6M5 steel are the formation of a diffusion layer of nitrogenous martensite, as well as the formation of fine particles of chromium nitride with the composition CrN in the surface layers;
3. It has been established that after electrolytic-plasma nitriding, the microhardness, wear resistance and resistance to abrasive wear of high-speed steels increase, depending on the nitriding mode. After electrolytic-plasma nitriding at a temperature of 550°C for 7 minutes, microhardness increases by 1.6 times, wear resistance increases by 2 times, and abrasive wear resistance increases by 1.3 times.

4. It has been established that the main mechanisms that ensure high wear resistance of the nitrided layer of high-speed steels obtained by electrolytic-plasma nitriding are, firstly, the formation of a modified layer consisting only of a diffusion zone (i.e., an internal nitriding zone), and secondly, precipitation in diffusion layer, preferably on dislocations of plate-like γ'-nitrides; thirdly, the formation of fine particles of chromium nitride with the composition CrN in the diffusion layer. It has been determined that the modified layer, consisting of a nitrided α'-phase with a fragmented substructure, M_6C and MC carbides, excess particles of the γ'-phase (Fe_4N) and finely dispersed chromium nitride, obtained after electrolytic-plasma nitriding at 550°C, is wear-resistant, hard, antifriction and has the property of extremely high wear resistance at high temperatures.

REFERENCES

Anel'nik, D.Ye. Polevoy, S.N., & Yevdokimov, V.D. (1981). Remont, vosstanovleniye i ispytaniye instrumenta i tekhnologicheskoy proverki: spravochnik. *Tekhnika*, 200. Kiyev.

Arshinger, I. (1982). Instrumental'nyye stali i ikh termicheskaya obrabotka: spravochnik. *Metallurgiya, 313.*

Artinger, I. (1982). Instrumental'nyye stali i ikh khimiko-termicheskaya obrabotka. *Spravochnik*, 312.

Arzamasov, B.N., Bratukhin, A.G., Yeliseyev, YU.S., & Panayoti, T.A. (1999). Ionnaya khimiko-termicheskaya obrabotka splavov. *MGTU im. Baumana*, 400.

Arzamasova, B.N., Mukhina, G.G. (2001). Materialovedeniye: uchebnik dlya vuzov. N.E. Baumana, (ed.) *Izd-vo MGTU im.* 648.

Astaf'yev A.A. (1997). Nekotoryye zakonomernosti vodorodnogo okhrupchivaniya konstruktsionnykh staley. *Metallovedeniye i termicheskaya obrabotka metallov*, (2), 5 -8.

Badisch, E., & Mitterer, C. (2003). Abrasive wear of high speed steels: Influence of abrasive particles and primary carbides on wear resistance. *Tribology International*, *36*(10), 765–770. doi:10.1016/S0301-679X(03)00058-6

Bakovets, V. V., Polyakov, O. V., & Dolgovesova, I. P. (1991). Plazmenno-elektroliticheskaya anodnaya obrabotka metallov. Novosibirsk, Nauka. Sib. otd.

Barabash R.I., Belobashskiy A.V., & Permyakov V.G. (1974). Tonkaya struktura i uprochneniye azotirovannykh volokon zheleza, legirovannogo khromom. Izvestiya vuzov. *Chernaya metallurgiya,* (10), 118–120.

Baranchikov V.I., Zharikov A.B., Yudina N.D., & Sadykhov A.I. (1990). Progressivnyye otvety i rezhimy rezaniya metallov, 400. *Mashinostroyeniye.*

Baranova L.V., & Demina E.L. (1986). Metallograficheskoye travleniye metallov i splavov: spravochnik, 256. *Metallurgiya.*

Belkin P.N. (2010). Anodnaya elektrokhimiko-termicheskaya modifikatsiya metallov i splavov. *Elektronnaya obrabotka materialov, 46*(6), 29–41.

Belkin P.N., Ganchar V.I., & Tovarkov A.K. (1986). Teploobmen mezhdu anodom i parogazovoy obolochkoy pri elektrolitnom nagreve. *Inzhenerno-fizicheskiy zhurnal. T. 51*(1), 154-155.

Belkin P.N., & Pasinkovskiy Ye.A. (1989). Termicheskaya i khimiko-termicheskaya obrabotka staley pri nagreve v rastvorakh elektrolitov. *Metallovedeniye i termicheskaya obrabotka metallov*, (5), 12 - 17.

Belkin P.P. (1997). Anodnyy nagrev v vodnykh rastvorakh. *Vestnik Kostromskogo gosudarstvennogo pedagogicheskogo universiteta*, (4), 55-58.

Belkin P.P., & Ganchar V.I. (1988). Prokhozhdeniye toka cherez parogazovuyu obolochku pri elektrolitnom elektrolitnom nagreve. *Elektronnaya obrabotka materialov*, (5), 59-62.

Belotskiy, A. V., Marchevskaya, E. I., & Permyakov, V. G. (1973). Osobennosti obrazovaniya azotistykh faz v sisteme Fe–V–N. Izv. AN SSSR. *Metally*, (3), 116–119.

Belotskiy A.V., & Permyakov V.G. (1972). O prirode vysokoprochnogo azotirovannogo sostoyaniya sloyev na legirovannom zheleze. *Zashchitnyye pokrytiya na metallakh*, (6), 83-86.

Belyy A.V., Karpenko G.D., & Myshkin N.K. (1991). *Struktura i metody formirovaniya iznosostoykikh sloyev,* 209. Mashinostroyeniye.

Boyarskaya YU.S., Grabko D.Z., & Kats M.S. (1986). *Fizika protsessov mikroidentirovaniya*, 295. Kishinev, Shtiinitsa.

Braun, E. D., Bushe, N. A., & Buyanovsky, I. A. (2001). Fundamentals of tribology (friction, wear, lubrication). In A.V. Chichinadze (ed.) Textbook for technical schools, 2nd edition. Center "Science and technology", 778.

Braun E.D., Bushe N.A., Chichinadze A.V., & Buyanovskiy I.A. (2001). Osnovy tribologii (treniye, iznos, smazka). *Uchebnik dlya tekhnicheskikh VUZov*, 2-ye izdaniye, 778. Tsentr «Nauka i tekhnika».

Budilov V.V., Agzamov R.D., & Ramazanov K.N. (2004). Sposobnost' vzryvnogo uprochneniya detaley. *Patent na izobreteniye Nº 2275433*, MPK 7, C21D1/09, C21D1/38. 3.

Budilov V.V., Agzamov R.D., & Ramazanov K.N. (2004). *Sposob azotirovaniya izdeliy v tleyushchem razryade s effektom pologo katoda.* Patent for invention No. 2276201, IPC 7, C23C8/36, C23C8/80.

Budilov, V. V., Agzamov, R. D., & Ramazanov, K. N. (2007). Tekhnologiya ionnogo azotirovaniya v tleyushchem razryade s polym katodom. *MiTOM*, (7), 25–29.

Chatterjee-Fisher, R., & Supova, A.B. (1990). Azotirovaniye i karbonitrirovaniye. 12. *Metallurgiya.*

Chaus, A. S. (2008). On the wear resistance of high-speed steels. *Journal of Friction and Wear*, *29*(1), 24–34. doi:10.3103/S1068366608010054

Chaus, A. S., & Hudakova, M. (2009). Wear resistance of high-speed steels and cutting performance of tool related to structural factors. In *17th International Conference on Wear of materials*. Las Vegas, USA. 10.1016/j.wear.2008.12.101

Chernyavskiy, K.S. (1977). Stereologiya v metallovedenii. 240. *Metallurgiya.*

Du, E. H., Marchie, V. V., Boerma, D. O., & Chechenin, N. C. (2002). Low-temperature extension of the Lehrer diagramand the iron-nitrogenp has ediagram. *Metallurgical and Materials Transactions*, (33), 34–39.

Duradzhi V. N., & Bryantsev I.V. (1979). O nekotorykh osobennostyakh issledovaniya magnitnogo polya na anodnyy i katodnyy protsessy. *Elektronnaya obrabotka materialov*, (5), 15 - 19.

Duradzhi V. N., & Parsadanyan A. S. (1988). *Nagrev metallov v elektrolitnoy plazme*. Kishinev, Shtiintsa.

Endryus K., Dayson D., & Kioun S. (1971). Elektronogrammy i ikh interpretatsiya. *Mir*, 56.

Fillipov, M. A., Fillipenkov, A. A., & Plotnikov, G. N. (2009). *Iznosostoykiye stali dlya otlivok*.

Geller, Y.A. (1983). *Instrumental'nyye stali*, 527. Metallurgiya.

Gerasimov, S.A., Kuksenova, L.I., & Lapteva, V.G. (2012). *Struktura i iznosostoykost' azotirovannykh staley i splavov*, 518.

Gnyusov S.F., Khazanov I.O., & Sovetchenko B.F. (2008). Primeneniye effekta sverkhplastichnosti staley v instrumental'nom proizvodstve. *Tomsk: Izd-vo nauch.-tekhn. lit.*, 237.

Gol'dshmidt KH.Dzh. (1971). Splavy potrebleniya. *Mir*, 424.

Gol'dshteyn M.I., Grachev S.V., & Veksler YU.G. (1985). Spetsial'nyye stali, 408. Metallurgiya.

Gorelik S.S., Skakov YU.A., & Rastorguyev L.N. (2002). Rentgenograficheskiy i elektronno-opticheskiy analiz. *MISIS*, 360.

GOST 9450-76. (n.d.). *Measurement of microhardness by intrusion of diamond tips* [Data set].

Grigor'yev, S. (2004). Kak povysit' nadezhnost' obnaruzheniya instrumenta. *TekhnoMIR.*, *21*(3), 53–57.

Grigorevich, V. K. (1976). Tverdost' i mikrotverdost' metallov. Nauka, 230.

Grigoriev, S. N., Tabakov, V. P., & Volosova, M. A. (2011). *Tekhnologicheskiye metody povysheniya iznosostoykosti kontaktnykh ploshchadok rezhushchego instrumenta*.

Guenzel, R., Matz, W., Ivanov, Yu. F., & Rothtein, V. P. (2000). *Pulsedelectron-beamtreatmentofhigh-speedsteelcurrenttools: struchire-phase transformation and wear resistance*. 1st International Congress on Radiation Physics, high current electronics, and modification of materials, Tomsk, Russia.

Gupta, P., Tenhundfeld, G., Daigle, E. O., & Ryabkov, D. (2007). Electrolytic plasma technology: Science and engineering – an overview. *Surface and Coatings Technology*, *201*(25), 87–96. doi:10.1016/j.surfcoat.2006.11.023

Guriev A. M., Ivanov S. G., Lygdenov B. D., Zemlyakov S. A., Vlasova O. A., Kosheleva E. A., & Guriev M.A. (2009). *MPK Sposob uprochneniya detaley iz konstruktsionnykh i instrumental'nykh staley*. Patent. 2345175 Russian Federation, N° 2007112368/02, (3), 9.

Guryev, A. M., Kozlov, E. V., & Popova, N. A. (2001). Izmeneniye fazovogo sostava i mekhanizm formirovaniya struktury perekhodnoy zony pri termotsiklicheskom borirovanii ferritno-perlitnoy stali. *Izv. Vuzov. Fizika*, (2), 58–63.

Guryev, A. M., Vlasova, O. A., & Lygdenov, B. D. (2007). Povysheniye prochnosti instrumental'nykh staley metodom termotsiklicheskogo borirovani. XVII Petersburg readings on the problems of strength, St. Petersburg, Russia.

Ivanov, Yu., Matz, W., Rotshtein, V., Günzel, R., & Shevchenko, N. (2002). Pulsed electronbeams melting of high-speed steel: Structural phase transformations and wear resistance. *Surface and Coatings Technology*, *150*(2-3), 188–198. doi:10.1016/S0257-8972(01)01542-0

Kardonina, N. I., Yurovskikh, A. S., & Kolpakov, A. S. (2010). Prevrashcheniya v sisteme Fe – N. *MiTOM*, (10), 5–15.

Kheyker D.M., & Zevin A.S. (1963). *Rentgenovskaya difraktometriya*. 540. M.:Fizmatgiz,

Khrushchov, M. M., & Babichev, M. A. (1970). *Abrazivnoye iznashivaniye*. Nauka.

Kittel, C. (1976). *Introduction to Solid State Physics*. John Wiley & Sons.

Kremnev L.S. (1985). Zaevtektoidnyye bystrorezhushchiye stali. *Metallovedeniye i termicheskaya obrabotka metallov*, (8), 24 – 25.

Kremnev, L. S. (2008). Teoriya legirovaniya i sozdaniya na yeye osnove teplostoykikh instrumental'nykh staley i splavov razlichnogo sostava. *Metallovedeniye i termicheskaya obrabotka metallov*, (11), 18–28.

Kuksenova, L. I., Lapteva, V. G., & Berezina, Ye. V. (2004). Struktura i iznosostoykost' azotirovannoy stali. *MiTOM*, (1), 31–34.

Lakhtin, Y. U. M. (1993). Sovremennoye sostoyaniye protsessa azotirovaniya. *Metallovedeniye i termicheskaya obrabotka metallov*, (JV27), 6–11.

Lakhtin, Y. U. M. (1995). Diffuzionnyye osnovy protsessa azotirovaniya. *MiTOM*, (7), 14–17.

Lakhtin, Y.M. (1995). Diffuzionnyye osnovy protsessa azotirovaniya. *Metallovedeniye termicheskaya obrabotka metallov*, (7),14-17.

Lakhtin, Y.M. (2009). *Materialovedeniye i termicheskaya obrabotka metallov: uchebnik dlya vuzov. 5-ye izd., pererab,* 448. TID Az – kniga.

Lakhtin, YU. M., & Arzamasov, B. N. (1985). Khimiko-termicheskaya obrabotka metallov: ucheb. posobiye dlya vuzov po spets. "Metallovedeniye, oborud. i tekhnologiya term, obrab. metallov. Metallurgiya, 256.

Lakhtin, YU.M., & Kogan, YA.D. (1982). Gazovoye azotirovaniye detaley mashin i instrumenta. Mashinostroyeniye, 60.

Lakhtin, Y.M., Kogan, YA.D., Shpis, G.Y., & Bemer, Z. (1991). Teoriya i tekhnologiya azotirovaniya. Metallurgiya, 320.

Lakhtin, Y.M., Neustroyev, G.N., & Ivanov, YU.P. (1973). Nizkotemperaturnoye tsianirovaniye instrumental'nykh staley. *Metallovedeniye i termicheskaya obrabotka metallov*, (12), 27-31.

Lakhtin, Y.M. & Silina, N.V. (1977). Kharakter vysokoy tverdosti legirovannogo ferrita posle azotirovaniya. *MiTOM*, 23-29.

Lazarenko, B.R., Duradzhi, V.N., & Bryantsev, I.V. (1979). O vliyanii povysheniya induktivnosti na kharakteristiki anodnogo i katodnogo protsessov. *Elektronnaya obrabotka materialov*, (5), 8 - 13.

Liu, Z.Y., Loh NH, Khor KA, & Tor S.B. Microstructure evolution during sintering of injection molded M2 high speed steel. *Mater SciEng*, -P. 46–55.

Loladze T.N. (1982). Prochnost' i iznosostoykost' rezhushchego instrumenta. *Mashinostroyeniye*, 320.

Maksakova, O. V., Plotnikov, S. V., Eskermesov, D. K., & Yurchenko, D. (2016). The effect of Si additions and deposition conditions on the structure and properties of the (Zr-Ti-Cr-Nb-Si)N coatings fabricated by vacuum-arc deposition. *Proceedings of the 6th International Conference Nanomaterials: Applications and Properties*. NAP.

Minkevich, A.N., & Suchevyanu, G. (1968). Khimiko-termicheskaya obrabotka staley v smesyakh soley, soderzhashchikh mochevinu. *Metallovedeniye i termicheskaya obrabotka metallov*, (10), 11-16.

Mirkin L.I. (1961). Spravochnik po rentgenostrukturnomu analizu polikristallov. *Gosizdat. Fiz. mat. literatury*, 863.

Moiseyev V.F., & Grigor'yev S.N. Instrumental'nyye materialy. *Izd. 2-ye*, 248. MGTUStankin.

Muras, V. S. (1955). *O svyazi mezhdu temperaturoy i napryazheniyem toka pri elektrolitnom nagreve metallov. Mashinostroitel' Belorussii*. Gosizdat BSSR.

Neustroyev G.P., Bogdanov V.V., & Ivanov YU.P. (1972). Nizkotemperaturnoye tsianirovaniye instrumental'nykh staley. *Progressivnyye metody termicheskoy i khimiko-termicheskoy obrabotki*, 89-91. Mashinostroyeniye.

Novikov, I. I. (1986). Teoriya termicheskoy obrabotki metallov. *Metallurgiya*, 480.

Parfenov, E. V., Yerokhin, A. L., & Mattews, A. (2007). Frequency response studies for the plasma electrolytic oxidation process. Surface and coatings technology. Science Direct. doi:10.1016/j.surfcoat.2007.04.044

Pastukh, I.M. (2006). *Teoriya i praktika bezvodorodnogo azotirovaniya v tleyushchem razryade*. Khar'kov, NNTS KHFTI, 364.

Pastukh, I. M., & Zdybel', A. S. (2008). Kharakteristiki obrazovaniya nitridov v stalyakh pri azotirovanii v tleyushchem razryade. OTTOM - 9: materialy mezhdunarodnoy konferentsii. Khar'kov, 62-68.

Permyakov V.G., Belotskiy A.V., & Barabash R.I. (1972). Struktura i svoystva diffuzionnoy zony pri azotirovanii khromistogo zheleza. Izvestiya vuzov. *Chernaya metallurgiya*, (40), 129–131.

Petrova L.G. (2001). Vnutrenneye azotirovaniye zharoprochnykh staley i splavov *MiTOM9*, (1), 10-17.

Petrova L.G. (2004). *Vysokotemperaturnoye azotirovaniye zharoprochnykh splavov*, (1), 18-24.

Petrova, L. G., Aleksandrov, V. A., & Zyuzin, D. M. (2003). Reguliruyemyye protsessy azotirovaniya korrozionnostoykikh staley. *Vestnik MADI*, (1), 20–26.

Petrova L.G. & Zyuzin D.M. (2005). Vysokotemperaturnoye azotirovaniye austenitnoy stali. *Uprochnyayushchiye tekhnologii pokrytiy*, (3), 29–36.

Petukhov, B. V. (2001). Statisticheskaya teoriya dvizheniya dislokatsiy pri nalichii spontannykh protsessov blokirovki-deblokirovki. *FTT*, *43*(5), 813–817.

Plotnikov, S. V., Pogrebnjak, A. D., Yerokhina, L. N., Yeskermessov, D. K., & Erdybaeva, N. K. (2016). Study of nanostructured (Ti-Zr-Nb)N coatings' physical- mechanical properties obtained by vacuum arc evaporation. *IOP Conference Series. Materials Science and Engineering*, *110*(1), 12–31. doi:10.1088/1757-899X/110/1/012031

Popilov L.YA., & Zaytseva L.P. (1963). Elektropolirovaniye i elektrotravleniye metallograficheskikh shlifov. *Metallurgiya*, 125.

Poznyak, L. A. (1996). Instrumental'nyye stali, 488. Kiyev: Naukova dumka.

Poznyak, L.A., Tishayev, S.I., & Skrynchenko, Y.M. (1977). Instrumental'nyye stali: spravochnik. *Metallurgiya*, 167.

Prokoshkin D.A. (1984). *Khimiko-termicheskaya obrabotka metallov karbonitratsiya*, Metallurgiya, mashinostroyeniye, 204.

Przhenosil B. (1969). Nitrotsementatsiya. *Mashinostroyeniye*, 212.

Rakhadilov, B. K., & Skakov, M. K. (2012). Issledovaniye tribologicheskikh svoystv stali bystrorezhushchey R6M5, uprochnennoy vysokoenergeticheskoy otsenki. *Vestnik VKGTU*, (4), 43–48.

Rakhadilov, B. K., & Skakov, M. K. (2014). Vysokoeffektivnaya tekhnologiya elektrolitno-plazmennogo azotirovaniya bystrorezhushchikh staley. *Vestnik VKGTU*, (1), 50–55.

Rakhadilov B.K., Skakov M.K., & Karipbayeva G.S. (2011). Elektrolitno-plazmennoye azotirovaniye bystrorezhushchey stali Sbornik trudov IV mezhdunarodnoy nauchno-prakticheskoy konferentsii s ispol'zovaniyem nauchnykh shkol dlya molodykh uchenykh *Innovatsionnyye tekhnologii i ekonomika mashinostroyeniya*, 151-155. Yurginskiy tekhnologicheskiy institut. Izd-vo Tomskogo politekhnicheskogo universiteta.

Rakhshtadt A.G., Bernshteyn L.M., Kurdyumov G.V., & Mes'kin V.S. (2005). Metallovedeniye i termicheskaya obrabotka stali i chuguna: spravochnik v 3 t.. *Intermet Inzhiniring,(*T.2), 528.

Ramazanov K.N., Budilov V.V., & Vafin R.K. (2010). Azotirovaniye stalirezhushchey R6M5 v tleyushchem razryade s nalozheniyem magnitnogo polya. *Uprochnyayushchiye bystrodeystvuyushchiye tekhnologii i pokrytiya*, (5), 39-42.

Ramazanov, K. N., & Vafin, R. K. (2011). Razrabotka khorosho ionno-stal'nogo azotirovaniya instrumental'noy KH12 v skreshchennykh elektricheskikh i magnitnykh polyakh. *Vestnik UGATU*, (41), 101–104.

Sablev, L. P., Andreyev, A. A., & Kunchenko, V. V. (2000). Plazmennoye azotirovaniye rezhushchego instrumenta iz bystrorezhushchey stali, 133-137. Trudy simpoziuma OTTOM, Khar'kov.

Sal'nikova S.S., & Rudman V.A. (1987). Opyt primeneniya ionnogo azotirovaniya v mashinostroyenii. *LDNTP, 20.*

Schastlivtsev V.M., Mirzayev D.A., & Yakovleva I.L. (1994). Struktura termicheski obrabotannoy stali. *Metallurgiya*, 288.

Shestopalova, L. P. (2011). Issledovaniye khimicheskogo sostava i sostava nitrid-oksidnoy zony oksiazotirovannogo sloya legirovannykh konstruktsionnykh staley. *Vestnik KHNADU*, (54), 78–81.

Skakov, M., Rakhadilov, B., & Scheffler, M. (2013). Wear-resistance increase of high speed steel surface layer. 5 World Tribological Congress. *Tribology of materials. –Wroclaw*, (3), 491.

Skakov, M. K., & Rakhadilov, B. K. (2012). Change of structure and wear-resistance of P6M5 steel for processing in electrolyte plasma. *32th All-Polish Tribology conference Autumnal school of Tribology 2012*. Wroclaw University of Technology Institute of Machine Design and Operation, Wroclaw, Poland.

Skakov, M. K., & Rakhadilov, B. K. (2013a). *Ustroystvo dlya nagrevaniya detaley v elektrolite: Patent na poleznuyu model' Respubliki Kazakhstan: MPK6 S21D 1/44 - N° 912*. Zayavl, Byul.

Skakov, M.K., & Rakhadilov, B.K. (2013b). Modifikatsiya poverkhnosti obnaruzheniya stali instrumentom iz bystrorezhushchey al'ternativno-plazmennoy obrabotki. *Sbornik tezisov V konferentsii Sovremennyye metody v teoreticheskoy i eksperimental'noy elektrokhimii i IV konferentsii Elektrokhimicheskiye i elektrolitno-plazmennyye metody obrabotki metallov*. Ples: Izd-vo Institut khimii rastvorov im. G.A. Krestova RAN – Ivanovo.

Skakov, M. K., Rakhadilov, B. K., Batyrbekov, E. G., & Manapbayeva, A. B. (2014). Povysheniye iznosostoykosti bystrorezhushchikh staleyel ektrolitno-plazmennym azotirovaniyem. *Vestnik KazNU*, (48), 44–52.

Skakov, M. K., Rakhadilov, B. K., Batyrbekov, E. G., Sheffler, M., Manapbayeva, A. B., Ayapbergenova, G. T., & Karipbayeva, G. S. (2014). Izmeneniye rezhimov elektrolitno-plazmennogo azotirovaniya na strukturno-fazovoye sostoyaniye i iznosostoykost' stali R6M5. *Vestnik KazNTU*, *103*(3), 65–71.

Skakov, M. K., Rakhadilov, B. K., & Karipbaeva, G. S. (2013). Specifics of microstructure and phase composition of high-speed steel R6M5. *Applied Mechanics and Materials*, *404*, 20–24. doi:10.4028/www.scientific.net/AMM.404.20

Skakov, M. K., Rakhadilov, B. K., Karipbayeva, G. S., & Manapbayeva, A. B. (2014). Strukturno-fazovoye sostoyaniye bystrorezhushchey stali R6M5 posle termicheskoy obrabotki. *Vestnik KazNU*, *48*(1), 53–59.

Skakov, M. K., Rakhadilov, B. K., & Rakhadilov, M. K. (2014). Wear-resistance of nitrided W-Mo-high speed steel in abrasive wear conditions. *Key Engineering Materials*, 1117–1121.

Slovetskiy, D. I. (1991). Mekhanizmy neravnovesnykh plazmokhimicheskikh reaktsiy, Novosibirsk, Nauka. Khimiya plazmy. T.3. Nizkotemperaturnaya plazma, (3), 94-100.

Slovetskiy D.I., Terent'yev S.D., & Plekhanov V.G. (1986). Mekhanizm plazmenno-elektrolitnogo nagreva. *Teplofizika vysokoy temperatury*, (2), 353-363.

Solntseva Yu.P. (1988). Metallovedeniye i tekhnologiya metallov. *Metallurgiya*, 491.

Suminov I.V., Belkin P.N., & Epel'fel'd A.V. (2011). Plazmenno-elektrolitnoye modifitsirovaniye poverkhnosti metallov i splavov. *Tekhnosfera*, (T.1), 464.

Suminov I.V., Epel'fel'd A.V., Lyudin V.B., Krit B.L., & Borisov A.M. (2005). Mikrodugovoye oksidirovaniye, teoriya tekhnologii, oborudovaniye. *EKOMET*, 368.

Seriya. (1975). Tekhnologicheskiy protsess poverkhnostnogo uprochneniya rezhushchego instrumenta. *Informatsionnyy listok*, 10-09.

Tomilov A.P. (1968). Elektrokhimiya pishchevykh soyedineniy. Izd-vo *Khimiya*, 115.

Usmanov, K. B., & Yakunin, G. I. (1984). Vliyaniye vneshney sredy na iznos i stoykost' obnaruzheniya instrumentov. Tashkent, 158.

Ustinovshchikov, Y. U. I. (1988). *Vydeleniye vtoroy fazy v tverdykh rastvorakh*. Nauka, 172.

Utevskiy L.M. (1973). Difraktsionnaya elektronnaya mikroskopiya v metallosoderzhanii. *Metallurgiya*, 584.

Vanin, V. S. (1967). Nagrev metallov v elektrolite. *Elektrotermiya*, (55), 18–19.

Veynberga F. (1973). Pribory i metody fizicheskogo metallovedeniya, (1), 427.

Vorob'yeva, G. A., & Skladnova, Ye. Ye. (2003). Instrumental'nyye stal'nyye materialy: Instrumental'nyye i splavy. *SPb.*, (1), 100.

Yasnorodskiy I.Z. (1971). Elektrolitnyy nagrev metallov. Elektrokhimicheskaya i elektromekhanicheskaya obrabotka metallov. *Mashinostroyeniye*, 117-168.

Yelyutina O.P. (1979). Prakticheskiye voprosy ispytaniy metallov. *Metallurgiya*, 280.

Yerokhin, A. L., Nie, X., Leyland, A., Matthews, A., & Dowey, S. J. (1999). Plasma electrolysis for surface engineering. *Surface and Coatings Technology*, *122*(2-3), 73–93. doi:10.1016/S0257-8972(99)00441-7

Younesi, Y., Bahrololoom, M. E., & Fooladfar, H. (2010). Development of Wear Resistant NFSS-HA Novel Biocomposites and Study of their Tribological Properties for Orthopedic Applications. *Journal of the Mechanical Behavior of Biomedical Materials*, *3*(3), 178–188. doi:10.1016/j.jmbbm.2009.08.003 PMID:20129417

Zubtsov M.E., & Korsakov V.D. (1971). Stoykost' shtampov. *Mashinostroenie*, 200.

Chapter 3
Changes in the Structural-Phase State of Ferrite-Perlite Steel Under the Action of Electrolyte-Plasma Surface Quenching:
Electrolytic-Plasma Surface Quenching of Steel Grade Two

Yerkezhan Tabiyeva
https://orcid.org/0000-0002-9726-7187
D. Serikbayev East Kazakhstan Technical University, Kazakhstan

Bauyrzhan Rakhadilov
PlasmaScience LLP, Kazakhstan

Gulzhaz Uazyrkhanova
D. Serikbayev East Kazakhstan Technical University, Kazakhstan

Didar Yeskermessov
https://orcid.org/0000-0002-2206-8132
D. Serikbayev East Kazakhstan Technical University, Kazakhstan

Zarina Aringozhina
https://orcid.org/0009-0001-8428-4033
D. Serikbayev East Kazakhstan Technical University, Kazakhstan

Waqar Ahmed
University of Lincoln, UK

ABSTRACT

This chapter presents an analytical review of literature sources devoted to research on the rapid heating of steels using various heat sources; in particular, on plasma surface hardening of materials. The results of studying the effect of a hardened surface layer during plasma hardening in the cathodic mode on the mechanical and tribological characteristics of grade two wheel steel are presented. Based on the experimental data obtained, a method for electrolytic-plasma surface hardening of wheel steel was developed and the optimal mode was determined. A description of the phase composition and morphology of grade two steel in the initial state and after electrolytic-plasma surface hardening is given. These descriptions are accompanied by images obtained by transmission electron microscopy. The authors present bright-field images and microdiffraction patterns obtained from these areas, as well as their dark-field patterns.

DOI: 10.4018/978-1-6684-6830-2.ch003

INTRODUCTION

The development of modern mechanical engineering is associated with the development of new materials, the introduction of new technologies and equipment, the invention of new ways to improve the performance of machine parts and mechanisms. The wear resistance, strength and corrosion resistance of machine parts are significantly affected by the condition of their surface layer, which is dictated by the manufacturing technology. Because the destruction of the surface during wear leads to the failure of 60-80 percent of mechanical engineering products (Makarov, 2009). Wear of working surfaces and fatigue of joints are key factors limiting the strength and reliability of various components of railway transport, construction, road and agricultural machinery, metallurgical and pressing equipment. Wear leads to a change in the shape, size and condition of the surface of the parts, which leads to a loss of functional qualities and operability of technical means of production, as well as to an increase in the probability of failure. Most mechanical engineering products are still made of steel or cast iron. Therefore, increasing the wear resistance of iron alloys is one of the most important and urgent tasks of technical physics.

There are several technical solutions to improve the quality of component surfaces. The most common electroplating and chemical methods of coating, surfacing, spraying, chemical and thermal treatment, surface hardening with high frequency currents, surface hardening with concentrated energy flows (plasma, electron beam, laser), etc. (Vyacheslav, 2018). Surface hardening by focused energy flows is the most promising method of surface hardening of iron alloys. However, the widespread introduction of some methods of surface hardening with a concentrated flow of energy, such as laser and electron beam, is hindered by the high cost and complexity of the equipment, its insufficient reliability and performance, the need for vacuum, special rooms with special requirements, the need for qualified maintenance and high operating costs (Lahtin & Arzamasov, 1985). As a consequence, according to its technical and economic indicators, as well as the results of a comparative study, a method of plasma surface hardening, devoid of the listed disadvantages, is proposed for widespread use (Troyannikov, 2021). Plasma surface hardening has been successfully developing in recent years and is currently widely used in various industries, primarily for the heat treatment of railway transport parts (the center plates of the freight wagons, the spring beam and the side frame of the trolley, the bandage, etc.) (Dyachenko, 1981).

Being one of the varieties of plasma quenching – electrolyte-plasma surface quenching (EPSQ) has recently been developing and intensively studied. This method is characterized by lower energy consumption, simplicity of technological equipment and large size of the hardened zone (Zaides, 2020). This method has several advantages, including a sufficiently high performance and the possibility of quenching parts of a larger and more complex profile, and the degree of hardening is equivalent to plasma quenching. It is well known (Tyurin, 1999) that the resource of steel parts is determined by the indicators of their operational characteristics (starting from the chemical composition of the material and ending with the conditions in which one or more parts should be operated), most of which are directly dependent on the structural and phase state of the material. There are very few publications devoted to the study of structural-phase transitions in wheel steel during plasma surface quenching, and they do not sufficiently cover the problems of structure formation. At the same time, there is no data in the literature on the treatment of wheel steels by the method of electrolytic-plasma hardening, which requires additional experimental work to assess the prospects for the applicability of the method for strengthening wheelset bands.

In connection with the above, this work focuses on the development of a method for hardening wheel steel, as well as the study of the formation of the structure, phase composition and tribological properties of the hardened surface layer of grade two steel by quenching during electrolytic plasma surface treatment.

1. PLASMA SURFACE HARDENING OF MATERIALS

In this section are presented analytical review of literature sources devoted to research on high-speed heating of steels using various heat sources, in particular, on plasma surface hardening of materials, is presented. The results of studies of the experimental work carried out on surface hardening of the working surfaces of steel parts using various methods and technologies used during manufacture or during their operation are considered.

1.1 The Status of Existing Methods of Surface Heat Treatment of Steels

Currently, methods of surface hardening of materials by forming nanoscale phases in them are poorly understood, although they are very promising, especially when modifying surface layers by plasma quenching, since nanostructured states are achieved especially effectively at high heating rates and with a short exposure time to high temperatures. Studies aimed at establishing the regularities of the formation of the main types of structures in materials allow us to purposefully apply certain modes of surface hardening to ensure the necessary complex of operational properties, therefore they are relevant (Tyurin, 1999).

Solving the issues of increasing the reliability and durability of critical products operating under extreme conditions under variable contact, shock, static loads require the use of modern high-strength materials and structural solutions. It is also possible to provide the necessary parameters of the product by changing the state of the surface layer of steel. There are large number of traditional technological processes for processing (hardening) surfaces, which have their own characteristics, advantages and disadvantages, but in the last decade much attention has been paid to high-energy methods for improving the surface of steels and, despite numerous publications, the mechanisms of hardening of metals and alloys cannot be considered fully elucidated (Majorov, 2009). The development of promising technologies aimed at improving traditional and creating new methods of influencing the surface to give it the properties required by operating conditions is undoubtedly one of the urgent tasks of modern science and technology (Paulmier, et al., 2008). Traditional technologies of thermal and chemical-thermal treatment of machine parts and tools are designed for volumetric surface treatment and do not allow local treatment of rapidly wearing areas (Balanovskij, 2006).

Along with metallurgical methods and heat treatment in the conditions of manufacturing plants, local surface hardening of wear surfaces using various technologies used both during manufacture and during operation is also considered to increase the service life of parts. At the same time, plasma and magnetoplasma surface hardening are recommended for widespread implementation from all existing methods of hardening (gas-flame, high-frequency currents, plasma, electron beam and laser) according to their technical and economic indicators and the results of comparative analysis (Leshinskij et al., 1990).

However, plasma and magnetoplasmic surface hardening technologies require the use of special expensive equipment, the creation of special processing sites and are economically feasible only for mass, large-scale production. Along with the above methods, one of the promising ones is electrolyte-plasma surface hardening (EPSH), which is the most economical and productive method. This technology is characterized by lower energy consumption, simplicity of technological equipment and large size of the hardened zone (Skakov et al., 2012). The advantages of the method are sufficiently high process productivity and the ability to harden parts of large mass and complex profile, and the degree of hardening is comparable to plasma hardening (Kondrat'ev et al., 2014). Functional qualities of steel parts, such as chemical composition and conditions of use, which are directly related to the structural and phase state of the part, are used to predict their service life (Grechneva, 1992).

The works devoted to the engineering of surface layers of wheel steel in the literature over the past few decades (since 1970) are few (Balanovskij, & Nesterenko, 1992; Sadovsky & Malyshev, 1984; Skripkin, et al., 1992; Van Gyui, 2015) and they do not fully cover the issues of structure formation. At the same time, despite the available amount of work on plasma quenching, the issue of electrolyte-plasma surface quenching in an aqueous electrolyte solution has not been sufficiently studied, and the quantitative characteristics of the fine structure of grade two ferritic-perlite class wheel steel have not been studied, which make this work relevant for use in the engineering, railway and manufacturing industries of our country. Currently, electrolyte-plasma technologies are intensively studied in many scientific and production teams, which indicates the relevance of research and the demand for their results by industry (Tyuftyaev, 2002).

Modern technologies of surface hardening of metals and alloys using concentrated energy flows (CEF), such as laser beam (Chudina, 2003; Majorov, 2009), electron beam (Poletika, et al., 2006), plasma arc (Balanovskij, 2015; Berdnikov, et al., ; Gulyaev, 1986), are characterized by high heating and cooling rates, short duration and locality of impact on the metal. The main requirements on the part of industrial enterprises when using new strengthening technologies based on CEF are associated with increasing the strength, wear and corrosion resistance of machine parts and tools (Zhang et al., 2009). The information available in the literature provides contradictory information about the prospects for the introduction of hardening technologies in various industries (Balanovskij, & Grechneva, 2015; Richard, 2007). Each of the authors of a particular technology gives a subjective assessment of the prospects of its development, finding shortcomings in other methods of hardening. In addition, the theory and practice of processing metals with concentrated energy flows are still far from perfect.

One of the first works on concentrated energy flows, in particular, on plasma surface hardening, were carried out at the Irkutsk Polytechnic Institute under the supervision of Professor V.N. Matkhanov. In the first works (Grigoriants, & Safonov, 1987; Lakhtin & Arzamasov, 1985; Zhang et al., 2005), the peculiarities of heating and cooling metals and alloys by concentrated energy flows were only stated without further investigation. Fundamentally new structures and phases in metals at extreme heating and cooling rates compared to furnace heat treatment, quenching with high frequency currents, welding were not detected. The resulting structures in the impact zone did not differ externally from the traditional ones observed in optical and electron microscopes (Balanovskij, 2016; Samotugin, et al., 1997), according to the authors, in general, everything looked consistent with the classical theory of phase and structural transformations.

At the same time, there was a high hardness and degree of dispersion of the resulting structures, heterogeneity of their distribution over the width and depth of the surface layer (Balanovskij, & Pletnikov, 1999; Balanovskij, & Pletnikov, 1999). In (Balanovskij, 2006; Stalinskij, & Rudyuk, 2007), new hypotheses of phase transformations were put forward from the perspective of nonequilibrium thermodynamics and the results of relevant experimental studies were presented, as well as an attempt was made to build a physical basis for the theory and practice of plasma surface hardening by the authors of the works. In (Medvedev, et al., 2015; Tyurin, 1999), general issues of plasma hardening were formulated for the first time, research goals were set and tasks requiring solutions for the successful introduction of real plasma hardening technologies into production were outlined. In these works, an important fact is noted that the surface hardening of the CEF of metals is not a classical volumetric heat treatment with stable critical points Ac_1 and A_{C3}, the stage of homogenization, critical cooling rates, thermokinetic diagrams. The theoretical concepts adopted for the classical heat treatment of metals are used in the analysis of the processes of structure formation in the surface layer of metal only as a guide for research and further interpretation of the results (Balanovskii, 2016; Balanovskij, & Pletnikov, 1999).

Surface hardening of metals and alloys by concentrated energy flows is an interdisciplinary branch of scientific research that has, on the one hand, much in common with the main processes noted, and on the other hand, fundamental differences and patterns (Balanovskij, 2006). A number of works on plasma hardening are aimed at solving the technological problem of increasing the hardness of a particular steel grade from which a particular part is made (Pogrebnjak, & Tyurin, 2000; Skakov, Bayatanova, & Scheffler, 2013; Skakov, et al., 2012). Most of the research in this area has been carried out on equipment with fundamental design and technological features: plasma torches with interelectrode inserts (Tyuirin, & Pogrebnjak, 2001), different plasma–forming gases and media (Leshchinsky & Samotugin, 2001; Menin et al., 1981; Pogrebnjak, et al., 2003), different technological modes - a jet or an arc of direct or reverse polarity (Medvedovskaya & Shur, 1983; Polyak, 1995), a three-phase arc, a vacuum arc with different execution techniques, with magnetic or mechanical expansion of the jet (arc), a creeping or reflected plasma jet (Balanovskij, 2015; Balanovskij, 2016).

A separate area of work in the field of plasma hardening of metals and alloys is surface modification and alloying using plasma-forming media and elements. The use of liquid media as carriers of alloying elements in plasma surface hardening is considered in (Belkin, 2005). By conducting plasma hardening through a liquid medium, it is possible to saturate the surface layer with various alloying elements (nitrogen, carbon, boron, silicon, etc.), as well as reduce residual deformations and stresses. Plasma cyclic hardening is considered in (Tyurin, 2000), where studies are given on changes in the structure in the hardened layer depending on the number of heating cycles for carbon and alloy steels. The results of tests on the wear resistance of a large number of grades of steel and cast iron are presented. In this work, the mechanism of structure formation in the hardened layer is revealed in the process of multi-cycle heating and cooling, which contributes to obtaining structures of complete hardening in the hardened layer. The issue of structure formation during plasma hardening is considered in more detail in (Grechneva, 1992; Yerokhin, et al., 1999; Yushenko, et al., 1994) and the authors draw conclusions about the parameters of the technological process according to the final structure of the hardened layer.

In (Bayatanova, et al., 2012), the kinetics of austenite growth and cooling during plasma heating is considered and it is argued that the heating rate significantly affects the size of the austenitic grain: the higher the plasma heating rate, the smaller the size of the austenitic grain when heated, therefore, highly dispersed martensite is obtained during cooling. In this regard, the proposed technological thermograms make it possible to regulate the growth of austenite grains. The authors of (Kong et al., 2012), investigating the structure of hardened carbon steels after plasma quenching, fix a large amount of residual austenite in the hardened layer. In their opinion, this is due to the incompleteness of the hardening process. The results of the work show that cyclic treatment should be used to regulate the proportion of residual austenite (Balanovskij, 2016). Technologies of electron beam hardening of metals using carbon-containing pastes and coatings are widely used in industry (Korotkov, 2016). In contrast to this method, plasma surface hardening is possible both in the micro-melting mode of the surface (Ivanov, et al., 2011) and without it, which is unattainable for electron beam alloying methods. Visualization of

Figure 1. Comparison of martensite microstructures in a hardened layer of 080A67 steel a) with electronic quenching; b) laser and plasma quenching; c) in arc mode (the upper picture in the corner of the SEM, the general picture of the TEM)
Source: Grechneva (2017)

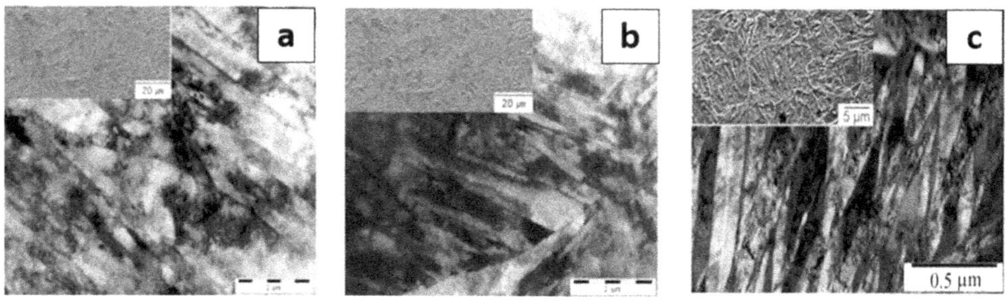

the plasma surface hardening process in the arc mode showed that for each steel grade there is a range of modes when surface melting does not occur (Balanovskij, 2016).

The presented information shows the competitiveness of the technology of plasma surface hardening of metals and alloys in comparison with other sources of concentrated energy. At the same time, the heterogeneity of technologies and equipment for plasma processing leads to different information about the processes of structure formation in the surface layer of metal. The results of a comparative evaluation of various methods of surface hardening summarized in (Grechneva, 2017) are shown in Figure 1.

The considered results of theoretical and experimental work in the field of plasma hardening represent the foundation for specific technologies of surface hardening of machine parts and tools. In particular, in the works (Korotkov et al., 1991), the issues of hardening of specific parts made of 5140, 1045, 080A67 steels are considered. For the first time, the effect of nanostructuring of the surface layer after plasma hardening was discovered during the development of plasma hardening technology for rail steel and was considered in detail in (Nesterov, & Sapozhkov, 1989). It is shown that the nanostructuring of the metal surface layer during plasma hardening depends on the modes used (Jin et al., 2013). In the works (Belkin, 2005; Kidin, 1978), new technological methods of heating machine parts and tools are proposed, which allow structuring the surface layer of metal at the nanoscale. In the development of this topic, a complex of works and studies has been carried out in order to develop a specific technology for plasma hardening of the bandage ridge (Tyuftyaev, 2002).

The results of the research carried out in the listed works with different types and technologies of surface hardening show the undeniable advantages of plasma surface hardening in comparison with the rest, both from the point of view of the economic component and from the side of resource conservation.

1.2 Physical Fundamentals of the Technology of Plasma Treatment of the Steel Surface in an Aqueous Electrolyte Solution

The development of scientific and technological progress in various spheres of the economy requires the widespread introduction into industry of new efficient and economical technological processes for strengthening materials based on the achievements of modern science and technology (Erkinbekkyzy et al.,). One of the promising areas that significantly expand the technological capabilities of the process of deep processing of metal products by strengthening heat treatment is the use of highly concentrated energy flows (laser, plasma, electron beam, etc.). However, the widespread introduction of most of the known methods of hardening heat treatment with a highly concentrated energy flow, namely, laser, electron beam, cathode-ion, is constrained by the high cost and complexity of the equipment, its insufficient reliability and performance, the need to use vacuum, special rooms with special requirements, the need for qualified maintenance, high operating costs.

Unlike the processes listed above, the technology of surface electrolyte-plasma surface hardening of metals is devoid of the above disadvantages, has been successfully developing in recent years and is increasingly being used in various industries. Practice shows that surface plasma hardening is rational in terms of versatility, accessibility, environmental friendliness and economic efficiency in order to extend the operational life of heavily loaded parts and assemblies. Without changing the surface roughness parameters, such a hardening heat treatment is easily integrated into the technological process of restoring parts, is low-cost, sufficiently productive and allows to effectively increase their operational durability (Skakov, Kurbanbekov, Tabieva, et al, 2013). In the electrolyte-plasma technology, the processed product is an anode or cathode, a corresponding positive or negative potential from the power source is supplied to it, and the working bath serves as the opposite electrode. Depending on the applied voltage when an electric current passes through an aqueous electrolyte solution (Balanovskii, 2016).

The available experimental data indicate that the classical laws of electrochemistry are fulfilled only at low values of current and voltage densities on the electrodes (Moon, & Jeong, 2009). A more general physical picture is when, after exceeding the critical values of the electrical parameters, a gas-plasma cloud forms around the metal electrode, pushing the electrolyte away from the electrode surface. In this case, a multiphase metal–plasma–gas–electrolyte system arises, in which not only ions, but also electrons serve as charge carriers, and the phenomena occurring in the near-electrode region do not fit into the framework of classical electrochemistry. The passage of an electric current through an electrolyte cell containing two metal electrodes, one of which has a much smaller surface than the second, with a gradual increase in voltage at the electrodes leads to different phases or modes of the process, which can be traced to the volt-ampere characteristic of the active electrode–electrolyte system (Figure 2).

Figure 2. Volt-ampere characteristic of the active electrode-electrolyte system
Source: Kulikov et al. (2010)

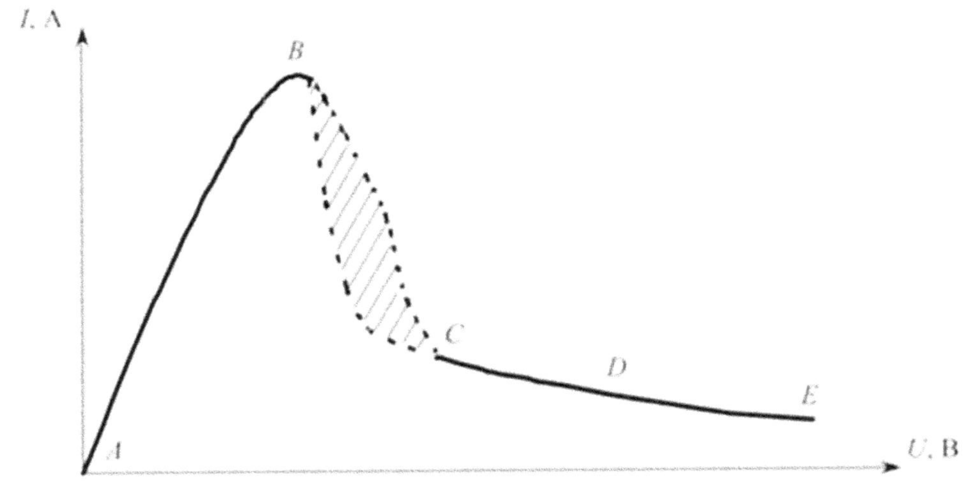

The AB section in Figure 2 is a conventional electrolysis process in which metal ions are transferred and gas release is observed depending on the composition of the electrolyte and the electrode material, and is described by classical electrochemistry. This section on the volt-ampere characteristic has a linear dependence of current on voltage, which means a constant value of the electrical resistance of the interelectrode gap.

With an increase in the voltage on the electrodes to 60-70 V, a transient or switching mode is established (a section of the BC), when a vapor-plasma shell is formed periodically around the active electrode, with a frequency of about 100 Hz, leading to current locking for 10^{-4} seconds (Kulikov et al., 2010). Due to the continuous change in the electrical resistance of the near-electrode region, each voltage value corresponds to a range of possible current values. At the moment of opening the circuit, a voltage pulse occurs according to the law of self-induction. The magnitude of the voltage in the switching mode may be variable. The resulting neutral gas shell periodically breaks through the resulting voltage pulse. After the breakdown, the active electrode comes into contact with the liquid electrolyte again and the process repeats. At a voltage of more than 200 V, the CD section indicated in Figure 2 forms a stable vapor-plasma shell around the active electrode, characterized by small current fluctuations at constant voltage. In this voltage region, the process of electrolyte-plasma treatment takes place. The falling character of the volt-ampere characteristic in this section can be considered as an increase in the resistance of the circuit with an increase in voltage. The increase in resistance is explained by an increase in the thickness of the vapor-gas shell. The solid vapor-plasma shell has a thickness of about 50 microns and constantly changes its shape. Electric current flows through a complex metal–plasma–gas–electrolyte system (Duradzhi & Parsadanyan, 1988).

With a further increase in voltage, the intensity of the glow of electric discharges increases, which leads to an increase in the thickness of the vapor plasma shell and separation over the entire surface of the active electrode, which is accompanied by a decrease in the current value by 2-2.5 times. An electrohydrodynamic mode is established in which the temperature of the active electrode drops to 100ºC with and the effect of the vapor-plasma shell disappears. Studies have shown (Kulikov et al., 2010) that the onset of the electrohydrodynamic regime depends on the concentration of the electrolyte. With an increase in the concentration of the solution, this mode is set at a lower voltage.

The current density is the main parameter that must be controlled during the electrolytic plasma treatment (EPT) process. Usually, to achieve the necessary conditions for plasma electrolysis, the current density is set within 0.01–0.3 A/cm^2. According to Faraday's first law, this determines the growth rate of the coating. As the coating grows, the voltage increases rapidly at first, then slowly, until constant plasma conditions are established. The critical rate of voltage change corresponds to the establishment of a spark discharge on the surface of the electrode. This value is very dependent on the characteristics of the metal–electrolyte combination and is usually located in the range of 120-350 V. At the first stage of the process, sparking is observed in the form of uniform white light surrounding the electrode. As the coating grows, it changes and appears as individual yellow sparks that move rapidly through the surface. Gradually, the density of sparks decreases, but their power increases. Finally, several red spots appear, slowly moving through the surface. From time to time, these powerful arcs lead to current fluctuations and damage to the coating. Therefore, as soon as such arcs are detected, the process is immediately terminated. EPT can be carried out using both anodic and cathodic polarization of the product (Kulikov et al., 2010).

1.3 Structural-Phase Transformations of Ferrite-Pearlite Steels Under Plasma Surface Hardening

Despite the difference in the physical processes underlying a particular method of surface hardening of metals (plasma, laser, electron beam, etc.), a common feature is characteristic of all - phase and structural transformations occur on the working hardened metal surface (Dunleavy et al., 2009). Let's consider the main, general points of the theory of hardening of ferrite-pearlite steels. As is known, the strength characteristics of any types of steel depend on the properties of the structural components of this steel, in our case ferrite and perlite as the main component of the structure of wheel steels.

Ferrite, depending on the degree of its alloying with other elements such as silicon, manganese, phosphorus, may have a different degree of hardening. To date, it can be said that such non-carbide-forming elements as phosphorus, silicon and nickel are entirely part of ferrite, molybdenum and chromium are partially dissolved in ferrite, whereas manganese can completely dissolve in ferrite. The most hardening effect is associated with carbon dissolved in the α-Fe lattice. The strength of the structural component also depends on the grain size and the Hall-Petch dependence is determined:

$$\sigma T = \sigma i + Kd^{-1/2}, \quad (1)$$

Value: σ_i – in ferrite-perlite steels, it is characterized by the friction stress of the α-Fe lattice, solid-solution hardening, hardening due to the formation of perlite, deformation hardening, dispersion hardening. $K_d^{-1/2}$ – grain boundary hardening.

The effect of these parameters on the yield strength of ferritic-pearlite steels is linearly additive, so it can be summed up. For a deeper understanding of the processes occurring during structure formation, we further use primary designations such as ferrite and iron carbide (cementite) Fe_3C, which are the main components in wheel steels.

In the first approximation, the contribution of the lattice is $2 \cdot 10^{-4}G$, where G is the shear modulus. The contribution of the lattice can be estimated by the ratio:

$$G_{dis} = \alpha 1 \cdot M \cdot G \cdot b \cdot \rho v^{1/2}, \quad (2)$$

where:

α_1 – coefficient depending on the nature of the distribution and interaction of the dislocation;
M – numerical coefficient;
b – the Brugers vector ~ 0,25 nm;
ρ_v – the density of dislocations in the grain volume.

Transmission electron microscopy or X-ray diffraction analysis of the material under study is used to determine the density of lattice dislocations.

Iron carbide (cementite) has a complex rhombic lattice, the unit cell of which consists of 72 Fe atoms and 24 C atoms, with a tightly packed structure where iron atoms are located in the interstices by carbon atoms, which provides a good orientation and structural correspondence between Fe_3C and γ-Fe, the same lattice correspondence is observed for Fe_3C and α-Fe. When martensite disintegrates, cementite is easily generated in it, which is known to have a high hardness.

Figure 3. Diagram of the "iron-carbon" state

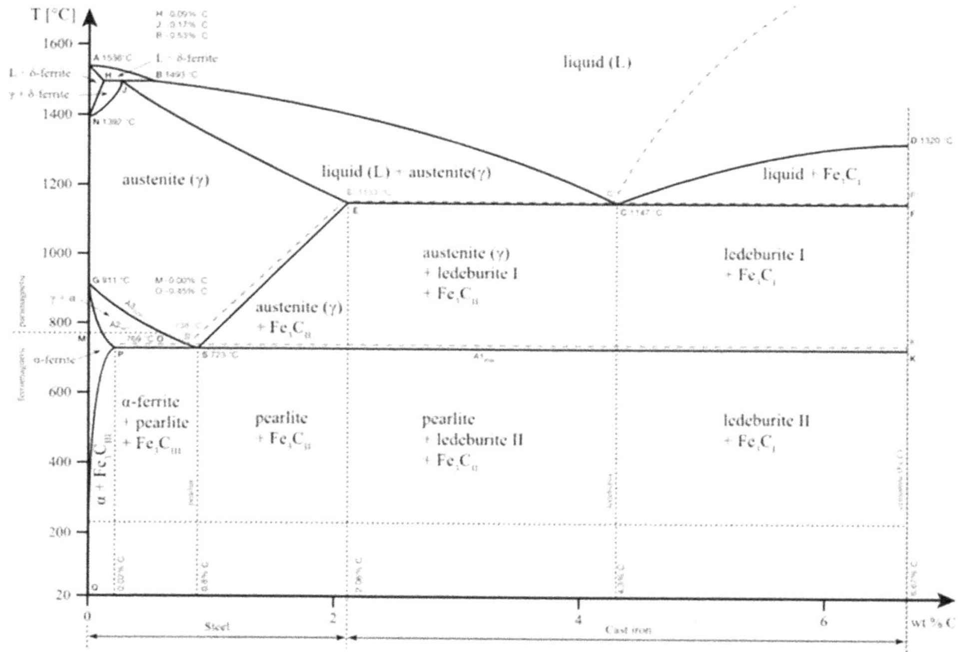

During the process of structuring, the iron atoms in cementite are easily replaced by atoms of some alloying elements that are part of the wheel steel. Thus, it was found that (Martin, et al., 2013), the basic element of wheel steel – manganese, is infinitely soluble in cementite, due to the fact that it forms a carbide similar to iron carbide. When analyzing the processes of structure formation during plasma hardening of wheel steel, it is important to take into account the structural properties of austenite formed in the high-temperature region in the "iron-carbon" transformation diagram (Figure 3).

Austenite, as a stable structure, exists at temperatures of 700-1400°C (##NO_NAME##, 2012), and metastable in the form of residual austenite after quenching. It is this structure that, when cooled, forms the final structural components such as perlite, martensite, bainite, troostite. Thus, a preliminary analysis of previously available works on the theory of hardening applied to wheel steels shows that during plasma hardening, dislocation, dispersion and deformation hardening occupy the first rows, which depend on the cycle, the duration of heating and cooling of the surface layer, which ultimately determine the strength of the material. The plasma surface hardening process consists of the following stages: heating, phase transformation, partial homogenization and cooling. When surface quenching with melting of the surface, the melting process is turned on as another stage-melting. The main issues of austenite formation and the kinetics of grain growth under the influence of various parameters and factors affecting it were considered in detail in the works of S.Dyachenko (Dyachenko, 1981), V.Sadovsky, K.Malyshev (Sadovsky & Malyshev, 1984), A.Gulyaev (Gulyaev, 1986), and other authors. In the works of L.Leshchinsky, S.Samotugin, I.Pirch (Leshinskij et al., 1990) detailed comparative technical and economic assessments of various methods of surface hardening were carried out, which makes it possible not to consider this aspect in this study.

It is known (Wang et al., 2013) that the performance characteristics of steel parts vary from the chemical composition of the material to the conditions in which a particular part should be used, most of which directly depend on the structural and phase state of the part. Despite the available amount of work on plasma quenching, the issue of electrolyte-plasma surface quenching in an aqueous electrolyte solution has not been sufficiently studied, and the quantitative characteristics of the fine structure of grade two ferritic-pearlite grade wheel steel have not been studied, which make this work relevant for use in the engineering, railway and manufacturing industries of our country.

Conclusions on the Section

The analysis of the literature data shows that:

1. The combination of the existing classical technologies of heat treatment and surface hardening by local treatment of their working surface will allow to obtain high performance properties of steel products.
2. Electrolyte-plasma treatment is one of the promising methods tested by numerous authors of works on the modification of the structural-phase state of the surface layers of steels and alloys. This method is largely devoid of the disadvantages inherent in traditional mechanical and electrochemical processing methods, and additionally allows you to save material resources. Its advantages are high productivity and efficiency, compliance with environmental cleanliness, high quality and speed of operations performed. However, experimental data and information are not sufficient to conduct surface hardening of bandage steel in an aqueous electrolyte solution, as well as information about structural-phase transformations and features of the formation of a fine structure in the surface layers of grade two wheel steel during surface hardening in an aqueous electrolyte solution, which makes this study relevant.

2. MATERIAL AND METHODS OF RESEARCH

2.1 Research Material and Sample Preparation

In accordance with the tasks set, grade two bandage steel was chosen as the research material. This steel is carbon steel, used in the field of metallurgy and mechanical engineering, for the manufacture of bandages of freight, passenger and shunting locomotives, motor cars and diesel trains, subway cars and wheel bandages for pairs of tram cars. The chemical composition of grade two steel, according to GOST 398-96 is given in Table 1.

Table 1. Chemical composition, % (GOST 398-96)

<table>
<tr><th colspan="7">Mass Fraction of Elements</th></tr>
<tr><th>C</th><th>Si</th><th>Mn</th><th>V</th><th>S</th><th>P</th><th>Fe</th></tr>
<tr><td rowspan="2">0.57-0.65</td><td rowspan="2">0.22-0.45</td><td rowspan="2">0.60-0.90</td><td colspan="3">no more</td><td rowspan="2">other</td></tr>
<tr><td>0.15</td><td>0.04</td><td>0.035</td></tr>
<tr><td colspan="7">Note: according to GOST 398-96, it is allowed: Mo £ 0.08%, Ni £ 0.25, Cr £ 0.20%, Cu £ 0.30%.</td></tr>
</table>

grade two steel in its initial state was a material that had been quenched from a temperature of 890°C (2-2.5 hours) with cooling in warm (30-60°C) water and subsequent tempering at 580°C (2.5-3 hours).

Blanks of steel samples of grade two for the study were cut out of a bandage in the form of a parallelepiped with a size of 15×15×10 mm^3. At low cutting speeds and low load, the sample does not experience deformation and thermal effects. Preparation of metallographic grinds of steel samples was carried out according to the methods.

2.2 Optical Metallography

Metallographic analysis is widely used in industry to control the structure of metal, semi-finished products and finished products to ensure their quality. To study the general nature of the structure, an optical microscope "ALTAMI-MET-1M" was used. The metallographic method of analysis has been introduced into a number of existing standards. According to the microstructure, the grain size was controlled [GOST 5639-82], the volume fraction of the material that has undergone decay and the volume fraction of the α-phase in grade two steel. The grain size of the microstructure of alloys was determined from photographs by the method of arbitrary secant, using the equation:

$$< d >= \frac{L \cdot p \cdot 10^3}{z \cdot B} \tag{3}$$

where:

L – secants length;
p – number of secants;
z – the number of grains crossed by straight lines;
B – magnification of the image in the photograph.

The slots for metallographic studies were prepared according to a standard technique, including mechanical grinding and mechanical polishing (Baranova, & Demina, 1986). Polishing was carried out on a grinding disc, covered with felt, pre-washed and soaked in water for 1-2 hours. Further, in order to identify the structure, the slots were etched. A 4% alcohol solution of nitric acid was used for etching. Qualitative and quantitative metallographic analysis was carried out according to the recommendations given in (Gorelik, et al., 2002; Rahadilov, 2014).

2.3 X-Ray Diffraction Analysis

Qualitative and quantitative phase analysis of the structure of steel samples was carried out on a Rengen diffractometer "X'pertPro" of the company "PANanalytical", using Cu-K_α radiation (λ=2.2897 A^0) at a voltage of 40 kV and a current of 30 mA. The samples were taken for reflection according to the Bragg–Brentano focusing scheme in cobalt and copper K_α - radiation. In order to exclude β-lines on diffractograms, a selectively absorbing β-filter made of iron was used when shooting on cobalt radiation, and a nickel filter was used when shooting on copper. When shooting on copper radiation, to exclude the influence of fluorescent radiation on the diffraction pattern of the samples, an Aluminium foil with a thickness of 0.05 mm was used on the X'pertPro diffractometer. The decoding of diffractograms was carried out manually using standard techniques and the PDF-4 database, and quantitative analysis was performed using the Powder Cell program.

Sample preparation, selection of shooting modes and calculation of diffractograms were carried out according to the methods. The interplane distances d_i were determined from the Wulf-Bragg formula:

$$\frac{d_i}{n} d_{HKL} = \frac{\lambda}{2\sin\theta} \tag{4}$$

where:

λ – the wavelength of the characteristic X-ray radiation;
n – the order of reflection;
θ – diffraction angle.

The interference indices and the type of crystal lattice of the phases were determined by the series Q found for each phase from the expression:

$$Q = \sin^2\theta i / \sin 2^{\theta} 1 \tag{5}$$

According to the found HKL indices and the value of the interplane distance for the line in the area of angles 77-112°, the periods of the unit cell of the lattices present in the sample phases were determined from the ratio:

$$a = \frac{\lambda}{2\sin\theta} \cdot \sqrt{H^2 + K^2 + L^2} \tag{6}$$

The error in determining the lattice parameter for cubic syngony crystals is found by the formula:

$$\Delta a = a \bullet ctg\theta \bullet \Delta\theta \tag{7}$$

Where Δθ – error in determining peak angles on the diffractogram.

The average size of the crystallites was determined using the Debye-Scherrer equation for broadening diffraction maxima, taking into account the instrumental broadening and the software "Powder Cell 2.4". The instrumental broadening was determined at the half-width of the maximum of the standard silicon powder and was 0.14.

The study of the phase composition was carried out on the basis of the fact that each crystalline substance (phase) has a certain crystal lattice characterized by a set of interplanar distances d_{HKL}. The latter are found from the Wolf-Bragg equation, which determines the direction of the diffraction maxima of elastically scattered X-ray radiation on the crystal:

$$2d_{HKL}\sin\theta = n\lambda \tag{8}$$

Where:

d_{HKL} – interplanar distance;
θ – sliding angle (Bragg angle);
λ – wavelength;
n – the order of the diffraction maximum.

The indexing of the radiograph involves the determination of all interference indices (HKL) of each line of the radiograph. In this case, the interference indices are equal to the product of the plane indices (hkl) by the order of reflection n. The indexing of crystal radiographs, depending on the syngony, is based on different ratios. So, for a cubic lattice:

$$a = \frac{\lambda}{2\sin\theta}\sqrt{H_i^2 + K_i^2 + L_i^2} \tag{9}$$

To establish the interference indices of each line, the ratio is used:

$$Q_i = \frac{\sin^2\theta_i}{\sin^2\theta_1} = \frac{H_i^2 + K_i^2 + L_i^2}{H_1^2 + K_1^2 + L_1^2} = \frac{d^2_{H_iK_iL_i}}{d^2_{H_1K_1L_1}} \tag{10}$$

Where Q_i – the line at the smallest angle θ.

Thus, by determining a number of Q_i ratios for all the lines of the X-ray in ascending order, it is possible to determine the type of lattice, interference indices and the period of the unit cell.

One of the methods of qualitative phase analysis is to compare the values of interplane distances and intensities of the studied sample with the reference one. At the same time, the JCPDS-ASTM card file is the most widely used. The calculation took into account both measurement errors of the instruments used and errors due to deviations from the intended ideal solution during the experiment. The error calculation data were compared with the data of statistical processing of the results of the study.

2.4 Scanning Element Analysis

The elemental composition of the sample after electrolytic plasma surface quenching was studied on a scanning electron microscope JSM-6390LV - JEOL (Japan), with the prefix of energy dispersive microanalysis INCA Energy by OXFORD Instruments. When electrons interact with matter, response signals of various shapes arise (reflected and secondary electrons, Auger electrons, X-rays, absorbed current, etc.), which are used to synchronously construct an image on the monitor screen.

Figure 4 shows a diagram of the formation of secondary signals under the influence of an electronic probe. An electron-optical system is not used for image formation, image scaling is carried out by radio engineering means. A scanning electron microscope is a vacuum device, since at normal atmospheric pressures the electron beam is strongly scattered and absorbed, which makes it impossible to focus it.

Therefore, the working vacuum in the microscope chamber should be 10^{-5} Torr or better. The movement of the probe over the surface of the sample should be carried out with very high accuracy. The accuracy of the movement and the size of the probe determine the resolution of the microscope. As a result of the interaction of the electron beam with the sample surface, a response occurs, which is recorded by the corresponding sensors. An electron beam from an electron source is formed by a special condenser system in the form of a well-formed probe and passes through a system of control electrodes or electromagnets that move the beam along the sample surface along a trajectory forming a raster. The signal recorded by the sensors is then used to modulate the brightness of the electron beam in the cathode ray tube of the monitor. The magnitude of this signal will depend on the physical properties of the sample surface and may vary from point to point. As a result, an image of the sample surface is formed on the monitor screen, displaying the topography of the corresponding physical property of the sample under study.

Figure 4. Scheme of formation of secondary signals during the interaction of probe electrons with a sample

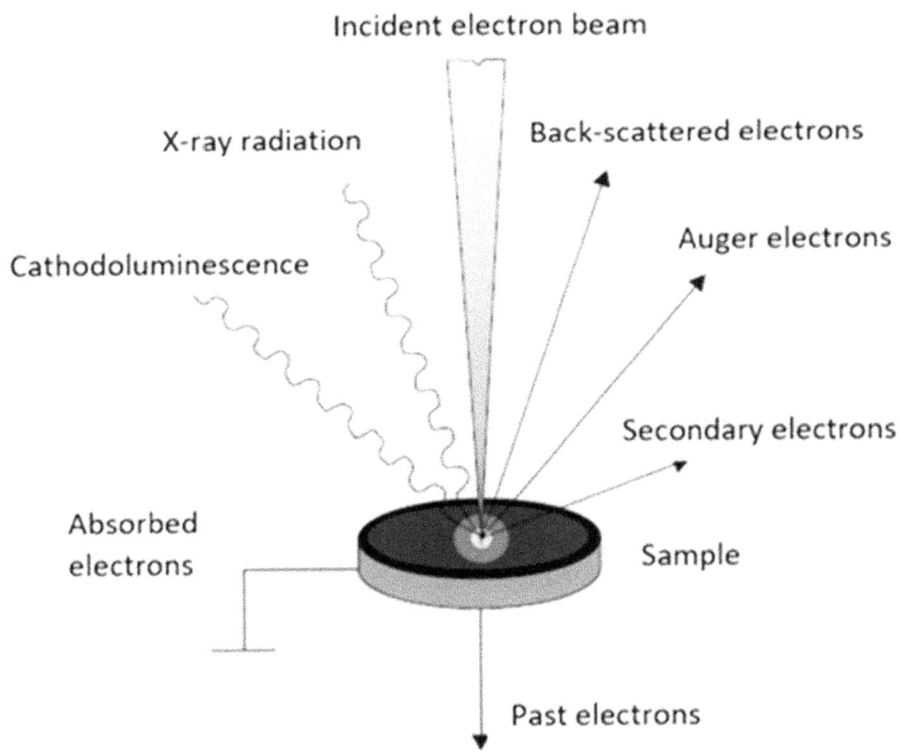

Thus, it is possible to study the topography of inhomogeneities of defects and surface conditions: for example, the topology of the surface (grain boundaries, pores, cracks, inhomogeneities of composition, etc.) – in reflected or secondary electrons. First of all, a scanning electron microscope (SEM) is used to study the structure of the surface. The possibility of studying not only the content of a particular chemical element in the sample, but also the distribution of selected chemical elements over the sample surface makes the scanning electron microscopy method unique.

The energy dispersion method is based on the registration of the so-called characteristic radiation resulting from electron transitions between the internal energy levels of atoms. Depending on the radius of the level, as well as on the binding energy, several types of spectra are distinguished: K, L, M, while only a few X-ray emission lines (K_α, K_β и L- series) are important for materials science (Utevskij, 1973). The use of an energy-dispersive semiconductor detector based on *Si* doped with *Li* in the energy range from 0.7 to 10 keV makes it possible to identify unknown elements. Quantitative elemental analysis consists in finding the concentration of the desired element using the measured intensity of continuous radiation. The mass fraction of an element can be determined by the following formula (Saltykov, 1976):

$$C_i = \frac{I_i \cdot I_{(B)} \cdot C_{(i)} \cdot G_i}{I_B \cdot I_{(i)} \cdot G_{(i)}} \quad (11)$$

Where:

C_i and $C_{(i)}$ – the mass fractions of element i in the analyzed micro-volume of the sample and the standard;
I_i – measured intensity of characteristic X-ray radiation;
I_B – measured intensity of continuous radiation.

The ratio of G-factors takes into account the fact that different intensity of continuous radiation per unit mass in a single volume can be generated in the standard and the sample.

2.5 Transmission Electron Microscopy

Structural studies were carried out using electron diffraction microscopy on thin foils on an EM-125 electron microscope at an accelerating voltage of 125 kV at Tomsk State University of Architecture and Civil Engineering (Tomsk, Russia). The working magnification in the microscope column was 25,000 times. Simplified diagram of a transmission electron microscope showed by Figure 5. The images of the fine structure of steel obtained by viewing in an electron microscope were used, firstly, to classify morphological features of the structure, secondly, to determine the size, volume fractions of the α- and γ-phases, as well as carbide phases and their localization, and, thirdly, to determine the parameters of the fine structure material.

The size and volume fraction of the carbide phases, as well as α- and γ-phases were determined from images confirmed by microdifraction patterns and dark-field images obtained in the reflexes of the corresponding phases. Phase identification was carried out according to the methods described in (Glagolev, 1941). For this purpose, schemes of microdifraction patterns calculated from tabular values of the parameters of crystal lattices were used. For each specific place under study, a large area of foil was viewed on the sample and up to 30-40 micrographs (light-field and dark-field images) and micro-difraction patterns were taken.

Preparation of the foil for viewing in an electron microscope was carried out as follows. Plates with a thickness of ~300 microns were cut out on an electric spark machine parallel to the surface of a grade two steel sample that underwent electrolyte-plasma surface hardening (measurements were carried out with a micrometer with an accuracy of 0.01 microns). It should be noted that the electric spark cutting mode was selected in such a way that it did not introduce additional distortions into the structure of the material.

Figure 5. Simplified diagram of a transmission electron microscope

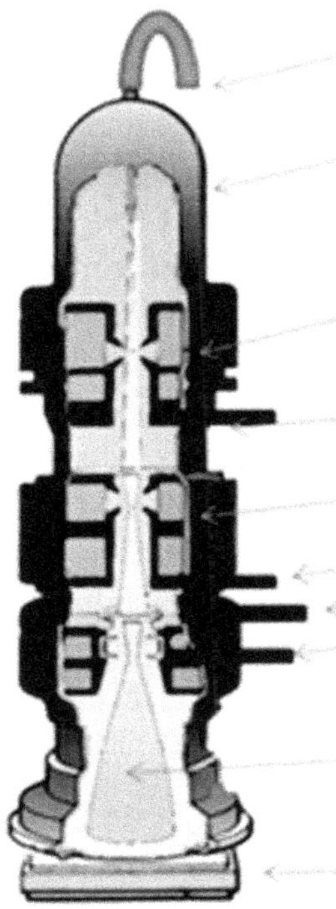

The cut-out plate was cut into two parts. On each part of the plate, one side corresponded to the surface of the processed sample, the opposite side was the side of the cut. A foil corresponding to the near–surface zone of the hardened sample was prepared from one part, and the other part was prepared at a distance of 100 microns from the sample surface.

The foil corresponding to the near-surface zone of the sample was prepared as follows. The surface of the hardened sample was varnished. Then, from the side of the cut, thinning was carried out. The thinning was carried out chemically in a solution of hydrogen peroxide (H_2O_2) with the addition of several drops of hydrofluoric acid (HF). Then, electrolytic polishing was also carried out on the cutting side in a foam of a supersaturated solution of chromium anhydride in orthophosphoric acid at a temperature of 60-80°C. The final stage is the removal of the film from the surface and the final polishing of the sample. Thus, the surface itself was only electrolytically polished for a very short time. Earlier (Malinovskaya, 2006), it was found by the nuclear gamma resonance method that after processing a sample in endothermic atmospheres, films of various iron oxides are always observed in surface zones up to 1 micron deep. Since we were interested in changes in the structure and phase composition of steel occurring in the material at much greater distances from the treated surface, the iron oxides formed on the surface of the sample do not affect the structure of the subsurface layers. Therefore, in order to remove the oxides and not obscure the overall picture, the final electrolytic polishing of the entire sample, i.e. the surface of the sample treated with surface hardening, was carried out for a very short time (~3 seconds). The foil corresponding to the distance from the surface of the sample equal to 100 microns was initially prepared in the same way, i.e. the surface of the hardened sample was also varnished and then thinning was carried out from the cutting side, first chemically, then electrolytically. But the thinning was carried out only to a thickness equal to ~ 30 microns. After that, the film was removed from the surface of the sample and the final electrolytic polishing was carried out on both sides.

2.6 Microhardness Measurements

Hardness, along with fatigue strength and wear resistance, is one of the most widely used mechanical characteristics of metals and alloys. At the same time, the hardness test differs from other methods for determining the mechanical properties of materials by its high sensitivity and short test time. It is carried out without destroying samples, requires a minimum of costs in their preparation. Measurements of the microhardness of alloy samples were carried out on the PMT-3M device in accordance with GOST 9450-76, with loads on the indenter P = 1H (100g) and the holding time at this load of 10 seconds (Figure 6). In a number of works there is quite complete information about the method of determining the microhardness of materials, about microstructural changes that occur during tests, about the physical nature of microidentification processes. As an indenter for measuring microhardness, a regular four-sided diamond pyramid with an angle at the apex between opposite faces of 136° was used, similar to the determination of hardness by Vickers. Prints were placed on the sample every 100 microns. The number of measurements per sample was at least 30. The microhardness of H_{μ} was determined in accordance with GOST 9450-76 by the formula:

$$H_{¼} = \frac{P}{S} = \frac{1854 \cdot P}{d^2}\left[\text{MPa}\right] \quad (12)$$

Figure 6. General view of PMT-3M

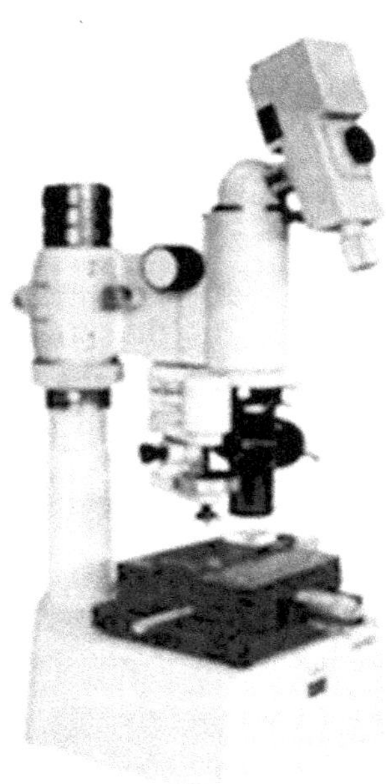

Where d – the diagonal of the print,

$$S = d^2 / 1854 \tag{13}$$

the area of the lateral surface of the resulting pyramidal print.

The tests were carried out on a PMT-3M microhardometer with a LOMO 92063 lens (F=6.3) and with an eyepiece FOM-2-16. The measurement error in accordance with the passport data of the device is 10%. When conducting tests in accordance with the recommendations of GOST 9450-76, the following rules were observed:

- The distance from the center of the print to the edge of the object must be at least twice the size of the print;
- The distance between the centers of neighboring prints should exceed the size of the print by more than 3 times;
- The minimum thickness of the object or layer must exceed the depth of the print by at least 10 times.

Standard computer programs of statistics and correlation analysis were used to process the results of microhardness measurements.

2.7 Tribological Tests

Wear resistance tests were carried out at the installation of the Scientific Research Center "Surface Engineering and Tribology" at the Sarsen Amanzholov East Kazakhstan University for testing materials for abrasive wear (Skakov & Rakhadilov, 2012). The wear resistance of the tested material was evaluated by comparing its wear with the wear of the reference sample (steel 45) according to GOST-23.208-79. The relative wear resistance of the tested steel was calculated by the formula:

$$K_{tm} = \frac{g_{rs} \cdot \rho_{tm \cdot N_{tm}}}{g_{tm} \cdot \rho_{rs} \cdot N_{rs}} \tag{14}$$

Where g_{rs}, g_{tm} – the values of mass loss during testing of the reference sample and samples of the test material, g; ρ_{rs}, ρ_{tm} – density of the reference and test material, g/cm^2; N_{rs}, N_{tm} – the number of revolutions of the roller when testing the reference and test materials.

Tribological dry friction tests were carried out at the center for collective use of TSU, Tomsk, Russia, where a high temperature tribometer (TNT, CSMInstruments, Switzerland) was used, as well as at the TRIBOMETER TRB^3 (Anton Paar) installation at the Scientific Research Center "Surface Engineering and Tribology" at the Sarsen Amanzholov East Kazakhstan University.The tests using the tribometer were based on the measurement by the LVDT sensor of the friction force that occurs when a spherical fixed counterbody pressed with a given force against the sample slides. Tribological measurements were carried out in air (30±5 ° C) in dry friction mode according to the "ball-disc" scheme (Figure 7) in accordance with international standards ASTMG-99 using a counterbody (ball with a diameter of 6 mm) made of Al_2O_3. The load on the counterbody was 5 N. The linear velocity was 2 cm/s. The friction path

Figure 7. Scheme of testing samples for wear resistance using the "ball-disc" method

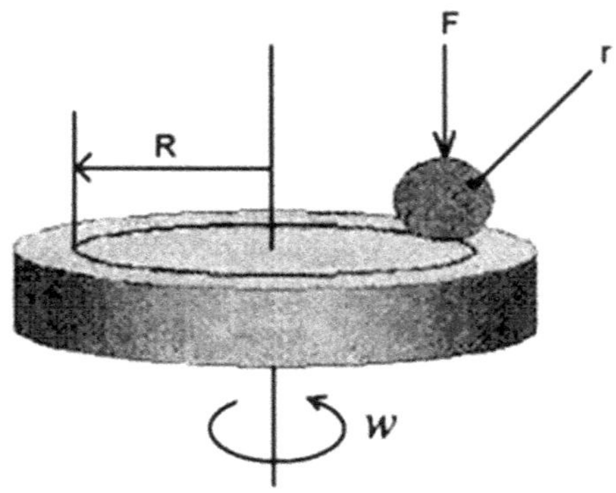

is 31.4 m. In the course of the study, the analysis of wear products, as well as the contact area on the counterbody, was carried out by examining the structures of wear grooves on the surface of the coatings under study, as well as changes in wear spots on the ball. Measurements of the vertical section of the wear grooves were carried out on a profilometer in four diametrically and orthogonally opposite areas, after which the average value of the groove depth and cross-sectional area was calculated.

The surface profile was measured using a profilometer device by measuring the vertical deflection of a diamond tip (probe) moving at a constant speed under conditions of mechanical contact with the sample. The device has a multifunctional RS-232 port, with which it is possible to transfer data to a printer for printing or to a computer for subsequent analysis using additional Talyprofile software. The software allows you to calculate parameters, set calculation modes in full compliance with international standards.

2.8 The Method of Electrolytic-Plasma Surface Hardening

Electrolyte-plasma surface quenching in cathode mode was carried out at the electrolyte-plasma treatment plant at the Scientific Research Center "Surface Engineering and Tribology" at the Sarsen Amanzholov East Kazakhstan University (Tabieva, Rahadilov, Uazyrhanova, et al, 2020).

The installation for conducting electrolyte-plasma surface hardening structurally consists of a power source, a material processing chamber and a personal computer. Table 2 shows the main technical characteristics of the installation.

A schematic representation of the installation (Figure 8) for electrolyte-plasma surface quenching (Tabieva, Rahadilov, Uazyrhanova, et al, 2020), and the processing process (Figure 8b) is shown in Figure 8. A rectifier was used as a power source for the EPSH installation, which gives a maximum output of 360 V /60 A in the form of direct current. Considering that the output of power supplies with the necessary electrical parameters is limited, the power supply of the installation for conducting electrolyte-plasma treatment was produced by a powerful rectifier manufactured by the authors of the works (Rahadilov, 2014). The manufactured DC power supply consists of a welded frame reinforced on the base of a transformer for lowering with a power of 40 kW. The power supply of this source is carried out from a three-phase AC network with a frequency of 50 Hz and a voltage with a maximum value of 380 V. The power supply is controlled by the buttons marked "Start" and "Stop" which are located on the front panel of the housing of this source. A fuse is installed to protect the circuits from short circuits.

Figure 8. Schematic representation (b) of the installation and processing of the sample mounted on the holder with electrolyte plasma in an aqueous electrolyte solution (a)
1– samples; 2 – cone–shaped stainless-steel cell; 3– pallet; 4–pump; 5- heat exchanger; 6– bath with electrolyte.

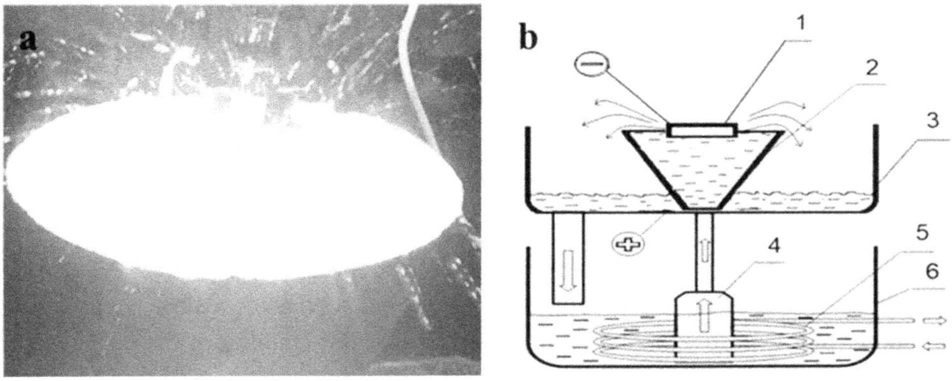

Table 2. Technical characteristics of the installation

Name of the Parameter	Meaning
Supply network voltage, V	380
Supply network current frequency, Hz	50
Maximum power consumption not more than, kW	40
Output voltage, V	0 to 350
Output current, A	0 to 60
Overall dimensions of the power supply, not more than, mm	700×440×600
Overall dimensions of the installation chamber, not more than, mm	660×660×1520
Weight of the power supply, not more than, kg	150
Weight of the installation chamber without solution, not more than, kg	40

Adjustable voltage for the electrolyte-plasma treatment process within the values U = from 0 to 360 Volts, the current value is up to I = 100 Amperes (Rahadilov, 2014).

For EPSH, we selected the following electrolytes containing carbamide and sodium carbonate with different concentrations indicated in Table 3. becouse for surface hardening of steels, aqueous solutions of sodium carbonate are most widely used. As is known (Rahadilov, 2014), for the cathode process, the most optimal component providing a stable discharge is sodium carbonate. In this regard, an electrolyte based on sodium carbonate was chosen for surface hardening. In order to prevent the decarburization process, carbamide was added to the electrolyte. The choice of urea is due to the fact that this substance is a source of carbon, and is also characterized by cheapness and environmental safety. It is also important to note that urea is highly soluble in water: at 20 ° C, the solubility of urea is 51.83% (by weight). In addition, sodium carbonate interacts well with urea. In this regard, sodium carbonate was chosen as the component providing the optimal value of the electrolyte's electrical conductivity, and carbamide was chosen to prevent decarburization of the surface.

Table 3. Working composition of water-based electrolytes

Electrolyte No.	Na_2CO_3, %	$(NH_2)_2CO$, %	H_2O, %
1	10	10	80
2	20	10	70
3	15	15	70
4	10	20	70
5	20	20	60

In general, the hardening process is carried out as follows: before starting work, the working bath is filled with electrolyte. Then the electrolyte enters the electrolytic cell using a pump installed at the bottom of the working bath. In this case, the electrolyte exits through the opening of the cone-shaped partition in the form of a jet and fills the electrolytic cell. Then the electrolyte is drained through the edge of the electrolytic cell into the pan, and then back into the working bath. Thus, the electrolyte is in circulation mode. For conducting electrolytic-plasma surface hardening, the test sample made of grade two steel is installed and clamped (Figure 8b) using a clamping mechanism (1) over a cone-shaped electrolytic cell (2) made of stainless steel, for conducting electrolytic-plasma surface hardening (Tabiyeva et al., 2019). A pump (4) is also used, which has a liquid electrolyte supply regulator, thereby ensuring that the electrolyte temperature is cooled to the required value (maintained at 25 ± 1°C). And the used aqueous electrolyte solution from the tray (3) is fed into the working bath (6) thereby ensuring circulation of the working electrolyte. Measurements of temperature values and temperature control of the active electrode and the working electrolyte are carried out by a digital multimeter "UNI-T" and UT-70B with built-in thermocouples (Rahadilov, 2014).

The very process of electrolytic-plasma surface quenching is controlled by a personal computer, which smoothly and with sufficiently high accuracy allows you to adjust the main energy parameters of the working plasma jet, as well as the dynamics of movement of the studied part (sample) undergoing plasma quenching. The depth of immersion of the sample or the working surface of the part in the working electrolyte can be adjusted by means of a clamp (1) moving up and down.

Conclusions on the Section

In the total volume of steel production, carbon steels make up the majority, which are widely used due to the relatively simple method of production and low cost. Parts made of this class of materials determine the performance of the product as a whole. In accordance with the tasks set, grade two bandage steel used in the field of metallurgy and mechanical engineering was selected as the research material for the manufacture of bandages for freight, passenger and shunting locomotives, subway cars and wheel bandages for pairs of tram cars that are subject to wear during operation and require hardening of their wear surface.

The considered types of surface hardening, according to their final results of the structural-phase state, hardness and wear resistance of the hardened layer, are equivalent, but differ in labor intensity, equipment cost and other technical indicators that determine the cost of processing. One of the promising methods of quenching with highly concentrated heating sources is electrolyte-plasma surface quenching, characterized by relatively lower energy consumption, simplicity of technological equipment and large size of the hardened zone. Phase and structural transformations in carbon steels during surface hardening play an important role, since the formation of defects and wear during operation largely depend on the structural components of the surface layer. The properties of various structures of the surface layer depend on the mechanism of formation, shape, size, density of dislocations of the hardening phase, which are formed during electrolyte-plasma surface quenching. To elucidate the physical mechanisms of the formation of structural-phase states and fine structure during electrolyte-plasma surface quenching, which provide an increase in hardness and wear resistance, their analysis by metallographic research using optical, scanning and transmission electron microscopy of high resolution, as well as X-ray diffraction analysis is necessary.

3. CHANGE IN THE MECHANICAL PROPERTIES OF THE SURFACE OF GRADE TWO STEEL AS A RESULT OF ELECTROLYTE-PLASMA SURFACE HARDENING

In this section are presented, the results of studying the effect of a hardened surface layer during plasma quenching in the cathode mode on the mechanical and tribological characteristics of grade two wheel steel.

3.1 Change in the Surface Hardness of Steel grade two During Electrolytic-Plasma Surface Quenching

Being one of the varieties of plasma quenching – electrolyte-plasma surface quenching (EPSQ) has recently been developing and intensively studied. This method is characterized by lower energy consumption, simplicity of technological equipment and large size of the hardened zone. One of the most important properties of the working, surface layer of parts, which significantly affects the quality of wear resistance of this part, is hardness. In this paper, studies have been carried out to study changes in the microhardness values of the surface layer of grade two wheel steel after EPSQ in an electrolyte from an aqueous solution containing 20% sodium carbonate and 10% urea (Rakhadilov, et al., 2019).

Figure 9. Microstructure of grade two steel before (a) and after EPSQ (b)

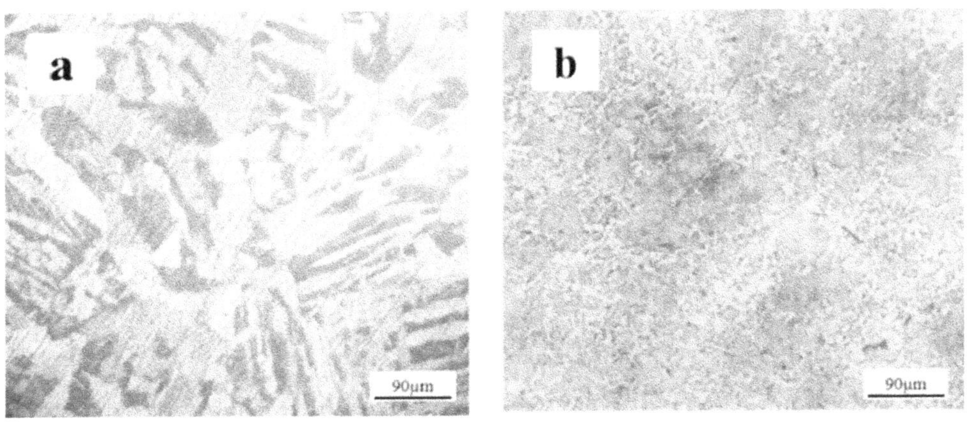

Figure 9 shows the surface structure of the initial and processed samples. Metallographic analysis showed that in the initial state, grade two steel has a ferrite-pearlite structure of carbon steel. The amount of ferrite is 35% (Rakhadilov, et al.,). After EPSQ with a heating duration of 2 s, the microstructure changes, a quenching effect is observed, and martensite is formed on the surface. The formation of a cementite mesh in the structure of grade two steel as a result of the EPSQ, which is located around the ferrite, significantly increases the microhardness of grade two steel. According to (Tabiyeva et al., 2020) the results obtained (Figure 10), the average values of microhardness in the surface hardening zone is ~ 420 HV, in the zone of thermal influence is ~ 260 HV and, accordingly, in the matrix of grade two steel, the microhardness value remained unchanged ~ 140 HV.

Figure 10. Microhardness of the cross-section of samples steel of grade two after EPSQ

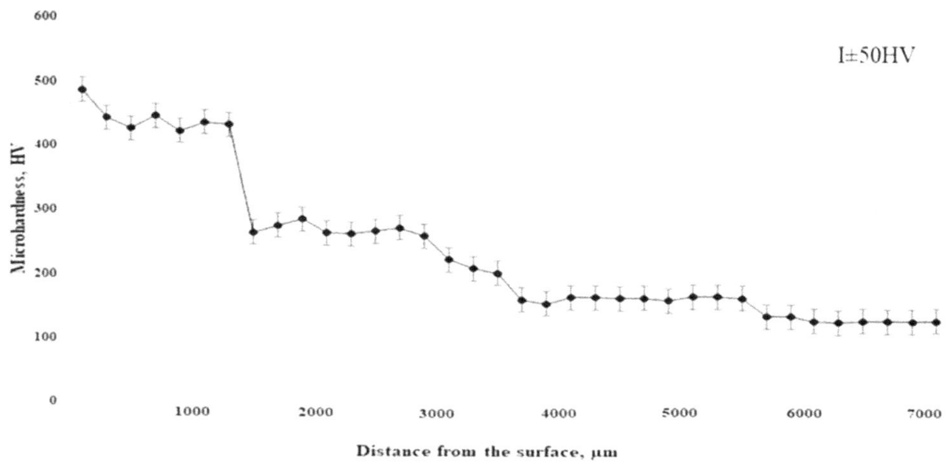

Figure 11. X-ray phase analysis of steel grade two

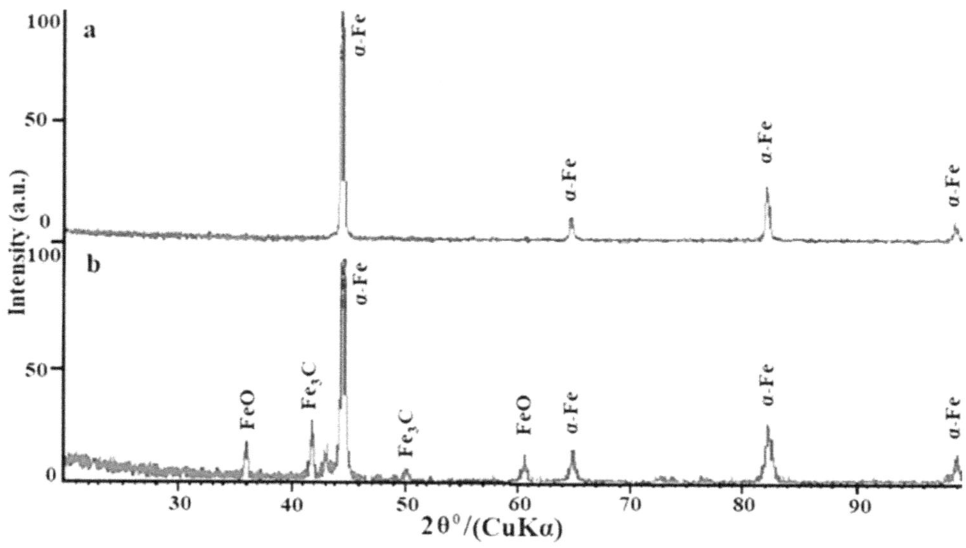

Figure 12. Cross-section microstructure images of grade two steel after EPSQ a) hardened layer; b) transition layer; c) uncured layer; d) general appearance.

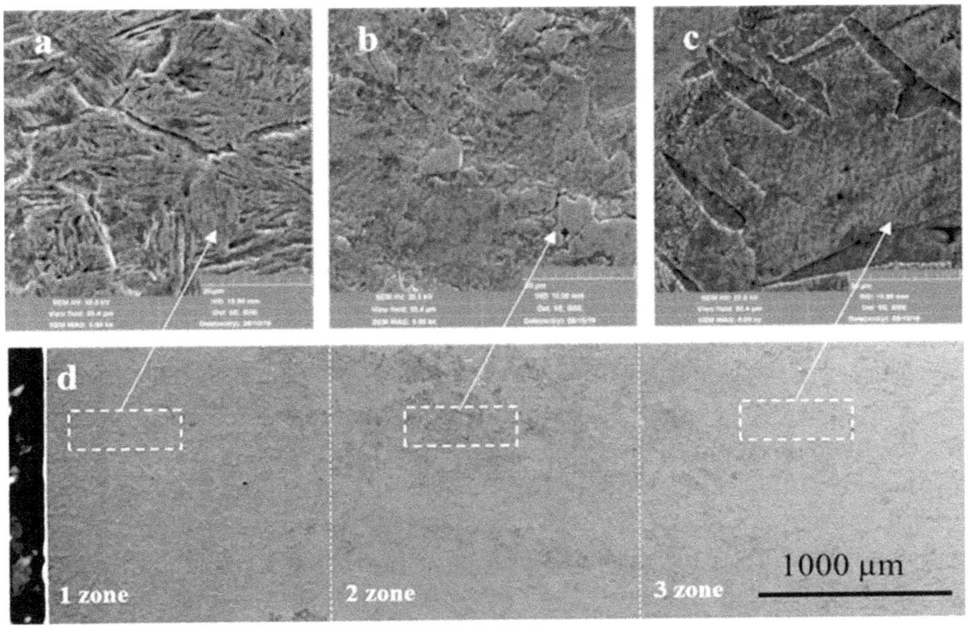

Structural changes were also detected during the study of the microstructure of the near-surface layers of grade two steel samples subjected to electrolyte-plasma surface quenching in electrolyte plasma.

Figure 11 shows X-ray diffractograms of grade two steel before and after the EPSQ. The obtained diffractograms show that after EPSQ, martensite, cementite, and iron oxide are formed on the surface of the steel. I would like to note (Bayatanova, 2014) that X-ray diffraction analysis determines the presence of any phase with its content in an amount of at least 2%. To identify the phases formed in the process of EPSQ in smaller proportions, the method of transmission electron microscopy is used.

Figure 12 (a-b) shows a snapshot of the microstructure of the cross-section of steel after treatment in electrolyte. As can be seen from Figure 12, electrolyte-plasma surface hardening led to a change in the microstructure of the cross-section of the sample, where the zoning of structures typical of electrolyte-plasma treatment is visible (Uazyrkhanova et al., 2020). To be more precise, the cross-section structure consists of 3 zones:

Zone 1 – surface hardening zone with a thickness of 1000-1500 μm,
Zone 2 is a zone of thermal influence,
Zone 3 is a matrix (initial structure).

Also, only using the method of electron transmission microscopy it is possible to obtain experimental data on the fine structure of the material, which will allow us to study the changes that have occurred both on a qualitative and quantitative level, which was done with samples of grade two steel, the results of which are described in detail and given in section 4 of this chapter. The microstructure of the surface of a hardened steel sample of grade two is a fine-grained ferritic structure. The formation of a modified layer of fine-grained ferrite with cementite particles in the surface layers will positively affect the operational properties of the parts.

3.2 The Effect of Electrolyte-Plasma Surface Quenching on the Tribological Properties of Wheel Steel

The mechanism of wear of parts during operation is complex and includes abrasive, adhesive and diffusion wear. Wear resistance tests were carried out according to the "ball – disk" and "flat surface – rotating disk" schemes.

Tribological properties of samples before and after EPSQ were investigated. Experimental curves of the dependence of the coefficient of friction on the friction time of the initial and processed samples are shown in Figure 13. The test was carried out according to the "ball-disc" scheme, the run length was 31 m, the speed was 2 cm/ s, the load was 5 N. It can be seen from Figure 13 that the coefficient of friction of steel samples of grade two before and after the EPSQ with a heating duration of 2 seconds increases from 0.12 to 0.42, then decreases to 0.4-0.338 and remains at this level until the end of friction. This is due to the roughness of the surface, as well as the unevenness of the surface layer in depth, i.e. the fragmented ferritic structure with the cementite mesh formed on the surface smoothly transitions into a ferrite-pearlite structure. The experimental curves of the dependence of the coefficient of friction on the length of the run are shown in Figure 14. The test results showed that the average coefficients of friction at a normal force of 5 N and the speed of mutual displacement of 2 cm/s are in the range from 0.27 to 0.40, which possibly indicates the destruction of the surface layer of the samples under study. All processed indicators have a good degree of improvement in tribological properties.

The tribological characteristics of the samples before and after EPSQ were characterized by the intensity of wear, the average values of which were $4.83 \cdot 10^{-4} mm^3/N \cdot m$ before quenching and $2.04 \cdot 10^{-4}$ $mm^3/N \cdot m$ after quenching. After EPSQ samples, there is an increase in wear resistance under dry friction conditions. With the help of a profilometer, images of the contact zone of the initial and processed EPSQ steel samples of grade two were taken (Figure 15).

Figure 13. Friction coefficient of steel grade two before (a) and after (b) EPSQ

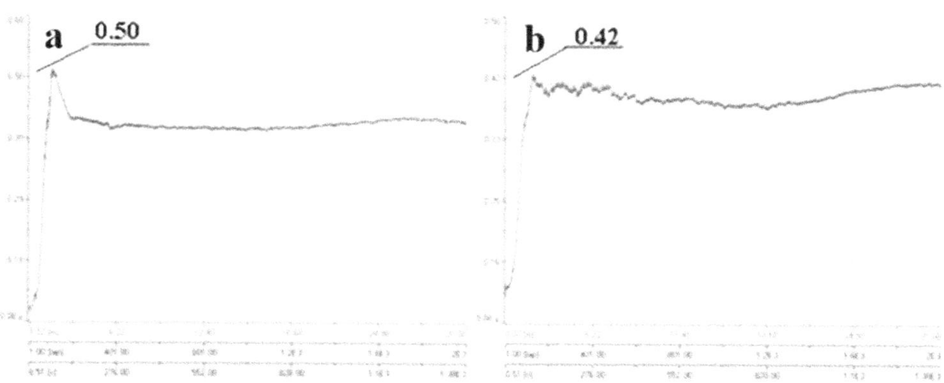

Figure 14. Friction coefficient of steel grade two before and after EPSQ: (b) initial; (c) 2 s; (d) 3 s; (e) 4 s

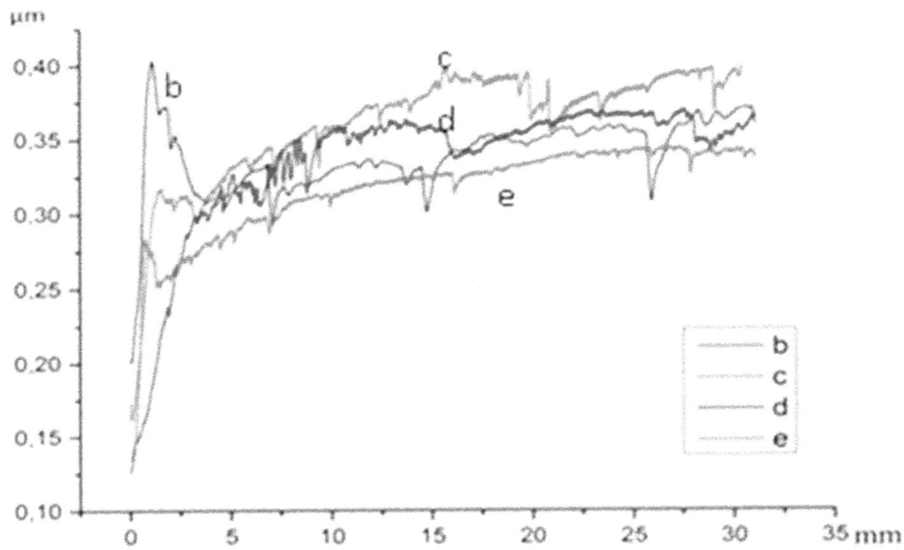

Figure 15. Profiles of the tracks of the initial (a) and hardened EPSQ (b) samples of steel grade two

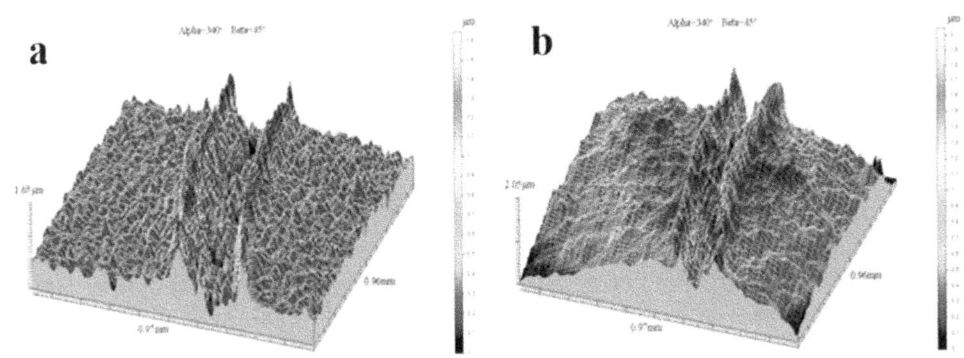

The results of the tests of the samples for abrasive wear were evaluated by the loss of mass of the samples after the test. The mass loss values of grade two steel samples before and after EPSQ were 0.0796 mg±0.01mg and 0.0231 mg±0.01mg, respectively. A significant increase in wear resistance is observed on all samples subjected to EPSQ.

To clarify the structural factors affecting the wear resistance, hardness and other characteristics of steels, the structural and phase states of the hardened surface layers of grade two steel were studied. Based on the X-ray phase results obtained, it can be said that during electrolytic plasma surface quenching, the surface layer of steel is saturated with carbon and martensite, cementite and oxides are formed on the surface layer iron. The thickness of the modified layer is 0.6-1 mm.

Conclusions on the Section

Based on the analysis of the results of the experimental work, a method for processing wheel steel by electrolytic-plasma surface quenching in an electrolyte from an aqueous solution containing 10% carbamide $(NH_2)_2CO$ and 20% sodium carbonate Na_2CO_3 has been developed.

The formation of hardening phases present on X-ray diffraction images, as a result of electrolytic-plasma surface hardening of grade two steel, led to an increase in the microhardness of the sample surface, the average values of which in the surface hardening zone are ~ 420 HV, while in the steel matrix the microhardness value remained unchanged ~ 140 HV.

It is established that the developed method of electrolyte-plasma quenching makes it possible to obtain a modified layer with a thickness of 1000-1500 microns with increased hardness, corrosion resistance and wear resistance by 2.5 – 3.8 times due to the formation of a fragmented structure of packet-plate martensite and carbide particles of the second phases.

4. STRUCTURAL AND PHASE STATE OF GRADE TWO WHEEL STEEL BEFORE AND AFTER ELECTROLYTIC PLASMA SURFACE QUENCHING

This section of the paper describes the phase composition and morphology of grade two steel in its initial state and after electrolytic-plasma surface quenching. These descriptions are accompanied by images obtained by transmission electron microscopy. The light-field images and micro-diffraction patterns obtained from these sites, as well as their dark-field patterns, are given.

Figure 16. Crystal lattices •– iron atom, ○– carbon atom, ○ – octahedral (a) and α – tetrahedral (b) internodes, some of which may be occupied by carbon atoms, (c) – cementite (Fe3C iron carbide)
Source: Uazyrkhanova et al. (2020)

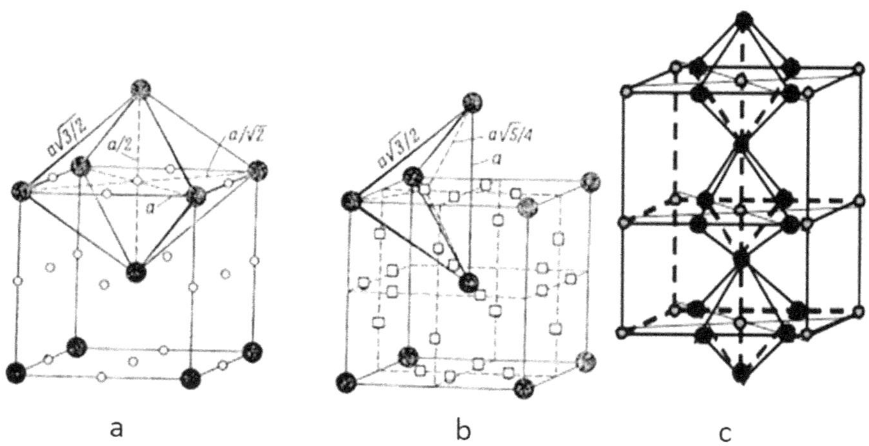

4.1 Initial Structure and Phase Composition of Grade Two Steel

The study was carried out both on a qualitative and quantitative level using the method of transmission electron diffraction microscopy on thin foils. All the data obtained were processed statistically. Phase identification was carried out according to the methods discussed in (Ivanov, & Kozlov, 1991; Ivanov & Kozlov, 2002; Kozlov, et al., 2004).

For this purpose, schemes of microdifraction patterns calculated from tabular values of the parameters of crystal lattices were used. The conducted studies have shown that in the initial state, the steel matrix is a α-phase – a solid solution of carbon and alloying elements in α-Fe. The α-phase has a body-centered cubic (BCC) lattice (Figure 16) and can be solid solutions based on iron atoms of introduction (C, N, B, S, P, etc.) and substitution (Si, Mn, Ni, Cr, Mo, V, W, etc.) simultaneously.

The morphological components of the α-phase are lamellar perlite (LP) and ferrite. Lamellar perlite, almost perfect, i.e. it is a conglomerate of alternating parallel plates of ferrite and cementite (Abuali Galedari, & Mousavi Khoei, 2013). Ferrite (α-phase) in perlite has a body-centered cubic (BCC) crystal lattice.

Cementite, which is a chemical compound of carbon with iron (Fe_3C iron carbide), has an orthorhombic crystal lattice. The carbon content in cementite is 6.67 wt.%. Cementite in perlite can be alloyed, i.e. iron atoms can be replaced by nonmetal atoms: nitrogen, oxygen, boron, and iron atoms by metals: manganese (in unlimited quantities), chromium (up to 20%), molybdenum (up to 2%), vanadium (tenths of %), etc. Pure carbonaceous cementite is a metastable phase and the entire $Fe–Fe_3C$ diagram is also metastable. This fact has a noticeable effect on the properties of steels of this class. The mutual parallelism of the plates means that:

- different plates of the same phase within the colony have the same orientation;
- the mutual orientation of the two phases (their orientation ratio) ensures the best coupling of the two crystal lattices along the habitus surface of the plates.

It has been established (Suhomlin,) that the orientation relation of Bagaryatsky is observed between ferrite and cementite in almost every pearlite colony. It should be noted that the effect of steel alloying on the crystallographic characteristics in perlite has not been established (Popova, et al., 2020). The electron microscopic image of lamellar perlite (LP) obtained during the study is shown in Figure 17.

Figure 17 shows that in the initial state, grade two steel is a ferrite-perlite mixture and that cementite particles have a lamellar shape, are located almost parallel to each other, and there are clear reflexes on the microdiffraction pattern (Figure 17b), which are identified (Figure 17c) like cementite reflexes. The black arrow (Figure 17a) indicates the coinciding directions $[011]_\alpha \,||\, [21\bar{1}]$ at the same time $\left(13\bar{3}\right)_\alpha \,||\, \left(1\bar{2}0\right)_C$ – Bagaryatsky ratio. The volume fraction of lamellar perlite is 35% in the volume of the material. Note that to determine the volume fraction, a planimetric method was used, which boils down to measuring the total cross-sectional area of this structural component on a certain area of the foil. It can be seen from the presented figures that the dislocation structure in ferrite plates represents dense dislocation grids. The average scalar dislocation density ρ was calculated by the rectangular grid method (Kozlov, et al., 2014). The calculations performed showed that the average scalar dislocation density in ferrite plates is equal to 2.1×10^{10} cm^{-2}. Dislocations are not allowed in cementite plates. As

Figure 17. Electron microscopic image of the fine structure of the initial state of the grade two wheel steel (FF – fragmented ferrite; LP – lamellar perlite) a – is a light–field image; b – is a micro–diffraction pattern obtained from the area marked on (a) with a circle; c – is its indexed scheme containing the reflexes of the α-phase and cementite.

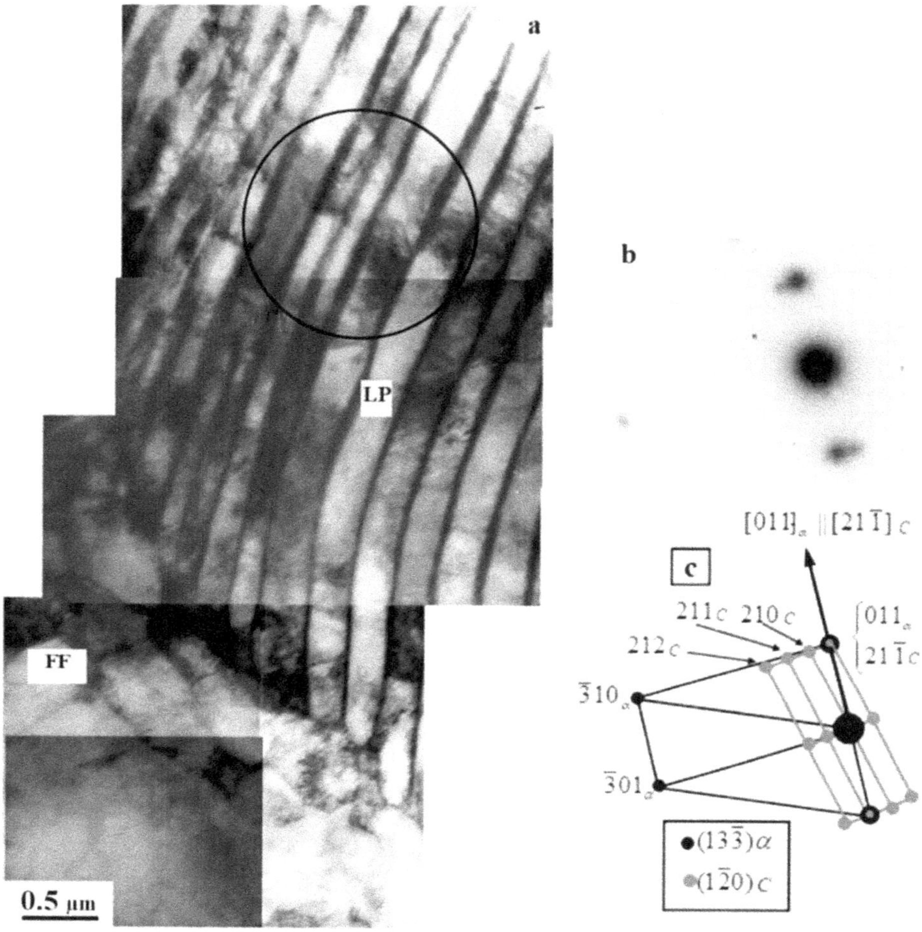

shown by the studies carried out on a transmission electron microscope, ferrite is present in the initial state of grade two wheel steel in the form of unfragmented ferrite (UF) and fragmented ferrite (FF).

Unfragmented ferrite – in the studied wheel steel, these are individual grains that are shown in Figures 18 and 20. The dislocation structure of unfragmented ferrite, as in perlite, is represented by dislocation grids, the average scalar dislocation density is 2.57×10^{10} cm^{-2}.

According to the conducted studies of the initial state of the grade two wheel steel, no carbide phase was detected either inside the grains or at their boundaries on a transmission electron microscope. This is evidenced by microdifraction patterns obtained from grains of unfragmented ferrite. One such example is shown in Figure 18 and Figure 20.

Figure 18. An image of the fine structure of the initial state of grade two steel obtained using a transmission electron microscope (UF – unfragmented ferrite)

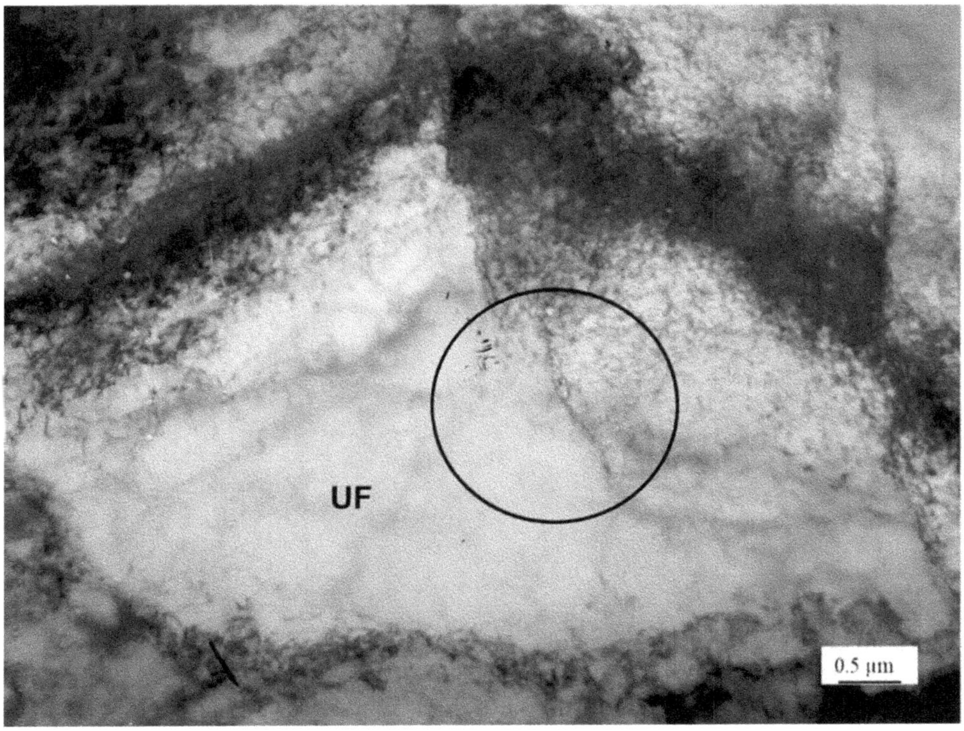

Figure 19. Microdifraction pattern (a) obtained from a grain section of unfragmented ferrite and its indexed scheme (b)

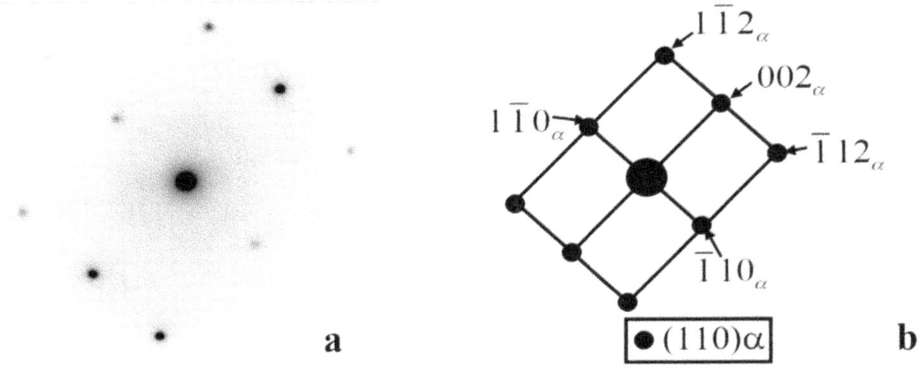

Figure 19 shows a microdifraction pattern (a) obtained from a grain section of unfragmented ferrite highlighted with a black circle in Figure 19 and its indicated scheme (b), where it is seen that reflexes belonging only to the α-phase are present in the initial state. According to the calculations performed, the volume fraction of unfragmented ferrite in the material is ~ 10%.

Figure 20. An image of the fine structure of the initial state of grade two steel obtained using a transmission electron microscope (UF – unfragmented ferrite; FF – fragmented ferrite)

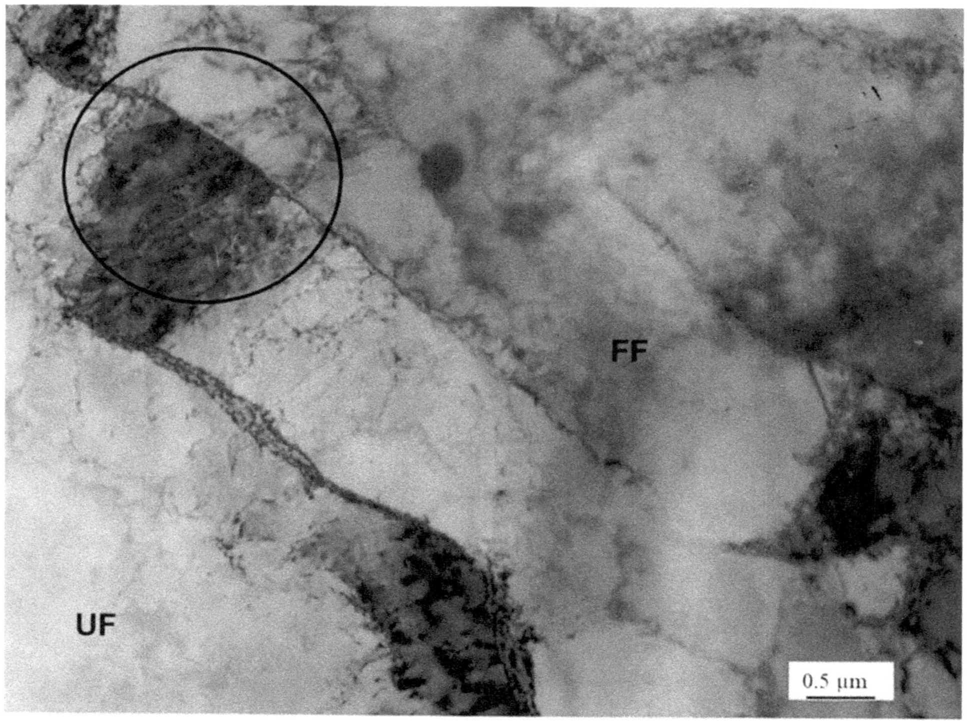

Figure 21. Microdifraction pattern (a) obtained from a grain section of fragmented ferrite and its indexed scheme (b)

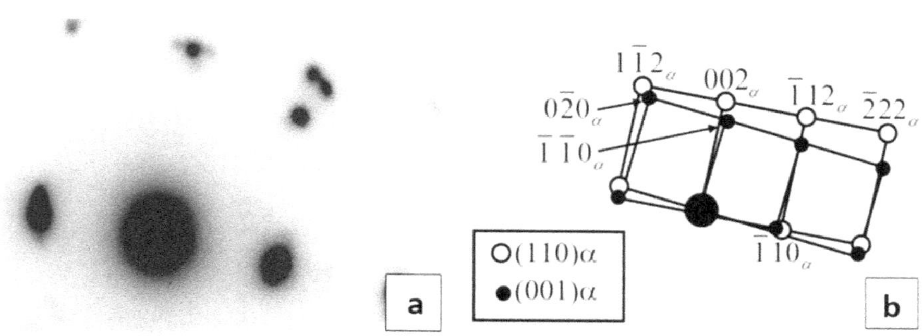

Fragmented ferrite is a substructure that consists of dislocation sub–boundaries (fragment walls) and an internal space containing or not containing dislocations. According to the calculations performed, the volume fraction of fragmented ferrite was 55% (Popova, et al., 2020). The dislocation structure inside the fragments is also mesh. The average scalar density in fragmented ferrite is almost the same as in unfragmented and amounted to 2.89×10^{10} cm^{-2}.

Figure 22. Electron microscopic image of the fine structure of the initial state of grade two steel (P – perlite; UF – unfragmented ferrite; FF – fragmented ferrite) a – ferrite-pearlite mixture; b – enlarged image of the area marked with a black circle on (a).

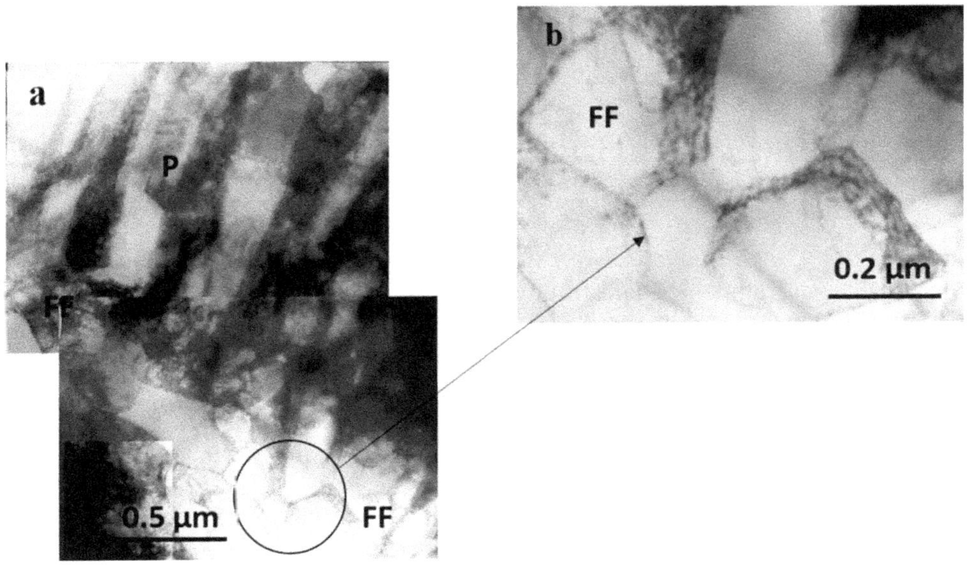

The boundaries of the fragments are free of carbides, and the carbide phase is also not detected inside the fragments. The proof of this is the microdifraction pattern shown in Figure 21a, obtained from a section of fragmented ferrite highlighted by a black circle in Figure 20 and its indexed scheme shown in Figure 21b – shows that there are reflexes belonging only to the α-phase.

According to the calculations carried out, the average size of fragments of fragmented ferrite is ~1 microns (Figure 22) (Tabieva, Popova, Nikonenko, et al, 2020). As studies have shown, at the joints of "perlite – ferrite" grains, there are often either groups of small (~0.3 microns) fragments, or chains of very small (0.08 microns) fragments, which is clearly visible in Figures 18-22, the area marked with a black circle, an enlarged image of which is shown in Figure 22b. This indicates that the fragmentation of ferrite began from the grain boundaries of "perlite – ferrite" (Rakhadilov, et al., 2020).

4.2 Quantitative Parameters of the Fine Structure of Grade Two Steel Before Electrolytic-Plasma Surface Hardening

The dislocation structure in both perlite and ferrite is polarized. This is indicated by the presence of flexural extinction contours in the entire material. We emphasize that such contours are not the result of deformation of the sample during the manufacture of thin foil from it, because the technique used in the preparation of foils almost completely eliminates the introduction of artifacts. The presence of extinction contours is the result of bending of the crystal lattice, or bending of the crystal α-matrix. At the same time, only a small part of the crystal is in the exact reflecting position (contour). The smaller the contour width, the more curved the crystal is. By measuring and calculating the average thickness (the value of the transverse dimension) of the extinction contour in any part of the material, it is possible to determine the χ-amplitude of the curvature-torsion of the crystal lattice. For each morphological

Table 4. Average quantitative parameters of the morphological components of the fine structure and in general for the material in the initial state of grade two steel

Morphological Component	Average Quantitative Parameters of Fine Structure					
	P_V	ρ, cm^{-2}	$\rho_{\pm}$, cm^{-2}	χ, cm^{-1}	σ_{Sh}, MPa	σ_{∂}, MPa
Perlite	35%	2.10×10^{10}	1.44×10^{10}	360	285	240
Unfragmented ferrite	10%	2.57×10^{10}	1.46×10^{10}	365	320	240
Fragmented ferrite	55%	2.89×10^{10}	2.76×10^{10}	690	340	330
In general, according to the material	100%	2.54×10^{10}	2.10×10^{10}	525	315	285

component: in perlite, unfragmented and fragmented ferrite, as well as for the whole material in the initial state of grade two steel, the following quantitative parameters of the fine structure were calculated according to the scalar dislocation density - ρ and the size of the bending extinction contours: χ, $\rho_{\pm}$, σ_{∂} and σ_{Sh}, the average values of which are shown in Table 4.

Analyzing the data indicated in Table 4, it can be argued that:

- the highest quantitative parameters were in fragmented ferrite, the lowest – in perlite.
- the conditions are met in all the material: $\rho > \rho_{\pm}$ and $\sigma_{Sh} > \sigma_{\partial}$.

This means that the bending-torsion (distortion) of the crystal lattice of the grade two wheel steel in its initial state is purely plastic in nature, which does not lead to the formation of microcracks in the material.

4.3 Structure and Phase Composition of the Near-Surface Zone of Wheel Steel After Electrolytic-Plasma Surface Quenching

The conducted studies have shown that the surface hardening of grade two steel has led to significant qualitative and quantitative changes, namely, to a change in the phase composition and the list of phases present, as well as their morphology. Thus, both the phase composition and the fine structure of the steel after surface hardening differ significantly from the initial state.

The surface hardening carried out led to the formation of packet-plate martensite. Electron microscopic images of the resulting structure are shown in Figures 23-27. It is known that martensite is a supersaturated solid solution of carbon in α-iron with the same concentration of dissolved carbon as the original austenite. It is also known that in high-carbon, low-alloy steels, batch (or rack-and-pinion), lamellar low-temperature and lamellar high-temperature martensite is formed. It was to this structure that the surface hardening of grade two steel led. Batch (or rack-and-pinion) martensite is a structural formation consisting of a set of elongated crystals (slats) almost parallel to each other, forming a package. A separate package has a certain shape and size. The shape of the cross-section of packages with a foil plane can be any: from triangular to polygonal. The calculations carried out on the obtained electron microscopic images showed that the average transverse size of a separate rail in different packages differs significantly from each other:

- There are packages (Figures 23-24) in which the average size is 0.15 microns;
- whereas in other packages (Figures 25-27) the average size of the rails is 0.50 microns.

Figure 23. Electron microscopic image of the fine structure of the surface of a grade two steel sample after electrolytic plasma surface quenching: packet-plate martensite, the boundary of two grains (P – packet martensite, LLM – Lamellar low–temperature martensite)

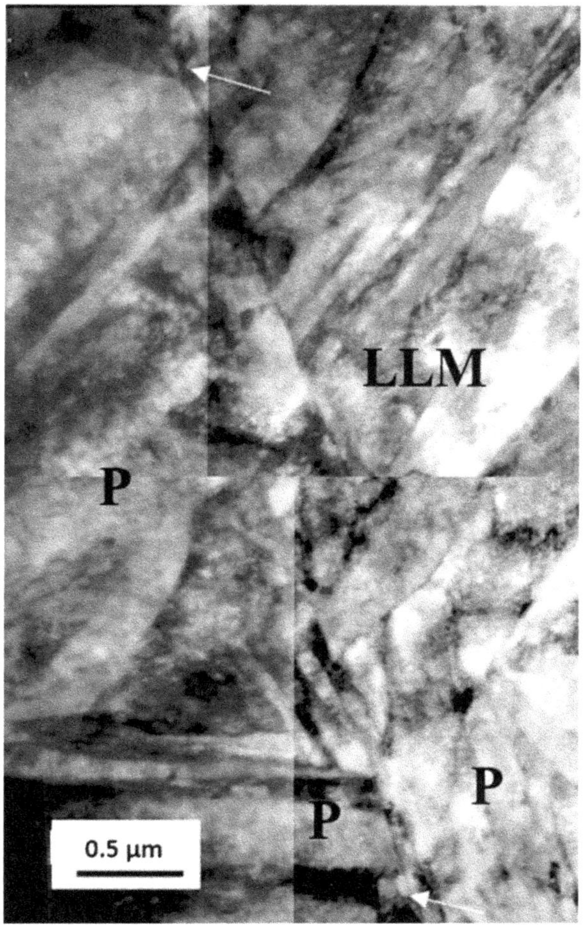

We noted above that in the initial state of steel in fragmented ferrite there are fragments that are sharply different in size: in the center of the grain – large, near the boundaries of "perlite – ferrite" – small. Apparently, packages with narrow slats were formed in sections of material with small fragments, packages with wide slats - in sections of material with large fragments. The fact that the formation of batch martensite occurred from fragmented ferrite is indicated by the fact that the volume fraction of batch martensite, calculated according to the formulas specified in the second section, is 60%, i.e. almost as much as there was fragmented ferrite in the initial state.

Lamellar low-temperature martensite (LLM) is a fairly large, or separately located, martensite crystals (plates) with a dislocation structure (Figure 23), or two or three plates arranged in parallel. According to the calculations, the volume fraction of lamellar low-temperature martensite is 10%. Comparison with the initial state of steel shows that lamellar low-temperature martensite was most likely formed from unfragmented ferrite, the volume fraction of which was also 10%.

Lamellar high–temperature martensite (LHM) is large, separately located martensite crystals (plates), often extending through the entire grain, as well as crystals of arbitrary shape that do not have a clear cut and do not have their own interface boundaries. It is well known that martensitic transformation is almost always incomplete. This leads to the presence of a certain amount of residual austenite (γ-phase) in the material.

The residual austenite (γ-phase) has a face-centered cubic (FCC) crystal lattice and is, like the α-phase, a solid solution based on iron atoms of introduction (C, N, etc.) and substitution (Cr, Ni, Mo, V, etc.) simultaneously. The main nodes, as in the α-phase crystal lattice, are occupied by iron atoms and substitution elements, tetrahedral (mostly) internodes are partially occupied by carbon atoms. The residual austenite (γ-phase) is located in the form of thin layers (~ 20-50 nm) along the boundaries of crystals (rails and plates) of martensite, as well as inside the plates in the form of "needles" or "islands". The crystal lattices of the residual austenite and the α-phase, regardless of the type and location of the residual austenite, are always interconnected by the Kurdyumov-Sachs orientation ratio.

Figure 24. Electron microscopic image of the fine structure of the surface of a grade two steel sample after EPSQ a – is a light–field image; b – is a micro–diffraction pattern; c – is its indexed scheme containing α- and γ-phase reflexes.

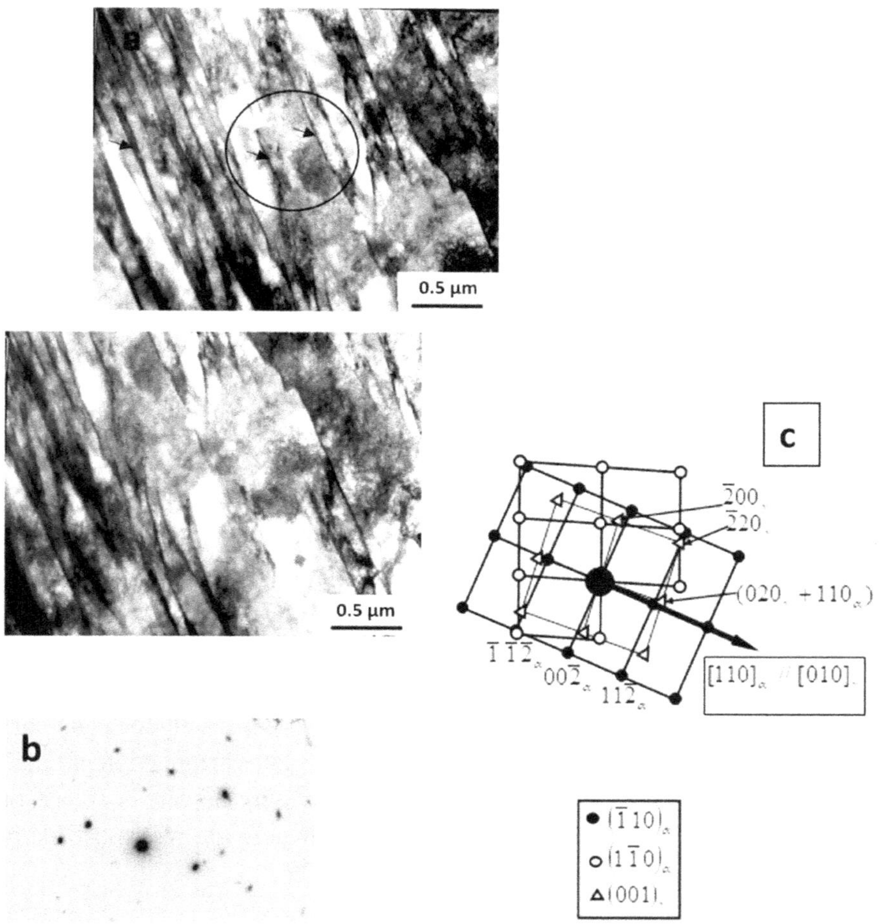

Studies have shown that in grade two steel after surface hardening, residual austenite (γ-phase) in batch martensite is also present at the boundaries of martensitic rails in the form of long thin layers. It should be noted that the γ-phase interlayers are present both in packages with narrow (Figure 24) and wide (Figure 25) slats.

As a confirmation, we will analyze this in more detail using the example of Figure 24. Residual austenite (γ-phase) is present at the boundaries of narrow martensitic slats in the form of thin layers (Figure 24a). The presence of the γ-phase is confirmed by the microdifraction pattern obtained from this site (Figure 24b) and its indexed scheme (Figure 24c). As can be seen from the microdifraction pattern and its indexed scheme, the direction (Tabieva, Popova, Nikonenko, et al, 2020)$_\alpha$ coincides with the direction (Balanovskij, 2006)$_\gamma$. This is also confirmed by the solution of matrix equations:

for planes $$\begin{pmatrix} u \\ v \\ w \end{pmatrix}_\alpha = \begin{bmatrix} 1\,\bar{1}\,0 \\ 1\,1\,0 \\ 0\,0\,2 \end{bmatrix} \cdot \begin{pmatrix} 0 \\ 0 \\ 1 \end{pmatrix}_\gamma = \left(\bar{1}10\right)_\alpha;$$

for directions $$\begin{bmatrix} h \\ k \\ l \end{bmatrix}_\alpha = \begin{bmatrix} 1\,\bar{1}\,0 \\ 1\,1\,0 \\ 0\,0\,1 \end{bmatrix} \cdot \begin{bmatrix} 0 \\ 1 \\ 0 \end{bmatrix}_\gamma = \left[110\right]_\alpha.$$

Thus, from the solution of matrix equations it turns out: (Tabieva, Popova, Nikonenko, et al, 2020)$_\alpha$. (Balanovskij, 2006)$_\gamma$ and $(\bar{1}10)_\alpha$. $(001)_\gamma$, This is observed in the microdifraction pattern (Figure 24b). The black arrow indicates the coinciding directions (Tabieva, Popova, Nikonenko, et al, 2020)$_\alpha$ and (Balanovskij, 2006)$_\gamma$, with the $\left(\bar{1}10\right)_\alpha$ *II* $\left(001\right)_\gamma$ – Kurdyumov-Sachs ratio.

Figure 25 shows an electron microscopic image of the fine structure of the surface of a sample of grade two steel after electrolytic-plasma surface hardening obtained on a TEM, which shows laths of packet martensite, and along the borders of laths there are interlayers of residual austenite (γ) with precipitates of special $M_{23}C_6$ carbides. Figure 25 shows a package consisting of a group of wide martensite laths, at the boundaries of which there are interlayers of the γ-phase. The presence of the γ-phase is confirmed by the dark-field image (Figure 25c) obtained in the reflection $[\bar{2}20]$of the γ-phase, the microdiffraction pattern obtained from this area (Figure 25b), and its indicated scheme (Figure 25d). As can be seen from the microdiffraction pattern and its indicated scheme, the direction $[0\bar{2}1]_\alpha$ coincides with the direction $[\bar{1}\bar{1}1]_\gamma$ (coincident directions in Figure 25d are marked with an arrow).

This is also confirmed by solving the matrix equations:

for planes $$\begin{pmatrix} u \\ v \\ w \end{pmatrix}_\alpha = \begin{bmatrix} 1\,\bar{1}\,0 \\ 1\,1\,0 \\ 0\,0\,2 \end{bmatrix} \cdot \begin{pmatrix} 1 \\ 1 \\ 2 \end{pmatrix}_\gamma = \left(012\right)_\alpha;$$

Figure 25. Electron microscopic image of the fine structure of the surface of a sample of steel grade two after EPSQ a – bright-field image; b – microdiffraction pattern; c – dark-field image obtained in closely spaced reflections [ ]γ and [ ]C; d – is its indicated scheme

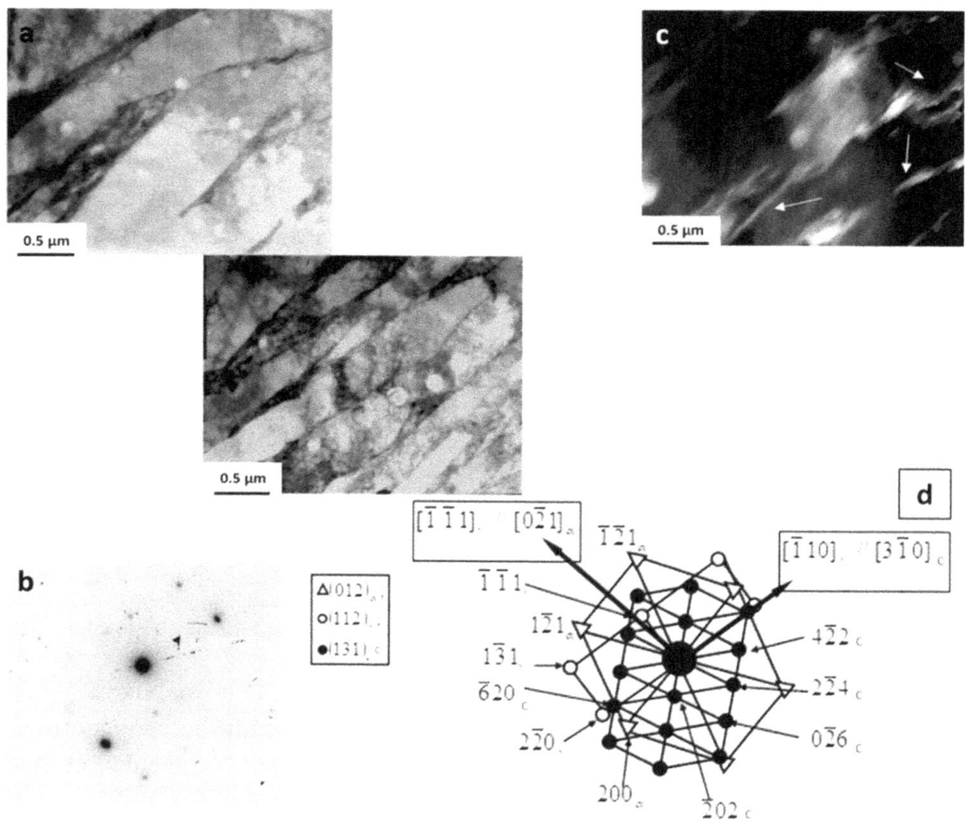

for destinations $\begin{bmatrix} h \\ k \\ l \end{bmatrix}_\alpha = \begin{bmatrix} 1\,\bar{1}\,0 \\ 1\,1\,0 \\ 0\,0\,1 \end{bmatrix} \cdot \begin{bmatrix} \bar{1} \\ \bar{1} \\ 1 \end{bmatrix}_\gamma = \left[0\bar{2}1\right]_\alpha$.

Thus, from the solution of matrix equations it turns out: $(012)_\alpha$. $(112)_\gamma$ and $[0\bar{2}1]_\alpha\ //[\bar{1}\bar{1}1]_\gamma$, which is observed in the microdiffraction pattern (Figure 25b) and indicated scheme (Figure 25d).

In Figure 25, the arrows indicate:

1) coinciding directions $[\bar{1}10]_\gamma$ and $[3\bar{1}0]_K$, which means $[\bar{1}10]_\gamma\ //\ [3\bar{1}0]_K$, at the same time $(112)_\gamma\ II\ (131)_K$ - the ratio "cube-cube";
2) coinciding directions $[0\bar{2}1]_\alpha$ and $[\bar{1}\bar{1}1]_\gamma$, which means $[0\bar{2}1]_\alpha\ //\ [\bar{1}\bar{1}1]_\gamma$, at the same time $(012)_\alpha\ //\ (112)_\gamma$) – Kurdyumov-Sachs ratio. In the diagram (d), the letter "K" denotes $M_{23}C_6$ carbide. In (c), white arrows mark interlayers of the γ-phase containing carbide particles.

Thus, the presented figures 24 and 25 indicate that residual austenite is present at the boundaries of all laths, regardless of their size. The volume fraction of retained austenite (γ-phase) in packet martensite is 6.5%.

An electron microscopic image of the fine structure of the surface of a sample of steel grade two after electrolytic-plasma hardening of the surface showed that along the boundaries of laths of packet martensite there are interlayers of residual austenite, which are marked with black arrows in Figures 26a and 27a. Inside the laths of packaged martensite, there are particles of cementite which are marked with white arrows in Figure 26a.

It was also revealed that there are reflections belonging to the α- and γ-phases, and carbides: cementite and carbides of a special type $M_{23}C_6$ that are in the γ-phase, marked with the letter "K" on the indicated scheme (Figure 27c) of the obtained microdiffraction pattern (Figure 27b) the area shown in Figure 27a.

Figure 26. Electron microscopic image of the fine structure of the surface of a sample of steel grade two after EPSQ a – bright-field image; b – microdiffraction pattern; c – its indicated scheme containing reflections of the α- and γ-phases

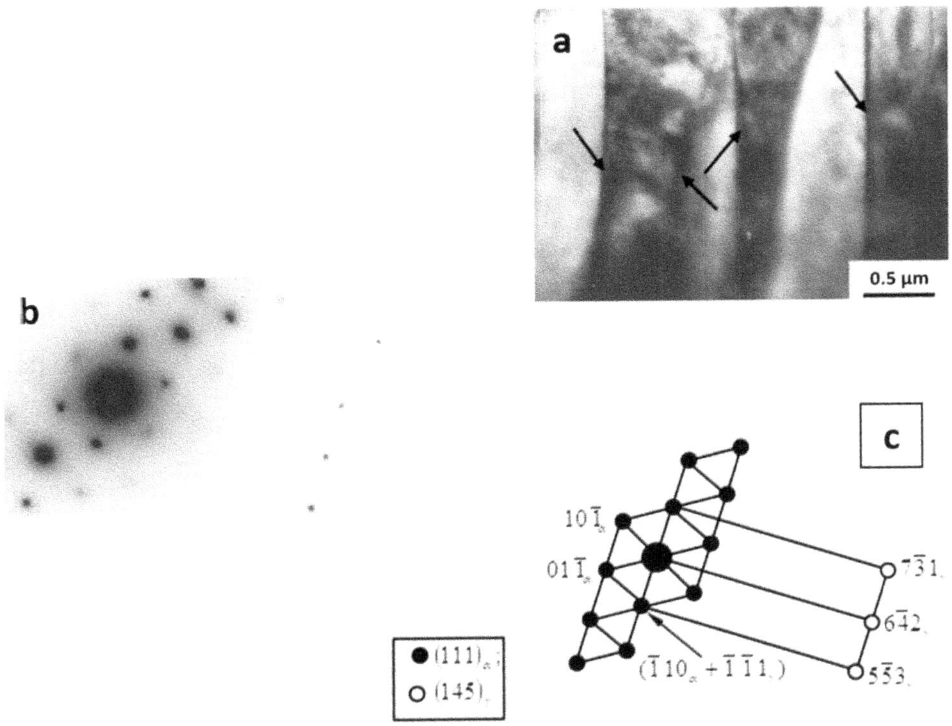

The volume fraction of the γ-phase in lamellar low-temperature martensite at the plate boundaries is 2%, inside the plates – 5.7%. In all martensite crystals, cementite particles are thin elongated plates (needles) located in two and three directions relative to the axis of the martensite crystal. The average particle size in martensitic slats is 12 × 40 nm, in plates of low–temperature martensite – 16 × 80 nm, in plates of high-temperature martensite - 32 × 84 nm. This means that the smallest particles of cementite are formed in the batch "self-released" martensite, the largest – in the lamellar high-temperature. Note that the distribution density of cementite particles in batch and lamellar low-temperature martensite is the same and higher than in lamellar high-temperature martensite, i.e. the distance between cementite particles in batch and lamellar low-temperature martensite (r = 100 nm) is less than in lamellar high-temperature (r = 140 nm) martensite.

The calculations performed showed that the volume fraction of cementite in batch (rack) martensite is 0.27%, in low–temperature lamellar – 0.95%, in high-temperature lamellar - 2%. Thus, the electrolyte-plasma surface hardening of grade two steel led not only to martensitic transformation, but also to the "self-release" of steel. The conducted studies have shown that in addition to cementite, there are also particles of carbides of a special type $M_{23}C_6$ in the structure of steel (Tabieva, Uazyrhanova, Popova, et al, 2020).

Carbides of the $M_{23}C_6$ type have a cubic crystal lattice (space group Fd3m). The unit cell of this carbide contains more than 100 atoms occupying various non-equivalent positions. Such carbides include Fe_4Mo_2C, $Fe_6Mo_6C_6$, $Fe_{21}Mo_2C_6$, $Fe_{21}Cr_2C_6$. These carbides are found at the boundaries of martensitic

Figure 27. Electron microscopic image of the fine structure of the surface of a sample of steel grade two after EPSQ a – bright-field image; b – microdiffraction pattern of area (a); c – its indicated scheme

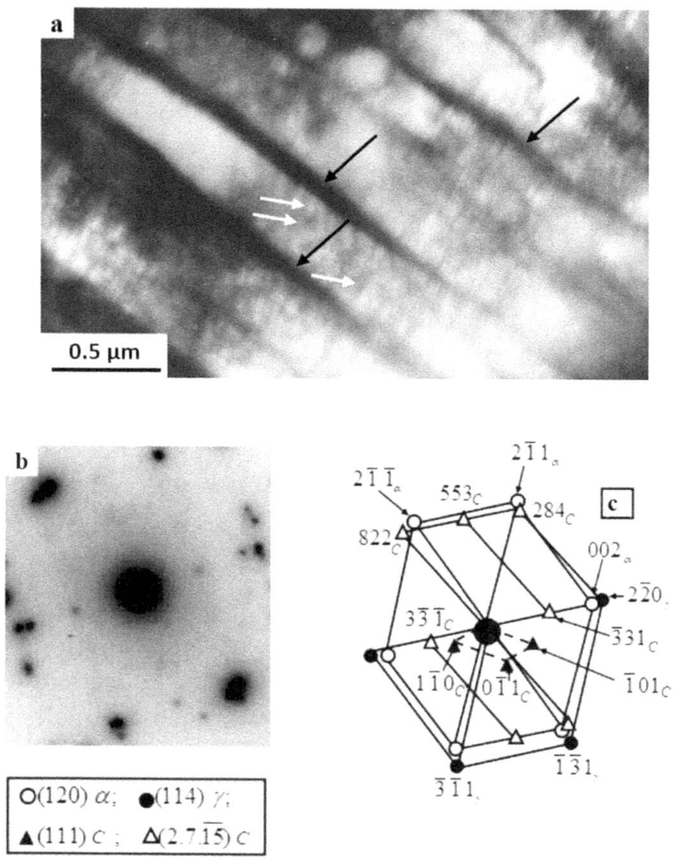

lath, as well as inside plates of low-temperature and high-temperature martensite. Microdifraction patterns and their indexed schemes obtained from these sites serve as proof of their presence. Let's explain this in more detail.

We noted above that surface hardening led to the "self-release" of steel, which is similar to ordinary tempering. It is well known that tempering of hardened steels leads to three main processes:

1) to the evolution of the defective structure, which consists in such a restructuring of the defective structure in which the internal stress fields decrease;
2) to reduce the density of defects and the energy of the defective structure;
3) to the decomposition of a solid solution with the formation of particles of carbide phases.

Due to the strong interaction of dislocations, sub-boundaries and carbon atoms, these processes are interconnected.

In addition, the "self-release" of steel means that the decomposition of the solid solution in the α- and γ-phases already begins during the surface quenching process and during the "self-release" a diffusion γ ® α transformation already begins to occur with the release of special type carbide particles (in particular, $M_{23}C_6$).

The evolution of the defective structure of steel during "self-release" and especially the change in its quantitative parameters have not been investigated in previously published works. Only a change in the phase composition of the hardened steel was noted, i.e. the presence of cementite (Ababkov, et al., 2021; Abuali Galedari, & Mousavi Khoei, 2013; Lee & Polycarpou, 2005; Zhang, 2002; Zhang, 2002; Zhang, 2002).

Based on the conducted studies, it can be argued that $M_{23}C_6$ carbide particles located at the boundaries of martensitic crystals are isolated from the γ-phase. The proof of this is that they are observed inside the layers of residual austenite: in microdifraction patterns obtained from the boundaries of martensitic crystals (see Figures 24b, 27b), along with $M_{23}C_6$ phase reflexes, γ-phase reflexes are always present.

The feasibility of this ratio is confirmed by the indexing of microdifraction patterns obtained from such sections of the structure (Figure 25). The particles of carbides $M_{23}C_6$, located in the layers of residual austenite, have a rounded shape. Their average size, regardless of the location of the particle, is ~ 5 nm. Due to the small size of the particles, they are detected only on dark-field images and at high magnification. However, they are reliably detected by microdifraction patterns when they are indexed.

In the near-surface zone of the grade two steel sample, after surface quenching, $M_{23}C_6$ carbide particles unrelated to the γ-phase were also found. These are particles with a rounded shape, the average size of which is ~ 80 nm. They were found in plates of high-temperature martensite on dislocations of the α-phase.

Thus, it can be argued that the release of special carbides of the $M_{23}C_6$ phase is due to:

- The decay of residual austenite and martensite,
- Partial dissolution of cementite,
- The departure of carbon from dislocations and crystal boundaries of the α-phase.

That is, in all cases, carbon from residual austenite, α-solid solution, cementite particles and crystal lattice defects goes to the formation of special carbides.

It should be noted that the intensity of the processes of decomposition of solid solutions of α and γ and the formation of a carbide phase in various structural components of the α-phase (in batch martensite, inside plates of low-temperature and high-temperature martensite) is different. It is also different within each structural component. It depends on many factors. Such factors, in particular, include the state of the source material (including the degree of fragmentation of the material, the type of fragments formed).

The calculations performed showed that the volume fraction of $M_{23}C_6$ carbides in lamellar high-temperature martensite is 2.7%. It should be noted that due to the small size of the $M_{23}C_6$ particles in the γ-phase, their volume fraction in each structural component of martensite does not exceed 0.01% and therefore is not taken into account in quantitative calculations in the future.

Thus, electrolytic plasma surface hardening of grade two steel in the near-surface zone of the sample resulted in:

1) to martensitic transformation;
2) to the “self-release” of steel;
3) to a diffusive γ ® α transformation with the release of carbide particles of a special type $M_{23}C_6$.

4.5 Quantitative Parameters of the Structural-Phase State of Grade Two Steel After Surface Quenching

Based on the conducted studies, it can be argued that in the near-surface zone of the grade two steel sample after the EPSQ, the quantitative phase composition in each structural component of martensite and in general for the material has the form shown in Table 5.

Table 5. Phase composition of grade two steel in the near-surface zone after electrolyte-plasma surface quenching

Phase Composition Parameters		Morphological Components of Martensite			In General, According to the Material
		Batch Martensite	Lamellar Martensite		
			Low Temperature	High Temperature	
Volume fraction (P_V)		60%	10%	30%	100%
Fraction γ-phases	at the borders	6.5%	2%	0	4.1%
	inside	0	5.7%	8.5%	3.1%
	sum	6.5%	7.7%	8.5%	7.2%
Cementite	δ_C	0.27%	0.95%	2.0%	0.86%
	dimensions, nm	12 × 40	16 × 80	32 × 84	18 × 60
	r, nm	100	100	140	110
$M_{23}C_6$ Carbides	δ_K	0	0	2.7%	0.8%
	dimensions, nm	–	–	80	80
	r, nm	–	–	240	70

Table 6. Average quantitative parameters of fine structure in batch, lamellar low-temperature and high-temperature martensite, as well as in general for the material in grade two steel after EPSQ

Morphological Component	Average Quantitative Parameters of Fine Structure					
	P_V	ρ, cm^{-2}	$\rho_{\pm}$, cm^{-2}	χ, cm^{-1}	σ_{Sh}, MPa	σ_{∂}, MPa
Batch martensite	60%	3.31×10^{10}	2.60×10^{10}	650	365	320
Lamellar low-temperature martensite	10%	2.52×10^{10}	1.86×10^{10}	465	315	270
Lamellar high-temperature martensite	30%	2.33×10^{10}	1.64×10^{10}	410	305	255
In general, according to the material	100%	2.94×10^{10}	2.24×10^{10}	560	340	295

The dislocation structure, as can be seen from the figures presented above, represents dense dislocation grids in all structural components of martensite. The calculations performed showed that in batch martensite the average scalar dislocation density is 3.31×10^{10} cm^{-2}, in plate low-temperature martensite the value of ρ is slightly lower and is equal to 2.52×10^{10} cm^{-2}, in plate high-temperature martensite the value of ρ is even lower (2.33×10^{10} cm^{-2}).

The dislocation structure in all structural components of martensite is polarized. This is indicated by the presence of flexural extinction contours in all martensitic crystals. For each morphological component (in batch, lamellar low-temperature and high-temperature martensite), as well as for the whole material in the near-surface zone of the grade two steel sample after surface quenching, the following quantitative parameters of the fine structure χ, $\rho_{\pm}$, σ_{∂} and σ_{Sh} were calculated according to the data ρ and the dimensions of the bending extinction contours, the average values of which presented in Table 6.

As can be seen from the calculations performed on the surface data of samples after EPSQ, the results of which are shown in Table 6:

- The highest quantitative parameters were found in batch martensite, the lowest – in lamellar high-temperature martensite.
- The conditions are met in all the material: $\rho > \rho_{\pm}$ and $\sigma_{Sh} > \sigma_{\partial}$. This means that the bending-torsion (distortion) of the crystal lattice of grade two steel in the near-surface zone of the sample after surface hardening is purely plastic in nature, which will not lead to the formation of micro-cracks in the material.

Conclusions on the Section

Thus, based on the analysis of the obtained results of studies of grade two wheel steel before and after electrolyte-plasma surface hardening, the following conclusions can be drawn:

1. The conducted studies on a transmission electron microscope have shown that, in the initial state, grade two steel:
    - Ferrite is present in the form of unfragmented and fragmented, with volume fractions of ~10% and ~55%. The volume fraction of lamellar perlite in the volume of the material is 35%;
    - The calculations performed showed that the average scalar density of dislocations in pearlite plates is equal to 2.1×10^{10} cm^{-2}, unfragmented and fragmented ferrite, as in perlite, is represented by dislocation grids, the average scalar density of dislocations is 2.57×10^{10} cm^{-2} and 2.89×10^{10} cm^{-2} ;
    - The boundaries of the fragments are free of carbides, the carbide phase is also not detected inside the fragments, the average size of fragments of fragmented ferrite is ~ 1 microns;
    - The conditions are met in all the material: $\rho > \rho_{\pm}$ and $\sigma_{Sh} > \sigma_{\partial}$. The bending-torsion (distortion) of the crystal lattice of the grade two wheel steel in its initial state is purely plastic in nature, which does not lead to the formation of microcracks in the material.
2. The conducted electrolyte-plasma surface hardening led to:
    - Formation of batch-plate martensite, as well as the presence of reflexes belonging to the α- and γ-phases, and carbides: cementite and carbides of a special type $M_{23}C_6$ in the γ-phase.
    - The presence of residual austenite (γ-phase) in the batch martensite at the boundaries of martensitic rails in the form of long thin layers. The volume fraction of residual austenite (γ-phase) in batch martensite is 6.5%.
    - Formation of particles of $M_{23}C_6$ carbides not associated with the γ-phase, having a rounded shape, an average size of ~80 nm, found in plates of high-temperature martensite on dislocations of the α-phase.
    - In lamellar low-temperature martensite, the volume fraction of the γ-phase at the plate boundaries is 2%, inside the plates – 5.7%. The volume fraction of lamellar low-temperature martensite is 10%, residual austenite is present not only at the plate boundaries in the form of long thin layers, but also inside the plates in the form of "needles" or twin-type colonies.
    - In lamellar high-temperature martensite, residual austenite is present only inside the plates in the form of "needles" or twin-type colonies, the volume fraction of the γ-phase in lamellar high-temperature martensite is 8.5%, the volume fraction of $M_{23}C_6$ carbides in lamellar high-temperature martensite is 2.7%;
3. The results of studies of electron microscopic pictures showed that:
    - The average size of cementite particles in martensitic slats is 12 × 40 nm, in plates of low–temperature martensite – 16 × 80 nm, in plates of high-temperature martensite - 32 × 84 nm. The volume fraction of cementite in batch (rack) martensite is 0.27%, in low–temperature lamellar – 0.95%, in high-temperature lamellar - 2%.
    - Bending-torsion (distortion) of the crystal lattice of grade two steel after surface hardening is purely plastic in nature, which will not lead to the formation of microcracks in the material, since the following conditions are met in the entire material: $\rho > \rho_{\pm}$ and $\sigma_{Sh} > \sigma_{\partial}$.

CONCLUSION

In this work, the structural-phase states, tribological and mechanical properties, features of the formation of the fine structure of grade two wheel steel before and after electrolytic-plasma surface hardening were studied, and the volume fractions of structural-morphological components, particles of the second phases, scalar excess dislocation densities were quantified, which allow us to formulate the following conclusions:

1. A method and an optimal mode of electrolyte-plasma surface hardening of the material of a railway bandage made of grade two wheel steel has been developed, providing a modified layer with a thickness of 1000-1500 microns with increased hardness and wear resistance due to the formation of a fragmented structure of packet-plate martensite and carbide particles of the second phases.
2. By transmission electron microscopy, it was found that:
    - in the initial state, grade two steel consists of a α-phase: ferrite is present in the form of unfragmented and fragmented, with volume fractions of ~ 10% and ~ 55%, respectively. The volume fraction of lamellar perlite is 35% in the volume of the material. The boundaries of the fragments are free of carbides, the carbide phase is also not detected inside the fragments, the average size of fragments of fragmented ferrite is ~ 1 microns.
    - electrolyte-plasma surface hardening of grade two steel leads to the formation of a batch-plate martensite with carbides of a special type $M_{23}C_6$, which are in the γ-phase. The release of special carbides $M_{23}C_6$ is due to the decay of residual austenite and martensite, partial dissolution of cementite, and the departure of carbon from dislocations and crystal boundaries of the α-phase.
    - the volume fractions of batch, lamellar low-temperature and lamellar high-temperature martensite are ~ 60%, ~ 10% and ~ 30%, respectively. It is shown that in lamellar high-temperature martensite, residual austenite is present only inside the plates in the form of "needles" or twin-type colonies with an average volume fraction of 8.5%.
3. The dislocation structure of grade two steel was studied before and after electrolytic plasma surface quenching. It is established that:
    - in the initial state, both in ferrite and perlite, the dislocation structure is polarized. This is indicated by the presence of flexural extinction contours throughout the material. Calculations of the average scalar dislocation density in pearlite plates, which are 2.1×10^{10} cm^{-2}, in unfragmented and fragmented ferrite – 2.57×10^{10} cm^{-2} and 2.89×10^{10} cm^{-2}, respectively, were performed;
    - after electrolytic-plasma surface hardening of grade two steel, the dislocation structure in all structural components of martensite is polarized, as indicated by the presence of flexural extinction contours in all martensite crystals. The calculations performed for the first time showed that in batch martensite the average scalar dislocation density is 3.31×10^{10} cm^{-2}, in lamellar low-temperature martensite and lamellar high-temperature martensite is equal to 2.52×10^{10} cm^{-2} and 2.33×10^{10} cm^{-2}, respectively.

The development of promising technologies aimed at improving traditional and creating new methods of impact on the surface to give it the properties required by operating conditions is undoubtedly one of the urgent tasks of modern science and technology. This method is characterized by lower energy consumption, simplicity of technological equipment, and large dimensions of the hardened zone. The advantages of the method are a rather high productivity of the process and the ability to harden parts of large mass and complex profile. Thus, the studies performed show that the developed method of electrolytic-plasma surface hardening of steels in an aqueous electrolyte solution leads to changes in the structural-phase state of the surface layers and the formation of hardening phases in order to obtain products with increased mechanical and tribological characteristics.

ACKNOWLEDGMENT

In conclusion, the author expresses his great gratitude:

To the team of LLP Plasma Science for the opportunity and support provided when performing experimental and analytical work.

To the staff of the Scientific Research Center "Surface Engineering and Tribology" for the opportunity to participate in the research and support provided in the performance of experimental and analytical work.

To the staff of the laboratory of Tomsk State University of Architecture and Civil Engineering for assistance in conducting research on a transmission electron microscope.

To the management and employees of the D.Serikbayev East Kazakhstan Technical University for the opportunity and support provided.

REFERENCES

Ababkov, N., Smirnov, A., Danilov, V., Popova, N., & Nikonenko, E. (2021). Structural-phase state, mechanical properties, acoustic and magnetic characteristics in the sustainable deformation localization zones of power equipment made of structural and heat resistant steels. *Metals, 11*(10), 1638.

Abuali Galedari, S., Mousavi Khoei, S.M. (2013). Effect of pulse frequency on microstructure and surface properties of Ck45 steel treated by plasma electrolysis method. *Journal of Alloys and Compounds, 551*, 415-421.

Balanovskii, A.E. (2016). Visualization of the process of metal heating and melting in the anode zone in an arc discharge with a tungsten electrode. *High Temperature, 54*, 627-631.

Balanovskij, A.E. (2006). Plazmennoe poverhnostnoe uprochnenie metallov. *Irkutsk: Izdatelstvo IrGTU*, 180.

Balanovskij, A.E. (2015). Ocenka zerna austenita pri plazmennom poverhnostnom uprochnenii sredneuglerodistyh stalej. *Uprochnyayushie tehnologii i pokrytiya, 6*, 27-32.

Balanovskij, A.E. (2016). Osnovnye voprosy teorii plazmennogo poverhnostnogo uprochneniya metallov (Obzor. Chast 3). *Uprochnyayushie tehnologii i pokrytiya, 2,* 20-30.

Balanovskij, A.E. (2006). Rezultaty vnedreniya tehnologii plazmennogo uprochneniya na VSZhD. *Zheleznodorozhnyj transport, 4,* 28-32.

Balanovskij, A.E. (2015). Osnovnye voprosy teorii plazmennogo poverhnostnogo uprochneniya metallov (Obzor. Chast 1). *Uprochnyayushie tehnologii i pokrytiya, 12,* 18-30.

Balanovskij A.E. (2016). Osnovnye voprosy teorii plazmennogo poverhnostnogo uprochneniya metallov (Obzor. Chast 2). *Uprochnyayushie tehnologii i pokrytiya, 1,* 25-34.

Balanovskij, A.E. (2016). Vizualizaciya processa nagreva i plavleniya metalla v anodnoj oblasti pri dugovom razryade s neplavyashimsya elektrodom. *Teplofizika vysokih temperature, 54,* 1-8.

Balanovskij, A.E. (2016). Vozmozhnosti cifrovoj vizualizacii processa nagreva i plavleniya metalla pri dugovom razryade s neplavyashimsya elektrodom. *Svarochnoe proizvodstvo, 6,* 31-39.

Balanovskij, A.E., & Grechneva, M.V., Vu Van, G. (2015). Issledovanie struktury relsovoj stali posle plazmennogo poverhnostnogo uprochneniya. *Uprochnyayushie tehnologii i pokrytiya, 11,* 23-32.

Balanovskij, A.E., & Nesterenko, H.A. (1992). Plazmennoe ciklicheskoe uprochnenie stalej. *Svarochnoe proizvodstvo, 11,* 19-20.

Balanovskij, A.E. & Pletnikov, I.A. (1999). Kompleksnaya ocenka kachestva tehnologii plazmennogo poverhnostnogo uprochneniya bandazhej lokomotivov. *Svarka i diagnostika, 3,* 45-50.

Balanovskij, A.E., & Pletnikov, I.A. (1999). Kompleksnaya ocenka kachestva tehnologii plazmennogo poverhnostnogo uprochneniya bandazhej lokomotivov. Chast 2. *Svarka i diagnostika, 4,* 46-50.

Balanovskij, A.E., & Pletnikov, I.A. (1999). Plazmennoe poverhnostnoe uprochnenie specialnyh stalej. *Metallurgiya mashinostroeniya, 5,* 24-30.

Baranova, L.V., & Demina, E.L. (1986). *Metallograficheskoe travlenie metallov i splavov*. Spravochnik.

Bayatanova, L.B. (2014). *Formirovanie modificirovannyh sloev na poverhnosti nizkouglerodistoj stali metodom vozdejstviya elektrolitnoj plazmoj: dissertaciya na soiskanie uchenoj stepeni doktora filosofii.* Vostochno-Kazahstanskij gosudarstvennyj tehnicheskij universitet im. D.Serikbaeva.

Bayatanova, L.B., Sapataev, E.E., & Tleukenov E.O. (2012). Uprochnenie poverhnosti podshipnikovoj dorozhki burovogo instrumenta elektrolitno-plazmennoj obrabotkoj. *Materialy III Mezhdunarodnoj nauchno-prakticheskoj konferencii s elementami nauchnoj shkoly dlya molodyh uchenyh «Innovacionnye tehnologii i ekonomika v mashinostroenii.* YuTI TPU.

Belkin, P.N. (2005). *Elektrohimiko-termicheskaya obrabotka metallov i splavov*. P. N. Belkin *Mir.*

Belkin, P.N. (2005). Stacionarnaya temperatura anoda, nagrevaemogo v vodnyh elektrolitah / P. N. Belkin, A. B. Belihov, *Inzhenerno-fizicheskij zhurnal, 75*(6), 19-24.

Berdnikov, A.A., Filippov, M.A., & Studenok, E.S. (2007). Struktura zakalennyh uglerodistyh stalej posle plazmennogo poverhnostnogo nagreva. *Metallovedenie i termicheskaya obrabotka metallov, 6,* 2-4.

Chudina O.V. (2003). Kombinirovannye metody poverhnostnogo uprochneniya stalej s primeneniem lazernogo nagreva. *Teoriya i tehnologiya*. M.: Izdatelstvo MADI (GTU).

Colm, G. (n.d.). Simplified diagram of a transmission electron microscope. Drawing by Graham Colm, courtesy of Wikimedia Commons. https://www.ccber.ucsb.edu/ucsb-natural-history-collections-botanical-plantanatomy / transmission-electron-microscope.

Dunleavy C.S., Golosnoy I.O., Curran J.A., Clyne T.W. (2009). Characterisation of discharge events during plasma electrolytic oxidation. *Surface & Coatings Technology, 203*, 3410-3419.

Duradzhi V. N., Parsadanyan A. S. (1988). *Nagrev v elektrolite*. Kishinev: Shtinca.

Dyachenko, S. S. (1981). *Formation of austenite in carbon alloys. M.* Metallurgy.

Erkinbekkyzy T.E., Zhurerova L.G, Daryn B. (2020). Influence of electrolyte-plasma hardening technological parameters on the structure and properties of banding steel 2. *Key Engineering Materials, 839*.

Glagolev, A. A. (1941). Geometricheskie metody kolichestvennogo analiza agregatov pod mikroskopom. Lvov: Gosgeolizdat.

Gorelik, S.S., Skakov, Y.A., & Rastorguev, L.N. (2002). *Rentgenograficheskij i elektronno-opticheskij analiz.* M.: MISIS.

Grechneva, M. V. (1992). Plasma hardening of metals in liquid media. Welding, 7, 18–22.

Grechneva, M.V. (1992). Plazmennoe uprochnenie metallov v zhidkih sredah. *Svarochnoe proizvodstvo, 7,* 8-12.

Grechneva, M. V. (2017). Kratkij analiz sostoyaniya tehnologij plazmennogo poverhnostnogo uprochneniya stalej i splavov v RF. *International conference on modern researches in science and technology. Conference Proceedings*. Scientific public organization.

Grigoriants, A.G., Safonov, A.N. (1987). *Methods of surface laser treatment*. Higher School Publ.

Gulyaev, A. P. (1986). *Metallurgy. M.* Metallurgy.

Ivanov, Y.F., Gromov, V.E., & Vorobev, S.V. (2011). Fazovyj sostav i defektnaya substruktura stali 20H13, obrabotannoj elektronnym puchkom v rezhime oplavleniya poverhnosti. *Fundamentalnye problemy sovremennogo materialovedeniya*.

Ivanov, Y.F. & Kozlov, E.V. (1991). Electron microscopic analysis of the martensitic phase of steel 38KHN3MFA. *Ferrous metallurgy, 8,* 38-41.

Ivanov Yu.F., Kozlov E.V. (2002). Volumetric and surface hardening of structural steel morphological analysis of the structure. Physics. - 2002. No. 3. pp. 5-23.

Jin, X., Wang, B., Xue, W., Du, J., Wu, X., & Wu, J. (2013). Characterization of wear-resistant coatings on 304 stainless steel fabricated by 148 cathodic plasma electrolytic oxidation. *Surface & Coatings Technology, 236*, 22-28.

Kidin I.N. (1978). Elektrohimiko-termicheskaya obrabotka metallov i splavov Tekst. *Metallurgiya*, 320.

Ko, Y., Lee, K., & Shin, D. (2013). Effect of ammonium metavanadate on surface characteristics of oxide layer formed on Mg alloy via plasma electrolytic oxidation. *Surface & Coatings Technology, 236*, 70-74.

Kondrat'ev V.V., Balanovskii A.E., Ivanov N.A., Ershov V.A., Kornyakov M.V. (2014). Evaluation of the Effect of Modifier Composition with Nanostructured Additives on Grey Cast Iron Properties. *Metallurgist, 58,* 377-387.

Kong, J. H., Takeda, T., Okumiya, M., Tsunekawa, Y., Yoshid, M., & Kim, S. G. (2012). The Study about Surface Modification of Steel by Water Plasma. *Proc. of 13th International Conference on Plasma Surface Engineering*, Garmisch-Partenkirchen.

Korotkov, V.A. (2016). The 15th Anniversary of the Manual Plasma Quenching. Ferrous Metallurgy. *Bulletin of Scientific, Technical and Economic Information, 10*, 62-68.

Korotkov, V.A., Baskakov, L.B., & Tolstov, I.A. (1991). Vosstanovitelnaya naplavka i uprochnenie rolikov rolgangov. *Svarochnoe proizvodstvo, 3*, 31-33.

Kozlov, E., Skakov, M., Uazyrkhanova, G., & Popova N. (2014). Redistribution of second phase particles in surface layers of deformed roll. *Advanced Materials Research, 1013*, 158-165.

Kozlov, E.V., Popova, N.A., & Koneva, N.A. (2004). Fragmented substructure formed in BCC steels during deformation. *RAS. The series is physical, 68*(10), 1419-1428.

Kulikov, I.S., Vashenko, S.V., & Kamenev A.Ya. (2010). *Elektrolitno-plazmennaya obrabotka materialov.* Minsk: Belarus.navuka.

Lahtin Y.M., & Arzamasov B.N. (1985). *Himiko-termicheskaya obrabotka metallov: uchebnoe posobie dlya vuzov.* M.: Metallurgiya.

Lakhtin Y.M., & Arzamasov B.N. (1985). *Chemical and thermal treatment of metals.* Metallurgy Publ.

Lee K.M., & Polycarpou A.A. (2005). Wear of conventional pearlitic and improved bainitic rail steels. *Wear, 259*, 391-399.

Leshchinsky L. K., & Samotugin S. S. (2001). *Mechanical properties of plasma-hardened 5% chromium tool steel deposited by arc welding. Welding,* 25-30.

Leshinskij, L. K., Samotugin, S. S., Pirch, I. I., & Komar, V. I. (1990). Plazmennoe poverhnostnoe uprochnenie. Kiev: Tehnika.

Majorov, V. S. (2009). *Lazernoe uprochnenie metallov. V kn.: Lazernye tehnologii obrabotki materialov: sovremennye problemy fundamentalnyh issledovanij i prikladnyh razrabotok.* M. Fizmatlit.

Majorov V.S. (2009). *Lazernye tehnologii obrabotki materialov: sovremennye problemy fundamentalnyh issledovanij i prikladnyh razrabotok.* M.: Fizmatlit.

Makarov A.V. (2009). Povyshenie iznosostojkosti splavov zheleza za schet sozdaniya metastabilnyh i nanokristallicheskih struktur: avtoref. ...dokt.teh.nauk. *Institut mashinovedeniya Uralskogo otdeleniya RAN.* Chelyabinsk.

Malinovskaya V.A. (2006). *Raspredelenie azota i ugleroda, fazovaya struktura gradientnyh sloev i mehanicheskie svojstva stali 20H2N4A posle nitrocementacii: dis... kand. tehn.* Tomsk.

Martin, J., Melhem, A., Shchedrina, I., Duchanoy, T., Nominé, A., Henrion, G., Czerwiec, T., & Belmonte T. (2013). Effects of electrical parameters on plasma electrolytic oxidation. *Surface & Coatings Technology, 221.*

Medvedev, S.I., Nezhivlyak, A.E., Grechneva, M.V., Balanovsky, A.E., & Ivakin, V.L. (2015). Optimization of plasma hardening conditions of the side surface of rails in PUR-1 experimental equipment. *Welding International, 29,* 643-649.

Medvedovskaya L. A., & Shur N. F. (1983). Oborudovanie i tehnologiya lazernoj termicheskoj obrabotki (obzor). *Metallovedenie i term.*

Menin, R., Ramos, E., & Magrini, M. (1981). Laser surface treatment of carbon and alloy steels. *Proc. 4 Inter. conf. of rapid quench. metals.* Sendai.

Moon, S., & Jeong, Y. (2009). Generation mechanism of microdischarges during plasma electrolytic oxidation of Al in aqueous solutions. *Corrosion Science, 51,* 1506-1512.

Nesterov, D.K., & Sapozhkov, V.E. (1989). i dr. Uprochnenie relsov iz zaevtektoidnoj stali kombinirovannoj termoobrabotkoj. *MiTom, 12,* 2-5.

Patent na poleznuyu model Nº 5365 Sposob uprochneniya zheleznodorozhnyh koles. Data registracii v Gosudarstvennom reestre poleznyh modelej Respubliki Kazahstan.

Paulmier, T., Bell, J.M., Fredericks, P.M. (2008). Development of a novel cathodic plasma-electrolytic deposition technique. Part 2: Physico-chemical analysis of the plasma discharge. *Journal of materials processing technology, 208,* 117-123.

Pogrebnjak, A. D., & Tyurin, Yu. (2000). Preparation and investigation of the structure and properties of Al_2O_3 plasma-detonation coatings. *Techn. Phys. Lett., 26*(11), 960-964.

Pogrebnjak, A. D., Tyurin, Y. N., & Kobzev, A. P. (2003). Mass transfer and doping during electrolyte-plasma treatment of cast iron. *Tech. Phys. Lett., 29*(4), 312-315.

Poletika, I.M., Golkovskij, M.G., & Perovskaya, M.B. (2006). Elektronno-luchevaya zakalka poverhnostnogo sloya stali vne vakuuma. *Fizicheskaya mezomehanika, 9,* 181-184.

Polyak M.S. (1995). Tehnologiya uprochneniya. *Tehnologicheskie metody uprochneniya: spravochnik v 2 t.* Mashinostroenie.

Popova, N.A., Nikonenko, E.L, Tabieva, E.E., & Uazyrkhanova, G.K. (2020). Structure and phase composition of ferritic-perlitic steel surface after electrolytic plasma quenching. *Russian Physics Journal, 63*(5), 791-796.

Rahadilov, B.K. (2014). *Elektrolitno-plazmennoe azotirovanie poverhnostnyh sloev bystrorezhushih stalej: dissertaciya na soiskanie uchenoj stepeni doktora filosofii* [Dissertation, Vostochno-Kazahstanskij gosudarstvennyj tehnicheskij universitet im. D.Serikbaeva].

Rakhadilov, B.K., Tabiyeva, Y.Y., Uazyrkhanova, G.K., Zhurerova, L.G., & Baizhan, D. (2020). Inñuence of electrolytic-plasma surface quenching on the structure and strength properties of ferritic-pearlite class wheel steel. *Eurasian Journal of Physics and Functional Materials, 4,* 167-173.

Rakhadilov, B.K., Tabiyeva, Y.Y., Uazyrkhanova, G.K., Zhurerova, L.G., Maulit, A., & Baizhan, D. (2019). Surface modification of steel mark 2 electrolytic-plasma exposure. *Eurasian Journal of Physics and Functional Materials, 3*, 355-362.

Rakhadilov, B.K., Tabiyeva, Y.Y., Uazyrkhanova, G.K., Zhurerova, L.G., & Popova N.A. (2020). Effect of electrolyte-plasma surface hardening on structure wheel steel 2. *Bulletin of Karaganda State University, 2* (98), 68-74.

Richard, R. (2007). *Track Demonstration of Laser-Treated Rail to Reduce Friction and WearTransportation Technology Center / R.* Richard. Inc.

Sadovsky, V. D., & Malyshev, K. A. (1984). *Phase and structural transformations during heating of steel.* Metallurgizdat.

Saltykov S.A. (1976). Stereometricheskaya metallografiya. *Metallurgiya.*

Samotugin, S.S., Muratov, V.A, & Kovalchuk, A.V. (1997). Plazmennoe uprochnenie instrumentov kolcevoj formy. *Metallovedenie i termicheskaya obrabotka metallov.*

Skakov, M., Bayatanova, L., & Sheffler, M. (2012). The drilling tool electrolyte-plasma material hardening technology planning and research. *The 7th International Forum on Strategic Technology.* Tomsk.

Skakov, M., Kurbanbekov, S., Tabieva, Y., & Zamanbekuly, E. (2013). Nitriding and carbonitriding influence on stainless steels surface layers changes. *Applied Mechanics and Materials, 379*, 105-109.

Skakov, M., & Rakhadilov, B. (2012). Change of structure and wear-resistance of P6M5 steel for processing in electrolyte plasma. *32th All-Polish Tribology conference Autumnal school of Tribology 2012.* Wroclaw University of Technology Institute of Machine Design and Operation.

Skakov, M.K., Bayatanova, L.B., & Scheffler, M. (2013). Changes of Structural-Phase Condition In *18CrNi3MoA-Sh Steel After Elektrolyte-Plasma Processing.* Advanced Materials Research.

Skakov, M.K., Bayatanova, L.B., & Scheffler, M. (2012). *Research of Surface Treatment on the Microstructure, Microhardness and Wear Resistance of 18CrNi3MoA-Sh Steel.* Tribologia.

Skripkin, A.A., Necvetaev, V.A., Sherbakov, V.E., & Minenko, N.Y. (1992). Poluchenie teplostojkih sloev na stali 20 s ispolzovaniem plazmennogo nagreva. *Svarochnoe proizvodstvo, 11*, 15-17.

Stalinskij, D.V., & Rudyuk, A.S. (2007). Sferidizaciya karbidnoj fazy v zaevtektoidnoj stali i ee vliyanie na svojstva relsov. *Metallurgicheskaya i gornorudnaya promyshlennost, 2*, 48-54.

Suhomlin, G.D. (1974). Elektronno-difrakcionnoe issledovanie orientacionnyh sootnoshenij ferrit-cementit v perlite. *FMM, 38*(4), 878-880.

Tabieva, E.E., Popova, N.A., Nikonenko, E.L., Uazyrkhanova, G.K., & Gromov, V.E. (2020). Influence of surface quenching on morphology and phase composition of ferritic-pearlitic steel. *Izvestiya Ferrous Metallurgy, 63*(11-12), 915–921.

Tabieva, E.E., Rahadilov, B.K., Uazyrhanova, G.K., & Bajzhan, D. (2020). Struktura i mehanicheskie svojstva kolesnoj stali do i posle elektrolitno-plazmennoj poverhnostnoj zakalki. *Vestnik VKTU, 3.*

Tabieva, E.E., Uazyrhanova, G.K., Popova, N.A., & Nikonenko E.L. (2020). Vliyanie poverhnostnoj zakalki na izmenenie strukturno-fazovogo sostoyaniya st2. *Fundamentalnye problemy sovremennogo materialovedeniya, 2*, 169-175.

Tabiyeva, E., Rahadilov, B., Uazyrkhanova, G., Satbaeva, Z., & Kengesbekov A. (2019). Influence of electrolytic-plasma surface hardening on the surface properties of bandage steel. *TRIBOLOGIA, 4*, 105-111.

Tabiyeva, Y. Y., Popova, N. A., & Nikonenko, E. L. (2020). Influence of electrolytic-plasma treatment on the morphological components of the matrix of steels of ferrite-perlite class. *International Scientific and Technical Conference PRFN-2020*. Tomsk.

Tabiyeva, Y. Y., Rakhadilov, B. K., Uazyrkhanova, G. K., Zhurerova, L. G., Sagdoldina, Zh. B., & Kurbanbekov, S. K. (2019). *Research of change in tribological properties of steel grade two after electrolytic-plasma exposure*. 11th International Scientific Conference "Chaos and Structures in Nonlinear Systems, Karaganda..

Troyannikov, A.Y. (2021). Sravnitelnyj analiz vysokoenergeticheskih metodov poverhnostnogo uprochneniya stali. *Molodoj uchenyj, 19*, (361).

Tyuftyaev, A.C. (2002). *Zakonomernosti strukturoobrazovaniya pri plazmennoj obrabotke stali 60 G i promyshlennoe osvoenie tehnologii poverhnostnogo uprochneniya grebnej zheleznodorozhnyh kolesnyh par.* OIVTRAN.

Tyuftyaev, A. S. (2002). *Zakonomernosti strukturoobrazovaniya pri plazmennoj obrabotke stali 60G i promyshlennoe osvoenie tehnologii poverhnostnogo uprochneniya grebnej zheleznodorozhnyh kolesnyh par*. FGUP CNIIchermet im. I.P.Bardina. Moskva.

Tyuirin, Y.N. & Pogrebnjak, A.D. (2001). *Electric Heating Using a Liquid Electrode*, 142-144. Surf. And Coat. Tech.

Tyurin, Y. (1999). *Technology and equipment electrolytic-plasma quenching*. 3rd Intern. conf. on quenching and control of distortion. Prague.

Tyurin, Y. (2000). Features and advantages of surface modification of metallic pieces using electrolytic plasma thermal cycling hardening. *Plasma Surface Eng*, 1230-1233.

Tyurin, Y.I. (1999). *Osobennosti elektrolitno-plazmennoj zakalki. 3-ya mezhdunarodnaya konferenciya. Vzaimodejstvie izluchenij s tverdym telom*. Minsk.

Uazyrkhanova, G. K., Rakhadilov, B. K., Tabieva, E. E., & Baizhan, D. (2020). The influence of electrolyte-plasma surface hardening on the strength characteristics of grade two steel. *VI International Scientific and Practical Conference "Science and education in the modern world: challenges of the XXI century"*. Nur-Sultan.

Utevskij, L.M. (1973). *Difrakcionnaya elektronnaya mikroskopiya v metallovedenii*. Metallurgiya.

Van Gyui, V. (2015). Plazmennaya cementaciya uglerodistyh stalej s ispolzovaniem past v tverdoj faze. *Vestnik nauki i obrazovaniya Severo-Zapada Rossii, 1*, 205-211.

Van Gyui, V. (2015). Parovodyanaya plazmennaya cementaciya v tverdoj faze na ustanovke MULTIPLAZ 3500. *Innovacii v nauke, 51* (1), 95-102.

Vyacheslav, F.B. & Szczerek, M. (2018). The influence of surface layer quality on the performance characteristics of machine parts under machining. *Tribologia, 271*(1), 23-29.

Wang, B., Jin, X., Xue, W., Wu, Z., Du, J., & Wu. J. (2013). High temperature tribological behaviors of plasma electrolytic borocarburized Q235 low-carbon steel. *Surface & Coatings Technology, 232*, 142-149.

Weinberg, F. (1973). *Pribory i metody fizicheskogo metallovedeniya, 1.*

Yerokhin, A.L., Nie, X., Leyland, A., Matthews, A., & Dowey S.J. (1999). Plasma electrolysis for surface engineering. *Surface and Coatings Technology, 122*, 73-93.

Yushenko, K.A., Borisov, Y.S., & Tyurin, Y.N. (1994). *Teoriya i praktika impulsno-plazmennogo modificirovaniya poverhnosti detalej mashin i instrumenta.* Svarochnoe proizvodstvo. Kiev.

Zaides S.A., Van Huy, V., & Van, D. (2020). Developing installation to increase cylindrical part surface hardness. *Proceedings of Irkutsk State Technical University, 24*(2), 262-274.

Zhang P., Nie X., & Northwood D.O. (2009). Influence of coating thickness on the galvanic corrosion properties of Mg oxide in an engine coolant. *Surface & Coatings Technology, 203*, 3271-3277.

Zhang, W. (2002). Integrated modeling of thermal cycles, austenite formation, grain growth and decomposition in the heat affected zone of carbon steel. *Science and Technology of Welding and Joining, 5*, 574-582.

Zhang W. (2002). Kinetics of Ferrite to Austenite Phase Transformation during Welding of 1005 Steel. *Scripta Materialia, 46*, 753-757.

Zhang, W. (2002). Modeling and Real Time Mapping of Phases during GTA Welding of 1005 Steel. *Materials Science and Engineering A, 333,* 320-335.

Zhang W., Elmer J.W., & Debroy T. (2005). Integrated modeling of thermal cycles, austenite formation, grain growth and decomposition in the heat affected zone of carbon steel. *Science and Technology of Welding and Joining, 10* (5), 574-582.

Chapter 4
Hard and Superhard Multicomponent Nitride Coatings Deposited Using Vacuum-Arc Evaporation:
The Effects of Cr and Si on the Structure and Properties of the Nanostructured (Zr-Ti-Nb)N Coatings

Didar Yeskermessov
https://orcid.org/0000-0002-2206-8132
D. Serikbayev East Kazakhstan Technical University, Kazakhstan

Zhanerke Toleukhanova
https://orcid.org/0009-0002-9254-3901
D. Serikbayev East Kazakhstan Technical University, Kazakhstan

Zarina Aringozhina
https://orcid.org/0009-0001-8428-4033
D. Serikbayev East Kazakhstan Technical University, Kazakhstan

Yerkezhan Tabiyeva
https://orcid.org/0000-0002-9726-7187
D. Serikbayev East Kazakhstan Technical University, Kazakhstan

Waqar Ahmed
University of Lincoln, UK

ABSTRACT

This chapter is devoted to experimental research on the investigated of the structural-phase, morphology, elemental composition, physical-mechanical, and tribological properties (friction, wear, and adhesion) of hard and superhard micro- and nanostructured coatings of systems based on (Zr-Ti-Nb)N, (Zr-Ti-Cr-Nb)N, and (Zr-Ti-Cr-Nb-Si)N were fabricated by vacuum-arc deposition in the nitrogen atmosphere. The

DOI: 10.4018/978-1-6684-6830-2.ch004

authors in the work used advanced proven experimental research methods (SEM-EDX, TEM-EDS, XRD, SIMS, GDMS, PIXE, XPS, AFM, hardness measurements, and adhesion testing), as well as theoretical methods for analysing the results. Based on the results obtained, according to the experimental data, an optimal mode was selected that provides an increase in mechanical and tribological characteristics, and a method for vacuum-arc surface hardening was also developed. Such coatings seem to have prospects as protective ones for couples of friction and cutting tools.

1. INTRODUCTION

High-tech industries such as: rocket and space, aviation, petroleum and gas production, as well as various alternative energy industries today pay great attention to improving the reliability and efficiency of products that are operated in various environments. The emphasis in modern materials analysis is usually directed at the structure and composition of the surface and outer layers of materials from several tens to hundreds of nanometers.

In particular, products based on titanium (Ti) and its alloys, which are used in various fields of industry, are of considerable interest due to their properties. The literature data indicate that various methods of surface treatment are used to improve the physical and mechanical properties of titanium and alloys based on it. The emphasis is on the realization that the surface and near-surface areas control many mechanical and chemical properties of solids: corrosion, friction, wear, adhesion and destruction. In addition, the composition and structure of the outer layers can be adapted using directed energy processes using lasers or electron and ion beams, as well as using more traditional methods such as oxidation and diffusion. The research of nanostructured objects is the fastest developing in modern materials science, since the ultrafine dispersed structure causes a significant improvement, and in some cases, a radical change in the properties of the material (Gavaleiro & De Hosson, 2006; Voevodin et al., 2004).

The investigation of ultrafine-grained materials has shown that a decrease in the size of crystals below a certain threshold value can lead to a significant change in properties. Dimensional effects are manifested when the average size of crystalline grains does not exceed 100 nm, and is most clearly observed when their size approaches 10 nm, and the intercrystalline (intergranular layer) is units of nanometres, consisting as a rule of an amorphous phase like nitrides, oxides and carbides (Gavaleiro & De Hosson, 2006; Nordin et al., 1998; Nordin et al., 1999; Pogrebnjak et al., 2015; Voevodin et al., 2004). From a physical point of view, the transition to the nanostate is associated with the appearance of dimensional effects, which should be understood as a complex of phenomena associated with changes in the properties of matter due to the coincidence of the size of the microstructure block and some critical length, characterizing the phenomenon as the free path lengths of electrons and phonons, the wall thickness of domains, the critical radius of the dislocation loop, etc (Morris, 1998). Conventionally, these coatings can be divided into conditionally hard (< 40 GPa) and superhard (>40 GPa), obtained using CVD (chemical vapor deposition), PVD (physical vapor deposition), magnetron sputtering and ion-assisted deposition methods (Kim & Cha, 2005; Lee et al., 2007; Leng et al., 2001; Rizzo et al., 2006; Shin et al., 2002).

Based on this, nanostructured nitride coatings based on multicomponent alloys containing at least 3-4 constituent elements are of scientific interest, and their synthesis and intensive research is an urgent task of materials science. Thus, the most diverse combination of constituent elements and changes in the physical parameters of deposition (working gas pressure and substrate bias voltage) will make it possible to change the structural state (grain size, texture, residual stresses) and properties of condensates in a wide range. Proceeding from the foregoing, the synthesis and study of nitride coatings of transition metals based on multicomponent alloys by vacuum-arc deposition is an urgent task. The analysis of the regularities of the formation of nanocrystalline coatings, the investigation of the correlation of their structural-phase state and properties depending on the deposition modes are of scientific and practical interest.

Therefore, the objective of this work is to conduct comprehensive research of the microstructure, elemental and phase compositions, surface morphology, physical-mechanical and tribological characteristics of the coating based on (Zr-Ti-Nb)N obtained by vacuum-arc deposition of the cathode, when the deposition and alloying parameters of the coating with chromium (Cr) and silicon (Si) elements change.

And for this reason, this chapter is generally devoted to experimental study on the investigation of the structural-phase, elemental composition, as well as the physical-mechanical and tribological properties of solid and superhard micro- and nanostructured coatings of systems based on Ti, Zr, Cr, Nb, Si, N elements. So, in section 2, the objects of research are discussed, as well as the principles, methods of obtaining, means of testing and determining the physical-mechanical and tribological properties of coatings are described in detail. The description of the equipment used in the work is given. Section 3 presents and analyzes the results of experimental and theoretical studies of the morphology, elemental and phase composition of coatings based on (Zr-Ti-Nb), (Zr-Ti-Nb-Cr) and (Zr-Ti-Nb-Cr-Si) obtained by vacuum-arc deposition. In Section 4, the relationship between mechanical properties and deposition parameters of multicomponent nitride coatings based on Zr, Ti, Nb, Cr, and Si was studied, and the tribological characteristics were also analyzed.

2. EQUIPMENT AND METHODS FOR RESEARCHING THE STRUCTURE AND PROPERTIES OF MULTI-COMPONENT NANOSTRUCTURED COATINGS

2.1. Materials and Equipment for Coating Deposition

Previously, the development of new technologies and principles for the design of new materials was based on the use of one to three elements as the main ones, for example: steel (Fe), heat-resistant alloys (Mn, Zr), bulk metallic glasses (Ni, Co), protective coatings based on (Ti). However, such compounds did not always meet the basic requirements for coatings. Further development of materials science led to the formation of 3- and 4-nitride systems, which were characterized by a significant increase in functional properties compared to binary nitrides. Recently, the so-called multicomponent (multi-element) alloys with variable chemical composition, which are characterized by universalization of physical and chemical properties, have been widely used.

It is known that, according to the configurational model of matter, the properties of nitride coatings (hardness, adhesive activity (strength), melting point, etc.) depend on the statistical weight of stable atomic sp- and d-configurations (SWSC) (Azarenkov et al., 2012). If titanium nitride (TiN) is considered as the basis, then its stability can be increased by doping with elements containing a large number of electrons at the d-level. Mostly stable d-configurations have elements of groups IV, V and VI of the periodic system. In this case, those alloying elements have the greatest influence on the properties of alloys, the atoms of which differ to the greatest extent from the atoms of the base in their external structure. The most common alloying elements are transition metals of groups IV and VI of the periodic system. Such elements have a high SWSC sp- и d- configurations at room temperature (Ti – 43%, Zr – 52%, Hf – 55%,

Table 1. Characteristics of elements used to produce nitride coatings

Element	Atomic Mass (g/mol)	Atomic Radius (pm)	Density (g/cm^3)	Melting Temperature (K)	Enthalpy of Formation (kJ/mol)
Ti	47.867	147	4.54	1933	336.3
Nb	92.906	146	8.57	2741	237.8
Zr	91.224	160	6.506	2125	355.6
Cr	51.996	130	7.19	2130	342.1
Si	28.086	132	2.33	1688	383.1
N	14.007	92	0.00125	63.29	-

V – 63%, Nb – 76%, Si – 81%, Cr – 73%, Mo – 84%, W – 96%) (Samsonov, 1969). According to the configurational model of a substance, the higher their total SWSC, the lower the adhesive interaction of the contacting metals. Thus, an increase in the properties of the TiN compound (for example: adhesive activity) is possible due to doping with elements that, on the one hand, will increase the SWSC of the d-configuration, and on the other hand, the elements must have high thermal stability. This circumstance suggests that alloying the TiN compound with atoms of Hf-, V-, Nb-, W-, Cr-, Si-elements will make it possible to obtain a compound with a unique set of properties (high hardness, heat resistance, melting point, etc.). However, taking into account the high degree of chemical affinity, it is advisable to use Zr, Nb, Cr and Si as constituent elements (Table 1), since the use of elements with a lower chemical affinity with nitrogen makes it difficult to grow a mononitride film.

If we compare the enthalpy of formation (-ΔH, kJ/mol), then the most stable nitrides with a high heat of formation are ZrN – 355.6 kJ/mol, TiN – 336.6 kJ/mol, Nb_2N – 255.8 kJ/mol, NbN – 237.8 kJ/mol, Cr_2N – 135.1 kJ/mol, CrN – 123.4 kJ/mol (Azarenkov et al., 2012). A class of compounds such as Cr_2N and CrN although it has the lowest heat of formation, however, has high functional properties (resistance to oxidation and wear). It should be noted that the Nb and Cr elements have a body-centered cubic (BCC)-lattice at room temperature, while Zr, Ti, and Si face-centered cubic (FCC) have an face-centered tightly packed (FCTP) lattice at room temperature and a BCC lattice at high temperatures. Thus, the high degree of chemical affinity of the constituent elements with nitrogen, as well as the stabilization of the BCC-lattice due to the inclusion of Nb and Cr, will make it possible to obtain a nitride phase based on a high-entropy single-phase alloy. An analysis of literature sources has shown that alloying titanium nitride with zirconium atoms favourably affects the functional properties, in particular, wear resistance increases due to the formation of a thin layer of zirconium oxide on the coating surface. Additional modification with Cr, Nb additives improves the thermal stability, oxidation and corrosion resistance of the protective coating, reduces the friction coefficient, and increases the level of hardness. It should also be noted that with the transition of metal nitrides from group VI (CrN, Cr_2N) to group IV (ZrN, NbN) resistance to high-temperature oxidation increases. Based on the above, it can be assumed that the compound based on Ti, Zr, Nb, Cr, Si and N will have a unique set of properties: high hardness, high oxidation resistance, excellent tribological characteristics and other functional properties.

Coatings were obtained by vacuum-arc evaporation at the "Bulat-6" installation, which consisted of a vacuum chamber, a control rack, a high-voltage rectifier, and three power sources for arc evaporators. (Shulayev et al., 2006). The main deposition parameters are given in Table 2.

Table 2. Deposition parameters of nitride coatings

Series (Sample) No.	Deposited Material	Arc current I_C, A	Nitrogen Pressure P_N, Pa	Bias Voltage U_b, V
1	(Zr-Ti-Nb)N	95	0.04	-100
2			0.5	-100
1	(Zr-Ti-Cr-Nb)N	110	0.3	-100
2			0.7	-100
3			0.3	-200
4			0.7	-200
1	(Zr-Ti-Cr-Nb-Si)N	110	0.3	-100
2			0.3	-200

Coatings (Zr-Ti-Nb)N, (Zr-Ti-Cr-Nb)N and (Zr-Ti-Cr-Nb-Si)N were deposited on a polished rotating substrate (material-steel 12Cr18Ni9Ti, A 570 Grade 36 steel and Si) 15×15×2.5 mm^3 in size with initial surface roughness R_a » 0.09 μm. The evaporated materials were solid cathodes Zr-Ti-Nb (Zr – 35 at%, Nb – 35 at%, Ti – 30 at%), Zr-Ti-Cr-Nb (Cr – 37.39 at%, Zr –27.99 at%, Nb – 22.30 at%, Ti – 12.32 at%) and Zr-Ti-Cr-Nb-Si (Cr –17.08 at%, Zr – 30.19 at%, Nb – 9.67 at%, Ti – 39.96 at%, Si – 3.1 at%). Cathodes from multicomponent alloys of the Zr-Ti-Nb, Zr-Ti-Cr-Nb and Zr-Ti-Cr-Nb-Si systems were fabricated by vacuum-arc melting in a high-purity argon atmosphere. Melting was carried out with a non-consumable tungsten electrode into a water-cooled copper container (hearth). The obtained ingots were remelted 6-7 times with a cooling rate of about 50 K/s for the final homogenization of the composition. The deposited (Zr-Ti-Nb)N, (Zr-Ti-Cr-Nb)N and (Zr-Ti-Cr-Nb-Si)N coatings were made in a molecular N_2 atmosphere, the pressure of the working gas varied from 0.04 to 0.7 Pa. The substrate bias was chosen as the control parameter, which varied from -95 to -200 V. The substrates were heated to 450°C before deposition. The distance between the substrates and the cathode was 250 mm. In practice, it is considered expedient to obtain thin coatings (up to 10 μm) due to a decrease in the influence of the scale factor on the properties of the substrate and the likelihood of destructive stresses in the coating itself. Following these, the thickness of the coatings (Zr-Ti-Nb)N, (Zr-Ti-Cr-Nb)N and (Zr-Ti-Cr-Nb-Si)N is 6.7, 10 and 6.2÷6.8 μm, respectively. The block diagram of the "Bulat-6" installation is shown in Figure 1.

The vacuum chamber is a cylindrical vessel with an inner diameter of 500 mm and a length of 500 mm with a horizontal axis (Figure 2). Vacuum-arc evaporators are fixed in the centre of the covers with flanges located at both ends of the chamber, as well as in the branch pipe located in the upper part of the chamber. In the lower part of the chamber there is a similar branch pipe with a flange, where a rotary device is fixed, on which the substrate holder is located. The rotary device is electrically isolated from the vacuum chamber and provides a rotation speed of 8 rpm. The chamber is evacuated using a steam-oil and mechanical vacuum pumps. (Shulayev et al., 2006).

The vacuum-arc evaporator (plasma source) consists of two main parts: the first is the cathode, containing the evaporated cathode with a magnetic system for holding the cathode spot on the evaporated surface of the cathode and an arc discharge ignition device, the second is the anode, which is a branch pipe with magnetic coils for focusing the plasma flow. In the case of using the cathode part as an independent evaporator without the anode part, the vacuum chamber body acts as the anode. Due

Figure 1. Installation block diagram "Bulat-6"

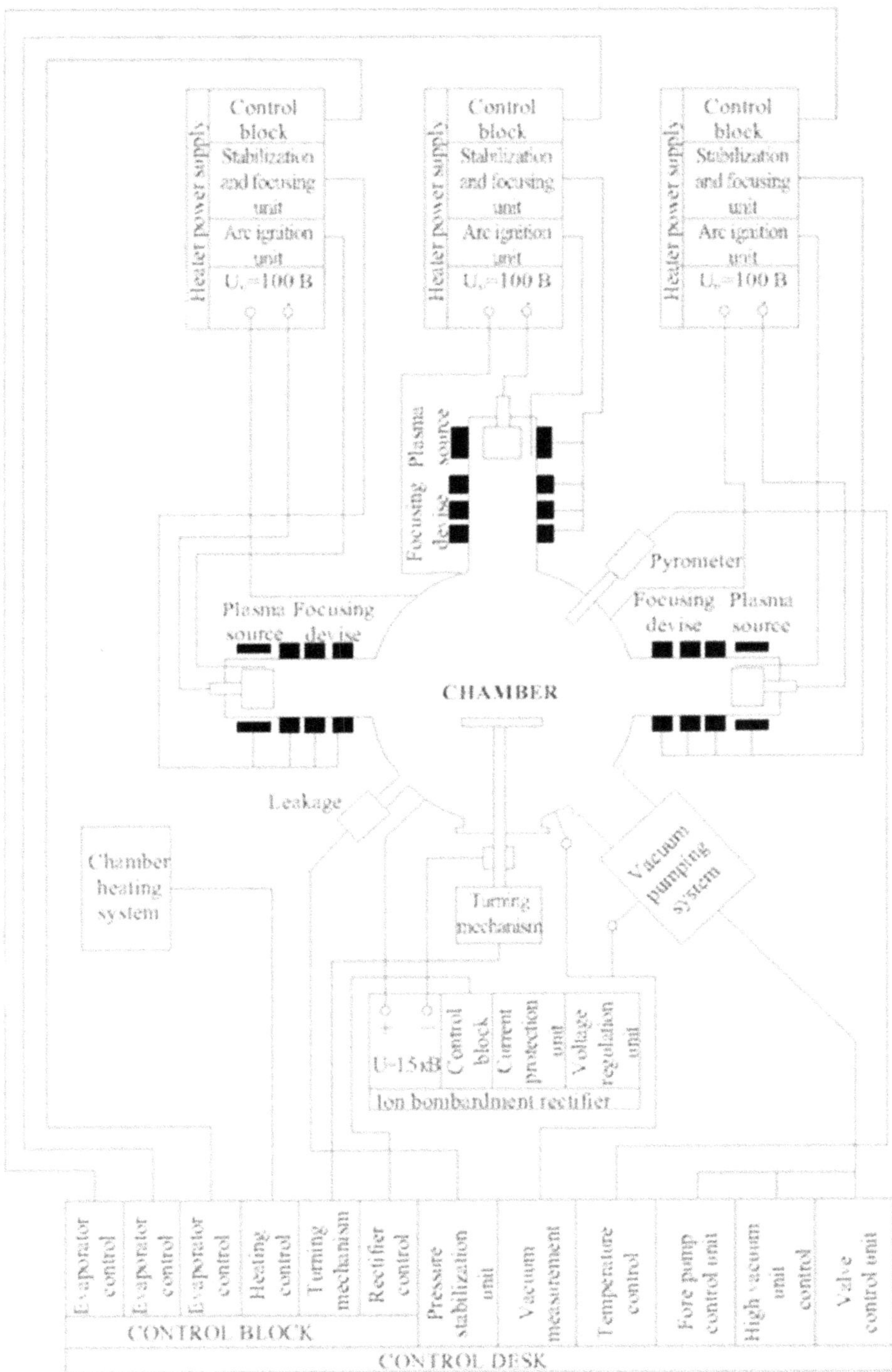

Figure 2. Vacuum-arc installation "Bulat-6": a – appearance; b – scheme

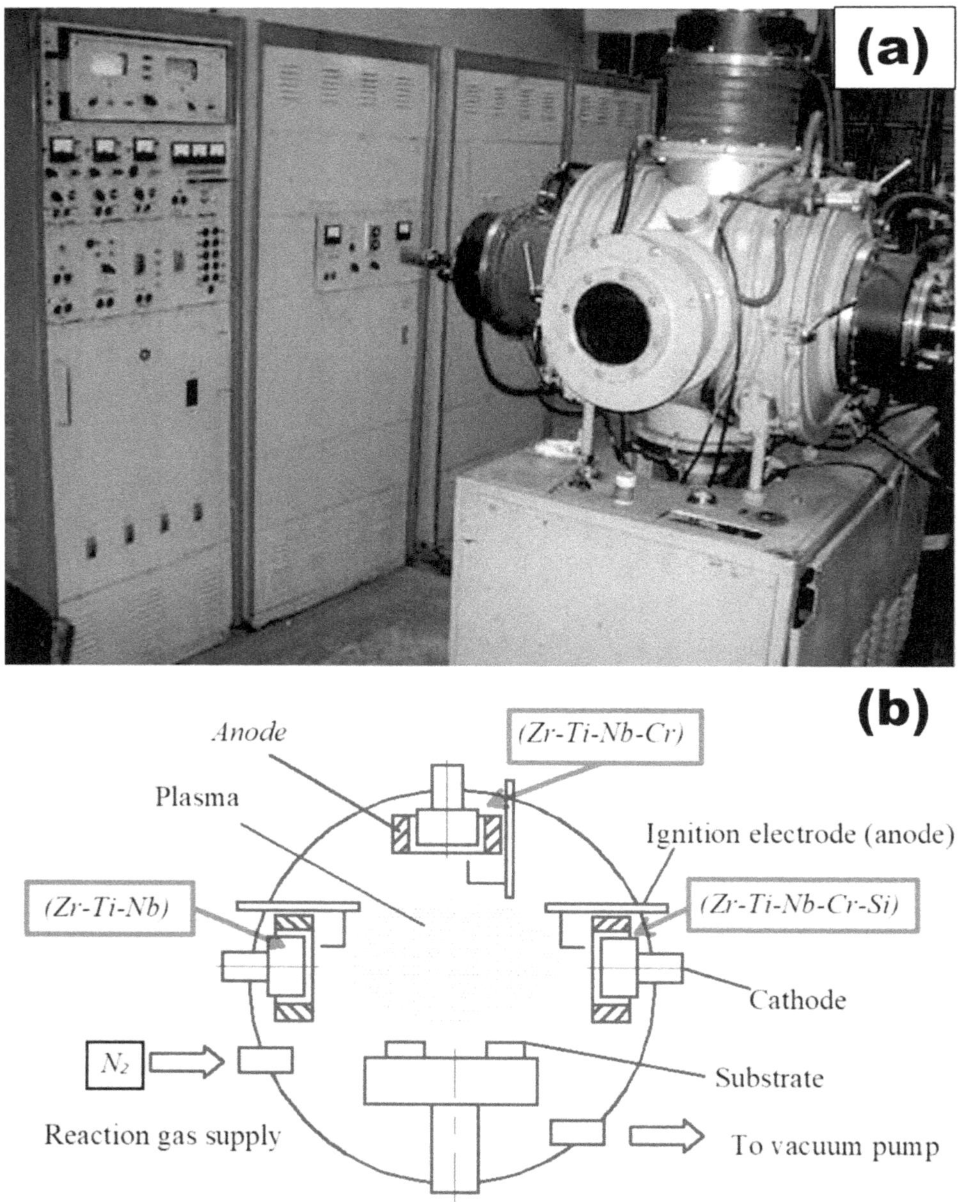

to this, the rate of deposition of coatings in the centre of the chamber increases by about one and a half to two times. The high voltage rectifier provides a voltage of 0.1 to 1.7 kV with a maximum current of 15 A. Power supplies for arc evaporators provide smooth current control from 90 to 150 A and an open circuit voltage of 100 V. The unit is equipped with a gas supply system with automatic maintenance of its pressure in the chamber in the range from 0.01 to 15 Pa (Shulayev et al., 2006).

2.2. Methods for Investigating the Structure and Properties of Multicomponent Nitride Coatings

Modern diagnostics of a substance implies the use of complementary methods for studying the physical and chemical parameters of structures, which provides complete information about the structure and physical-chemical properties. At the same time, the analysis of the substance consists in interpreting the signals recorded by the instruments, which can be the energy and intensity of the energy spectrum of absorbed or emitted electromagnetic radiation, electrons, ions, changes in the strength of the interaction of the probe with the surface under study, etc. (Troyan et al., 2008). This section provides a brief description of the research methods of nanostructured coatings used in the work.

2.2.1. Methods for Studying the Surface Morphology of Coatings

The surface morphology of nitride multicomponent coatings based on (Zr-Ti-Nb)N, (Zr-Ti-Cr-Nb)N and (Zr-Ti-Cr-Nb-Si)N was studied using scanning electron microscopy on JEOL-7000F, JSM-6390LV microscopes with energy-dispersive X-ray spectroscopy (EDS/EDX) analysers and JEOL-6010LA In-TouchScope. The use of such a technique makes it possible to study the sample area or micro-volume by irradiating it with a thin focused electron beam. Changing the angle of incidence of the primary beam makes it possible to study in detail the topography of various microparticles of the sample. Secondary, reflected and absorbed electrons are used to obtain an image of the surface. Of greatest interest are the images obtained using secondary and reflected electrons, which have a relatively high resolution (1.2 nm at 30 kV – JEOL-7000F), and also provide information about the sample areas located in the depth of a complex relief (Engel, 1986). The study was carried out at an accelerating voltage of 20 kV in real time (163.13 s.), while the scanning area was perpendicular to the main optical axis.

An atomic-force microscope (AFM) study was also conducted for a more in-depth detailed analysis of the surface morphology of the obtained nitride coatings. In this method, a change in the interaction force of the tip of the probe (needle) with the surface under study is recorded. The needle is located at the end of a cantilever beam with a known rigidity capable of bending under the action of small forces arising between the surface of the sample and the tip of the tip. These forces in a number of variants of the method can be van der Waals (molecular), electrostatic or magnetic. A beam with a needle is called a cantilever. The deformation of the cantilever is measured by the deflection of the laser beam incident on its back surface, or by the piezoresistive effect that occurs in the cantilever material during bending.

2.2.2. Elemental Analysis of Coatings

The study of the elemental composition of the coating was carried out using a scanning electron microscope JSM-6390LV – by JEOL (Japan), with the prefix of energy dispersive microanalysis INCA Energy by OXFORD Instruments. When electrons interact with matter, response signals of various forms arise (reflected and secondary electrons, Auger electrons, X-rays, absorbed current, etc.), which are used to synchronously construct an image on the monitor screen. Figure 3 shows a diagram of the formation of secondary signals under the influence of an electronic probe. An electron-optical system is not used for image formation, image scaling is carried out by radio-technical means. A scanning electron microscope is a vacuum device, since at normal atmospheric pressures the electron beam is strongly scattered and absorbed, which makes it impossible to focus it. Therefore, the working vacuum in the microscope cham-

Figure 3. Scheme of formation of secondary signals during the interaction of probe electrons with a sample

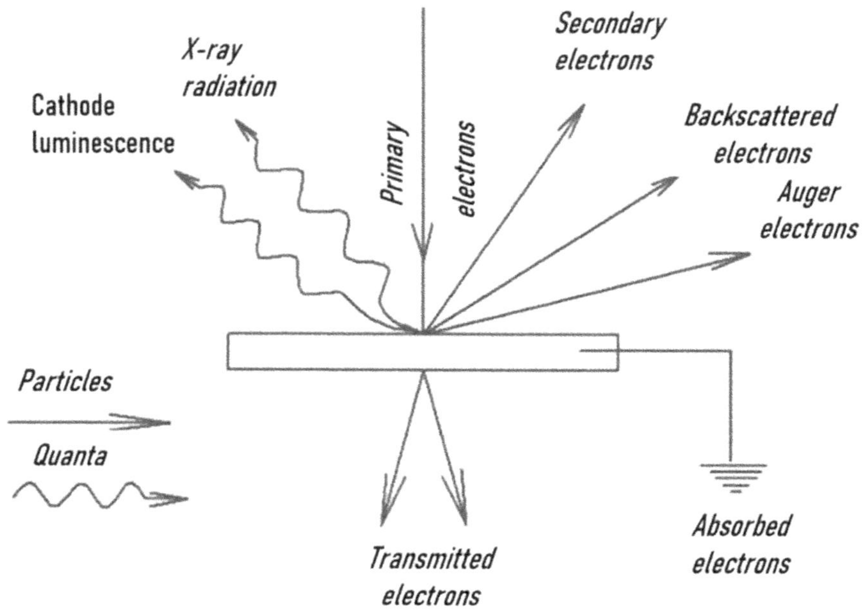

ber should be 10^{-5} Torr or better. The movement of the probe over the surface of the sample should be carried out with very high accuracy. The accuracy of the movement and the size of the probe determine the resolution of the microscope.

As a result of the interaction of the electron beam with the sample surface, a response occurs, which is recorded by the corresponding sensors. The appearance and diagram of the main nodes of the scanning microscope is shown in Figure 4. An electron beam from an electron source by a special condenser system is formed in the form of a well-formed probe and passes through a system of control electrodes or electromagnets that move the beam along the sample surface along a trajectory forming a raster. The signal recorded by the sensors is then used to modulate the brightness of the electron beam in the cathode ray tube of the monitor. The magnitude of this signal will depend on the physical properties of the sample surface and may vary from point to point. As a result, an image of the sample surface is formed on the monitor screen, displaying the topography of the corresponding physical property of the sample under study. Thus, it is possible to study the topography of inhomogeneities of defects and surface conditions: for example, the topology of the surface (grain boundaries, pores, cracks, inhomogeneities of composition, etc.) – in reflected or secondary electrons.

First of all, a scanning electron microscope (SEM) is used to study the structure of the surface. Currently manufactured devices have an increase from tens to several hundred thousand times. The possibility of studying not only the content of a particular chemical element in the sample, but also the distribution of selected chemical elements over the sample surface makes the scanning electron microscopy method unique. The energy dispersion method is based on the registration of the so-called characteristic radiation resulting from electron transitions between the internal energy levels of atoms. Depending on the

Figure 4. Scanning electron microscope (SEM): a – appearance; b – block diagram

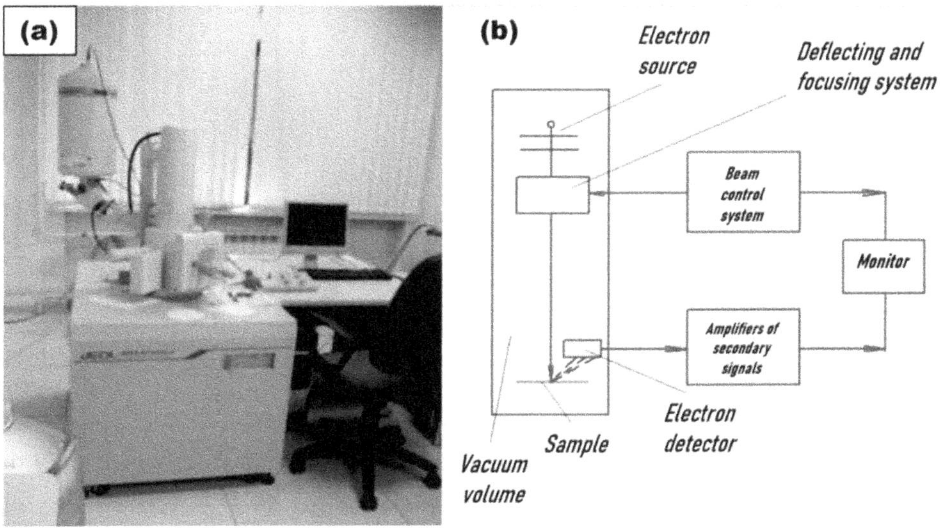

radius of the level, as well as on the binding energy, several types of spectra are distinguished: K, L, M, while only a few X-ray emission lines (K_{α}, K_{β} and L-series) are important for materials science (Uglov et al., 2007). The use of an energy-dispersive semiconductor detector based on Si doped with Li in the energy range from 0.7 to 10 keV makes it possible to identify unknown elements.

Quantitative elemental analysis consists in finding the concentration of the desired element using the measured intensity of continuous radiation. The mass fraction of an element can be determined by the following formula (1) (Gouldsteyn et al., 1984):

$$C_i = \frac{I_i \cdot I_{(E)} \cdot C_{(i)} \cdot G_i}{I_B \cdot I_{(i)} \cdot G_{(i)}} \tag{1}$$

where C_i и $C_{(i)}$ – the mass fractions of element i in the analysed micro-volume of the sample and the standard, respectively; I_i – measured intensity of characteristic X-ray radiation; I_B – the measured intensity of continuous radiation. The ratio of G-factors takes into account the fact that the standard and the sample may generate unequal intensity of continuous radiation per unit mass in a single volume (Gouldsteyn et al., 1984).

One of the additional methods of quantitative investigation of the elemental composition of coatings is Rutherford backscattering (RBS). One of the advantages of this technique is the indestructibility of the sample structure during the study. The RBS method is based on irradiation of the sample surface with an ion beam with an energy from 1 to 3 MeV. The resulting RBS spectrum is a graph along the abscissa axis of which an energy channel (n_i) is deposited, into which a scattered ion with a certain energy enters, and along the ordinate axis is the number of ions (H_i) that have fallen into the n_i channel (Uglov et al., 2007). The graph itself is a set of surface peaks, the intensity analysis of which provides information about the structure of the surface. One of the many possibilities of RBS is the study of the distribution of elements by depth (or the construction of concentration profiles). The proportion of the desired element can be found by the formula (2), (3):

$$C_i = \frac{N_i / N_m}{\sum_i N_i / N_m + 1} \tag{2}$$

$$\frac{N_i}{N_m} = \frac{H_i}{H_m}\left(\frac{Z_m}{Z_i}\right)^2 \frac{\left[\varepsilon(E)\right]_i^{ef}}{\left[\varepsilon(E)\right]_m^{ef}} \tag{3}$$

where H_i – the height of the energy peak for the i-th element; H_m – the height of the energy peak for the main element (the highest concentration in the sample); Z_i и Z_m – atomic numbers of the i-th and main element; $\left[\varepsilon(E)\right]_i^{ef}$ and $\left[\varepsilon(E)\right]_m^{ef}$ – the factor of the effective braking cross-section for the i–th and the main element (the upper index means the braking medium, and the lower one means the scattering atom of the element on which the scattering occurs). The studies were carried out on He^+ ions with an energy of 1.7 MeV (scattering angle $\theta = 170°$) with a normal incidence of probing ions on coated samples. At the same time, the energy resolution of the detector was 16 keV, and the dose of helium ions was 5 μCi. Standard software was used to process RBS spectra and obtain profiles of the distribution of elements over the depth of coverage.

Recently, there has been a tendency to use RBS to determine the depth distribution profiles of elements in combination with ion etching methods: secondary mass spectrometry (SIMS) and glow discharge mass spectrometry (GDMS). These methods are destructive in nature and involve spraying the surface (Ganeyev et al., 2012). The fundamental difference between these methods is the mechanism of formation of ions of the studied elements. Thus, in the SIMS method, the formation of ions occurs due to the simultaneous knocking out and ionization of the sample atoms by a beam of primary ions, whereas in the GDMS method, electrons emitted and accelerated in an electric field cause excitation and ionization of the atoms of the working gas (often argon). Further, argon ions bombard the cathode, cause not only ion-electron emission, but also knock out atoms of the cathode material. The atomized atoms of the sample, in turn, participate in various collisions in the plasma, which also lead to excitation and ionization (Ganeyev et al., 2012). The SIMS method is characterized by low detection limits, good spatial resolution (1 microns), allows detecting the presence of any element. The minimum concentrations that can be determined depend on the element, the matrix, the energy of the primary electrons and can vary from 10^{13} to 10^{18} cm^{-3}. The use of the GDMS method also leads to erosion of the coating surface, however, one of the main advantages of the glow discharge is its speed and the ability to study sufficiently thick coatings (several mm).

The study using the SIMS method was carried out using a SAJW-05 analyser equipped with a Physical Electronics 06-350E ion cannon and a QMA-410 Balzers quadrupole mass analyser with a rod diameter of 16 mm. The depth analysis profile was obtained using Ar^+ ions with an energy of 3 keV, the ion beam current was 1.5 μA, the angle of incidence was – 45°. For GDMS analysis, an SMWJ-01 analyser equipped with an SRS-300 quadrupole mass analyser with a rod diameter of 6 mm was used. In this case, the alternating current voltage was 1.8 kV, and the pressure of the working gas (Ar) was 0.2 Torr.

PIXE (particle induced X-ray emission) method is one of the widely used elemental analysis methods based on detection of characteristic X-ray radiation induced by a charged particle beam. One of the main advantages of the PIXE method is the ability to perform analysis in air without placing the sample in a vacuum setup. This method is characterized by high sensitivity due to the use of high-energy ions (0.1-5 MeV). An analysis of the intensity and wavelength of the characteristic X-ray radiation makes it possible to carry out a qualitative and quantitative analysis of the elemental composition of the material under study (Pogrebnyak et al., 2012). To register the emitted X-ray quanta in the PIXE technique, a semiconductor silicon-lithium detector is used. This method is primarily used to determine extremely low impurity concentrations in a coating. The PIXE method can be combined with micrometre beamforming to produce highly accurate maps of the horizontal location of negligible elemental concentrations. Such beams can be used to study biological and vacuum-degrading samples in air (Uglov et al., 2007). For the study, a microbeam of protons (μ-PIXE) with an initial energy of 1.4 MeV was used on the basis of the electrostatic accelerator of the "IPF NANU" (Sumy) with a beam size of up to 0.4 μm (charge 3×10^{-10} C/pixel, raster 50×50, scan step 0.5). As a result, maps of the distribution of the elements Ti, Nb, Cr, Zr, and Si in the coatings immediately after deposition were obtained. The chemical bonding state of the coatings was analysed by X-ray photoelectron spectroscopy (XPS, EC 2401) using MgK_α radiation (E=1253.6 eV). Prior to XPS measurements, the samples were subjected to sputter etching in argon plasma for 5 min.

2.2.3. Investigation of the Phase Structure of Coatings

The structure of the coating was studied using transmission electron microscopy (JEOL 2100F) in the dark field mode with an accelerating voltage of 200 and 250 keV. The study of the structure and phase composition was carried out on X-ray diffractometers DRON-4, D8 ADVANCE and Shimadzu XRD-7000 in filtered Cu-K_α radiation using a graphite monochromator in the secondary beam. The diffraction spectra were obtained in a point-by-point mode with a scanning step 2θ = 0.02-0.2°, X-ray diffraction was also used using a grazing beam in Cr-radiation at an angle of 3°. The average crystallite size was determined using the Debye-Scherrer equation from the broadening of diffraction peaks, taking into account instrumental broadening and the "Powder Cell 2.4" software. The instrumental broadening was determined at the half-width of the maximum of the standard silicon powder and amounted to 0.14.

The study of the phase composition was carried out on the basis that each crystalline substance (phase) has a certain crystal lattice, characterized by a set of interplanar distances d_{HKL} (Anishik et al., 2011; Umanskiy et al., 1982). The latter are found from the Wulf-Bragg equation:

$$2d_{HKL}\sin\theta = n\lambda \tag{4}$$

where θ – slip angle; λ – wavelength; n – order. Indexing a radiograph involves determining all interference indices (HKL) of each radiograph line. In this case, the interference indices are equal to the product of the plane indices (hkl) and the reflection order n. Indexing of X-ray patterns of crystals depending on syngony is based on different ratios. So, for a cubic lattice:

$$a = \frac{\lambda}{2\sin\theta_i}\sqrt{H_i^2 + K_i^2 + L_i^2} \tag{5}$$

To establish the interference indices of each line, the relation is used:

$$Q_i = \frac{sin^2\theta_i}{sin^2\theta_1} = \frac{H_i^2 + K_i^2 + L_i^2}{H_1^2 + K_1^2 + L_1^2} = \frac{d^2_{H_iK_iL_i}}{d^2_{H_1K_1L_1}} \tag{6}$$

где Q_i – line at the smallest angle θ. Thus, having determined a series of ratios Q_i for all lines of the X-ray diffraction pattern in ascending order, one can determine the grating type, interference indices, and unit cell period. One of the methods of qualitative phase analysis is the comparison of the values of interplanar distances and intensities of the studied sample with the reference one. At the same time, the JCPDS-ASTM card index file is the most widely used. One of the methods of studying the stressed state of a solid is the X-ray diffraction method. This technique allows you to determine the stresses at each point of the object surface at any stress distribution. The X-ray deformation measurement allows the assumption of a plane-stressed state, since the deformation refers to a surface layer with a thickness of ~ 10 microns. There are various ways to determine macro-stresses (principal stress sum method, single oblique survey method, multiple oblique survey method), but all of them can be combined within the so-called $sin^2\psi$-method (Gladkikh et al., 2006; Palatnik et al., 1972). This technique allows the assumption of a random orientation of crystallites, which allows one to write equations relating strain and stress for a homogeneous isotropic medium. To visualize the representation of the deformation of the body, it is convenient to use the so-called deformation ellipsoid - the surface into which the sphere of unit radius, selected in the original undeformed body, passes after deformation of this body. Under the assumption of a plane stress state, a biaxial stress state is realized in the X-ray method for studying residual stresses. It should be noted that in this case the main plane of the deformation ellipsoid coincides with the sample surface, which means that the coordinate axes also coincide. As a result, the following equation holds for the biaxial stress relation (Gladkikh et al., 2006):

$$\varepsilon_{\psi,\varphi} = \frac{1+\nu}{E}\sigma_\varphi sin^2\psi + \varepsilon_3 \tag{7}$$

where $\varepsilon\psi_{,\varphi}$ – deformation in the direction specified by the angles ψ and φ, *al*ong which it is measured (Figure 5); ν – *P*oisson's ratio; $\sigma\varphi$ – s$_{tr}$ess in the direction of the angle φ; E – Yo*u*ng's modulus.

For equation (7), there should be a linear relationship between the measured strain and и sin2ψ. ^{W}*h*en using X-ray analysis of residual macro-stresses, the interplanar distance of a family of crystallographic planes (hkl) *is* used as a scale in determining the deformation. Then the quantities $\varepsilon\psi,\varphi$ a$_{nd\ \varepsilon 3}$ can be represented as (Gladkikh et al., 2006):

$$\varepsilon_3 = \frac{d_3 - d_0}{d_0} = \frac{d_\perp - d_0}{d_0} \tag{8}$$

$$\varepsilon_{\psi,\varphi} = \frac{d_{\psi,\varphi} - d_0}{d_0} \tag{9}$$

Figure 5. Geometry of diffraction in the X-ray method for determining macro-stresses: N – normal to the sample surface; σ_1, σ_2 – main voltages; ε_1, ε_2, ε_3 – main deformations; s_0, s_1 – primary and reflected beams

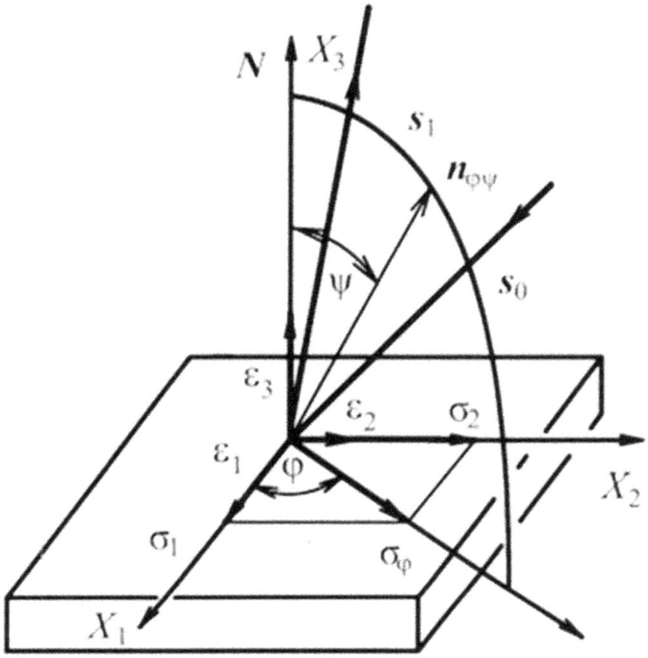

where d0 – is the interplanar distance of the planes (hkl) *in* the unstressed state; d3 ($d_\perp$) – is the interplanar distance in the direction of the normal to the sample surface (ψ=0); $d\psi,\varphi$ $_{-in}$terplanar distance of planes (hkl) *in* a stressed state. Taking into account relations (8) and (9), equation (7) takes the form:

$$\frac{d_{\psi,\varphi}-d_0}{d_0}=\frac{1+\upsilon}{E}\sigma_\varphi sin^2\psi+\frac{d_\perp-d_0}{d_0} \tag{10}$$

Equation (10) is the main relation of the sin2ψ method. To improve the measurement accuracy, it is necessary to carry out several surveys at different angles ψ. As a result, equation (10) can be rewritten as:

$$\frac{d_{\psi_i}-d_0}{d_0}=\frac{1+\upsilon}{E}\sigma_\varphi sin^2\psi_i+\frac{d_\perp-d_0}{d_0} \tag{11}$$

where $\sigma\varphi$ $_{de}$fined as:

$$\sigma_\varphi=\sigma_1 cos^2\varphi+\sigma_2 sin^2\varphi \tag{12}$$

Based on equation (12), the radiographic method allows you to determine the stresses separately σ_1 and σ_2, so when $\varphi = 0°$ $\sigma_\varphi = \sigma_1$, by $\varphi = 90°$ $\sigma_\varphi = \sigma_2$. As a result, the linear dependence of $\frac{d_{\psi_i} - d_0}{d_0}$ from is a $sin^2\psi$- graph. The component of normal stresses σ_φ is found from the angle of inclination of the $sin^2\psi$- graph according to the ratio (Gladkikh et al., 2006):

$$\sigma_\varphi = \frac{E}{1+\mu} tg\alpha \tag{13}$$

One of the most promising and most significant methods for studying the defect structure of solids is the positron annihilation method (Pogrebnyak et al., 2012). To analyse defects of the vacancy type in the coating, we used a slow positron microbeam (SPB) from Halle, Germany, which was used to measure the S-parameter depending on the energy of the incident positron beam (up to 30 keV), i.e. from the depth of analysis. To study structural vacancies, the method of positron annihilation was used, namely, the method of Doppler broadening of the gamma-ray line. The Doppler broadening spectrum can be quantified using the characteristic S-parameter of the line shape. This parameter is the ratio of the area of the central part of the spectrum with low momentum (A_s) to the area over the entire curve (A_0) (Pogrebnyak et al., 2012).

2.2.4. Methods of Investigation of Mechanical and Tribological Properties of Coatings

Hardness, along with fatigue strength and wear resistance, is one of the most widely used mechanical characteristics of metals and alloys. At the same time, the hardness test differs from other methods for determining the mechanical properties of materials by its high sensitivity and short test time. It is carried out without destroying samples, requires a minimum of costs in their preparation.

The study of the hardness of nitride coatings based on (Zr-Ti-Nb)N, (Zr-Ti-Cr-Nb)N and (Zr-Ti-Cr-Nb-Si)N was carried out using microhardmeters manufactured by "CSM Systems AG" (Switzerland). The microhardness value was determined using the Oliver-Pharr technique (Oliver & Pharr, 1992). Measurement of Vickers microhardness was carried out using a digital automated microhardness tester "DM8-B" and "Durascan-20" by the method of pressing a tetrahedral pyramid into the sample (see Figure 6). The loads used in the measurement were 0.05–1 N. The exposure of all samples under load was 10 s. At least 10 measurements were carried out on each sample. The value of microhardness was determined by the formula (14) (Geller & Rakhshadt, 1989):

$$H = 1.854 \cdot \frac{P}{d^2} \tag{14}$$

where P – load; d – arithmetic mean of the length of both diagonals of the imprint after unloading, mm. The microhardness values varied in a certain range due to the internal heterogeneity of the coatings (see Figure 6). The method is used to determine the hardness of parts of small thickness and thin surface layers having high hardness. The thinner the material, the less the load should be. The Vickers hardness

Figure 6. Traces of a diamond tip on the surface of nitride coatings (Zr-Ti-Cr-Nb)N

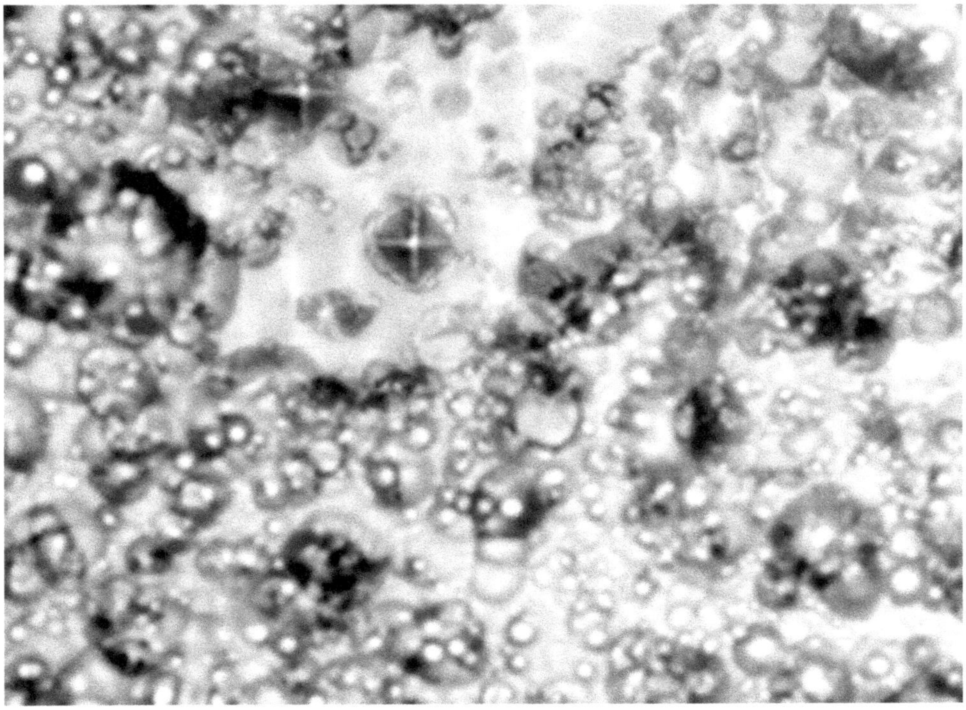

number *HV* is determined by special tables according to the measured value d (diagonal of the print in millimetres).

Additionally, the surface of the coatings was tested for nanohardness based on the use of a triangular pyramid (Berkovich's triangular pyramidal indenter). It avoids the problem of reducing four faces to one point, characteristic of Vickers indentors, and to obtain a radius of rounding of the vertex less than 100 nm. Probing with a pyramidal indenter leads to the fact that, unlike macro-tests, the deformable volume of the sample does not remain unchanged during loading but grows by many orders of magnitude. In addition, this method makes it possible to study physical and mechanical properties in nanoscale and even at the atomic level, which is not available to conventional mechanical testing technologies (Golovin, 2008). Nanohardness measurements were carried out by a diamond triangular Berkovich indenter on the "Hysitron TI 950 TriboIndenter/Nanoindenter" nanohardometer using the Oliver and Far technique (Oliver & Pharr, 1992). The maximum load was 10,000 μN. With the help of a computer program, test parameters are set – load, loading speed, holding time, unloading speed. During the tests, the dependence of the indenter vertex movement on the load was recorded with high accuracy using a capacitive sensor (Figure 7).

The device contains a loading unit of 1 m, a precision sensor 2 for registering the movement of the indenter 3 on soft springs 4, structurally combined into one measuring head 5, a controller unit 6 and a computer 7 with a software package for controlling all operating cycles of the device, collecting, processing and storing data. An optical microscope is used to select the injection site, and a two– or three-coordinate table 9 is used for positioning and moving the sample 8. In the most advanced devices, the table is motorized and also controlled by a computer. The set of nodes, their functions and relationships

Figure 7. Block diagram of nanoindentation and scratch testing: a – nanoindometer device, 1 – power cell; 2 – sensor for registering the movement of a movable rod with an indenter (3); 4 – rod suspension springs; 5 – measuring head housing; 6 – controller unit; 7 – computer; 8 – sample; 9 – slide table; b – atomic force microscope, 5 – measuring head housing; 6 – controller unit; 7 – computer; 8 – sample; 9 – slide table; 10 – probe; 11 – piezoelectric actuator; 12 – cantilever micro-beam (quantileever); 13 – four-window photodetector (recorder of probe movements); 14 – laser

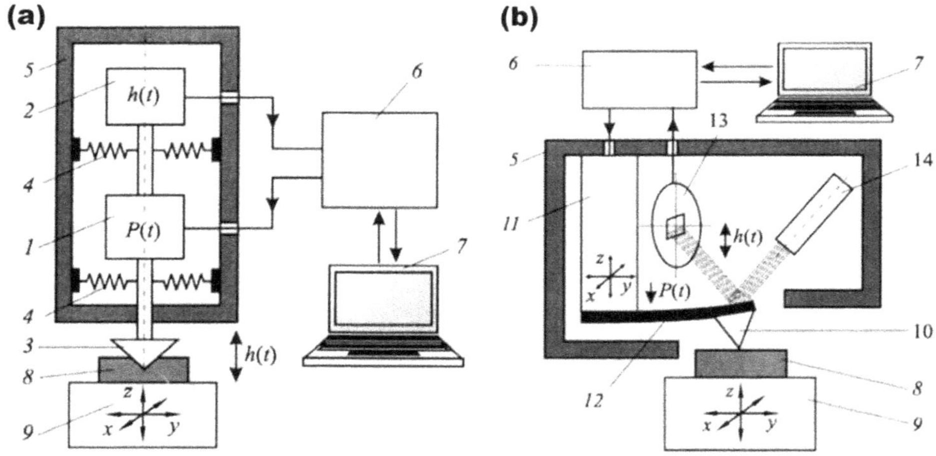

in nanoindentometers and atomic force microscopes (AFM) are similar (Figure 7b), and they developed almost in parallel and simultaneously. The resolution of the measuring path of the probe movement in them is also comparable and can be hundredths of nm. Therefore, they are often combined in one complex or even in one head, which makes it possible to expand the capabilities of probe methods and make them the most popular in modern nanotechnology. The combination of AFM and nanoindentation methods makes it possible to carry out 2D studies of the surface in normal and lateral modes and 3D characterization of mechanical properties at a given depth (from units to thousands of nm).

Figure 8 shows a general view of the nanohardometer "Hysitron TI 950 TriboIndenter/Nanoindenter". In indentation studies, a kinetic diagram is recorded, and at the same time the hardness and Young's modulus are determined, and their ratio allows you to compare the relative hardness of different materials, including in different structural states. This method makes it possible to study the micromechanical behaviour and structural sensitivity of the mechanical properties of thin near-surface layers. The sensor is formed by two glass plates located at a distance of 200 microns from each other.

The third plate is fixed on the rod of the indenter. When the depth of the print changes, the sensor capacity changes, which allows you to continuously monitor the movement of the top of the indenter. A triangular Berkovich indenter is fixed at the lower end of the rod, and an inductor is located at the upper end. The coil is placed in a permanent magnet, while when a current is passed through the inductor, the rod with the indenter is pushed out of the magnet and a load is applied to the indenter. In the range from 0 to 150 mN, the connection between current and load is linear, which allows you to control the force applied to the indenter. During the tests, the device allows you to record three parameters – the load, the movement of the top of the indenter and the time. During the tests, the dependence of the movement of the vertex of the Berkovich indenter on the load was recorded with high accuracy. The accuracy of

Figure 8. Optical, scanning probe microscopy and instrumental indentation: a – appearance; b – measuring techniques

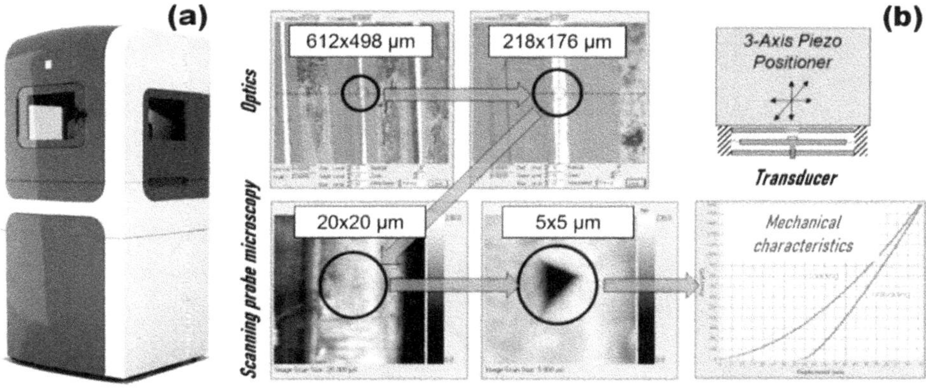

measuring the depth of the print is ± 0.04 nm, the load on the indenter is ± 75 nN. The device performs about three load and displacement measurements in 1 second. To reduce vibrations, the device is mounted on a vibration-isolating table. The rate of introduction of the indenter is 5 nm/s. During each test, the indenter was loaded/unloaded three times, each time for a higher load, which did not exceed 10 mN at a depth of 150 nm. To reduce the temperature difference between the sample and the indenter, the sample is placed in the device for 12 hours before the start of the tests. The room temperature is kept constant with an accuracy of ± 0.5 C. The tests do not start if the thermal expansion rate of the indenter rod is higher than 0.05 nm/s. During unloading, for each test, the rate of thermal expansion of the indenter was measured again and as a result, an appropriate correction was made. When an indenter is introduced near the contact area, a complex stress state is created, close to comprehensive compression, and the deformation spreading deep into the material has both elastic (reversible) and plastic (irreversible) components. Due to this, during nanoindentation, it is possible to obtain information about both hardness and Young's modulus, as well as to estimate the proportion of the elastic component in the total deformation or elastic recovery. Figure 9 shows a typical experimental curve of continuous indentation, the dependence of the load on the depth of indentation. The value of the elastic recovery W_e of the surface layer was calculated by the curves "loading – unloading" according to the formula (15):

$$W_e = \frac{h_{max} - h_r}{h_{max}} \tag{15}$$

where h_{max} - maximum penetration depth; h_r - the residual depth after removing the load, which consists in selecting the parameters of a power function describing the experimental dependence of the loading depth on the applied load, and calculating the hardness and Young's modulus from these data. The hardness value was calculated as the ratio of the maximum load to the projection area of the unrecoverable print, and the modulus of elasticity was calculated based on the projection area of the print, contact stiffness, defined as the slope of the upper third of the loading curve, and the specified Poisson's ratio.

To determine the tribotechnical characteristics, a universal "SMC-2" installation was used. The obtained coatings and uncoated samples were tested for wear resistance according to the "plane cylinder"

Figure 9. Dependence of the load on the depth of indentation

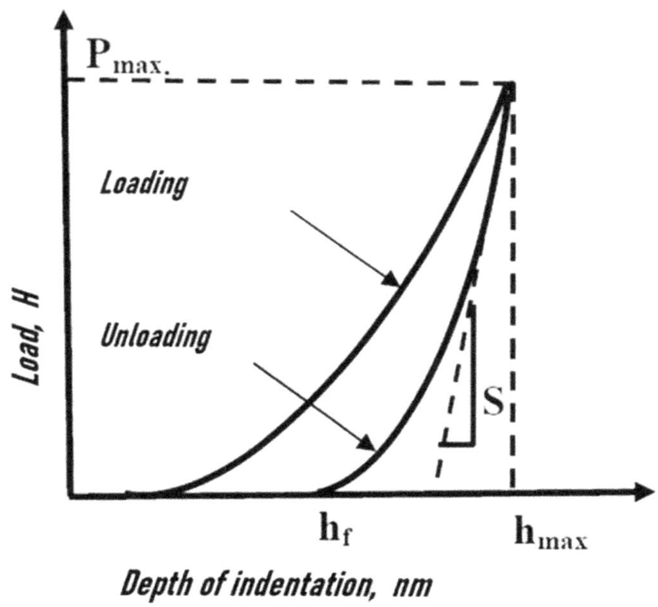

scheme using technical vaseline. During the entire test, the sample was not removed from the friction machine. The width of the groove and its length in the wear zone were carried out using a Brinell "MPB-2" microscope, which provides measurement accuracy of ± 0.025 mm. The coefficient of friction and the wear rate of nitride coatings and counterbody were determined at different temperatures on an automated friction machine (PC-Operated High-Temperature Tribometer, CSM Instruments, Switzerland), in air according to the "ball-disc" scheme (Figure 10). The coatings were applied to discs made of 45 steel (HRC = 55) with a diameter of d = 50 mm and a thickness of h = 5 mm. The load during the tests was 3.0 N, and the sliding speed was 10 cm/s. The sample was installed in a holder, a rod was fixed perpendicular to the sample plane, at the end of which there was a ceramic ball with a diameter of 6 mm made of Al_2O_3, SiN, SiC. By adjusting the displacement sensor, the wear radius of curvature was selected, another sensor compensated for the friction force and allowed the value of the friction coefficient to be set at a certain point in time.

In the course of the study, the analysis of wear products, as well as the contact area on the counterbody, was carried out by examining the structures of wear grooves on the surface of the coatings under study, as well as changes in wear spots on the ball. Measurements of the vertical section of the wear grooves were carried out on a profilometer in four diametrically and orthogonally opposite areas, after which the average value of the groove depth and cross-sectional area was calculated. The sample is in the form of a disk with a hardened surface (coating), rotates at a given frequency in the heating zone. The counterbody in the form of a certified ball is fixed immobile in the holder, which transmits a given load to it and is connected to the sensors of the friction force and the depth of immersion of the counterbody. Such a contact scheme also allows for the mathematical calculation of the initial contact stresses (Hertz stresses). The tests are carried out according to ASTM G99-959 and DIN50324 standards. The following temperatures were selected as temperatures for tribotechnical tests: 300°C and 500°C.

Figure 10. Scheme of testing samples of nitride coatings for wear resistance using the "ball-disc" method

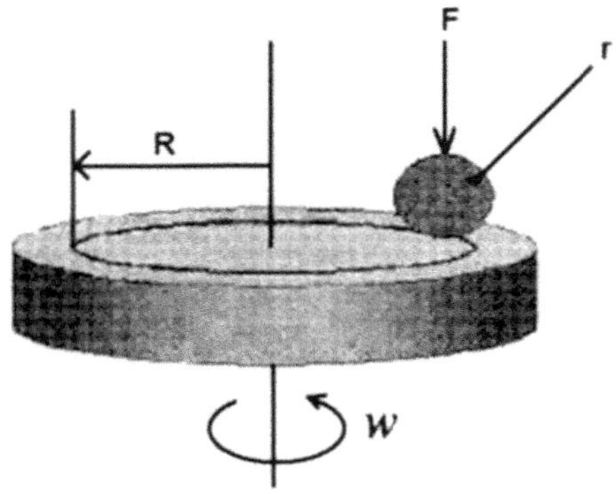

The surface profile was measured using a "Surtronic 25" profilometer device by measuring the vertical deflection of a diamond tip (probe) moving at a constant speed under conditions of mechanical contact with the sample. Surtronic-25 has a multifunctional RS-232 port, with which it is possible to transfer data to a printer for printing or to a computer for subsequent analysis using additional Talyprofile software. The software allows you to calculate parameters, set calculation modes in full compliance with international standards. Special functions allow you to get a vertical/horizontal reflection of the profile, enlarge individual sections for a more detailed examination, get an inverted profile, and also calculate the undulation and roughness separately. The structure of the coating wear grooves and wear spots on the balls were studied using an Olympus GX 51 optical inverted microscope and a Quanta 200 3D scanning ion electron microscope. As a result of the tests, the wear factor (Vershinin et al., 2010) of the coated sample and the statistical partner (ball) were evaluated according to the method (Randall, 2002):

$$W = \frac{V}{Pl} \tag{16}$$

where W – wear factor [mm³/H·m]; V – volume of removed material [mm³]; P – load [N]; l – friction path [m]. The wear diameter of the ball was determined using an "Olympus GX 71" optical inverted microscope, the volume of the removed material on the ball was calculated as

$$V_{\varnothing} = \pi h^2 \left(r - \frac{1}{3} h \right) \tag{17}$$

where $h = r - \sqrt{r^2 - \left(d/2 \right)^2}$ segment height; d – wear diameter; r – ball radius. The volume of removed coating material is equal to $V_{п} = s l_{п}$, where $l_{п}$ circumference, s – wear track cross-sectional area.

The study of adhesive-cohesive strength, scratch resistance and determination of the mechanism of destruction of the coating were carried out using the "REVETEST" scratch tester (CSM Instruments). Scratches were applied to the surface with a diamond spherical indenter of the "Rockwell C" type with

Figure 11. Scheme and appearance of the scratch tester

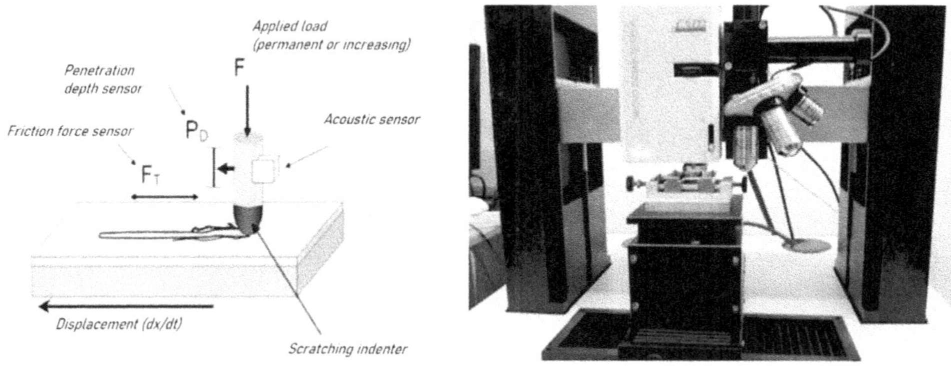

a radius of 200 microns with a continuously increasing load (Figure 11), after which physical parameters were recorded: acoustic emission, coefficient of friction and depth of penetration of the indenter.

The tests were carried out under the following conditions: the load on the indenter was increased from 0.9 to 70 N, the speed - 1 mm/min, the scratch length - 5 mm, the loading speed - 6.91 N/min, the signal sampling rate - 60, the acoustic emission - 9. The registration of various parameters during the testing process made it possible to fix the stages of coating destruction. During the movement of the indenter at a given speed and with a continuously increasing load, the readings of several sensors are recorded on the computer, namely: loading force, acoustic emission intensity, friction force, friction coefficient, scratch depth. The moment of adhesive or cohesive destruction of the coating is recorded after testing visually using an optical microscope equipped with a digital camera, as well as by analysing the shape of the "property-load" curves, by changing one of five parameters: acoustic emission, friction force, friction coefficient, indenter penetration depth and residual scratch depth. The data is stored on the hard disk of the control computer. The combination of various parameters recorded during the testing process increases the reliability of the methodology and the accuracy of determining the critical load. Additional advantages of the scratch testing method include the ability to study the type of material failure (cohesive or adhesive) and evaluate the crack resistance of materials. The sclerometric head includes a device for measuring the normal and tangential load on the indenter (a diamond pyramid of the "Rockwell C" type) when scratching, a device for measuring the magnitude of the normal movement of the indenter, a device for lifting and lowering the head, a device for lifting and lowering the indenter, a drive for moving the indenter in the tangential direction. The loading of the indenter is provided by a spring squeezed by a screw mechanism. In this case, the spring presses on the indenter through a dynamometric semicircle with a load cell to measure the load applied to the indenter. The loaded indenter is lowered and embedded in the surface. The movement and speed of the indenter insertion is measured by an inductive sensor. The scratch resistance force is measured using strain gauges.

During the test, a scratch is formed not only due to the destruction of the material, but also as a result of plastic deformation. The process of material deformation by a diamond indenter consists of the following main stages:

- preliminary incline during vertical insertion of the indenter to a given depth;

- the introduction of a rigid indenter into a solid leads to the appearance of a characteristic "crown" at the base of the print;
- at the initial moment of the scratch formation, the subsequent deformation of the riveted material occurs, caused by the movement of the lateral surface of the pyramid;
- a slice of the "crown" of the main print is transformed into an influx;
- with further movement ahead of the indenter, the influx increases and reaches the maximum height with maximum scratching force.

At this moment, local destruction of the material occurs at the tip of the influx, resulting in a reduced scratching force. As a result of the tests, it is possible to determine the minimum (critical) load (L_C), which leads to the destruction of the coating. It should be noted that not all recorded events related to the destruction of the coating describe the actual adhesion of the coating to the substrate. By changing the values of the coefficient of friction and the acoustic emission signal with an increase in the scribing load, the characteristic values of the critical load L_C were determined: The critical load responsible for the moment of the appearance of the first chevron cracks in the lower part of the scratches was determined as L_{C1}; L_{C2} – the time of the appearance of multiple chevron cracks in the lower part of the scratches; L_{C3} – when the destruction has an adhesive-cohesive character; L_{C4} – local peeling of the coating areas; L_{C5} – plastic abrasion of the coating on the substrate, loss of adhesion strength. The combination of various parameters recorded during the testing process increases the reliability of the methodology and the accuracy of determining the critical load. This technique complies with the international standard ISO DIS 20502.

Conclusions for Section 2

1. Based on the literature sources, the expediency of choosing the elements Ti, Zr, Cr, Nb, and Si as components of the nitride coating was determined.
2. Taking into account the features of the constituent elements, the optimal physical parameters were selected when using the method of vacuum-arc deposition of nitride coatings based on (Zr-Ti-Nb), (Zr-Ti-Nb-Cr) and (Zr-Ti-Nb-Cr-Si).
3. A set of complementary research methods has been selected that allows analysing the change in the morphology of the coating surface, the elemental composition and its redistribution in depth, structural and phase transformations depending on the deposition modes (change in bias voltage and pressure of working gas).
4. To determine the physical and mechanical properties, a technique has been developed for determining microhardness and nanohardness, as well as testing protective coatings for physical, mechanical and tribological tests.

3. INFLUENCE OF DEPOSITION PARAMETERS AND DOPING PROCESS ON THE ELEMENTAL AND PHASE COMPOSITION OF MULTICOMPONENT COATINGS

3.1. Morphology and Elemental Composition of the Surface of the Investigated Coatings of the System (Zr-Ti-Nb)N, (Zr-Ti-Nb-Cr)N and (Zr-Ti-Nb-Cr-Si)N

A review of the literature has shown that ion-vacuum coating technologies are the most promising (Thornton, 1982). Nitride multicomponent coatings obtained by magnetron sputtering have an amorphous structure (in the absence of nitrogen). With the addition of reaction gas (Ar+N), the structure of the coating changes from amorphous to crystalline (FCC lattice of the NaCl type).

A characteristic feature of the coatings obtained by vacuum arc deposition is the presence of a preferred orientation and high compressive stresses. The phase composition and structural state of coatings obtained by vacuum arc deposition vary widely depending on the ion energy of the deposited flow and the pressure of the working gas. At the same time, the presence of nonequilibrium condensation conditions (the influence of impurities, the radiation effect of the deposited flow, etc.) determines the metastability of the properties of the coatings obtained (Azarenkov et al., 2012). Thus, changing the energy of the deposited particles allows you to adjust the level of residual compressive stresses and the size of the crystallites. Adjusting the pressure of the working gas (nitrogen) provides a wide variation of the microstructure by changing the stress-strain state, grain size, crystallographic orientation, etc.

The chemical composition and morphology of the coatings were studied using the PEGASUS X-ray energy dispersive spectrometric system, scanning electron microscopy/energy dispersive X-ray spectroscopy (SEM/EDX) JEOL JSM-6610 LV and JEOL 7001TTLS with a voltage of 15-20 kV. The chemical bond state of the coatings was analysed by X-ray photoelectron spectroscopy (XPS, EC 2401) using MgK_α radiation (E = 1253.6 eV). Before XPS measurements, the samples were etched by sputtering in argon plasma for 5 min. According to the results of scanning electron microscopy, vacuum-arc condensation of a multicomponent system has a number of features in the formation of surface morphology.

Figures 12 and 13 show images of the coating surface (Zr-Ti-Nb)N. The study of surface morphology showed that deposition of coatings leads to an increase in surface roughness (from 0.09 to 0.42 μm) due to the droplet component of the plasma flow. Figure 12b clearly shows the presence of a clear boundary between the coating and the substrate without transition layers. The thickness of the coating (Zr-Ti-Nb)N is d » 6.7-10 μm, which is typical for nitride coatings obtained by vacuum-arc deposition. In the cross-sectional image in Figure 13, the coating showed a columnar structure, which is characterized for materials obtained by deposition in a vacuum arc. It should be noted that there are small voids and gaps in the columns. The concentration of various elements in the coatings (Zr-Ti-Nb)N, (Zr-Ti-Nb-Cr)N and (Zr-Ti-Nb-Cr-Si)N at different deposition parameters is shown in Table 3. The content of elements in all coatings was close to the concentration of elements in the sprayed target.

Figure 14 shows a SEM image of one of the obtained coatings of the system (Ti-Zr-Cr-Nb)N. The main difference between vacuum arc deposition, in particular, in the cathode spot arc mode, is the production of molten droplets that can be incorporated into the coating in the form of microparticles. As can be seen from Figure 14, the coating (Ti-Zr-Cr-Nb)N contains inclusions of droplet fractions of various sizes (up to 10 μm in diameter). In our case, the microdroplets have the shape of an ellipsoid, which indicates that the droplets fly almost parallel to the plane of the substrate. An increase in the temperature of the deposition flow using a bias voltage (U_C = -200 V) significantly reduces the con-

Figure 12. SEM-images of coatings based on (Zr-Ti-Nb)N at P_N=0.5 Pa: a – surface morphology; b – cross-section of coatings

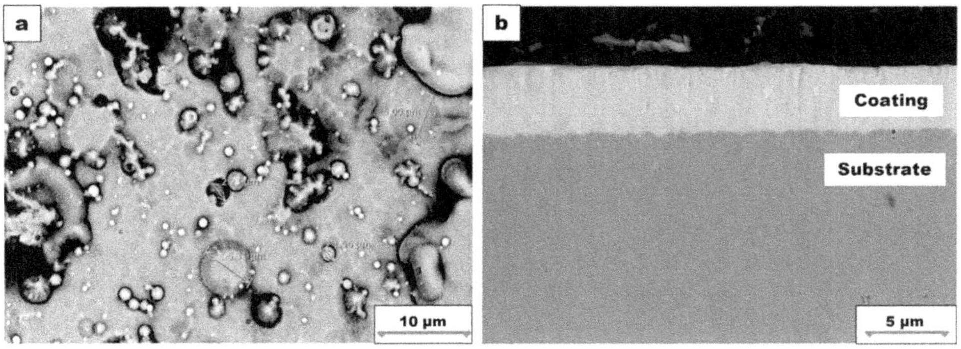

Figure 13. SEM-images of crack structures of (Ti-Zr-Nb)N coatings obtained at P_N = 0.5 Pa: coating fracture fractogram and magnified view (columnar structure)

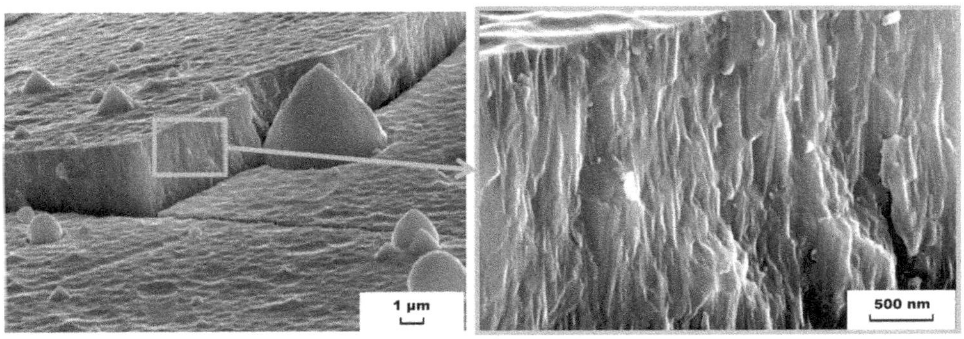

centration of the droplet fraction on the surface (Figure 14b). A reduced droplet content may occur as a result of melting processes on the coating surface. It is likely that focusing a beam of charged particles activates the process of cleaning the surface from smaller fractions, and its heating allows the formation of a homogeneous protective layer.

As morphological studies show (Figure 14b), the obtained at higher values of the displacement potential and the pressure of the working gas is a fairly flat surface with minimal droplet presence, whereas the use of reduced values of physical deposition parameters leads to the formation of micron-sized droplets of molten material on the surface (Figure 14a). According to the work (Boxman & Zhitomirsky,), the use of high pressure of the working gas contributes to the activation of the formation of chemical compounds on the cathode surface, and the content of droplets in the plasma stream depends on the thermophysical properties of the cathode, namely, on the melting temperature. As a result of the entry of a chemically active gas into the chamber, a nitride layer of constituent elements forms on the cathode surface (they often have a high melting point), which leads to a decrease in its erosion in the droplet phase. The reduction of the droplet component in the coating can also be achieved by increasing the energy of the sprayed ions. Thus, with an increase in the displacement potential of the substrate, the condensation

Figure 14. SEM-images of the surface of (Zr-Ti-Cr-Nb)N coatings: a – PN = 0.3 Pa, UC = -100 V; b – PN = 0.7 Pa, UC = -200 V

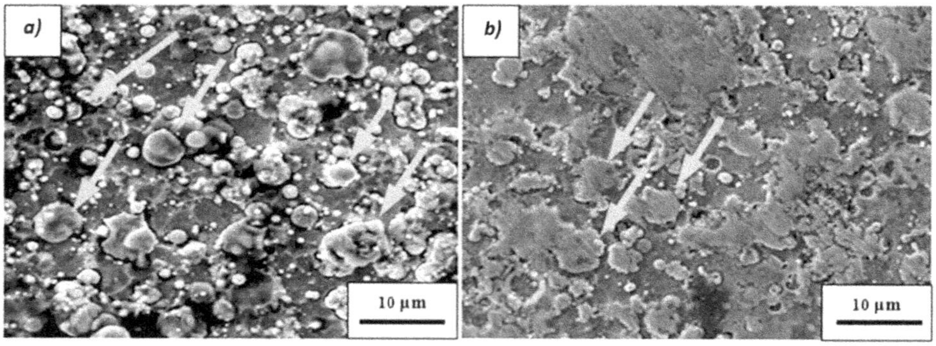

Figure 15. SEM/EDX integral elemental composition of (Zr-Ti-Cr-Nb)N coatings

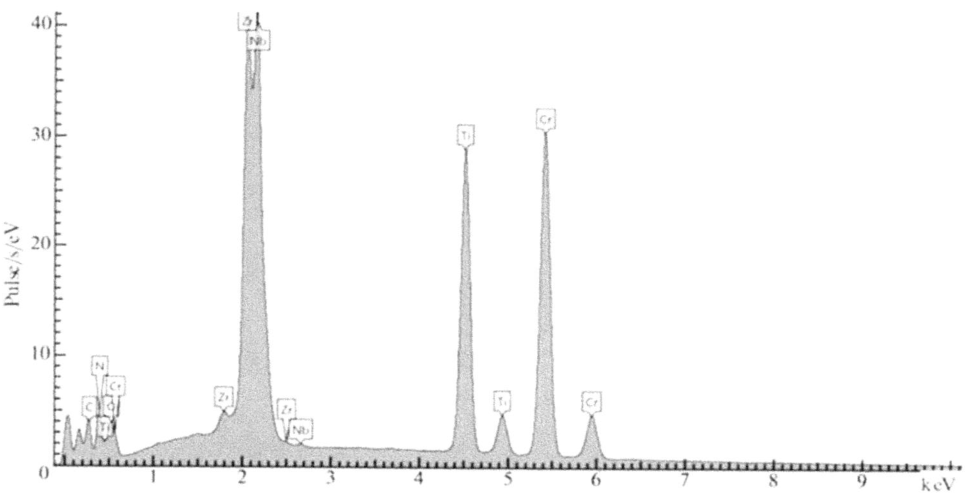

surface temperature increases, which leads to partial melting of small droplets. As a result, the number of droplets decreases due to spraying with high-energy ions, as well as due to evaporation processes.

Integral element analysis showed that the matrix of the near-surface area of the coating consists of Zr, Ti, Cr and Nb (Figure 15 and Table 3). The presence of nitrogen peaks in the spectra is associated with the composition of the gas atmosphere of the vacuum-arc source.

The high carbon content on the sample surface can be explained by the difficulty in recognizing elements such as C, N, O by energy dispersive X-ray spectroscopy due to the similarity of their electronic structure. By calculating the energy E of the K_α radiation for some elements, determining the difference between them and comparing them with the resolution of the analyser, it is possible to draw conclusions about the accuracy of the results obtained. It follows from Moseley 's law:

$$E=10,\ 2eV\bullet(Z-1)^2 \quad (18)$$

$$\Delta E = 10,2eV \cdot \left((Z_A - 1)^2 - (Z_B - 1)^2 \right) = 10,2eV \cdot (Z_A + Z_B - 2)(Z_A - Z_B) \quad (19)$$

where Z – the ordinal number of an atom in the periodic table of chemical elements of Mendeleev. According to calculations, the difference between the radiation energies ΔE when determining an element using energy-dispersive X-ray spectroscopy is 112.2 eV for C and N, 132.6 eV for N and O, 805.8 eV for Zr and Nb. Since the difference between the radiation energies from elements C, N and N, O is less than the resolution of the analyser, it is impossible to accurately determine the ratio between these elements in the coating. It is likely that oxygen is practically absent in the coating, and the processes of diffusion of oxygen atoms into the surface composition from the air atmosphere can be considered the reason for its appearance on the spectrum. The nitrogen concentration should be slightly higher due to a decrease in the carbon concentration.

Based on the heterogeneous contrast of inclusions on the SEM-images of the surface (Figure 14), the droplet fractions have different elemental composition in stoichiometry. According to the obtained maps of the distribution of elements (Figure 16), the basis of these inclusions are zirconium, niobium and nitrogen. Note also that Nb and Zr practically duplicate their location on the map of elements and, basically, make up a matrix of inclusions of droplet fractions. The reason for this distribution is the high melting point of both elements and the almost identical work of the electron output from the metal surface (A_{out}(Nb) = 3.99 eV, A_{out}(Zr) = 3.96–4.16 eV). According to calculations, the difference in the radiation energies of zirconium and niobium is several times greater than the resolution threshold of the EDS analyser, which makes it possible to accurately recognize the mentioned elements. Therefore, the presence of droplet fractions from Nb and Zr is quite possible (light balls in Figure 14a). Titanium and nitrogen have an equally uniform distribution over the entire surface of the coating (Figure 16), and the chromium concentration is significantly lower in areas where niobium predominates. It is also seen that

Figure 16. SEM/EDX-maps of the distribution elements on the surface of the sample series 1

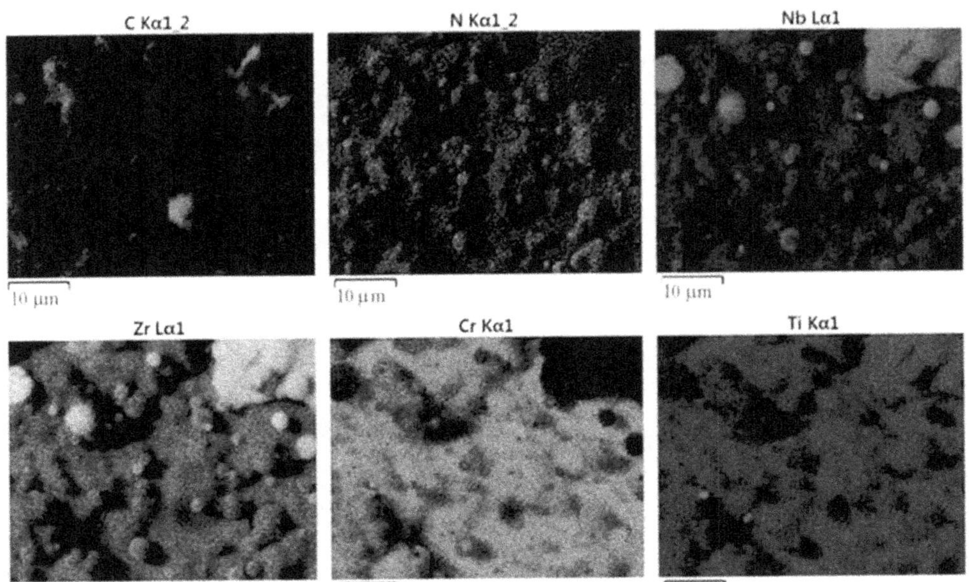

the distribution of carbon does not repeat the distribution of nitrogen over the entire surface, therefore, this is not a measurement error.

In all obtained coatings, the nature of the distribution of elements over concentrations in the surface composition has a similar pattern $C_N > C_{Cr} > C_{Ti} > C_{Zr} > C_{Nb} > C_C/C_O$. This is clearly illustrated by the pie charts presented in Figure 17. We also note that an increase in the bias voltage to –200 V leads to a noticeable decrease in the concentration of titanium and chromium atoms in the surface composition. This means that in the process of coating deposition, particles of the coating surface with the lowest atomic mass are sputtered.

From the graphs shown in Figure 18, it can be seen that an increase in nitrogen pressure in the chamber from 0.3 to 0.7 Pa leads to a decrease in the carbon concentration in the surface composition and an increase in the concentration of nitrogen, niobium and zirconium atoms. The same was observed with an increase in the negative potential of the substrate from -100 to -200 V (table 3).

The chemical bonding state and elemental composition of the (Zr-Ti-Cr-Nb)N coating were examined by the XPS method (Figure 19) with peaks associated with Ti 2p, Cr 2p, Zr 3d, Nb 3d, N 1s. There was also an oxygen peak (O 1s). According to the existing data (Morris, 1998), the Ti 2p peaks at 455.8 and 458.7 eV correspond to the Ti-O and Ti-N bonds, respectively; Cr 2p peak at 576.1 eV to the Cr-N bond; Zr 3d at 180.1 eV and 182.2 eV to Zr-N and Zr-O bonds respectively; Nb 3d at 203.8 eV and 207.5 eV,

Figure 17. SEM/EDX-elemental composition of (Zr-Ti-Cr-Nb)N coatings: a – series (sample) 1; b – series (sample) 2; c – series (sample) 3; d – series (sample) 4

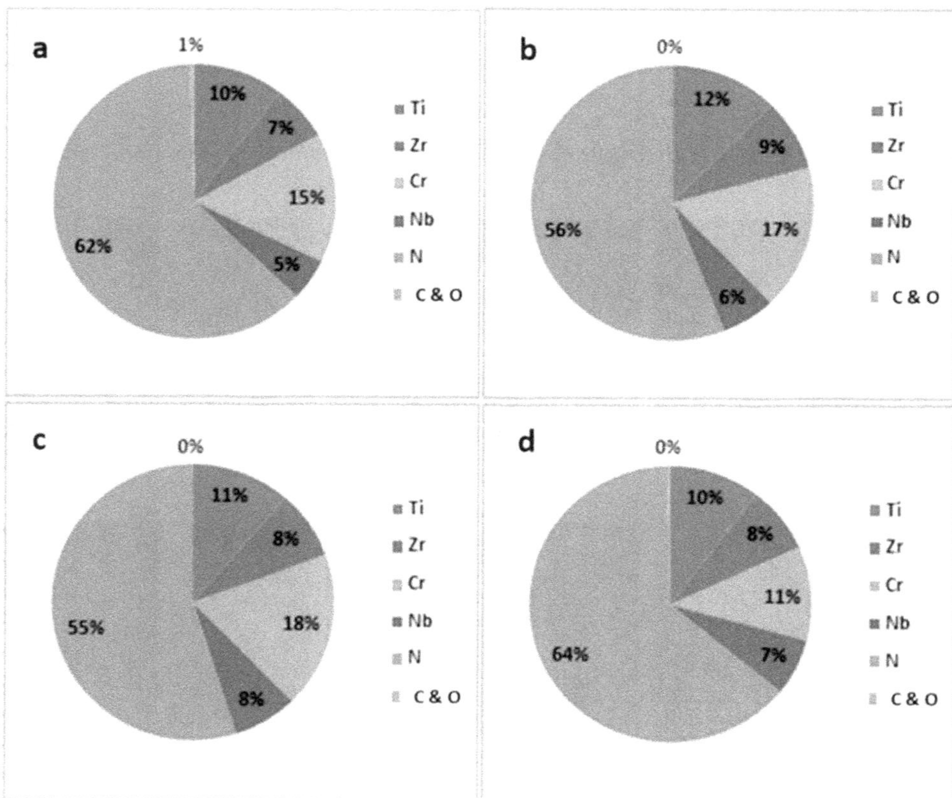

Figure 18. Change in the concentration of elements in (Zr-Ti-Cr-Nb)N coatings depending on the deposition parameters: a – P_N = 0.3 Pa; b – U_b = -100 V

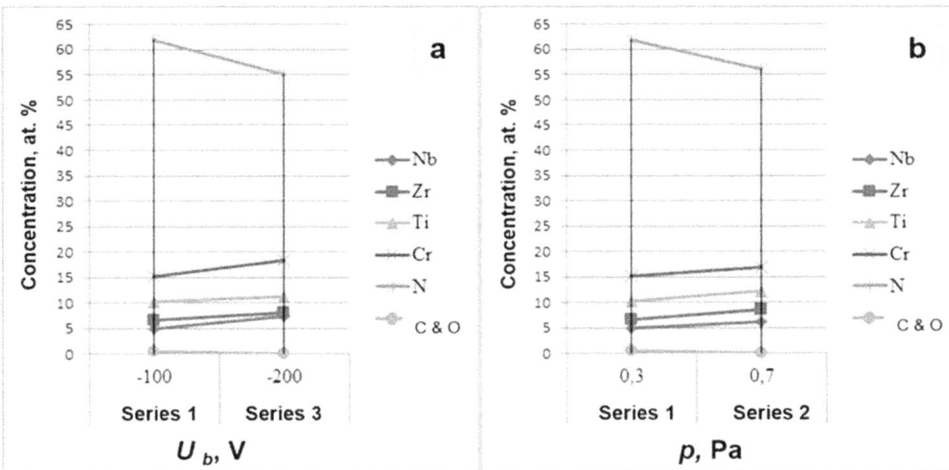

Table 3. Elemental analysis of (Zr-Ti-Nb)N, (Zr-Ti-Cr-Nb)N and (Zr-Ti-Cr-Nb-Si)N coatings

Series (Sample) N°	Concentration, at.%						
	Ti	Zr	Cr	Nb	N	Si	Impurities (C and O)
(Zr-Ti-Nb) N							
1	20.91	20.38	-	19.99	39.00	-	-
2	20.52	19.36	-	19.26	41.00	-	-
(Zr-Ti-Cr-Nb)N							
1	10.21	06.63	15.22	04.96	62.00	-	0.57
2	12.30	08.48	16.92	06.17	56.00	-	-
3	11.27	08.03	18.23	07.48	55.00	-	-
4	10.40	07.81	11.00	06.73	64.00	-	0.39
(Zr-Ti-Cr-Nb-Si)N							
1	34.50	11.58	11.00	03.30	37.00	02.10	-
2	32.30	13.28	07.95	03.62	40.00	03.50	-

respectively, to the Nb-N and Nb-O bonds; N 1s peak at 397.4 eV to Cr-N bonds; O 1s at 531.3 eV to Nb-O bonds. These results show that the (Zr-Ti-Cr-Nb)N coating is composed of Ti-N, Ti-O, Cr-N, Zr-N, Zr-O, Nb-N, and Nb-O bonds, which can be are attributed to the TiN, TiO_2, Cr_2N, ZrN, ZrO_2, NbN and Nb_2O_5 phases, respectively. Thus, it can be assumed that the multiphase material (Ti, Zr, Nb)N-Cr is formed in the coatings.

The results of the RBS analysis of the elemental composition of nitride coatings based on (Zr-Ti-Cr-Nb)N are shown in Figure 20. As can be seen from Figure 20a, all peaks correspond to all the constituent elements of the coating, which corresponds to the chemical composition of the sputtered cathodes. From the element distribution profiles shown in Figure 20b, it is possible to judge the homogeneous distribu-

Figure 19. XPS-peaks of the (Zr-Ti-Cr-Nb)N coating deposited at P_N = 0.3 Pa, U_b = 200 V

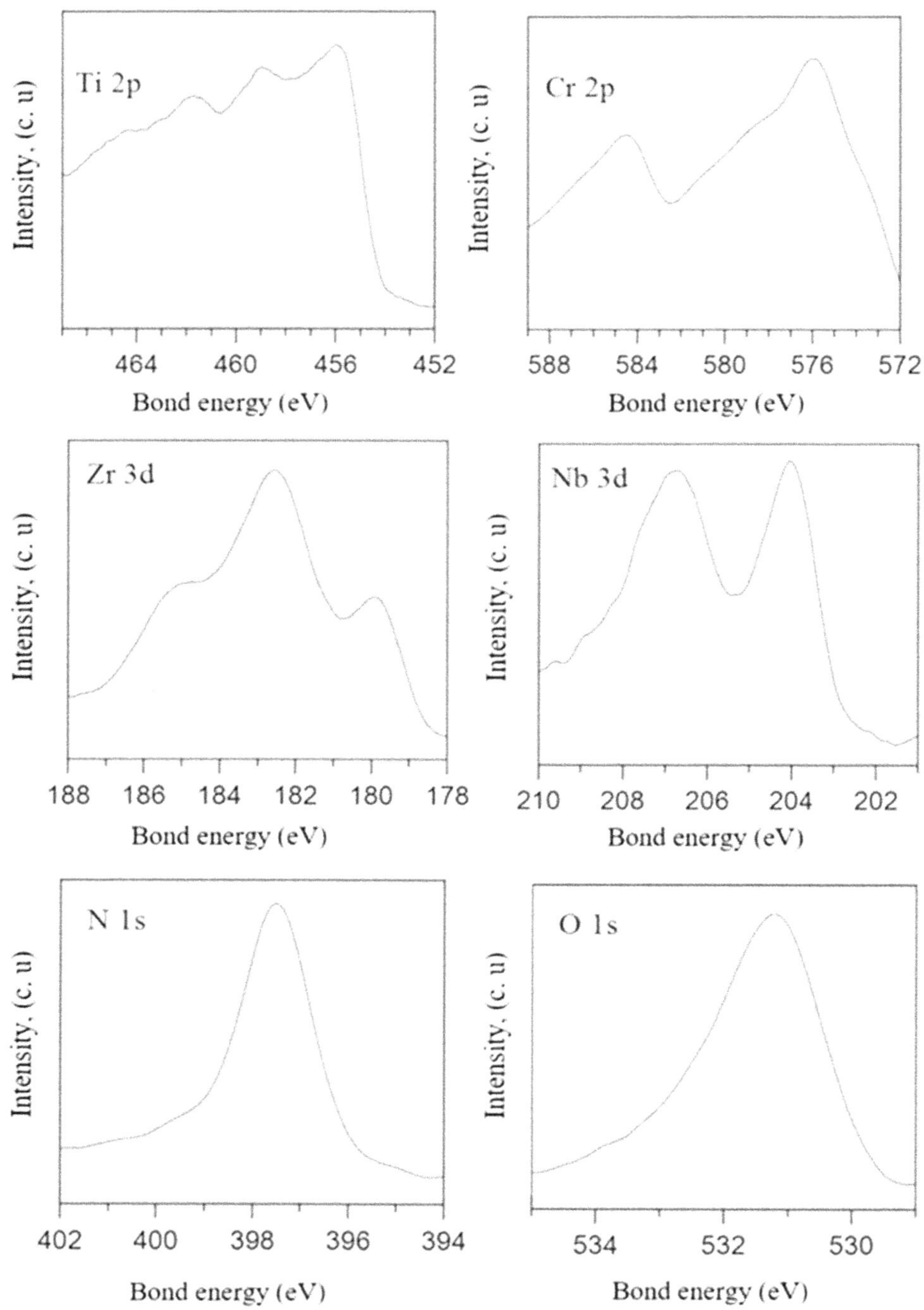

Figure 20. RBS-analysis of (Zr-Ti-Cr-Nb)N nitride coatings at P_N = 0.3 Pa and U_b = 200 V: a – energy spectrum (channeling); b – element distribution profile in depth

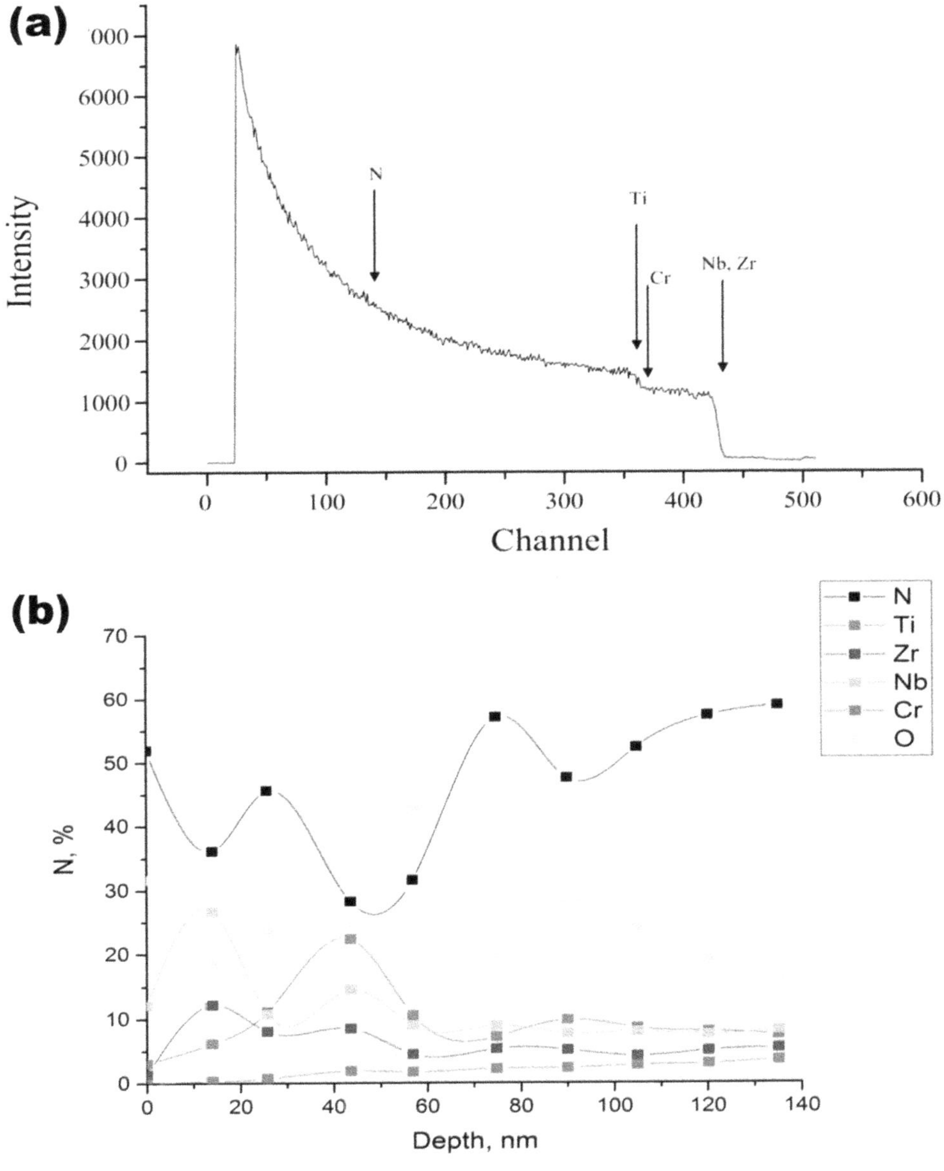

tion of the constituent elements in the near-surface layer, while the surface layer is enriched with N, O and Nb atoms, which have the highest concentration, which correlates well with the results of EDX/ EDS analysis. In the case of coatings (Zr-Ti-Cr-Nb-Si)N with a high concentration of nitride-forming elements (Ti and Zr), the carbon content was controlled at a low level of about 2 at.%.

It is well known that the RBS method is a reference method for measuring the concentration of heavy elements and for determining the thickness of the coating with high accuracy (up to 0.1 nm). At the same time, peaks corresponding to light elements (N, C and O), which have a significant effect on the properties of thin films, are almost impossible to separate from the general background of heavy elements due to the large atomic number of the substrate. In addition, as can be seen from Figure 20, the peaks corresponding to elements with a close atomic radius (located side by side in the periodic table) are superimposed on each other (Ti and Cr, Zr and Nb).

Figure 21 shows the initial state of the coating surface after deposition. The images were obtained using AFM.

The presence of micro-spherical and flat formations of various dispersity on the surface of condensed layers of titanium, chromium, zirconium, and niobium indicates the presence of a droplet phase in plasma flows emitted at high velocities from cathode spots towards the substrates. The use of a negative bias potential applied to the substrate, -100 and -200 V, affects the surface morphology of the multicomponent condensate and promotes the formation of denser layers. It is known that the efficiency of plasma-chemical reactions in the synthesis of compounds is determined, in particular, by the degree and nature of excitation of molecules, molecular ions, as well as by the ion kinetic energy E_i, which varies depending on U_b. These factors contribute to the dissociation of active gas molecules in the condensation zone and subsequent reactions of synthesis of compounds of metal and non-metal atoms. An increase in the energy of ions and fast atoms due to RF pulse technology leads to an increase in the surface diffusion mobility of atoms, causing the predominant growth of crystallites in the condensation plane, as well as to a redistribution of the components of the growing condensate (implanted atoms, vacancies, their complexes, second phases) over the depth of radiation exposure bombarding particles, as a result of which the microlayered structure of the coating is formed.

Studies of the features of the formation of the structural-phase state of multicomponent coatings based on Zr, Ti, Nb, Cr, Si and N were prepared from one evaporating Zr-Ti-Cr-Nb-Si cathode (Cr -17.08 at%, Zr - 30, 19 at%, Nb - 9.67 at%, Ti - 39.96 at%, Si - 3.1 at%). To study the features of the formation of the structural-phase state of the coatings, two series of coating samples were formed and at least five samples from each series were studied (see Figures 19 and 20).

Figures 22 and 23 show images of the coating surface (P_N = 0.3 Pa и U_b = -100 V). From which it can be seen that during the deposition process in the coating there are areas with a drop fraction up to 10 μm in size. However, no cracks were found, which indicates the good quality of the coating obtained

Figure 21. Surface morphology of the (Zr-Ti-Cr-Nb)N coating obtained using AFM: a – 2D-image; b – 3D-image

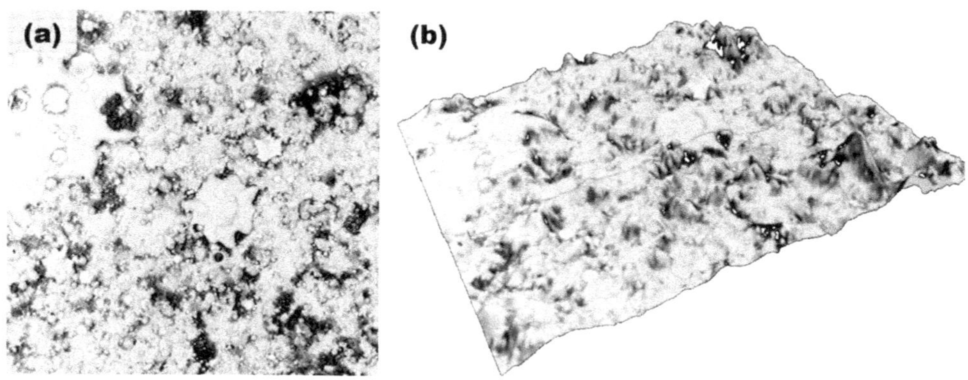

Figure 22. SEM-images of the surface morphology of coatings of the (Zr-Ti-Nb-Cr-Si)N system, obtained at P_N = 0.3 Pa and U_b = -100 V: a – general view, x2750; b – enlarged view, x6000

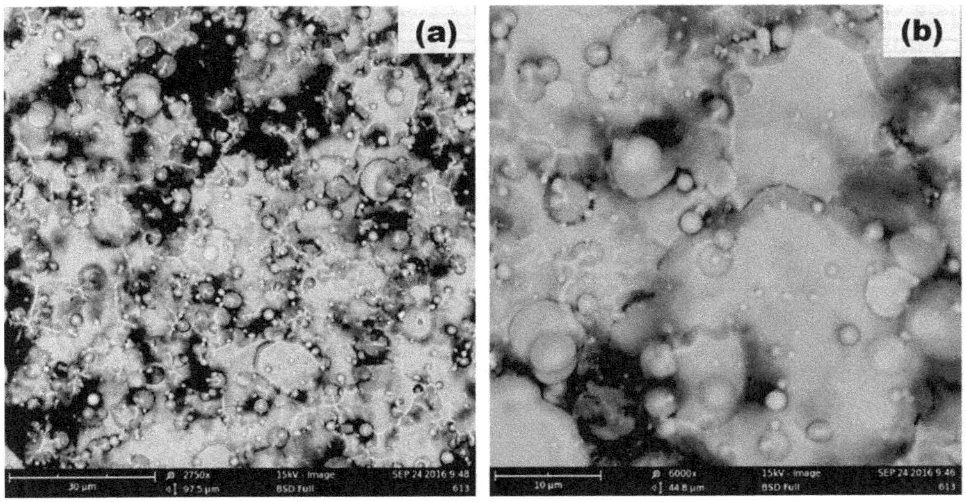

Figure 23. SEM-images of the surface morphology of coatings of the (Zr-Ti-Nb-Cr-Si)N system, obtained at P_N = 0.3 Pa and U_b = -200 V: a – general view, x1350; b – enlarged view, x18000

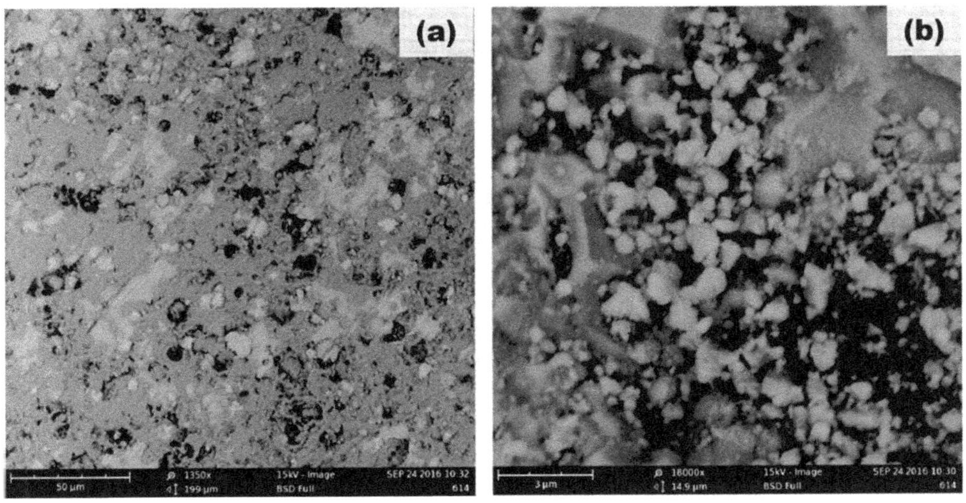

using pulsed RF stimulation of coating deposition. An increase in the bias potential changes the surface topography (Figure 23) and leads to a decrease in the silicon component in the coating.

More generalized data on the coverage area are given by the results of energy dispersive microanalysis, which are shown in Figures 24 and 25.

Energy dispersion analysis data, resulting mean values of elemental microanalysis of system coatings based on (Zr-Ti-Nb)N, (Zr-Ti-Cr-Nb)N and (Zr-Ti-Cr-Nb-Si)N show a good correspondence of the composition of the coating material to the composition of the atomized cathode for heavy elements. It should be noted that the weak diffusion characteristic of a multicomponent coating due to the heteroge-

Figure 24. SEM/EDX-elemental composition of coatings based on (Zr-Ti-Nb-Cr-Si)N, obtained by microanalysis of the surface of the first series of samples, at P_N = 0.3 Pa and U_b = -100 V: a – areas for EDX/EDS-analysis; b – EDX/EDS-spectrum

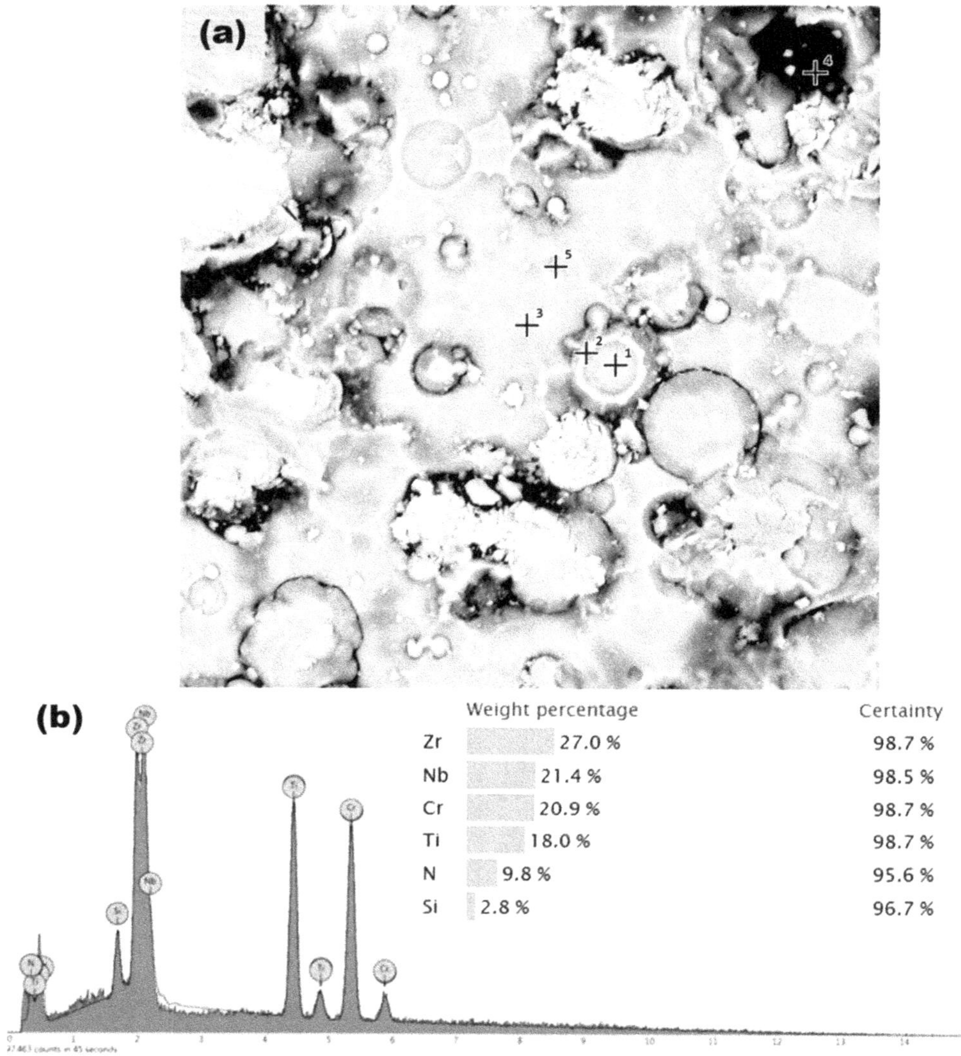

neity of the constituent elements that reduce the effective diffusion distance also leads to more difficult grain growth.

A distinctive feature of multicomponent coatings is the local redistribution of elements in the periodic lattice with the need to maintain one type of solid solution. At the same time, strong fluctuations in the chemical composition cannot but affect the distortions of the crystal lattice and the physical-mechanical properties of the coating. At the same time, it should be noted that with thermal exposure, this process occurs more intensively. As can be seen from the distribution of elements along the profile, in the initial state, the elements in the coating are distributed almost evenly over the surface and depth (Figure 22). From the presented data (Figure 25) it can be seen that the profiles corresponding to the elements Zr,

Figure 25. SEM/EDX-analysis, the profile of the distribution of elements over the depth of (Zr-Ti-Cr-Nb) N coatings

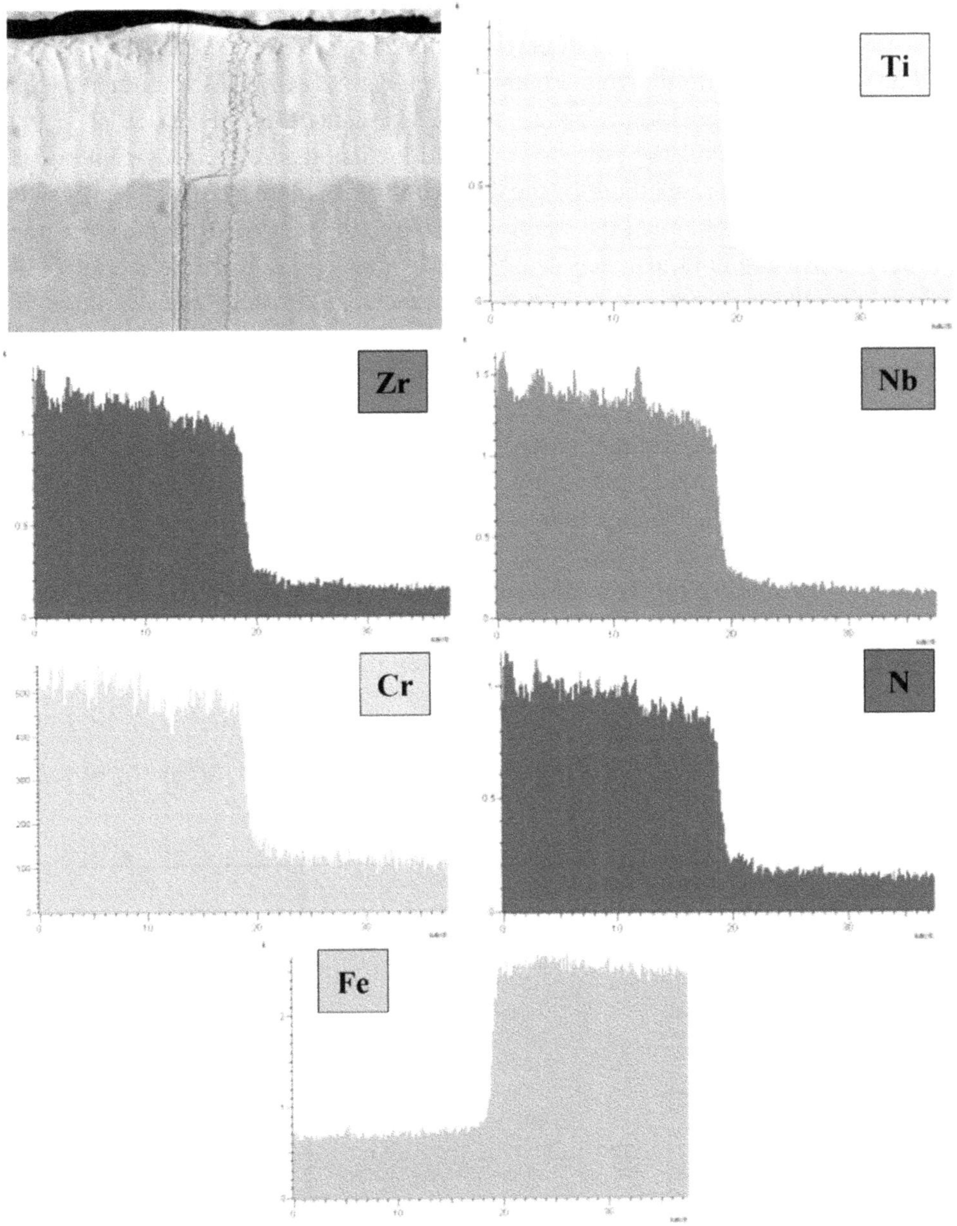

Ti, Cr, Nb and N indicate a homogeneous distribution of elements in depth. As can be seen from the profile of iron, the increase in intensity begins after the end of the coating matrix.

Thus, the use of complementary methods for studying surface morphology and elemental analysis, such as SEM, XPS, EDS, RBS and AFM, made it possible to study the surfaces of coatings and determine the chemical composition of the surface layer, to establish patterns of distribution of elements in depth, as well as to identify uncontrolled impurities.

3.2. Phase Composition of Multicomponent Nitride Coatings Based on (Zr-Ti-Nb)N, (Zr-Ti-Cr-Nb)N and (Zr-Ti-Cr-Nb-Si)N

The results of X-ray diffraction analysis for all samples obtained under various conditions of deposition of coatings based on (Zr-Ti-Nb)N are shown in Figure 26. Analysis of X-ray diffractometer spectra showed that the determining phase composition is a phase with a face-centered cubic lattice (FCC phase). A low peak intensity at $2\theta=38°$ indicates the presence of small inclusions with a body-centered cubic lattice (BCC), typical of the vacuum-arc method for the droplet phase (Figures 12a and 13a). A characteristic feature of the increase in the pressure of the reaction gas (from 0.05 to 0.5 Pa) is the strengthening of the peaks of the family of planes {111}, which is determined by an increase in the perfection of the preferred orientation of the growth of crystallites with the axis perpendicular to the surface plane. It was found that the size of the crystallites increases from 10 nm at the lowest pressure of 0.05 Pa and 63 nm at the maximum operating pressure of 0.5 Pa.

Phase analysis of (Zr-Ti-Cr-Nb)N nitride coatings indicates the presence of the FCC TiN-phase (a=0.243 nm, $a_{tab.}$=0,244 nm) and trigonal modification Cr_2N (space group P31m, a = 0.4800 nm, c = 0.472 nm) (Figure 27). From the X-ray diffraction patterns, it was established that the substrate displacement potential strongly affects the change in the (111) orientation, which is preferable for mixed (111) and (200), and the appearance of (112) and (300) reflections at large angles (trigonal Cr_2N phase). The structural analysis of (Zr-Ti-Cr-Nb)N nitride coatings is consistent with XPS results, while the coatings undergo phase separation. Table 4 shows the grain sizes and lattice parameters of (Zr-Ti-Cr-Nb)N coatings, which were additionally estimated from X-ray diffraction patterns. This table shows that

Figure 26. Areas of XRD-patterns of (Zr-Ti-Nb)N coatings obtained at different partial pressures of nitrogen: 1 – P_N = 0.05 Pa; 2 – P_N = 0.5 Pa

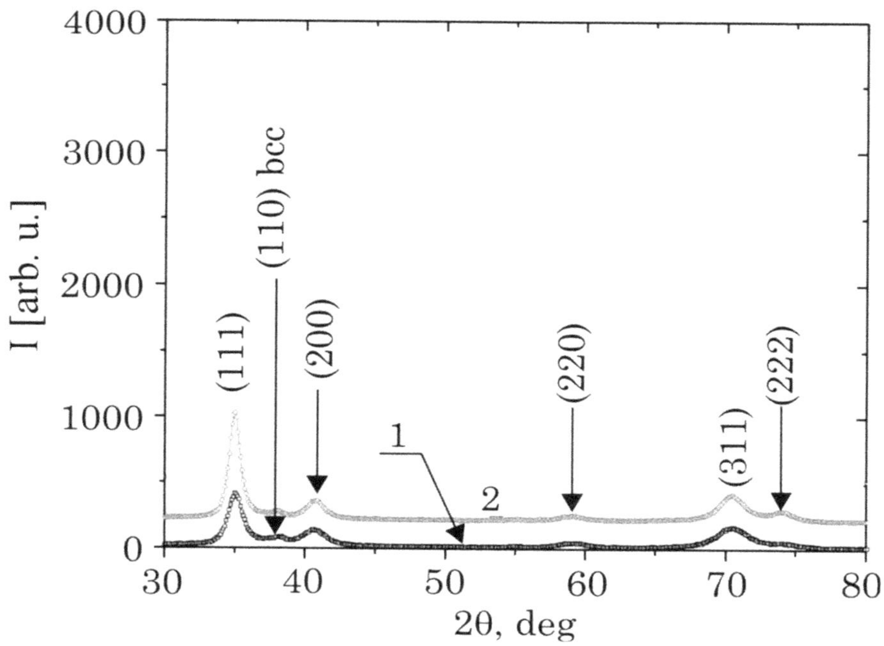

Figure 27. XRD-patterns of (Zr-Ti-Cr-Nb)N coatings deposited at different values P_N and U_b

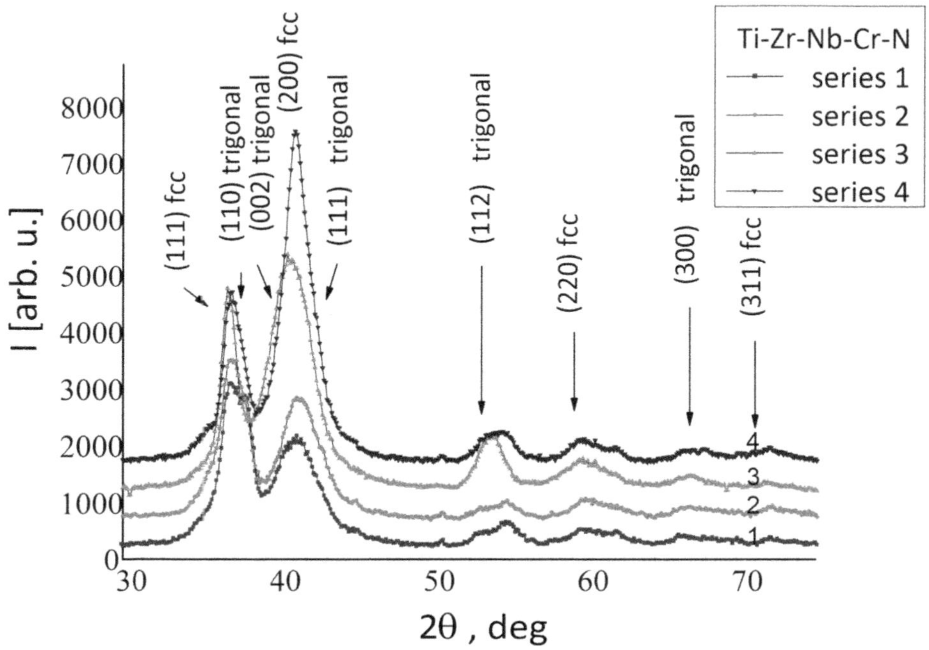

Table 4. Crystallite size and lattice period of (Zr-Ti-Cr-Nb-Si)N and (Zr-Ti-Cr-Nb)N coatings evaluated by X-ray diffraction analysis

N° Series (Sample)	Nitride Coating	Lattice Period, nm	Crystallite Size, nm
1	(Zr-Ti-Cr-Nb)N	0.4365	5.2
2		0.4359	4.5
3		0.4410	5.1
4		0.4381	6.9
1	(Zr-Ti-Cr-Nb-Si)N	0.4332	11.5
2		0.4337	9.7

the lattice constant increases with U_b. This is reasonable since the energy bombardment results in the incorporation of nitrogen atoms into spaces in the growing film that are smaller than the usual atomic volume (the “atomic shot blasting effect”). In addition, the grain size of (Zr-Ti-Cr-Nb)N nitride coatings is about 5.5 nm for all possible biases, which means that the films have very fine-grained nanocrystalline structures. This statement is confirmed by the results of electron microscopic and diffraction studies. As can be seen from Figure 28, the transition metals of the Zr-Ti-Cr-Nb system form a coating with a finely dispersed structure during the condensation process. The electron diffraction patterns show rings that belong to the FCC-phase.

Figure 28. TEM-image of (Zr-Ti-Cr-Nb)N coatings: a – brightfield image; b – micro-diffraction pattern

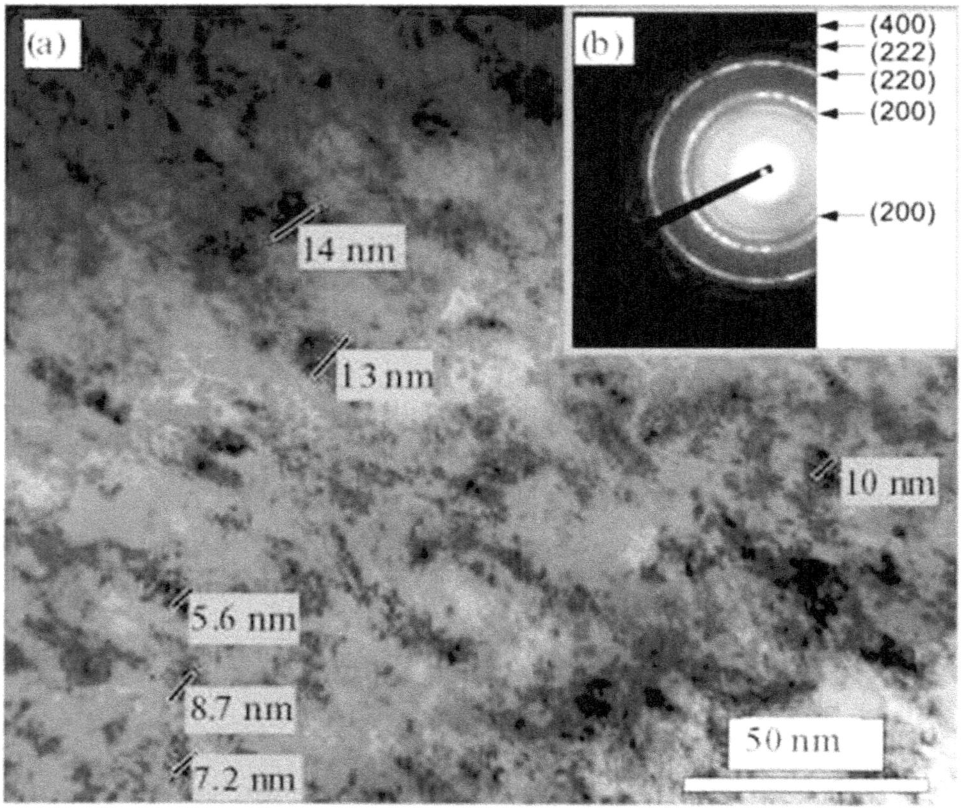

X-ray patterns of (Zr-Ti-Nb-Cr-Si)N coatings deposited at P_N = 0.3 Pa and U_b = 100 and 200 V are shown in Figure 29. As we can see, the substrate displacement has a great influence on the X-ray patterns of (Zr-Ti-Cr-Nb-Si)N coatings. For nitride coatings, only the XRD-lines related to the (111), (200), (220), (222), and (311) peaks from the FCC structure of the B1 phase (NaCl type) are confirmed. It is well known that nanocomposite coatings such as TiSiN are characterized by the formation of amorphous (Si_3N_4)/ crystalline (TiN) composite structures (Wagner, 1995). The non-FCC nitride phase of Si_3N_4 has low solubility in the FCC structure, and thermodynamically conditioned (spinodal) segregation may occur during precipitation. It should be noted that the appropriate Si content to achieve a single layer (Si_3N_4) around each TiN grain is between 5 and 12 at.%, although the exact amount varies depending on the average grain size (Hsieh et al., 2013; Martin et al., 2005). In our case, although the present systems contain several immiscible nitrides, such as TiN and Si_3N_4 (in low concentration), only one fcc phase in the coating was clearly identified (see figure 28). The formation of the nitride of the FCC phase is associated with the high mixing entropy effect and limited diffusion kinetics, in which the atoms cannot reach their stable configuration. (Holleck, 1988).

It was found that with an increase in the displacement potential on the substrate, the positions of the XRD peaks are shifted towards smaller angles (Figure 29). One explanation for this effect is an increase in the lattice parameters of (Zr-Ti-Cr-Nb-Si)N coatings due to the formation of Frenkel pairs and anti-Schottky defects. This shift can also be explained by the fact that a slight compression of the NaCl-type

Figure 29. XRD-patterns of (Ti-Zr-Nb-Cr-Si)N coatings, at P_N = 0.3 Pa and different values of U_b (-100 and -200 V)

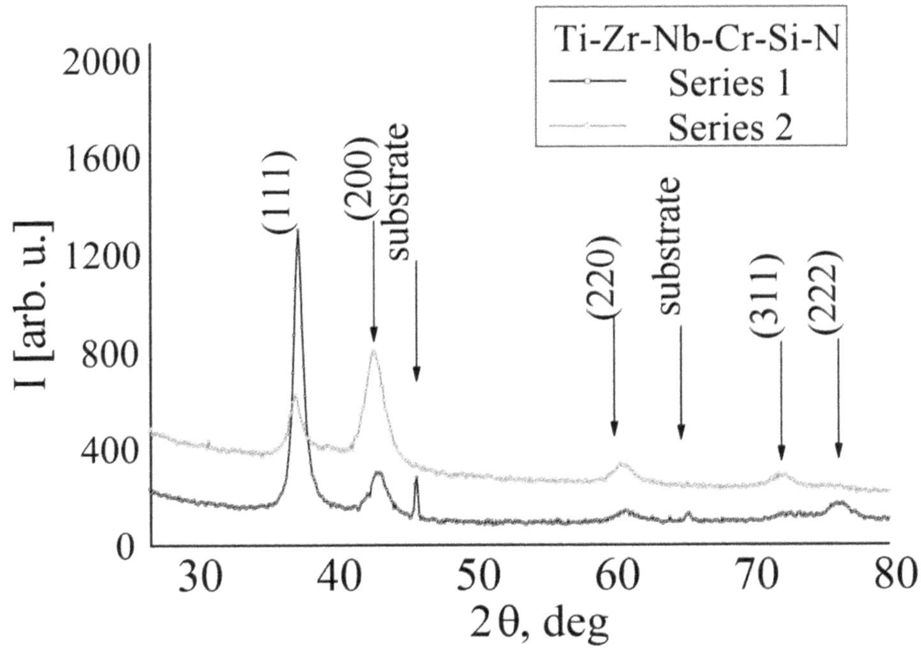

lattice with the presence of Si may be due to a small inclusion of Si in the lattice and the formation of short Me-Si bonds. Moreover, an increase in the Si concentration probably causes a shift of the XRD-peaks, which indicates a possible change in the lattice constant (Kim et al., 2002; Pilloud et al., 2004). Lattice parameters for deposited (Zr-Ti-Cr-Nb-Si)N coatings gradually increases from 0.4332 nm at a negative displacement of the substrate -100 V, 0.4337 nm at a displacement of -200 V (Table 5). The average grain sizes of the films were calculated from X-ray images using the Scherrer formula, and it was found that they strongly depend on an increase in the displacement potential on the substrate. The size of the crystallites in the coating of series 1 (U_b = -100 V) averaged 11.5 nm, for series 2-9.7 nm (U_b = -200 V). It is well known that high-energy ion bombardment prevents the migration of grain boundaries due to an increase in the number of preferred nucleation centers, which reduces the grain size. As can be seen from the results in Table 5, when the Si content increased from zero to 3.5, the size of the crystallites increased from 4.5 to 11.5 nm, while the lattice parameter decreased. Similar results were obtained in various papers devoted to the study of Si-containing nitride systems (Kim et al., 2002; Pilloud et al., 2004). From the XRD analysis, as shown in Figures 26, 27 and 29, it was found that on all (Zr-Ti-Nb)N and (Zr-Ti-Cr-Nb-Si)N coatings, only a single solid nitride phase was formed, and not separate nitrides coexisting with each other. Although some factors, such as the nonequilibrium process of physical deposition, from the vapor phase, high entropy of mixing (for multicomponent coatings) and a small difference in atomic radii, may contribute to the formation of crystal structures of this nature. Similar results were also obtained as precipitated nitrides, such as $Zr_xTi_{1-x}N$, $Cr_xTi_{1-x}N$, $Nb_xTi_{1-x}N$, (AlCrTaTiZr)N, (CrTaTiVZr)N, (TiHfVNbZr)N (Bagdasaryan et al., 2014; Cecchini et al., 2011).

Figure 30. Calculated mixing energy (E_{mix}) of $Ti_{1-x}Nb_xN$, $Ti_{1-x}Zr_xN$ u $Zr_{1-x}Nb_xN$ alloys depending on composition x

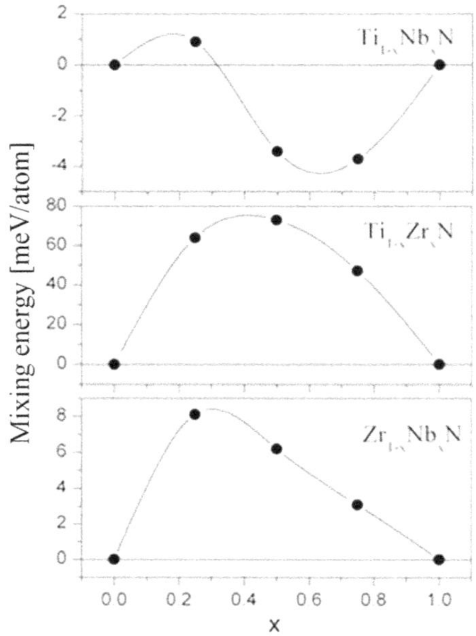

Figure 30 shows the Gibbs free energy of mixing of the studied alloys calculated at T = 0 K (i.e., the energy of formation) depending on the composition of *x*. The positive energy of formation means that $Ti_{1-x}Zr_xN$ alloys are unstable and will decompose into TiN and ZrN with chemical driving force (E_{mix}). However, all alloys can be stabilized in a certain range of composition, depending on temperature, since the configuration entropy is always positive and contributes to a decrease in the Gibbs free energy. Since the E_{mix} values for $Ti_{1-x}Nb_xN$ and $Zr_{1-x}Nb_xN$ are very small, and some of them are negative, these alloys can be stabilized as solid solutions at moderate temperatures. These data confirm the possibility of formation of solid FCC solutions based on TiN, ZrN and NbN in thin coatings. Since we observed only solid solutions based on these nitrides in our coatings, it can be assumed that the presence of NbN in the films excludes the separation of TiN and ZrN from $Ti_{1-x}Zr_xN$ alloys. Nevertheless, for the coatings of the (Zr-Ti-Cr-Nb)N system, experimental results indicate the existence of a two-phase structure: the FCC phase TiN (a = 0.243 nm, a_{tab} = 0.244 nm) and the trigonal modification Cr_2N (spatial group P31m, a = 0.4800 nm, c = 0.4472 nm). The formation of a two-phase structure is obviously associated with the presence of high concentrations of Cr and Nb elements, which have a low enthalpy of nitride formation (Table 5).

Table 5. The enthalpy of formation of nitrides of the constituent elements of the studied nitride coatings

-	TiN	ZrN	NbN	CrN	Cr_2N	Si_3N_4
ΔH, kJ/mol	-337.7	-365.3	-234.7	-117.2	-129.0	-106.0

Figure 31. Triangular map of the distribution elements, obtained using EDX (TEM): a – solid solution Nb-Zr-N; b – solid solution Zr-Ti-N; c – solid solution Cr-Zr-N; d – solid solution Nb-Ti-N

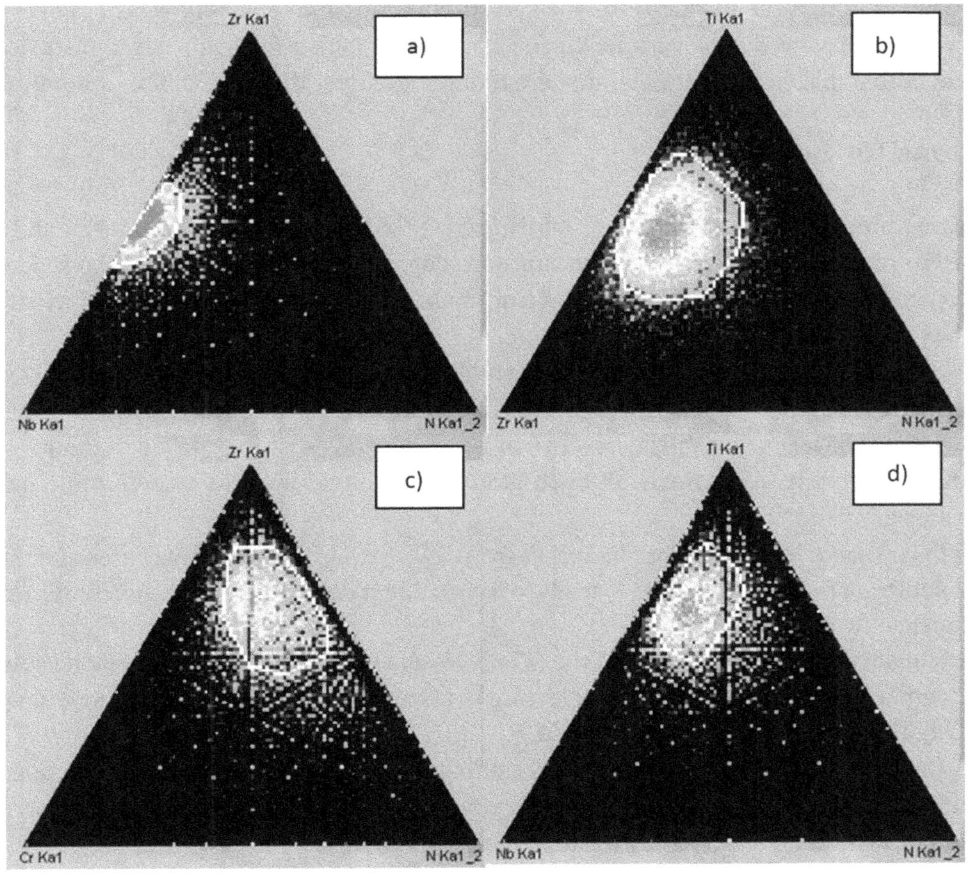

Cecchini R. et al. (Cecchini et al., 2011) showed that with increasing nitrogen pressure, the crystalline phase of the coatings passes from Cr_2N (main phase) + CrN to CrN (main phase) + Cr_2N. These results are confirmed by the presentation of the Gibbs-Rosenbaum triangle (Figure 31). The vertices of an equilateral triangle correspond to the 100% content of each component of the system. The elemental composition determination method is based on the fact that the sum of perpendiculars from any point inside an equilateral triangle on each side is equal to the height of the triangle. Rosenbaum proposed the use of three segments of straight lines parallel to the sides of the triangle and extending from that point to intersect with each side of the triangle. The sum of the three-line segments for each point inside an equilateral triangle is equal to the length of its sides. As can be seen from Figure 31, the percentage of Ti-N, Zr-N systems is higher than that of Cr-N and Nb-N, thus resulting in the formation of a solid solution with a high TiN and ZrN content.

It is well known that the application of a bias potential on a substrate can significantly change the structure of coatings and their properties. When a substrate bias potential is applied, reactive gas and metal ions bombard the substrate. With an increase in the bias potential on the substrate, the energetic ion bombardment becomes more intense. Due to the enhanced bombardment effect, the densification of the coating is increased by reducing void areas in the microstructure.It should be noted that in all coatings, the predominant (111) orientation was observed at U_b = –100 V. When the substrate bias potential was increased to U_b = -200 V, the nitride films showed strong (200) and relatively weak (111) orientations. This phenomenon may be due to various kinetic constraints (anisotropy in surface diffusion, adatom mobility, and collisional cascade effects) that are thought to affect the preferential orientation.

Conclusions for Section 3

1. Nanostructured coatings of the (Zr-Ti-Nb)N, (Zr-Ti-Cr-Nb)N and (Zr-Ti-Cr-Nb)N and (Zr-Ti-Cr-Nb-Si)N systems were obtained by vacuum-arc deposition of a solid-cast cathode in a nitrogen reaction gas medium, the main matrix has a pronounced columnar structure and consists of droplet fractions up to 10 μm.
2. The use of complementary methods of elemental analysis (SEM, TEM, XPS, EDX/EDS, RBS and AFM) made it possible to study the surfaces of coatings and determine the chemical composition of the surface layer, to establish patterns of homogeneous distribution of titanium, chromium, zirconium, niobium, silicon and nitrogen atoms in depth, as well as to determine uncontrolled impurities as C and O.
3. When the pressure in the chamber increases from 0.05 to 0.3-0.7 Pa in some cases and the potential on the substrate from -100 to -200 V leads to a significant decrease in the concentrations of droplet fractions on the surface.
4. It is established that in the coating of the (Zr-Ti-Nb)N system, the main phase is the solid solution phase with a FCC lattice with a growth texture (111) and a small volume fraction of the phase with a BCC lattice with a growth texture (110).
5. When Cr and Si elements are added to the cathode in series, a TiN-based FCC phase and a phase with a Cr_2N trigonal lattice are formed in the coatings, for the first case, and for the second, a solid solution of FCC is also formed by a lattice and a growth texture (111).

4. MECHANICAL AND TRIBOLOGICAL CHARACTERISTICS OF MULTICOMPONENT NITRIDE COATINGS BASED ON Zr, Ti, Nb, Cr, AND Si

4.1. Mechanical Properties of Multicomponent Nitride (Zr-Ti-Nb)N, (Zr-Ti-Nb-Cr)N and (Zr-Ti-Nb-Cr-Si)N Coatings

The research and development of nanostructured materials is a promising area of materials science, promising the creation of fundamentally new protective coatings with excellent physical and mechanical properties. The uniqueness of such materials lies in the presence of structural elements with a size of less than 100 nm, which allows them to acquire physical and mechanical properties different from bulk bodies. Special attention should be paid to nanocrystalline multicomponent coatings based on nitrides, borides, oxides, etc. A feature of such coatings is the formation of a grain structure due to phase segregation, in which nanocrystals (solid phase) are separated by an interlayer of an amorphous phase. Studies show that coatings with a similar structure have excellent performance characteristics (superhardness > 40 GPa, elastic recovery > 70%, etc.) (Veprek, 2013).

A review of the literature showed that the main factors that increase the hardness of coatings are: the adhesion forces between atoms, compressive stresses, plastic deformation with the dominant role of dislocations, and nanostructure. The maximum hardness can be obtained with grain sizes d_c ~ 10 nm. The region near H_{max} corresponds to a continuous transition from microscopic processes of nucleation and motion of dislocations at $d > d_c$, described by the well-known Hall-Petch law ($H \sim d^{-1/2}$) for ordinary polycrystalline materials, to intercrystalline processes of local slip along grain and phase boundaries at $d \sim d_c$. The increase in the hardness of the coating is due to the interaction of all factors, which greatly complicates the determination of the real origin of hardness. The literature review showed the existence of two ways to increase the strength of the coating, namely: energy ion bombardment and the creation of nanocomposite coatings. Thus, in the case of an increase in the energy of the sprayed ions, the grain boundaries are compacted, the grain size is reduced, and high compressive stresses are formed. The creation of a nanocomposite structure in the coating leads to inhibition of the propagation of deformations such as cracks and dislocations, which has a beneficial effect on strength characteristics. In the case of nitride coatings based on multicomponent alloys, special attention should be paid to solid-solution hardening. A wide choice of coating components can lead to serious distortions of the crystal lattice due to the discrepancy of atomic radii. According to the dislocation theory, the strength increase occurs due to the inhibition of dislocations in the fields of elastic stresses caused by distortions of the crystal lattice. As a result, dislocations appear to be blocked by the atoms of the dissolved substance. An analysis of the periodic literature (Liang et al., 2011; Tsai et al., 2010) devoted to the study of the mechanical characteristics of nitride coatings based on multicomponent alloys has shown that the main factors affecting the hardness value are: the number of nitride-forming constituent elements, as well as the parameters of the deposition condition.

The study of the mechanical characteristics of nitride coatings deposited at various deposition parameters was carried out using microhardness testers "DM8-B" and "Durascan-20" manufactured by "CSM Systems AG" (Switzerland). The results of measurements of mechanical characteristics, in particular, the hardness of all obtained coatings, are shown in Table 6.

Table 6. Average hardness values of nitride coatings

Series (Sample) No.	Hardness, GPa
(Zr-Ti-Nb)N $HV_{0.025}$ GPa	
1	37.2
2	44.5
(Zr-Ti-Cr-Nb)N $HV_{0.1}$ GPa	
1	30.9
2	34.7
3	38.8
4	43.9
(Zr-Ti-Cr-Nb-Si)N $HV_{0.1}$ GPa	
1	29.0
2	24.0

As can be seen from Table 6, the maximum hardness value (44.5 GPa) for the (Zr-Ti-Nb)N nitride coating is achieved at a reaction gas pressure of P_N = 0.5 Pa. As we can see, the hardness of (Zr-Ti-Cr-Nb)N coatings deposited at P_N = 0.3 Pa and U_b = -100 V is 30.9 GPa and changes up to 43.9 GPa with increasing bias voltage on the substrate up to -200 V. It should be noted that coatings applied at a lower nitrogen content have low hardness. When P_N increases to 0.7 Pa, the increase in hardness is explained by the formation of more strong Me/N bonds present in the films. Moreover, the high hardness may be due to the high content of the Cr_2N phase since the hardness of the Cr_2N coating is higher than that of the CrN coating. For (Zr-Ti-Cr-Nb-Si)N nitride coatings deposited at P_N = 0.3 Pa and U_b = -100 V, the hardness is 29 GPa and the elastic modulus is 291 GPa. As the bias voltage on the substrate increased to –200 V, the hardness and elastic modulus decreased to 24 and 254 GPa, respectively. One explanation for the decrease in hardness is excessive ion bombardment, resulting in less textured coatings, a transition from a predominantly (111) high-intensity orientation to a high FCC orientation, including additional (111) and (200) orientations. The nanohardness and elastic modulus of (Zr-Ti-Cr-Nb-Si)N coatings were measured using a “Hysitron TI 950 TriboIndenter” nanoindenter with a Berkovich diamond indenter, the maximum load was 10,000 μN. The hardness value and modulus of elasticity were evaluated from indenter load-unload curves using the Oliver and Farr methods. From nanoindentation tests of load-unload curves of (Zr-Ti-Cr-Nb-Si)N coatings, the nanohardness and modulus of elasticity of the films deposited on the substrate at various bias voltages were determined and shown in Figure 32. Table 7 shows the results of tests for adhesion of (Zr-Ti-Nb)N, (Zr-Ti-Cr-Nb-Si)N and (Zr-Ti-Cr-Nb)N coatings

Figure 32. Dependences of hardness and elastic modulus on the contact depth for (Zr-Ti-Vr-Nb-Si)N coatings, at P_N = 0.3 Pa and different values: U_b = -100 V, a – hardness [GPa]; b – Young' modulus (Elastic modules) [GPa]; U_b = -200 V, c – hardness [GPa]; d – Young' modulus (Elastic modules) [GPa]

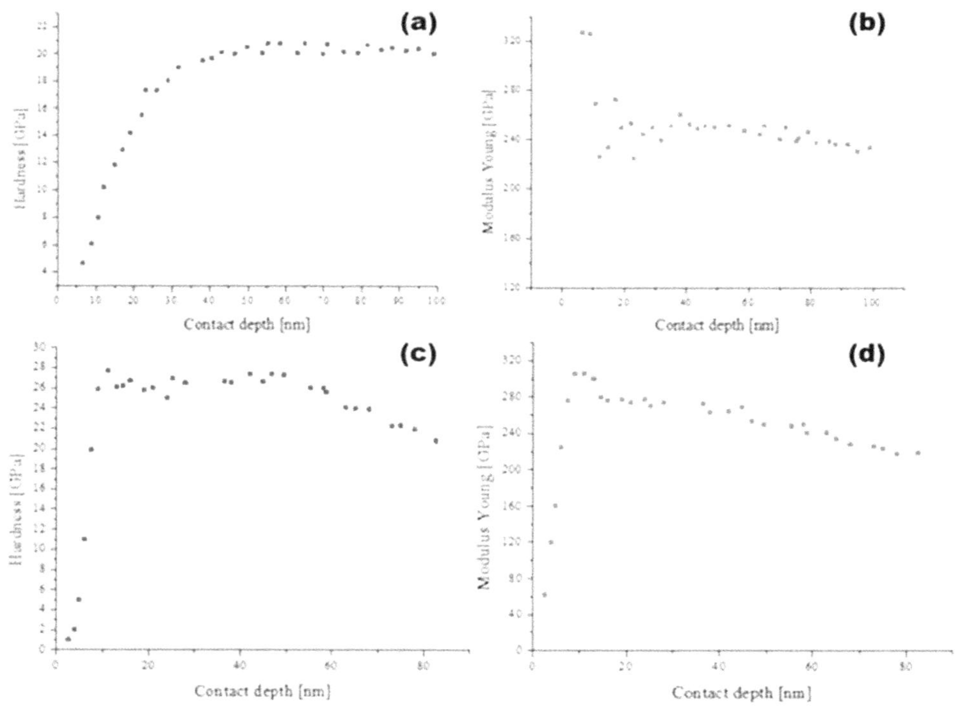

Table 7. Comparative results of adhesion tests for (Zr-Ti-Nb)N, (Zr-Ti-Cr-Nb)N and (Zr-Ti-Cr-Nb-Si)N coatings

Critical Loads [N]	Series (Sample) No.			
	No.1	No.2	No.3	No.4
(Zr-Ti-Nb)N				
L_{C1}	02.91	09.89	-	-
L_{C2}	29.04	20.62	-	-
L_{C3}	43.18	36.43	-	-
L_{C4} (L_{C5})	59.26	66.77	-	-
(Zr-Ti-Cr-Nb)N				
L_{C1}	10.94	11.80	10.35	15.21
L_{C2}	18.69	20.93	18.42	24.29
L_{C3}	26.95	30.35	23.12	33.45
L_{C4}	39.15	45.94	45.12	40.97
L_{C5}	49.09	56.17	61.08	62.06
(Zr-Ti-Cr-Nb-Si)N				
L_{C1}	09.54	11.28	-	-
L_{C2}	12.48	14.04	-	-
L_{C3}	18.36	24.00	-	-
L_{C4}	29.86	34.09	-	-
L_{C5}	45.33	45.57	-	-

Figure 33. Micrographs of the contact area of a diamond indenter coated with (Zr-Ti-Nb)N (P_N=0.5 Pa), at different stages of scratch testing: a – L_{C1}; b – L_{C2}; c – L_{C3}; d – L_{C4} (L_{C5})

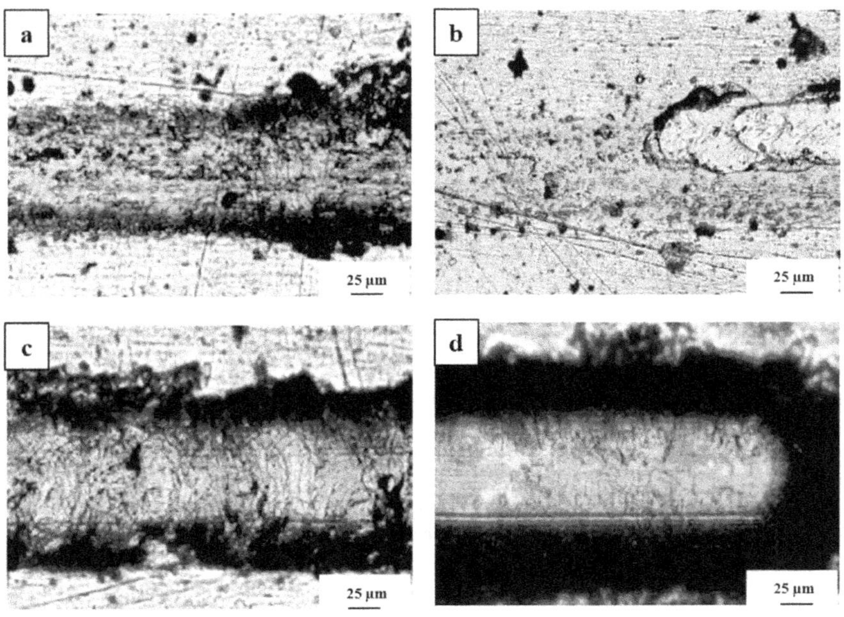

Figure 34. Dependence of the friction coefficient and the acoustic emission signal on the applied load during scratch testing of the (Zr-Ti-Nb)N coating obtained at P_N=0.5 Pa

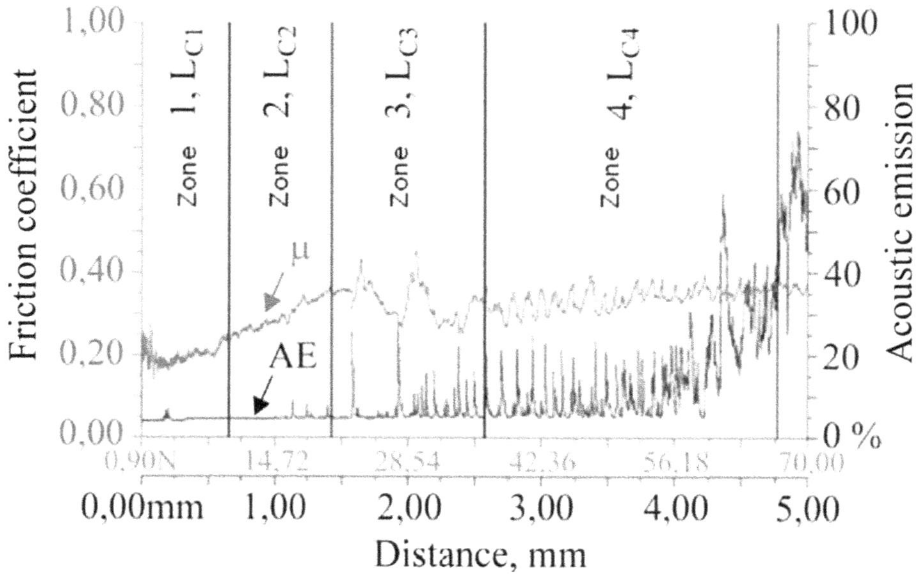

obtained at various deposition process parameters. The results of the study of adhesive-cohesive strength and scratch resistance of coatings are shown in Figures 33 and 34. In accordance with these criteria, the process of coating destruction during scratching with an indenter can be conditionally divided into four stages. In the load range from F=0.9 to 9.89 N, monotonous penetration of the indenter into the coating occurs, while the friction coefficient slightly increases, and the acoustic emission signal remains unchanged. At a load F=15.81 N, the indenter is completely immersed in the coating, and the diamond indenter slides over the coating with a friction coefficient of 0.35. With an increase in load (F=20.6-36.4 N), the material is squeezed out in front of the indenter in the form of tubercles and the penetration depth of the indenter increases.

Comparative analysis shows that (Zr-Ti-Cr-Nb)N coatings wear off but do not peel off during scratches, i.e. are destroyed due to the cohesive mechanism of plastic deformation and the formation of fatigue cracks in the coating material. Figure 35 shows the curve of change in the coefficient of friction (μ) when moving the diamond indenter on the surface of the coating of the (Zr-Ti-Cr-Nb)N system (sample 4), the curve of change in acoustic emission (AE) and the image of the remaining fragments of the coating at the bottom of the scratch after impact of a diamond indenter.

As we can see, monotonous penetration of the indenter into the coating occurs, and the first cracks appear (load up to 15.21 N); the coefficient of friction (μ) increases, but the acoustic emission signal remains unchanged. Subsequently, with an increase in the load, chevron and diagonal cracks appear, which leads to an increase in the friction coefficient to a value of 0.3. At a load of up to 14 N, the amplitude of the acoustic emission signal sharply increases, and its value remains at the same level until the end of the test. After that, with an increase in the load to 62 N, local abrasion of the coating occurs down to the substrate material. The dependence of the change in the friction coefficient and acoustic emission on the applied load on the (Zr-Ti-Cr-Nb-Si)N coating samples for series 1 and 2 is shown in

Figure 35. Dependence of the coefficient of friction and acoustic emission signal on the applied load during scratch testing of the (Zr-Ti-Cr-Nb)N coating series sample 4

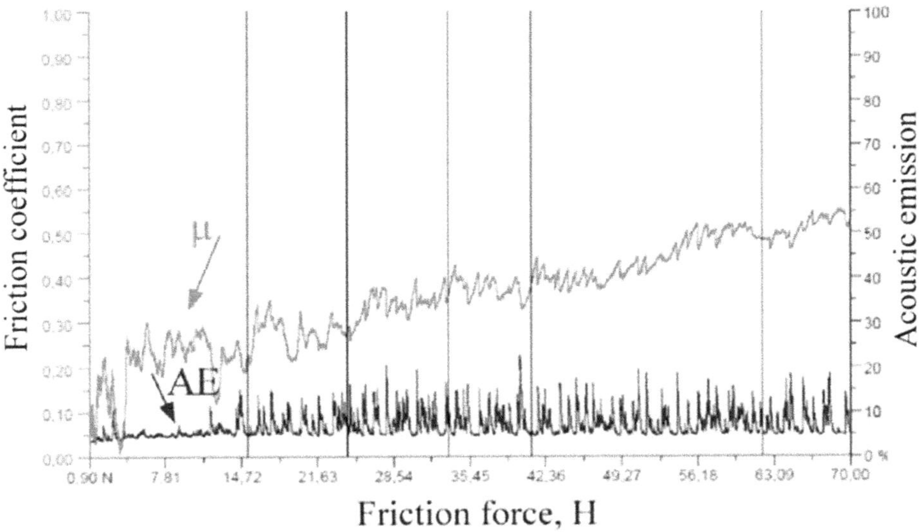

Figure 36. As we can see, at a load of 9 N (L_{C1} for series 1) and 11 N (L_{C1} for series 2) correspond to the appearance of the first cracks and fragments (figure 37 and 38). This is confirmed by the beginning of the increase in the amplitude of acoustic emission (Figure 36). The initial abrasion of coatings occurs at loads L_{C3}=18 N (series 1) and L_{C3}=24 N (series 2) (Figures 37 and 38), which leads to an increase in the friction coefficient to a value of 0.49 for all coatings. Local abrasion of coatings down to the substrate material occurs when the load reaches 45 and 46 N. In this case, the friction coefficient increases to a maximum value of 0.63.

Figure 39 shows the changes in the chemical composition of (Zr-Ti-Cr-Nb-Si)N coatings' during the scratch-test. A sharp increase in the iron concentration (at a distance of 3 mm) corresponds to the appearance of the first cracks and the beginning of the initial abrasion of the coatings.

In addition, it should be noted that the iron (Fe) content in the (Zr-Ti-Cr-Nb-Si)N coatings increases more intensively at a bias voltage U_b = -100 V. This statement is confirmed by a lower value of the critical load ((L_{C3}): 18.36 and 24 N, respectively), which corresponds to the beginning of the initial abrasion of the coatings (Table 7). The chemical composition of the (Zr-Ti-Cr-Nb-Si)N coatings at local abrasion is practically the same, but the nitrogen concentration in series 2 is higher than in series 1, which is consistent with the EDX result. For multicomponent (Zr-Ti-Cr-Nb-Si)N coatings for series 1, the adhesion strength at different intervals of the experiment differs significantly from the strength of coatings (samples) of series 2. As the load increases, a noticeable crack propagation on the coating surface does not occur. Based on the nature of the failure, the main contribution appears to be related to shear stresses. Cohesive failure occurs during cracking in a plane perpendicular to the direction of coating growth. The first signs of coating abrasion were recorded at a load of 18 N. The results of the adhesive strength tests show that the coatings were worn during scratches but did not peel off, i.e., the failure occurred due to the cohesive mechanism associated with plastic deformation and the formation of fatigue cracks in the coating material. A comparative analysis shows (Table 7) that the difference in

Figure 36. Dependence of the friction coefficient and acoustic emission signal on the applied load during scratch testing of the (Zr-Ti-Cr-Nb-Si)N coating, a – series (sample) 1: U_b = -100 V; b – series (sample) 2: U_b = -200 V

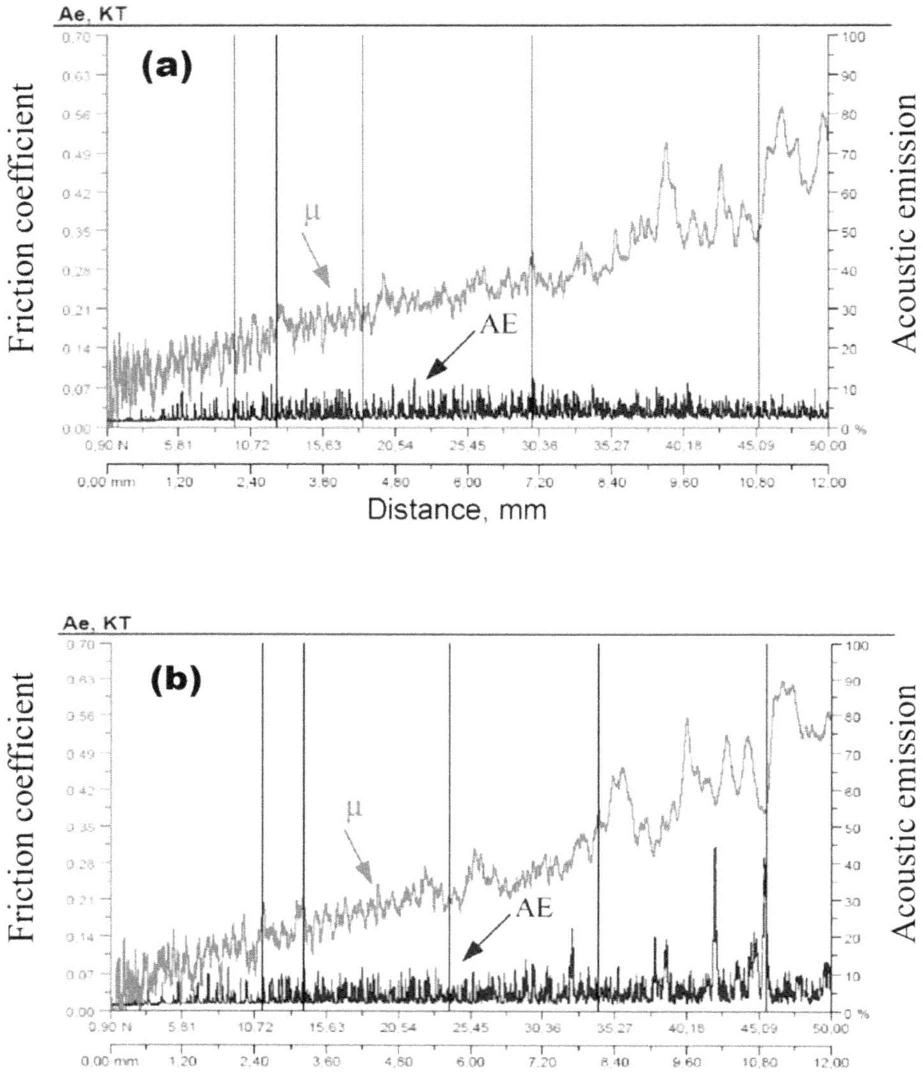

the adhesion strength (adhesion) of the coatings is apparently due to differences in their structure and mechanical properties. As shown in Table 7, as the substrate displacement increases, the adhesion strength between the coating and the substrate increases. Increased adhesion may be associated with increased resistance to cracking and plastic deformation. Loads (L_{C5}) for (Zr-Ti-Cr-Nb)N and (Zr-Ti-Cr-Nb-Si)N coatings increases, with increasing substrate displacement (from -100 to -200 V), their maximum value was 62.06 and 45.57 N, respectively. In addition, it should be noted that coatings obtained with a high displacement potential have a higher hardness and are more stable (Pogrebnjak, Yakushchenko, Bondar et al, 2016; Shypylenko et al., 2016). Nitride coatings (Zr-Ti-Cr-Nb)N and (Zr-Ti-Cr-Nb-Si)N with

Figure 37. Microphotographs of the contact area of a diamond indenter coated with (Zr-Ti-Cr-Nb-Si) N series (sample) 1, at different loads on the indenter: a – 9.54 N (L_{C1}); b – 12.48 N (L_{C2}); c – 18.36 N (L_{C3}); d – 29.86 N (L_{C4}); e – 45.33 N (L_{C5})

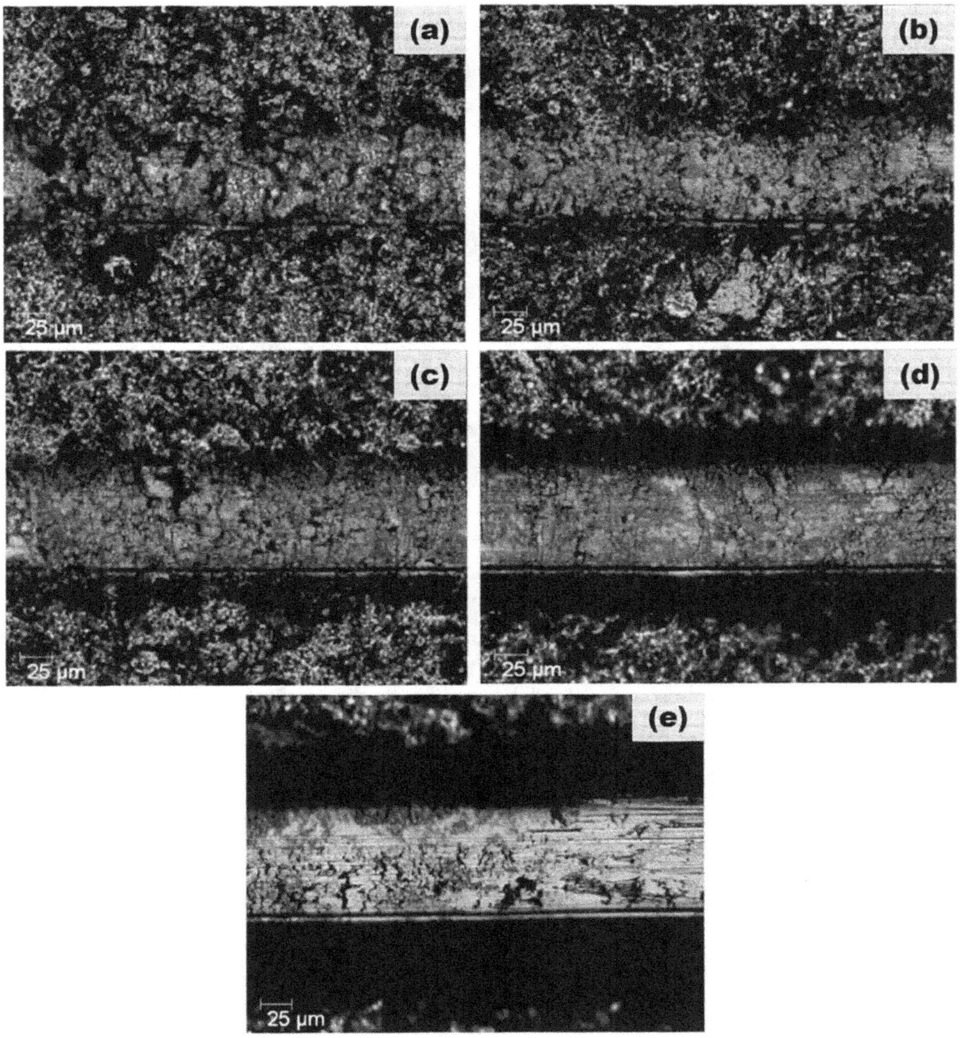

good adhesive strength demonstrate high hardness values of 43.9 and 29 GPa, respectively. Comparing the results from Tables 6 and 7, it can be seen that coatings with high hardness have good adhesive strength. It was previously reported that the application of the displacement potential on the substrate during deposition can cause collision mixing and, consequently, lead to improved adhesion (Hones et al., 1998; Yamamoto et al., 2003).

Figure 38. Microphotographs of the contact area of a diamond indenter coated with (Zr-Ti-Cr-Nb-Si)N series (sample) 2, at different loads on the indenter: a – 11.28 N (L_{C1}); b – 14.04 N (L_{C2}); c – 24 N (L_{C3}); d – 34.09 N (L_{C4}); e – 45.57 N (L_{C5})

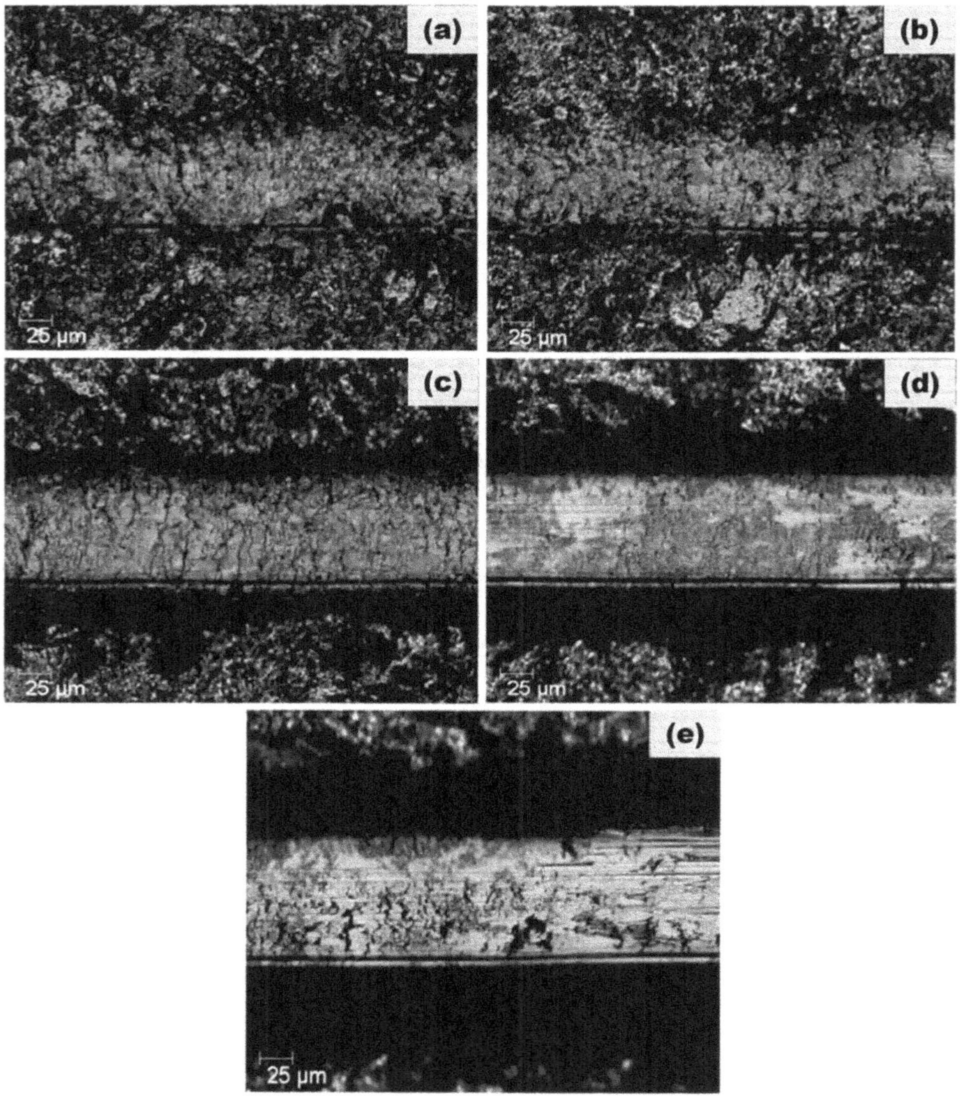

4.2. Tribological Characteristics of Coatings Based on Zr, Ti, Nb, Cr and Si Elements

One of the tasks of materials science and industry is to solve the problems of wear of machine parts, tools, mechanisms, etc. Thus, the concept of creating tribological coatings seeks to optimize two main characteristics of nanostructured coatings, namely, an increase in hardness *H* with a simultaneous decrease in the elastic modulus *E*, in other words, it is necessary to increase the ratio *H/E*. This ratio plays a key role for coating surfaces, since the behaviour of the latter under load must be consistent with the char-

Figure 39. EDX-chemical composition of scratches on (Zr-Ti-Cr-Nb-Si)N coatings after testing, a – series (sample) 1: U_b = -100 V; b – series (sample) 2: U_b = -200 V

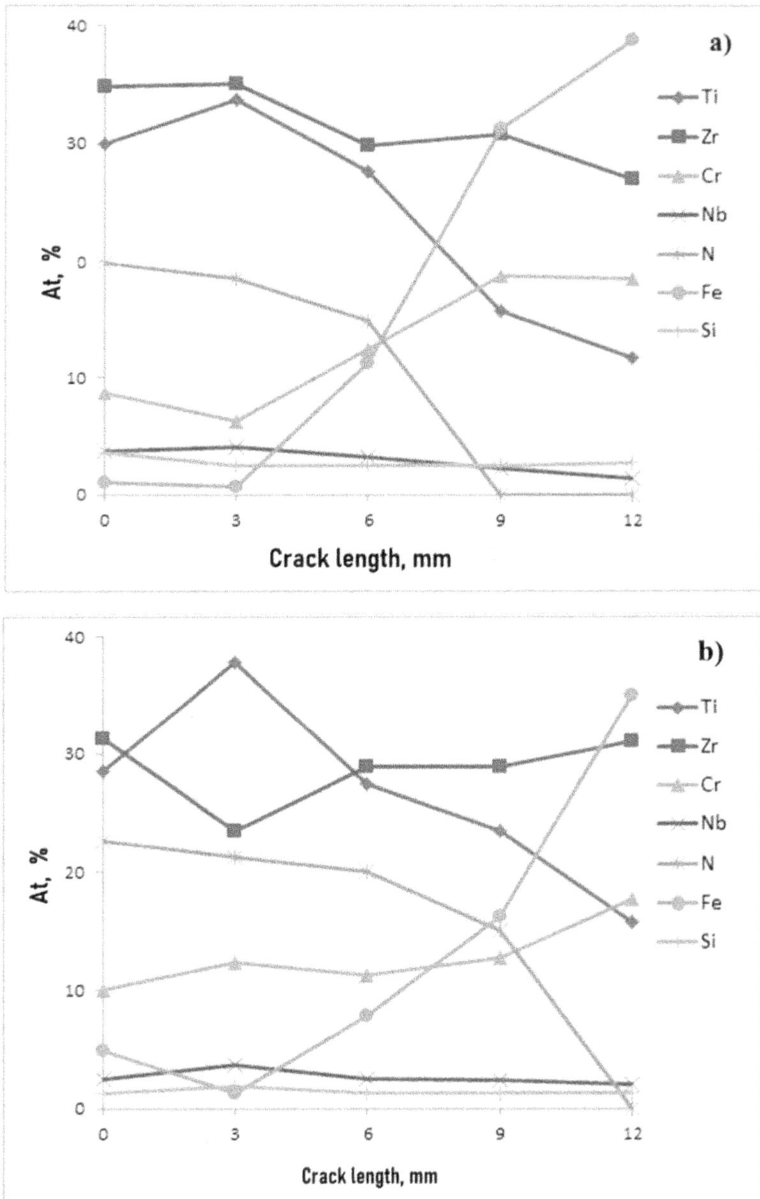

acteristics of the substrate. It follows from this that the elastic modulus of the substrate and the coating should be as close as possible in value, this allows to reduce the peak stress values at the contact points, as well as to reduce the stress values at the interface between the hard coating and the substrate (Godet et al., 1991; Ramalingam & Zheng, 1995).

The development of general ideas and ideas for creating wear-resistant coatings is primarily related to the works of Holmberg (Holmberg & Matthews, 1994), where the classification of interactions in the process of friction of surfaces is given, namely:

a) macromechanical;
b) micromechanical;
c) substance transfer processes;
d) tribomechanical;
e) nanomechanical.

The first class (macromechanics) includes macroscopic processes that occur under the action of macroscopic loads that cause stresses and deformations in the near-surface layer (Gavaleiro & De Hosson, 2006). The mechanisms of processes at the micro level are primarily associated with the origin and development of cracks, the formation of fragments, etc. The transfer of matter and tribomechanical reactions of the surface are related to the characteristics of materials and their phase composition. One of the main methods of applying protective coatings to the inside of parts is vacuum arc deposition. The specific features of this method make it possible to obtain coatings that are superior to ceramic in many respects (adhesion, hardness, fracture toughness, etc.) (Gavaleiro & De Hosson, 2006). Coatings obtained by vacuum arc deposition are often a compound of the type of embedding phases, however, the use of alloying elements with low solubility can lead to the formation of substitution phases, which is a key condition for the creation of a nanocrystalline phase, as well as glassy films.

First of all, the selection of constituent elements for the creation of wear-resistant coatings comes from the ability of the elements to form nitrides. In our case, Zr, Ti, Nb, Cr and Si elements have a high heat of nitride formation (see Table 1). At the same time, the melting point of the mentioned elements exceeds 1700°C, which allows them to be attributed to refractory materials. It is also worth noting that the atoms of the selected elements have similar geometric dimensions and electronic compatibility, which significantly expands the area of their mutual solubility. As a result, the driving thermodynamic forces of crystallization from the melt are stronger than amorphization. Thus, the use of materials that have their own high temperature resistance should be the basis for creating wear-resistant coatings. It is known that the main disadvantage of vacuum arc deposition is the presence of a micron-sized droplet fraction, which significantly affects the degree of wear of the protective coating.

As a sample for tribological tests, nitride multicomponent (Zr-Ti-Nb)N, (Zr-Ti-Nb-Cr)N and (Zr-Ti-Nb-Cr-Si)N coatings obtained by vacuum-arc deposition at the previously mentioned displacement potentials and working gas pressure were selected (see table 2). According to (Boxman & Zhitomirsky,), high values of technological parameters of deposition contribute to a decrease in the droplet fraction on the surface of coatings. The morphology of the surface of the test sample is shown in Figure 40. As can be seen from Figure 40, the coating has a homogeneous surface with minimal inclusion of microparticles. The presence of small craters is obviously associated with sputtering processes due to the high energy of the sprayed ions (Beresnev et al., 2016; Maksakova et al., 2017; Plotnikov et al., 2016; Pogrebnjak, Maksakova, Kozak et al, 2016). The study of surface morphology shows that deposition of coatings leads to an increase in surface roughness (0.088-0.42 μm for (Zr-Ti-Nb)N coatings and 1.02-1.82 μm for (Zr-Ti-Cr-Nb-Si)N) coatings due to the molten components of the droplets, which can be incorporated into the coating in the form of macroparticles. This is a typical phenomenon in the process of vacuum-arc deposition. Tribological tests were carried out on an automated friction machine of the brand "Tribometer"

(CSM Instruments) in air according to the "ball–disc" scheme at a temperature of 20°C. The discs on which the coatings were applied were made of 45 steel (HRC = 55) with a diameter of 42 mm, a height of 5 mm. A ball with a diameter of 6.0 mm, made of sintered certified material – Al_2O_3. was used as a counterbody. The load was 3.0 N, the sliding speed was 10 cm/s. The test conditions conformed to the international standards ASTM G99-959, DIN50324 and ISO 20808 (Kantay et al., 2021). Profilograms of the surface of the steel disc before and after the test are shown in Figures 41 and 42.

In the absence of coating, the average disc roughness is Ra = 0.088 microns. The application of a coating (Zr-Ti-Nb)N (6.7 μm thick) in a nitrogen reaction gas medium by vacuum-arc deposition on the polished surface of a steel disc leads to an increase in roughness (Figure 42) up to 0.42 μm. A noticeable increase in the roughness of the steel disc is obviously due to both the presence of a flow of macroparticles and the formation of craters during deposition of the coating. The results of tribological tests of the steel disc after coating are shown in Figure 43 and in Table 8.

An important parameter determining the performance of the coating is also its tribological characteristics (friction coefficient and wear factor). The coefficient of friction μ determines the adhesion strength of the rubbing materials, and the wear factor determines the resistance to wear (the smaller the wear factor, the higher the wear resistance). From the data presented in Figure 42 and in Table 8, it can be seen that the application of a nitride (Zr-Ti-Nb)N coating on steel 45 leads to an increase in the coefficient of friction from 0.674 to 1.95, which is consistent with the results of the study of the surface roughness of the steel disc.

Figure 40. SEM-images of the surface morphology of (Zr-Ti-Nb)N coatings obtained at Ub = -100 V: a – PN = 0.04 Pa, Ra » 1.17 μm; b – PN = 0.5 Pa, Ra » 0.42 μm

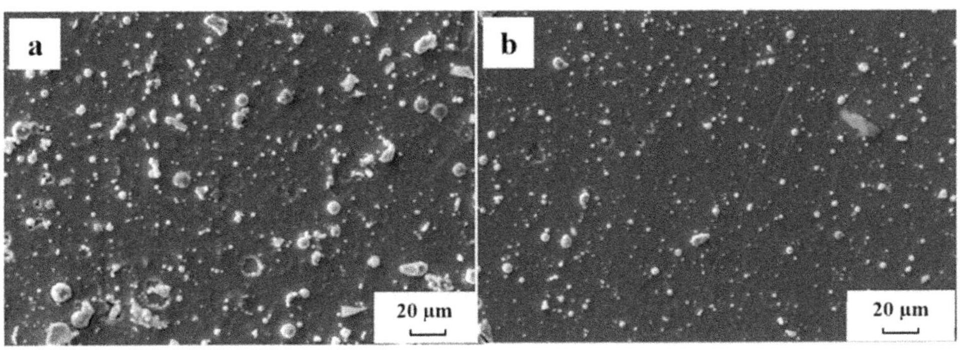

Figure 41. Surface roughness profile of the steel disc after polishing (R_a » 0.088 μm)

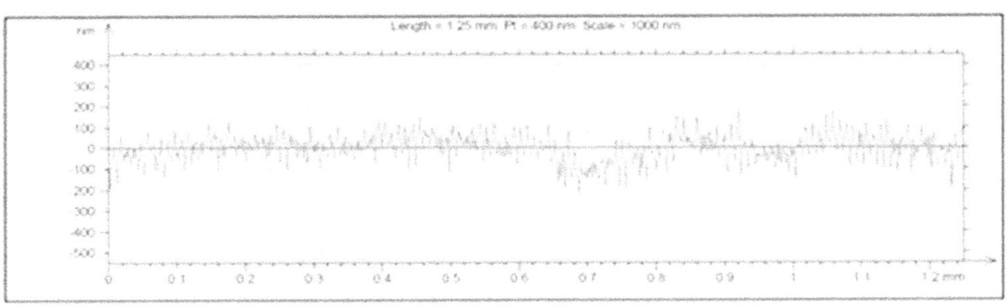

Figure 42. Profile of the surface roughness of the steel disk after deposition of the (Zr-Ti-Nb)N coating (Ra » 0.42 μm)

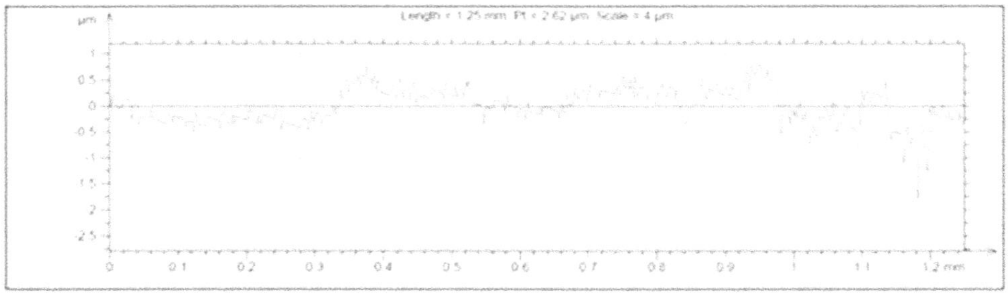

Figure 43. Results of tribological tests of a steel disk after (Zr-Ti-Nb)N coating

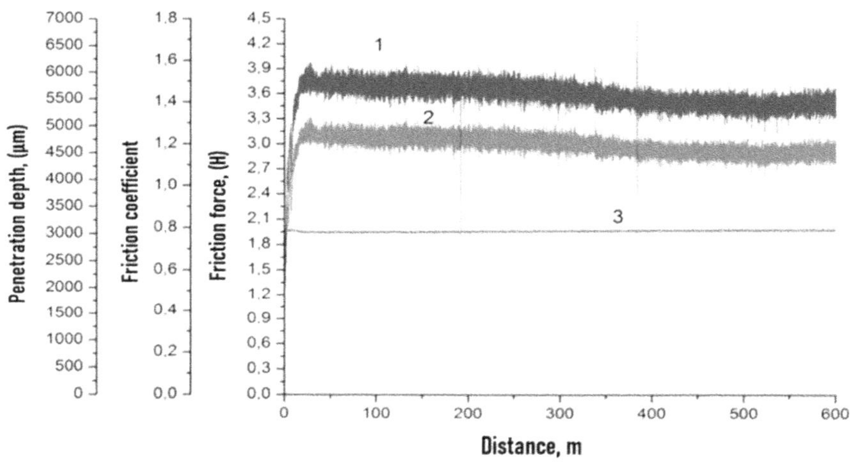

The friction coefficient takes on values from initial (during the first contact) to stationary (column "during tests" in Table 8) - when reaching constant values during tests. On all coated samples (series 1, 2), the friction coefficient was higher than 1.0. Such high values can be explained by the high roughness (Figure 40) associated with the presence of a droplet fraction on the surface and in the coating, which is formed during vacuum-arc deposition. The appearance of a solid droplet component, as well as the formation of wear products in the form of particles consisting of solid nitrides during the destruction of the coating, leads to abrasive wear of the coating. Reducing the roughness reduces the friction coefficient from 1.95 to 1.05. With an increase in hardness (see table 6) of the coating, the wear factor *W* of the coating decreases, while the counterbody increases (table 8). Of considerable interest is the study of friction tracks, as a result of which it is possible to draw a conclusion about the wear mechanism. It is known that the main mechanisms of wear of coatings during friction are: 1) adhesive wear, in which the coating material adheres to the counterbody; 2) abrasive wear, in which the formation of grooves on the surface of the material by a harder material of the counterbody occurs; 3) fatigue wear - the process of removing coating particles; 4) plastic deformation of the coating material.

Table 8. Tribological characteristics of the system "(Zr-Ti-Nb)N coating - Al_2O_3"

(Zr-Ti-Nb)N Coating, Series (Samples)	Friction Coefficient, μ		Wear Factor *W*, 10^{-5} mm³/N·m	
	Initial	During Testing	Counterbody	Coated Sample
1	0.61	1.95	0.391	9.69
2	0.491	1.05	3.21	2.4
Steel 45 (substrate)	0.204	0.674	0.269	35.36

Figure 44 (a) and (b) show photographs of the friction tracks of the (Zr-Ti-Nb)N coating obtained by vacuum-arc deposition. Analysing the images of the friction tracks of the coating under study, we can conclude that the presence of scratch marks, as well as particles of similar wear products (see Figure 44) indicate the abrasive nature of the wear of the coating. From the figures shown (Figure 43) it can be seen that the groove has a smooth surface. The presence of such a groove is obviously associated with a large number of defects in the coating. Figures 44b and 44d show the results of the analysis of wear products along the formed groove. The micrograph of the wear products clearly shows the presence of the counterbody material on the surface. The energy dispersive spectrum (Figures 44b and 44d) indicates the presence of aluminium oxide on the surface of the coating, which is obviously associated with the transfer of the counterbody material. This circumstance is confirmed by the data presented in Table 8, namely by comparing the wear factors of the counterbody and the test sample.

Figure 44. SEM-surface of the system (Zr-Ti-Nb)N coating after tribotechnical tests (a, c) and EDS-energy-dispersive spectra of friction tracks (b, d): a, b – sample series 1; c, d – sample series 2

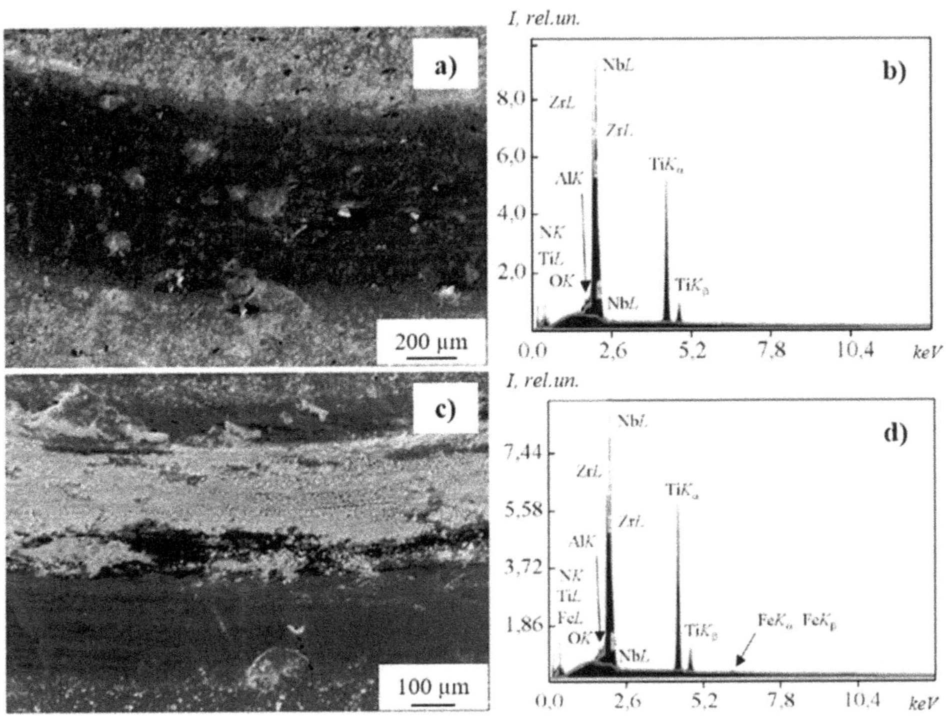

The results of processing profilograms indicate that the average value of the substrate roughness for non-deposited (Zr-Ti-Cr-Nb)N coatings is R_a = 0.088 μm, and it changes after coating deposition (Figure 45).

Figure 45. Surface roughness profile of (Zr-Ti-Cr-Nb)N coatings: a – series (sample) 1 (R_a » 1.88 μm); b – series (sample) 4 (R_a » 1.55 μm)

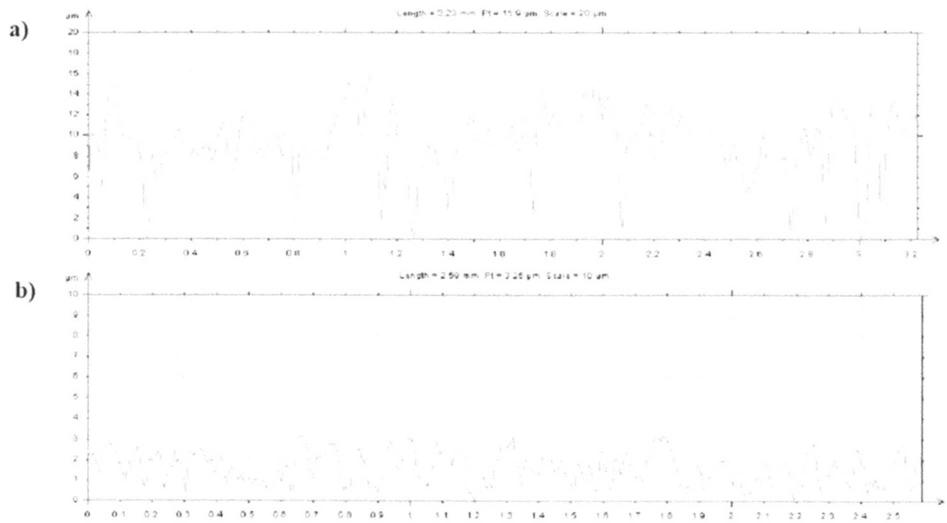

Figure 46. AFM-three-dimensional images of the (Zr-Ti-Cr-Nb)N coating surface: a – PN=0.3 Pa, Ub=-100 V, Ra=1.88 μm; b – PN=0,7 Pa, Ub=-100 V, Ra=1.83 μm; c – PN=0.3 Pa, Ub=-200 V, Ra=1.69 μm; d – PN=0.7 Pa, Ub=-200 V, Ra=1.55 μm

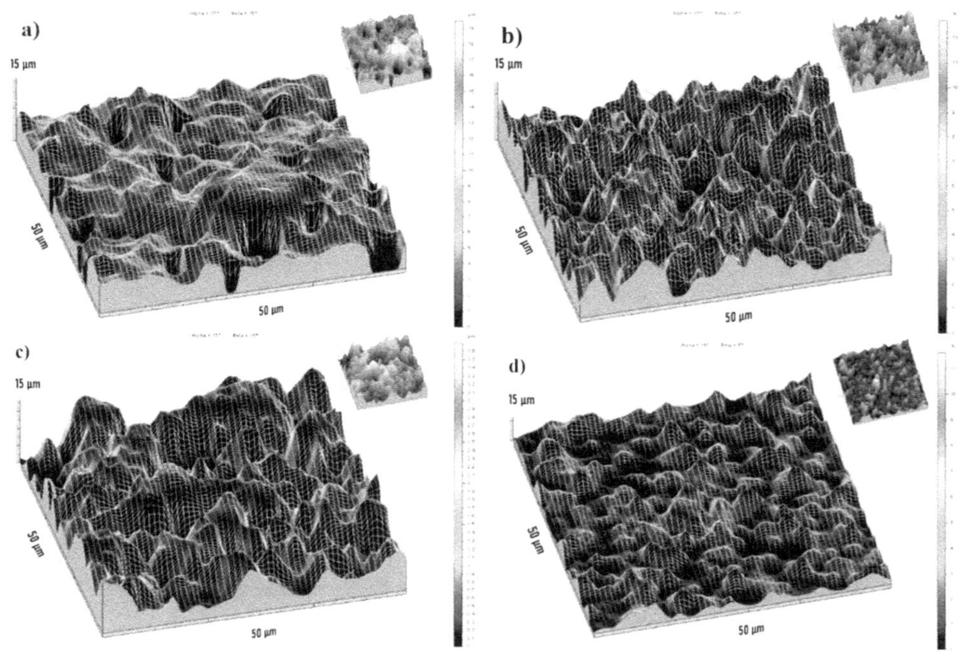

Figure 46 shows AFM-images of the topography surface of the coating of the 4th series of the (Zr-Ti-Cr-Nb)N system.

Comparison of the obtained data shows that the change in the roughness of the (Zr-Ti-Cr-Nb)N coating directly depends on the pressure of the nitrogen reaction gas and the mixing voltage. At a pressure of $P_N = 0.7$ Pa, the surface of the coating is more homogeneous, with clearly expressed droplet inclusions of small sizes. At a pressure of $P_N = 0.3$ Pa, incompletely melted compounds of the elements Ti, Zr, Cr, Nb, with nitrogen are formed on the surface of the coating.

Conclusions for Section 4

1. The relationship between the structural-phase state and the mechanical characteristics of the coatings has been studied. It has been established that the hardness of (Zr-Ti-Nb)N nitride coatings, depending on the physical parameters of deposition, varies from 37 to 45 GPa.
2. An analysis of experiments on measuring microhardness indicates that the hardness for (Zr-Ti-Cr-Nb)N coatings, depending on the physical parameters of deposition, is in the range from 31 to 44 GPa.
3. The influence of physical-technological parameters of deposition on the hardness and modulus of elasticity for (Zr-Ti-Cr-Nb-Si)N nitride coatings deposited at $P_N = 0.3$ Pa and $U_b = -100$ V, the hardness is 29 GPa, and the modulus elasticity is 291 GPa. With an increase in the bias potential on the substrate to -200 V, the hardness and elastic modulus decreased to 24 and 254 GPa, respectively.
4. It is shown that the application of a (Zr-Ti-Nb)N coating on a steel disk leads to an increase in the friction coefficient from 0.674 to 1.95. The study of wear resistance (wear factor) of the analysed coating showed a noticeable increase in the resistance of the sample to wear (0.039×10^{-5} $mm^3\times N^{-1}\times mm^{-1}$ and 35.36×10^{-5} $mm^3\times N^{-1}\times mm^{-1}$).
5. It was found that the adhesion of the studied coatings was improved by using higher deposition parameters (bias voltage on the substrate and nitrogen working gas pressure). Local wear of (Zr-Ti-Cr-Nb-Si)N coatings to the substrate material occurs when the load reaches the maximum value of 46N, for (Zr-Ti-Cr-Nb)N coatings - 62N and for (Zr-Ti- Nb)N - 66N.

CONCLUSION

The research results obtained in the course of this work made it possible to solve a scientific and technical problem, which consisted in the development of physical and technological foundations for the creation of multicomponent hard and superhard nanostructured coatings obtained by vacuum-arc deposition and the establishment of their structural-phase, physical-mechanical and tribological-technical characteristics. Also, in this work, the authors considered the role of Cr and Si addition on the properties of (Zr-Ti-Nb)N coatings, including morphology, relative hardness and adhesion. Based on the results of the research work obtained for the first time, the following conclusions are formulated:

1. Using complementary methods of analysis, the influence of the physical parameters of deposition (the pressure of the nitrogen working gas and the bias voltage of the substrate) on the elemental composition and phase structure of multicomponent nanostructured nitride (Zr-Ti-Nb)N, (Zr-Ti-Nb-Cr)N and (Zr-Ti-Nb-Cr-Si)N coatings obtained by vacuum-arc deposition, in which the formation of nanocrystallites with sizes ranging from 5.2 nm to 63 nm of various phases was found.

2. It has been found that the main phase in triple coatings is a solid solution phase with an FCC-lattice with a (111) growth texture and a small volume fraction of a phase with a BCC-lattice with a (110) growth texture.
3. It has been found that, as a result of adding Cr and Si elements to the cathode in succession, the TiN-based FCC-phase and the Cr_2N trigonal lattice phase are formed in the coatings in the first case, while in the second case, a solid solution of the FCC-lattice and (111) growth texture is also formed.
4. The relationship between the structural-phase state and the mechanical characteristics of the coatings is determined. An analysis of experiments on measuring hardness indicates that the highest hardness values were obtained in the (Zr-Ti-Nb)N coating - 37-45 GPa and (Zr-Ti-Nb-Cr)N - 31-44 GPa, at the same time, alloying with Si leads to a decrease in hardness to 24–29 GPa. The increase in the hardness of the coatings was due to the formation of a preferred orientation of crystallite growth with the (111) axis.
5. It is shown that with an increase in pressure in the chamber from 0.05 to 0.3-0.7 Pa, in some cases and an increase in the potential to the substrate from -100 to -200 V, the adhesion of coatings to the substrate increases. Local abrasion of coatings (Zr-Ti-Cr-Nb-Si)N to the substrate material occurs when the load on the indenter reaches 46N, for coatings (Zr-Ti-Nb-Cr)N - 62N, and for coatings from (Zr-Ti-Nb)N - 66N (i.e. the largest value).
6. It has been found that coating (Zr-Ti-Nb)N on a steel disk results in an increase in the friction coefficient from 0.674 to 1.95. A study of the wear resistance of the analyzed coating showed a noticeable decrease in the wear factor of the sample (0.391×10^{-5} $mm^3/N\cdot m$ and 35.36×10^{-5} $mm^3/N\cdot m$).

Summing up, these results represent a new step in solving the problem of creating protective coatings based on multicomponent systems. A change in the structural and phase properties that affect the increase in the operational characteristics of various products operating at high temperatures, loads and wear rates.

REFERENCES

Anishik, V. M., Ponaryadov, V. V., & Uglov, V. V. (2011). *Difraktsionnyy analiz.*

Azarenkov, N. A., Sobol, O. V., & Pogrebnjak, A. D. (2012). Materyalovedenye neravnovesnogo sostoyania modificirovannoy poverhnosti. Sumy, Sumy State University.

Bagdasaryan A.A., Smirnova E., Konarski P., Misnik M. & Zawada A. (2014). The analysis of elemental composition and depth profiles of nitride nanostructures coating based on the TiHfVNbZr high-entropy alloy. *JNEP*, (6), 1–5.

Beresnev, V. M., Sobol, O. V., Grankin, S. S., Nemchenko, U. S., Novikov, V. Y., Bondar, O. V., Belovol, K. O., Maksakova, O. V., & Eskermesov, D. K. (2016). Physical and mechanical properties of (Ti–Zr–Nb)N coatings fabricated by vacuum-arc deposition. *Inorganic Materials: Applied Research*, *7*(7), 388–394. doi:10.1134/S2075113316030047

Boxman R.L. & Zhitomirsky V.N. Vacuum-arc deposition devices. *Review Science Instruments* (77), 1 – 15.

Cecchini, R., Fabrizi, A., Cabibbo, M., Paternoster, C., Mavrin, B. N., Denisov, V. N., Novikova, N. N., & Haidopoulo, M. (2011). Mechanical, microstructural and oxidation properties of reactively sputtered thin Cr/N coatings on steel. *Thin Solid Films*, *519*(19), 6515–6521. doi:10.1016/j.tsf.2011.04.115

Engel, L. (1986). *Rastrovaya electronnaya microskopia*. Metallurgy.

Ganeyev, A. A., Gubal', A. R., Uskov, K. N., & Potapov, S. V. (2012). Analiticheskaya mass-spektrometriya s tleyushchim razryadom. Izvestiya Akademii Nauk. *Seriya Khimicheskaya*, (4), 1–17.

Gavaleiro, A., & De Hosson, J. T. (2006). *Nanostructured Coating*. Springer-Verlag. doi:10.1007/978-0-387-48756-4

Geller, Yu. A., & Rakhshadt, A. G. (1989). *Materialovedeniye. Metody analiza, laboratornyye raboty i zadachi*. Metallurgiya.

Gladkikh, A. I., Malykhin, S. V., & Pugachev, A. T. (2006). *Difraktsionnyye metody analiza vnutrennikh napryazheniy, Teoriya i eksperiment*. NTU HPI.

Godet, M., Berthier, Y., Vincent, L., & Flamand, L. (1991). Hard coatings for tribological applications: A pluridisciplinary approach. *Surface and Coatings Technology*, *45*(1-3), 1–8. doi:10.1016/0257-8972(91)90199-7

Golovin, Yu. I. (2008). Nanoindentirovaniye i mekhanicheskiye svoystva tverdykh tel v submikroob"yemakh, tonkikh pripoverkhnostnykh sloyakh i plenkakh. *FTT*, *50*(12), 2113–2140.

Gouldsteyn, Dj., N'yuberi, D., & Echlin, P. (1984). Rastrovaya elektronnaya mikroskopiya i rentgenovskiy mikroanaliz. *Mir, Moscow*, (2), 348.

Holleck, H. (1988). Metastable coatings – Prediction of composition and structure. *Surface and Coatings Technology*, *36*(1-2), 151–159. doi:10.1016/0257-8972(88)90145-4

Holmberg, K., & Matthews, A. (1994). Coating's tribology: A concept, critical aspects and future directions. *Thin Solid Films*, *253*(1-2), 173–178. doi:10.1016/0040-6090(94)90315-8

Hones, P., Sanjine, R., & Levy, F. (1998). Sputter deposited chromium nitride based ternary compounds for hard coatings. *Thin Solid Films*, *332*(1-2), 240–246. doi:10.1016/S0040-6090(98)00992-4

Hsieh, M. H., Tsai, M. H., Shen, W. J., & Yeh, J. W. (2013). Structure and properties of two Al-Cr-Nb-Si-Ti high-entropy nitride coatings. *Surface and Coatings Technology*, *221*, 118–123. doi:10.1016/j.surfcoat.2013.01.036

Kantay N., Rakhadilov B., Kurbanbekov Sh., Yeskermessov D., Yerbolatova G. & Apsezhanova A. (2021). Influence of detonation-spraying parameters on the phase composition and tribological properties of Al2O3 coatings. *Coatings, 11*(7), 793, 1-10.

Kim, H., Choi, S. R., & Yoon, S. Y. (2002). Superhard Ti-Si-N coatings by a hybrid system of arc ion plating and sputtering techniques. *Surface and Coatings Technology*, *161*(2-3), 243–248. doi:10.1016/S0257-8972(02)00499-1

Kim, S. K., & Cha, B. C. (2005). Deposition of tantalum nitride thin films by D.C. magnetron sputtering. *Thin Solid Films*, *475*(1-2), 202–207. doi:10.1016/j.tsf.2004.08.059

Lee, G. R., Kim, H., Choi, H. S., & Lee, J. J. (2007). Superhard tantalum-nitride films formed by inductively coupled plasma-assisted sputtering. *Surface and Coatings Technology*, *201*(9-11), 5207–5210. doi:10.1016/j.surfcoat.2006.07.207

Leng, Y. X., Sun, H., Yang, P., Chen, J. Y., Wang, J., Wan, G. J., Huang, N., Tian, X. B., Wang, L. P., & Chu, P. K. (2001). Biomedical properties of tantalum nitride films synthesized by reactive magnetron sputtering. *Thin Solid Films*, *398-399*, 471–475. doi:10.1016/S0040-6090(01)01448-1

Liang, S. C., Chang, Z. C., Tsai, D. C., Lin, Y.-C., Sung, H.-S., Deng, M.-J., & Shieu, F.-S. (2011). Effects of substrate temperature on the structure and mechanical properties of (TiVCrZrHf)N coatings. *Applied Surface Science*, *257*(17), 7709–7713. doi:10.1016/j.apsusc.2011.04.014

Maksakova, O. V., Beresnev, V. M., & Eskermesov, D. K. (2017). Effects of Cr and Si additions under the various deposition conditions on the mechanical properties of the (Zr-Ti-Nb)N coatings. *Proceedings of the 2017 IEEE 7th International Conference on Nanomaterials: Applications & Properties*, (1), 1-6.

Martin, P. J., Bendavid, A., Cairney, J. M., & Hoffman, M. (2005). Nanocomposite Ti-Si-N, Zr-Si-N, Ti-Al-Si-N, Ti-Al-V-Si-N thin film coatings deposited by vacuum-arc deposition. *Surface and Coatings Technology*, *200*(7), 2228–2235. doi:10.1016/j.surfcoat.2004.06.012

Morris, D. G. (1998). Material. Science foundation, Trans Technical publication LVD.

Nordin, M., Larsson, M., & Hogmark, S. (1998). Mechanical and tribological properties of multilayered PVD TiN/CrN, TiN/MoN, TiN/NbN and TiN/TaN coatings on cemented carbide. *Surface and Coatings Technology*, *106*(2-3), 234–241. doi:10.1016/S0257-8972(98)00544-1

Nordin, M., Larsson, M., & Hogmark, S. (1999). Mechanical and tribological properties of multilayered PVD TiN/CrN. *Wear*, *232*(2), 221–225. doi:10.1016/S0043-1648(99)00149-0

Oliver, W. C., & Pharr, G. M. (1992). An improved technique for determining hardness and elastic modulus using load and displacement sensing indentation experiments. *Journal of Materials Research*, *7*(7), 1564–1583. doi:10.1557/JMR.1992.1564

Palatnik, L. S., Fuks, M. Y. A., & Kosevich, V. M. (1972). *Mekhanizm obrazovaniya i substruktura kondensirovannykh plenok*. Nauka.

Pilloud, D., Pierson, J. F., Marques, A. P., & Cavaleiro, A. (2004). Structural changes in Zr-Si-N films vs. their silicon content. *Surface and Coatings Technology*, *180-181*(180), 352–356. doi:10.1016/j.surfcoat.2003.10.087

Plotnikov, S. V., Pogrebnjak, A. D., Yerokhina, L. N., Yeskermessov, D. K., & Erdybaeva, N. K. (2016). Study of nanostructured (Ti-Zr-Nb)N coatings' physical- mechanical properties obtained by vacuum arc evaporation. *IOP Conference Series: Material Science and Engineering, 110*(1), 1-6 (012031). 10.1088/1757-899X/110/1/012031

Pogrebnjak, A. D., Maksakova, O., Kozak, C., Koltunowicz, T. N., Grankin, S., Bondar, O., Eskermesov, D. K., Drozdenko, A., Petrov, S., & Erdybaeva, N. (2016). *Przegląd Elektrotechniczny*, (8), 180–183.

Pogrebnjak, A. D., Yakushchenko, I. V., Bondar, O. V., Beresnev, V. M., Oyoshi, K., Ivasishin, O. M., Amekura, H., Takeda, Y., Opielak, M., & Kozak, C. (2016). Irradiation resistance, microstructure and mechanical properties of nanostructured (TiZrHfVNbTa)N coatings. *Journal of Alloys and Compounds*, *679*, 155–163. doi:10.1016/j.jallcom.2016.04.064

Pogrebnjak, A. D., Yakushchenko, I. V., Bondar, O. V., Sobol', O. V., Beresnev, V. M., Oyoshi, K., Amekura, H., & Takeda, Y. (2015). Influence of Implantation of Au- Ions on the Microstructure and Mechanical Properties of the Nanostructured Multielement (TiZrHfVNbTa)N Coating. *Physics of the Solid State*, *57*(8), 1559–1564. doi:10.1134/S1063783415080259

Pogrebnyak, A. D., Ponomarev, A. G., Shpak, A. P., & Kunitskiy, Yu. A. (2012). Primeneniye mikro- i nanozondov dlya analiza malorazmernykh 3D materialov, nanosistem i nanoobyektov. *UFN*, *182*(3), 287–321.

Ramalingam, S., & Zheng, L. (1995). Film-substrate interface stresses and their role in the tribological performance on surface coatings. *Tribology International*, *28*(3), 145–161. doi:10.1016/0301-679X(95)98963-E

Randall, N. (2002). Overview of mechanical testing standards. *CSM Instruments Applications Bulletin*, (18), 3.

Rizzo, A., Signore, M. A., Penza, M., Tagliente, M. A., De Riccardis, F., & Serra, E. (2006). RF sputtering deposition of alternate TiN/ZrN multilayer hard coatings. *Thin Solid Films*, *515*(2), 500–504. doi:10.1016/j.tsf.2005.12.279

Samsonov G.V. (1969). Nitridy. *Scientific Opinion*, 380.

Shin, C. S., Kim, Y. W., Gall, D., Greene, J. E., & Petrov, I. (2002). Phase composition and microstructure of polycrystalline and epitaxial TaNx layers grown on oxidized Si(001) and MgO(001) by reactive magnetron sputter deposition. *Thin Solid Films*, *402*(1-2), 172–182. doi:10.1016/S0040-6090(01)01618-2

Shulayev, V. M., Anreyev, A. A., & Rudenko, V. P. (2006). Moderniacia vacuumno-dugovih ustanovok dlya sinteza pokriti i azotirovania metodom ionnoy implantacii i osajdenia. *FIP*, *4*(3-4), 136–142.

Shypylenko, A., Pshyk, A. V., Grzeskowiak, B., Medjanik, K., Peplinska, B., Oyoshi, K., Pogrebnjak, A., Jurga, S., & Coy, E. (2016). Effect of ion implantation on the physical and mechanical properties of Ti-Si-N multifunctional coatings for biomedical applications. *Materials & Design*, *110*, 821–829. doi:10.1016/j.matdes.2016.08.050

Thornton, J. A. (1982). *Deposition technologies for films and coating*. Noyes Publications.

Troyan, V. I., Pushkin, M. A., Borman, V. D., & Tronin, V. N. (2008). *Fizicheskie osnovi metodov issledovania nanostruktur i poverhnosti tverdogo tela*. MIFI.

Tsai, D. C., Huang, Y. L., Lin, S. R., Liang, S.-C., & Shieu, F.-S. (2010). Effect of nitrogen flow ratios on the structure and mechanical properties of (TiVCrZrY)N coatings prepared by reactive magnetron sputtering. *Applied Surface Science*, *257*(4), 1361–1367. doi:10.1016/j.apsusc.2010.08.078

Uglov, V. V., Chereda, N. N., & Anishin, V. M. (2007). *Metodi analiza elementnogo sostava poverhnostnih sloev. Learning guide*. Belarusian State University.

Umanskiy, Ya. S., Skakov, Yu. A., Ivanov, A. N., & Rastorguyev, L. N. (1982). *Kristallografiya, rentgenografiya i elektronnaya mikroskopiya*. Metallurgiya.

Veprek S. (2013). Recent search of new superhard materials: Go nano. *Journal Vacuum Science Technology A*(31), 1-33.

Vershinin D.S., Smolyakova M.Yu., Manokhin S.S., Druchinina O.A. & Akhmadeyev Yu.H. (2010). Issledovaniye tribologicheskikh svoystv azotirovannogo titanovogo splava VT16. Zavodskaya laboratoriya. *Diagnostika materialov, 76*(12), 45-49.

Voevodin, A. A., Shtansky, D. V., Levashov, E. A., & Moore, J. J. (2004). *Nanostructured Thin Films and Nanodispersion Strengherred Coatings*. Kluger Academic. doi:10.1007/1-4020-2222-0

Wagner, C. D. (1995). Handbook of XPS. Physical Electronics.

Yamamoto, K., Sato, T., Takahara, K., & Hanaguri, K. (2003). Properties of (Ti,Cr,Al)N coatings with high Al content deposited by new plasma enhanced arc-cathode. *Surface and Coatings Technology, 174-175*(174), 620–626. doi:10.1016/S0257-8972(03)00580-2

Chapter 5
Surface Hardening on Wheel Steel Using Electrolytic Plasma:
Formation of Fine Structure and Phase Composition of Wheel Steel Due to Plasma Hardening

Yerkezhan Tabiyeva
https://orcid.org/0000-0002-9726-7187
D. Serikbayev East Kazakhstan Technical University, Kazakhstan

Bauyrzhan Rakhadilov
LLP PlasmaScience, Kazakhstan

Gulzhaz Uazyrkhanova
D. Serikbayev East Kazakhstan Technical University, Kazakhstan

Waqar Ahmed
University of Lincoln, UK

ABSTRACT

Transmission electron microscopy (TEM) investigations of the structure and phase composition of ferritic-pearlitic wheel steel mark two surface after electrolyte-plasma surface hardening are presented. Initially the morphology of the steel matrix consists of lamellar perlite and non-fragmented and fragmented ferrite. Electrolytic plasma quenching of the steel surface results in the martensite transformation, steel self-tempering, and the formation of cementite particles in all martensite crystals. This treatment also leads to the diffusion transformation of gamma to alpha phases, the release of residual austenite along the low-temperature martensite laths and lamellas and in all crystals of lamellar martensite, the formation of M23C6 special carbides and, finally, to the enhancement of all parameters of the steel fine structure.

DOI: 10.4018/978-1-6684-6830-2.ch005

INTRODUCTION

Industrial progress is determined implementation of new technologies that improve the quality of products. Efficient production of machinery is impossible without using new technologies that provide the required combination of strength and plastic properties of constructional steels. Hence, machinery life and performance depend on the wear resistance of parts, which undergo considerable reversing impact loads. The surface layer must possess high strength and hardness combined with sufficient core viscosity. This can be achieved using a variety of the surface hardening techniques (Kondrat'ev et al., 2014; Nguyen et al., 2022; Paulmier et al., 2008; Zhang et al., 2009). An effective method of surface hardening of iron-carbon steels is electrolytic plasma treatment. High-speed heating and cooling in the thermal influence zone, structural changes due to phase transitions provide the operational characteristics of the working surface (Rakhadilov et al., 2016; Rakhadilov et al., 2018). An important feature of plasma hardening is its effective application for additional surface hardening after conventional volumetric thermal treatment. The life and economical use of railway products is related to the strength and tribological properties of products made from wheel steels (Konstantinova et al., 2019; Wang et al., 2003). The pace of economic and social development of society largely depends on the efficiency and quality of railway transport. There is a need to equip railways with modern traction rolling stock with increased reliability of its components, one of which is wheel sets, the tires of which are made of grade 2 wheel steel. At the moment, the problem of wear of wheel sets of traction rolling stock has become especially acute. In this regard, it is effective to increase the reliability and extend the uptime of railway equipment, which, in turn, contributes to the safety of train traffic. To improve the properties of wheel steels, it is necessary to study the influence of surface hardening on the steel structure to achieve the desired mechanical characteristics. This study focuses on phase composition, fine structure and mechanical properties of wheel steel mark 2 before and after electrolyte-plasma surface hardening.

MATERIAL AND METHODS OF RESEARCH

Wheel steel mark 2 was used to fabricate railway wheel sets. In its initial state wheel steel mark 2 represents material exposed to hardening at 890^{0}C for 2-2.5 hours with cooling in warm water (30-60^{0}C) followed by temperingat 580^{0}C for 2.5-3 hours. Steel samples were cut out of the wheel as parallelepiped in size 15×15×10 mm^3. The samples were free from deformation and thermal effect under slow cutting speed and low load. According to GOST 398-96, chemical composition of steel (in %) is C-0.57-0.65; Mn-0.50-0.90; Si-0.22-0.45; V - not exceeded 0.10; S- not exceeded 0.030 and P-not exceeded 0.035 respectively.

Electrolyte-plasma surface hardening in cathode mode was carried out at the electrolyte-plasma treatment plant at the Scientific Research Center "Surface Engineering and Tribology" at the S. Amanzholov East Kazakhstan University (Rakhadilov, Tabieva, & Zhurerova, 2019; Rakhadilov, Tabiyeva, Uazyrkhanova et al, 2019; Tabieva et al., 2020).

Steel structure and phase composition on thin foils were investigated using the EM-125 TEM at 125 kV accelerating voltage, before and after electrolyte-plasma surface hardening. The phase identification was carried out using TEM, microdiffraction patterns and dark-field images obtained in the respective reflections. TEM images of the fine steel structure were used for the structure morphology classification, phase identification and localization, and determination of such parameters as scalar and excess disloca-

Figure 1. Process of processing of an electrolyte plasma sample (a) and functional diagram (b) of the machine 1– in-process part; 2 – conic stainless steel electrolytic cell; 3– bottom plate; 4–pump; 5– heat exchanger; 6– bath filled with electrolyte

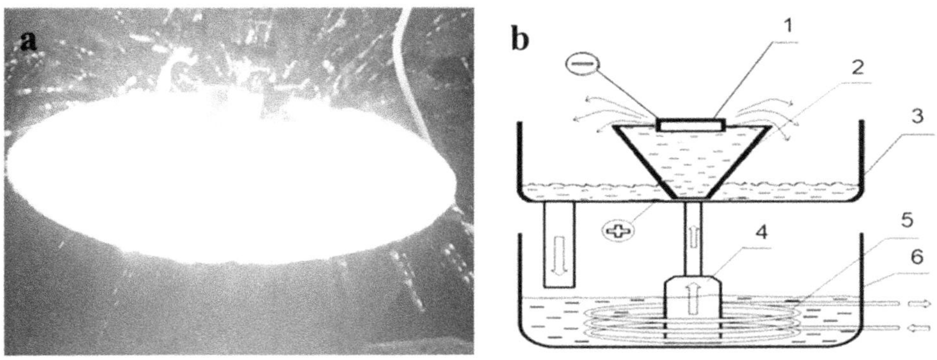

tion density, bending-torsion amplitudes of the crystal lattice, and internal stress amplitude for the fine steel structure. All the parameters were calculated using standard procedures. The surface morphology was studied on the scanning electron microscope JSM-6390LV, equipped with the energy-dispersive analysis prefix INCA Energy Penta FET X3. The microstructure was studied using an Altami MET-1M optical metallographic microscope in reflected light at a light field. The microhardness was measured in PMT-3M machine in accordance with GOST 9450-76, with loads of 100g and holding time for 10 seconds. Electrolyte-plasma surface hardening (EPSH) of the steel was carried out in the cathode mode in electrolyte-plasma treatment machine (Rakhadilov et al.,), scheme (Figure 1a) and processing (Figure 1b). High-current rectifier with output power of 360V/60A as was used as a power source. The samples were processed by rapid heating for 2 seconds followed by cooling in a flowing electrolyte. The process used the following parameters: electrolyte composition (%, mass): 10% urea $(NH_2)_2CO$ + 20% sodium carbonate Na_2CO_3+70% water, processing lasted for 2 seconds, T_{max} = 850-900°C; U= 320V; I=40A.

RESULTS AND DISCUSSION

Microstructural of the near-surface layers of wheel steel mark 2 samples subjected to electrolyte-plasma surface quenching in electrolyte plasma showed structural changes. Figure 2 shows the microstructures

Figure 2. Microstructure of Wheel Steel Mark 2 before and after electrolyte-plasma surface hardening obtained by Optical metallographic microscope a) initial state; after electrolyte-plasma surface hardening: b) 1 s, c) 2 s, d) 3 s, e) 4 s

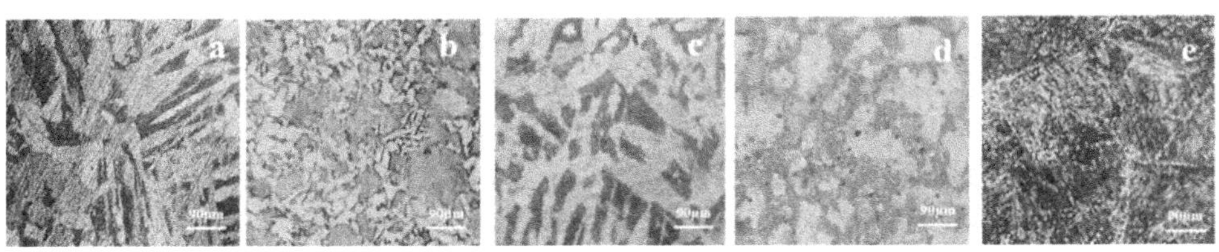

Figure 3. X-ray phase analysis of Wheel Steel Mark 2 a) initial state, b) after electrolyte-plasma surface hardening

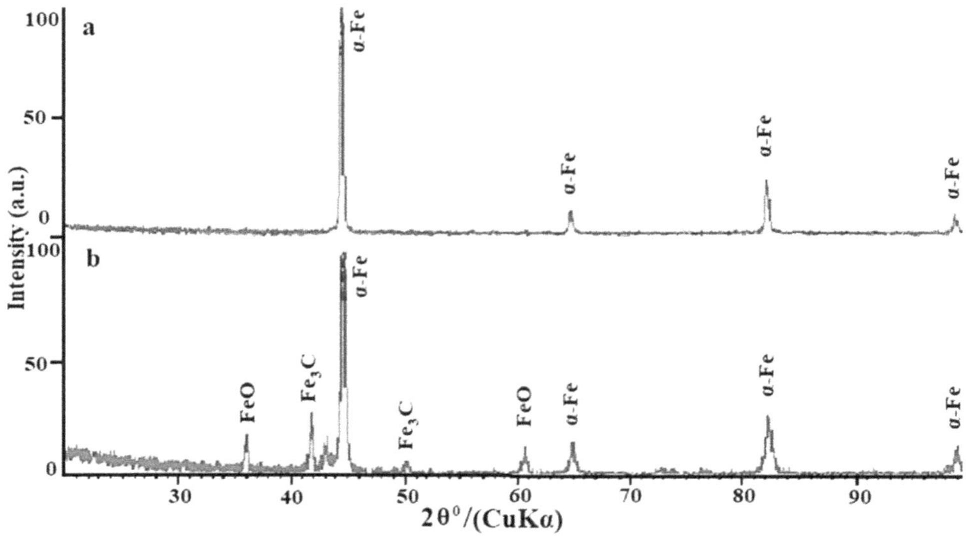

of the surface layer of steel before and after hardening at processing time of 1,2,3,4 seconds at a temperature of 850-900°C.

After electrolyte-plasma surface hardening for 3-4 seconds, the growth of the ferritic structure was observed, while the grains of perlite decrease. After electrolyte-plasma surface hardening with heating for 2 seconds, the formation of a cementite mesh is observed in the structure of wheel steel mark 2, which is located around the ferrite. X-ray diffraction analysis in Figure 3 shows that iron oxide was found along with cementite. This is explained by the fact that the oxygen concentration on the surface is not high enough, as well as the content of carbamide in the composition of the electrolyte prevents the formation of a continuous oxide layer, so it is more correct to say the surface layer contains iron oxide, which does not affect the further characteristics of the steel after EPSH. The grains of the perlite phases decrease gradually compared to the microstructure of wheel steel mark 2 before EPSH. Thus, the microstructure of the surface of the hardened steel sample is a fine-grained ferrite structure.

Tribological properties of the samples before and after electrolyte-plasma hardening were also investigated.

Experimental curves of the dependence of the coefficient of friction on the length of the runare shown (Figure 4). The test was carried out according to the “ball-disc” scheme, the length of the run was 31m, the speedwas 2cm/s, theloadwas 5N.

Test results showed that the average friction coefficients at a normal force of 5N and a mutual displacement velocity of 2cm/s are in the range from 0.27 to 0.40, which possibly indicates the destruction of the surface layer of the samples. All processed samples have a good degree of improvement in tribological properties. The results of experimental studies on structural-phase state of wheel steel mark 2 showed that in its initial state the steel matrix represents α-phase - solid solution of carbon and alloying elements in α-Fe with BCC matrix. Lamellar pearlite and ferrite are morphological components of α-phase. Lamellar perlite, almost ideal, i.e., it is a conglomerate of alternating parallel plates ferrite and cementite (Kozlov et al., 2014). Ferrite (α-phase) in the pearlite has a volumetric-centered cubic (BCC) crystal lattice. Mutual

Figure 4. Coefficient of friction of Wheel Steel Mark 2 before and after electrolyte-plasma surface hardening: (b) initial; (c) 2 s; (d) 3 s; (e) 4 s

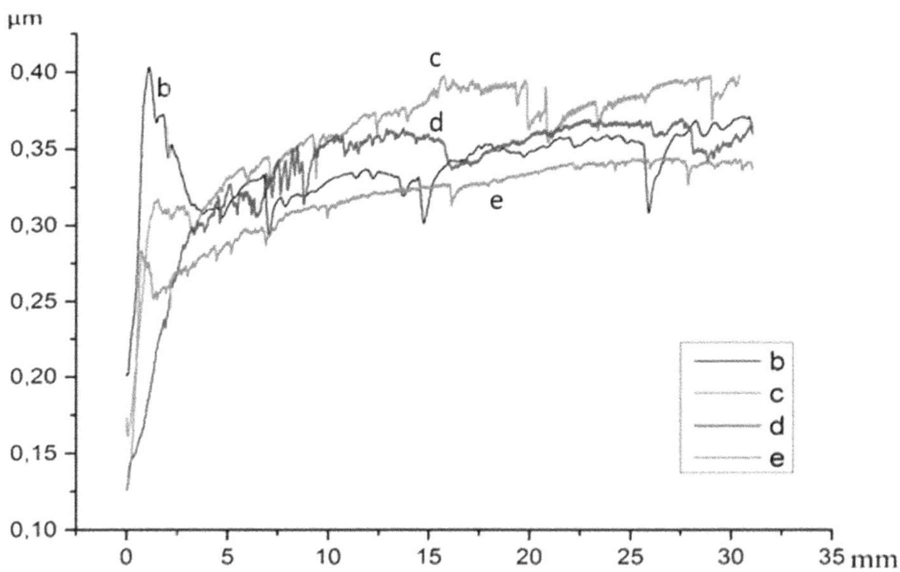

parallelism of the lamellas (Abuali & Mousavi, 2013) means that, firstly, different lamellas of the same phase within the colony have the same orientation and, secondly, mutual orientation of two phases (their orientation ratio) provides best coupling of two crystal lattices along the habit surface of the lamellas. It is established (Skakov et al., 2013) that Bagaryatsky orientation relationship is observed between ferrite and cementite in nearly every pearlite colony, which is not related to the carbon content in the steel. It should be noted that steel alloying does not affect the crystallographic characteristics within the pearlite (Rakhadilov et al., 2020). The volume ratio of lamellar pearlite is 35% along the material (Table 1). It should be noted that the planning method was used to determine the volume fraction, which is reduced to the measurement of the total area of sections of this structural component on a certain area of the foil. The conclusion of this method is based on the principle of Cavalieri-Anker-Glagolev. It postulates the relationship between the Area (P_S) and Volume (P_V) shares:

$$P_V = P_S, \tag{1}$$

This is one of the fundamental relationships of stereology was postulated by S.A. Saltykov [17;51]. The formula has the following form:

$$P_S = \frac{S}{L^2} = \frac{V}{L^3} = P_V, \tag{2}$$

Where S and V are the area and volume occupied by the corresponding structural component in the sample element representing the L-rib cube.

Ferrite in the initial state of the wheel steel mark 2 is present in the form of non-segmented and fragmented ferrites. The volume share of non-segmented ferrite is ~10%. The volume of fragmented ferrite is 55%. The surface hardening carried out resulted in the formation of packet-plate martensite. The volume share of pacts of martensite is 60%, plate low-temperature martensite is 10%, plate high-temperature martensite is 30%. Martensite transformation does not take place completely due to residual austenite (γ phase). The crystal lattices of residual austenite and the α-phase, regardless of the type and location of residual austenite, are always interconnected by the Kurdyumov–Sachs orientation relation. The volume fraction of residual austenite (γ-phase) in lath martensite is 6.5%. Studies have shown that in wheel steel mark 2 after surface hardening inside all crystals of martensite's there are particles of cementite. Crystalline grates of cement and α-phase are connected among themselves by the orientation ratio of Bagaryatskiy. The volume fraction of cementite in lath martensite is 0.27%, in low–temperature lamellar martensite is 0.95% and in high-temperature lamellar martensite is 2%. (Table 1). The volume fraction of cement (δ_{FeC}) was calculated using:

$$\delta_{Fe_3C} = \frac{V_{Fe_3C}}{t \cdot r^2}, \qquad (3)$$

Where V_{Fe_3C} - average volume of one particle of cement, t - thickness of the foil, r - average distance between particles.

Figure 5 shows that in the initial state, wheel steel mark 2 is a ferrite-perlite mixture and that cementite particles have a lamellar shape, are located almost parallel. There are clear reflexes on the microdiffraction pattern (Figure 5b), which are identified (Figure 5c) like cementite reflexes. The black arrow (Figure 5) indicates the coinciding directions $[011]_\alpha \,||\, [21\bar{1}]$ at the same time $(13\bar{3})_\alpha \,||\, (1\bar{2}0)_C$ – the Bagaryatsky ratio. The dislocation structure in perlite and ferrite represents close dense dislocation networks. Table 1 shows the highest scalar dislocation density belongs to fragmented ferrite, and the lowest – to the α-phase of lamellar perlite. The mean value of scalar dislocation density in steel is $2.5 \cdot 10^{10}$ cm^{-2}. The dislocation structure of all the morphological components is polarized, i.e., dislocations in the material are mostly excess (Popova, Nikonenko, Tabieva, Uazyrkhanova et al, 2020). This is indicated by the presence of bend extinction contours in the bulk material. Based on the size of the bend extinction contours in each morphological component and in the bulk material, the following quantitative parameters are calculated for the fine structure: bending-torsion amplitude of the crystal lattice, excess dislocation density and amplitude of long-range internal stresses caused by the excess dislocation density. Using the scalar dislocation density, we can calculate the amplitude of internal shear stresses induced by the dislocation structure known as forest dislocations. Table 1 shows that fragmented ferrite is characterized by the highest quantitative parameters, while perlite possess the lowest. The conditions of $\rho > \rho_\pm$ and $\sigma_{sh} > \sigma_l$ are satisfied for the whole material. This means that the lattice bending-torsion (distortion) of the wheel steel in the initial state is plastic and does not lead to the steel microcracking (Popova, Nikonenko, Tabieva, & Uazyrkhanova, 2020).

Electrolyte-plasma quenching results in martensite transformation, i.e., the formation of lath and lamellar (low- and high-temperature martensite) (Lobanov et al., 2016). Lath martensite (Figure 6a) is a structural formation comprising a number of elongated laths parallel to each other, which form a packet. Lath martensite probably generates from fragmented ferrite. This is supported by the 60% volume fraction of lath martensite i.e. it is the same as the volume fraction of fragmented ferrite in the initial state.

Table 1. Mean quantitative parameters of perlite and ferrite fine structure and bulk material in the initial state

Morphological Components	Mean Quantitative Parameters of Fine Structure					
	P_V, %	ρ, cm^{-2}	$\rho_{\pm}$, cm^{-2}	χ, cm^{-1}	σ_{sh}, MPa	σ_l, MPa
Perlite	35	2.1×10^{10}	1.4×10^{10}	360	285	240
Nonfragmented ferrite	10	2.6×10^{10}	1.5×10^{10}	365	320	240
Fragmented ferrite	55	2.9×10^{10}	2.8×10^{10}	690	340	330
Bulk material	100	2.5×10^{10}	2.1×10^{10}	525	315	285

Notation: P_V – volume fraction, ρ – scalar dislocation density, $\rho_{\pm}$ – excess dislocation density, χ– bending-torsion amplitudes of the crystal lattice, σ_{sh}– shear stress, σ_l – long-range stress.

Figure 5. Electron microscopic image of the fine structure of the initial state of Wheel Steel Mark 2 obtained by Transmission Electron Microscope a – is a light-field image (LP- lamellar pearlite, FF - fragmented ferrite); b – is a micro-diffraction pattern obtained from the area marked on (a) with a circle; c – is its indexed scheme containing the reflexes of the x-phase and cementite

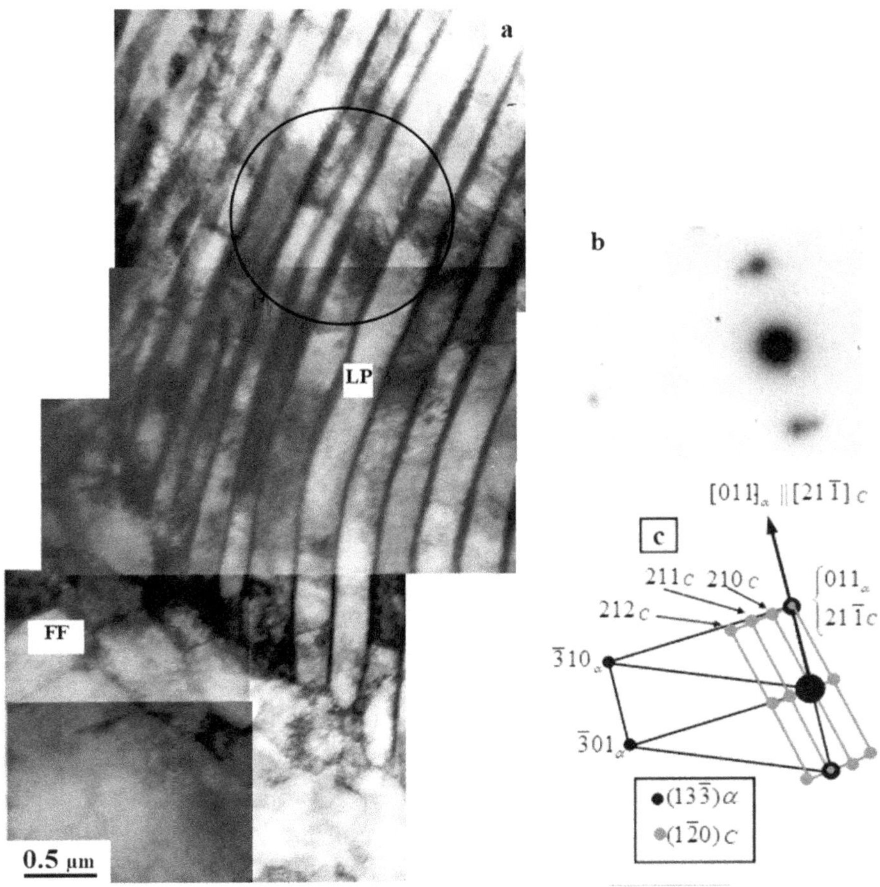

Figure 6. Electron microscopic image of the fine structure of the sample surface after electrolyte-plasma surface hardening obtained by transmission electron microscope

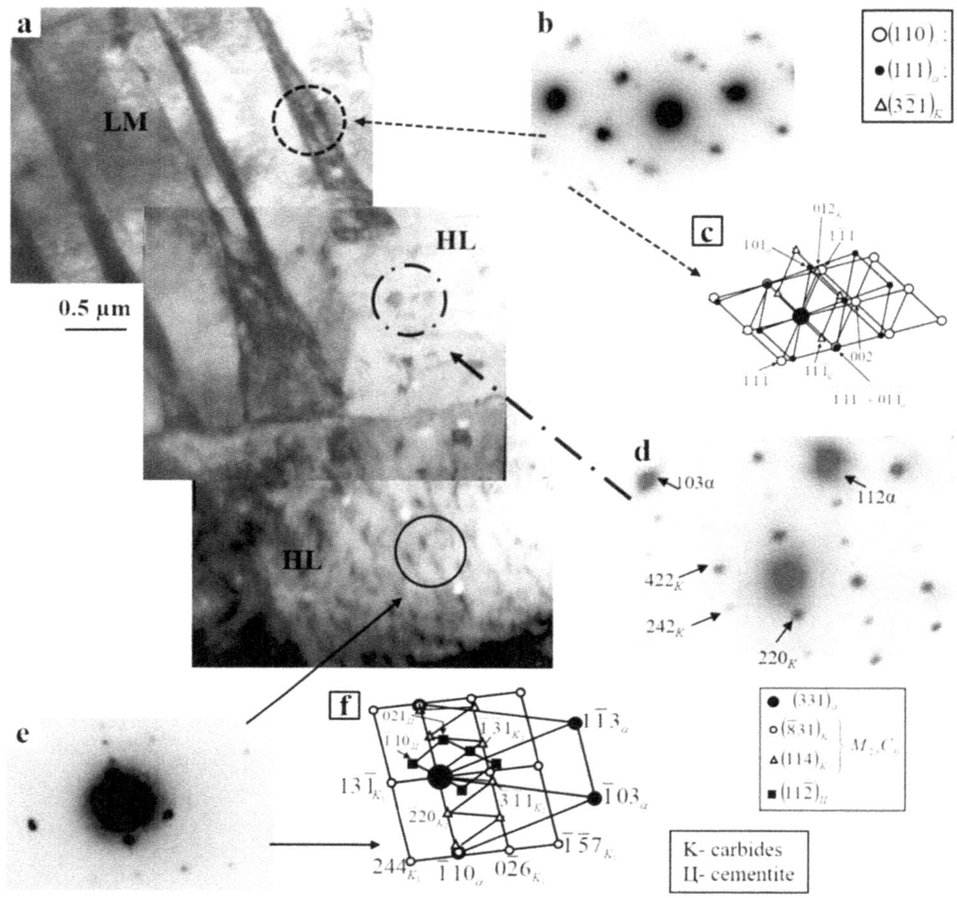

Figure 6 shows a electron microscopic image of the fine structure of the surface of wheel steel mark 2 after electrolyte-plasma surface hardening. In this section, an image of a lath-lamellar or packet-plate martensite is visible: a – a light–field image (LM- lath (packet) martensite, HL-high-temperature lamellar martensite); b, c – microdifraction pattern and its scheme obtained from the area marked on (a) with a dashed line circle; d - the indicated microdifraction pattern obtained from the area marked on (a) with a dashed line with dots circle; e, f – microdifraction pattern and its scheme obtained from the area marked on (a) with a black circle.

Low-temperature lamellar martensite presented in Figure 6a, has the volume fraction of 10%; it is rather large, separately located martensite crystals (laths) having the dislocation structure. In comparison with the initial steel state, low-temperature lamellar martensite probably generates from non-fragmented ferrite, the amount of which is also 10%. High-temperature lamellar martensite presented in Figure 6a, has the volume fraction of 30%; it is large, separately located martensite crystals (laths) propagating through the whole grain and crystals of an arbitrary shape, without a well-defined faceting and grain boundaries. This martensite type probably results from lath perlite. Our findings show the surface hardening results not only in the martensite transformation, but also in the formation of residual austenite (α-phase) having a face centered cubic crystal system. In lath martensite, residual austenite is present as long thin layers on the boundaries of the martensite laths. In low-temperature lamellar martensite, it represents both long thin layers on the boundaries and aciculae inside the lamellas which constitute the twin-type colonies. In high-temperature lamellar martensite, residual austenite is present in the form of aciculae also forming the twin-type colonies. Our early research (Kashchenko et al., 2012) showed that in lath and lamellar martensite, both low- and high-temperature, Fe_3C particles (self-tempering martensite) appear already during the quenching process. This leads to the formation of carbide particles, viz. cementite inside martensite crystals. Cementite particles generating in self-tempering, are similar to those releasing in steel tempering, namely in shape, localization and crystal geometry. The cementite particle size, the volume fraction and mutual arrangement depend on the martensite type (Rahadilov et al., 2021).

According to our studies, cementite particles observed in all martensite crystals of the wheel steel mark 2 after electrolyte-plasma surface hardening, have a thin lamellar structure and arrange in two and three directions relative to the martensite crystal axis. It is found that the fine cementite particles form in self-tempered lath martensite, while the coarsest ones appear in high-temperature lamellar martensite. The surface quenching process thus leads to martensite transformation and steel self-tempering. Besides cementite, the steel structure contains special carbide $M_{23}C_6$ particles having a cubic crystal lattice. These carbide particles are observed on the boundaries of martensite laths and inside low- and high-temperature lamellar martensite. The analysis shows that special carbide $M_{23}C_6$ particles locating on the boundaries of the martensite crystals, result from the γ-phase. This is confirmed by the fact that the particles are observed inside the residual austenite layers, and the XRD patterns of the martensite crystal boundaries demonstrate the γ -phase reflections along with the reflections of $M_{23}C_6$ carbide. According to the XRD patterns, inside low- and high-temperature lamellar martensite with the twin-type colonies, $M_{23}C_6$ carbide particles are also present in the steel structure. After the surface hardening, $M_{23}C_6$ carbide particles which are not associated with the γ -phase are also found in high-temperature lamellar martensite, on α-phase dislocations. Thus, special carbide $M_{23}C_6$ particles appear due to firstly, decomposition of residual austenite and martensite, secondly, partial dissolution of cementite and, thirdly, carbon moving from α-phase crystal dislocations and boundaries. Thus, in all cases, carbon provides the formation of special carbide $M_{23}C_6$ particles from residual austenite, α-phase solid solution, cementite particles and defects of the crystal lattice. It should be noted that the decomposition intensity of α- and γ-phase solid solutions and the formation of the carbide phase in various structural components of the α -phase (lath martensite, inside low- and high-temperature martensite) are different. They also differ within each structural component. This difference depends on many factors, in particular, the initial state of the material (including the material fragmentation and the type of fragments formed) (Bayatanova et al., 2021).

The dislocation structure of all structural martensite components represents dense dislocation networks. According to Table 2, the highest scalar dislocation density belongs to lath martensite, and the lowest – to high-temperature lamellar martensite. This also confirms that lath martensite results from fragmented ferrite, in which the mean value of scalar dislocation density is the highest, whereas high-temperature lamellar martensite releases from lamellar perlite, in which the mean value of scalar dislocation density is the lowest. The dislocation structure in all structural components of martensite is polarized. This is indicated by the presence of bend extinction contours in all martensite crystals. For each morphological component (lath and low- and high-temperature lamellar martensite) and for the bulk material, the quantitative parameters of the fine structure are calculated before and after the surface quenching, using

Table 2. Mean quantitative parameters of fine structure after electrolytic plasma quenching surface of wheel steel mark two

Morphological Components	Mean Quantitative Parameters of Fine Structure					
	P_V, %	ρ, cm^{-2}	$ρ_±$, cm^{-2}	χ, cm^{-1}	$σ_{sh}$, MPa	$σ_l$, MPa
Lath martensite	60	3.3×10^{10}	2.6×10^{10}	650	365	320
Low-temperature lamellar martensite	10	2.5×10^{10}	1.9×10^{10}	465	315	270
High-temperature lamellar martensite	30	2.3×10^{10}	1.6×10^{10}	410	305	255
Bulk material	100	2.9×10^{10}	2.2×10^{10}	560	340	295
Notation: P_V – volume fraction, ρ – scalar dislocation density, $ρ_±$ – excess dislocation density, χ– bending-torsion amplitudes of the crystal lattice, $σ_{sh}$ – shear stress, $σ_l$ – long-range stress.						

the values of scalar dislocation density and the size of bend extinction contours. These parameters are summarized in Table 2.

As can be seen from Table 2, the highest quantitative parameters belong to lath martensite, and the lowest belong to high-temperature lamellar martensite. Electrolytic plasma quenching of the steel surface improves all the parameters of the fine structure. Nevertheless, the conditions of $ρ>ρ_±$ and$σ_{sh}>σ_l$ are satisfied in the whole material. This means that after the wheel steel mark surface treatment, the lattice distortion is purely plastic, as in the initial state, and does not lead to the microcrack formation in the material.

Figure 7. Microstructure along the cross section of Wheel Steel Mark 2 after electrolyte-plasma surface hardening obtained by scanning electron microscope a) a hardened layer; b) a transition layer; c) non-hardened layer; d) general view

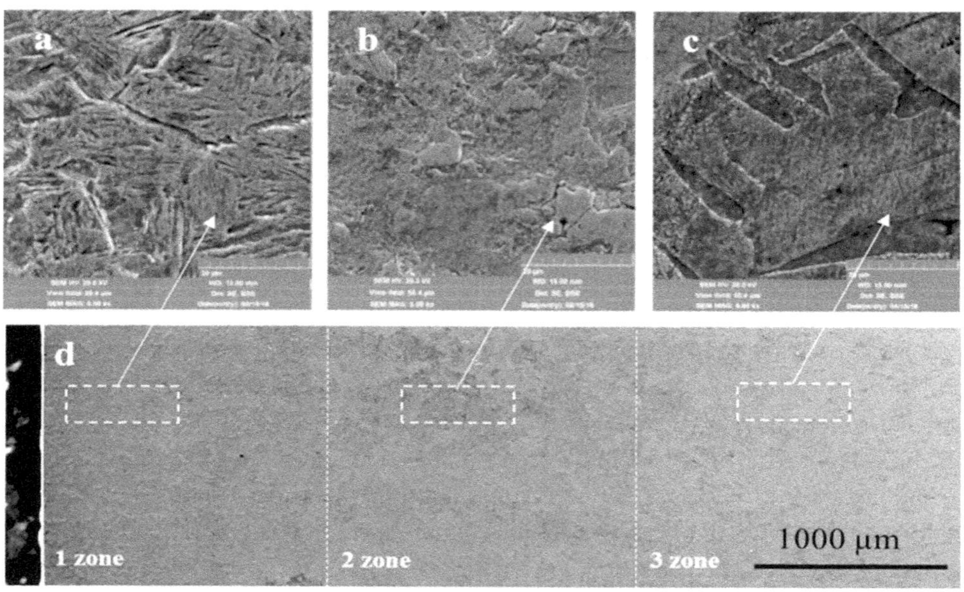

Figure 8. Microhardness over the cross section of wheel steel mark two after electrolytic plasma surface hardening

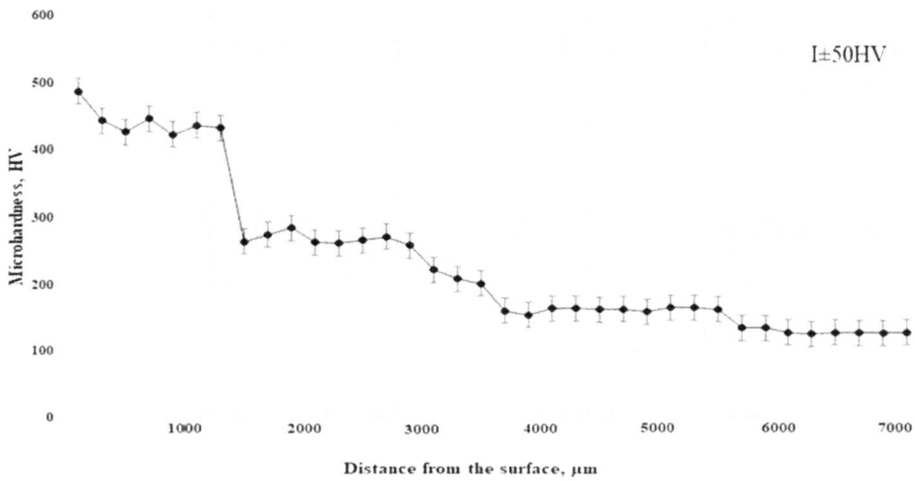

The research results show that electrolytic plasma quenching of the steel surface leads to the formation of the structure consisting of α- and γ-phases and $M_{23}C_6$ carbides. The wheel steel mark 2 represents a mixture of ferrite and carbide. However, according to TEM observations, this structure always contains the boundaries inherited from the martensite structure. Therefore, this structure can also be called lath and lamellar tempered martensite.

Figure 7 shows that the electrolyte-plasma surface hardening led to a change in the microstructure of the cross section, where the zoning of structures typical of electrolyte-plasma treatment is visible. The cross-sectional structure consists of 3 zones: 1 zone - a zone of surface hardening with a thickness of 1000-1500 µm, 2 zone - a zone of thermal influence, 3 zone - a matrix.

As is known, one of the most important properties of the surface layer, which significantly affects the strength characteristics, is microhardness, the value of which in the initial state (matrix) of steel mark 2 is ~ 140 HV. In this work, we studied the changes in the microhardness of a specimen of steel mark 2 over a cross section after electrolytic plasma surface hardening. According to the results obtained, the average microhardness in the surface hardening zone is ~ 420 HV, in the heat-affected zone it is ~ 260 HV and, accordingly, in the steel mark 2 matrix, the microhardness remains unchanged.

The development of promising technologies aimed at improving traditional and creating new methods of impact on the surface to give it the properties required by operating conditions is undoubtedly one of the urgent tasks of modern science and technology. Electrolytic plasma surface hardening, which is one of the varieties of plasma hardening, has recently been developed and intensively studied. This method is characterized by lower energy consumption, simplicity of technological equipment, and large dimensions of the hardened zone. The advantages of the method are a rather high productivity of the process and the ability to harden parts of large mass and complex profile, and the degree of hardening is comparable to plasma hardening. Practice shows that in order to extend the service life of heavily loaded parts and assemblies, surface plasma hardening is rational in terms of versatility, accessibility, environmental friendliness and economic efficiency.

CONCLUSION

This study results of phase composition, fine structure, and properties of wheel steel mark 2 before and after electrolyte-plasma surface hardening has shown that:

- In its initial state, matrix of wheel steel mark 2 represents α-phase the volume ratio of unfragmented ferrite is ~10% and 55% of fragmented one respectively and pearlite with a volume of ~ 35%.
- The morphological components of the structure of wheel steel mark 2 after EPSH are: martensite in the form of packet martensite with a volume fraction of 60%, lamellar low-temperature and lamellar high-temperature martensite's with volume fractions of ~ 10% and ~ 30%, respectively.

Electrolyte-plasma surface hardening of wheel steel led to:

- Martensite transformation, steel self-tempering (formation of cementite particles inside all martensite crystals), diffusion transformation of γ®α phases and release of residual austenite (γ-phase) along low-temperature martensite laths and lamellas and in all crystals of lamellar martensite.
- Formation of special carbide $M_{23}C_6$ particles due to decomposition of residual austenite and martensite, partial dissolution of cementite and carbon removal from α-phase crystal dislocations and boundaries, increase in all the parameters of the steel fine structure.

It was determined that electrolyte-plasma surface hardening leads to a change and hardening of the surface layer of wheel steel mark 2, the thickness of the hardened layer is ~ 1000-1500 μm, and the microhardness increases by ~ 3 times.

REFERENCES

Abuali, G. S., & Mousavi, S. M. (2013). Effect of pulse frequency on microstructure and surface properties of Ck45 steel treated by plasma electrolysis method. *Journal of Alloys and Compounds*, *551*, 415–421. doi:10.1016/j.jallcom.2012.10.176

Bayatanova, L., Rakhadilov, B., Kurbanbekov, S., Skakov, M., & Popova, N. (2021). Fine structure of low-carbon steel after electrolytic plasma treatment. *Materials Testing*, *63*(9), 842–847. doi:10.1515/mt-2020-0119

Grishunin, V. A. (2014). *Increasing the fatigue life of rail steel by electron beam processing*. [Dissertation of Candidate of Technical Sciences. Siberian State Industrial University, Novokuznetsk].

Kashchenko, M. P., Dzhemilev, K. N., & Chashchina, V. G. (2012). The connection of various habituses with variants of orientation relations under γ-α martensitic transformation in dynamic theory. Fundamental problems of modern. *Materials Science*, (1), 50–56.

Kondrat'ev, V. V., Balanovskii, A. E., Ivanov, N. A., Ershov, V. A., & Kornyakov, M. V. (2014). Evaluation of the Effect of Modifier Composition with Nanostructured Additives on Grey Cast Iron Properties. *Metallurgist*, *58*(5-6), 377–387. doi:10.100711015-014-9919-x

Konstantinova, M. V., Balanovskiy, A. E., Gozbenko, V. E., Kargapoltsev, S. K., Karlina, A. I., Shtayger, M. G., & Kuznetsov, B. O. (2019). Application of plasma surface quenching to reduce rail side wear. *IOP Conference Series. Materials Science and Engineering*, *560*(1), 012146. doi:10.1088/1757-899X/560/1/012146

Kozlov, E., Skakov, M., Uazyrkhanova, G., & Popova, N. (2014). Redistribution of second phase particles in surface layers of deformed roll. *Advanced Materials Research*, *1013*, 158–165. doi:10.4028/www.scientific.net/AMR.1013.158

Lobanov, M. L., Rusakov, G. M., Redikul'tsev, A. A., Belikov, S. V., Karabanalov, M. S., Struina, E. R., & Gervas'ev, A. M. (2016). Investigation of Special Misorientations in Lath Martensite of Low Carbon Steel Using the Method of Orientation Microscopy. *The Physics of Metals and Metallography*, *117*(3), 254–259. doi:10.1134/S0031918X1603008X

Nguyen, V. T., Astafeva, N. A., Tikhonov, A. G., Balanovskiy, A. E., & Vu, V. H. (2022). Evaluation of the Hardness and Wear Resistance of Alloyed Coatings From Fastening CuSn/CrxCy Mixture Hardened by Plasma and Laser. *Tribology in Industry*, *44*(3), 518–527. doi:10.24874/ti.1074.03.21.07

Paulmier, T., Bell, J. M., & Fredericks, P. M. (2008). Development of a novel cathodic plasma-electrolytic deposition technique. Part 2: Physico-chemical analysis of the plasma discharge. *Journal of Materials Processing Technology*, (208), 117–123. doi:10.1016/j.jmatprotec.2007.12.101

Popova, N. A., Nikonenko, E. L., Tabieva, E. E., & Uazyrkhanova, G. K. (2020). Structure and phase composition of ferritic-perlitic steel surface after electrolytic plasma quenching. *Russian Physics Journal*, *63*(5), 791–796. doi:10.100711182-020-02099-z

Popova, N. A., Nikonenko, E. L., Tabieva, E. E., Uazyrkhanova, G. K., & Gromov, V. E. (2020). Influence of surface quenching on morphology and phase composition of ferritic-pearlitic steel. *Izvestiya Ferrous Metallurgy*, *63*(11-12), 915–921. doi:10.17073/0368-0797-2020-11-12-915-921

Rahadilov, B. K., Zhurerova, L. G., Sagdoldina, Z. B., Kenesbekov, A. B., & Bayatanova, L. B. (2021). Morphological Changes in the Dislocation Structure of Structural Steel 20GL after Electrolytic-Plasma Hardening of the Surface. *Journal of Surface Investigation. X-ray, Synchrotron and Neutron Techniques*, *15*(2), 408–413. doi:10.1134/S1027451021020300

Rakhadilov B. K., Baizhan D. R., Uazyrkhanova G. K. & Tabieva Y.Y. (2020). *Patent for utility model No. 5365.* State Register of Utility Models of the Republic of Kazakhstan.

Rakhadilov, B. K., Buranich, V. V., Satbayeva, Z. A., Kozhanova, R. S., & Pogrebnjak, A. D. (2020). The cathodic electrolytic plasma hardening of the 20Cr2Ni4A chromium-nickel steel. *Journal of Materials Research and Technology*, *9*(4), 6969–6976. doi:10.1016/j.jmrt.2020.05.020

Rakhadilov, B. K., Tabieva, E. E., & Zhurerova, L. G. (2019). Plasma-electrolytic Nitriding of 0.3Cr-1Mn-1Si-Fe Construction Steel. *METAL 2019 - 28th International Conference on Metallurgy and Materials, Conference Proceedings*, (pp. 1174-1179). Research Gate.

Rakhadilov, B. K., Tabiyeva, Y. Y., Uazyrkhanova, G. K., Zhurerova, L. G., Maulit, A., & Baizhan, D. (2019). Surface modification of steel mark 2 electrolytic-plasma exposure. *Eurasian Journal of Physics and Functional Materials*, (3), 355–362.

Rakhadilov, B. K., Zhurerova, L. G., & Pavlov, A. V. (2016). Electrolyte-plasma surface hardening of 65G and 20GL low-alloy steels. *Bulletin of the University of Karaganda*, *4*(84), 8–13.

Rakhadilov, B. K., Zhurerova, L. G., Scheffler, M., & Khassenov, A. K. (2018). Change in high-temperature wear resistance of high-speed steel by plasma nitriding. *Bulletin of the University of Karaganda*, *3*(91), 59–65.

Skakov, M. K., Uazyrkhanova, G. K., & Popova, N. A. (2013). Electron microscopic analysis of 30CrMnSiA steel surface layers after hot deformation. *Applied Mechanics and Materials*, *395-396*(396), 336–340. doi:10.4028/www.scientific.net/AMM.395-396.336

Tabieva, E. E., Zhurerova, L. G., & Baizhan, D. (2020). Influence of Electrolyte-Plasma Hardening Technological Parameters on the Structure and Properties of Banding Steel 2. *Key Engineering Materials*, *839*, 57–62. doi:10.4028/www.scientific.net/KEM.839.57

Wang, L., Pyzalla, A., Stadlbauer, W., & Werner, E. A. (2003). Microstructure features on rolling surfaces of railway rails subjected. *Materials Science and Engineering A*, *359*(1-2, A359), 31–43. doi:10.1016/S0921-5093(03)00327-7

Zhang, P., Nie, X., & Northwood, D. O. (2009). Influence of coating thickness on the galvanic corrosion properties of Mg oxide in an engine coolant. *Surface and Coatings Technology*, *203*(20-21), 3271–3277. doi:10.1016/j.surfcoat.2009.04.012

Chapter 6

Electrolytic Plasma Treatment of Turkurite-Phase in High-Speed Steels:

High Temperature Testing of Nitrided Layers High-Speed Steel

Zarina Aringozhina
https://orcid.org/0009-0001-8428-4033
D. Serikbayev East Kazakhstan Technical University, Kazakhstan

Yerkezhan Tabiyeva
https://orcid.org/0000-0002-9726-7187
D. Serikbayev East Kazakhstan Technical University, Kazakhstan

Bauyrzhan Rakhadilov
PlasmaScience LLP, Kazakhstan

Waqar Ahmed
University of Lincoln, UK

Didar Yeskermessov
https://orcid.org/0000-0002-2206-8132
D. Serikbayev East Kazakhstan Technical University, Kazakhstan

ABSTRACT

Scanning electron microscopy (SEM), EBSD-analysis, and X-ray structure (XRD) analysis were used to investigate the microstructure, morphology, elemental composition, phase composition, and crystal structure of M2 high speed steel after standard thermal treatment. It has been shown that the microstructure of the M2 high speed steel after hardening and three-time tempering consists of tempered martensite and solid carbide M6C-type and MC spherical shape. The volume fraction of carbides and their distribution have been defined. The main carbides in the study of steel after heat treatment – M6C and MC carbides who have a complex FCC lattice and space group Fd3m, have been established by X-ray structural analysis. Carbides are homogeneous and single crystal. EBSD-analysis method with the support of X-ray structural analysis has established that the bright carbides spherical shape M6C correspond to the composition Fe3W3C, and grey carbides spherical shape MC correspond to the composition of the VC.

DOI: 10.4018/978-1-6684-6830-2.ch006

INTRODUCTION

This study focused on the changes in the structure, phase composition of the surface layers of M2 high speed steels resulting from electrolytic-plasma nitriding treatment. It is well known (Parfenov et al., 2007) that treatment by electrolytic-plasma action causes significant changes in the structural-phase states of the material in thin surface layers occur due to the physical action of low-temperature plasma ions and electric discharge. Structural phase transformations occur under conditions far from thermodynamically equilibrium states and make it possible to obtain modified surface layers with a unique set of physical and mechanical properties (Gupta et al., 2007; Pogrebnjak et al., 2016).

High cutting properties of high-speed steels (high heat resistance in the presence of high hardness and wear resistance) are achieved through special alloying and complex heat treatment with a certain phase composition (Plotnikov et al., 2016). The main alloying elements of high-speed steels are carbon, tungsten, molybdenum, vanadium, and chromium. These elements under certain temperature-time conditions form particles in the steel carbide phase strengthening the material (Suminov,, 2005). This leads to a reduction in the carbon content in martensite and formation of ultramicroscopic carbides. These carbides play an important role in mechanical properties of steel such as hardness, wear resistance and heat resistance. Microstructure analysis, of steel after quenching followed by tempering, can be used to evaluate its properties under operatingconditions. Studies linking the microstructure of a material with its physical and mechanical properties have been carried out widely. However, little work has been done on the carbide phases, which play an important role in mechanical properties of M2 high-speed steel. Therefore, microstructures of high-speed steel and its carbide phases is very important. This study of changes in the structural-phase state of high-speed steels due to electrolytic-plasma nitriding is of significant scientific and practical interest.

MATERIALS, EQUIPMENT, AND RESEARCH METHODS

M2 high-speed steel was used as the material in this study. The use of high-speed steels for cutting tools makes it possible to increase the cutting speed by several times, and tool life by dozens of times (Petrova, 2001). The main distinguishing feature of high-speed steels is their high heat resistance or red hardness (600-700°C) in the presence of high hardness (63-70 HRC) and tool wear resistance. The unique properties of high-speed steels are achieved through special alloying and complex heat treatment, providing a certain phase composition (Moiseev, & Grigor'ev,, 2004). Table 1 shows the chemical composition of M2 high speed steels.

The choice of research materials is also justified by the fact that M2 high-speed steels are common in metalworking, typical high-speed steels of moderate heat resistance. Currently, to reduce the consumption of expensive and scarce elements, especially tungsten, sparingly alloyed steels are used. Of these,

Table 1. Chemical composition of M2 high speed steel (GOST 19265-73)

Grade Steel	C	Mn	Si	Cr	W	V	Co	Mo	Ni	S	P
M2	0.82-0.9	0,20 - 0.50	0,20 - 0.50	3.80 - 4.40	5.50 - 6.50	1.70-2.10	before 0.50	4.80-5.30	before 0.60	before 0.025	before 0.03

M2 steel has the widest application. Blanks of steel samples for research in the form of parallelepipeds with dimensions10×30×30 mm^3 cut out from bars of M2 steel were used. The samples were subjected to standard heat treatment: quenching from 1230°C in oil and subsequent tempering at 560°C three times for a duration of 1 hour each and cooling in the air (Petrova, 2004).

Optical metallography was used to determine the structure. For metallographic analysis, an optical light microscope "ALTAMI-MET-1M" was used. X-ray diffraction studies of steel samples were performed using known methods of XRD on X'PertPRO diffractometers. Diffraction patterns were taken using CuK$_\alpha$ radiation (λ=1.540598 Å) at a voltage of 35 kV. In the D8 ADVANCE diffractometer, reflection from a flat pyrographite monochromator was used in front of the detector to isolate a narrow portion of the spectrum (monochromatization), and K$_\alpha$ radiation with certain wavelengths was selected into the detection unit.The interpretation of the diffraction patterns was carried out manually using standard methods and the PDF-4 database, and the quantitative analysis was performed using the Powder Cell program. The morphology and elemental composition of the sample treated in electrolytic plasma was studied in the engineering laboratory "VERITAS" of the VKTU named after D. Serikbaev on a JSM-6390LV scanning electron microscope - JEOL (Japan), with an INCAEnergy energy-dispersive microanalysis attachment from OXFOR D Instruments. Studies of the phase composition of carbide phases and their sizes were carried out by EBSD analysis (analysis of backscattered electrons) on a system with electron and focused ion beams Quanta 200 3D.

RESULTS AND DISCUSSION OF RESEARCH

To establish the main regularities of structural-phase transformations in high-speed steels during electrolytic-plasma nitriding, the structural-phase states of M2 high speed steel samples of high-speed steels in the initial state were studied.

Figure 1 shows the microstructures of M2 steel i.e. after standard heat treatment. The microstructures of M2 steels consist of tempered martensite and special carbides (Figure1). In previous studies (Skakov, Rakhadilov, and Sheffler, 2013; Skakov, Rakhadilov, and Karipbayeva, 2013), we found that martensite and special carbides are present in the steel structure after standard heat treatment. No residual austenite was observed in the matrix. The preliminary estimate was confirmed by X-ray diffraction phase analysis.

Figure 1. Microstructure of M2 high-speed steel before (a) and after (b) of electrolytic-plasma nitriding

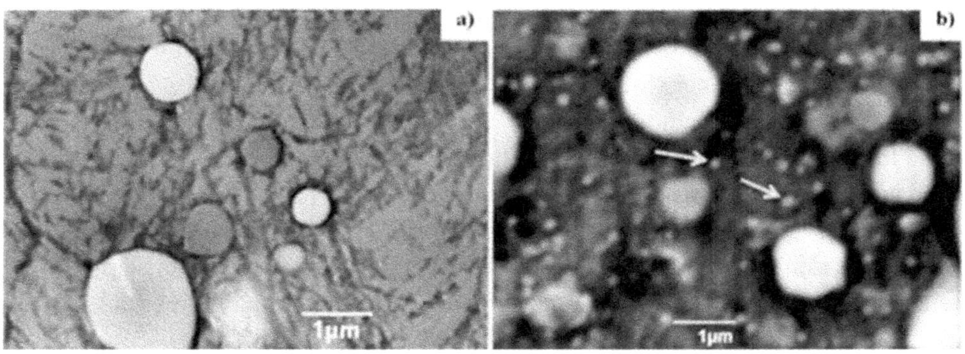

Figure 2. Surface microstructure and distribution map of alloying elements of M2 high speed steel before nitriding

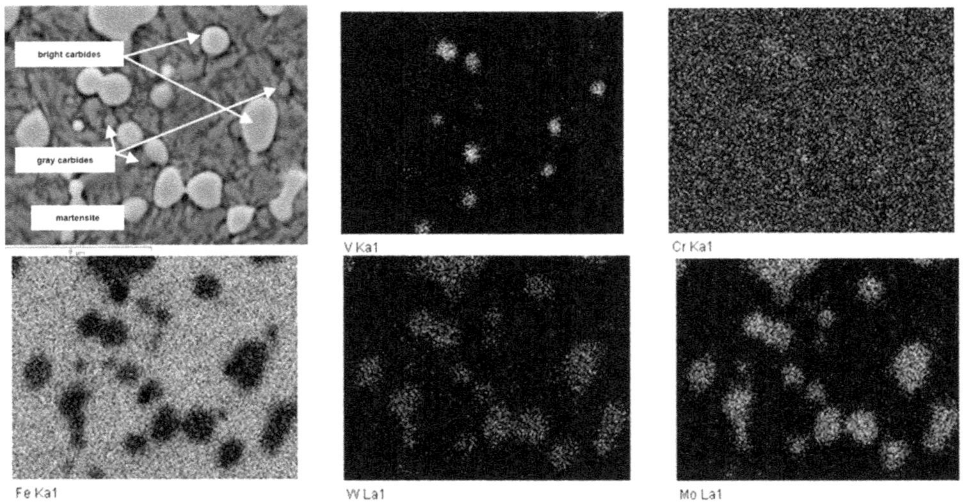

To reveal the composition of carbides and their distribution, a map of the distribution of alloying elements in the steel structure was obtained. The general nature of the distribution of alloying elements in the structure of M2 steels is shown (Figures 1). Light spherical carbides contain tungsten and molybdenum, and gray carbides is enriched with vanadium. The maps of the distribution of alloying elements confirmed that two types of carbides are present in M2 steels - light and dark. It should be kept in mind that V, W, Mo and Cr are carbide-forming elements. The carbides of these metals have a high binding energy and stability (Erkinbekkyzy et al., 2020; Kremnev,, 2008; Skakov, Kurbanbekov, Tabieva et al, 2013; Skakov, Rakhadilov, & Karipbaeva, 2013). Hence, the most of the alloying elements are in carbides, and not in solid solution.

To determine the elemental composition of particles of precipitated carbides and matrix (martensite), microprobe analysis was carried out (Figure 2). Table 1 shows alloying elements in carbides and matrix for steel M2. Tungsten, molybdenum, vanadium, and chromium form special carbides in steel: M_6C based on tungsten and molybdenum, MC (special carbides) based on vanadium and $M_{23}C_6$ based on chromium. The results of mapping and microprobe analysis show that the structure of M2 steel after standard heat treatment contains M_6C and MC carbides and no $M_{23}C_6$ carbides, which is in good agreement with the literature data (Kremnev,, 2008; Vorob'yeva, & Skladnova,, 2003). However, some studies (Guenzel et al., 2000; Ivanov et al., 2002) show that after the standard heat treatment, only M_6C type carbide particles are present in the structure of M2 steel. This could be due to the small volume fraction of particles of MC type carbides and the similarity of these particles with the matrix, which makes it impossible to detect them. The methods used in these works have limitations in the detection of carbide particles with a low concentration. Therefore, in this study, along with X-ray diffraction phase analysis, special methods of scanning electron microscopy are used.

The supposed observed carbide M_6C is between the formulas $Fe_3(W, Mo)_3C$ - $Fe_3(W, Mo)_2C$. In other words, together with tungsten and molybdenum atoms, M_6 carbide can contain up to 2/3 iron atoms out of the total number of metal atoms. In addition, chromium and vanadium atoms can be dissolved,

Table 2. The content of alloying elements in the structural components of high speed steels after heat treatment

Structural Components	Content of Elements, % (wt.)				
	V	Cr	Fe	Mo	W
M2 high speed steel					
Bright carbides	3.42	3.31	30.85	26.05	36.37
Gray carbides	26.47	4.45	30.82	16.62	21.64
Martensite	1.33	4.62	84.50	4.25	5.30

Figure 3. X-ray diffraction patterns of high-speed steels in the initial state (after heat treatment)

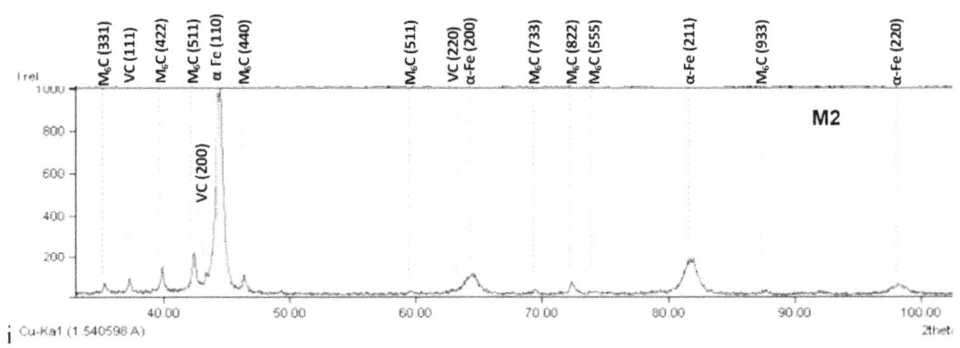

which replace iron atoms. According to the results of microprobe analysis, it can be assumed that the gray carbide particles are vanadium-based MC carbides.

Figure 3 shows the diffraction patterns of M2 steel. XRD analysis showed that in the initial state, i.e., after heat treatment, the structure of M2 steel contains an α-phase and carbides M_6C, MC. Thus, XRD confirmed that the main carbides in the steel samples are M_6C and MC carbides. It has been shown that M_6C type carbides, which have a complex FCC crystal lattice and Fd3m space group, correspond to the Fe_3W_3C composition. The MC type carbides, have a cubic crystal lattice and Fm3m space group, corresponding to the VC (vanadium carbides) composition. In this case, it should benoted that the M_6C type carbide can have both Fe_3W_3C and Fe_3Mo_3C forms. One of the advantages of XRD is that the peak position and lattice parameters can be determined accurately. Nevertheless, when studying individual carbides, it is advisable to apply the EBSD analysis. Therefore, to confirm the results of X-ray diffraction analysis, the crystal structure of M_6C and MC carbides was studied by EBSD analysis using a reflected electron detector on a scanning microscope (system) with electron and focused ion beams.

Figure 4 shows the results of an EBSD analysis of the surface of M2 steel. The EBSD analysis showed that tungsten-rich M_6C carbides are most optimally combined with the Fe_3W_3C cubic phase, while MC-type carbides correspond to the VC phase. However, it is worth noting that in this case Fe_3W_3C other carbide-forming elements are present in the form of M_6C carbides.

In this study we have characterised the structures and carbide phases of M2 high-speed steel in the initial state, i.e. after standard heat treatment. The study of structural-phase states before a certain treatment is necessary in terms of identifying patterns of structure changes and its effect on properties. Since

Figure 4. Results of EBSD analysis of M2 steel

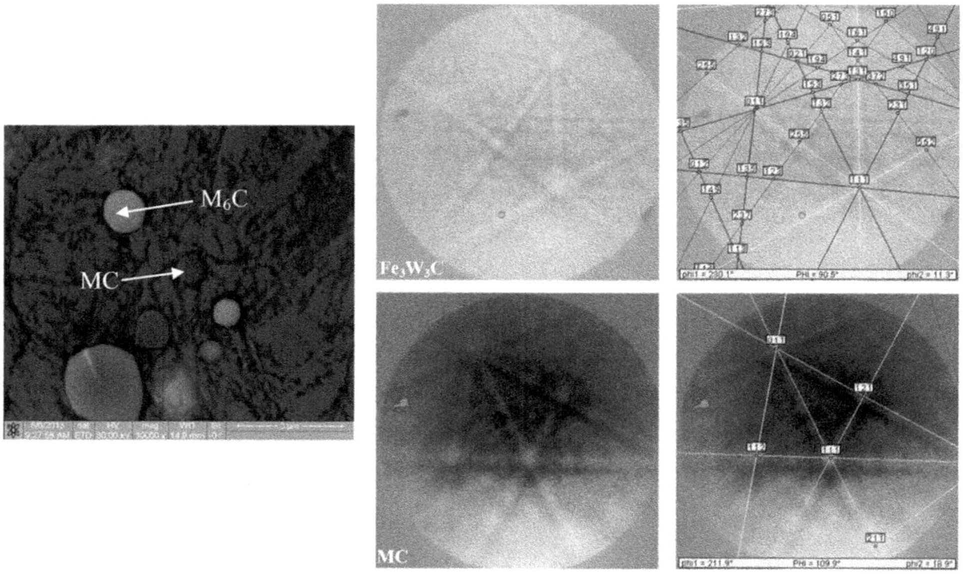

Figure 5. Microstructure of the modified layer of high-speed steels after nitriding a) M2, at 450°C; b) M2, at 500°C; c) M2, at 550°C

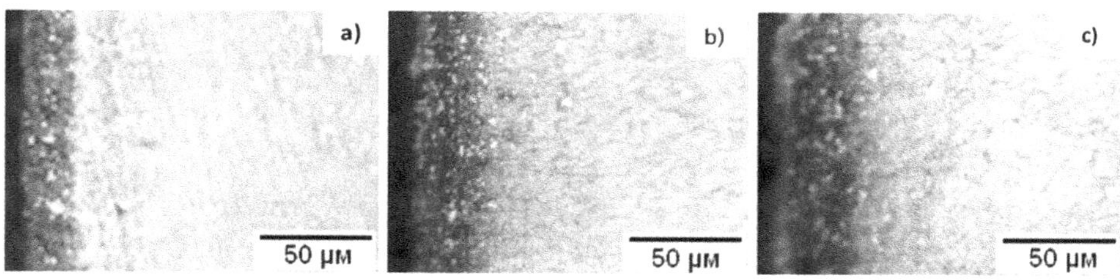

the physical and mechanical properties of high-speed steel are largely determined by the structure and state of the carbide phases and their shape, size, distribution in volume.

Figure 5 shows the microstructures of the modified surface layer of samples of M2 high-speed steels nitrided for 7 minutes. After nitriding, a dark-etched nitrided layer is observed on the surface, which is a zone of internal nitriding. In this case, the dark-etched zone smoothly passes into the base. As expected, after nitriding at temperatures of 450°C, 500°C and 550°C, the thickness of the nitrided layer of M2 steel is 25–40 μm and increases with temperature.

Metallographic analysis has shown that during nitriding by electrolytic-plasma action, diffusion saturation of steel with nitrogen proceeds with accelerated formation of a modified layer consisting only of a well-developed zone of internal nitriding (diffusion layer).

The dependence of the thickness of the diffusion layer of the M2 steel on the exposure time at various nitriding temperatures is evident. Figure 6 shows that the thickness of the modified layer increases with nitriding temperature and holding time. A modified layer is obtained withthe thickness controlled

Figure 6. Growth kinetics of nitrided layers of M2 steel at different temperatures depending on the holding time 1 -450°C; 2 -500°C; 3 -550°C

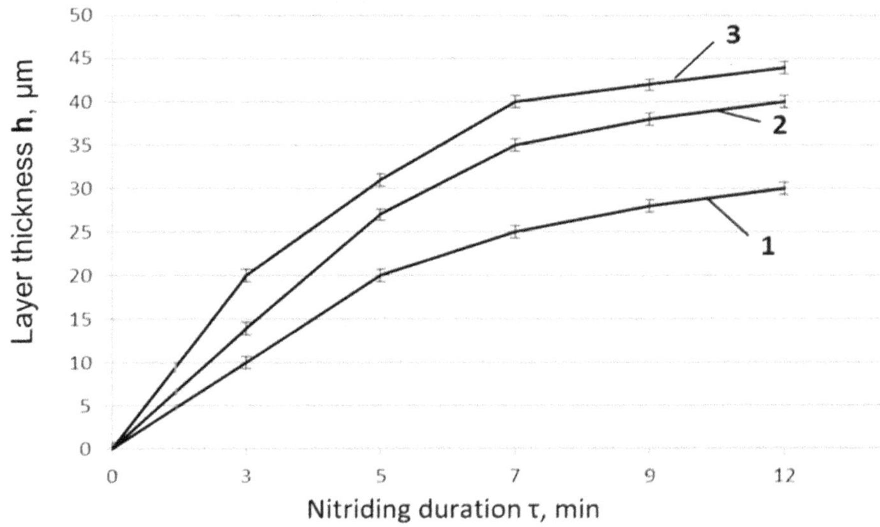

by varying the temperature and nitriding time. Studies (Artinger, 1982; Lakhtin,, 2009), for a cutting tool reveal that the value of h (layer thickness) should not exceed 50–60 μm. Therefore, the modes of nitriding studies are suitable for hardening cutting tools. The optimal time for electrolytic-plasma nitriding for high-speed steel is 7 minutes. Hence, only samples nitride for 7 minutes were studied further.

Figures 7 show SEM images of the surface of M2 steels before and after nitriding for 7 minutes. SEM showed that after electrolytic-plasma nitriding at 550°C, fine particles are formed on the surface of M2 steel (Figure 7 d, fine particles are marked by arrows). And when nitriding at 450°C and 500°C, they are not observed (Figure 7. b, c). The formation of fine nitrides of alloying elements at a temperature of 550°C is possible due to the fact that this temperature corresponds to the tempering temperature of this steel. Since, during the tempering of M2 steel at a temperature of 550–560°C, precipitation hardening occurs as a result of the partial decomposition of martensite and the precipitation of finely dispersed inclusions of strengthening phases (Artinger, 1982).

Figures 8, 9 and 10 show the results of surface mapping of M2 steel after nitriding at 450°C, 500°C and 550°C. The mapping results showed that after nitriding, the surface of M2 steel is saturated with nitrogen. In this case, after nitriding at 500°C and 550°C, areas are observed in the surface where the nitrogen content is higher. These areas are most consistent with the matrix. However, mapping does not allow one to accurately determine which structural component is more saturated with nitrogen, i.e. where nitrides are located. Therefore, to identify the nitrogen content in the structural components and changes in the chemical composition of the surface as a whole, it is necessary to conduct a quantitative energy dispersive analysis of the surface. In order to identify changes in the elemental composition of the surface after nitriding, an energy dispersive analysis of the surface of M2 steel samples nitrided at 450°C, 500°C, and 550°C was carried out.

Figures 11, 12 and 13 show the results of the energy dispersive analysis.

Figure 7. Microstructure of M2 high-speed steel before (a) and after electrolytic-plasma nitriding at temperatures of 450°C (b), 500°C (c), 550°C (d)

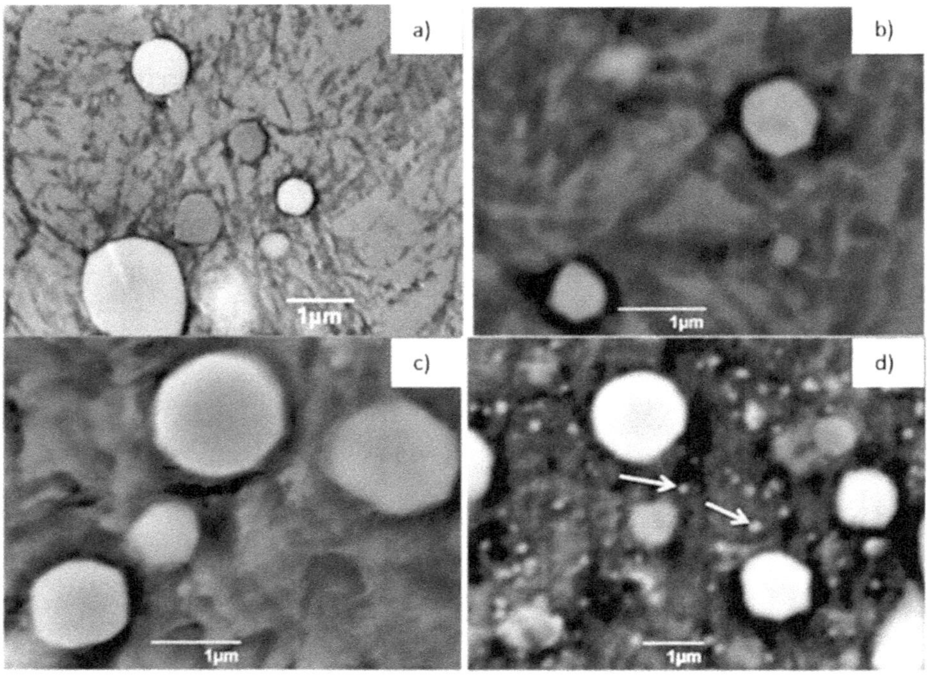

Figure 8. Surface microstructure and distribution map of alloying elements of M2 steel after nitriding at 450°C

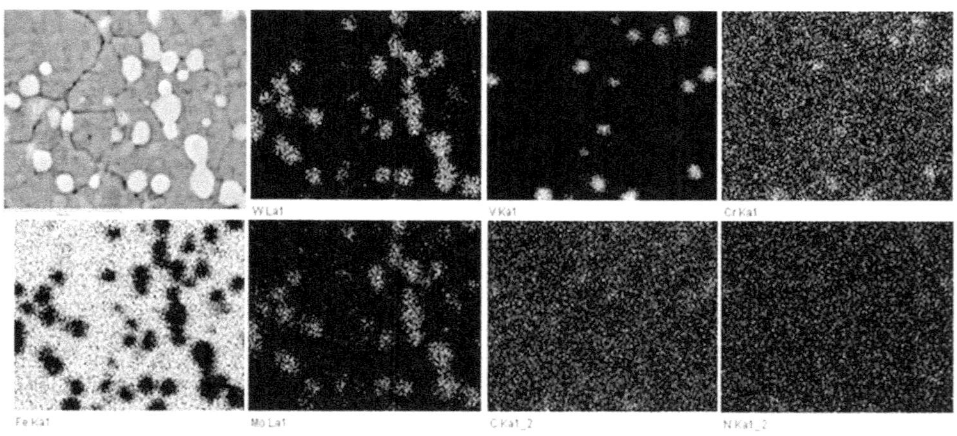

The analysis showed that nitrogen is observed only in the matrix, as expected from the mapping results. The nitrogen content in the surface of M2 steel reaches up to 7% (in weight%). The results of mapping and energy dispersive analysis showed a more uniform distribution of nitrogen over the matrix of the nitrided layer. The observed phenomenon of uniform adsorption and uniform growth of the nitrided layer should be explained by the appearance of a special defective substructure in the grain volume

Figure 9. Surface microstructure and distribution map of alloying elements of M2 steel after nitriding at 500°C

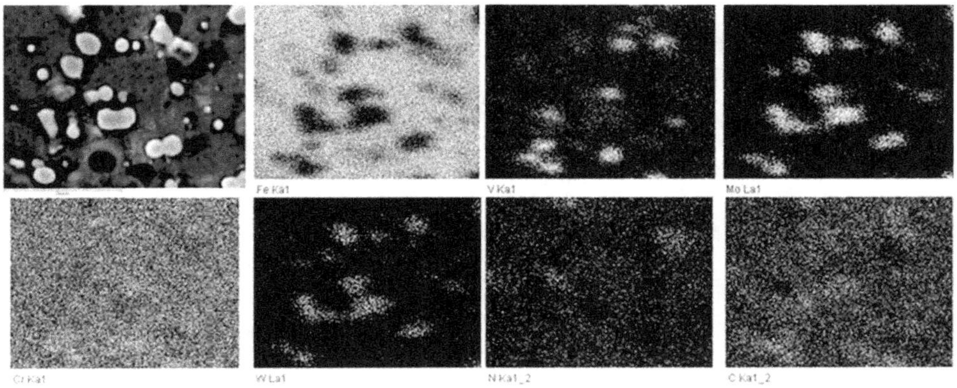

Figure 10. Surface microstructure and distribution map of alloying elements of M2 steel after nitriding at 550°

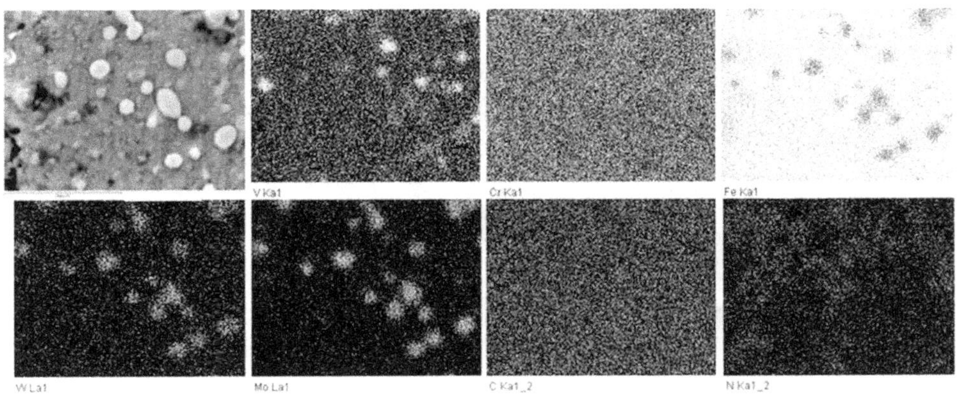

(dislocation-disclination structure of grains in the near-surface zone). The resulting imperfection, apparently, approaches the defectiveness of the grain boundaries. Since, electrolytic-plasma nitriding is carried out under conditions of excessive excitation of the metal surface and subsurface layers. An important factor affecting the growth rate of the nitrided layer and its structure in this process is the exclusion of the predominant role of boundary diffusion. Plasma, accelerating the directed mass transfer of ions to the surface of the sample, creates conditions for the uniform adsorption of nitrogen atoms over the entire surface of the metal, and not selectively along the grain boundaries, as is observed during conventional nitriding by traditional methods.

Thus, the conducted studies have shown that after electrolytic-plasma nitriding, a modified layer is formed on the surface layer of high-speed steels, consisting of a diffusion layer, which is a nitrogenous martensite with fine nitrides of alloying elements. It is determined that the surface is saturated with nitrogen up to 7% (in weight %). Nevertheless, in order to verify the results obtained and to establish the regularities of the formation of the modified layer, it is necessary to study the phase transformations in the surface layers of high-speed steel during nitriding by the X-ray diffraction analysis.

Figure 11. SEM image of M2 steel after nitriding at 450°C (a), characteristic radiation spectra (b) and microanalysis results

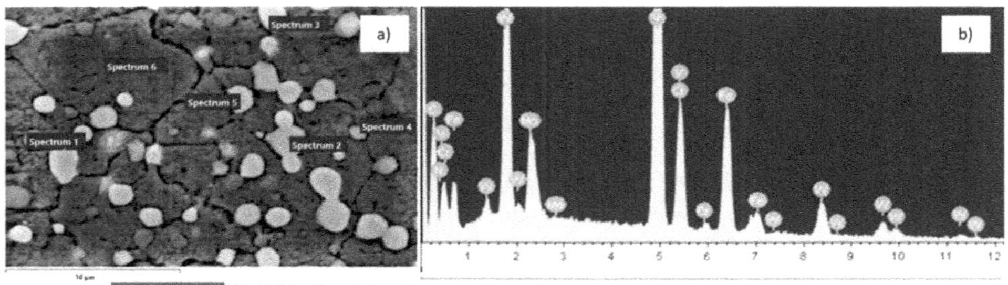

Spectrum	N	V	Cr	Fe	Mo	W	Total
Spectrum 1	0.00	39.66	5.78	17.06	9.87	27.63	100.00
Spectrum 2	0.00	2.31	3.40	26.73	13.74	53.82	100.00
Spectrum 3	0.00	2.42	3.25	26.60	13.55	54.18	100.00
Spectrum 4	0.06	4.06	4.69	78.71	3.33	9.15	100.00
Spectrum 5	0.12	12.33	5.05	65.71	3.91	12.88	100.00
Spectrum 6	0.40	1.03	4.44	85.09	1.97	7.07	100.00

c)

Figure 12. SEM image of M2 steel after nitriding at 500°C (a), characteristic radiation spectra (b) and microanalysis results (c)

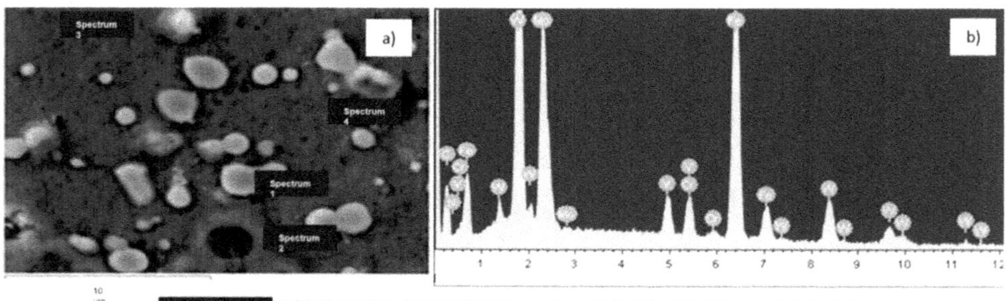

Spectrum	N	V	Cr	Fe	Mo	W	Total
Spectrum 1	0.00	3.25	3.51	32.04	25.91	35.29	100.00
Spectrum 2	6.36	9.86	6.14	62.59	7.12	7.93	100.00
Spectrum 3	5.87	1.45	4.70	80.41	3.60	3.97	100.00
Spectrum 4	6.00	1.56	4.91	78.45	3.97	5.11	100.00

c)

Figure 13. SEM image of M2 steel after nitriding at 550°C (a), characteristic radiation spectra (b) and microanalysis results (c)

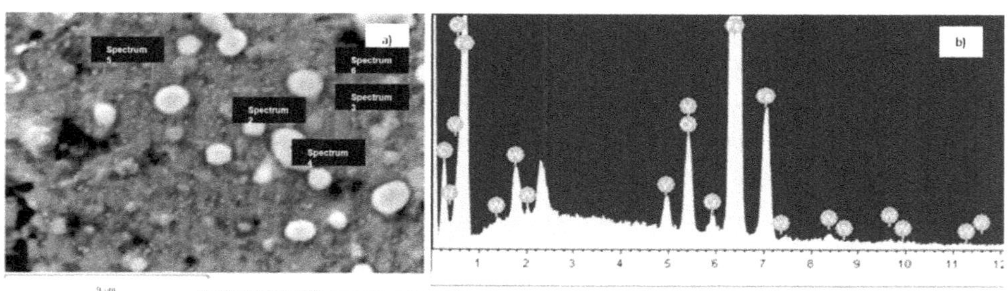

Spectrum	N	V	Cr	Fe	Mo	W	Total
Soectrum 2	0.14	10.46	5.19	72.12	5.29	6.80	100.00
Spectrum 3	7.08	1.82	5.12	77.05	4.47	4.46	100.00
Spectrum 4	0.00	1.47	4.53	33.93	24.26	35.81	100.00
Spectrum 5	5.91	2.47	5.12	78.02	3.94	4.54	100.00
Spectrum 6	3.58	2.59	4.78	81.23	3.21	4.61	100.00

c)

Figure 14. XRD patterns of samples of M2 high-speed steel before (a) and after nitriding at temperatures of 450°C (b), 500°C (c) and 550°C (d)

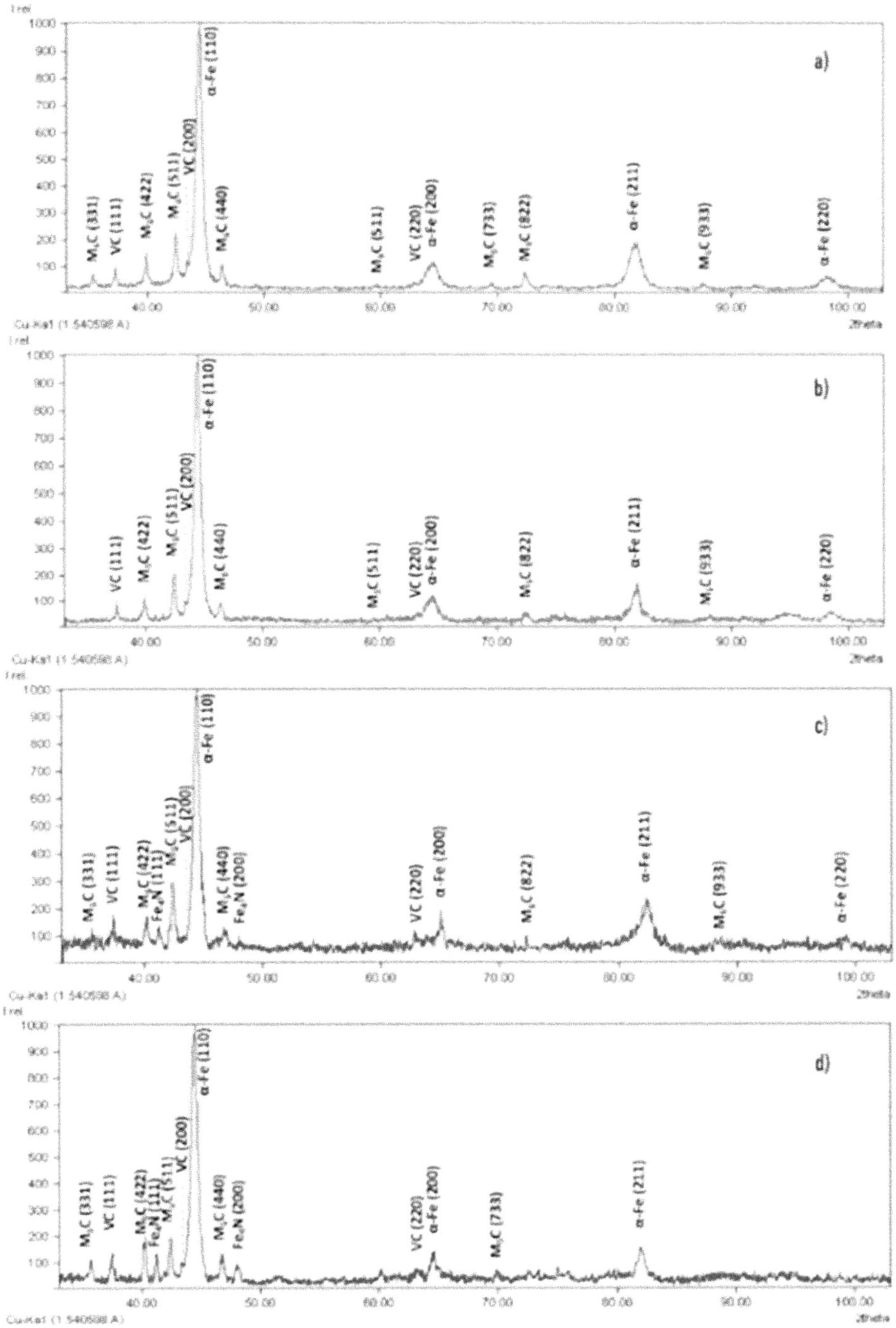

One of the most effective ways to improve the physical and mechanical properties of the surface layer of high-speed steels is the development of optimal nitriding modes. It is practically impossible to choose the necessary nitriding regimes without a detailed study of phase transformations in the material during nitriding.

Figure 14 shows XRD patterns of M2 steel before and after nitriding. XRD analysis show that in the initial state, after standard heat treatment, the M2 steel structure contains martensite (α-phase) and M_6C, MC carbides. After nitriding, broadening, a decrease in intensity, and a shift towards smaller Bragg angles of the interference line (110) of the α-phase are observed. These indicate the formation of a solid solution of nitrogen in iron, i.e., internal nitriding zone (Kuksenova, et al., 2004). On the diffraction patterns of M2 steel samples nitrided at 500°C and 550°C, interference lines of the Fe_4N phase were observed.

After nitriding the martensite structure is more fragmented and characterised by a defective substructure. The resulting imperfection, apparently, approaches the defectiveness of the grain boundaries. The formation of such a substructure of grains during electrolytic-plasma nitriding may be due to the increased energy in the surface and subsurface layers bombarded with ions and neutral atoms of low-temperature plasma during treatment.

More work is needed to study fine structure of steels and its evolution during chemical-thermal treatment, particularly high-speed steels. The formation of secondary phases in the surface layers of high-speed steels during plasma nitriding is of interest.

As noted earlier, the structure of high-speed steel in the initial (heat-treated) state is tempered martensite with hard carbides. XRD analysis established that Fe_4N nitride is formed in the surface layer after nitriding. SEM analysisshowed that after nitriding at 550°C, fine nitrides of alloying elements are formed in the surface of high-speed steel. However, XRD did not show the presence of nitrides of alloying elements, possibly due to the low concentration and shallow depth of their formation (Skakov, Rakhadilov, Karipbayeva et al, 2014; Skakov, Rakhadilov, & Rakhadilov, 2014). Therefore, to reveal the crystal structures of the secondary phases formed during nitriding, the fine structure of M2 steel samples before and after nitriding at 550°C for 7 min were investigated using transmission electron microscopy (TEM).

Previously, X-ray diffraction studies, the results of which are discussed in (Skakov, Rakhadilov, and Sheffler, 2013; Skakov, Rakhadilov, and Karipbayeva, 2013), showed that the structure of M2 steel in the initial state, i.e. after standard heat treatment, consists of α'-phase based on iron with special carbides M6C and MC. The results of TEM studies confirm these data. The fine structure of M2 steel, observed by electron microscopy, is shown in Figure 15. It is clearly seen that the main phase is the α′-phase (martensite) and that carbide particles of globular morphology occupy a significant part of the volume. Morphologically, the carbide phases are more clearly faceted. It should be noted that the α′-phase has a body-centered cubic bcc crystal lattice and can be iron-based solid solutions of interstitial atoms (C, N, B, S, P, etc.) and substitution (Si, Mn, Ni, Cr, Mo, V, W, etc.) at the same time. In the case of the steel under study, the α′-phase is possibly a substitutional solid solution of Cr, W and Mo and an interstitial solid solution of carbon.

After nitriding, the martensite structure is more fragmented and is characterized by a defective substructure. The resulting imperfection, apparently, approaches the defectiveness of the grain boundaries. The possibility of forming such a substructure of grains during electrolytic-plasma nitriding is provided by the increased energy state of the surface and subsurface layers, which are bombarded with ions and neutral atoms of low-temperature plasma throughout the treatment.

Figure 15. Electron microscopic image of the fine structure of M2 steel before (a) and after (b) nitriding at 550⁰C

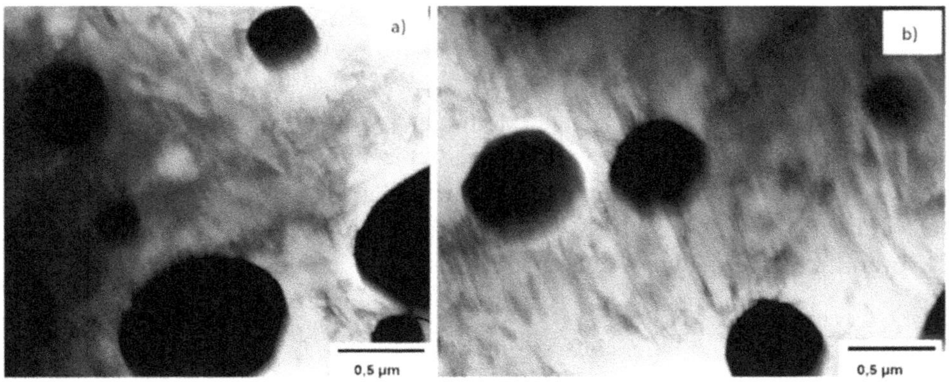

Figure 16. Electron microscopic images of the fine structure of steel M2 before nitriding and their microdiffraction patterns obtained with a matrix (a) and carbides M_6S (b) and MC (c) and schemes for their indexing by Transmission Electron Microscope

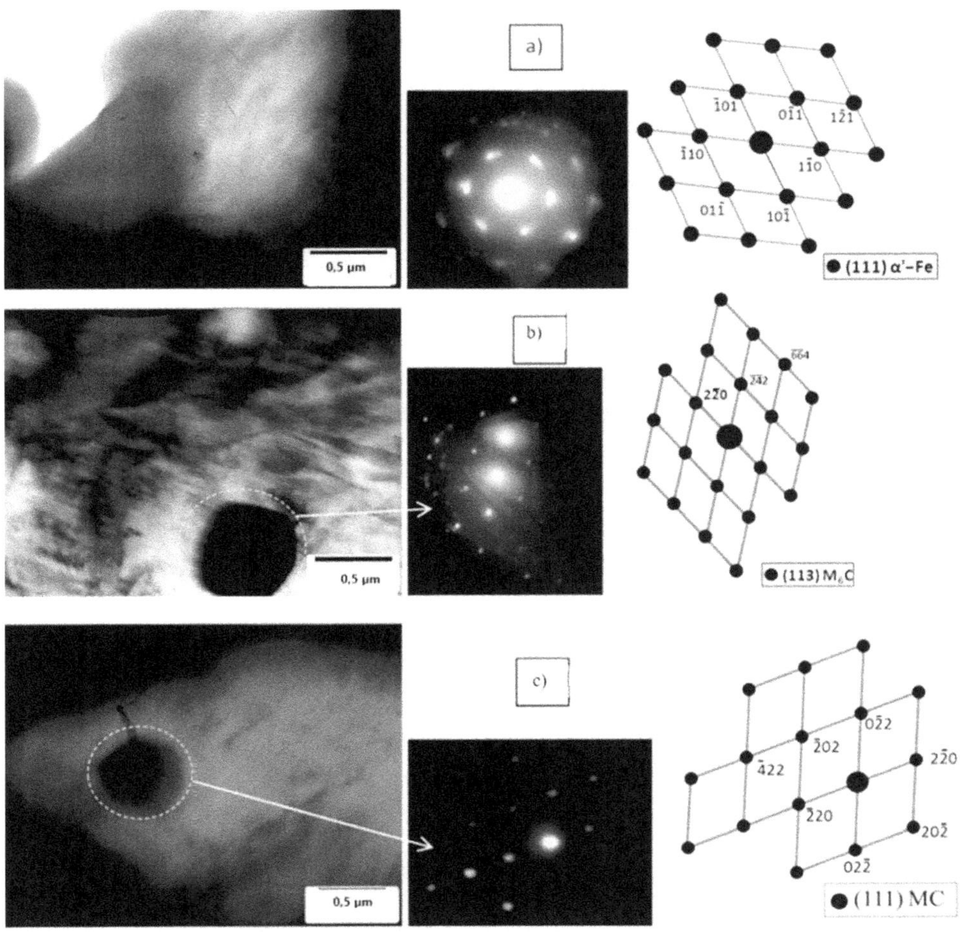

Figure 16 shows TEM images of the fine structure of M2 steel before nitriding and their microdiffraction patterns obtained from the matrix and M_6C and MC carbides and their indexing schemes. Microdiffraction studies have confirmed that the main carbides in steel are M_6C carbide which has a complex FCC crystal lattice and space group Fd3m, and MC carbide, which has an FCC lattice of the NaCl type and space group Fm3m.

Figure 17 shows electron microscopic images of the fine structure of M2 steel after nitriding at 550^{0}C. After nitriding, particles are formed in the surface layer in the form of thin interlayers (Figure 17 a, indicated by arrows). Finely dispersed precipitates of irregular shape (Figure 17 b, indicated by arrows), randomly located in the martensitic matrix. Thus, the formation of fine precipitates of secondary phases in the surface layer after nitriding was confirmed by the electron microscopic method.

Figure 17. Electron microscopic images of the fine structure of M2 steel after nitriding at 5500C (arrows mark particles of the γ′-phase (a) and fine precipitates of secondary phases (b)) by transmission electron microscope

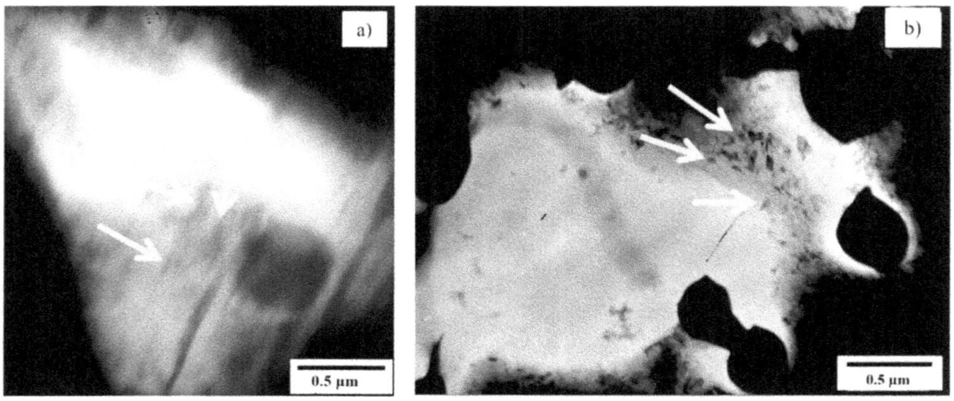

Electron microscopic images of the fine structure of M2 steel after nitriding and their microdiffraction patterns obtained from carbide, matrix and fine nitrides and their indexing schemes are shown in Figure 18.

X-ray diffraction analysis determined that after nitriding at 500 and 550^{0}C, γ′-nitride is formed in the structure of high-speed steels. From the electron microscopic image obtained from the surface of M2 steel nitrided at 550°C, one can see particles of γ′-nitrides in the form of thin layers. The indexing of the microelectron diffraction pattern shown in Figure 18 showed that the formed thin layers (plate-shaped particles) are iron nitrides of the composition Fe_4N (γ′-phase). Iron nitrides of composition Fe_4N (γ′-phase) are formed upon cooling. Thus, in the process of cooling to room temperature, due to the low solubility of nitrogen in the α-phase at low temperatures, a strongly supersaturated solid solution appears. And this leads to the precipitation, preferably on dislocations, of α′-nitride particles of a lamellar shape with a certain orientation, which are subsequently transformed into γ′-nitrides (Pastukh,, 2006). The mechanism corresponds to the release of carbon and the formation of carbides during the tempering of hardened steel.

Electron microscopic studies have shown that after nitriding, fine precipitates with a size of 50–100 nm are formed on the surface of M2 steel, which are in a nanocrystalline state.

Figure 18. Electron microscopic images of the fine structure of the surface layer of M2 steel formed after nitriding and their microdiffraction patterns obtained from carbide (a), matrix (b) and fine nitrides (c) and schemes for their indexing by transmission electron microscope

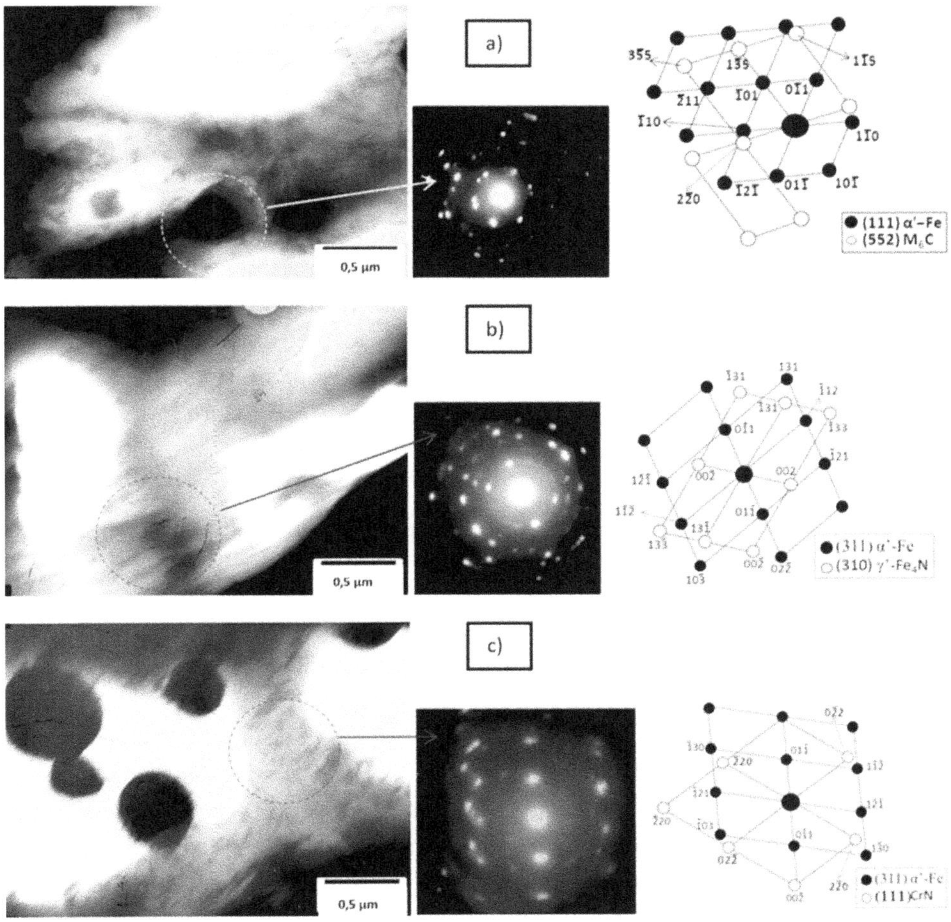

Microdiffraction analysis also showed that the crystal structure of these fine precipitates corresponds to chromium nitride with the composition CrN, which has an FCC lattice of the NaCl type and space group Fm3m. The formation of such CrN nitrides in the surface layers of M2 tool steels after nitriding was observed earlier in (Ramazanov & Vafin, 2011; Shestopalova, 2011). The formation of CrN nitride in high-speed steels is possible in the presence of fluctuations with a sufficiently high concentration (Permyakov, et al., 1972). The presence of concentration fluctuations in solutions in general and in solid solutions in particular is beyond doubt and has been experimentally confirmed in many systems. Under conditions when fluctuations of chromium occur in a medium supersaturated with carbon, they are stabilized at the initial stage due to interaction with carbon (Barabash et al., 1974). This is all the more likely that chromium is a carbide-forming element. Stabilization of fluctuations by carbon creates prerequisites during nitriding for the transformation of the chromium carbon cluster into the nitride phase. The latter is natural in connection with the higher affinity of chromium for nitrogen compared with the affinity for carbon.

Thus, we have characterised all secondary phases formed after nitriding in the surface layers of M2 high-speed steel. These studies are of practical significance because the mechanical properties of high-speed steel are largely determined by the structure and composition of the secondary phases (Gerasimov et al., 2012). It is assumed that the formation of the γ'-phase (Fe_4N) and finely dispersed CrN nitride leads to an increase in the hardness and wear resistance of M2 steel.

CONCLUSION

It has been established that during electrolytic-plasma nitriding of high-speed steels, the process of diffusion saturation of steel with nitrogen occurs with accelerated formation of a modified layer, consisting only of a well-developed zone of internal nitriding, i.e. diffusion layer. It was found that after electrolytic-plasma nitriding at a temperature of 450°C, a modified layer is formed on the surface of high-speed steel, consisting of α′-phase ($Fe_{\alpha(N)}$) and carbides (M_6C and MC), with an increase in the nitriding temperature from 450⁰C to 500⁰C in the modified layer particles of the γ′-phase (Fe_4N) are formed, and at a nitriding temperature of 550⁰C, finely dispersed particles of chromium nitride of composition CrN with an FCC lattice of the NaCl type are formed.

As a result of electrolytic-plasma nitriding at 550°C, a selection of high-speed steels, there was an increase in surface microhardness by 1.6 times. It has been experimentally established that the main increase in the hardness of the diffusion layer occurs after nitriding at 550°C, which is the source of precipitation of finely dispersed nitrides in the supersaturated α-phase, which distort the crystal α-lattice. It has been determined that the important factors influencing the increase in the microhardness of steel is the formation of a diffusion layer from nitrogenous martensite.

It has been established that the main mechanisms that ensure high wear resistance of the nitrided layer of high-speed steels obtained by electrolytic-plasma nitriding are, firstly, the formation of a modified layer consisting only of a diffusion zone (i.e., an internal nitriding zone), and secondly, precipitation in diffusion layer, preferably on dislocations of plate-like γ'-nitrides; thirdly, the formation of fine particles of chromium nitride with the composition CrN in the diffusion layer. It has been determined that the modified layer, consisting of a nitrided α′-phase with a fragmented substructure, M_6C and MC carbides, excess particles of the γ′-phase (Fe_4N) and finely dispersed chromium nitride, obtained after electrolytic-plasma nitriding at 550°C, is wear-resistant, hard, antifriction and has the property of extremely high wear resistance at high temperatures.

The development of engineering technologies in a wide range depends on the technical level of tool production. Thus, one of the problems of environmental protection in mechanical engineering is the high wear resistance of metalworking tools under various conditions of the loading process when machining parts by cutting. Tool life depends not only on the properties of the material, but also on the increase in surface properties. A high role in checking the properties of properties is constantly increasing, which is observed, high efficiency using the methods of chemical-thermal treatment, the scale and development of new areas – engineering of surfaces, methods of energy and physical chemical effects. The implementation of this study when choosing a material significantly improves the performance properties of the tool, as well as reducing the consumption of natural materials. So, in recent applications for calculating emissions and the manifestation of hardening, the increasing use and production of high speed steels, which measures emissions by emissions of hard alloys. An important role in this is played by the use of protection against detection and explosive hardening, which is associated with the use of resource-saving technologies that help reduce resource costs and increase labor productivity. The most promising method for carrying out methodical hardening is chemical-thermal treatment in an electrolyte plasma, which makes it possible to intensify the saturation process. Reducing the duration of saturation of the process is the subject of close attention to the development of this process.

REFERENCES

Artinger I. (1982). *Instrumental'nye stali i ih termicheskaja obrabotka*, 312.

Barabash R.I., Belobashskiy A.V., & Permyakov V.G. (1974). Tonkaya struktura i uprochneniye azotirovannykh volokon zheleza, legirovannogo khromom. Izvestiya vuzov. *Chernaya metallurgiya,* (10), 118–120.

Erkinbekkyzy, T. E., Zhurerova, L. G., & Daryn, B. (2020). Influence of electrolyte-plasma hardening technological parameters on the structure and properties of banding steel 2. *Key Engineering Materials*, (839), 57–62.

Gerasimov S.A., Kuksenova L.I., & Lapteva V.G. (2012). *Struktura I iznosostoykost' azotirovannykh staley i splavov*, 518. APAN

Guenzel, R., Matz, W., Ivanov, Yu. F., & Rothtein, V. P. (2000). Pulsed electron-beam treatment of high-speed steel current tools: struchire-phase transformation and wear resistance. *1st International Congress on Radiation Physics, high current electronics, and modification of materials*, (3), 303-307.

Gupta, P., Tenhundfeld, G., Daigle, E. O., & Ryabkov, D. (2007). Electrolytic plasma technology: Science and engineering – an overview. *Surface and Coatings Technology*, *201*(25), 87–96. doi:10.1016/j.surfcoat.2006.11.023

Ivanov, Yu., Matz, W., Rotshtein, V., Gunzel, R., & Shevchenko, N. (2002). Pulsed electronbeams melting of high-speed steel: Structural phase transformations and wear resistance. *Surface and Coatings Technology*, *150*(2-3), 188–198. doi:10.1016/S0257-8972(01)01542-0

Kremnev, L.S. (2008) Teoriya legirovaniya i sozdaniye na yeye osnove teplostoykikh instrumental'nykh staley i splavov optimal'nogo sostava. *Metallovedeniye i termicheskaya obrabotka metallov*, (11),18-28.

Kremnev, L.S. (2008). Teoriya legirovaniya i sozdaniye na yeye osnove teplostoykikh instrumental'nykh staley i splavov optimal'nogo sostava. *Metallovedeniye i termicheskaya obrabotka metallov*, (641), 18-28.

Kuksenova, L.I., Lapteva, V.G., & Berezina, Y.V. (2004). *Struktura i iznosostoykost' azotirovannoystali*, (1), 31-34.

Lakhtin, Y.M. (2009). *Materialovedeniye i termicheskaya obrabotka metallov*, (5),448.

Moiseev, V.F., & Grigor'ev, S.N. (2004). *Instrumental'nye materialy*, 248.

Parfenov, E. V., Yerokhin, A. L., & Mattews, A. (2007). Frequency response studies for the plasma electrolytic oxidation process. *Surface and Coatings Technology*, *201*(21), 8661–8670. doi:10.1016/j.surfcoat.2007.04.044

Pastukh, I.M. (2006). *Teoriya i praktika bezvodorodnogo azotirovaniya v tleyushchem razryade*, 364.

Permyakov, V.G., Belotskiy, A.V., & Barabash, R.I. (1972). Struktura i svoystva diffuzionnoy zony pri azotirovanii khromistogo zheleza. Izvestiya vuzov. *Chernaya metallurgiya*, (4), 129–131.

Petrova L.G. (2001). *Vnutrenneye azotirovaniye zharoprochnykh staley i splavov*, (1), 10-17.

Petrova L.G. (2004). *Vysokotemperaturnoye azotirovaniye zharoprochnykh splavov*, (1),18-24.

Plotnikov, S. V., Pogrebnjak, A. D., Yerokhina, L. N., Yeskermessov, D. K., & Erdybaeva, N. K. (2016). Study of nanostructured (Ti-Zr-Nb)N coatings' physical-mechanical properties obtained by vacuum arc evaporation. *IOP Conference Series: Material Science and Engineering, 110*(1), 1-6. (012031). 10.1088/1757-899X/110/1/012031

Pogrebnjak, A. D., Maksakova, O., Kozak, C., Koltunowicz, T. N., Grankin, S., Bondar, O., Eskermesov, D. K., Drozdenko, A., Petrov, S., & Erdybaeva, N. (2016). *Przegląd Elektrotechniczny*, (8), 180–183.

Ramazanov K. N., & Vafin R. K. (2011). *Razrabotka khorosho ionno-stal'nogo azotirovaniya instrumental'noy KH12 v skreshchennykh elektricheskikh i magnitnykh polyakh*, (41), 101-104.

Shestopalova L.P. (2011). *Issledovaniye khimicheskogo sostava i sostava nitrid-oksidnoy zony oksiaz-otirovannogo sloya legirovannykh konstruktsionnykh staley,* (54), 78-81.

Skakov, M., Kurbanbekov, S., Tabieva, Y., & Zamanbekuly, E. (2013). Nitriding and carbonitriding influence on stainless steels surface layers changes. *Applied Mechanics and Materials*, *379*, 105–109. doi:10.4028/www.scientific.net/AMM.379.105

Skakov, M. K., Rakhadilov, B. K., & Karipbaeva, G. S. (2013). Specifics of microstructure and phase composition of high-speed steel R6M5. *Applied Mechanics and Materials*, *404*, 20–24. doi:10.4028/www.scientific.net/AMM.404.20

Skakov M.K., Rakhadilov B.K., Karipbayeva G.S., & Manapbayeva A.B. (2014). *Structural-phase state of R6M5 high-speed steel after thermal processing*, (48), 53-59

Skakov M.K., Rakhadilov B.K., & Rakhadilov M.K. (2014). Wear-resistance of nitrided W-Mo-high speed steel in abrasive wear conditions. *Key engineering materials*, (594 – 595).

Skakov, M. K., Rakhadilov, B. K., & Sheffler, M. (2013). Modification of structure and properties of steel R6M5 at electrolyteplasma treatment. *Advanced Materials Research*, (601), 64–68.

Suminov, I.V. (2005). Mikrodugovoye oksidirovaniye teoriya, tekhnologiya oborudovaniye. *EKOMET*, 368.

Vorob'yeva, G.A., & Skladnova, Y.Y. (2003). *Instrumental'nyye materialy: instrumental'nyye stali i splavy,* 100.

Chapter 7
Detonation-Gun Spraying Technology for Multilayer Coatings:
Obtaining Multilayer Coatings by the Detonation Spraying Method

Didar Yeskermessov
D. Serikbayev East Kazakhstan Technical University, Kazakhstan

Nurgamit Kantay
Sarsen Amanzholov East Kazakhstan University, Kazakhstan

Dauir Kakimzhanov
D. Serikbayev East Kazakhstan Technical University, Kazakhstan

Bauyrzhan Rakhadilov
PlasmaScience LLP, Kazakhstan

Akbota Apsezhanova
D. Serikbayev East Kazakhstan Technical University, Kazakhstan

Waqar Ahmed
University of Lincoln, UK

ABSTRACT

This study investigated the structural-phase, morphology, elemental composition, and tribological properties of coatings systems based on Al2O3, NiCr, and NiCr-Al2O3 fabricated by detonation-gas spraying method in a nitrogen/propane atmosphere. The coatings were characterised using SEM-EDX, XRD, Raman spectroscopy, and hardness measurements, and results analysed are presented. According to the experimental data, an optimal mode was selected that provides an increase in mechanical and tribological characteristics, and a method for detonation spraying surface hardening was also developed. Tribology tests have shown that the wear rates and friction coefficients of the coatings are highly dependent on the degree of filling of the gun.

DOI: 10.4018/978-1-6684-6830-2.ch007

INTRODUCTION

Engineering materials are often exposed to mechanical influences, thermal radiation and aggressive chemical environments. The loss of performance often occurs due to wear, erosion and corrosion of the surface. The life of machine parts and mechanisms can be increased by applying protective coatings or modifying the surface layer. A promising technology is thermal spraying. Thermal spraying combines all the methods in which the coating on the surface of the substrate is formed from heated particles hitting it at high speed. This heating process involves the dispersion and transfer of condensed particles of the sprayed material by a gas or plasma stream to form a layer of the desired material on the substrate. At the same time, the three most important parameters: particle size, particle temperature and their velocity can vary greatly, depending on the spraying method. Under the general name thermal spraying (TS), the following methods are combined: gas flame spraying (wire and powder), supersonic gas flame spraying (wire and powder), cold powder spraying, detonation powder spraying, electric arc wire spraying and plasma powder spraying (atmospheric and vacuum). In essence, gas thermal spraying is very similar to welding, the difference lies in the functional purpose of the transferred material. The purpose of welding is to connect the construction elements of buildings, the purpose of thermal spraying is to protect the surface from corrosion, wear, etc. Gas-thermal methods have made it possible to obtain high-quality, durable coatings (Bannier et al., 2014; Dhakar et al., 2018; Ulianitsky et al., 2018; Zhang et al., 2008). For example, detonation spraying (DS) provides excellent adhesive strength, low porosity and a strong, dense and wear-resistant coating (Gallyamov et al., 2018; Thirumalaikumarasamy et al., 2014).

DS is the most promising and is carried out using a special detonation gun filled with an explosive gas mixture. Powder sprayed materials are used to form the coatings. The detonation process accelerates the powder particles to high speeds of up to 1300 m/s which melt and deposit on the sprayed surface. The advantage of DS over other gas-thermal methods (plasma and flame spraying, HVOF, HVAF) is the low porosity of the coatings obtained, high adhesion strength to the base of the workpiece (high adhesion), low thermal impact, which avoids unwanted thermal stresses, in addition, the sprayed parts are not deformed (Rakhadilov et al., 2019).

Detonation coatings are used in various fields of experimental science and technology. For example, for the oil and gas industry, the DS method has developed a technology for electrical insulation of threaded connections of drilling column elements. DS has proven to be effective in hardfacing applications, so it is used in the aircraft engine industry to treat the edges of the anti-vibration shelves of gas turbine blades in order to increase wear resistance. The high melting point (2044°C) at DS, combined with the specified strength characteristics and low porosity, makes it possible to apply heat-resistant and fire-resistant coatings made of Al_2O_3 to various metals, as well as to carbon and fiberglass. Also, DS technology is of interest in using in the production of ceramic coatings to ensure high dielectric strength, wear resistance of electrical insulating coatings (Algatti et al., 2017; Kaya et al., 2020; Ulianitsky et al., 2011; Zhu et al., 2020).

A combination of high values of adhesion strength, wear resistance, corrosion resistance, etc. are needed to improve the durability of detonation coatings operating under severe conditions. Multilayer coatings are very effective in enhancing performance particularly multilayer with a gradient structure which produces a smooth change in the chemical composition, structure and properties (physical, mechanical, etc.) along the coating thickness (Rakhadilov et al., 2020). A multilayer structure can be an effective method for improving the corrosion resistance of coatings. Interfaces between layers can act as protective barriers in aggressive environments (Pogrebnjak et al., 2009; Pogrebnjak & Tyurin, 2005).

Hence, the aim of this work is to study the effect of process parameters on the properties of NiCr and Al_2O_3 coatings and graded multilayer coatings of NiCr/NiCr-Al_2O_3/Al_2O_3.

MATERIALS, EQUIPMENT, AND RESEARCH METHODS

Powder of NiCr, Al_2O_3 and NiCr-Al_2O_3 (50/50) with a dispersion of 20-45 μm was used as the sputtering material. The coatings were applied to samples made of AISI321 (12Ch18N10T) steel by detonation spraying on a "CCDS2000" (Computer Controlled Detonation Spraying 2000) detonation machine. Figure 1 shows a general view and schematic diagram of the "CCDS2000" detonation facility. The injector was placed on a "CCDS2000" detonation unit with a diameter of 20 mm and a length of 800 mm. The coating is applied with a detonation gun, the barrel of which is filled with an explosive gas mixture, a dosed portion of the powder is thrown into the barrel, and an electric spark excites detonation. The detonation products (heated to 3500-4500 K) heat the powder particles until they melt and throw them at high velocity onto a part mounted in front of the cannon barrel. The reaction time of gases with scattered particles is 2-5ms. The particle velocity can reach 800-1200 ms^{-1}. On impact, micro-welding occurs, and the powder is firmly (on a molecular level) bonded to the surface of the part. After each shot, the barrel is purged with nitrogen to clean the detonation products residue. The required thickness is built up by a series of successive series of shots, during which the object can be moved using a manipulator (Buitkenov et al., 2020). The distance between the barrel and the substrate was 250 mm. The surface of the substrate was subjected to sandblasting and dry cleaning for 5–7 min.

Coatings based on NiCr and Al_2O_3 were obtained at different values of the barrel filling volume. As the amount of explosive mixture increases from 25 to 60% of the barrel volume, the temperature inside the barrel increases. Process parameters NiCr, Al_2O_3 and NiCr-Al_2O_3 coatings are shown in Table 1. The multilayer coating on the bases of NiCr/NiCr-Al_2O_3/Al_2O_3 was also obtained by successive spraying of three kinds of powder: 11-NiCr, 12-mixture of NiCr powders (50%) and Al_2O_3 (50%), 13-Al_2O_3 (Table 1). An acetylene-oxygen mixture was used as a combustible gas, which detonates the process of combustion powder in the barrel. Optimal ratio was chosen of O_2/C_2H_2 = 1.856 to ensure complete melting of the powder. The consumption of working gases at the average frequency of shots of 4 Hz is: acetylene 4–7; propane butane mixture 2–3, 5; oxygen 10–12; nitrogen 10–15 m^3/h. Nitrogen was used as a carrier gas (see Figure 1b).

The dimensions of the steel samples were 75×50×5 mm^3 and the surface roughness was R_a=0.088 μm. The surface was cleaned with electrocorundum (Contracor Eco140S, Wuppertal, Germany) with a grain size of about 20–80 μm, and the roughness of the substrate was about 4–4.5 μm. The surface was chemically wiped for 5–7 minutes and the substrate was dried for 20–30 min, then coated powders. Table 2 shows the chemical composition of AISI321 steel.

The thickness of the coatings was 180 μm. The average size of the formed coating in a single shot was 8–10 μm. Each sample was shot 20 times. Depending on the volume of gas filling the barrel in the detonation device, the degree of explosion will be different, i.e. the larger the gas filling volume, the higher the explosion power at the same gas ratios. The power of the explosion also depends on the type of gases used and their ratio. Experimentally, we have found that when the filling volume is below 40%, the sprayed powder does not have a sufficient speed and temperature. During spraying, with a filling volume of 30%–40%, some of the powders are not applied and, hitting to the surface of the substrate, are pushed back. In order for spraying to occur, the powders must partially or completely melt and have

Table 1. Detonation spraying process parameters of coatings

Samples No.	Powder Name	Barrel Filling, %	Shots No.
Al_2O_3			
1	Al_2O_3	53	20
2	Al_2O_3	58	20
3	Al_2O_3	63	20
4	Al_2O_3	68	20
NiCr-Al_2O_3			
5	Al_2O_3	45	20
6	Al_2O_3	54	20
7	Al_2O_3	63	20
8	NiCr	54	20
9	NiCr	58	20
10	NiCr	65	20
NiCr/NiCr-Al_2O_3/Al_2O_3 (Multilayer)			
11	NiCr	54	20
12	NiCr-Al_2O_3	54	20
13	Al_2O_3	63	20

Figure 1. "CCDS2000" complex detonation sprayer general view (a) and schematic diagram of the installation (b): 1-control computer, 2-gas distributor, 3-mixing-ignition chamber, 4-spark plug, 5-barrel valve, 6-fuel line, 7-oxygen line, 8-gas valves, 9-gas supply unit, 10-breech block, 11-powder batcher, 12-workpiece, 13-manipulator, 14-muzzle of the barrel
Source: Rakhadilov, Sagdoldina, Baizhan et al. (2021)

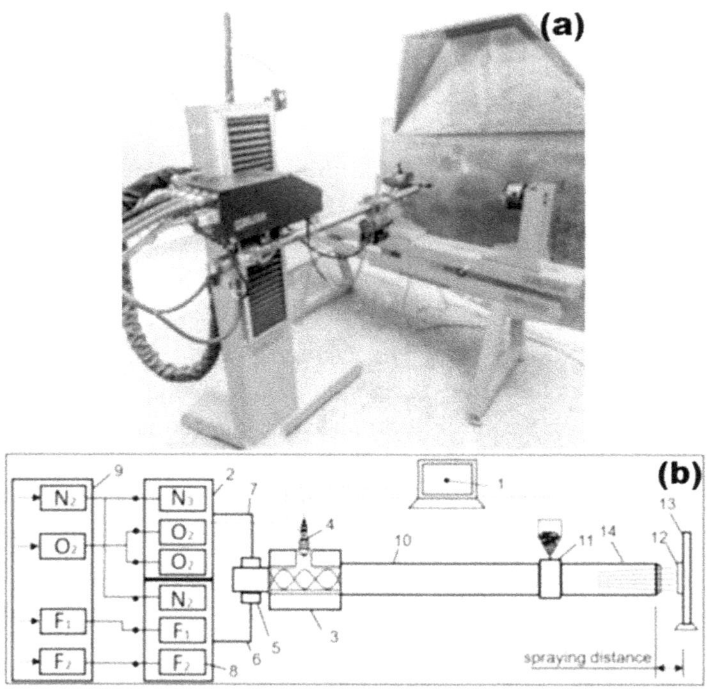

Table 2. Chemical composition of substrate material: AISI321 steel (12Ch18N10T)

C	Si	Mn	P	S	Cr	Mo	Ni	V	Ti	Cu	W	Fe
<0.12	<0.8	<2.0	<0.035	<0.02	17.0-19.0	<0.5	9.0-11.0	<0.2	<0.8	<0.4	<0.2	The rest

a fixed speed depending on the size of the powder. Taking into account the above and the technical characteristics of the detonation installation (maximum filling volume up to 70%), We carried out experiments with filling volumes from 45% to 68%.

The phase composition of the coatings was determined by X-ray diffraction (XRD) on an X'Pert PRO diffractometer (Philips Corporation, Netherlands) with Cu-K_α radiation ($\lambda = 1.541$ Å) at a voltage of 40 kV and a current of 30 mA. Diffraction patterns were decoded using the "High Score" software; measurements were carried out in the 2θ range, equal to 10°–90°, with a step of 0.02 and a counting time of 0.5 s/step. The surface roughness was evaluated in accordance with ASTM D7127–05 by the R_a parameter using a profilometer model 130 (JSC Plant PROTON). The surface of the coatings was examined at optical magnification of ×20 using Altami MET 5S metallographic microscope (Altami LLC). Mechanical properties of the coatings (Young's modulus, nanohardness) were investigated using nanohardness meter NanoScan-4Dcompact. Nanoindentation of the coatings was performed by the Oliver and Farr method using a Berkovich indenter at a load of 100 Mn (ASTM E2546-07). The cross-sectional morphology of the samples was examined by scanning electron microscopy (SEM) using backscattered electrons (BSE) on a JSM-6390LV scanning electron microscope at accelerated stress.

Tribological properties were assessed in dry slip tests using TRB-3 high temperature tribometer (Anton Paar Srl, Switzerland) by a standard ball-disc technique in accordance with ASTM G 133–95 and ASTM G99 (ASTM G133–95, 1995; ASTM G99–05, 2005). The sliding pair was a stationary ball with a diameter of 6 mm and a hardness of 62 ± 2 HRC, made of 100Cr6 steel. Pressed against a rotating disk made of AISI321 steel, respectively, without a coating and with a sprayed coating of Al_2O_3. The contact load was 10N and a sliding speed was 0.2 m/s. The cycle time was 60 min. The tribological properties of the coatings were characterised by the wear rate and the coefficient of friction. To obtain reliable research results tests were carried out on three samples from each batch.

RESULTS AND DISCUSSION

a. Results for Single Layer DS

Figure 2 shows the structure of the substrate surface and Al_2O_3 powder after detonation injection using an optical microscope. Figure 3 shows the microstructure of the surface layer of the Al_2O_3 coating by SEM. The granules in the surface layer of AISI321steel are large/coarse grained (figure 2) and the surface layer of the Al_2O_3 coating is small/fine grained.

Figure 4 shows the microstructure and surface roughness, taken from the cross-section of the Al_2O_3 coating, depending on the filled barrel volume with gas 53%, 58%, 63% and 68%. R_a values ranged from 0.439 to 1.442 μm. This degree of barrel filling does not appear to be greatly affect the roughness of Al_2O_3 coatings. When analyzing the results of the surface roughness of the obtained coatings, it can be seen

Figure 2. Surface structure of the substrate (a) and after detonation spraying Al_2O_3 coating (b)

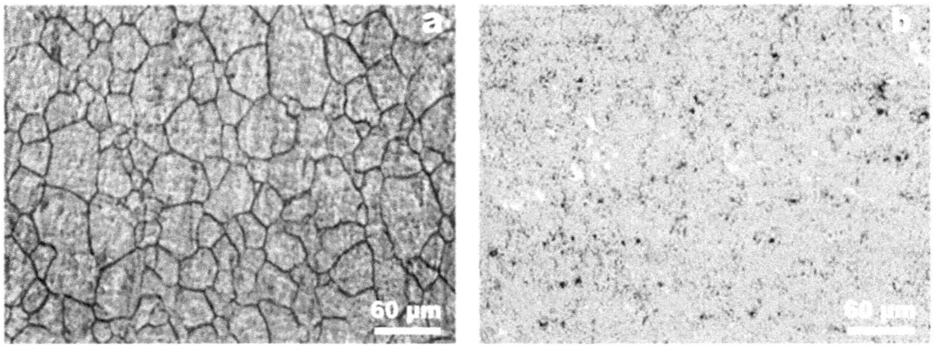

Figure 3. SEM-images of surface (a) and of fracture fractograph cross-section (b) of the Al_2O_3 coatings

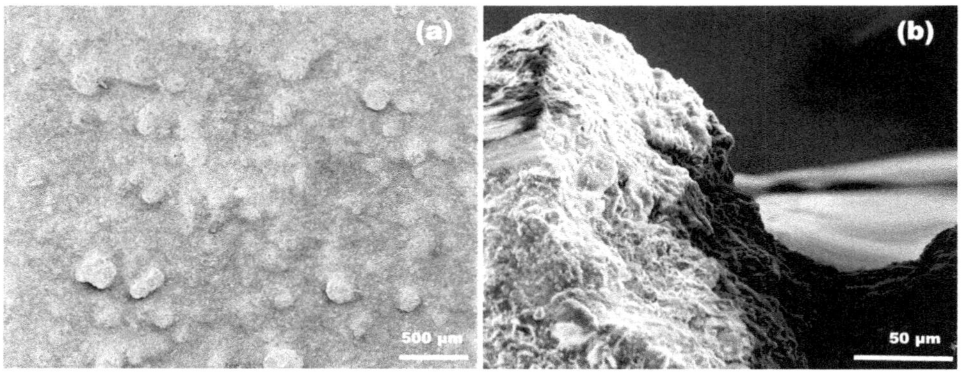

that the filling of the barrel by 53% and 68% has lower Ra compared to 58% and 63%. This difference between the percentage of barrel filling can be characterized by the formation of various Al_2O_3-phases and a specific location on the substrate surface, as well as an increase in the detonation temperature and the impact of a shock wave when a powder mixture collides with the substrate (see figure 7).

Figure 5 shows SEM-image taken from the cross section of the Al_2O_3 coating when 53% of the barrel is filled with gas. Figure 6 shows the results of spectral microanalysis. According to Figure 6, the coating is dense and uniform, as well as the composition of the coating in the spectrum 001–005 consists of pure Al and O atoms, as well as in the spectrum 006 taken from the boundary of the substrate and the coating, it was observed that the substrate material contains Fe, Cr, Ni atoms and the coating contains Al and O atoms.

The SEM image shows that the cross-section of the sprayed coating is not flat and continuous. It follows from the image analysis that the coating structure has an inhomogeneous structure with pores, bubbles, cracks, nodules (brighter spherical inclusions) and a typical layered, wave-like arrangement of structural components. Significant pores (dark area) are visible in the cross-sectional image of the coating. The border between the coating and the base has a characteristic zigzag shape.

Figure 7 shows the diffraction patterns of Al_2O_3 coatings obtained at 53, 58, 63, and 68% of barrel filling respectively. The coatings after detonation spraying are composed of α-Al_2O_3 (ICDD/JCPDS No.

Figure 4. Optical images taken from the cross section of the Al_2O_3 coating and the surface roughness curve: 53% (a), 58% (b), 63% (c), and 68% (d)

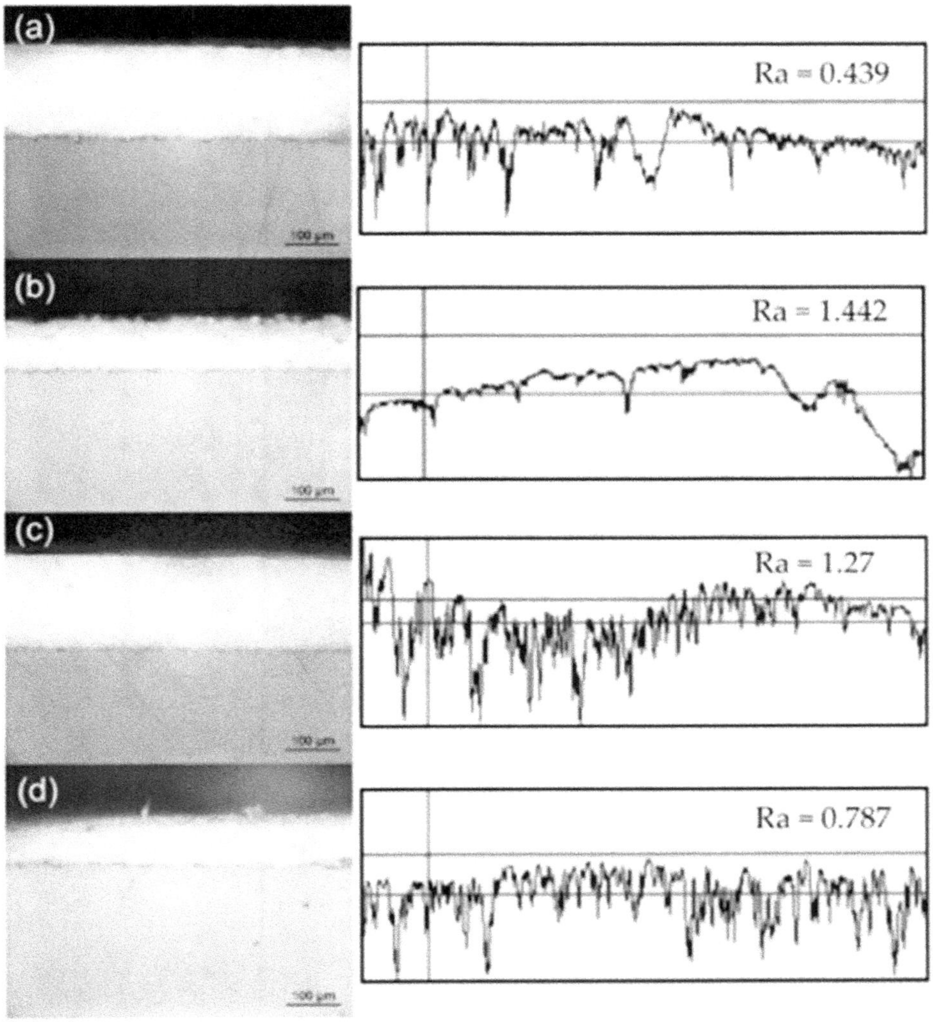

96–230–0376) and γ-Al_2O_3 (ICDD/JCPDS No. 96–154–1583). Corundum α-Al_2O_3 was used as a powder for spraying. The resulting coating mainly consists of γ-Al_2O_3. This is because the powder of Al_2O_3 can undergo several modifications under the influence of high temperatures (Kantay et al., 2020).

Figure 7(a-e) shows that alumina oxide in powder form consists of a single α-Al_2O_3 phase. Figure 7 shows that with an increase in the filling of the barrel with gas from 53% to 68%, the content of γ-Al_2O_3 phase in the coating increases, while the content of α-Al_2O_3 phase decreases. In our opinion, it can be assumed that the γ-Al_2O_3 phase is formed when the barrel is filled by 68% due to an increase in the explosion temperature and the impact of a shock wave during flight along the barrel and subsequent spraying onto the surface. Since, it is known that when the temperature rises to 950–1250°C, γ-Al_2O_3 turns into

Figure 5. SEM-image taken from the cross section of the Al_2O_3 coating when 53% of the barrel is filled with gas

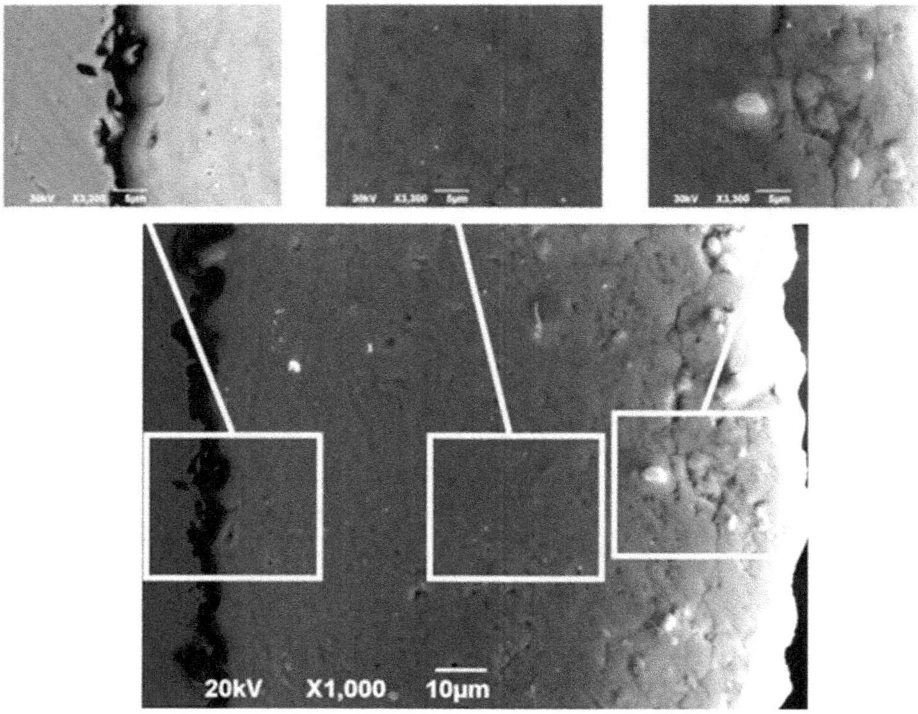

Figure 6. SEM-EDX microanalysis of the cross section Al_2O_3 coating at 53%-barrel gas filling

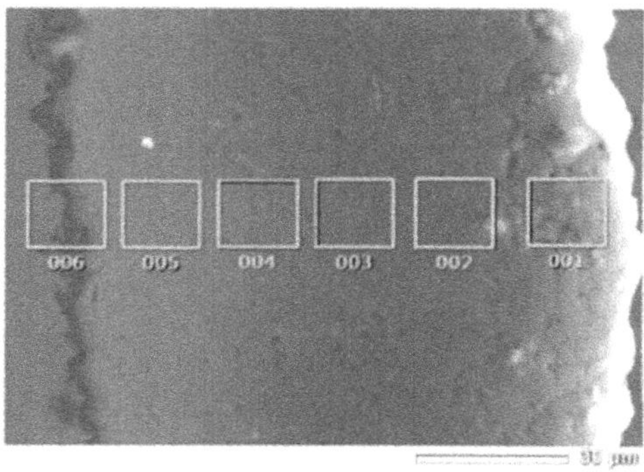

Atom, %						
spectrum	O	Al	Cr	Fe	Ni	Total
001	57.13	42.87				100.00
002	55.90	44.10				100.00
003	54.67	45.33				100.00
004	54.12	45.88				100.00
005	52.98	47.02				100.00
006	22.17	36.91	8.76	28.91	3.25	100.00

thermodynamically stable α-Al_2O_3. This transformation can proceed directly or through intermediate phases such as δ- and θ-Al_2O_3 (Shoja et al., 2021).

Figure 8 shows the tribological tests of Al_2O_3 coatings by the ball-disk method. The wear resistance of the coatings was characterised by the volume of wear. The samples have different values of the volume of wear depending on the degree of barrel filling. Wear gradually increases with an increased filling. The

Figure 7. XRD-pattern of Al_2O_3 coating depending on the filling of the barrel: Al_2O_3 powder (a), 53% (b), 58% (c), 63% (d), and 68% (e)

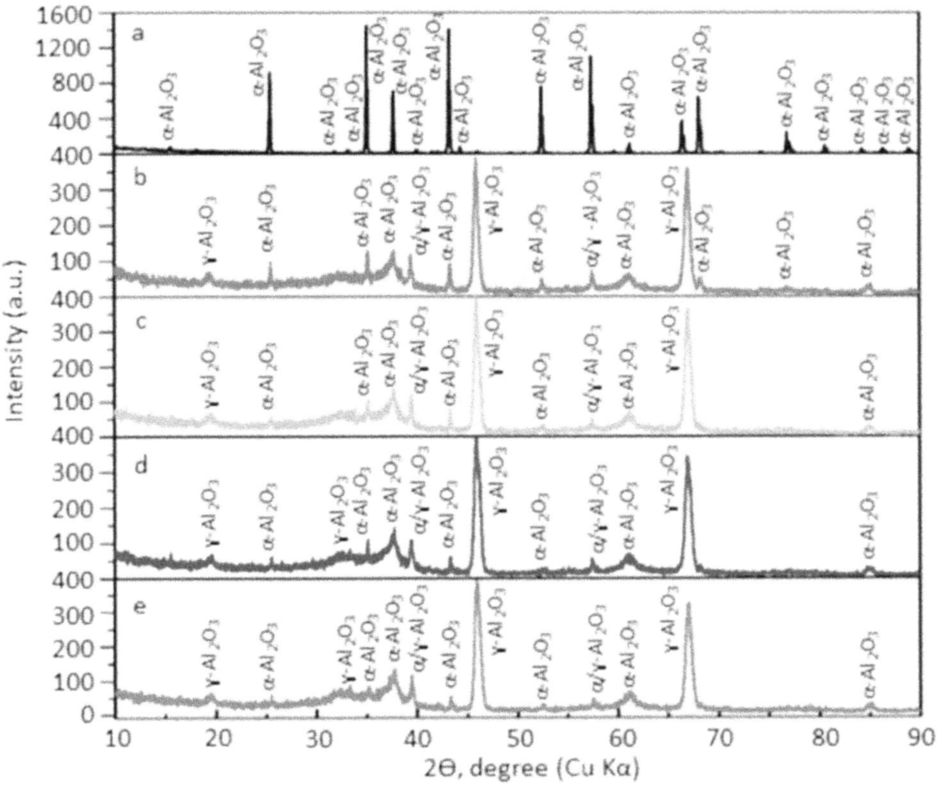

coating obtained when the barrel is filled to 53% has a high wear resistance. Based on X-ray diffraction analysis, it can be assumed that an increase in the volume fraction of α-Al_2O_3 phase in coatings leads to an increase in wear resistance. This is because α-Al_2O_3 has several features, including low density, relatively high melting point, excellent corrosion and wear resistance, and high strength at high temperatures.

Figure 9 shows the friction coefficients of coatings obtained with the barrel filled 53%, 58%, 63% and 68%. In (Deng et al., 2017) the tribological properties of the Al_2O_3 coating obtained by the plasma method using ZrO_2, Si_3N_4, Al_2O_3 and stainless steel balls (1Cr18Ni9Ti) with a diameter of 6 mm. The coefficients of friction were μ»0.55–0.6 with ZrO_2, with Si_3N_4 μ»0.5–0.55, with Al_2O_3 ball μ»0.6–0.65, with stainless steel ball (1Cr18Ni9Ti) μ»0.7–0.8. In (Ling et al., 2009). A steel ball 100Cr6 was used to study the friction and wear resistance of ZrO_2–Al_2O_3-based coatings obtained by air-plasma injection. In this study, a steel ball 100Cr6 was used and the coefficient of friction was in the range μ»0.52–0.59. The coefficient of friction was observed when filling the barrel with gas at least 53%. The coefficients of friction of all coatings obtained when filling the barrel by 53%, 58% and 68% have the same values and are $\mu=0.52\pm0.01$. The coating obtained when filling the barrel by 63% has a high coefficient of friction $\mu=0.59\pm0.01$ in comparison with other samples due to the roughness and porosity of the coatings.

Figure 10 shows the change in the micro-hardness of the coating. Modifications of α-and γ-phases have different values of physical and mechanical properties. Modifications of α-Al_2O_3 have high hard-

Figure 8. Tribological tests Al_2O_3 coatings according to the "ball-disk" scheme: 53% (1), 58% (2), 63% (3), and 68% (4)

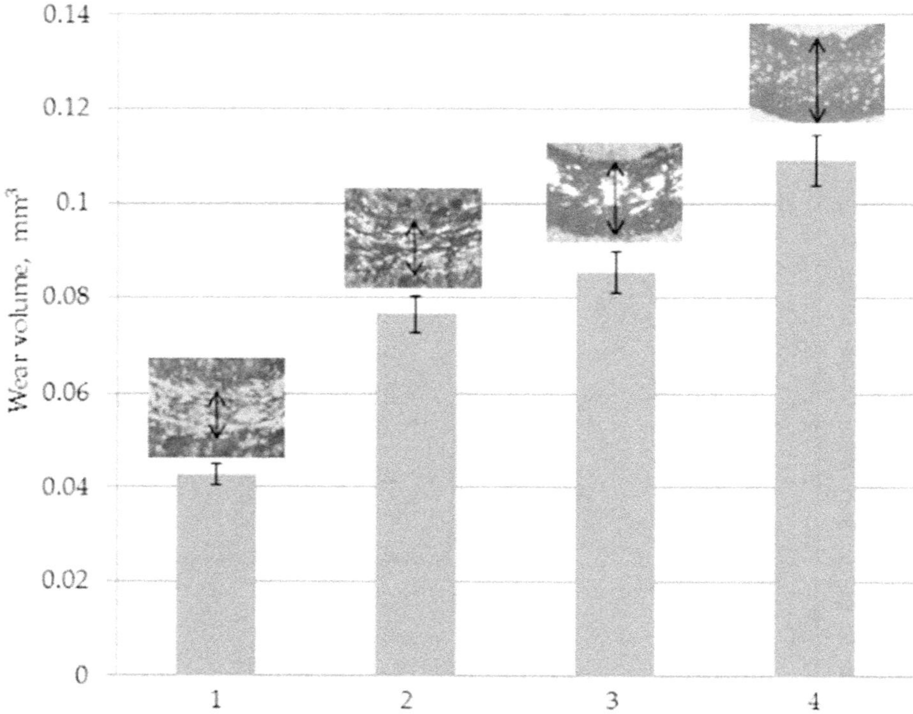

Figure 9. Tribological studies Al_2O_3 coatings depending on the filling of the barrel with gas: 53% (1), 58% (2), 63% (3), and 68% (4)

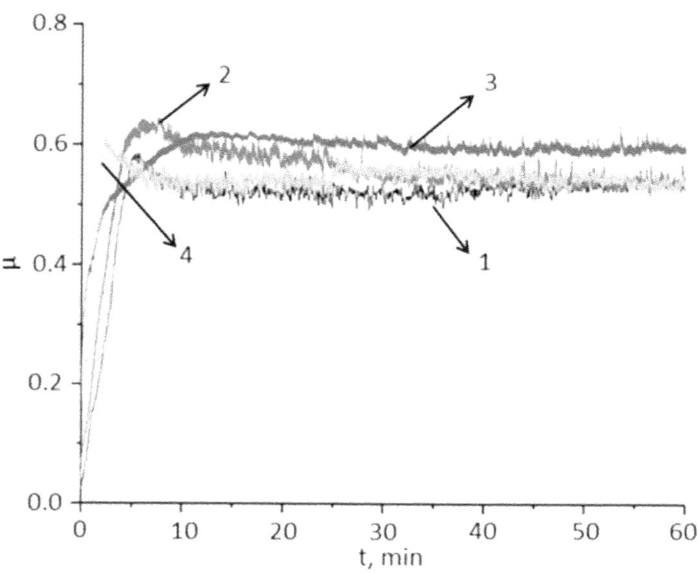

ness and wear resistance, while γ-Al_2O_3 is relatively elastic and provides high adhesion to the substrate. Analysis of the effect of phase composition on the physicochemical and mechanical characteristics of plasma detonation coatings made of aluminium oxide shows that an increase in the stable γ-phase, as well as other (metastable) phases reduces the hardness, wear resistance and corrosion resistance of the coating. In addition, the maximum hardness of the coating obtained by plasma detonation is 1.98 GPa, Thus, the use of plasma-detonation technology for spraying coatings made of Al_2O_3 allows to obtain dense multi-phase coatings with good adhesion to the substrate (due to damping of low-temperature stable and metastable modifications) and high physical and mechanical properties (due to high α-phase composition) (Pogrebnyak et al., 2000). Based on these analyses, based on the results of our study, we conclude that the mechanical and tribological properties increased due to the increase in α-Al_2O_3, and the adhesion strength may improve due to the γ-Al_2O_3 phase. Reference (ASTM G99–05, 2005) reported that the hardness of the Al_2O_3 coating obtained increased from 10.87 to 16.33 GPa by reducing the delay time between shots during a detonation explosion from 1 to 0.25 s. Reference (Rakhadilov, Kantay, Sagdoldina et al, 2021) reported that the hardness of the Al_2O_3 coating increases from 7.86 to 11.08 GPa as the thickness increases from 250 to 1100 µm. In (Cheng et al., 2020) reported that the microhardness of ZrO_2–Al_2O_3-based coatings obtained by air-plasma injection was in the range of 8.06–9.11 GPa. The micro-hardness of the obtained coatings, depending on the filling with gas, was at 53%–20.6±0.3 GPa, 58%–16.3±0.15 GPa, 63%–17.9±0.25 GPa, and at 68%–18.16±0.27 GPa. At the same time, the highest value of micro-hardness was observed at 53% filling. An increase in the micro-hardness of coatings is associated with an increase in the α-Al_2O_3 phase. γ-Al_2O_3 has a lower hardness and mechanical resistance than α-Al_2O_3.

Figure 10. Microhardness of the surface Al_2O_3 coatings

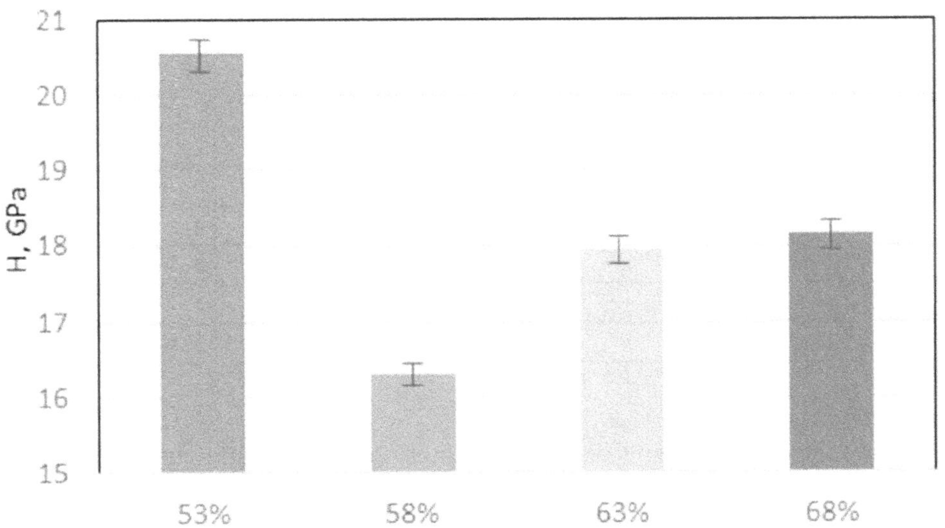

b. Results for Multilayer DS

Figure 11 shows the SEM-image of the cross-section of multilayer coating based on NiCr/NiCr-Al_2O_3/Al_2O_3. The elemental composition of the coatings was determined by using EDX-analysis. From the data, the coatings consist of a lower metal layer, an upper ceramic layer and a transitional layer of a mechanical mixture of metal and ceramics. This confirms the EDX-analysis. The results of the EDX-analysis are shown in Table 3.

Figure 11. SEM cross-section images of multilayer NiCr/NiCr-Al_2O_3/Al_2O_3 coating

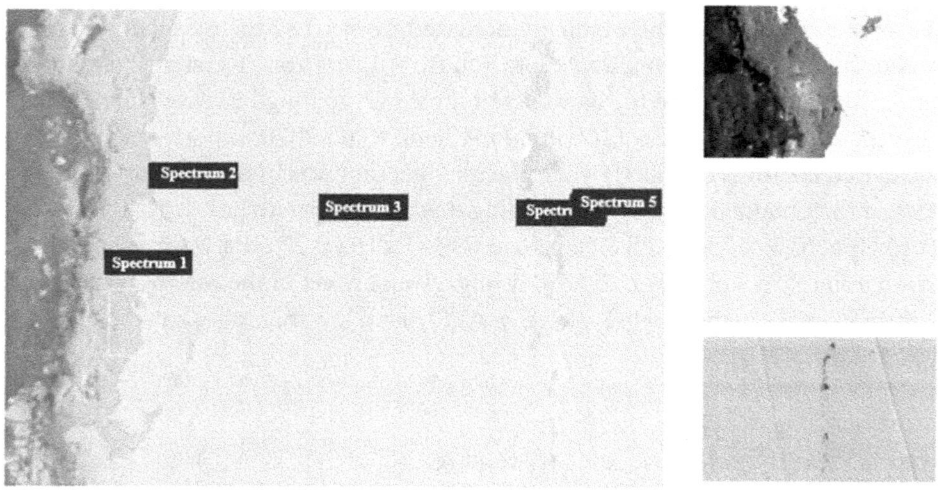

Table 3. EDX analysis of multilayer coating

Spectrum No.	Al, %	O, %	Ni, %	Cr, %	Fe, %	Total, %
Spectrum 1	39.78	31.36	13.88	14.98	0	100
Spectrum 2	29.63	26.74	24.71	18.92	0	100
Spectrum 3	19.56	17.42	39.56	23.46	0	100
Spectrum 4	7.53	4.66	49.68	38.13	0	100
Spectrum 5	1.25	1.67	23.97	15.06	58.09	100

Figure 12 shows the diffractograms of Al_2O_3 coatings obtained at different degrees of barrel filling volume. The phase composition of Al_2O_3 coatings changes depending on the degree of barrel filling. The intensity of α-Al_2O_3-phase peaks increases at barrel is filled by 63% and 54%, indicating an increase in the concentration of α-Al_2O_3-phase. At filling of the shaft by 45% the coating consists mainly of γ-Al_2O_3-phase. The change in the phase composition of Al_2O_3 coatings is related to the temperature of powder heating in the spraying process. Since, the powder temperature changes depending on the volume of barrel filling.

Figure 12. Areas of the diffraction spectra of Al_2O_3 coatings obtained at different barrel filling volume: powder 63% (a), 54% (b), and 45% (c)

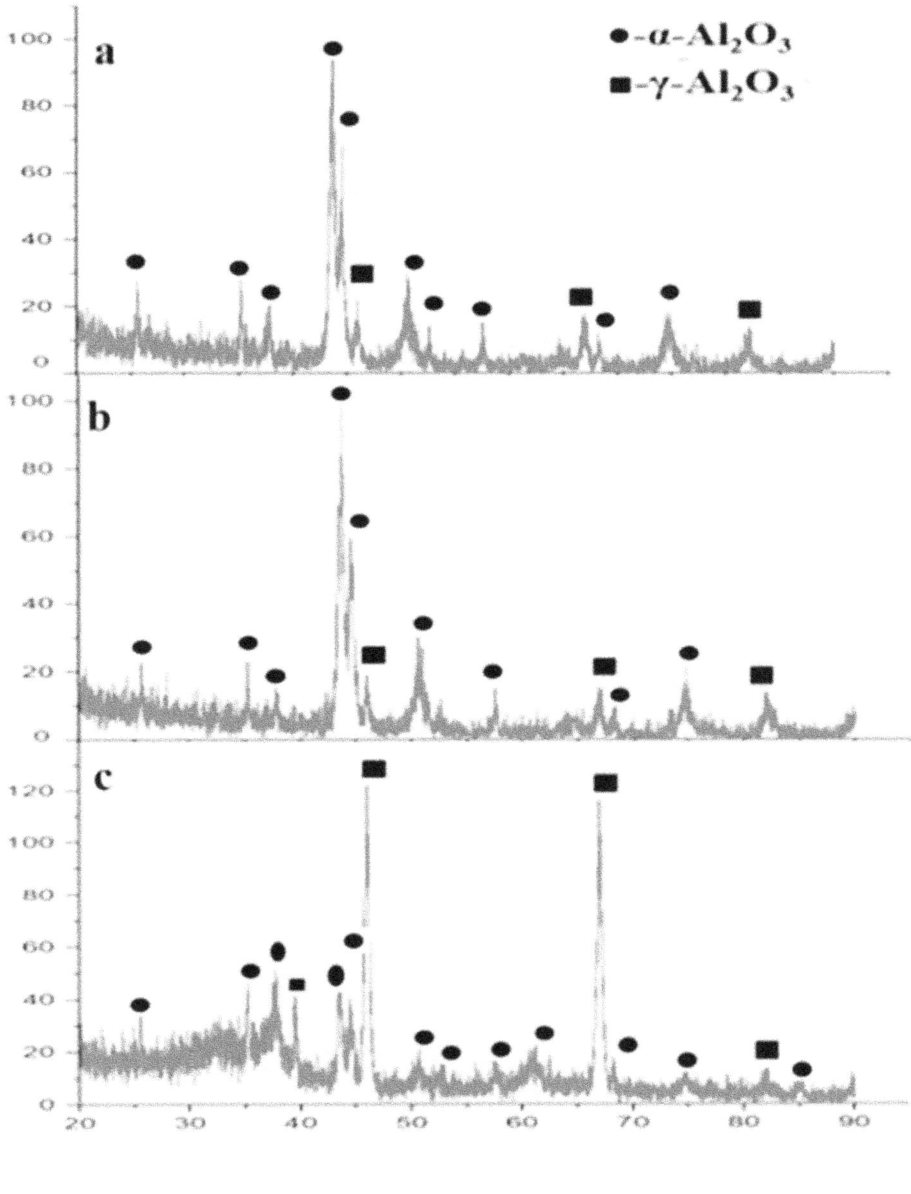

Diffractograms on the bases of NiCr coatings obtained at different degrees of barrel filling are shown in Figure 13. All coatings consist of $CrNi_3$ and an increase in the reflex intensity (020) is observed when the barrel is filled to 58%.

Table 4 shows the values of nanohardness and modulus of elasticity of the obtained coatings. The results showed that the hardness of the Al_2O_3 coating increases as a function of its α-Al_2O_3 content with a maximum nanohardness of 16.42 GPa when the borehole is filled with 63%. The increase in hardness of Al_2O_3 coatings with increasing α-Al_2O_3-phase in it is quite natural. Since, α-Al_2O_3 has a higher hardness and wear resistance than γ-Al_2O_3. The nanohardness of the NiCr coating has maximum values when the

Figure 13. Diffraction spectra of NiCr coatings obtained at different barrel filling volume: powder 58% (a), 54% (b), and 65% (c)

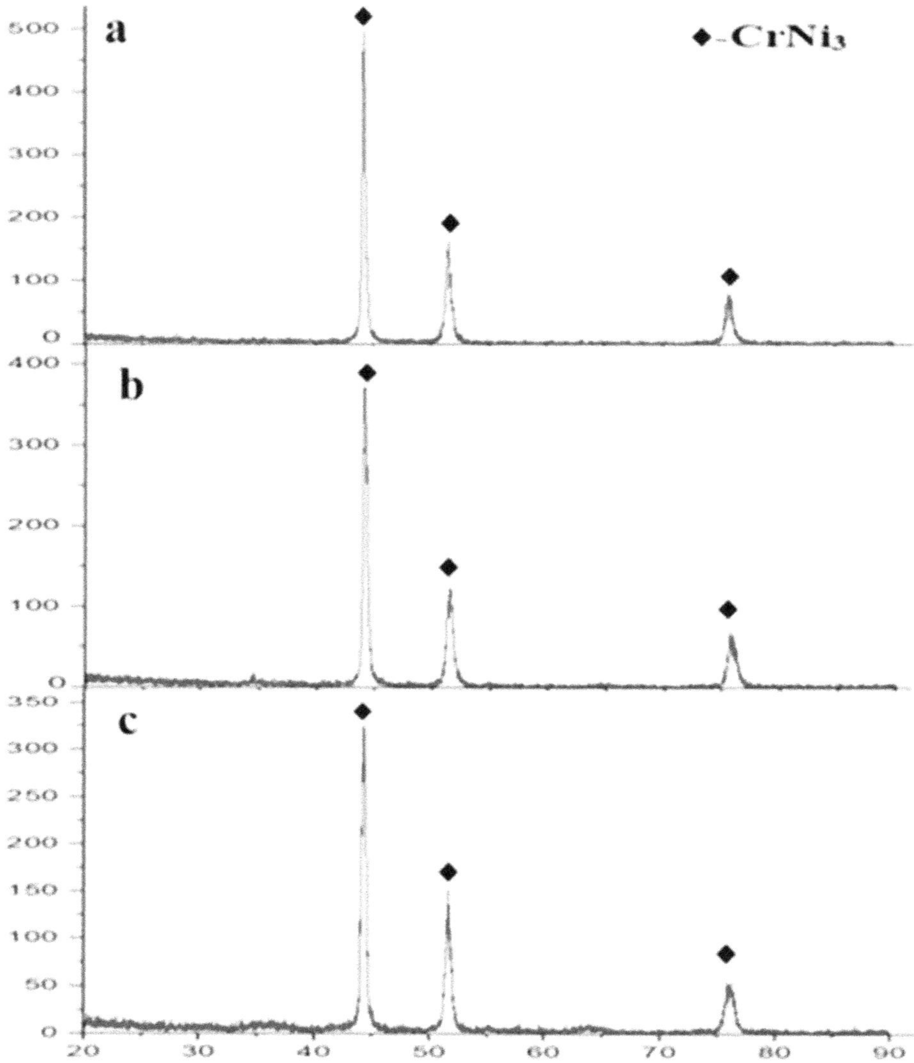

Table 4. Nanoindentation and modulus of elasticity of NiCr-Al_2O_3 coatings

Sample Number	Powder Name	Nanohardness, GPa	Young's Modulus, GPa
1	Al_2O_3	13.54	256.13
2	Al_2O_3	14.73	261.31
3	Al_2O_3	16.42	286.12
4	NiCr	6.92	165.32
5	NiCr	8.02	176.23
6	NiCr	7.63	169.12

barrel is filled to 58% and is 8.02 GPa. The small difference in hardness of NiCr coatings is apparently due to the porosity of the coating, which changes depending on the spraying mode. The table shows that the modulus of elasticity of all coatings varies with hardness. The sputtering mode influences the structure and properties of NiCr and Al_2O_3 coatings and hence multilayer coatings NiCr/NiCr-Al_2O_3/Al_2O_3 were made.

Thus, summing up the results of this section and continuing this topic, additional tests will be introduced to determine adhesion and corrosion, depending on the technological modes with varying powder composition and obtaining gradient coatings with specified characteristics.

CONCLUSION

On the bases of experimental obtained data and their analysis, the following conclusions were made: It was determined that the phase composition and properties of detonation coatings strongly depend on the technological parameters of spraying. When spraying powders based on Al_2O_3, the chemical and phase composition of the coatings obtained strongly depend on the degree of powder supply to the cylinder with a gas mixture, which detonates using a spark plug located in the barrel itself, which allows you to control the structure and composition of coatings and, thus, obtain coatings with unique characteristics. When spraying the powder with the DS method, special preparation of the substrate surface does not affect the formation or formation of specific phases. But it significantly affects the adhesive properties of the resulting coating. For example, the treatment of the substrate surface with electrocorundum before spraying improves adhesion depending on the type and size of the particles of the powder itself. In addition, one of the uniqueness of this method is that the chemical and phase composition of the substrate used and the resulting coating are strictly separated by boundaries, while maintaining high adhesive properties between the substrate and the coating. The method of obtaining multilayer coatings based on NiCr/NiCr-Al_2O_3/Al_2O_3 which consists of the lower metal layer, the upper ceramic layer and the transition layer of a mechanical mixture of metal and ceramics on a detonation machine was developed. The study of the elemental composition by the SEM-EDX method showed that multilayer coating on the bases of NiCr/NiCr-Al_2O_3/Al_2O_3 has a gradient structure, as there is an increase in the concentration of ceramics from the substrate to the coating surface.

REFERENCES

Algatti, M. A., Santos, C. N., Mota, R. P., De Campos, E., Lucena, E. F., Machida, M., & Melo, F. C. (2017). Characterization of SiC and Al2O3 ceramics exposed to nitrogen ions from inverse-Z pinch plasma discharge. *Proceedings of the 16th Latin American Workshop on Plasma Physics (LAWPP)*, 5–11. 10.1109/LAWPP.2017.8692198

ASTM G133–95. (1995). *Standard Test Method for Linearly Reciprocating Ball-on-Flat Sliding Wear.* ASTM International.

ASTM G99–05. (2005). *Standard Test Method for Wear Testing with a Pin-on-Disk Apparatus.* ASTM International.

Bannier, E., Vincent, M., Rayon, E., Benavente, R., Salvador, M. D., & Sanchez, E. (2014). Effect of TiO2 addition on the microstructure and nanomechanical properties of Al2O3 suspension plasma sprayed coatings. *Applied Surface Science*, *316*, 141–146. doi:10.1016/j.apsusc.2014.07.168

Buitkenov, D., Rakhadilov, B., Erbolatuly, D., & Sagdoldina, Z. (2020). Influence of Heat Treatment on the Phase Composition and Microhardness of Coatings Based on Ti_3SiC_2/TiC. *Key Engineering Materials*, *839*, 137–143. doi:10.4028/www.scientific.net/KEM.839.137

Cheng, J., Ge, Y., Wang, B., Zhang, L., Hu, X., Hong, S., Liang, X., & Zhang, X. (2020). Microstructure and Tribocorrosion behavior of Al2O3/Al composite coatings: Role of Al2O3 addition. *Journal Thermal Spray Technology*, *29*(7), 1741–1751. doi:10.100711666-020-01062-1

Deng, W., Shuangjian, L., Guoliang, H., Xia, L., Xiaoqin, Z., Yulong, A., Huidi, Z., & Jianmin, C. (2017). Comparative study on wear behavior of plasma sprayed Al2O3 coatings sliding against different counterparts. *Ceramics International*, *43*(9), 6976–6986. doi:10.1016/j.ceramint.2017.02.122

Dhakar, B., Namdeo, A., Chatterjee, S., & Sabiruddin, K. (2018). Heat treatment of plasma sprayed alumina-chromia composite coatings. *Surface Engineering*, *34*(10), 737–746. doi:10.1080/02670844.2017.1316029

Gallyamov, A. A., Nenashev, M. V., Ibatullin, I. D., & Murzin, A. Y. (2018). Application detonation coatings to design a new metal cutting tool. *Transactions of the IMF*, *96*(6), 290–294. 10.1080/00202967.2018.1477267

Kantay, N., Kasmamytov, N., Rakhadilov, B., Plotnikov, S., Paszkowski, M., & Kurbanbekov, S. (2020). Influence of temperature on structural-phase changes and physical properties of ceramics on the basis of aluminum oxide and silicon. *Materials Testing*, *62*(7), 716–720. doi:10.3139/120.111537

Kaya, G., Gunhan, B., Ozer, I. O., & Poyraz, H. B. (2020). Production of TiO2 coated α-Al2O3 platelets by flame spray pyrolysis and their characterization. *Ceramics International*, *46*(16), 25512–25519. doi:10.1016/j.ceramint.2020.07.022

Ling, B., Zhang, G., Liao, H., Coddet, C., & Ding, C. (2009). Friction and wear behavior of ZrO2–Al2O3 composite coatings deposited by air plasma spraying: Correlation with physical and mechanical properties. *Surface and Coatings Technology*, *203*(20, 21), 3235–3242. doi:10.1016/j.surfcoat.2009.03.056

Pogrebnjak, A. D., Shpak, A. P., Azarenkov, N. A., & Beresnev, V. M. (2009). Structure and properties of hard and superhard nanocomposite coatings. *Physics Uspekhi*, *52*(1), 29–54. doi:10.3367/UFNe.0179.200901b.0035

Pogrebnjak, A. D., & Tyurin, Y. N. (2005). Modification of material properties and coating deposition using plasma jets. *Physics Uspekhi*, (48), 515–544.

Pogrebnyak, A. D., Tyurin, Y. N., Ivanov, Y. F., Kobzev, A. P., Kulmentieva, O. P., & Ilyashenko, M. I. (2000). Obtaining and investigation of the structure and properties of plasma-detonation coatings from Al2O3. *Letter Journal Technology Physics*, (21), 53–60.

Rakhadilov, B., Kantay, N., Sagdoldina, Z., Erbolatuly, D., Bektasova, G., & Paszkowski, M. (2021). Experimental investigations of Al2O3 - and ZrO2 -based coatings deposited by detonation spraying. *Materials Research Express*, *8*(5), 1–12. doi:10.1088/2053-1591/abfbb7

Rakhadilov, B. K., Buitkenov, D. B., Tuyakbaev, B. T., Sagdoldina, Zh. B., & Kenesbekov, A. B. (2019). Structure and properties of detonation coatings based on titanium carbosilicide. *Key Engineering Materials*, (821), 301–306.

Rakhadilov, B. K., Buytkenov, D. B., Kakimzhanov, D., Kozhanova, R. S., & Bektasova, G. S. (2020). The effect of detonation spraying on the phase composition and hardness of Al2O3 coatings. *Eurasian Journal of Physics and Functional Materials*, *4*(2), 160–166. doi:10.29317/ejpfm.2020040207

Rakhadilov, B. K., Sagdoldina, Zh. B., Baizhan, D. R., Zhurerova, L. G., Yeskermessov, D. K., Kalitova, A. A., & Smaiylova, M. (2021). Obtaining of hydroxyapatite coatings on a titanium substrate by detonation-gas spraying. *Eurasian Physical Technical Journal*, *3*(37), 30–36. doi:10.31489/2021No3/30-36

Shoja, S., Alm, O., Norgren, S., Andrén, H. O., & Halvarsson, M. (2021). Calculated and experimental Schmid factors for chip flow deformation of textured CVD α-alumina coatings. *Surface and Coatings Technology*, *412*, 126991. doi:10.1016/j.surfcoat.2021.126991

Thirumalaikumarasamy, D., Shanmugam, K., & Balasubramanian, V. (2014). Corrosion performance of atmospheric plasma sprayed alumina coatings on AZ31B magnesium alloy under immersion environment. *Journal of Asian Ceramic Societies*, *2*(4), 403–415. doi:10.1016/j.jascer.2014.08.006

Ulianitsky, V., Shtertser, A., Zlobin, S., & Smurov, I. (2011). Computer-Controlled Detonation Spraying: From Process Fundamentals Toward Advanced Applications. *Journal of Thermal Spray Technology*, *20*(4), 791–801. doi:10.100711666-011-9649-6

Ulianitsky, V. Y., Batraev, I. S., Shtertser, A. A., Dudina, D. V., Bulina, N. V., & Smurov, I. (2018). Detonation spraying behaviour of refractory metals: Case studies for moand Ta-based powders. *Advanced Powder Technology*, *29*(8), 1859–1864. doi:10.1016/j.apt.2018.04.023

Zhang, S. H., Cho, T. Y., Yoon, J. H., Fang, W., Song, K. O., Li, M. X., Joo, Y. K., & Lee, C. G. (2008). Characterization of microstructure and surface properties of hybrid coatings of WC–CoCr prepared by laser heat treatment and high velocity oxygen fuel spraying. *Materials Characterization*, *59*(10), 1412–1418. doi:10.1016/j.matchar.2008.01.003

Zhu, C., Zhang, W., Wang, L., Ning, Z., Feng, Y., Feng, K., Liao, J., Yang, Y., Liu, N., & Yang, J. (2020). Effect of thermal cycles on structure and deuterium permeation of Al2O3 coating prepared by MOD method. *Fusion Engineering and Design*, *159*, 111750. doi:10.1016/j.fusengdes.2020.111750

Chapter 8
The Photocatalytic Treatment of Industrial Wastewater Using Semiconductor-Based Materials Fabrication and Modification Methods

Athule Ngqalakwezi
https://orcid.org/0000-0003-2024-0389
Mintek, South Africa

Sivuyisiwe Mapukata
Mintek, South Africa

Mpfunzeni Raphulu
Mintek, South Africa

Lizzie Mampane
Wits University, South Africa

ABSTRACT

Water scarcity has been exacerbated in recent years due to factors including the expansion of industrialization and agricultural activities. The contaminants released by industries into surrounding water bodies include toxic aromatic molecules, dyes, and heavy metals, which upon consumption pose a threat to human life. The development of water treatment technologies and catalysts targeting these pollutants has thus emerged as a critical area of research and development. This chapter therefore focuses on providing comprehensive information on the application of metal oxide semiconductor photocatalysis for the treatment of a range of industrial pollutants. With focus mainly being on titanium dioxide (TiO2), zinc oxide (ZnO), tin oxide (SnO2), cadmium oxide (CdO), and tungsten trioxide (WO3), the synthesis and modification methods for enhancement of the photocatalytic efficiencies of the semiconductor materials is discussed. Their physical, morphological, and optical properties are compared, and the mechanisms behind their activities are reported.

DOI: 10.4018/978-1-6684-6830-2.ch008

1. INTRODUCTION

Although industrialization plays an important role in the global socio-economic development, it often has a detrimental environmental impact. Most industries including mining, pharmaceutical, textile as well as the tannery industry release toxic contaminants into surrounding water streams (da Silva-Rêgo et al., 2022; Lellis et al., 2019; Nassiri Koopaei & Abdollahi, 2017; Saxena, 2005.). These contaminants include amongst others, heavy metals, organic and inorganic compounds which consist of sulphates and pesticides (Briffa et al., 2020; Hussain & Asi, 2008; Radwan et al., 2017; Torres-Martínez et al., 2021).

The stability, toxicity and accumulation of these contaminants impose various repercussions including the diminishing of aquatic life, devastating health effects and deteriorating soil quality (Banerjee, 1986; Bashir et al., 2020; Kasem Mahmoud & Mohamed Ghoneim, 2016). Industrialisation together with natural calamities such as drought have therefore contributed to the decrease in the global water supply (T. Ahmed et al., 2020). On the other hand, the population growth, economic development and changing consumption pattern has heightened the need for potable water (Okello et al., 2015). This has therefore led to the increased research efforts in finding efficient ways of treating and thus recycling the contaminated water.

With the aim of creating a sustainable environment and thriving eco-system, numerous researchers have gravitated towards semiconductor photocatalysis for environmental and wastewater remediation (Belver et al., 2018; D. Chen et al., 2000; Gonzalez-Perez et al., 2018). Although other methods like filtration, adsorption and coagulation have been implemented for the treatment of water pollutants(Precious Sibiya et al., 2021; Rathi & Kumar, 2021; Utari & Herdiansyah, 2020), these treatment technologies often possess limitations such as consuming large amounts of energy, high cost, low efficiency and even toxic by-product formation etc. (Kar et al., 2021; Sun et al., 2022). Semiconductor based photocatalysis on the other hand can attribute its growing popularity to its high efficiency, low cost as well as complete degradation of pollutants with no added pollution (Deng et al., 2018). Upon generation of reactive oxygen species in the presence of light, this process can efficiently degrade organic pollutants into water, carbon dioxide or other small molecules, and reduce or oxidize inorganic pollutants to harmless substances (Ren et al., 2021). In addition, semiconductors possess good electrical conductivity as well as sufficient stability to be applied in large scale practical applications (Dey et al., 2015).

This chapter focuses on providing detailed information on the application of various metal oxide semiconductor photocatalysts (TiO_2, ZnO, SnO_2, WO_3 and CdO) for the treatment of a range of industrial pollutants. Unlike the covalent semiconductors like silicon, metal oxide semiconductors are particularly unique due to their electronic charge transport properties (H. He, 2020).

Figure 1 demonstrates the mechanism of activity of semiconductor photocatalysis by depicting the manner in which highly active oxidizing and reducing agents (h^+ and e^-) are generated. Upon exposure to the UV light, electrons are transferred from the valence band to the conduction band, thereby leaving behind photo-generated holes in the former (W. J. Ong et al., 2014). The generated positive holes are direct oxidants and thus react with the adsorbed water molecules on the surface of the photocatalyst to generate hydroxyl radicals. The electrons on the other hand reduce dissolved oxygen molecules into superoxide anion radicals (George et al., 2016). It is these reactive oxygen species (ROS) generated from both the conduction and valence bands that facilitate the activity of the semiconductors against various pollutants. The fast recombination of the electron–hole pairs however suppresses the photocatalytic process (Gao et al., 2019). Several approaches have since been developed to alleviate this and hence increase the photocatalytic activity of semiconductors. These include doping with metals, non-metals

Figure 1. Basic mechanism of semiconductor based photocatalysis

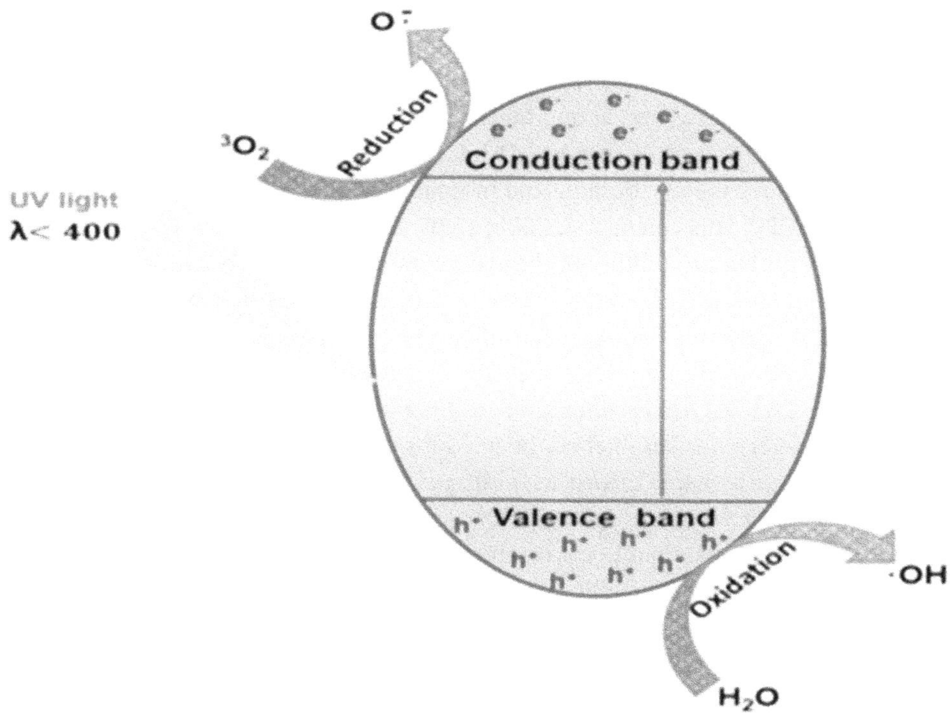

and coupling with carbonaceous nanoscale materials amongst others (Kuo et al., 2021; Malekshoar et al., 2014; Vargas Hernández et al., 2017).

The direct band gap energies (the minimum energy that is needed to excite an electron into the conduction band) of the semiconductor materials of interest are listed in Table 1. They show the photo-activation potentials of the semiconductor materials, although some of their large values (>2 eV) limit their application for real life applications as the UV regime is only a small fraction of the sun's energy (<10%) (Y. Wang et al., 2016).

Another limitation on the practical applications of these semiconductor photocatalysts is that they are usually applied in powdered form, meaning they often require filtration post-application, a process that can be rather tedious. In order to eliminate this, the immobilization of the semiconductor nanoparticles on high surface area substrates is often implemented (Lei et al., 2012). Some of the supporting materials interact synergistically with the semiconductors due to their unique pore structure, adsorption efficiency and electronic properties which enhances their photocatalytic activities (P. Yao et al., 2018). Other solutions include utilising synthetic methods that produce retrievable and thus reusable semiconductor photocatalysts such as electrospinning for the fabrication of semiconductor nanofibers (Saidin et al., 2017). Conjugation of semiconductor photocatalysts to magnetic materials also aids in their efficient retrieval and thus recyclability post-application (Álvarez et al., 2010).

The fabrication methods of the semiconductor materials are therefore discussed as well as the various modification strategies that can be employed for reducing their band gap energies and thus enhancing their respective photocatalytic activities in a wider range of the spectrum. The commonly treated pollutants and the mechanisms behind their treatment are also highlighted.

Table 1. Direct bandgap energies of various semiconductor materials

Semiconductor	Band Gap Energy (eV)	Ref
TiO_2 (anatase)	3.20	(Dette et al., 2014)
TiO_2 (rutile)	3.00	(Y. zhen Fang et al., 2018)
ZnO	3.37	(Z. L. Wang, 2004)
SnO_2	3.60	(Doyan et al., 2021)
WO_3	2.60-3.00	(González-Borrero et al., 2010)
CdO	2.20	(Przeździecka et al., 2021)

1.1 Titanium Dioxide (TiO_2)

Being the most popular photocatalyst, TiO_2 owes its diverse applications to its strong oxidizing power, high chemical stability as well as affordability (Zhu & Zhou, 2019). It is also biologically benign and can thus be applied for water treatment with no significant additional toxicity being introduced to the water (George et al., 2016; J. L. Wang & Xu, 2012). Lastly, TiO_2 is attractive as it can be mass-produced at a low cost in different sizes, shapes and polymorphs which ultimately influence its photoactivity (Liao & Liao, 2007). Titanium dioxide occurs as three polymorphs, the stable rutile and metastable anatase and brookite with the former two being the most common (Allen et al., 2018; de Matteis et al., 2016). These polymorphs exhibit different properties and consequently different photocatalytic performances (Luttrell et al., 2015). The most common commercial TiO_2 photocatalyst is the Degussa P-25, a powder which consists of both rutile and anatase crystallites (Ohno et al., 2001). The phase mixture of the different polymorphs is known to have synergistic effects and an increased photocatalytic activity is observed compared to pure phases (Paul & Choudhury, 2014). However, for pure phases, the anatase generally exhibits a higher photocatalytic efficiency compared to the rutile TiO_2 (J. Zhang et al., 2014). In addition to the crystalline phase, various other factors including the size, specific surface area, pore volume, pore structure and structural dimensionality can affect the photocatalytic performance of TiO_2 photocatalysts (Nakata & Fujishima, 2012). Various studies have thus been conducted to fabricate TiO_2 nano- and micro-structures including nanosheets, spheres, hierarchical architectures, nanotubes and fibers (Cheng et al., 2010; Y. Chen et al., 2011; H. Li et al., 2007; Macak et al., 2007; Tiwari et al., 2018) amongst others, with the aim of obtaining the best photocatalytic activities.

Increased attention on the photocatalytic properties of TiO_2 has heightened interest in the photocatalytic applications of related metal oxides for catalyst comparison and optimisation. The high photosensitivity and thermal stability, as well as their low cost and negligible toxicity have thus made ZnO and SnO_2 valuable alternatives as photocatalysts (Al-Hamdi et al., 2017). Interest has also been drawn to WO_3 and CdO, which just like previously mentioned oxides, often possess octahedral sites in their structures. An octahedral network is considered a prerequisite for high photocatalytic activity as it increases the mobility of electron–hole pairs, thereby influencing the probability of reactions occurring on the surfaces of the photocatalysts (Abe et al., 2006).

1.2 Zinc Oxide (ZnO)

Due to the cost of ZnO nanoparticles being up to 75% lower than that of TiO_2 (S. Liang et al., 2012; C. B. Ong et al., 2018), its applications have significantly expanded and diversified in recent years. Interestingly, ZnO almost has the same band gap energy as TiO_2 but exhibits much higher absorption efficiency across a large fraction of the solar spectrum when compared to TiO_2 (C. B. Ong et al., 2018; Qiu et al., 2008). This is attributed to the position of the valence band of ZnO, which is lower than that of TiO_2. As a result, the oxidation potential of hydroxyl radicals generated by ZnO is higher than that of the hydroxyl radicals generated by TiO_2 (Qi et al., 2017). Thus, the photocatalytic performance of ZnO in degrading pollutants is often higher than that of TiO_2 (Gnanaprakasam et al., 2015; Sakthivel et al., 2003). This is however not always the case as photocorrosion of ZnO frequently occurs when it is illuminated with UV light, thereby decreasing its photocatalytic activity in aqueous solutions (Hariharan, 2006).

The major crystal forms of ZnO are rocksalt, zinc blende and wurtzite (Espitia et al., 2012). Due to its high thermodynamic stability at ambient conditions and easy synthesis, the wurtzite phase has attracted the most attention (T. Li et al., 2020). In addition to the crystal structure, the most common parameter that has been varied to optimise the ZnO properties is the morphology. As a result it has been fabricated in a range of structures including nanopyramids, nanotubes and microspeheres (M. Z. Ahmad et al., 2013; Ali et al., 2020; X. Chen et al., 2013) to name a few.

1.3 Tin Oxide (SnO_2)

Tin (VI) oxide (also commonly known as stannic oxide) is an n-type semiconductor with potential in a variety of water treatment applications including photooxidation of dyes and bacterial photoinactivation (Foletto et al., 2012), (Erkan et al., 2006). However, SnO_2 is rarely used on its own, it is commonly used as a component of composite photocatalysts such as SnO_2/TiO_2 and SnO_2/ZnO or in the presence of a dopant (*3*, n.d.; Karunakaran et al., 2013; Uddin et al., 2012). This is because SnO_2 alone is incapable of reducing oxygen molecules due to its conduction band position (Kim et al., 2016). It is however a good electron acceptor due to its positive conduction band edge and is thus easily coupled with other semiconductors with suitable matching of band levels to form heterostructures. Simultaneous formation of Sn^{2+} and Sn^0, causing mixed phases of SnO_2 and SnO or Sn in the photocatalyst has also been reported as a possible factor for its poor performance (Toloman et al., 2020).

Under normal conditions, the crystal structure of SnO_2 is rutile (Q. Zhang et al., 2015) and it has been developed in various morphologies including nanowires, nanobelts, nanofibers, nanospheres, and nanofilms (Hu & Yuan, 2017; Kolmakov et al., 2005; Mali et al., 2018; Rameshkumar et al., 2020) amongst others.

1.4 Tungsten Trioxide (WO_3)

In recent years, tungsten oxides (WO_X, $X \leq 3$) have received increasing interest due to their easy preparation, tuneable structures, strong photocatalytic ability, Earth-abundance and suitable band gaps for visible light absorption (Quan et al., 2020). Furthermore, WO_3 in particular has outstanding photo-corrosion resistance and chemical stability when in contact with solar light irradiation (Samuel et al., 2022). The crystal phases of pure WO_3 have five phase transitions at temperatures ranging between 180 and 900 oC (Y. C. Liang & Chang, 2019). The phases are monoclinic I (γ-WO_3), monoclinic II (ε-WO_3), triclinic

(δ-WO_3), orthorhombic (β-WO_3) and tetragonal (α-WO_3) (Faudoa-Arzate et al., 2017). The monoclinic I (γ-WO_3) phase is the most investigated among the five phases due to its greater stability at room temperature (Zheng et al., 2014).

Their easy fabrication methods have led to the development of various WO_3 morphologies including nanocubes, nanoflowers and nanorods (W. Fang et al., 2016; Khan et al., 2016; L. Wang et al., 2019) which have been effectively applied in various photocatalytic applications.

1.5 Cadmium Oxide (CdO)

Cadmium Oxide (CdO) is an n-type semiconductor with high electrical conductivity and a moderate refractive index (Omran Alkhayatt et al., 2014; Zargar et al., 2016). In addition to its great stability, wide availability and cost effectivity, CdO has interstitial cadmium atoms as well as oxygen vacancies which are necessary for photocatalytic applications. It can also absorb light in the visible region with high charge carrier mobility and can thus serve as an effective visible-light-driven photocatalyst (Y. N. Rane et al., 2019). Having a rock-salt crystal structure (G. Yao et al., 2013), several unique nanostructures of CdO have been developed over the years such as nanowalls, nanoneedles, nanowires, nanoclusters, nanotubes and nanocrystals (Chang, 2015; Guo et al., 2008; X. Liu et al., 2003; Lu et al., 2008; Radi et al., 2006; Yakuphanoglu, 2010). These nanomaterials together with those of the other semiconductor oxide photocatalysts are all fabricated using a range of techniques and methods as elaborated in the next section.

2. FABRICATION METHODS OF SEMICONDUCTOR PHOTOCATALYSTS

The different fabrication techniques influence the semiconductor morphologies and sizes which can directly affect their photocatalytic properties. For instance, when the size of semiconductor materials is reduced to nanoscale, their physical and chemical properties change drastically, often resulting in enhanced properties due to their large surface area or quantum size effect (Suresh, 2013). It has also been reported that cylindrical structures like nanorods and nanotubes have shown higher photocatalytic activity compared to the commercially available Degussa P-25 nanoparticles due to their reduced charge recombination (Reghunath et al., 2021). Enhanced photocatalytic activity has also been observed for semiconductor photocatalysts with a hierarchical architecture (Xu et al., 2011). This is because they have an interconnected porous network which can facilitate the movement of reactants towards the active sites located on the pore walls, thus enabling better diffusion (G. Wang & Coppens, 2010).

As depicted in Figure 2, fabrication methods are distinctly differentiated by their phases, with the most commonly used method being the liquid phase hence it is the focus of this section.

2.1 Hydrothermal Synthesis Method

The hydrothermal method is the most commonly used and reported method of synthesizing semiconductor materials for photocatalytic applications. The ease of scaling up and accurate control of product microstructure has made it the "go to method" for large scale applications. In the hydrothermal method, pressure, temperature and concentration are the key parameters that control and determine the success of the synthesis. Water is the commonly utilized solvent and its dielectric constant and density are highly dependent on the pressure and temperature. An increase in the temperature and decrease in pressure

Figure 2. Different fabrication methods for semiconductor materials

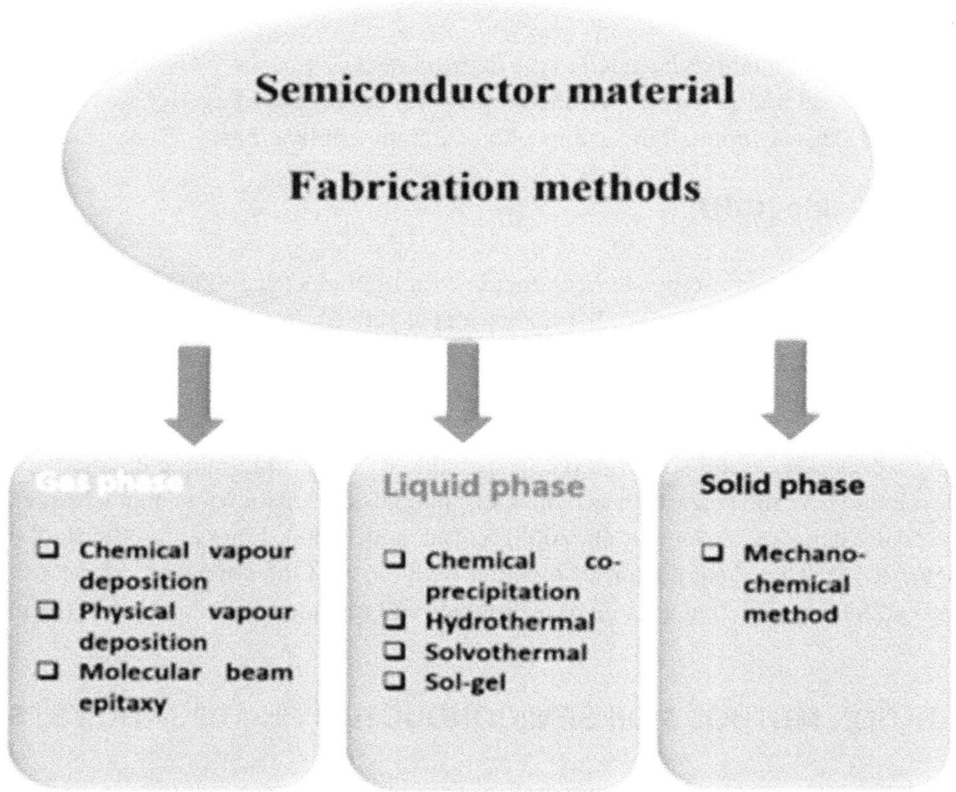

results in the decrease of the dielectric water constant. Subsequently, this results in the facilitation of nucleation growth of the crystals due to the enhanced reaction rate after the dielectric constant decreases.

Vu et al. synthesized TiO_2 nanotubes using the one-step robust hydrothermal method from the commercial Degussa P-25 powder (Vu et al., 2014). Various parameters that affect the morphology of the TiO_2 nanotubes such as reaction temperature, NaOH concentration and the reaction time and temperature were evaluated (Vu et al., 2014). Characterization revealed that stirring minimizes the reaction time it takes for the particles to form nanotubes. Furthermore, the nanotubes were formed when the temperature reached 130 °C. Yu et al. synthesized ZnO using the one step/one pot hydrothermal synthesis method for the photodegradation of rhodamine B (Yu & Yu, 2008). Glucose precursor and $ZnCl_2$ mixture were used at 180 °C for 24 hrs and then annealed at different temperatures for 4 h. The ZnO was characterised and tested for photocatalytic activity. The results showed the structural properties of ZnO (shell thickness, specific surface area, crystalline size and pore structure) were controlled by varying the molar ratio of zinc ions to glucose (Yu & Yu, 2008).

Viet et al. synthesized SnO_2 nanoparticles at low temperature using the hydrothermal method (Viet et al., 2016). The formation of the SnO_2 nanoparticles was confirmed through characterization. The photocatalytic activity of the nanoparticles was tested through the photodegradation of methylene blue wherein 88.88% of it was degraded in 120 minutes using UV irradiation and 90 minutes under sunlight illumination (Viet et al., 2016). Adhikari et al., synthesized hierarchical WO_3 structures using ethylenediaminetetraacetic acid (EDTA) mediated microwave hydrothermal synthesis method (Matalkeh et al., 2022). EDTA is a complexing agent that determines the morphology of the material. In the method, titanium dihydrate solution was mixed with EDTA solution and HCl solution. The resultant solution was transferred to a Teflon lined vessel and microwaved at 200°C for 2 h at 200W power. The resultant precipitate was washed and dried overnight at 70°C. The results showed that EDTA improved the crystallinity of WO_3 by 2.6 times (Matalkeh et al., 2022). CdO nanoparticles were synthesized using the hydrothermal method by Somasundaram (Somasundaram et al., 2019). Cadmium nitrate and ammonium hydroxide were used as precursor material and after the hydrothermal reaction, the CdO was annealed at 400 °C. The CdO nanoparticles were then tested for the treatment of methylene blue under UV light wherein and up to 80.41% of it was degraded.

2.2 Solvothermal Method

The solvothermal synthesis method is very similar to the hydrothermal method, the only difference is the precursor solution. In this method, non-aqueous solvents such as dimethylformamide are used. Similar to the hydrothermal method, the solvothermal method is based on heating both the solvent and precursor in an autoclave reactor at high pressure and temperature, however the temperatures are usually higher than the boiling point of the precursor solvent. The uncommon behaviour of the solvent under supercritical conditions exerts various effects on the precursor resulting in the generation of the desired material. The downfall of this method is that the solvents used are often not environmentally friendly and as such, the hydrothermal method is preferred over this method. Though, this is the case, a lot of semiconductor materials have been synthesized using this method. Nam et al., for instance synthesized TiO_2 nanostructures using a highly alkaline ketone/water solution with Degussa P-25 powder as a precursor material (Nam et al., 2013). The morphologies and photocatalytic activities of the TiO_2 nanostructures were tested. The solvent enhanced the nanostructure properties and thus resulted in a high photocatalytic activity of the as-prepared TiO_2 materials. Using the solvothermal method depicted in Figure 3, Wang et al. synthesized ZnO nanoparticles for the degradation of methyl orange and p-nitrophenol (Y. Wang et al., 2021) Zinc acetate dihydrate, a precursor material, was mixed with NaOH and ethanol followed by stirring for 2 h. A precipitate gradually formed and was transferred to a Teflon liner and sealed in an autoclave at 180 °C for 12 h. The resultant precipitate was dried and annealed at 400 °C. The obtained ZnO nanoparticles were characterized for morphological and photocatalytic properties for the degradation of methyl orange and p-nitrophenol. Figure 3 is the schematic diagram of the synthesis process.

The morphological analysis of the ZnO nanoparticles through Scanning Electron Microscopy (SEM) are shown in **Figure 4** and they demonstrate that the particle sizes are uniform and the structure between the particles is loose. The loose structure has nanoparticles that are capable of improving the photocatalytic performance of ZnO (Tian et al., 2008; Y. Wang et al., 2021).

The study of the photocatalytic degradation of the methyl orange and p-nitrophenol showed that both methyl orange and p-nitrophenol depicted higher degradation efficiency under UV light (Y. Wang et al., 2021).

He et al. synthesized SnO_2 nanoparticles for the degradation of rhodamine B using the microwave solvothermal method and tin chloride dihydrate as a precursor material (Z. He & Zhou, 2013). The material was characterized and tested for photocatalytic activity under UV light. The SnO_2 depicted a high photocatalytic activity in the degradation of rhodamine B and a red shift was observed in the UV-vis absorption. This was attributed to the unique structure obtained. Balta et al. synthesized a WO_3 composite

Figure 3. Solvothermal synthesis method using zinc acetate dihydrate as a precursor material for the degradation of methyl orange and p-nitrophenol
Source: Reprinted with permission from Y. Wang et al. (2021) MDPI Open Access

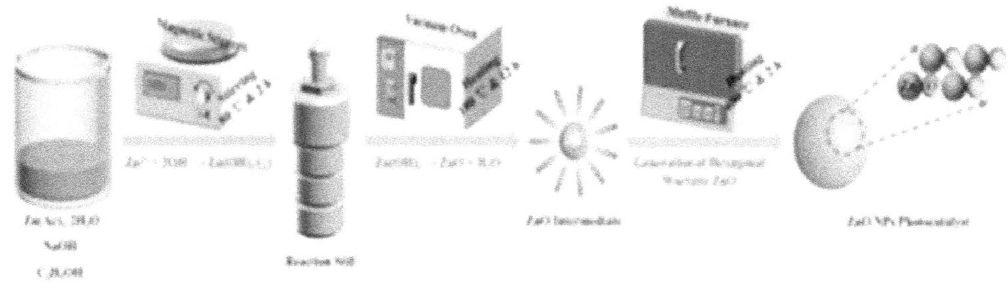

Figure 4. SEM images of ZnO nanoparticles: a) shows the agglomerated nanoparticles, b) the uniformity of the nanoparticles can be observe.
Source: Reprinted with permission from (Y. Wang et al., 2021) MDPI Open Access

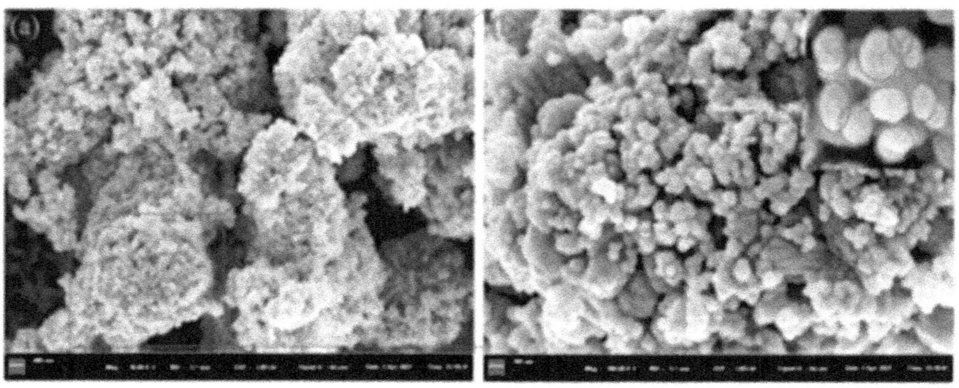

using the solvothermal synthesis method to enhance the photocatalytic activity of the material under sunlight illumination (Balta et al., 2019). Orange II azo dye was used to test the degradation capability of the composite wherein a high degradation performance 85% was obtained. The high performance of the composite catalyst was attributed to the reduced band gap in the composite relative the individual constituents of the composite.

2.3 Sol-Gel Method

The sol-gel method is regarded as one of the most appropriate methods to synthesize different nano-structures, more particularly metal oxide nanoparticles (Bokov et al., 2021). There are basic principles with which sol-gel synthesis methods are carried out:

1) The chemistry of the precursors in organic solvents
2) The colloidal chemistry in aqueous media (Guizard, 1996).

Figure 5. Sol gel process and the SEM image of a dried aerogel
Source: Reprinted with permission from Bokov et al. (2021) Hindawi Open Access

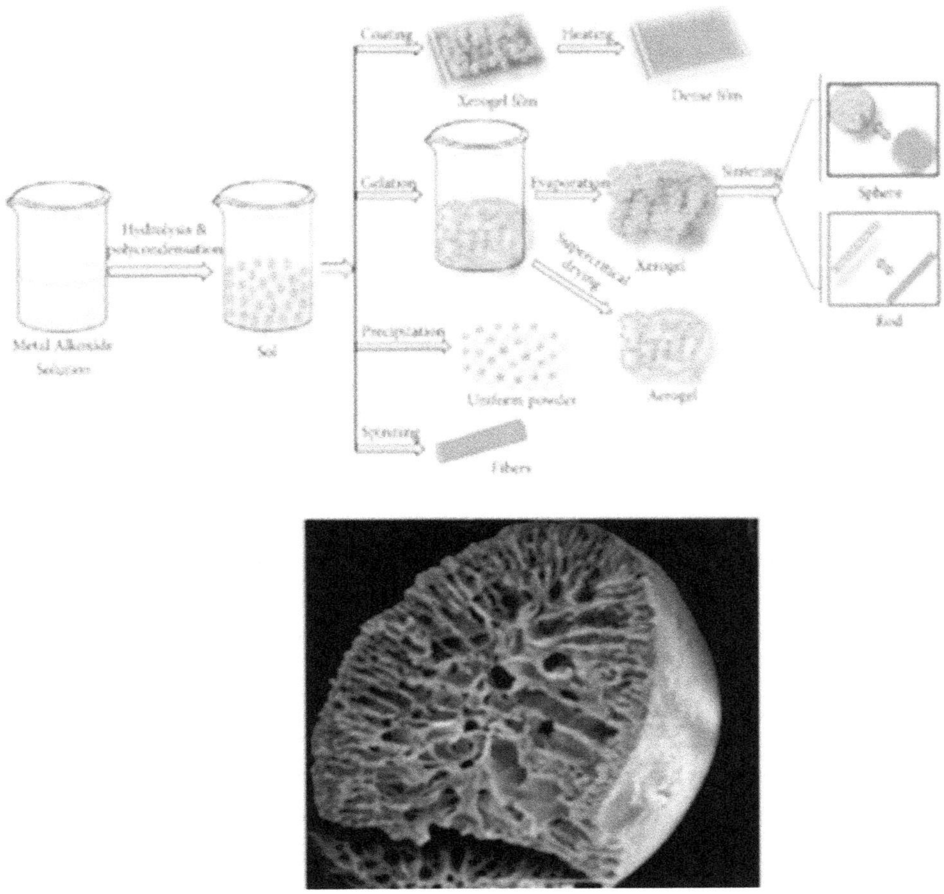

These affect the surface morphology of the resultant nanoparticles. In a typical sol gel method, a metal alkoxides precursor is dissolved in either alcohol or water because the metal alkoxides precursors used in this method usually have weaker ligands compared to water. The hydrolysed metal alkoxides then condense to produce small colloidal nanoparticles suspended in sol (the liquid formed). If the sol particles undergo further polycondensation then an extended oxo bridged network of polymeric oxide based material will be formed (Dave & Lockwood, 2016). The formed gel obtained is a biphasic material that has a gel network with a lot of water in it. Figure 5 depicts an overview of the sol-gel method and the types of processes that can be made and the products of each process. The SEM image in the figure is a product of the gelation method which has been super critically dried.

In addition to the sol-gel method having the efficiency to produce highly pure and homogeneous composites (Kajihara, 2013; *Scholar*, n.d.), it is very cheap and the low reaction temperature allows for managed control of the chemical composition of the products (Bokov et al., 2021).

Mushtaq et al. synthesized TiO_2 nanoparticles for the degradation of the 2nd generation of ofloxacin using the sol-gel method (Mushtaq et al., 2020). Titanium iso-propoxide was used as the precursor material and was calcined at different temperatures. The material was characterized wherein anatase TiO_2

was found to have been formed at low temperature. Degradation rates of 64, 76, 87 & 91% were attained from TiO_2 annealed at 450, 400, 350 and 300 °C, respectively (Mushtaq et al., 2020). These results showed that the increase in particle size and lower surface area due to increasing annealing temperature decreased the degradation rate of ofloxacin (Mushtaq et al., 2020). Cherepivska et al. synthesized WO_3 and ZnO nanoparticles using the sol-gel method for the degradation of phenol (Cherepivska & Prihod'Ko, 2014). In this work, sodium tungstate dehydrate was used a precursor tungsten source. The precursor was dissolved in water and nitric acid to the solution and it was aged for 24 h, at ambient temperature. The precipitate was filtered and calcined at different temperatures for 2 h and was subsequently characterized and tested. Synthesis of the ZnO nanoparticles was achieved by preparing an aqueous solution of zinc nitrate and urea which was heated and stirred for 24 h. The resultant precipitate was washed, dried, calcined, characterized and tested. The investigation of phenol degradation showed that the ZnO nanoparticles degraded phenol better than the WO_3 nanoparticles.

Benhebal et al. on the other hand synthesized crystallized pure SnO_2 powders using the sol-gel method for the degradation of phenol (Benhebal et al., 2013). The photocatalytic degradation of phenol was done in a glass reactor with a side external UV lamp. The phenol was degraded to about 60% after 120 min. The annealing temperature had an influence on the photocatalytic degradation of phenol. Khan et al. synthesized CdO nanoparticles using the sol gel method for the degradation of the methyl orange in wastewater. Ethanol was used as a solvent and the as prepared CdO was characterized and tested for its morphological, structural, optical and photocatalytic properties. The photocatalytic properties were tested under visible light wherein optimum results were obtained 45 min into reaction.

2.4 Chemical Co-Precipitation Method

The chemical co-precipitation synthesis method entails precipitation of a metal from a salt precursor in the form of hydroxide using a base solvent. In this method, nucleation, growth, coarsening and the agglomeration processes occur simultaneously (A. V. Rane et al., 2018). This synthesis method can be conducted using various variations such as:

1) Oxides produced from non-aqueous and aqueous solutions
2) Sonication/microwave assisted co-precipitation
3) Formation of metal from aqueous solutions either by reducing non-aqueous solutions, electrochemical reduction or decomposing metalorganic precursors (A. V. Rane et al., 2018).

Chemical co-precipitation has been preferred over other methods because it is easy to manipulate the particle composition and size. The method is also fast and simple, it is energy efficient and reaction temperatures are very low (A. V. Rane et al., 2018). In this method, materials with varying chemical components are mixed to form the precursor precipitate. The precursor precipitate is dried and calcined to produce nanoparticles. The formation of the precipitate is determined by the particles formed i.e. when the particle diameter is about 1 μm, a precipitate forms. This particle size depends on the supersaturation of the solution and the solubility of the precipitate. Generally, photocatalysts that have been synthesized using the chemical co-precipitation method have enhanced separation of electron-hole pairs thereby improving their photoactivity and they also possess good light absorption capability.

The preparation of TiO_2 photocatalysts for the degradation of methylene blue with the chemical co-precipitation has been implemented before using $TiCl_3$ and NH_4OH as precursor materials (Yudoyono et

Figure 6. SEM images of SnO_2 synthesized via chemical co-precipitation: a) Pure SnO_2, b) doped SnO_2
Source: Reprinted with permission from Letifi et al. (2021) MDPI Open Access

al., 2016). The semiconductor material was annealed at different temperatures to test the effects on the photocatalytic activity of the material. The characterization results showed that TiO_2 only had a single phase i.e. anatase. The photocatalytic activity results showed that all the samples annealed at different temperatures exhibited some activity due to the reduced band gap (Yudoyono et al., 2016). This method has also aided in the synthesis of large surface area and highly efficient titanium doped SnO_2 photocatalysts (Letifi et al., 2021). Tin chloride dihydrate and titanium isopropoxide were used as precursor materials for tin and titanium respectively. The tin chloride dihydrate was dissolved in deionised water and stirred vigorously. Titanium isopropoxide was then added and stirred for an hour. The pH was adjusted, heated, cooled and centrifuged. The resultant precipitate was then dried for 12 h at 80 °C and annealed for 4 h at 600 °C. The optical, morphological, microstructural and photocatalytic properties were tested using various characterization techniques. The SEM images of the pure SnO_2 and doped SnO_2 are shown in Figure 6. The images reveal that pure SnO_2 is marked with the occurrence of non-homogeneous large isolate aggregates and spherical particles whereas the doped SnO_2 shows a decrease in size and a more uniform dispersion of spherical particles (Letifi et al., 2021). The photocatalytic test showed that undoped SnO_2 particles depicted a higher photocatalytic activity for the degradation of rhodamine B under UV illumination. The incorporation of titanium however improved the photodegradation of rhodamine B under visible light where 52% degradation percentage was obtained within 120 min (Letifi et al., 2021).

Cruz et al. synthesized WO_3 nanoparticles for the degradation of indigo carmine using the chemical co-precipitation method (Cruz et al., 2010). Ammonium tungstate hydrate was used as a starting material and it was dissolved in deionised water at 80 °C. Concentrated nitric acid was then added dropwise and the solution was stirred at 800 rpm until a precipitate started forming. The resultant precipitate was allowed to settle, dried and then calcined at different temperatures over varying times (Cruz et al., 2010). The photocatalytic activity tests showed that WO_3 degraded indigo carmine substantially. The morphology of the nanoparticles played an important role on the degradation of indigo carmine (Cruz et al., 2010).

3. MODIFICATION OF SEMICONDUCTOR PHOTOCATALYSTS

As mentioned in the introduction, semiconductor materials are often modified so as to enhance their photocatalytic activities over a larger range of the spectrum. The methods that are commonly implemented for their modification are discussed next and they include; doping, construction of heterojunctions and dye photosensitization.

3.1 Doping

Doping; the introduction of a foreign element in the lattice of another, is a very common modification method used to enhance the photocatalytic ability of many semiconductor materials. The use of metals as dopants is the most common mode of semiconductor modification. The metal is incorporated in a metal ion state where it forms an energy or band level in the forbidden energy area and this increases the probability of the material in absorbing visible light irradiation and subsequently increasing the overall quantum efficiency. Metallic elements such as manganese (Mn), palladium (Pd), copper (Cu), silver (Ag) and tin (Sn) have been used to improve photocatalytic activity of semiconductor photocatlysts through the improvement of surface properties (S. A. Ahmed, 2017; Chakhtouna et al., 2021; Gonzalez-Hernandez et al., 2012; F. Huang et al., 2016; Jiang et al., 2021; Serpone & Emeline, 2012; Thi Quyen et al., 2021). Matalkeh et al. on the other hand synthesized WO_3 nanoparticles doped with silver (Ag) for the degradation of methylene blue (MB) and inhibition of bacteria under visible light using the deposition precipitation method. The High-resolution Transmission Electron Microscope (HRTEM) images of the nanomaterials are shown in Figure 7 and they depict that the WO_3 nanoparticles possess a sheet-like morphology with irregular sizes, with the largest particle size ranging from 500 nm. The images further reveal that the Ag nanoparticles with a size range of 3-10 nm are highly dispersed on the surface of the WO_3 nanosheets.

Figure 7. High-resolution transmission electron microscope of Ag/WO_3 and the equivalent periodicity at various level of magnifications (2nm – 100nm)

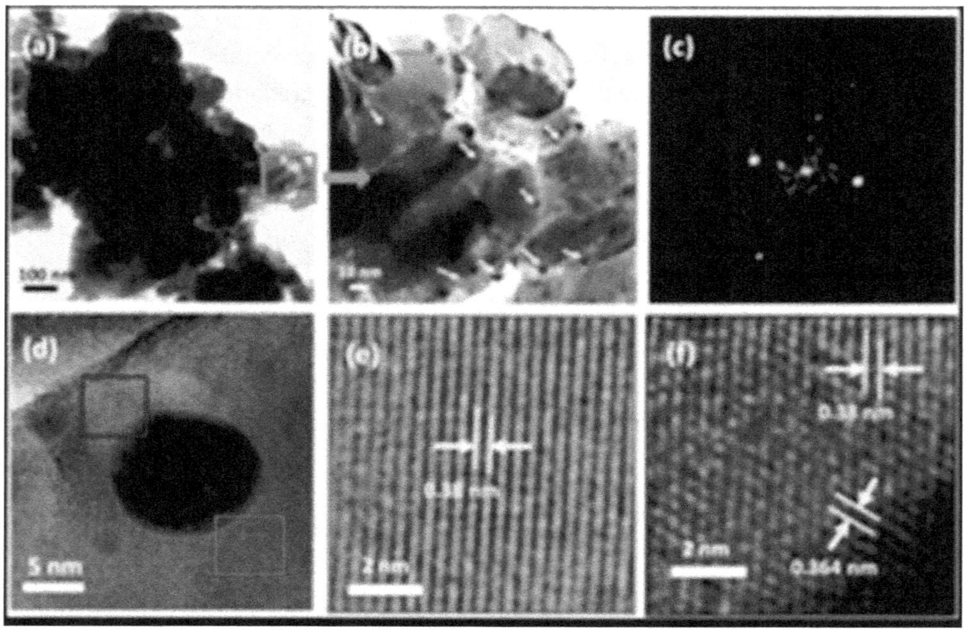

As listed **in Table 2, various** other metallic dopants have been used to modify semiconductor materials of various morphologies which were fabricated using different methods. Interestingly, they all showed enhanced photocatalytic activity against their treated pollutants post-doping relative to when the bare materials were used. Doping semiconductor materials with non-metals like nitrogen (N) and iodine (I) and boron (B) has however also been reported to enhance the photocatalytic efficiencies of semiconductor photocatalysts (Kabir et al., 2020; Szkoda et al., 2016).

Table 2. Doped semiconductor photocatalysts applied for the treatment of various water pollutants

Type of Semiconductor	Material Type	Dopant	Preparation Method	Treated Pollutants	Ref
TiO_2	Nanoparticles	Cobalt	Chemical precipitation	Rhodamine B Methylene Blue	(Noonuruk & Wattanawikkam, 2020)
	Nanotubes	Tungstate disulfide	Hydrothermal	Rhodamine B	(Wu et al., 2019)
	Membranes	Zicornium	Electrospinning	Methylene Blue	(J. Song et al., 2017)
	Nanoparticles	Gadolinium	Sol-gel	Methylene Blue	(ben Chobba et al., 2019)
ZnO	Nanomaterials	Silver	Hydrothermal	Phenols	(Sampaio et al., 2017)
	Nanoparticles	Cerium	Chemical precipitation	Methylene Blue	(Rezaei & Habibi-Yangjeh, 2013)
	Nanoparticles	Nickel	Facile electrochemical	Indigo Carmine	(Uma et al., 2019)
	Nanopowders	Aluminium	Combustion	Methyl Orange	(M. Ahmad et al., 2013)
SnO_2	Nanoparticles	Nickel	Co-precipitation	Rhodamine B, Congo red, Direct red	(Kandasamy et al., 2018)
	Nanoparticles	Vanadium	Co-precipitation	Rhodamine B	(Letifi et al., 2019)
	Nanoparticles	Zinc	Hydrothermal	Methylene Blue	(Baig et al., n.d.)
WO_3	Nanostructures	Iron	Hydrothermal	Rhodamine B	(H. Song et al., 2015)
	Nanoparticles	Zinc	Co-precipitation	Methylene Blue	(Arshad et al., 2020)
	Nanowires	Molybdenum	Hydrothermal	Methylene Blue	(X. C. Song et al., 2010)
	Nanoparticles	Europium	Sol-gel	Rhodamine B	(C. Wang & Cao, 2011)
	Nanorods	Chromium	Microwave irradiation	Methylene Blue	(Parthibavarman et al., 2018)
CdO	Nanowires	Chromium Cobalt	Co-precipitation	Methylene Blue Methyl Orange, Rhodamine B	(Munawar et al., 2021)
	Nanoparticles	Zinc	Co-precipitation	Atrazine	(Gupta et al., 2017)
	Nanoparticles	Cerium	Chemical precipitation	Methylene Blue	(Elayaraja et al., 2020)
	Nanostructures	Gadolinium Yttrium	Wet Chemical method	Methylene Blue	(Kannan et al., 2021)

3.2 Heterojunctions

The formation of heterojunctions using a semiconductor photocatalysts with another material (co-catalyst or secondary semiconductor) has the potential of broadening light harvesting properties of the formed composite material. Construction of heterojunctions also enhances the chemical stability and photo-excited charge separation of the semiconductor materials. This is generally because the heterojunction-based photocatalysts generally allow spatial separation of photogenerated charge carriers at the interface between a semiconductor and the secondary material, leading to enhanced utilization of those photogenerated charge carriers (Z. Liu et al., 2022).

Another important aspect considered when modifying semiconductor materials to improve their efficiency, is the crystal lattice structure at the point of interface. A discrepancy in the lattice spacing of two semiconductor materials may result in reduced electron-hole fusion because of the lattice disparity. As such, it is vital to select semiconductor material, paying specific attention to their band edge positions. The ideal characteristics to look for when synthesizing nanocomposites are: a) high absorption capacity of light, to use the solar spectrum efficiently, b) Enough active sites and large surface area, c) efficient segregation of charge carriers in order to manufacture sufficient active radicals for the degradation of pollutants. Numerous researchers have explored the use of carbon materials as co-catalysts in heterojunction development, including Rodríguez et al. who fabricated graphene oxide–ZnO nanocomposites the treatment of acid mine drainage (Rodríguez et al., 2020). Kumar et al. on the other hand synthesized CdO nanoparticles in-situ, on reduced graphene for the photocatalytic degradation of methylene blue under UV irradiation as depicted in **Figure 8** (Kumar et al., 2016).

Figure 8. Photocatalytic degradation of methylene blue using CdO nanoparticles/rGO
Source: Adapted from Kumar et al. (2016)

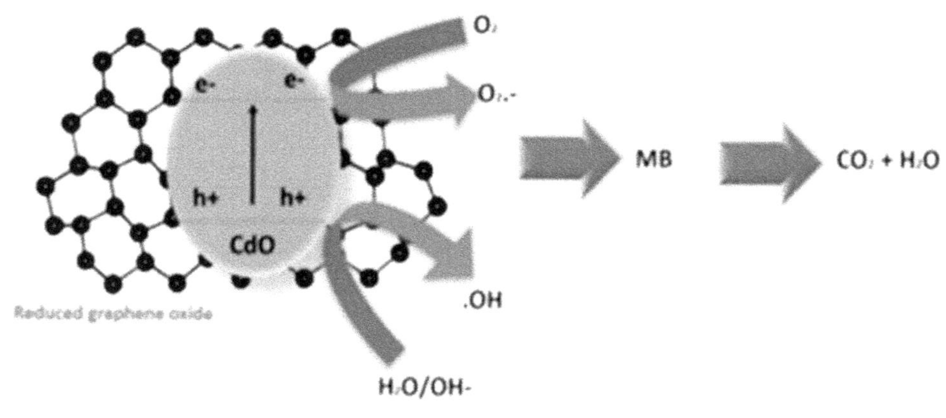

In this work, one pot in-situ hydrothermal method was used because it offers an easy and facile approach to decorate ferromagnetic CdO nanoparticles on reduced graphene sheets (Kumar et al., 2016). The photocatalytic measurement of this nanocomposite showed that approximately 80% of methylene blue dye was degraded. This was achieved because of the delay in the recombination time of electron-hole pairs due to the incorporated reduced graphene sheets. Various binary semiconductor heterojunctions have also been developed. Kanjwal et al. synthesized a TiO_2/CdO composite for the degradation of methyl orange using the electrospinning method (Kanjwal et al., 2010). The photoreaction was carried out in a glass reactor covered with aluminium foil and a ultra-violet lamp emitting light at a wavelength of 365 nm. The obtained results showed that the composite nanofibers could eliminate all of the methyl orange dye within about 75 min, whereas the pristine TiO_2 nanofibers could not eliminate more than 50% even after 180 min.

3.3 Dye Photosensitization

Another approach to improving the efficiency of semiconductor photocatalysts under sunlight is by dye photosensitization (Bayrak et al., 2016) The conjugation of dyes with semiconductor materials yields photocatalysts with two light absorbers, wherein the dye absorbs visible light, which is beneficial as most of the solar irradiation is in the visible region. The synergistic interaction of a dye with a semiconductor photocatalyst is depicted in Figure 9. The dye absorbs visible light, and an electron is promoted from the highest occupied molecular orbital (HOMO) to the lowest unoccupied molecular orbital (LUMO). The excited dye then injects its electrons into the conduction band of the excited semiconductor material (Z. Huang et al., 2014). The increased flux of electrons in the conduction band of the semiconductor enhance their reaction with molecular oxygen to generate reactive superoxide anion radicals. The holes in the valence band of semiconductor react with the adsorbed water molecules on the surface of the photocatalyst to generate hydroxyl radicals. The generated species from the interaction of the dye and semiconductor photocatalyst are strong oxidizing agents that can decompose a range of pollutants.

Numerous researchers have implemented dye photosensitization in modifying semiconductor materials for photocatalytic applications. Keşir et al. for instance sensitized TiO_2 nanoparticles with a metallophthalocyanine for the photoreduction of Cr(VI) and photodegradation of Antibiotic Amoxicillin (Keşir et al., 2021). Using the in situ hydrothermal method, Zhang et al. fabricated metallophthalocyanine

Figure 9. Mechanism of synergistic interaction of dye photosensitized semiconductor photocatalysts

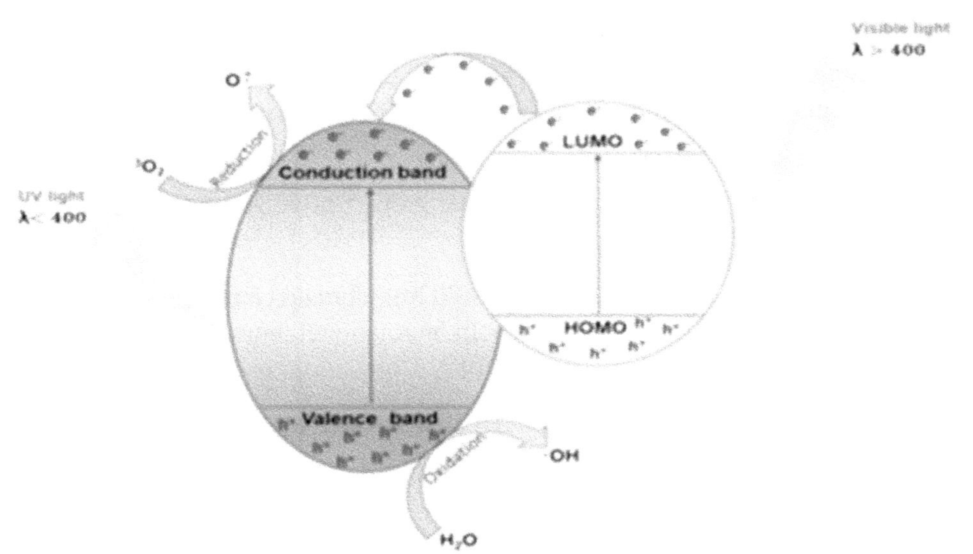

sensitized SnO_2 nanoparticles for the degradation of thiophene (G. Zhang et al., 2018). Comparison of the pure SnO_2, with its dye sensitized counterpart showed that the latter had enhanced photocatalytic activity under visible irradiation relative to the former.

All the discussed modification techniques have proven to effectively enhance the electrical, optical and photocatalytic properties of semiconductor materials. An understanding of the desired properties of a potential photocatalytic systems is thus an important step towards the tailored design and modification of semiconductor photocatalysts.

CONCLUSION

Wastewater treatment is an essential part of water systems management that works towards saving the environment and preserving life. In this chapter, the synthesis and modification methods of semiconductors for photocatalytic treatment of industrial wastewater were discussed. Liquid synthesis methods such as, hydrothermal, solvothermal, chemical co-precipitation & sol-gel were discussed in detail, mentioning recent work done in each of these methods. Furthermore, modification methods that improve the photocatalytic activity of the semiconductor material such as doping, heterojunctions and dye sensitization were discussed. More work in improving the photocatalytic activity of semiconductor material under visible light still needs to be done.

REFERENCES

Abe, R., Higashi, M., Sayama, K., Abe, Y., & Sugihara, H. (2006). Photocatalytic activity of R_3MO_7 and $R_2Ti_2O_7$ (R = Y, Gd, La; M = Nb, Ta) for water splitting into H_2 and O_2. *The Journal of Physical Chemistry B*, *110*(5), 2219–2226. doi:10.1021/jp0552933 PMID:16471807

Ahmad, M., Ahmed, E., Zhang, Y., Khalid, N. R., Xu, J., Ullah, M., & Hong, Z. (2013). Preparation of highly efficient Al-doped ZnO photocatalyst by combustion synthesis. *Current Applied Physics*, *13*(4), 697–704. doi:10.1016/j.cap.2012.11.008

Ahmad, M. Z., Chang, J., Ahmad, M. S., Waclawik, E. R., & Wlodarski, W. (2013). Non-aqueous synthesis of hexagonal ZnO nanopyramids: Gas sensing properties. *Sensors and Actuators. B, Chemical*, *177*, 286–294. doi:10.1016/j.snb.2012.11.013

Ahmed, S. A. (2017). Structural, optical, and magnetic properties of Mn-doped ZnO samples. *Results in Physics*, *7*, 604–610. doi:10.1016/j.rinp.2017.01.018

Ahmed, T., Zounemat-Kermani, M., & Scholz, M. (2020). Climate change, water quality and water-related challenges: A review with focus on Pakistan. In International Journal of Environmental Research and Public Health (Vol. 17, Issue 22, pp. 1–22). MDPI AG. https://doi.org/ doi:10.3390/ijerph17228518

Al-Hamdi, A. M., Rinner, U., & Sillanpää, M. (2017). Tin dioxide as a photocatalyst for water treatment: A review. In Process Safety and Environmental Protection (Vol. 107, pp. 190–205). Institution of Chemical Engineers. doi:10.1016/j.psep.2017.01.022

Ali, M., Amrane, N., & Tit, N. (2020). Relevance of defects in ZnO nanotubes for selective adsorption of H_2S and CO_2 gas molecules: Ab-initio investigation. *Results in Physics*, *16*, 102907. Advance online publication. doi:10.1016/j.rinp.2019.102907

Allen, N. S., Mahdjoub, N., Vishnyakov, V., Kelly, P. J., & Kriek, R. J. (2018). The effect of crystalline phase (anatase, brookite and rutile) and size on the photocatalytic activity of calcined polymorphic titanium dioxide (TiO_2). *Polymer Degradation & Stability*, *150*, 31–36. doi:10.1016/j.polymdegradstab.2018.02.008

Álvarez, P. M., Jaramillo, J., López-Piñero, F., & Plucinski, P. K. (2010). Preparation and characterization of magnetic TiO_2 nanoparticles and their utilization for the degradation of emerging pollutants in water. *Applied Catalysis B: Environmental*, *100*(1–2), 338–345. doi:10.1016/j.apcatb.2010.08.010

Arshad, M., Ehtisham-Ul-Haque, S., Bilal, M., Ahmad, N., Ahmad, A., Abbas, M., Nisar, J., Khan, M. I., Nazir, A., Ghaffar, A., & Iqbal, M. (2020). Synthesis and characterization of Zn doped WO_3 nanoparticles: Photocatalytic, antifungal and antibacterial activities evaluation. *Materials Research Express*, *7*(1), 015407. Advance online publication. doi:10.1088/2053-1591/ab6380

Baig, A., Baig, A., Rathinam, V., & Ramya, V. (n.d.). *Facile fabrication of Zn-doped SnO_2 nanoparticles for enhanced photocatalytic dye degradation performance under visible light exposure*. https://doi.org/doi:10.1007/s42114-020-00195-9

Balta, Z., Bilgin Simsek, E., & Berek, D. (2019). Solvothermal synthesis of WO_3/TiO_2/carbon fiber composite photocatalysts for enhanced performance under sunlight illumination. *Photochemistry and Photobiology*, *95*(6), 1331–1338. doi:10.1111/php.13117 PMID:31074885

Banerjee, D. K. (1986). Water pollution and human health. *Water and Sanitation at Mid-Decade: Proceedings of the 12th WEDC Conference*, 47–50. 10.4066/2529-8046.100020

Bashir, I., Lone, F. A., Bhat, R. A., Mir, S. A., Dar, Z. A., & Dar, S. A. (2020). Concerns and threats of contamination on aquatic ecosystems. In Bioremediation and Biotechnology: Sustainable Approaches to Pollution Degradation (pp. 1–26). Springer International Publishing. doi:10.1007/978-3-030-35691-0_1

Bayrak, R., Albay, C., Koç, M., Altin, I., Değirmencioğlu, I., & Sökmen, M. (2016). Preparation of phthalocyanine/TiO_2 nanocomposites for photocatalytic removal of toxic Cr(VI) ions. *Process Safety and Environmental Protection*, *102*, 294–302. doi:10.1016/j.psep.2016.03.023

Belver, C., Bedia, J., Gómez-Avilés, A., Peñas-Garzón, M., & Rodriguez, J. J. (2018). Semiconductor Photocatalysis for Water Purification. In *Nanoscale Materials in Water Purification* (pp. 581–651). Elsevier. doi:10.1016/B978-0-12-813926-4.00028-8

ben Chobba, M., Messaoud, M., Weththimuni, M. L., Bouaziz, J., Licchelli, M., de Leo, F., & Urzì, C. (2019). Preparation and characterization of photocatalytic Gd-doped TiO_2 nanoparticles for water treatment. *Environmental Science and Pollution Research*, *26*(32), 32734–32745. doi:10.1007/s11356-019-04680-7

Benhebal, H., Chaib, M., Léonard, A., Tasseroul, L., & Paèz, C., Salmon, T., Poelman, D., Crine, M., Heinrichs, B., & Pirard, J.-P. (2013). Sol-gel preparation and characterisation of SnO_2 powders employed as catalyst for phenol photodegradation. *Scientia Iranica*, *20*(6), 1891–1898.

Bokov, D., Turki Jalil, A., Chupradit, S., Suksatan, W., Javed Ansari, M., Shewael, I. H., Valiev, G. H., & Kianfar, E. (2021). Nanomaterial by Sol-Gel Method: Synthesis and Application. *Advances in Materials Science and Engineering*, *5102014*, 1–21. Advance online publication. doi:10.1155/2021/5102014

Briffa, J., Sinagra, E., & Blundell, R. (2020). Heavy metal pollution in the environment and their toxicological effects on humans. In Heliyon (Vol. 6, Issue 9). Elsevier Ltd. doi:10.1016/j.heliyon.2020.e04691

Chakhtouna, H., Benzeid, H., Zari, N., Qaiss, A., & Bouhfid, R. (2021). Recent progress on Ag/TiO_2 photocatalysts: Photocatalytic and bactericidal behaviors. *Environmental Science and Pollution Research International*, *28*(33), 44638–44666. doi:10.100711356-021-14996-y PMID:34212334

Chang, Y. C. (2015). Cadmium hydroxide and oxide nanoporous walls with high performance photocatalytic properties. *Journal of Alloys and Compounds*, *637*, 112–118. doi:10.1016/j.jallcom.2015.02.214

Chen, D., Sivakumar, M., & Ray, A. K. (2000). Heterogeneous Photocatalysis in Environmental Remedia tion. In *Dev* (Vol. 8). Chem. Eng. Mineral Process.

Chen, X., Jing, X., Wang, J., Liu, J., Song, D., & Liu, L. (2013). Self-assembly of ZnO nanoparticles into hollow microspheres via a facile solvothermal route and their application as gas sensor. *CrystEngComm*, *15*(36), 7243–7249. doi:10.1039/c3ce40654g

Chen, Y., Tian, G., Ren, Z., Tian, C., Pan, K., Zhou, W., & Fu, H. (2011). Solvothermal synthesis, characterization, and formation mechanism of a single-layer anatase TiO_2 nanosheet with a porous structure. *European Journal of Inorganic Chemistry*, *5*(5), 754–760. doi:10.1002/ejic.201000999

Cheng, Y., Huang, W., Zhang, Y., Zhu, L., Liu, Y., Fan, X., & Cao, X. (2010). Preparation of TiO_2 hollow nanofibers by electrospining combined with sol-gel process. *CrystEngComm*, *12*(7), 2256–2260. doi:10.1039/b922564a

Cherepivska, M. K., & Prihod'Ko, R. V. (2014). Sol-Gel Synthesized Semiconductor Oxides in Photocatalytic Degradation of Phenol. *ISRN Physical Chemistry*, *2014*, 1–7. Advance online publication. doi:10.1155/2014/724095

da Silva-Rêgo, L. L., de Almeida, L. A., & Gasparotto, J. (2022). Toxicological effects of mining hazard elements. In Energy Geoscience (Vol. 3, Issue 3, pp. 255–262). KeAi Communications Co. doi:10.1016/j.engeos.2022.03.003

Dave, B. C., & Lockwood, S. B. (2016). *Sol–Gel Method BT.* In B. Bhushan (Ed.), *Encyclopedia of Nanotechnology* (pp. 3748–3761). Springer Netherlands. doi:10.1007/978-94-017-9780-1_359

de Matteis, V., Cascione, M., Brunetti, V., Toma, C. C., & Rinaldi, R. (2016). Toxicity assessment of anatase and rutile titanium dioxide nanoparticles: The role of degradation in different pH conditions and light exposure. *Toxicology In Vitro*, *37*, 201–210. doi:10.1016/j.tiv.2016.09.010 PMID:27622577

Deng, Y., Feng, C., Tang, L., Zeng, G., Chen, Z., & Zhang, M. (2018). Nanohybrid Photocatalysts for Heavy Metal Pollutant Control. In *Nanohybrid and Nanoporous Materials for Aquatic Pollution Control* (pp. 125–153). Elsevier. doi:10.1016/B978-0-12-814154-0.00005-0

Dette, C., Pérez-Osorio, M. A., Kley, C. S., Punke, P., Patrick, C. E., Jacobson, P., Giustino, F., Jung, S. J., & Kern, K. (2014). TiO_2 anatase with a bandgap in the visible region. *Nano Letters*, *14*(11), 6533–6538. doi:10.1021/nl503131s PMID:25252265

Dey, A., Singh, A., Das, D., & Iyer, P. K. (2015). Organic semiconductors: A new future of nanodevices and applications. In *Thin Film Structures in Energy Applications* (pp. 97–128). Springer International Publishing. doi:10.1007/978-3-319-14774-1_4

Doyan, A. (1816). The effect of dopant material to optical properties: Energy band gap Tin Oxide thin film. *Journal of Physics: Conference Series*, (1). Advance online publication. doi:10.1088/1742-6596/1816/1/012114

Elayaraja, M., Punithavathy, I. K., Jothibas, M., Muthuvel, A., & Jeyakumar, S. J. (2020). Effect of rare-earth metal ion Ce^{3+} on the structural, optical and photocatalytic properties of CdO nanoparticles. *Nanotechnology for Environmental Engineering*, *5*(3), 25. Advance online publication. doi:10.100741204-020-00091-z

Erkan, A., Bakir, U., & Karakas, G. (2006). Photocatalytic microbial inactivation over Pd doped SnO_2 and TiO_2 thin films. *Journal of Photochemistry and Photobiology A Chemistry*, *184*(3), 313–321. doi:10.1016/j.jphotochem.2006.05.001

Espitia, P. J. P., Soares, N. de F. F., Coimbra, J. S. dos R., de Andrade, N. J., Cruz, R. S., & Medeiros, E. A. A. (2012). Zinc Oxide Nanoparticles: Synthesis, Antimicrobial Activity and Food Packaging Applications. In Food and Bioprocess Technology (Vol. 5, Issue 5, pp. 1447–1464). doi:10.100711947-012-0797-6

Fang, W., Yang, Y., Yu, H., Dong, X., Wang, T., Wang, J., Liu, Z., Zhao, B., & Yang, M. (2016). One-step synthesis of flower-shaped WO_3 nanostructures for a high-sensitivity room-temperature NOx gas sensor. *RSC Advances*, *6*(108), 106880–106886. doi:10.1039/C6RA21322G

Fang, Y., Kong, X., Wang, D., Cui, S., & Liu, J. (2018). Role of dopant Ga in tuning the band gap of rutile TiO_2 from first principles. *Zhongguo Wuli Xuekan*, *56*(4), 1370–1377. doi:10.1016/j.cjph.2018.04.011

Faudoa-Arzate, A., Arteaga-Durán, A., Saenz-Hernández, R. J., Botello-Zubiate, M. E., Realyvazquez-Guevara, P. R., & Matutes-Aquino, J. A. (2017). HRTEM microstructural characterization of β-WO_3 thin films deposited by reactive RF magnetron sputtering. *Materials (Basel)*, *10*(2), 200. Advance online publication. doi:10.3390/ma10020200 PMID:28772559

Foletto, E. L., Battiston, S., Collazzo, G. C., Bassaco, M. M., & Mazutti, M. A. (2012). Degradation of leather dye using CeO_2-SnO_2 nanocomposite as photocatalyst under sunlight. *Water, Air, and Soil Pollution*, *223*(9), 5773–5779. doi:10.100711270-012-1313-3

Gao, W., Lu, J., Zhang, S., Zhang, X., Wang, Z., Qin, W., Wang, J., Zhou, W., Liu, H., & Sang, Y. (2019). Suppressing Photoinduced Charge Recombination via the Lorentz Force in a Photocatalytic System. *Advancement of Science*, *6*(18), 1901244. Advance online publication. doi:10.1002/advs.201901244 PMID:31559139

George, R., Bahadur, N., Singh, N., Singh, R., Verma, A., & Shukla, A. K. (2016). Environmentally Benign TiO_2 Nanomaterials for Removal of Heavy Metal Ions with Interfering Ions Present in Tap Water. *Materials Today: Proceedings*, *3*(2), 162–166. doi:10.1016/j.matpr.2016.01.051

Gnanaprakasam, A., Sivakumar, V. M., Sivayogavalli, P. L., & Thirumarimurugan, M. (2015). Characterization of TiO_2 and ZnO nanoparticles and their applications in photocatalytic degradation of azodyes. *Ecotoxicology and Environmental Safety*, *121*, 121–125. doi:10.1016/j.ecoenv.2015.04.043 PMID:25937630

González-Borrero, P. P., Sato, F., Medina, A. N., Baesso, M. L., Bento, A. C., Baldissera, G., Persson, C., Niklasson, G. A., Granqvist, C. G., & Ferreira Da Silva, A. (2010). Optical band-gap determination of nanostructured WO3 film. *Applied Physics Letters*, *96*(6), 061909. Advance online publication. doi:10.1063/1.3313945

Gonzalez-Hernandez, R., Perez, W., & Rodriguez, M. J. (2012). Magnetic ordering and electronic properties of Pd-doped ZnO. *Physica Status Solidi. B, Basic Research*, *249*(1), 198–201. doi:10.1002/pssb.201147113

Gonzalez-Perez, A., Persson, K. M., & Samuelson, L. (2018). Semiconductor Eco-Materials for Water Treatment. In *Handbook of Ecomaterials* (pp. 1–27). Springer International Publishing. doi:10.1007/978-3-319-48281-1_170-1

Guizard, C. (1996). Chapter 7 Sol-gel chemistry and its application to porous membrane processing. In *Fundamentals of Inorganic Membrane Science and Technology* (Vol. 4, pp. 227–258). Elsevier. https://doi.org/https://doi.org/10.1016/S0927-5193(96)80010-6

Guo, Z., Li, M., & Liu, J. (2008). Highly porous CdO nanowires: Preparation based on hydroxy- and carbonate-containing cadmium compound precursor nanowires, gas sensing and optical properties. *Nanotechnology*, *19*(24), 245611. Advance online publication. doi:10.1088/0957-4484/19/24/245611 PMID:21825823

Gupta, V. K., Fakhri, A., Tahami, S., & Agarwal, S. (2017). Zn doped CdO nanoparticles: Structural, morphological, optical, photocatalytic and anti-bacterial properties. *Journal of Colloid and Interface Science*, *504*, 164–170. doi:10.1016/j.jcis.2017.05.026 PMID:28550746

Hariharan, C. (2006). Photocatalytic degradation of organic contaminants in water by ZnO nanoparticles: Revisited. *Applied Catalysis A, General*, *304*(1–2), 55–61. doi:10.1016/j.apcata.2006.02.020

He, H. (2020). Metal oxide semiconductors and conductors. In *Solution Processed Metal Oxide Thin Films for Electronic Applications* (pp. 7–30). Elsevier. doi:10.1016/B978-0-12-814930-0.00002-5

He, Z., & Zhou, J. (2013). Synthesis, characterization, and activity of tin oxide nanoparticles: Influence of solvothermal time on photocatalytic degradation of rhodamine B. *Mod. Res. Catal*, *2*(3), 13–18. doi:10.4236/mrc.2013.23A003

Hu, W., & Yuan, X. (2017). Facile Hydrothermal Synthesis of SnO_2 Nanospheres as Photocatalysts. *Journal of Nanomaterials*, *2017*, 1–6. Advance online publication. doi:10.1155/2017/6976203

Huang, F., Yan, A., & Zhao, H. (2016). *Influences of Doping on Photocatalytic Properties of TiO2 Photocatalyst*. doi:10.5772/63234

Huang, Z., Zheng, B., Zhu, S., Yao, Y., Ye, Y., Lu, W., & Chen, W. (2014). Photocatalytic activity of phthalocyanine-sensitized TiO_2-SiO_2 microparticles irradiated by visible light. *Materials Science in Semiconductor Processing*, *25*, 148–152. doi:10.1016/j.mssp.2013.10.014

Hussain, A., & Asi, M. R. (2008). Pesticides as water pollutants. In Groundwater for Sustainable Development: Problems, Perspectives and Challenges (pp. 95–101). CRC Press. https://doi.org/ doi:10.4018/978-1-5225-6111-8.ch001

Jiang, D., Otitoju, T. A., Ouyang, Y., Shoparwe, N. F., Wang, S., Zhang, A., & Li, S. (2021). A Review on Metal Ions Modified TiO2 for Photocatalytic Degradation of Organic Pollutants. In Catalysts (Vol. 11, Issue 9). doi:10.3390/catal11091039

Kabir, R., Saifullah, M. A. K., Ahmed, A. Z., Masum, S. M., & Molla, M. A. I. (2020). Synthesis of n-doped zno nanocomposites for sunlight photocatalytic degradation of textile dye pollutants. *Journal of Composites Science*, *4*(2), 49. Advance online publication. doi:10.3390/jcs4020049

Kajihara, K. (2013). Recent advances in sol–gel synthesis of monolithic silica and silica-based glasses. *Journal of Asian Ceramic Societies*, *1*(2), 121–133. doi:10.1016/j.jascer.2013.04.002

Kandasamy, M., Seetharaman, A., Sivasubramanian, D., Nithya, A., Jothivenkatachalam, K., Maheswari, N., Gopalan, M., Dillibabu, S., & Eftekhari, A. (2018). Ni-Doped SnO_2 Nanoparticles for Sensing and Photocatalysis. *ACS Applied Nano Materials*, *1*(10), 5823–5836. doi:10.1021/acsanm.8b01473

Kanjwal, M. A., Barakat, N. A. M., Sheikh, F. A., & Kim, H. Y. (2010). Electronic characterization and photocatalytic properties of TiO_2/CdO electrospun nanofibers. *Journal of Materials Science*, *45*(5), 1272–1279. doi:10.100710853-009-4078-3

Kannan, K., Radhika, D., Reddy, K. R., Raghu, A., Sadasivuni, K. K., Palani, G., & Gurushankar, K. (2021). $Gd^{3+}$and Y^{3+}co-doped mixed metal oxide nanohybrids for photocatalytic and antibacterial applications. *Nano Express*, *2*(1). Advance online publication. doi:10.1088/2632-959X/abdd87

Kar, P., Shukla, K., Jain, P., Sathiyan, G., & Gupta, R. K. (2021). Semiconductor based photocatalysts for detoxification of emerging pharmaceutical pollutants from aquatic systems: A critical review. *Nano Materials Science*, *3*(1), 25–46. doi:10.1016/j.nanoms.2020.11.001

Karunakaran, C., Sakthiraadha, S., & Gomathisankar, P. (2013). Photocatalytic and bactericidal activities of hydrothermally and sonochemically prepared Fe_2O_3-SnO_2 nanoparticles. *Materials Science in Semiconductor Processing*, *16*(3), 818–824. doi:10.1016/j.mssp.2012.12.030

Kasem Mahmoud, E., & Mohamed Ghoneim, A. (2016). Effect of polluted water on soil and plant contamination by heavy metals in El-Mahla El-Kobra, Egypt. *Solid Earth*, *7*(2), 703–711. doi:10.5194e-7-703-2016

Keşir, M. K., Sökmen, M., & Bıyıklıoğlu, Z. (2021). Photocatalytic efficiency of metallo phthalocyanine sensitized tio2 (Mpc/tio2) nanocomposites for Cr(vi) and antibiotic amoxicillin. *Water (Basel)*, *13*(16), 2174. Advance online publication. doi:10.3390/w13162174

Khan, M. E., Khan, M. M., & Cho, M. H. (2016). Fabrication of WO_3 nanorods on graphene nanosheets for improved visible light-induced photocapacitive and photocatalytic performance. *RSC Advances*, *6*(25), 20824–20833. doi:10.1039/C5RA24575C

Kianfar, E., & Suksatan, W. (2021). Nanomaterial by Sol-Gel Method: Synthesis and Application. *Advances in Materials Science and Engineering*, *2021*, 1–21. doi:10.1155/2021/5102014

Kim, S. P., Choi, M. Y., & Choi, H. C. (2016). Photocatalytic activity of SnO_2 nanoparticles in methylene blue degradation. *Materials Research Bulletin*, *74*, 85–89. doi:10.1016/j.materresbull.2015.10.024

Kolmakov, A., Klenov, D. O., Lilach, Y., Stemmer, S., & Moskovitst, M. (2005). Enhanced gas sensing by individual SnO_2 nanowires and nanobelts functionalized with Pd catalyst particles. *Nano Letters*, *5*(4), 667–673. doi:10.1021/nl050082v PMID:15826106

Kumar, S., Ojha, A. K., & Walkenfort, B. (2016). Cadmium oxide nanoparticles grown in situ on reduced graphene oxide for enhanced photocatalytic degradation of methylene blue dye under ultraviolet irradiation. *Journal of Photochemistry and Photobiology. B, Biology*, *159*, 111–119. doi:10.1016/j.jphotobiol.2016.03.025 PMID:27045279

Kuo, C. Y., Jheng, H. K., & Syu, S. E. (2021). Effect of non-metal doping on the photocatalytic activity of titanium dioxide on the photodegradation of aqueous bisphenol A. *Environmental Technology*, *42*(10), 1603–1611. doi:10.1080/09593330.2019.1674930 PMID:31580775

La Cruz, A. M.-D., Martínez, D. S., & Cuéllar, E. L. (2010). Synthesis and characterization of WO_3 nanoparticles prepared by the precipitation method: Evaluation of photocatalytic activity under vis-irradiation. *Solid State Sciences*, *12*(1), 88–94. doi:10.1016/j.solidstatesciences.2009.10.010

Lei, P., Wang, F., Gao, X., Ding, Y., Zhang, S., Zhao, J., Liu, S., & Yang, M. (2012). Immobilization of TiO_2 nanoparticles in polymeric substrates by chemical bonding for multi-cycle photodegradation of organic pollutants. *Journal of Hazardous Materials*, *227–228*, 185–194. doi:10.1016/j.jhazmat.2012.05.029 PMID:22658211

Lellis, B., Fávaro-Polonio, C. Z., Pamphile, J. A., & Polonio, J. C. (2019). Effects of textile dyes on health and the environment and bioremediation potential of living organisms. *Biotechnology Research and Innovation*, *3*(2), 275–290. doi:10.1016/j.biori.2019.09.001

Letifi, H., Dridi, D., Litaiem, Y., Ammar, S., Dimassi, W., & Chtourou, R. (2021). High Efficient and Cost Effective Titanium Doped Tin Dioxide Based Photocatalysts Synthesized via Co-precipitation Approach. In Catalysts (Vol. 11, Issue 7). doi:10.3390/catal11070803

Letifi, H., Litaiem, Y., Dridi, D., Ammar, S., & Chtourou, R. (2019). Enhanced Photocatalytic Activity of Vanadium-Doped SnO_2 Nanoparticles in Rhodamine B Degradation. *Advances in Condensed Matter Physics*, *2019*, 1–11. Advance online publication. doi:10.1155/2019/2157428

Li, H., Bian, Z., Zhu, J., Zhang, D., Li, G., Huo, Y., Li, H., & Lu, Y. (2007). Mesoporous titania spheres with tunable chamber stucture and enhanced photocatalytic activity. *Journal of the American Chemical Society*, *129*(27), 8406–8407. doi:10.1021/ja072191c PMID:17571890

Li, T., Cai, H., Li, C., Liu, X., & Huang, F. (2020). Rocksalt-Zincblende–Wurtzite Mixed-Phase ZnO Crystals With High Activity as Photocatalysts for Visible-Light-Driven Water Splitting. *Frontiers in Chemistry*, *8*, 351. Advance online publication. doi:10.3389/fchem.2020.00351 PMID:32411673

Liang, S., Xiao, K., Mo, Y., & Huang, X. (2012). A novel ZnO nanoparticle blended polyvinylidene fluoride membrane for anti-irreversible fouling. *Journal of Membrane Science*, *394–395*, 184–192. doi:10.1016/j.memsci.2011.12.040

Liang, Y. C., & Chang, C. W. (2019). Preparation of orthorhombic WO_3 thin films and their crystal quality-dependent dye photodegradation ability. *Coatings*, *9*(2), 90. Advance online publication. doi:10.3390/coatings9020090

Liao, D. L., & Liao, B. Q. (2007). Shape, size and photocatalytic activity control of TiO_2 nanoparticles with surfactants. *Journal of Photochemistry and Photobiology A Chemistry*, *187*(2–3), 363–369. doi:10.1016/j.jphotochem.2006.11.003

Liu, X., Li, C., Han, S., Han, J., & Zhou, C. (2003). Synthesis and electronic transport studies of CdO nanoneedles. *Applied Physics Letters*, *82*(12), 1950–1952. doi:10.1063/1.1562331

Liu, Z., Yu, Y., Zhu, X., Fang, J., Xu, W., Hu, X., Li, R., Yao, L., Qin, J., & Fang, Z. (2022). Semiconductor heterojunctions for photocatalytic hydrogen production and Cr(VI) Reduction: A review. In Materials Research Bulletin (Vol. 147). Elsevier Ltd. doi:10.1016/j.materresbull.2021.111636

Lu, H. B., Liao, L., Li, H., Tian, Y., Wang, D. F., Li, J. C., Fu, Q., Zhu, B. P., & Wu, Y. (2008). Fabrication of CdO nanotubes via simple thermal evaporation. *Materials Letters*, *62*(24), 3928–3930. doi:10.1016/j.matlet.2008.05.010

Luttrell, T., Halpegamage, S., Tao, J., Kramer, A., Sutter, E., & Batzill, M. (2015). Why is anatase a better photocatalyst than rutile? - Model studies on epitaxial TiO_2 films. *Scientific Reports*, *4*(1), 4043. Advance online publication. doi:10.1038rep04043 PMID:24509651

Macak, J. M., Zlamal, M., Krysa, J., & Schmuki, P. (2007). Self-organized TiO_2 nanotube layers as highly efficient photocatalysts. *Small*, *3*(2), 300–304. doi:10.1002mll.200600426 PMID:17230591

Malekshoar, G., Pal, K., He, Q., Yu, A., & Ray, A. K. (2014). Enhanced solar photocatalytic degradation of phenol with coupled graphene-based titanium dioxide and zinc oxide. *Industrial & Engineering Chemistry Research*, *53*(49), 18824–18832. doi:10.1021/ie501673v

Mali, S. S., Patil, J., Kim, H., & Hong, C. K. (2018). Synthesis of SnO_2 nanofibers and nanobelts electron transporting layer for efficient perovskite solar cells. *Nanoscale*, *10*(17), 8275–8284. doi:10.1039/C8NR00695D PMID:29687121

Matalkeh, M., Nasrallah, G. K., Shurrab, F. M., Al-Absi, E. S., Mohammed, W., Elzatahry, A., & Saoud, K. M. (2022). Visible Light Photocatalytic Activity of Ag/WO_3 Nanoparticles and its Antibacterial Activity Under Ambient Light and in The Dark. *Results in Engineering*, *13*, 100313. doi:10.1016/j.rineng.2021.100313

Messaadi, C., Ghrib, T., Jalali, J., Ghrib, M., Alyami, A., Gaidi, M., Manso-Silván, M., & Ezzaouia, H. (2018). Synthesis and Characterization of SnO2-TiO2 Nanocomposites Photocatalysts. *Current Nanoscience*, *14*. Advance online publication. doi:10.2174/1573413714666180927110912

Munawar, T., Rehman, M. N., Nadeem, M. S., Mukhtar, F., Manzoor, S., Ashiq, M. N., & Iqbal, F. (2021). Facile synthesis of Cr-Co co-doped CdO nanowires for photocatalytic, antimicrobial, and supercapacitor applications. *Journal of Alloys and Compounds*, *885*, 160885. Advance online publication. doi:10.1016/j.jallcom.2021.160885

Mushtaq, K., Saeed, M., Gul, W., Munir, M., Firdous, A., Yousaf, T., Khan, K., Sarwar, H. M. R., Riaz, M. A., & Zahid, S. (2020). Synthesis and characterization of TiO_2 via sol-gel method for efficient photocatalytic degradation of antibiotic ofloxacin. *Inorganic and Nano-Metal Chemistry*, *50*(7), 580–586. doi:10.1080/24701556.2020.1722695

Nakata, K., & Fujishima, A. (2012). TiO2 photocatalysis: Design and applications. In Journal of Photochemistry and Photobiology C: Photochemistry Reviews (Vol. 13, Issue 3, pp. 169–189). doi:10.1016/j.jphotochemrev.2012.06.001

Nam, C. T., Yang, W.-D., & Duc, L. M. (2013). Solvothermal Synthesis of TiO_2 Photocatalysts in Ketone Solvents with Low Boiling Points. *Journal of Nanomaterials*, *627385*, 1–11. Advance online publication. doi:10.1155/2013/627385

Nassiri Koopaei, N., & Abdollahi, M. (2017). Health risks associated with the pharmaceuticals in wastewater. In DARU, Journal of Pharmaceutical Sciences (Vol. 25, Issue 1). BioMed Central Ltd. doi:10.118640199-017-0176-y

Noonuruk, R., & Wattanawikkam, C. (2020). Visible-light-driven photodegradation of commercial dyes by the cooperation of co-doped TiO_2 material. *Current Applied Science and Technology*, 20(1), 43–51. doi:10.14456/cast.1477.4

Ohno, T., Sarukawa, K., Tokieda, K., & Matsumura, M. (2001). Morphology of a TiO_2 photocatalyst (Degussa, P-25) consisting of anatase and rutile crystalline phases. *Journal of Catalysis*, *203*(1), 82–86. doi:10.1006/jcat.2001.3316

Okello, C., Tomasello, B., Greggio, N., Wambiji, N., & Antonellini, M. (2015). Impact of population growth and climate change on the freshwater resources of Lamu Island, Kenya. *Water (Basel)*, *7*(3), 1264–1290. doi:10.3390/w7031264

Omran Alkhayatt, A. H., Al-Hussainy, I. A. D., & Al-Rikaby, O. A. C. (2014). *Annealing Effect on the Structural and Optical Properties of Sol-Gel Deposited Nanocrystalline CdO Thin Films* (Vol. 34). www.iiste.org

Ong, C. B., Ng, L. Y., & Mohammad, A. W. (2018). A review of ZnO nanoparticles as solar photocatalysts: Synthesis, mechanisms and applications. In *Renewable and Sustainable Energy Reviews* (Vol. 81, pp. 536–551). Elsevier Ltd. doi:10.1016/j.rser.2017.08.020

Ong, W. J., Tan, L. L., Chai, S. P., Yong, S. T., & Mohamed, A. R. (2014). Facet-dependent photocatalytic properties of TiO2-based composites for energy conversion and environmental remediation. In ChemSusChem (Vol. 7, Issue 3, pp. 690–719). doi:10.1002/cssc.201300924

Parthibavarman, M., Karthik, M., Sathishkumar, P., & Poonguzhali, R. (2018). Rapid synthesis of novel Cr-doped WO_3 nanorods: An efficient electrochemical and photocatalytic performance. *Journal of the Indian Chemical Society*, *15*(6), 1419–1430. doi:10.100713738-018-1342-y

Paul, S., & Choudhury, A. (2014). Investigation of the optical property and photocatalytic activity of mixed phase nanocrystalline titania. *Applied Nanoscience*, *4*(7), 839–847. doi:10.100713204-013-0264-3

Precious Sibiya, N., Rathilal, S., & Kweinor Tetteh, E. (2021). Coagulation treatment of wastewater: Kinetics and natural coagulant evaluation. *Molecules (Basel, Switzerland)*, *26*(3), 698. Advance online publication. doi:10.3390/molecules26030698 PMID:33572703

Przeździecka, E., Strąk, P., Wierzbicka, A., Adhikari, A., Lysak, A., Sybilski, P., Sajkowski, J. M., Seweryn, A., & Kozanecki, A. (2021). The Band-Gap Studies of Short-Period CdO/MgO Superlattices. *Nanoscale Research Letters*, *16*(1), 59. Advance online publication. doi:10.118611671-021-03517-y PMID:33835276

Qi, K., Cheng, B., Yu, J., & Ho, W. (2017). Review on the improvement of the photocatalytic and antibacterial activities of ZnO. In *Journal of Alloys and Compounds* (Vol. 727, pp. 792–820). Elsevier Ltd. doi:10.1016/j.jallcom.2017.08.142

Qiu, R., Zhang, D., Mo, Y., Song, L., Brewer, E., Huang, X., & Xiong, Y. (2008). Photocatalytic activity of polymer-modified ZnO under visible light irradiation. *Journal of Hazardous Materials*, *156*(1–3), 80–85. doi:10.1016/j.jhazmat.2007.11.114 PMID:18248886

Quan, H., Gao, Y., & Wang, W. (2020). Tungsten oxide-based visible light-driven photocatalysts: Crystal and electronic structures and strategies for photocatalytic efficiency enhancement. In Inorganic Chemistry Frontiers (Vol. 7, Issue 4, pp. 817–838). Royal Society of Chemistry. doi:10.1039/C9QI01516G

Radi, P. A., Brito-Madurro, A. G., Madurro, J. M., & Dantas, N. O. (2006). Characterization and Properties of CdO Nanocrystals Incorporated in Polyacrylamide. *Brazilian Journal of Physics*, *36*(2A), 412–414. doi:10.1590/S0103-97332006000300048

Radwan, E. H., Saber, M. A. K., Saber, M. E. K., & Fahmy, G. H. (2017). The Impact Of Some Organic And Inorganic Pollutants On Fresh Water (Rashid, River Nile), Egypt. *Journal of Advances in Biology*, *10*(2), 2133–2145. doi:10.24297/jab.v10i2.6481

Rameshkumar, C., Anderson, A., Ananth, D., Hinduja Mohan, T., & Subalakshmi, R. (2020). An investigation of SnO_2 nanofilm for solar cell application by spin coating technique. *Materials Today: Proceedings*, *45*, 6042–6045. doi:10.1016/j.matpr.2020.09.732

Rane, A. V., Kanny, K., Abitha, V. K., & Thomas, S. (2018). Chapter 5 - Methods for Synthesis of Nanoparticles and Fabrication of Nanocomposites. In *Micro and Nano Technologies* (pp. 121–139). Woodhead Publishing. https://doi.org/https://doi.org/10.1016/B978-0-08-101975-7.00005-1

Rane, Y. N., Shende, D. A., Raghuwanshi, M. G., Koli, R. R., Gosavi, S. R., & Deshpande, N. G. (2019). Visible-light assisted CdO nanowires photocatalyst for toxic dye degradation studies. *Optik (Stuttgart)*, *179*, 535–544. doi:10.1016/j.ijleo.2018.10.215

Rathi, B. S., & Kumar, P. S. (2021). Application of adsorption process for effective removal of emerging contaminants from water and wastewater. *Environmental Pollution*, *280*, 116995. Advance online publication. doi:10.1016/j.envpol.2021.116995 PMID:33789220

Reghunath, S., Pinheiro, D., & KR, S. D. (2021). A review of hierarchical nanostructures of TiO2: Advances and applications. In *Applied Surface Science Advances* (Vol. 3). Elsevier B.V. doi:10.1016/j.apsadv.2021.100063

Ren, G., Han, H., Wang, Y., Liu, S., Zhao, J., Meng, X., & Li, Z. (2021). Recent advances of photocatalytic application in water treatment: A review. In Nanomaterials (Vol. 11, Issue 7). MDPI AG. doi:10.3390/nano11071804

Rezaei, M., & Habibi-Yangjeh, A. (2013). Simple and large scale refluxing method for preparation of Ce-doped ZnO nanostructures as highly efficient photocatalyst. *Applied Surface Science*, *265*, 591–596. doi:10.1016/j.apsusc.2012.11.053

Rodríguez, C., Tapia, C., Leiva-Aravena, E., & Leiva, E. (2020). Graphene oxide–zno nanocomposites for removal of aluminum and copper ions from acid mine drainage wastewater. *International Journal of Environmental Research and Public Health*, *17*(18), 1–18. doi:10.3390/ijerph17186911 PMID:32967362

Saidin, N. U., Choo, T. F., Kok, K. Y., Yusof, M. R., & Ng, I. K. (2017). Fabrication and characterization of ZnO nanofibers by electrospinning. *Materials Science Forum, 888 MSF*, 309–313. https://doi.org/ doi:10.4028/www.scientific.net/MSF.888.309

Sakthivel, S., Neppolian, B., Shankar, M. V., Arabindoo, B., Palanichamy, M., & Murugesan, V. (2003). Solar photocatalytic degradation of azo dye: comparison of photocatalytic efficiency of ZnO and TiO2. In Solar Energy Materials & Solar Cells (Vol. 77). Academic Press.

Sampaio, M. J., Lima, M. J., Baptista, D. L., Silva, A. M. T., Silva, C. G., & Faria, J. L. (2017). Ag-loaded ZnO materials for photocatalytic water treatment. *Chemical Engineering Journal*, *318*, 95–102. doi:10.1016/j.cej.2016.05.105

Samuel, O., Othman, M. H. D., Kamaludin, R., Sinsamphanh, O., Abdullah, H., Puteh, M. H., & Kurniawan, T. A. (2022). WO3–based photocatalysts: A review on synthesis, performance enhancement and photocatalytic memory for environmental applications. In Ceramics International (Vol. 48, Issue 5, pp. 5845–5875). Elsevier Ltd. doi:10.1016/j.ceramint.2021.11.158

Saxena, V. K. (2005). *Impact of Pollution due to Tanneries on Groundwater Regime*. https://www.researchgate.net/publication/216087091

Serpone, N., & Emeline, A. V. (2012). Semiconductor Photocatalysis — Past, Present, and Future Outlook. *The Journal of Physical Chemistry Letters*, *3*(5), 673–677. doi:10.1021/jz300071j PMID:26286164

Somasundaram, G., Rajan, J., Sangaiya, P., & Dilip, R. (2019). Hydrothermal synthesis of CdO nanoparticles for photocatalytic and antimicrobial activities. *Results in Materials, 4*, 100044. https://doi.org/https://doi.org/10.1016/j.rinma.2019.100044

Song, H., Li, Y., Lou, Z., Xiao, M., Hu, L., Ye, Z., & Zhu, L. (2015). Synthesis of Fe-doped WO3 nanostructures with high visible-light-driven photocatalytic activities. *Applied Catalysis B: Environmental*, *166–167*, 112–120. doi:10.1016/j.apcatb.2014.11.020

Song, J., Wang, X., Yan, J., Yu, J., Sun, G., & Ding, B. (2017). Soft Zr-doped TiO_2 Nanofibrous Membranes with Enhanced Photocatalytic Activity for Water Purification. *Scientific Reports*, *7*(1), 1636. Advance online publication. doi:10.103841598-017-01969-w PMID:28487571

Song, X. C., Yang, E., Liu, G., Zhang, Y., Liu, Z. S., Chen, H. F., & Wang, Y. (2010). Preparation and photocatalytic activity of Mo-doped WO_3 nanowires. *Journal of Nanoparticle Research*, *12*(8), 2813–2819. doi:10.100711051-010-9859-8

Sun, X., Wei, D., Liu, J., & Du, Y. (2022). Free radical-enhanced formation of toxic byproduct benzoyl benzoquinone during the combined UV-chlorine treatment on BP-1. *Chemical Engineering Journal*, *433*, 134344. Advance online publication. doi:10.1016/j.cej.2021.134344

Suresh, S. (2013). Semiconductor Nanomaterials, Methods and Applications: A Review. *Nanoscience and Nanotechnology*, *3*(3), 62–74. doi:10.5923/j.nn.20130303.06

Szkoda, M., Siuzdak, K., & Lisowska-Oleksiak, A. (2016). Non-metal doped TiO_2 nanotube arrays for high efficiency photocatalytic decomposition of organic species in water. *Physica E, Low-Dimensional Systems and Nanostructures*, *84*, 141–145. doi:10.1016/j.physe.2016.06.004

Thi Quyen, V., Jitae, K., Thi Huong, P., Thi Thu Ha, L., My Thanh, D., Minh Viet, N., & Quang Thang, P. (2021). Copper doped titanium dioxide as a low-cost visible light photocatalyst for water splitting. *Solar Energy*, *218*, 150–156. doi:10.1016/j.solener.2021.02.036

Tian, L., Yang, X., Lu, P., Williams, I. D., Wang, C., Ou, S., Liang, C., & Wu, M. (2008). Hollow single-crystal spinel nanocubes: The case of zinc cobalt oxide grown by a unique Kirkendall effect. *Inorganic Chemistry*, *47*(13), 5522–5524. doi:10.1021/ic702457b PMID:18507455

Tiwari, A., Krishna, N. V., Giribabu, L., & Pal, U. (2018). Hierarchical Porous TiO_2 Embedded Unsymmetrical Zinc-Phthalocyanine Sensitizer for Visible-Light-Induced Photocatalytic H2 Production. *The Journal of Physical Chemistry C*, *122*(1), 495–502. doi:10.1021/acs.jpcc.7b09759

Toloman, D., Popa, A., Stefan, M., Silipas, T. D., Suciu, R. C., Barbu-Tudoran, L., & Pana, O. (2020). Enhanced photocatalytic activity of Co doped SnO_2 nanoparticles by controlling the oxygen vacancy states. *Optical Materials*, *110*, 110472. Advance online publication. doi:10.1016/j.optmat.2020.110472

Torres-Martínez, J. A., Mora, A., Mahlknecht, J., Kaown, D., & Barceló, D. (2021). Determining nitrate and sulfate pollution sources and transformations in a coastal aquifer impacted by seawater intrusion—A multi-isotopic approach combined with self-organizing maps and a Bayesian mixing model. *Journal of Hazardous Materials*, *417*, 126103. Advance online publication. doi:10.1016/j.jhazmat.2021.126103 PMID:34229392

Uddin, M. T., Nicolas, Y., Olivier, C., Toupance, T., Servant, L., Müller, M. M., Kleebe, H. J., Ziegler, J., & Jaegermann, W. (2012). Nanostructured SnO_2-ZnO heterojunction photocatalysts showing enhanced photocatalytic activity for the degradation of organic dyes. *Inorganic Chemistry*, *51*(14), 7764–7773. doi:10.1021/ic300794j PMID:22734686

Uma, H. B., Ananda, S., & Nandaprakash, M. B. (2019). High efficient photocatalytic treatment of textile dye and antibacterial activity via electrochemically synthesized Ni-doped ZnO nano photocatalysts. *Chemical Data Collections*, *24*, 100301. Advance online publication. doi:10.1016/j.cdc.2019.100301

Utari, A. W., & Herdiansyah, H. (2020). Using filtration as a technology to remove pollutants in domestic wastewater. *IOP Conference Series. Materials Science and Engineering*, *725*(1), 012025. Advance online publication. doi:10.1088/1757-899X/725/1/012025

Van Viet, P., Thi, C. M., & Van Hieu, L. (2016). The High Photocatalytic Activity of SnO_2 Nanoparticles Synthesized by Hydrothermal Method. *Journal of Nanomaterials*, *4231046*, 1–8. Advance online publication. doi:10.1155/2016/4231046

Vargas Hernández, J., Coste, S., García Murillo, A., Carrillo Romo, F., & Kassiba, A. (2017). Effects of metal doping (Cu, Ag, Eu) on the electronic and optical behavior of nanostructured TiO_2. *Journal of Alloys and Compounds*, *710*, 355–363. doi:10.1016/j.jallcom.2017.03.275

Vu, T. H. T., Au, H. T., Tran, L. T., Nguyen, T. M. T., Tran, T. T. T., Pham, M. T., Do, M. H., & Nguyen, D. L. (2014). Synthesis of titanium dioxide nanotubes via one-step dynamic hydrothermal process. *Journal of Materials Science*, *49*(16), 5617–5625. doi:10.100710853-014-8274-4

Wang, C., & Cao, L. (2011). Preparation, spectral characteristics and photocatalytic activity of Eu^{3+}-doped WO_3 nanoparticles. *Journal of Rare Earths*, *29*(8), 727–731. doi:10.1016/S1002-0721(10)60531-5

Wang, G., & Coppens, M. O. (2010). Rational design of hierarchically structured porous catalysts for autothermal reforming of methane. *Chemical Engineering Science*, *65*(7), 2344–2351. doi:10.1016/j.ces.2009.09.079

Wang, J. L., & Xu, L. J. (2012). Advanced oxidation processes for wastewater treatment: Formation of hydroxyl radical and application. *Critical Reviews in Environmental Science and Technology*, *42*(3), 251–325. doi:10.1080/10643389.2010.507698

Wang, L., Hu, H., Xu, J., Zhu, S., Ding, A., & Deng, C. (2019). WO_3 nanocubes: Hydrothermal synthesis, growth mechanism, and photocatalytic performance. *Journal of Materials Research*, *34*(17), 2955–2963. doi:10.1557/jmr.2019.189

Wang, Y., Sun, C., Zhao, X., Cui, B., Zeng, Z., Wang, A., Liu, G., & Cui, H. (2016). The Application of Nano-TiO2 Photo Semiconductors in Agriculture. In Nanoscale Research Letters (Vol. 11, Issue 1). Springer New York LLC. doi:10.118611671-016-1721-1

Wang, Y., Yang, C., Liu, Y., Fan, Y., Dang, F., Qiu, Y., Zhou, H., Wang, W., & Liu, Y. (2021). Solvothermal Synthesis of ZnO Nanoparticles for Photocatalytic Degradation of Methyl Orange and p-Nitrophenol. *Water (Basel)*, *13*(22), 3224. Advance online publication. doi:10.3390/w13223224

Wang, Z. L. (2004). Zinc oxide nanostructures: Growth, properties and applications. *Journal of Physics Condensed Matter*, *16*(25), R829–R858. Advance online publication. doi:10.1088/0953-8984/16/25/R01

Wu, Y., Liu, Z., Li, Y., Chen, J., Zhu, X., & Na, P. (2019). WS 2 nanodots-modified TiO_2 nanotubes to enhance visible-light photocatalytic activity. *Materials Letters*, *240*, 47–50. doi:10.1016/j.matlet.2018.12.056

Xu, F., Shen, Y., Sun, L., Zeng, H., & Lu, Y. (2011). Enhanced photocatalytic activity of hierarchical ZnO nanoplate-nanowire architecture as environmentally safe and facilely recyclable photocatalyst. *Nanoscale*, *3*(12), 5020–5025. doi:10.1039/c1nr11033k PMID:22057641

Yakuphanoglu, F. (2010). Nanocluster n-CdO thin film by sol-gel for solar cell applications. *Applied Surface Science*, *257*(5), 1413–1419. doi:10.1016/j.apsusc.2010.08.045

Yao, G., An, X., Lei, H., Fu, Y., & Wu, W. (2013). Electronic and Optical Properties of Rocksalt CdO: A first-Principles Density-Functional Theory Study. In *Modeling and Numerical Simulation of Material Science* (Issue 3). http://www.SciRP.org/journal/mnsms

Yao, P., Liu, H., Wang, D., Chen, J., Li, G., & An, T. (2018). Enhanced visible-light photocatalytic activity to volatile organic compounds degradation and deactivation resistance mechanism of titania confined inside a metal-organic framework. *Journal of Colloid and Interface Science*, *522*, 174–182. doi:10.1016/j.jcis.2018.03.075 PMID:29601959

Yu, J., & Yu, X. (2008). Hydrothermal Synthesis and Photocatalytic Activity of Zinc Oxide Hollow Spheres. *Environmental Science & Technology*, *42*(13), 4902–4907. doi:10.1021/es800036n PMID:18678024

Yudoyono, G., Ichzan, N., Zharvan, V., Daniyati, R., Santoso, H., Indarto, B., Pramono, Y. H., Zainuri, M., & Darminto. (2016). Effect of calcination temperature on the photocatalytic activity of TiO_2 powders prepared by co-precipitation of $TiCl_3$. *AIP Conference Proceedings*, *1725*(1), 20099. doi:10.1063/1.4945553

Zargar, R. A., Chackrabarti, S., Arora, M., & Hafiz, A. K. (2016). Synthesis, characterization and interpretation of screen-printed nanocrystalline CdO thick film for optoelectronic applications. *International Nano Letters*, *6*(2), 99–104. doi:10.100740089-015-0172-5

Zhang, G., Ren, J., Liu, B., Tian, M., Zhou, H., & Zhao, J. (2018). In situ hydrothermal preparation and photocatalytic desulfurization performance of metallophthalocyanine sensitized SnO_2. *Inorganica Chimica Acta*, *471*, 782–787. doi:10.1016/j.ica.2017.12.025

Zhang, J., Zhou, P., Liu, J., & Yu, J. (2014). New understanding of the difference of photocatalytic activity among anatase, rutile and brookite TiO_2. *Physical Chemistry Chemical Physics*, *16*(38), 20382–20386. doi:10.1039/C4CP02201G PMID:25144471

Zhang, Q., Liu, P., Miao, C., Chen, Z., Lawrence Wu, C. M., & Shek, C. H. (2015). Formation of orthorhombic SnO_2 originated from lattice distortion by Mn-doped tetragonal SnO2. *RSC Advances*, *5*(49), 39285–39290. doi:10.1039/C5RA04946F

Zheng, J. Y., Song, G., Hong, J., Van, T. K., Pawar, A. U., Kim, D. Y., Kim, C. W., Haider, Z., & Kang, Y. S. (2014). Facile fabrication of WO_3 nanoplates thin films with dominant crystal facet of (002) for water splitting. *Crystal Growth & Design*, *14*(11), 6057–6066. doi:10.1021/cg5012154

Zhu, D., & Zhou, Q. (2019). Action and mechanism of semiconductor photocatalysis on degradation of organic pollutants in water treatment: A review. In Environmental Nanotechnology, Monitoring and Management (Vol. 12). Elsevier B.V. https://doi.org/ doi:10.1016/j.enmm.2019.100255

Chapter 9
Nanotechnological Approach to Improve Water and Wastewater Treatment Processes:
Concept, Recent Advances, and Challenges

Jnanraj Borah
Gauhati University, India

Anupam Chetia
Indian Institute of Technology, Jodhpur, India

ABSTRACT

The issue of ensuring that people have access to water that is free from hazardous microbes and toxic chemicals presents a significant challenge that calls for a comprehensive investigation. Nanotechnology has enormous potential applications in water purification because it paves the way for the exploration and production of nanomaterials that have never-before-seen characteristics in the size range of one nanometer to one hundred nanometers. These nanomaterials make the process of water treatment more effective, which in turn reduces the amount of money, time, and effort that are required by conventional methods. This short review chapter focuses on the limitations of current methodologies for wastewater treatment, nanotechnology as an alternative for wastewater treatment, advancements made in nanomaterials for water treatment, and a brief look at potential challenges and future aspects of nanotechnology-based water treatment processes.

1. INTRODUCTION

Water is one of the vital natural resources and a necessity without which no life will survive on this planet. Even though water makes up over 70% of the Earth's surface, only 2.5% of it is really consumable. The different sources of freshwater are groundwater, river, glacier, and atmospheric water vapor. The sustainability of human life is dependent on the availability of water in its pure form. The pollutants

DOI: 10.4018/978-1-6684-6830-2.ch009

are added to the environment through human activities or industrial processes (Chaturvedi et al., 2019; Patanjali et al., 2019). The most crucial step is making sure everyone has access to safe, affordable water at the same time globally a grand challenge. Millions of people suffer from diseases mediated through contaminated water. According to a report by the World Health Organization (WHO), about 12 million individuals die every year from diseases associated with water. Around 90% of all diseases are caused by impure water in developing countries. Almost all the sources of natural freshwater are contaminated. Even the water rich countries are also inescapable from water scarcity due to the growing modern civilized industries, human population, and agricultural activities. Fresh, potable water is a global resource that is becoming difficult to obtain. Millions of lives have been lost every year due to unsafe water, and an estimated 1.2 billion people worldwide lack access to safe drinking water. (Montgomery & Elimelech, 2007; Shannon et al., 2008). The natural quality of the water has been compromised by the introduction of toxic pollutants from a variety of sources. The extensive use of chemically prepared pesticides in the farming fields is deteriorating groundwater quality. Due to different human activities, harmful contaminants are entering water supply systems worsening public health. It is anticipated that existing water problems will become much more severe over the next several years. Worldwide water shortage is occurring as a result of droughts, rising population, and increased urbanization. There are predictions that the global population will reach roughly nine billion by the year 2050, and that by the year 2075, nearly 75% will be affected by clean water shortages (Kargari & Shirazi, 2014; Tlili & Alkanhal, 2019). It is, therefore, of the highest significance to do an extensive study on the issue of providing people with access to clean and safe water. Intestinal parasitic infections and diarrheal have become major concerns regarding malnutrition as a result of poor digestion of food by people who have been sickened by water (Lima et al., 2000; Shannon et al., 2008). Bacteria and enteric viruses found in water are to blame for these illnesses. Human activities are responsible for the introduction of a significant number of pollutants into the municipal water supply systems of both developing countries and developed nations, which raises worries about both public health and the environment. In order to solve these widespread issues, significant research needs to be conducted into innovative methods of water treatment that offer financial advantages and require only a small amount of energy, while also minimising the consumption of chemicals and their impact on the surrounding environment. Existing water and wastewater treatment methods include liquid distillation, anion exchange, ultrafiltration, adsorption via activated carbon, reverse osmosis, deionization, and ultraviolet (UV) filtration. Such conventional processes involve high investment costs and low output-input ratio, and sometimes can be inefficient too. Nearly about 2-3% of the nation's total energy is consumed by water treatment processes. Biological treatment plants consume a major part of the total energy used in water treatment. To this end, there is a pressing want for more efficient, low-cost, technologically superior, and robust methods of water treatment with the involvement of minimal energy, and chemicals, and having low direct or indirect impact on the environment and human life (Chaturvedi et al., 2019; Patanjali et al., 2019; Qu et al., n.d.; Tlili & Alkanhal, 2019).

This chapter of the book contains a discussion on the current scenario of the declining quality of freshwater, which affects not only the health of people but also the ecosystem. The current water and wastewater treatment systems are examined in a comprehensive fashion for their shortcomings, and it is recognized that there is a pressing requirement for alternative water and wastewater treatment methods. A nanotechnological approach for water treatment processes is explored here as a better and more efficient alternative. In the final part of this article, we will talk about nanotechnology-based water and wastewater treatment methods, including the employment of a variety of different nanomaterials in these processes. In this short review chapter, we also take a brief look at potential challenges and future aspects of nanotechnology-based water treatment processes.

2. BACKGROUND

2.1 Clean or Pure Water and Contaminants

Clean or pure water is defined as water free from any form of contaminants, i.e., free from toxic chemicals and pathogens. According to SDWA (The Safe Drinking Water Act) contaminants are defined as any biological, chemical, physical, or radiological matter in water. Physical contaminants are organic materials or sediment suspended in different water sources like rivers, lakes, and streams caused by soil erosion. Chemical pollutants include things like pesticides, nitrogen, bleach, salts, metals, toxins that are generated by bacteria, and pharmaceuticals that have been used by humans or animals. Biological contaminants consisted of microbes or microbiological contaminants. Different radiological contaminants are unbalanced numbers of protons and neutrons, resulting in an unstable atom, which can emit ionizing radiation (Chaturvedi et al., 2019; *Types of Drinking Water Contaminants | US EPA*, n.d.).

2.2 Need for Alternative Water and Wastewater Technologies

In the last several years, an immediate rise in worldwide population, as well as overexploitation of groundwater, has resulted in moderate and serious water scarcity in a lot of areas of the planet. Different projections predict that by 2050, the global population will have increased to nine billion. The result is expected to be an increase in water contamination and the accumulation of trash in these systems. The nonexistence of treatment measures that are appropriate for wastewater control in many nations has given rise to enhanced pollution in currently limited freshwater reserves. As an outcome of greater use of water, and shortage of drinking water, high importance is going to be given on wastewater recycling. Particularly, it is going to be needed to treat water to really fulfill the global need for fresh water. Methods often used to treat wastewater include different physical, biological processes, and chemical. Higher investment expense, and in some cases poor treatment efficiency, might restrict the effects of the treatment. Because of this, new methods are always being developed to enhance or replace standard water purification practices. Taking into consideration the limits of these standard techniques, nanotechnology enters into the picture, which not just offers a toxic free ambiance, but additionally should be used for the preservation of resources

2.3 Significance of Nanotechnology in Water and Wastewater Treatment

In 1867, James Clerk Maxwell was the very first to bring up a few nano ideas in nanotechnology by way of a proposed experiment on a Maxwell's Demon, a little creature able to handle certain molecules. During the early twentieth century, the very first size measurements and observations of nanoparticles working with an ultramicroscope had been made possible in research of other nanomaterials and gold sols with sizes down to 10 nm and smaller than 10 nm (Zsigmondy 1914). Zsigmondy (1914) was the very first to characterize particle measurements utilizing the phrase nanometer and also he created the original system of distinction, based on particle dimensions in the nanometer span. Many developments in the area of nanomaterial characterization happened to be achievable with Blodgett and Langmuir (1920s) who unveiled the monolayer idea, along with Abrikosova and Derjaguin (1950s) who conducted the very first measurement of surface forces (Derjaguin, 1954). Professor Richard Feynman introduced a revolutionary idea to the world in 1959. From "There's Plenty of Room in the Bottom," he foresaw several beneficial applications of nanotechnology. The 1980s, and more specifically 1981, saw the creation of the first "scanning tunnelling microscope" (STM), which made his vision possible by allowing not only the viewing of individual molecules and atoms but also their building and manipulation to form a wide variety of nanostructures (*Maxwell's Demon: Nano Machine Of The Future Captures Great Scientist's Bold Vision -- ScienceDaily*, n.d.; Sharma, n.d.).

Nanotechnology provides a viable approach to address the issues related to disinfect and decontaminate water. Nanotechnology offers the ability to control matter at dimension comparable to or less than 100 nm (1-100 nm) to obtain materials termed as nanomaterials with specific novel properties. Nanomaterials in water treatment techniques can provide a new dimension to water purification, contaminant sensing, and detection technology. Nanomaterials' extraordinary features, such as their high surface to volume ratio, photosensitivity, optical, electro-chemical, magnetic, pore size tunability, etc., allow for the production of cutting-edge water purification methods. (Jeevanandam et al., 2018; Patanjali et al., 2019; Qu et al., n.d.). All of this theory hinges on a single observation: the phenomena that materials display entirely different behaviour at nanoscale compared to their properties at the macrostate. The

Figure 1. Schematic representation of top-down and bottom-up approach and list of some synthesis techniques

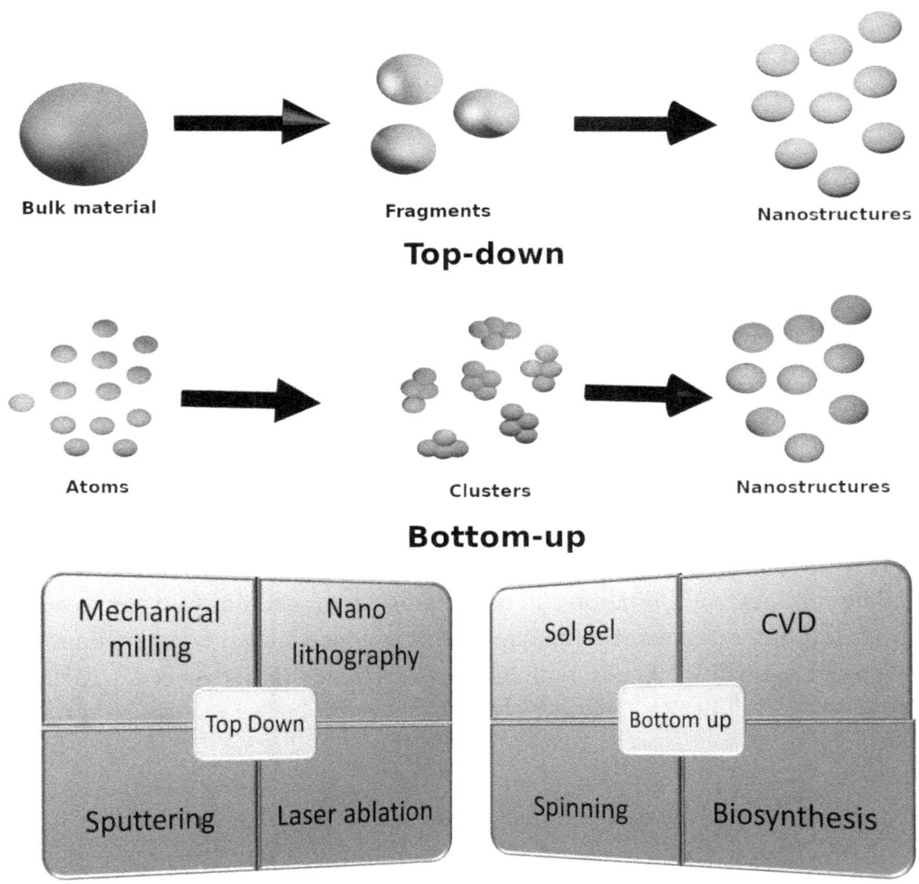

properties of nanomaterials also depend on the synthesis technique. The significant properties such as stability, adsorption capability, morphology, degree of catalyst are affected by the synthesis route taken for structuring the nanomaterial. To synthesise or even structure nanomaterials, one can use either a top-down or a bottom-up approach.

In the top-down approach, bulk materials are stably divided to get nanostructures. Sputtering, laser ablation, nanolithography are examples of the top-down approach. The bottom-up method involves creating nanostructures by forming chemical linkages between atoms. Bottom-up methods include techniques like Sol-gel, Chemical vapour deposition, spinning, and biosynthesis (Baig et al., 2021; Wagner Ferreira Sabará et al., 2017). Figure 1 is the schematic representation of top down and bottom up approach. The most crucial feature of nanomaterial for water treatment application is the high surface to volume ratio. Due to their high surface-to-volume ratio, nanomaterials may be used in a variety of water purification techniques, including adsorption, membrane process, photocatalysis, etc. As a result, nanoparticles are characterised by superior optical, electrical, and magnetic capabilities, enhanced reactivity with neighbouring molecules, and accelerated chemical reactions. Because of their unique properties, nanoparticles are a promising new method for removing impurities in water (Chaturvedi et al., 2019; Patanjali et al., 2019; Qu et al., n.d.). One may classify nanotechnology-based wastewater treatment into three broad categories: I treatment and remediation; (ii) sensing and detection; and (iii) pollution prevention. The process of treatment and remediation is given more priority (Zekić et al., 2018). The sensing and detection of pollutant molecules can provide an advantage to the water purification technology.

A number of other countries have expended significantly in nanotechnology analysis. Research into nanotechnology in the United States is state-of-the-art. Japan invests in a similar manner through its research programs dedicated for nanotechnology. South Africa developed its own National Strategy for Nanotechnology in 2006 and has innovation centres devoted to water purification. Brazil, China, India, Saudi Arabia, and other emerging countries are also members of this group. The goal is to provide water/wastewater treatment systems that are economically viable to confirm adequate hygiene issues over an extended period of time. (Zekić et al., 2018).

In developed countries about 2-3% of the total electrical consumption is used by water treatment plants. Out of these 2-3% of electrical energy consumption for municipal wastewater treatment, biological treatment alone consumes a heavy amount of energy which is close to 50-60% of the plant's total energy. Treatment of sludge, a semi solid residue produced following physical, chemical, along with biological treatment of wastewater, before disposal is extremely important as well as a challenging task. The sludge is polluted with assortment of impurities like heavy metals, pathogens, and chemicals. Sludge treatment plants are very expensive besides they occupy large area of space. Unfortunately, the market for nanotech-based wastewater solutions is still in its infancy, with only a handful of items now accessible. To fully use this technology's potential for environmental preservation, more in-depth study is required to address its current limitations. In order to use as an alternative water treatment technique nanotechnology-based water treatment techniques need to reduce energy consumption, produce less or no chemical by products, involve low cost with increase efficiency (Patanjali et al., 2019; Qu et al., n.d.; Zekić et al., 2018).

Figure 2. Advantages of using nanomaterials

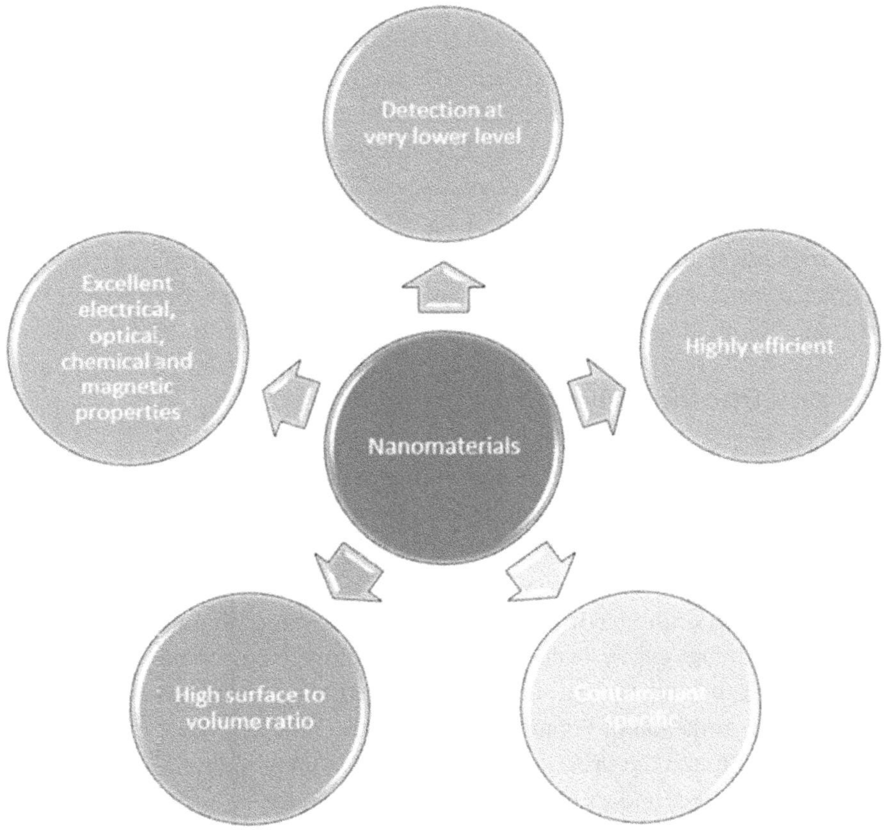

3. NANOTECHNOLOGY BASED WATER AND WASTEWATER TREATMENT TECHNIQUES

3.1 Adsorption

Nanomaterial based adsorbents or nanoadsorbents are now commonly utilized for eliminating contaminants from wastewater. Nanoparticles are very desirable as sorbents due to two distinct characteristics. Compared to bulk particles, their surface areas are large. To further enhance their affinity for target substances, nanoparticles can be functionalized with various chemical groups. Nanoadsorbents provide considerable improvements over traditional adsorbents with small intraparticle diffusion distance, and tunable surface chemistry and pore size. Strong adsorption capability may be attributed mainly to the material's large surface area. In addition, the nanoscale surface construction, which is dependent on both surface energy and size, may provide highly efficient adsorption sites, leading to increased surface-area-normalized adsorption capacity. Many different nanomaterials' surfaces might be functionalized to specifically target certain pollutants, leading to high levels of selectivity. Nanoadsorbents could be conveniently incorporated into current treatment systems like slurry reactors, filters, as well as adsorbers (Auffan et al., 2008; Qu et al., 2013).

3.2 Membrane and Membrane Process

Membrane and Membrane procedures such as ultrafiltration as (UF), nanofiltration (NF) and reverse osmosis (RO) are growing as crucial ingredients of sophisticated water purification and desalination technologies. It removes a wide variety of pollutants, makes use of non-traditional water sources (such brackish water, seawater, or wastewater), is highly automated, uses little space or chemicals, and can be scaled to accommodate a wide range of plant sizes. Nanomaterials are offering novel opportunities to produce much more effective and cost effective nanostructured as well as reactive membranes for drinking water purification and desalination. The material used in a membrane device is crucial to its performance, since it must strike a balance between allowing solvent permeability and blocking out solutes (Imran et al., 2018; Theron et al., 2008; Tlili & Alkanhal, 2019).

3.3 Catalysis and Photocatalysis

To combat the lethal consequences of untreated garbage, it is vital to design such chemical processes which are efficient and ecologically benign and to discover methods to treat trash originating from factories. Catalytic reactions account for roughly 85-90% of all chemical reactions. From the last few decades, environmental catalysis has been widely employed in many commercial and industrial sectors for the sake of conservation of the environment and for the development of the quality of our lives. Homogeneous and heterogeneous catalysts are efficiently employed to minimise water pollution, to oxidise organic particles, to limit the emission of nitrogen oxides from power plants, to remove waste and pesticides, and to regulate the emission of volatile organic compounds, etc. To speed up a chemical reaction, a catalyst is often used to make the process go by faster; this phenomenon is known as catalysis. In the end, the catalyst is regenerated, and its quantity does not change as a result. The homogeneous catalysis takes place in a single phase, where the catalyst is dispersed in an aqueous solution or gas combination with the reactants. On the other hand, heterogeneous catalysis takes place in more than one phase; for example, the reactants in this process are liquids, but the catalyst itself is solid (Zhang, 2020). These days, catalytic technologies are widely regarded as efficient methods for eliminating organic pollutants from the environment. The technique, in general, makes use of a combination of oxidation (O3, H_2O_2), irradiation, electron, and catalyst in order to generate hydroxyl radicals (.OH), superoxide radicals (O_2.), and hydroperoxyl (HO_2.) radicals, among other types of radicals, at temperatures and pressures that are typical of the environment. When it comes to boosting the breakdown and mineralization of refractory organic contaminants in water, the synergistic impact from numerous active species is far stronger than the solo effects of each species. The majority of the catalytic technologies consist of photocatalysis, electrochemical catalysis, photo-electrochemical catalysis, catalytic ozonation, and other similar processes. During the treatment of water, one of the most important procedures is the oxidation process, which can be biological, chemical, or physical. Biological oxidation is seen as being both practically and financially viable in a broad variety of contexts. Nevertheless, the utilisation of this technique can be rendered difficult due to the existence of hazardous contaminants in the treated water. Although it is a successful method, the destruction of compounds using high-temperature thermal processes is not economically practical. Chemical oxidation with the use of multiple different oxidants such as ozone, hydrogen peroxide, chlorine, or chlorine dioxide typically solves these challenges, despite the fact that in many instances, it does not result in the complete mineralization of water pollutants (LI & QU, 2009; Parvulescu et al., 2022).

Strong catalytic activity is demonstrated by nanomaterials, resulting in the oxidation of pollution molecules, leading to the production of less dangerous compounds or even the transformation of pollutants into ecologically preferable end products. Nanoparticles' unique properties—such as their high reactivity and large surface-to-volume ratio—are due mostly to their extremely small particle size. In advanced oxidation processes, nanocatalysts might be employed effectively for chemical substance oxidation of inorganic and organic contaminants in water. Extremely reactive radicals are created during these processes, and they react readily with pollution molecules. Photocatalysis is considered the most significant oxidation process. It is well recognised that advanced oxidation processes, often known as AOPs, are promising technologies for removing pollutants of rising concern from the effluents of wastewater treatment plants. For the treatment of wastewater, photocatalysis has received a lot of attention among AOPs (Parvulescu et al., 2022). In the process of photocatalysis, a substance known as a photocatalyst is present to absorb light energy, boost its energy level, and then transfer that energy to a reactive material. Upon excitation by radiation of energy greater than band gap energy of the photocatalyst material, an electron jumps to conduction band from valence leaving a hole in the valence band. The electron-hole pairs formed in the process are capable of forming highly reactive radicals and thus breaking contaminant pollutants into non-hazardous substances. In the past two decades, there has been a great increase in the use of photocatalysis on semiconductor materials as part of an effort to comprehend and duplicate natural photosynthesis, which is perhaps the most important natural chemical activity (Khan et al., 2015). The process of photosynthesis is the transformation of water and carbon dioxide into oxygen and carbohydrates through the utilisation of light from the sun. In a similar fashion, photocatalysts on semiconductor surfaces may transform water into oxygen and hydrogen through the process of photocatalysis, or they can transform carbon dioxide and water into solar fuels. Light is responsible for the creation of electron–hole pairs, which, once they reach the surface of the semiconductor nanoparticles, can cause chemical oxidation or reduction processes to take place. Multiple materials have been developed, described, and examined to this point. The production of semiconductors that have absorption in the visible range and systems that reduce the occurrence of electron–hole recombination is the primary requirement. This effect has a significant negative impact on the efficiency of the photocatalytic conversion, so minimising its occurrence is essential (Giamello & Pacchioni, 2020). In order to develop new classes of photocatalytic materials in a reasonable manner, one must first have an in-depth grasp of the fundamental physical principles that underlie the phenomena of photocatalysis. This necessitates the creation of cutting-edge and original experimental methodologies and theoretical processes, as well as their application. Important characteristics of the photocatalytic system include having the necessary band gap, having a proper shape, having a large surface area, being stable, and having the potential to be reused. Similar fundamental photocatalytic mechanisms, such as light absorption, cause a charge separation process, and result in the creation of positive holes that can oxidise organic substrates in metal oxides including vanadium, chromium, titanium, zinc, tin, and cerium. Other primary photocatalytic processes include the formation of electron holes, which are also capable of oxidising organic substrates. During this process, either ultraviolet light (UV) or visible light (or a mixture of the two) is used to activate a metal oxide. Consequently, an electron-hole pair (e-/h+) is formed when photoexcited electrons are shifted up to the conduction band. Adsorbed chemicals on the photocatalyst's surface can be reduced or oxidised depending on whether the photogenerated pair (e-/h+) is positive or negative. Two different processes are responsible for the photocatalytic activity of metal oxide: I the oxidation of OH anions result in the creation of OH radicals, and (ii) the reduction of oxygen results in the generation of O_2 radicals. Radicals and anions are both capable of interacting with pollutants and causing them to decay or otherwise be transformed into less hazardous byproducts (Khan et al., 2015). The process of photocatalysis involves the following steps

Photocatalyst + light → $e^- + h^+$ (1)

$e^-/h^+ + O_2/H_2O$ → Reactive oxygen species (2)

Reactive oxygen species + Pollutant → Degradation products (3)

Since Fujishima and Honda's 1972 discovery of water splitting on a TiO2 electrode under the irradiation of ultraviolet (UV) light, it has been proven that several semiconductors display photocatalytic characteristics. This was accomplished by exposing the TiO_2 electrode to UV light. Thanks to the work of scientists, semiconductor photocatalysts may now be used in a far wider variety of applications. The qualities of semiconductor photocatalysts are, unfortunately, generally poor for practical use. This is because of quick charge separation and restricted light absorption, both of which limit the amount of light that can be absorbed. Researchers have proposed a number of different methodologies, such as doping, dye-sensitization, and coupling, amongst others, in an effort to enhance the characteristics of semiconductor photocatalysts. However, each of these approaches comes with its own set of advantages and disadvantages. Combining semiconductor photocatalysts with metals, especially noble metals, can greatly increase their capabilities. The surface plasmon resonance (SPR) of metals and the increased charge separation at the metal–semiconductor contact makes this feasible. In the past, the fundamental disadvantage of the coupling technique was the lack of precise control over the process of fusing semiconductors and metals together. This was the case even if the price of noble metals was rather high. The coupling process, on the other hand, is now able to be sensitively regulated thanks to the advent of nanotechnology, and various novel structures have been created in order to improve the characteristics of metal/semiconductor (M/S) nanocomposite photocatalysts (Fu et al., 2011). One of the most innovative and all-encompassing strategies for overcoming the limitations of semiconductor materials is defect engineering of semiconductor photocatalyst. To change the photocatalytic activity of semiconductors, it is necessary to get a thorough understanding of their electrical characteristics, geometric properties, and chemical properties, as well as their synergistic interactions, a process known as defect engineering (Raizada et al., 2021). The vast majority of semiconductors can only take in light of the ultraviolet or visible spectrum when exposed to sunlight; as a result, only about half of the near-infrared spectrum has been used. Consequently, it is still a difficult task to commercialise photocatalysts that are capable of absorbing NIR light. Many different approaches have been taken to modify semiconductors in order to maximise the amount of solar energy that can be extracted from the sun. Some of these approaches include loading noble metals, doping nonmetal elements, combining with other semiconductors that have narrower bandgaps, developing semiconductor near IR plasmonic materials, Using upconversion (UC) luminous materials in conjunction with semiconductors and constructing Z-scheme photocatalysts (X. Li et al., 2019).

3.4 Disinfection and Microbial Control

While conventional disinfection techniques including chlorination as well as ozonation have extremely enhanced public health, the task to offer highly effective disinfection without developing dangerous disinfection byproducts (DBPs) requires scientific development. Microorganism contained in wastewater and water is able to trigger a range of diseases to people. Pathogens are defined as microorganisms that cause disease in humans and other animals. Disinfecting water is the process of eliminating disease-causing microorganisms. Disinfection is the phrase for removal of just pathogens, and it doesn't eliminate all of the microorganisms contained in water as some helpful microorganisms also contained in water. Several nanomaterials have been proven to exhibit strong antibacterial activity in addition to their remarkable adsorption and catalytic characteristics (Qu et al., 2013; Yaqoob et al., n.d.).

3.5 Sensing and Detection

Low concentrations of micropollutants, the complexity of water and wastewater matrices, and a dearth of rapid, cost-effective pathogen detection methods all contribute to the difficulty of water quality monitoring. Precise and rapid sensors capable to identify pollutants in the molecular level might improve the human power to safeguard the sustainability of the environment and human health. Successful integration of nanomaterials could deliver selective, sensitive, and fast sensors for water pollutants. With their unique electrochemical, optical, or magnetic properties, nanomaterials can increase sensor sensitivity and speed and enable multiplex target detection. Rapid identification of microorganisms is particularly important for diagnosis-based disinfection or biofilm management, in which treatment decisions are made based on data from specialized sensors to provide high-efficiency, responsive (on demand), and accurate treatment. This is often accomplished by integration of nanomaterials as well as recognition agents (e.g., carbohydrates, aptamers, antibodies, plus antimicrobial peptides) (Qu et al., 2013; Vikesland & Wigginton, 2010).

4. USE OF NANOMATERIALS IN WATER AND WASTEWATER TREATMENT PROCESSES

4.1 In Adsorption-Based Techniques

Nanomaterials possess distinctive mechanical, chemical, electrical, optical as well as numerous other attributes which let them have significantly improved absorption characteristics for many contaminants (heavy metals as well as organic pollutants). This is precisely why they are been given the name "material of the 21st century". The development of carbon nanotubes (CNTs) (Iijima, 1991) introduced the world to a brand-new carbon-based adsorption substance. As compared to the most widely known some material - activated carbon CNTs have around exactly the same high substantial specific surface, but the great advantage of theirs is based on the construction of nanomaterials and a far better configuration of carbon atoms. Due to their efficacy in eliminating both organic and metallic contaminants, carbon nanotubes (CNTs) are currently being considered as an alternative to activated carbon. The available binding sites as well as contaminant CNT interactions influence organic contaminant adsorption on CNTs. The interaction forces are π-π bonding, hydrophobic, covalent bonding, hydrogen bonding, and then electrostatic interactions. Binding sites on CNTs are definitely more accessible compared to the on activated carbon, that contains unavailable pores, particularly for bulky molecules like tetracycline. Their π electron abundant surface can function as either acceptor or electron donor for lots of polar aromatics like phenols and nitroaromatics. CNTs have an impressive adsorption proficiency for metallic ions, which are chiefly adsorbed on the surface functional groups. These can be reused by reversing with pH adjustments. Another potential alternative nanoadsorbent for metallic and organic pollutants is graphite oxide nanosheets, which might be produced economically by exfoliating graphite (Qu et al., 2013; K. Yang & Xing, 2010).

Metal-based nanoparticles, in addition to carbon nanotubes, have adsorption properties. Metal oxide nanomaterials, like nano-TiO_2 and nanomagnetite, are effective, inexpensive adsorbents (H. G. Yang et al., 2008). They are primarily used for adsorption of quite heavy metals as well as radionuclides (unstable nuclides). Their surface structure could be manipulated to maximize efficient adsorption sites for example corners, vacancies, edges, along with high energy crystallographic facets. The adsorption process utilizes the electrostatic interaction between nanoadsorbent surface and dissolved metal ions. Nanomagnetite possesses different superparamagnetic properties which allow simple separation from water in a magnetic field of low strength. Such magnetic characteristics allow a brand new category of core-shell system nanoparticles: the shell offers desired performance while the magnetic center provides effortless particle separation. An additional promising form of coreshell nanomaterials consists of a shell chemically tailored for fast, picky adsorption and a reactive center for degradation of adsorbed contaminants. Specialty nanoadsorbents may additionally be designed by using amphiphilic dendrimers with targeted binding sites (Qu et al., 2013).

4.2 In Membrane and Membrane Process

Aligned CNT and biomimetic membranes both make use of nanochannels (CNTs and protein channels known to as aquaporins, respectively) that successfully permit water particles to flow in a single file with extraordinary permeation rates (Holt et al., 2006a). The water selectivity is achieved through CNT diameter as well as the chemistry of the nanotube aperture, or even the peculiar hourglass shape, size, and chemical structure of aquaporins. Incorporation of vertically aligned CNTs and aquaporins straight into membrane matrices, even when it is at a minimal percent can possibly offer flux exceeding that of present industrial seawater reverse osmosis (RO) membranes. The bigger flux might not substantially improve the power usage of seawater desalination by RO, though it is able to significantly decrease the footprint as well as price of wastewater treatment by RO, whose theoretical power usage is significantly less. Excessive rejection of small contaminant molecules and salts stays difficult for aligned CNT membranes as a result of the absence of nanotubes with even subnanometer diameters; along with the opening gating possibly is vulnerable to charge screening or lessens the membrane permeability. Along with water channels, biological membranes offer good examples of picky ion channels. Nanozeolites are embedded in the polyamide active layer of thin film composite membranes to form TFNC membranes. The enhanced permeability was attributed to little, hydrophilic skin pores of zeolites which produce preferential paths for water while excluding hydrated ions,46 and perhaps defects in the zeolite polyamide interface. Nanozeolites also can function as carriers for antimicrobial agents; e.g., Ag^+, whose regeneration and release may be recognized by ion exchange. TFNC technological innovation has come to the first phase of commercialization (www.nanoh2o.com). Much studies have been dedicated to producing reactive membranes, usually comprising nano TiO_2, to concurrently separate and degrade contaminants.

Another essential use of nanomaterials is fouling management of other surfaces and membranes in water treatment process, storage space, and distribution methods (Holt et al., 2006b; Qu et al., 2013).

4.3 In Catalysis and Photocatalysis

Metallic nanoparticles as well as metal oxides have shown to be very good catalysts in oxidation responses. They display a strong catalytic exercise whereby pollution molecules are oxidized developing much less hazardous substances, or even changed into ecologically suitable ultimate products. Recently, different zero valent metallic nanoparticles, like Fe, Zn, Al, and also Ni, in water pollution remedy have pulled great research interest. As a result of the incredibly high reductive power, nano-zero-valent Al is thermodynamically unstable in the presence of water, which favours the development of oxides/hydroxides on the surface, impeding (completely) the transfer of electrons from the metallic surface on the contaminants. In contrast to Fe, Ni has a much less negative standard reduction potential, indicating a lower reducing capacity. With a reasonable standard reduction potential, nano-zero-valent Zn or Fe holds good potential to serve as reducing agents relative to several redox-labile contaminants. As a promising and emerging technology, photocatalytic degradation has attracted excellent interest after 1972 when Honda and Fujishima observed electrochemical photolysis of water on TiO_2 semiconductor electrode. Photocatalysis, utilizing nanostructures of metallic oxide semiconductors such as for instance zinc oxide (ZnO), titania (TiO_2), tungsten oxide (WO_3) as well as zinc stannate (Zn_2SnO_4) are excellent method of water filtration as it is effective at detatching chemical substance in addition to biological contaminants. Owing to the high photocatalytic activity of its, photostability, reasonable price, and biological and chemical stability TiO_2 has been most extensively examined in the past decades. The larger band gap energy (3.2 eV) of TiO_2 needs ultraviolet (UV) excitation to induce charge separation in the particles. Also, TiO2 NPs show minimal selectivity and therefore are ideal for the degradation of all sorts of contaminants. In the discipline of photocatalysis, aside from TiO_2 NPs, ZnO NPs have emerged as an additional effective prospect of wastewater and water treatment. the unique characteristics of ZnO, like wide and direct band gap in the near UV spectral region, powerful oxidation ability, and excellent photocatalytic property make it suitable candidate for water treatment. Intense exploration over the past years for the implementation of it's in the filtration of drinking water may be discovered in the literature (Fu et al., n.d.; Giamello & Pacchioni, 2020; Khan et al., 2015; LI & QU, 2009; X. Li et al., n.d.; Parvulescu et al., 2022; Raizada et al., 2021; Yaqoob et al., n.d.; Zhang, 2020).

4.4 In Disinfection and Microbial Control Process

Several nanomaterials have powerful antimicrobial properties, CNTs, nano-Ce_2O_4, nano-TiO_2, nano-ZnO, including nano-Ag, and fullerenes. These nanomaterials inactivate microorganisms by releasing poisonous metal ions compromising cellular membrane integrityupon immediate contact or even producing reactive oxygen species with a lot fewer tendencies to create harmful disinfection byproducts. Nano-Ag is a very common option for water treatment due to its good and wide spectrum antimicrobial activity and minimal toxicity to humans. The fibrous framework, antibacterial activity, and then conductivity of CNTs allow the use of theirs in antimicrobial filters. An actively pursued water treatment approach utilizes sunlight and additionally nanophotocatalysts to degrade biological contaminants and inactivate pathogens. TiO_2 photocatalysts effectively produce reactive oxygen species especially hydroxyl radicals and, superoxide, under UV A irradiation. Reactive oxygen species inactivate pathogens and degrade natural contaminants with little hazardous byproduct formation. Fullerenes as well as CNTs are photosensitive and will also generate reactive oxygen species in water (Klaine et al., 2008; Q. Li et al., n.d.; Qu et al., 2013).

4.5 In Sensing and Detection

CNTs and magnetic nanoparticles are being investigated for sample concentration and purification. CNTs, noble metallic nanoparticles, dye-doped nanoparticles, and quantum dots (QDs) are frequently employed in nanosensor technology. According to changes in the aggregation state of the noble metallic nanoparticles (such as nano Au and nano Ag) and local refractive index, stable localised area plasmon resonances of these particles enable optical pathogen identification (Qu et al., 2013; Vikesland & Wigginton, 2010). Additionally, Nobel metal nanoparticles improve surface enhanced Raman spectroscopy, obtaining amplification factors up to 10^{14} and enabling the detection of single molecules. (Nie & Emory, 1997).

5. ENVIRONMENTAL AND HUMAN HEALTH EFFECTS OF NANOMATERIAL USED IN WATER AND WASTEWATER TREATMENT

Contemporary nanotechnology exploration in the region of wastewater treatment shouldn't entirely be depending on the possibility for enhancing attributes of nanomaterials utilized in wastewater treatment, the efficiency of theirs, and standardization. A really crucial element which is extremely substantial for broader industrial use of these kinds of products in the future is the impact of theirs on the ecosystem and human health. In most of the forms of application of the technology, probably the most crucial component would be the presence of nanomaterials within wastewater, due to the potential environmental impact of theirs. Furthermore, when nanoparticles arrive at nature, the interaction of theirs with chemical substances in the surroundings can oftentimes have negative effects. A certain number of nanomaterials will undoubtedly be released into the environment through the usage of nanoparticles in the water treatment process. The behaviour of such nanomaterials in the environment, their ecotoxicity are some serious matter of concerns that requires extensive studies. Even though no specific information is still available, it's recognized that largest levels of nanomaterials which turn up in the earth will ultimately end up in soil, while smaller quantities of nanoparticles are contained in air and water. Among the biggest "sources" of nanomaterials in the dirt may be the sludge produced in wastewater treatment plants. Following wastewater treatment, complete pollution, including nanoparticles accumulates in sludge, and that will be transferred and also managed in different ways. The sludge will be used for different uses and it winds up in the planet just where it's possibly deposited for landfills, used as dirt improver for agriculture, etc. This domain of wastewater treatment needs extensive research efforts. Essentially, a number of studies on the effect of nanomaterials on the environment have thus far been made, and several good results are accomplished. Nevertheless, several nanotoxicological studies continue to be in the first phase along with a lot of effort is necessary to make precise conclusions about the consequences of some nanomaterials on the environment and human lives. Although encouraging water treatment results are obtained for different nanomaterials, this particular problem may be a restricting element for industrial use of certain promising nanotechnologies (Qu et al., 2013; Zekić et al., 2018).

6. CHALLENGES AND FUTURE ASPECTS

Even though nanotechnology enabled water/wastewater treatment procedures show great promise of laboratory studies, the readiness of theirs for commercialization varies widely. Significant barriers stand between several of these promises as well as the delivery of theirs. The future development of theirs as well as commercialization face an assortment of problems including technical hurdles, cost effectiveness, and potential environmental as well as human threat. Wide acceptance of novel water as well as wastewater treatment nanotechnologies will depend on each their affordability and performance. Regardless of the better functionality, the adoption of innovative technologies firmly relies on the cost effectiveness and also the possible risk concerned. The present price of nanomaterials is prohibitively substantial with some exceptions like nano TiO_2, polymeric nanofibers, and nanoscale ion oxide. One best way to address the cost issue is to retaining and reusing of nanomaterials. The issues of depleting freshwater sources with the passage of time and also boosting demands have guided a huge advancement toward finding greener and newer technologies for wastewater treatment. Whether nanotechnology may be incorporated as an environmentally friendly methods or otherwise is nonetheless a situation of debate among scientists (Riffat & Husnain, 2022; Zekić et al., 2018).

CONCLUSION

Nanomaterials possess excellent structural, surface, electro-optical, and magnetic properties and can be applied in various fields like physics, chemistry, medicine, biology, water treatment, etc. Due to their novel properties, nanomaterials possess the ability to remove the contaminants present in groundwater, surface water, and wastewater. The use of these nanomaterials possibly removes of different organic, inorganic, metal ions, microorganism pollutants, which are otherwise unmanageable or requires more cost or labor. Despite the advantages offered by the nanomaterials, their toxic effect on human health and the environment also requires significant attention and needs tremendous research for their practical employment in water treatment techniques.

REFERENCES

Auffan, M., Rose, J., Proux, O., Borschneck, D., Masion, A., Chaurand, P., Hazemann, J.-L., Chaneac, C., Jolivet, J.-P., Wiesner, M. R., van Geen, A., Bottero, J.-Y., Doherty, L., & Observatory, E. (2008). Enhanced adsorption of arsenic onto maghemites nanoparticles: As (III) as a probe of the surface structure and heterogeneity. *Langmuir*, *24*(7), 3215–3222. doi:10.1021/la702998x PMID:18266393

Baig, N., Kammakakam, I., Falath, W., & Kammakakam, I. (2021). Nanomaterials: A review of synthesis methods, properties, recent progress, and challenges. *Materials Advances*, *2*(6), 1821–1871. doi:10.1039/D0MA00807A

Chaturvedi, S., & Methodologies, P. D.-C. (2019). undefined. (2019). Water purification using nanotechnology an emerging opportunities. *Iranjournals.Nlai.Ir*, *3*, 115–144. doi:10.22034/CHEMM.2018.143461.1069

Fu, Y., & Li, J. (2019). *Metal/semiconductor nanocomposites for photocatalysis: Fundamentals, structures, applications and properties*. Retrieved July 29, 2022, from https://www.mdpi.com/421488

Giamello, E., & Pacchioni, G. (2020). Applied vs fundamental research in heterogeneous photocatalysis: Problems and perspectives. An introduction to "physical principles of photocatalysis." *Journal of Physics Condensed Matter*, *32*(36), 360301. Advance online publication. doi:10.1088/1361-648X/ab926b PMID:32396884

Holt, J. K., Park, H. G., Wang, Y., Stadermann, M., Artyukhin, A. B., Grigoropoulos, C. P., Noy, A., & Bakajin, O. (2006). Fast mass transport through sub-2-nanometer carbon nanotubes. *Science*, *312*(5776), 1034–1037. doi:10.1126cience.1126298 PMID:16709781

Imran, M. A., Miraj, F., Khan, I., & Tlili, I. (2018). MHD fractional Jeffrey's fluid flow in the presence of thermo diffusion, thermal radiation effects with first order chemical reaction and uniform heat flux. *Results in Physics*, *10*, 10–17. doi:10.1016/j.rinp.2018.04.008

Jeevanandam, J., Barhoum, A., Chan, Y. S., Dufresne, A., & Danquah, M. K. (2018). Review on nanoparticles and nanostructured materials: history, sources, toxicity and regulations. *Beilstein Journal of Nanotechnology, 9*(1), 1050–1074. doi:10.3762/bjnano.9.98

Kargari, A., & Shirazi, M. M. A. (2014). Water Desalination: Solar-Assisted Membrane Distillation. Encyclopedia of Energy Engineering and Technology, 2095–2109. doi:10.1081/E-EEE2-120051388

Khan, M. M., Adil, S. F., & Al-Mayouf, A. (2015). Metal oxides as photocatalysts. *Journal of Saudi Chemical Society*, *19*(5), 462–464. doi:10.1016/j.jscs.2015.04.003

Klaine, S. J., Alvarez, P. J. J., Batley, G. E., Fernandes, T. F., Handy, R. D., Lyon, D. Y., Mahendra, S., McLaughlin, M. J., & Lead, J. R. (2008). Nanomaterials in the environment: Behavior, fate, bioavailability, and effects. *Environmental Toxicology and Chemistry*, *27*(9), 1825–1851. doi:10.1897/08-090.1 PMID:19086204

Li, D., & Qu, J.LI. (2009). The progress of catalytic technologies in water purification: A review. *Journal of Environmental Sciences (China)*, *21*(6), 713–719. doi:10.1016/S1001-0742(08)62329-3 PMID:19803071

Li, Q., Mahendra, S., Lyon, D., Brunet, L., & Liga, M. (n.d.). *Antimicrobial nanomaterials for water disinfection and microbial control: potential applications and implications*. Elsevier. Retrieved July 29, 2022, from https://www.sciencedirect.com/science/article/pii/S0043135408003333?casa_token=mmxoSvxfy3AAAAAA:i05GpT1KbFBTdQCFDuXMLsQIM8Oi7_r0Pn32bDBKf-6InqxNHbBSs_68bjCZCj4inhzUNAXyzw

Li, X., Yang, K., Yu, C., Zhang, K., & Yang, S. (2019). *Broadband photocatalysis using a Z-scheme heterojunction of Au/NaYF 4: Yb, Er/WO 3· 0.33 H 2 OW 18 O 49 via a synergetic strategy of upconversion function and*. Retrieved July 29, 2022, from https://pubs.rsc.org/en/content/articlehtml/2019/qi/c9qi00823c

Lima, A. A. M., Moore, S. R., Barboza, M. S. Jr, Soares, A. M., Schleupner, M. A., Newman, R. D., Sears, C. L., Nataro, J. P., Fedorko, D. P., Wuhib, T., Schorling, J. B., & Guerrant, R. L. (2000). Persistent Diarrhea Signals a Critical Period of Increased Diarrhea Burdens and Nutritional Shortfalls: A Prospective Cohort Study among Children in Northeastern Brazil. *The Journal of Infectious Diseases, 181*(5), 1643–1651. doi:10.1086/315423 PMID:10823764

Maxwell's Demon: Nano Machine Of The Future Captures Great Scientist's Bold Vision -- ScienceDaily. (n.d.). Retrieved July 29, 2022, from https://www.sciencedaily.com/releases/2007/02/070201144706.htm

Montgomery, M. A., & Elimelech, M. (2007). Water and sanitation in developing countries: Including health in the equation - Millions suffer from preventable illnesses and die every year. *Environmental Science & Technology, 41*(1), 17–24. doi:10.1021/es072435t PMID:17265923

Nie, S., & Emory, S. R. (1997). Probing Single Molecules and Single Nanoparticles by Surface-Enhanced Raman Scattering. *Science, 275*(5303), 1102–1106. doi:10.1126cience.275.5303.1102 PMID:9027306

Parvulescu, V. I., Epron, F., Garcia, H., & Granger, P. (2022). Recent Progress and Prospects in Catalytic Water Treatment. *Chemical Reviews, 122*(3), 2981–3121. doi:10.1021/acs.chemrev.1c00527 PMID:34874709

Patanjali, P., Singh, R., Kumar, A., & Chaudhary, P. (2019). Nanotechnology for water treatment: A green approach. *Green Synthesis, Characterization and Applications of Nanoparticles*, 485–512. doi:10.1016/B978-0-08-102579-6.00021-6

Qu, X., & Alvarez, P. (n.d.). *Applications of nanotechnology in water and wastewater treatment*. Elsevier. Retrieved July 29, 2022, from https://www.sciencedirect.com/science/article/pii/S0043135413001772

Qu, X., Brame, J., Li, Q., & Alvarez, P. J. J. (2013). Nanotechnology for a safe and sustainable water supply: Enabling integrated water treatment and reuse. *Accounts of Chemical Research, 46*(3), 834–843. doi:10.1021/ar300029v PMID:22738389

Raizada, P., Soni, V., Kumar, A., Singh, P., Parwaz Khan, A. A., Asiri, A. M., Thakur, V. K., & Nguyen, V. H. (2021). Surface defect engineering of metal oxides photocatalyst for energy application and water treatment. *Journal of Materiomics, 7*(2), 388–418. doi:10.1016/j.jmat.2020.10.009

Riffat, R., & Husnain, T. (2022). *Fundamentals of Wastewater Treatment and Engineering*. doi:10.1201/9781003134374

Shannon, M. A., Bohn, P. W., Elimelech, M., Georgiadis, J. G., Mariñas, B. J., & Mayes, A. M. (2008). Science and technology for water purification in the coming decades. *Nature, 452*(7185), 301–310. doi:10.1038/nature06599

Sharma, N. (n.d.). *Nanotechnology in Stem Cell Research by Neha Sharma, Keshav Mahavidyalaya* : SSRN. Retrieved July 29, 2022, from https://papers.ssrn.com/sol3/papers.cfm?abstract_id=3958890

Theron, J., & Walker, J. (2008). *Nanotechnology and water treatment: applications and emerging opportunities*. Taylor & Francis. doi:10.1080/10408410701710442

Tlili, I., & Alkanhal, T. A. (2019). Nanotechnology for water purification: Electrospun nanofibrous membrane in water and wastewater treatment. *Journal of Water Reuse and Desalination*, *9*(3), 232–248. doi:10.2166/wrd.2019.057

Types of Drinking Water Contaminants | US EPA. (n.d.). Retrieved July 29, 2022, from https://www.epa.gov/ccl/types-drinking-water-contaminants

Vikesland, P. J., & Wigginton, K. R. (2010). Nanomaterial enabled biosensors for pathogen monitoring - A review. *Environmental Science & Technology*, *44*(10), 3656–3669. doi:10.1021/es903704z PMID:20405829

Wagner Ferreira Sabará, E., Pereira, V., Luiz Molisani, A., & Mary Ealias, A. (2017). A review on the classification, characterisation, synthesis of nanoparticles and their application. *IOP Conference Series. Materials Science and Engineering*, *263*(3), 032019. doi:10.1088/1757-899X/263/3/032019

Yang, H. G., Sun, C. H., Qiao, S. Z., Zou, J., Liu, G., Smith, S. C., Cheng, H. M., & Lu, G. Q. (2008). Anatase TiO2 single crystals with a large percentage of reactive facets. *Nature, 453*(7195), 638–641. doi:10.1038/nature06964

Yang, K., & Xing, B. (2010). Adsorption of organic compounds by carbon nanomaterials in aqueous phase: Polanyi theory and its application. *Chemical Reviews*, *110*(10), 5989–6008. doi:10.1021/cr100059s PMID:20518459

Yaqoob, A. A., Parveen, T., Umar, K., Nasir, M., & Ibrahim, M. (n.d.). Role of nanomaterials in the treatment of wastewater: A review. *Mdpi.Com*. doi:10.3390/w12020495

Zekić, E., Vuković, Ž., Halkijević, I., Zelić, E., Hidrokon doo, M., & Ivan Halkijević, A. (2018). Primjena nanotehnologije u pročišćavanju otpadnih voda. *Casopis-Gradjevinar.Hr, 4*, 315–323. doi:10.14256/JCE.2165.2017

Zhang, T. (2020). *Heterogeneous Catalytic Process for Wastewater Treatment*. Advanced Oxidation Processes - Applications, Trends, and Prospects. doi:10.5772/intechopen.90393

KEY TERMS AND DEFINITIONS

Adsorption: The term “adsorption” refers to the process through which atoms, ions, or molecules from a gas, liquid, or solid that has been dissolved adhere to a surface.

Bottom-Up Approach: The bottom-up method involves creating nanostructures by forming chemical linkages between atoms. Bottom-up methods include techniques like Sol-gel, Chemical vapour deposition, spinning, and biosynthesis.

Catalysis: To speed up a chemical reaction, a catalyst is often used to make the process go by faster; this phenomenon is known as catalysis.

Clean Water: Clean or pure water is defined as water free from any form of contaminants, i.e., free from toxic chemicals and pathogens. According to SDWA (The Safe Drinking Water Act) contaminants are defined as any biological, chemical, physical, or radiological matter in water.

Membrane: Membrane filters can either remove the particles that are polluting the water by acting as a barrier between the water and the pollutants, or they can remove the contaminants themselves.

Nanomaterials: Nanotechnology offers the ability to control matter at dimension comparable to or less than 100 nm (1-100 nm) to obtain materials termed as nanomaterials.

Top-Down Approach: In the top-down approach, bulk materials are stably divided to get nanostructures. Sputtering, laser ablation, nanolithography are examples of the top-down approach.

Chapter 10
Role of Two-Dimensional Nanomaterials in Electrodes for Rechargeable Batteries and Supercapacitors

Rafael Vargas-Bernal
https://orcid.org/0000-0003-4865-4575
Universidad de Guanajuato, Mexico

ABSTRACT

It has become urgently necessary to develop energy storage technologies and materials due to the rising need for advanced sustainable energies. Among the emerging advances in materials chemistry is the research and development of new electrode materials for rechargeable batteries and supercapacitors. To increase their energy density, service life, specific power, charge and/or discharge capacity, and environmental friendliness, electroactive electrode materials with high surface areas and redox-active mediators are essential. Two-dimensional materials such as MXenes, Xenes, phosphorene, silicene, siloxane, stanene, arsenene, plumbene, bismuthene, antimonene, metallenes, metal-organic frameworks (MOFs), and other materials are being investigated to optimize their performance for various implementations of rechargeable batteries and supercapacitors. This chapter discusses recent progress and future challenges in the design of two-dimensional materials for electrodes of electrochemical energy storage devices.

INTRODUCTION

The use of electricity produced by renewable energies like wind, solar, and water-power has become necessary due to the energy crisis and environmental damage (Tang, 2018). However, because these renewable energies are sporadic, it is crucial to create energy storage technologies for their better exploitation. It has become imperative to create rechargeable batteries and supercapacitors with improved electrochemical performance, especially since conventional energy storage technologies are unable to satisfy the energy market's present and future expectations (Zhang, 2022). Advanced electrode materi-

DOI: 10.4018/978-1-6684-6830-2.ch010

als with superior electrochemical performance, high-energy density, and high-power density are being developed with great effort for energy storage. Two-dimensional materials have mainly become in the last decade ideal candidates for energy applications due to their outstanding electrical, chemical, and physical properties (Fan, 2021). Among the advantages that these materials can offer are a high surface-volume surface rate, good transport properties, excellent physicochemical properties, and remarkable confinement effects. Additionally, the morphology of these materials allows controllable interfacial chemistry, high electrical conductivity, tunable layers structure, high porosity, modifiable by functional groups, controllable chemical composition, small strain expansion, etc. (Mao, 2021; Rasheed, 2022). Due to these characteristics, two-dimensional materials are being applied in energy storage, electrocatalysis, medicine, water purification, superconductivity, etc. (Guo, 2020). However, its practical applicability in energy storage systems is limited due to the low efficiency of material use, slow reaction kinetics, and limited active sites. To overcome these restatements, it is necessary to improve the electrochemical performance of these materials through rational microstructural design and synthesis. A synergistic effect between materials has been found when using 0D-2D composites, 1D-2D composites, 2D-2D composites, nanosheet-substrate composites, and core-shell composites (Oh, 2020).

In comparison to ordinary capacitors, which have a very high capacitance of up to 10,000 F, supercapacitors have a high-energy density and high-power density (Dongre, 2022). Supercapacitors (SCs) have received a lot of interest recently in energy storage, and research efforts are heavily reliant on the design and engineering of efficient electrode materials (Qin, 2020). Rechargeable batteries that use an aqueous electrolyte are categorized as Li^{+}-, Na^{+}-, K^{+}-, Zn^{2+}-, Ca^{2+}-, Mg^{2+}-, Al^{3+}-, or Cl^{-}-ion aqueous batteries depending on the kind of principal charge carrier (Choi, 2021). The development of advanced batteries has also been stimulated by the desire for greater efficiency and lower emissions following decades of the energy revolution. These batteries frequently have the advantages of improved cycling stability, high specific capacity, high energy density, and long service lives (Zhang, 2022). The anode must be both ionically and electrically conductive for a battery to operate at high rates.

Graphene, a two-dimensional carbon nanosheet, has remained the main source of scientific fascination for energy storage in the last decade (Kumar, 2018). Among the two-dimensional materials of the IV-A group used as electrodes are graphene (C), silicene (Si), germanene (Ge), stanene (Sn), and plumbene (Pb) (Guo, 2020). In the V-A group, two-dimensional materials that have been used as an electrode in energy storage devices are phosphorene or black-phosphorus (P) (Zhu, J.P. 2020), arsenene (As) (Benzidi, 2019), antimonene (Sb) (Liang, 2021), and bismuthene (Bi) (Zhou, 2019). In addition, there are compound two-dimensional materials such as metal-organic frameworks (MOFs) (Zhang, 2022), transition metal nitrides and carbides (MXenes) (Rasheed, 2022), transition metal dichalcogenides (TMDs) (Peng, 2016), and 2D metal oxides (Mendoza-Sánchez, 2016). There is another group of emerging two-dimensional materials that are also being used in energy storage devices such as electrodes such as InSe (Sun, 2019), metal thiophosphates (Zhu, M. 2020), spinel ferrites (Mao, 2021), perovskite oxides (Qin, 2020), and transition metals sulfides (Maksoud, 2021), etc.

In this review, the functions that two-dimensional materials have in the design of electrodes for the different types of rechargeable batteries and supercapacitors, are discussed. The characteristics, structure, and applications of two-dimensional materials in energy storage devices are analyzed. The use of these materials is specifically influenced by metal-sulfur batteries, alkali metal-ion batteries (Li^{+}, Na^{+}, K^{+}), alkaline earth metal-ion batteries (Mg^{2+}), zinc-ion batteries (Zn^{2+}), aluminum-ion batteries (Al^{3+}), multivalent metal-ion batteries, and supercapacitors (electrochemical double-layer capacitors (EDLCs), pseudocapacitors (PCs), asymmetric capacitors (ACs) and hybrid capacitors (HCs)). Despite the advances achieved by two-dimensional materials so far, their use in energy storage devices is expected to be feasible in the next decade, so this chapter also aims to point out the future research addresses those researchers in materials engineering will be developing.

The chapter is divided as follows: In the section entitled "Introduction to the Two-Dimensional Nanomaterials," a description of different two-dimensional materials that have been used in the design of electrodes for energy storage devices is made. In some cases, the description has been made in a group form because they belong to the same group of chemical elements in the periodic table. Cases deserve a separate specific description. Following the section entitled "Physical properties of two-dimensional materials used in rechargeable batteries and supercapacitors" a description of the outstanding properties that two-dimensional materials must be used in electrodes of energy storage devices is presented. Later, in the section entitled "Solutions and recommendations for rechargeable batteries and supercapacitors" some of the contributions that two-dimensional materials have had in the design of rechargeable batteries and supercapacitors are discussed. Next, a section entitled "Future research directions" aims to point out some of the possible activities that should be carried out to improve all the solutions proposed so far. Finally, the conclusions of the chapter are presented to summarize the achievements to date.

FUNDAMENTALS OF TWO-DIMENSIONAL MATERIALS

In the field of materials science, crystalline solids made of a single layer of atoms are referred to as single-layer materials or 2D materials. Although the research is still primarily on these materials, they show promise for various applications. The *-ene* suffix is typically added to the names of single-layer materials made from a single element, such as graphene, silicene, germanene, borophene, stanene, phosphorene, antimonene, bismuthene, and plumbene. Materials with the suffixes *-ane* or *-ide* are single-layer nanomaterials made of two or more elements. 2D materials can typically be divided into two categories: composites or 2D allotropes of different elements (consisting of two or more covalently bonding elements). Research on 2D materials is expanding quickly, and graphene is just one of several 2D materials that are being studied (Dongre, 2022). Several 2D materials have been successfully synthesized based on theoretical predictions, and it was shown that these materials had several structural characteristics to graphene despite having distinct compositions. Although the first two-dimensional material, graphene, was discovered in 2004 and is known to have higher electronic conductivity than silicon, it cannot be employed in all electronic applications since it is not a semiconductor. Following the discovery of graphene, 2D materials such as hexagonal boron nitride, graphitic carbon nitride, transition metal dichalcogenides, layered metal oxides and hydroxides, and MXenes have attracted a lot of attention in the last 10 years (Venkateshalu, 2022). Novel 2D nanomaterials that are comparable to graphene, like phosphorene, antimonene, silicene, siloxene, germanene, etc., were very recently found and are now posing a serious competitor to already existing two-dimensional materials. A summary of the different types of materials that have been and will be used in the emerging design of rechargeable batteries and supercapacitors is presented in Figure 1. This figure breaks down the main two-dimensional materials that can be used in electrochemical energy storage devices, which can be found in topologies based on monolayers, composite materials, hybrid materials, or heterostructures. Within these materials, you can find the crystalline materials produced by the chemical elements of groups IV-A, V-A, non-metallic elements, elements of the main metal groups, metals of transition elements, lanthanides, actinides, MXenes,

Figure 1. Two-dimensional materials used in energy storage devices

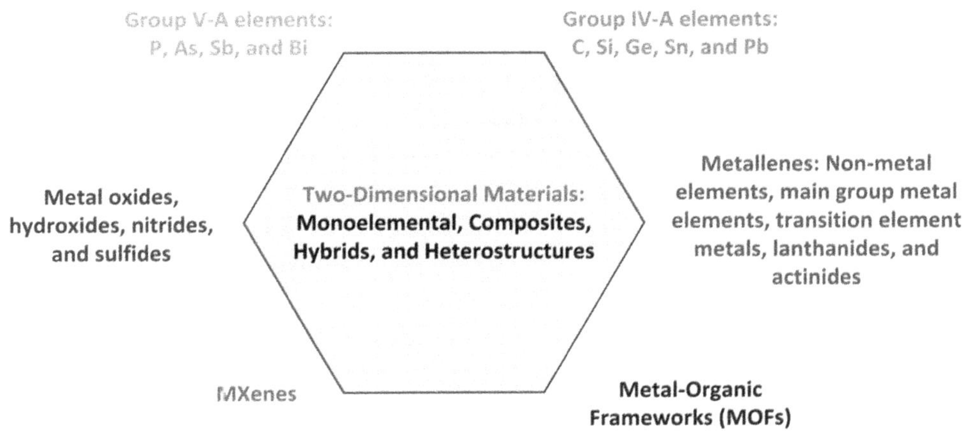

materials ceramics such as oxides, hydroxides, nitrides, and sulfides, as well as metal-organic frameworks (MOFs). Next, a set of subsections is used to describe each of them in a basic way.

Introduction to Graphene

Graphite was mechanically exfoliated to produce graphene, a single-atom-thick nanosheet with sp^2 hybridized carbon atoms (Wang, 2020). The number of layers of graphene has a significant impact on its properties and functions, which are necessary for further research. Utilizing its superior mechanical strength, exceptional electronic and thermal conductivities, high surface specific area, and other properties, graphene, a newly discovered fabric made of carbon atoms, has demonstrated its versatility in a variety of fields including electronics, optoelectronics, and thermoelectrics (Guo, 2020). Numerous initiatives have been made to improve its intriguing qualities and broaden the applications it can be used in, including batteries, supercapacitors, catalysis, solar water evaporation, and moist electric generators.

Introduction to MXenes

When *M* is an early transition metal (Mo, Ti, Nb), *A* is an element from group IIIA or group IVA (Al, Ga, Si, Ge, Sn), *X* is C or N, and $n = 1, 2$, and 3, MXenes with *MAX* phases ($M_{n+1}AX_n$ composition) are produced. The *A* element from the *MAX* phase is removed by etching to produce $M_{n+1}X_n$ or MXenes. This was accomplished for the first time by treating Ti_3AlC_2 with HF to create a multilayer $Ti_3C_2T_x$, which was then processed using sonication to create delaminated MXene sheets. On the delaminated MXene surfaces, this treatment generates the existence of chemical functionalities T_x, such as =O, -OH, and -F groups. These surface functions cause hydrophilicity and have an impact on the electrical characteristics of MXenes. MXenes often exhibit a hexagonal tight packing structure (Rasheed, 2022). The atomic ordering patterns used in M_2X, M_3X_2, and M_4X_3 are, respectively, ABABAB and ABCABC. The MXene sheets generally have a laterally aligned orientation, just like the parent *MAX* phases.

Introduction to Silicene, Siloxene, Germanene, Stanene, and Plumbene

Since its discovery in 2012, silicene has become a popular subject of study (Venkateshalu, 2022). These recently found 2D materials differ significantly from graphene in several aspects and are only comparable to it in terms of being 2D. First, unlike the exfoliating 3D layered parent of graphene, these materials do not have one. Only chemical processes or the molecular-beam epitaxy approach may create them. The structural and electrical characteristics of these 2D structures are significantly influenced by the substrate on which they are produced. The second difference is that group IV elements contain mixed sp^3 hybridization in their layers, which rises with an increase in atomic scale, whereas graphene is a flat substance with exclusively sp^2 hybridization. This causes the atoms' bonds to buckle, leading to layers that are not flat. Silicene's characteristics are strongly influenced by the substrate and the buckling that is present in it, allowing it to be employed in electronic devices. A comparable hexagonal honeycomb structure to graphene may be found in the atomically thick materials silicene, germanene, and stanene (Mortazavi, 2016).

The buckled honeycomb lattice structure of silicon's important allotrope, silicene, has been created in Si-based electronics. Wet chemical exfoliation is used to synthesize silicene in addition to epitaxial growth, which is widely used. This method offers sufficient opportunities and the fundamental building blocks for a variety of applications, ranging from energy storage and conversion devices to biomedical nanomedicines. Because group IVA elements (Si, Ge, Sn, and Pb)-derived 2D nanosheets have been effectively manufactured by physical and chemical means and have filled special and vital roles in smart electronics, energy storage, and conversion devices. Due to its potential uses in the topological insulator, batteries, capacitors, catalysis, and topological superconductivity, the success of additional two-dimensional materials after graphene has drawn attention to silicene, siloxene, germanene, stanene, and plumbene (Guo, 2020).

Germanene is a buckled honeycomb-shaped two-dimensional single-atomic-layer germanium (Ge) material. Ge is more favorable for sp^3 hybridization than Si due to its greater ionic radius, which suggests that the buckled structure of 2D Ge is more stable (Fan, 2021). To obtain stable 2D Ge structures, stabilization from substrates or surface modification might be useful.

Gray tin (α-Sn) with a face-centered cubic diamond structure and white tin (β-Sn) with a body-centered tetragonal structure are the two stable homogenous structures of bulk tin (Fan, 2021). The Sn allotrope with a single atomic layer that resembles graphene is called stanene. The diatomic layer of α-Sn(111) that makes up stanene has two triangular sublattices that stack on top of one another to create a buckled honeycomb lattice.

Plumbene is atomically thin lead (Pb) with a honeycomb lattice-like graphene (Fan, 2021). Due to the orbital electron structure of lead, previous theoretical simulations demonstrated that plumbene has the highest spin-orbit interaction and energy bandgap among the group IV monolayers. A brand-new family of 2D quantum spin Hall (QSH) insulators built on plumbene monolayers linked to halogens (PbX; X = H, F, Cl, Br, I) and featuring enormous bulk gaps ranging from 1.03 eV to 1.34 eV. These QSH insulators offer a viable platform for investigating fascinating topological phenomena and extending application possibilities at high temperatures. Plumbene has enticing qualities for use in batteries, electronics, wearing radiation protection, and other applications in addition to its desirable qualities as a topological insulator.

Through the topotactic deintercalation of calcium and liquid exfoliation from the Zintl phase of calcium silicide, siloxene is created ($CaSi_2$) (Venkateshalu, 2022). Due to its dual-band function location, 2D siloxene exhibits a squat buckled structure, which is different from the planar arrangement of graphene. Siloxene can be employed as electrode material in energy storage devices due to the presence of surface-terminating functional groups and a combination of sp^2 and sp^3 hybridization.

Introduction to Exfoliated Pnictogens

Any chemical element in group 15 of the periodic table is a pnictogen. Group 15 is often referred to as the nitrogen family or group (Venkateshalu, 2022). Nitrogen (N), phosphorus (P), arsenic (As), antimony (Sb), and bismuth (Bi) are the constituent chemical elements. The two-dimensional materials produced by these chemical elements are called nitrogene (N), phosphorene (P), arsenene (As), antimonene (Sb), and bismuthene (Bi), respectively.

In 2014, the mechanical exfoliation of 2D black phosphorous, known as phosphorene, from its parent material via the blue tape method was proposed (Venkateshalu, 2022). Due to its resemblance to graphene, this isolated single layer of black phosphorous was given the name phosphorene. Contrary to graphene, phosphorene has a puckered structure because of its in-plane bonding being sp^3 hybridized. Through Van der Waals forces, the layers in phosphorene are tenuously linked together.

Due to the poor contact between layers in bulk grey arsenic, arsenene is created experimentally by exfoliating from gray arsenic (Benzidi, 2019). With strong covalent connections between layers and stronger interlayer contacts, antimony, and arsenic form a rhombohedral structure (Dongre, 2022). Due to interactions between sp^2 and sp^3 hybrid bonds that are identical to those in graphene, the atoms are organized in a puckered honeycomb lattice. The bulk antimony (Sb) exists in a variety of allotropic forms, the most stable of which is the rhombohedral structure (β-phase), whose structure is like that of gray arsenic (β-phase). Arsenene and antimonene have uses in a variety of industries, including photocatalysis, electronics, optoelectronics, and electrochemical devices including supercapacitors and batteries (Dongre, 2022).

Bismuth is like graphite in both look and structure; it is flaky and gray with a puckered layered crystal structure made up of six-membered rings (Zhou, 2019). Group VA's heaviest element, bismuth, has a layered structure in its stable form (β-phase) (Fan, 2021). Bismuthene has demonstrated great application promise in energy storage, electrocatalysis, and photonics. Real van der Waals bound layered structures cannot be adopted by bulk Bi having a rhombohedral structure. Therefore, exfoliating bulk Bi into bismuthene using conventional ultrasonication or other mechanical processes is challenging. However, several top-down techniques, such as heat pressing and liquid exfoliation, can be employed to create bismuthene.

Introduction to Metallenes

Metallenes are two-dimensional materials based on non-metal elements, main group metal elements, transition element metals, lanthanides, and actinides (Liu, 2020). Metallenes include two-dimensional materials based on aluminum (aluminene), gallium (gallenene), indium (indienene), iodine (iodinene), hafnium (hafnene), molybdenum, iron, ruthenium, cobalt, rhodium, iridium, nickel, palladium, platinum, copper, silver, gold, zinc, cadmium, etc. and its alloys forming bimetallenes and trimetallenes. It is important to note that the suffix "*ene*" in metallene refers to monolayers in general and does not specifically relate to graphene-like honeycomb structures. Metallenes typically consist of a few layers that are less than 5 nm thick.

Introduction to Metal-Organic Frameworks (MOFs)

Metal-organic frameworks (MOFs) are 2D or 3D porous coordination polymers that are created when metal ions and organic ligands, which often have nitrogen or oxygen groups as part of their structure, self-assemble (Guo, 2021). MOFs can be created by choosing and controlling distinct organic building blocks at the molecular level, unlike conventional inorganic porous materials. Furthermore, MOFs have increased porosity, more specific surface area, and a variety of structural and functional variations. Planar 2D MOFs undergo extended 2D conjugation in comparison to 3D MOFs, and they have graphene-like structures that could provide one of the most conductive skeleton structures. Additionally, because of their lamellar structures, 2D MOFs have more open coordination and electroactive sites for molecular interaction due to their unique surfaces and surface areas, unique interlayers, larger specific surface areas, and more accessible active sites. This is because metal centers have both planar square coordination and octahedral coordination on their 2D surfaces.

Introduction to Other 2D Materials

The term "thiophosphate" refers to chemical substances and anions with the general chemical formula $PS_{4-x}O^{-3}_{x}$ ($x = 0$, 1, or 3), as well as related derivatives, where organic groups are linked to one or more O or S. Tetrahedral phosphorus(V) centers are present in thiophosphates. A class of common 2D materials called metal thiophosphates MPS_3 can be represented by M, a divalent transition metal ion, such as Co(II), Mn(II), Fe(II), Ni(II), Sn(II), or Zn(II) (Zhu, M. 2020).

Titanium-based compounds used in energy storage are titanium oxide (TiO_2), titanium-based MXenes, titanates, and layered dichalcogenides (Mao, 2021). Trivalent Ti ions are ineffective for storing energy and have a low cell voltage, but they are useful for improving structural stability and enabling high power performance.

Vanadium(V)-based nanocomposites are of particular interest, mainly due to V's cost-effectiveness and wide range of valence states (including V^{2+}, V^{3+}, V^{4+}, and V^{5+}) (Qin, 2020). As V-based materials, vanadium oxides, vanadium nitrides, vanadium sulfides, and mixed metal vanadates are mainly investigated.

The Crystalline Structure of Two-Dimensional Materials

Two-dimensional materials have a crystalline structure based on layers of single or few atoms in which interatomic interactions in the plane are much stronger than along the stacking direction. To analyze them, it is necessary to understand the pattern of atoms in cross-section and their surface view. The regular arrangement of atoms is based on distributing cations and anions of different chemical elements or of the same chemical element with stable and flexible physical structures to take full advantage of the surface properties. The crystalline structure of some two-dimensional materials that can be used to implement electrochemical storage devices is illustrated in Figure 2. Materials such as carbon-based graphene, phosphorous-based black phosphorous or phosphorene, transition metal dichalcogenides such as molybdenum disulfide (MoS_2), and MXene as titanium carbide (Ti_2C) are illustrated.

Figure 2. Crystal structure of some two-dimensional materials used in electrochemical storage devices

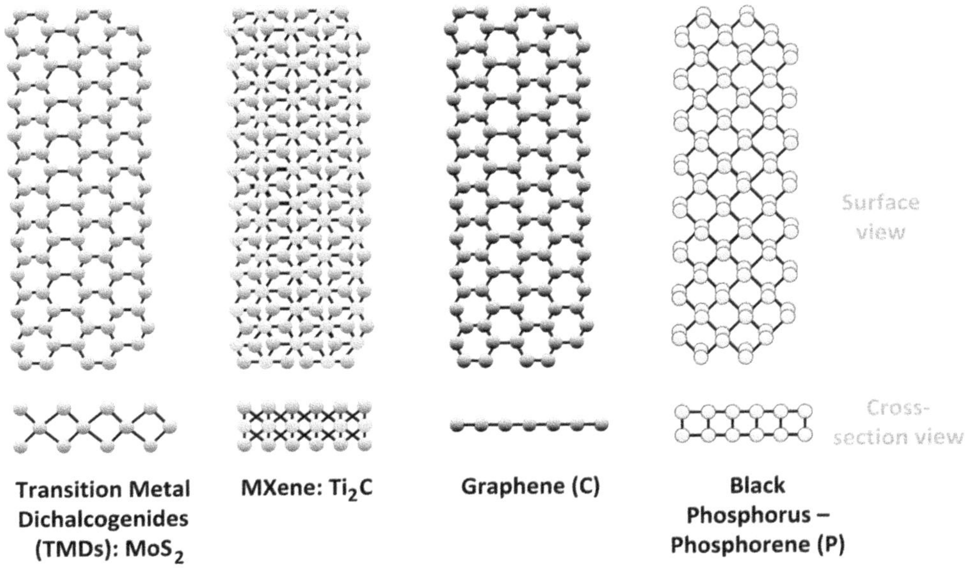

Fundamentals of the Synthesis of Two-Dimensional Materials

Electrochemical energy storage performance depends on the composition of the two-dimensional material, the texture, and morphology of the material, the configuration of the device, as well as the architecture of the electrode (Mendoza-Sánchez, 2016). Among the challenges to exploiting the maximum potential of two-dimensional materials for energy storage is the reduction of their stacking nature which would allow full use of surface area and electrolyte access to all electrode material. One of the possible strategies is to synthesize nanosheets with curved or wrinkled morphologies. Several synthesis methods have been proposed so far to implement the two-dimensional materials to satisfy the above requirements.

Two-dimensional material synthesis approaches make use of top-down and bottom-up methods (Cui, 2020). Top-down methods are mainly based on exfoliation based on mechanical processes, by intercalation or by etching. In the case of bottom-up methods, they are based on epitaxial growth and chemical vapor deposition (CVD). Through exfoliation methods, the bulk material is reduced to a layered material either under the action of a shear force to break the interaction between layers (mechanical exfoliation), insertion of ions or molecules between the layers to weaken the interaction (exfoliation by intercalation), and selective etching by acids that remove weak bonds from the bulk material (exfoliation by etching). The use of chemical vapor deposition (CVD) allows to assembly of atoms to form the crystalline structure of two-dimensional materials through convenient chemical reactions based on oxidation and/or reduction processes. Epitaxial growth develops when a substrate and an overlayer are lattice matched or when a chemical balance condition is met to form a stable two-dimensional crystalline structure.

The main methods of synthesis of two-dimensional materials have been outlined in Figure 3. These methods include mechanical etching, intercalated etching, etching exfoliation, epitaxial growth, and chemical vapor deposition (CVD). The obtained monolayers are illustrated as individual nanoflakes or deposited on a substrate.

Figure 3. Synthesis methods of two-dimensional materials used in electrochemical storage devices

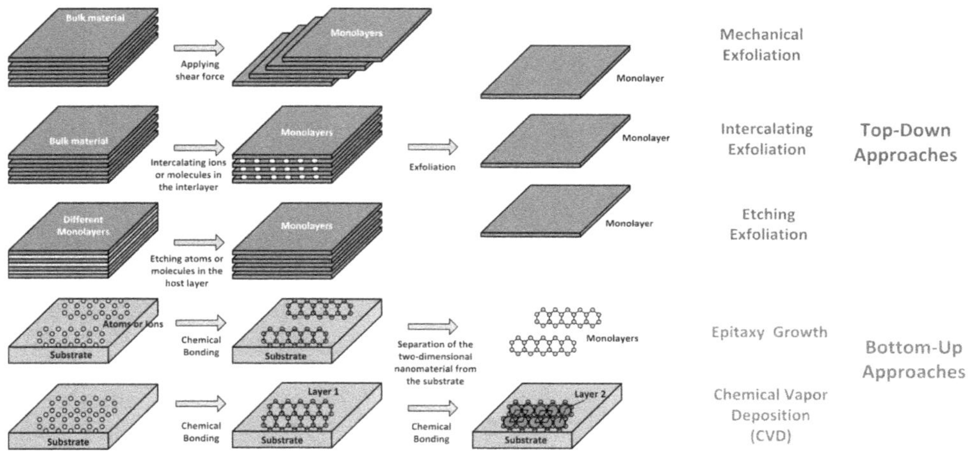

PHYSICAL PROPERTIES OF TWO-DIMENSIONAL MATERIALS USED IN RECHARGEABLE BATTERIES AND SUPERCAPACITORS

Two-dimensional (2D) materials with thicknesses of a few atoms have recently become more important as electrodes for rechargeable batteries and supercapacitors (Venkateshalu, 2022). Nanosheets made of a single atom are used to reduce the thickness of layered materials, giving them unusual and intriguing physical features (Guo, 2020). Larger surface areas, weak van der Waals forces that work between the layers to aid in the intercalation of electrolyte ions, and great mechanical strength all contribute to the advantages of 2D electrode materials, which also make them suitable for use as flexible electrodes. When compared to materials in other dimensions, 2D materials are distinct due to their strong quantum confinement, tunable thickness, electrical conductivity, and enormous surface area.

Advantages and Disadvantages of the Nanomaterials Used in Electrochemical Storage Devices

Nanomaterials have been one of the relevant material options to improve the performance of electrochemical storage devices (AbdelHamid, 2022). In the last three decades, researchers around the world have synthesized a wide diversity of nanomaterials to develop a broad set of applications. Zero-dimensional (OD) nanomaterials or nanoparticles have all their dimensions within the nanoscale. One-dimensional (1D) nanomaterials have one of their dimensions outside the nanoscale and have morphologies such as nanowires, nanotubes, or nanorods. Two-dimensional (2D) nanomaterials have two of their dimensions outside the nanoscale and have morphologies in the form of nanofilms, nanocoatings, and nanolayers. Three-dimensional (3D) nanomaterials have their three dimensions outside the nanoscale with diverse architectures such as multiple nanolayers, metal-organic frameworks, nanowire bundles, and nanoparticle assemblies. Hybrid nanomaterials are defined as unique chemical conjugates of organic and/or inorganic materials. That is, these are mixtures of two or more inorganic components, two or more organic components, or at least one of both types of components. The resulting material is not a simple mixture of its components but a synergistic material with properties and performance to develop applications with

Figure 4. Advantages and disadvantages of the materials used in electrochemical storage devices

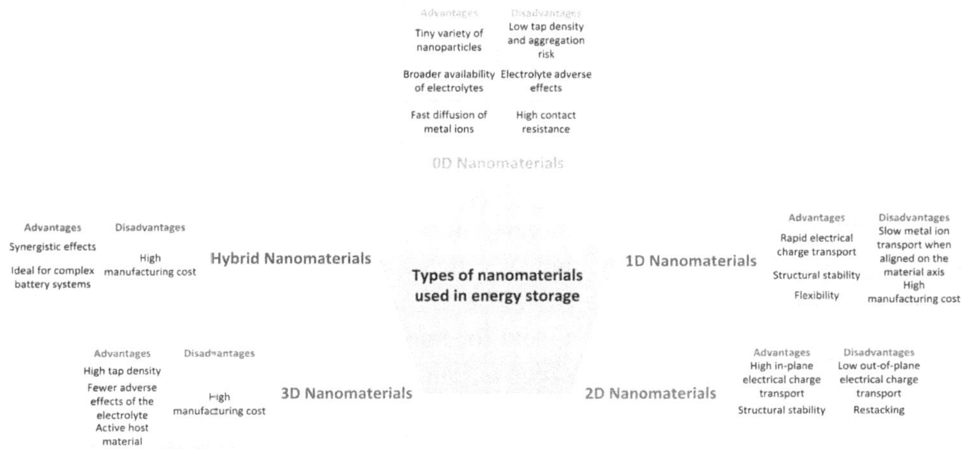

unique properties, which are determined by the interface of the components at the molecular or supramolecular level. Its functionality is associated with the improvement of physicochemical properties. All these materials have been used to implement different electrochemical storage devices such as rechargeable batteries and supercapacitors. In addition to being good materials for storing metal ions, they also optimize the performance of electrochemical storage devices through different mechanisms. A summary of these mechanisms has been compiled for each of these materials as advantages and disadvantages. The main advantages and disadvantages of the nanomaterials used in electrochemical storage devices have been compiled in Figure 4. Nanomaterials are classified as zero-dimensional, one-dimensional, two-dimensional, three-dimensional, and hybrid (i.e., using at least 2 of the above types). This last type is expected to be the most used by designers of this century to exploit multifunctional design to exploit the multiple properties of each of the types of nanomaterials.

Properties of the Graphene as Electrodes

Graphene is the first and most studied of the 2D materials in group IVA. It is a superior electronic and thermal conductor and has several intriguing properties, including ease of synthesis, high specific surface area, tunable electronic structure, and the ability to modify and functionalize its structure (Guo, 2020). This help to explain its broad range of applications.

Properties of the MXenes as Electrodes

MXenes' chemical capabilities can be used to increase energy storage even more (Mendoza-Sánchez, 2016). Due to their adequate electrical conductivity, hydrophilicity, tensile strength, and flexibility, free-standing films made from delaminated MXenes hold promise for flexible supercapacitor applications. Due to their strong metallic conductivity, binder-free MXenes can be utilized as current collector-free electrodes, reducing the number of inactive materials, and raising the device's energy density even more. Since MXenes can only function at negative potentials in aqueous solutions, finding a complementary positive electrode with high capacitance and conductivity is the main challenge in using them as pseudocapacitive materials. MXenes can be used in a range of applications due to their remarkable thermoelectric, electrical, plasmonic, and optical properties (Rasheed, 2022).

Properties of the Phosphorene and Borophene as Electrodes

Anisotropic corrugation is a characteristic of both borophene and phosphorene, however, borophene has a higher carrier density and lower mechanical stiffness than that phosphorene (Wang, 2020). The anisotropic structure of borophene gives it a very high valley-direction diffusivity. The calculations revealed that borophene has an incredibly high specific capacity of 1218 mAh/g, and good conductivity can be ensured by its metallic characteristic. The average open-circuit voltage for sodium-ion and sodium-oxygen batteries is 0.53 V, which is ideal for minimizing dendrite formation and maximizing energy density. The practical uses of phosphorene are severely constrained by its rapid air deterioration.

Properties of the Silicene, Siloxene, Germanene, and Stanene as Electrodes

Xenes are regarded as promising electrode materials in the field of electrochemical energy with big capacity, high rate, and high safety thanks to their amazing properties, including ultrathin folded structure, ultrahigh surface-volume ratio, superior mechanical strength, and flexibility (Wang, 2020). Due to their extremely high specific surface area, extremely thin thickness, strong carrier mobility, variable band gap, and in-plane anisotropy, Xenes have garnered a lot of interest. Different Xenes' chemical, electrical, and optical characteristics are closely related to their atomic configuration, elemental subgroups, surface/edge terminations, and interlayer forces.

IVA element monolayers are anticipated to display the typical electrical and structural properties of graphene (Fan, 2021). It's interesting to note that silicene has a similar conductivity as well as positive biological benefits for biocompatibility and biodegradation (Guo, 2020). These chemical elements have fascinating features that make them candidates for use in rechargeable batteries, including a large surface area, high mechanical flexibility, and higher electron mobility (Mortazavi, 2016). Since silicon is expected to have a stronger spin-orbit coupling than graphene and a nontrivial electronic structure, silicon-based quantum electronics may be made possible by accessible quantum spin Hall events at high temperatures (Fan, 2021). Graphene, silicene, germanene, stanene, and plumbene all have different energy bandgaps that are 10^{-3} meV, 8 meV, 23 meV, 300 meV, and 400 meV, respectively. Stanene has a high bandgap, making it an attractive material for spintronics and quantum electronics at ambient temperature. On a nanometer scale, the surface of the stanene sheet is not completely flat.

The conductivity of 2D siloxene sheets decreases when there are abundant oxygen functional groups on the edge or basal surface (Venkateshalu, 2022). To boost conductivity and electrochemical active sites for redox processes, O_2 groups must be removed.

Properties of the Exfoliated Pnictogens as Electrodes

Phosphorene has a band gap that is 1.5 eV larger than that of its bulk form (0.3 eV) (Venkateshalu, 2022). With hole mobility of 1000 cm^2/Vs and a great sensitivity to an external electric field, phosphorene is a semiconducting material. The fascinating aspect of phosphorene is that it has an adjustable band gap since its band gap shrinks with an increase in 2D layer spacing.

Due to its several outstanding qualities, arsenene, a single-layer arsenic nanosheet with a honeycomb shape, has recently gained more attention (Benzidi, 2019). Antimonene and arsenene have unique properties such as large surface area, optical, electrical, mechanical, and biocompatibility (Dongre, 2022). In bulk form, the chemical elements such as antimonene and arsenene are semimetals, but when exfoliated or thinned, they become semiconductors with indirect band gaps of 2.28 and 2.49 eV, respectively. Furthermore, by producing a little biaxial strain or by applying an external electric field, they can be changed into direct bandgaps. A recent development that has received a lot of interest is the few-layer antimony, which has improved air stability and an adjustable bandgap due to the small number of layers. By reducing the number of layers, As and Sb materials are transformed from metallic to semiconductors, along with the ensuing changes in electrical, electrocatalytic, and optical properties.

The last pnictogen element, bismuth, is a well-known inexpensive, non-toxic, and eco-friendly metal with strong electronic conductivity and chemical characteristics resembling those of its lighter homologs, phosphorus, and antimony (Zhou, 2019). Monolayer bismuth (bismuthene) is a viable contender for a room-temperature quantum spin Hall material because it presently has the highest experimentally determined bulk gap (0.8 eV) (Fan, 2021). The carrier mobilities for 2D Bi are also quite high for electrical applications. Bi has a huge capacity for storing K and is an environmentally acceptable material with high electronic conductivity and low cost, but it inevitably pulverizes in commonly used carbonate-based electrolytes, and its rechargeability is poor (Lei, 2018).

Properties of the Metallenes as Electrodes

Metallene is a class of 2D nanomaterials that is being investigated as a prospective candidate due to its outstanding physicochemical characteristics (Liu, 2020). Metallenes have unrivaled advantages for electrical, optoelectronic, and energy-related devices, such as excellent conductivity and extensive exposure to metal active sites. While being scaled down to a 2D scale, several types of metallenes can maintain the basic properties of metal materials with great air stability. Given these benefits, 2D metallenes are a strong contender for upcoming sophisticated energy devices. Due to their extremely high surface area, high surface energy, and unsaturated bonds, 2D metallenes may have exceptional catalytic activity.

Properties of the Metal-Organic Frameworks as Electrodes

Metal-organic frameworks (MOFs) in two dimensions (2D) are a new class of 2D materials that exhibit good catalytic performance, intriguing magnetic characteristics, and tolerable electrical conductivities in addition to uniform active sites and distinctive open channels (Guo, 2021). Due to their distinct shape, large specific surface area, functional connectivity, and metal active sites, MOFs provide good electrode materials for electrical storage devices (Zhang, 2022). These MOFs have quick charge and discharge capabilities, high power densities, and extremely long cycle life factors, making them particularly well suited for usage in applications in energy-related industries or portable devices.

Properties of the Other 2D Materials as Electrodes

In contrast to their applications, investigations on MPS_3 throughout the next two decades have mostly concentrated on their structures, conductivity, magnetism, optical characteristics, and intercalation behaviors (Zhu, M. 2020). Based on the constant-voltage two-terminal technique, MPS_3 materials can be split into two classes. While the electrical conductivities of $MnPS_3$, $NiPS_3$, and $ZnPS_3$ are lower than 10^{-9} S cm^{-1}, $FePS_3$, VPS_3, and $V_{0.78}PS_3$ exhibit greater electrical conductivities of about 10^{-5}, 10^{-2}, and 10^{-1} S cm^{-1}, respectively. Temperature, pressure, and lighting conditions can all have an impact on the conductivity of MPS_3 materials.

Due to its lightweight, exceptional mechanical strength, good corrosion and heat resistance, biocompatibility, etc., metallic titanium and its alloys have found extensive use in the domains of aviation, aerospace, navigational devices, chemical engineering, metallurgy, and other industries. Thanks to its low strain expansion, numerous ion diffusion pathways, high safety, and low cost, two-dimensional (2D) titanium (Ti)-based nanomaterials have been thoroughly studied. Such Ti-based nanosheets, however, often only show weak electrical conductivity and are readily stackable (Mao, 2021).

Fabricating porous structured materials improves the specific surface area and active sites (Qin, 2020). Increased contact between the material and the electrolyte, more active sites, more redox reactions, and shorter electron and ion transport routes are all benefits of a high specific surface area.

SOLUTIONS AND RECOMMENDATIONS FOR RECHARGEABLE BATTERIES AND SUPERCAPACITORS

The conventional lithium-ion rechargeable battery (LIB) provides stability, but it also has drawbacks such as low capacity, high cost, and slow charge or discharge; it also degrades quickly and has a finite number of life cycles (Benzidi, 2019). Because of their impressive use in several energy storage fields; electrode material development has rapidly accelerated. Several studies have been completed that have enhanced cathode, electrolyte, and anode performance. However, given that the positive electrode of a traditional rechargeable lithium-ion battery is made of oxide and the negative electrode of graphite, the bulk system still faces several inherent challenges. This results in slow charge-discharge operations, quick electrode failure, and rate performance limitations since the ions move slowly. Additionally, capacity reduction may be caused by phase changes that result in inactive phases and volume expansion. Phase changes and mechanical deterioration of electrode materials therefore fatally harm batteries and supercapacitors. The system must have quick ion and electron transport and maximize the number of ion intercalation sites to achieve high power density. One of the most effective strategies for raising the performance of rechargeable batteries and supercapacitors is the creation of novel electrode materials with a nanoscale scale. The processes used to store charges as well as the dynamics that take place at the electrode/electrolyte interface control the energy and power density of storage devices (Venkateshalu, 2022). Batteries are the most popular storage technology in this regard since they operate on the theory of reversible redox processes, which have a high-energy density and a low power density. Supercapacitors are another type of cutting-edge storage system, offering greater power density and cycle life. The choice of electrode material is crucial for enhancing the efficiency of supercapacitors and rechargeable batteries. Larger surface areas, weak van der Waals forces that work between the layers to aid in the intercalation of electrolyte ions, and great mechanical strength all contribute to the advantages of 2D electrode materials, which also make them suitable for use as flexible electrodes.

The main electrical performance parameters of a rechargeable battery are shown in Figure 5 (Gao, 2014). In rechargeable batteries, the main electrical parameters that must be monitored during design and their useful life are cycle life, energy density, charge and/or discharge rate, self-discharge rate, internal resistance, capacity, rated voltage, and discharge cut-off voltage. All these parameters must be considered

Figure 5. Main electrical performance parameters of a rechargeable battery

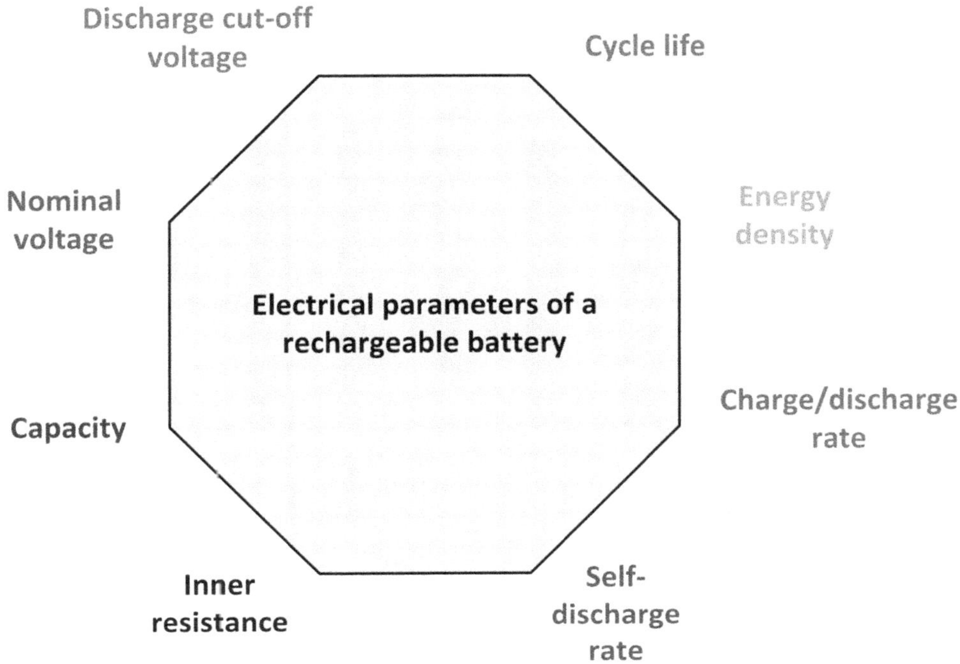

to achieve the best performance of the batteries depending on their application in their actual operation. These vary according to the lifetime of the supercapacitor, so they cannot be considered constant or are not influenced, for example, by the operating temperature or the use cycles of the device.

The theoretical specific capacities that can be achieved by different two-dimensional materials used in rechargeable batteries are illustrated in Figure 6 (data obtained from Mortazavi, 2016; Putungan, 2016; Shukla, 2017; Benzidi, 2019. Zhu, M. 2020; Liang, 2021). The specific capacity that two-dimensional materials can offer ranges from hundreds to thousands of mAh/g, allowing for a wide variety of practical implementation possibilities for rechargeable batteries. The higher the specific capacity, the better the charge and discharge ratio of the battery, and the greater the possibility of recharging or reusing the rechargeable battery.

A basic schematic of the use of two-dimensional electrode materials for rechargeable battery design is outlined in Figure 7. In the illustrated design the anode is based on a two-dimensional material and the cathode is on zinc. Metal ions are incorporated into the anode to increase the efficiency of the rechargeable battery. The electrolyte flows in different directions during the charging and discharging processes. Two-dimensional materials improve the electrical conductivity and electrochemical properties of rechargeable batteries mainly used in the electrode to improve gravimetric and volumetric capacities as well as the number of charge and/or discharge cycles.

The main electrical performance parameters of a supercapacitor are shown in Figure 8 (Zhang, 2015). In the design of your supercapacitor, the electrical parameters that determine its behavior are capacitance, operating voltage, series resistance, power density, energy density, and time constant. All these parameters must be considered to achieve the best performance of the supercapacitor depending on its application in its actual operation. These vary according to the lifetime of the supercapacitor, so

Figure 6. Theoretical specific capacities achieved by different two-dimensional materials used in rechargeable batteries

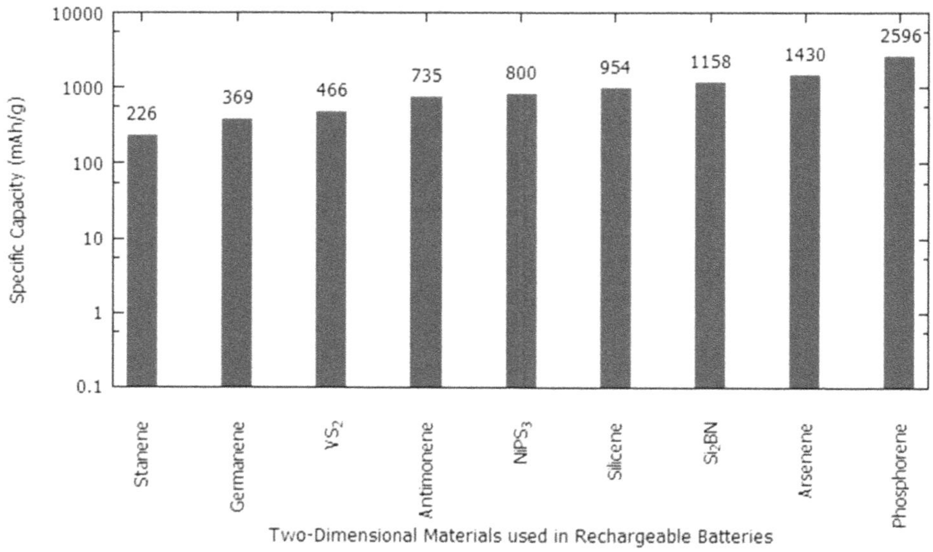

Figure 7. Rechargeable battery based on two-dimensional materials

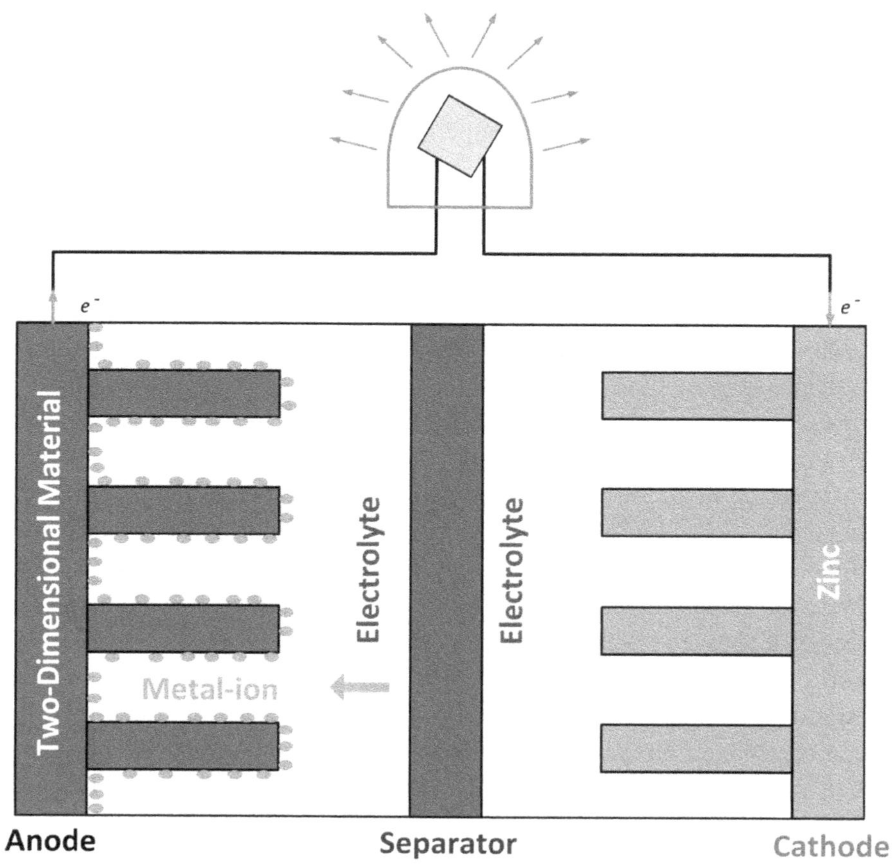

Figure 8. Main electrical performance parameters of a supercapacitor

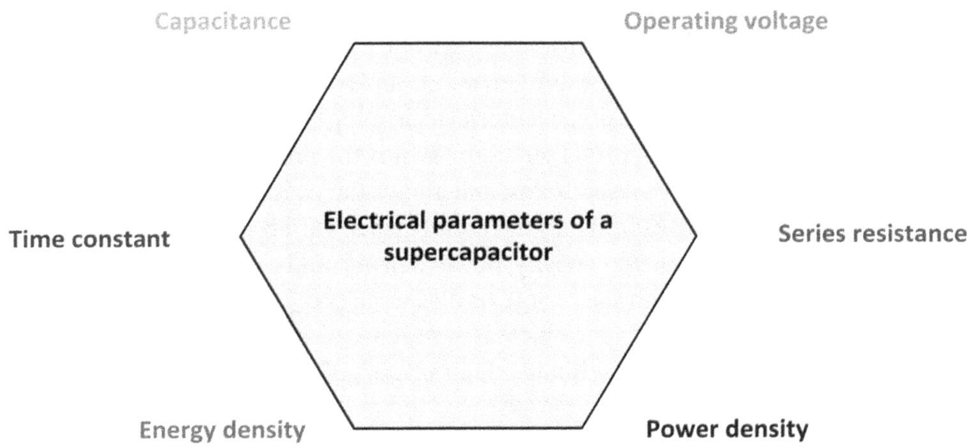

they cannot be considered constant or are not influenced, for example, by the operating temperature or the use cycles of the device.

The theoretical specific capacitance that can be achieved by different two-dimensional materials used in supercapacitors is illustrated in Figure 9 (data obtained from Oh, 2020; Qin, 2020; Guo, 2021; Dongre, 2022; Venkateshalu, 2022). The specific capacitance for these materials ranges from tens to thousands of F/g, therefore there are a variety of capacitance possibilities for different designs depending on the applications to be implemented. The higher the specific capacitance, the higher the energy density stored by the supercapacitor, and with it, the longer its operating life will be. Two-dimensional materials such

Figure 9. Theoretical specific capacitances achieved by different two-dimensional materials used in supercapacitors

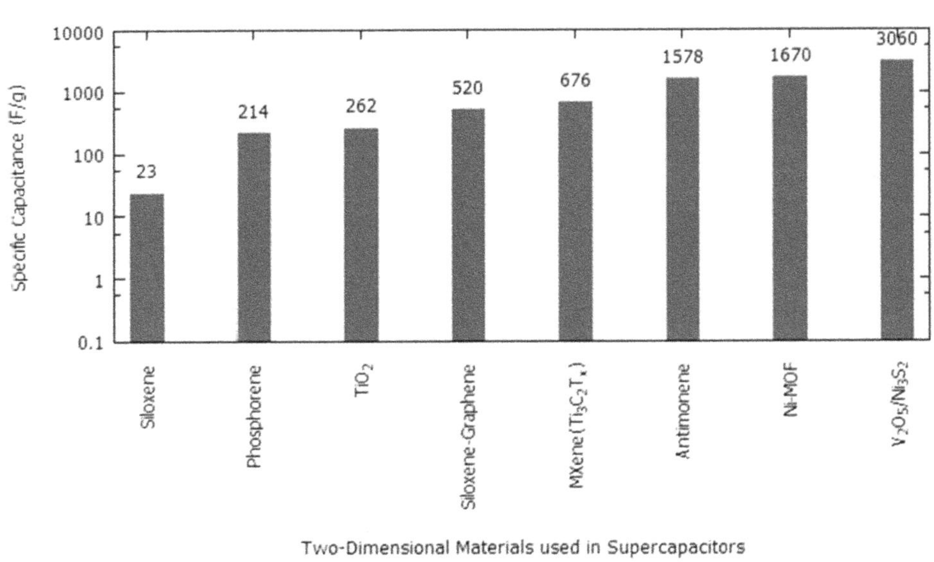

as electroactive materials through the specific capacitance allow balancing the charges stored on the two electrodes and the mass charge ratio of the anions and cations present in the supercapacitor.

A basic schematic of the use of two-dimensional electrode materials for supercapacitor design is outlined in Figure 10. The supercapacitor consists of an anode and a cathode implemented through a nickel electrode using some different two-dimensional material to develop the positive or negative charge buildup depending on the type of electrode. A porous separator avoids the production of short circuits, the storage of the electrolyte in the pores of the electrodes, and allows the passage of ions during the charge and discharge processes. The electrolyte establishes the performance variables of the supercapacitor such as safety, stability, speed of operation, energy density, power density, and capacity. The direction of the flow of anions and cations is associated with the charge or discharge process. The two-dimensional material present in the electrodes of the supercapacitor increases the charge and/or discharge rates, and the active surface area of the electrode, as well as facilitates the transport of electrolytes between the electrodes.

Figure 10. Supercapacitors based on two-dimensional materials

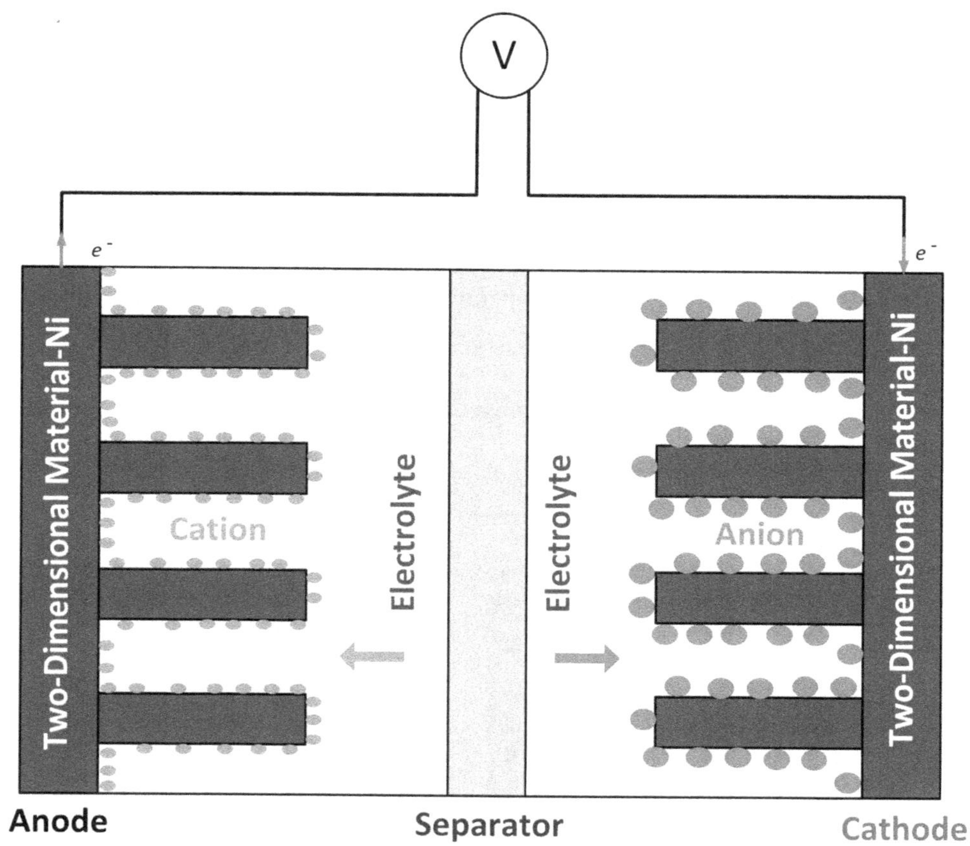

Applications of the MXenes in Energy Storage Devices

In microsized, binder-free, and mechanically robust electrodes, MXenes, a new class of 2D carbides with metallic conductivity, hydrophilicity, and the ability for ion intercalation, have demonstrated excellent volumetric capacitances (Mendoza-Sánchez, 2016). In addition to having a large storage capacity, MXenes also deliver it quickly. MXenes store charge by cation adsorption into swollen MXene galleries because of water and electrolyte ions intercalated in between MXene layers, which is accompanied and facilitated by rapid layer expansion and contraction, as opposed to storage processes in battery materials that are limited by Li-ion diffusion. MXene-based composites with improved interlayer spacing and superior electrical conductivity offer significant promise for high-power lithium-ion batteries, even though MXene electrodes deliver an inclined charge-discharge curve and very low initial Coulombic efficiency (Tang, 2018). Other multivalent ions, such as sodium and potassium, can also intercalate into the MXene interlayers in addition to lithium ions, suggesting the possibility of using MXenes in other battery systems (such as sodium-ion batteries, potassium-ion batteries, aluminum-ion batteries, and hybrid capacitors, among others).

Applications of the Silicene, Siloxene, Germanene, and Stanene in Energy Storage Devices

The following are some unique benefits of Xenes in electrochemical energy (Wang, 2020). First, an extremely high theoretical specific capacity is produced by the folded structure and extremely high surface volume ratio. Second, the reaction rate can be accelerated by using consecutive and condensed ion/electron channels equipped with ultrathin Xenes. Third, Xenes' outstanding flexibility and wide layer spacing can effectively buffer the reaction's volume growth. Fourth, using high mechanical strength Xenes as the battery separator material can reduce the development of metal dendrites and raise the batteries' safety level. Fifth, Xenes are simple to chemically functionalize due to the accessible surface lone pair electrons. Xenes have been widely used in the electrochemical energy domains, including rechargeable batteries, and supercapacitors, as well as in several innovative applications, due to their exceptional qualities.

Due to their high specific capacities of 4200 mAh/g ($Li_{22}Si_5$) for silicon bulk and 954 mAh/g for silicene, silicon-based anodes can be regarded as one of the promising options (Benzidi, 2019). With a specific capacity of 954 mAh/g, silicene is far more capable of storing sodium and lithium ions than germanene (369 mAh/g) and stanene (226 mAh/g) (Mortazavi, 2016). One of the crucial elements affecting an electrode material's potential for use is its electrical conductivity. The electrical conductivity of the electrode materials and the internal electronic resistance of a battery are tightly correlated. Additionally, the electronic conductivity of the active two-dimensional material is proportional to the ohmic heating produced during the charging or discharging operation. It was demonstrated that the quantum capacitance of silicene electrodes may be increased by doping (N, B, P, and S) vacancy defects into silicene (Yang, 2018). An increase in quantum capacitance was seen as the defect concentration rose. In contrast to P and N doped single vacancy silicene, S and B may form strong bonds with Si atoms at very low formation energies. These findings demonstrate the viability of using silicene as a supercapacitor electrode.

Layer-by-layer Xenes and other 2D layered materials combined to form 2D van der Waals heterojunctions (vdWHs) exhibit additional new properties that go far beyond those of their parts and can be used to design electrocatalytic and supercapacitor applications for high-performance energy-related devices (Wang, 2020). The vdWHs are created by weak vdW forces, which help maintain the electronic structures of the participating materials and, as a result, cannot sacrifice the original features of 2D block units. This is true even when there is a significant lattice mismatch in stacked materials. To create gate-tunable band structures for further enhancing electronic transport, vdWHs are applied to bandgap engineering considering the atomic thickness and exceptional flexibility of 2D materials. More crucially, the vdWHs allow for the efficient combination of several functionalities, the compensation of individual weaknesses, and even the endowment of new features, which considerably improves performance in applications using energy. Additionally, the vdWHs can prevent Xenes from stacking, thus enhancing their stability in electrochemical applications.

Siloxene, a kind of two-dimensional silicon (Si) that is used in supercapacitors and lithium-ion batteries, has recently received greater attention (Venkateshalu, 2022). In the organic electrolyte tetraethyl ammonium tetrafluoroborate (TEABF4), the pseudocapacitance behavior of 2D siloxene was assessed. With energy density and areal energy density of 5 Wh/Kg and 9.82 mJ/cm^2, respectively, the supercapacitor exhibits remarkable pseudocapacitive outcomes. Despite 10,000 charge-discharge cycles, the gadget shows a near 98 percent retention of capacitance.

Applications of the Phosphorene in Energy Storage Devices

The structure of phosphorene attracts study into the development of phosphorus-based lithium batteries because it offers a high-performance anode for building these devices (Benzidi, 2019). The black phosphorus has a large theoretical capacity of 2596 mAh/g; however, the volume expansion limits the cycle efficiency. As a result, the phosphorene monolayer exhibits an ultrafast diffusion along a zigzag direction of about (0,08 eV), but its theoretical capacity is restricted to 432,79 mAh/g because of structural degradation during the lithiation process.

The application of phosphorene in energy storage devices began in 2015 in lithium-ion batteries and sodium-ion batteries, 2016 it was used in lithium-sulfur batteries and supercapacitors, in 2017 in micro-supercapacitors, in 2018 in potassium-ion batteries, in 2019 in dual ion batteries, and in 2020 in hybrid metal ion capacitors (Venkateshalu, 2022). The electrical conductivity and quick ion diffusivity of phosphorene with a puckered structure make it appropriate for use as supercapacitor electrodes. Phosphorene has been used individually (Hao, 2016), in nanocomposites (Yang, 2017), or as hybrid materials (Xiao, 2017) to implement supercapacitors.

Applications of the Arsenene, Antimonene, and Bismuthene in Energy Storage Devices

The interaction of Li, Na, or Mg ions with the arsenene monolayer was examined and contrasted using density functional theory (DFT) computations. It is done on purpose for lithium, sodium, and magnesium ion rechargeable batteries to employ them as anodes (Benzidi, 2019). The findings show that Li, Na, and Mg adatoms preferred to adsorb on valley sites with negative adsorption energies of -2,55, -1,91, and -1,10 eV, respectively. This finding is crucial for battery applications because it introduces strong interaction between alkali-metal and arsenene monolayers. Additionally, it was discovered that arsenene has a theoretical specific capacity that is significantly higher than that of the commercially available graphite anode (372 mAh/g) in Li-ion batteries, reaching up to 1430 mAh/g for Li and Mg and 1073,18 mAh/g for Na. These ions' activation energy barriers exhibit isotropic behavior for the different pathways

(x, y, and diagonal directions), with the measured values for Li, Na, and Mg ions being 0,16; 0,05; and 0,016 eV, respectively. The arsenene is a good contender for use as an anode material for rechargeable batteries due to its high current capacity, low open-circuit voltage, and extremely high barrier diffusion.

Antimonene displayed a high specific capacitance of 597 F/g compared to ball-milled antimony (101 F/g), low charge transfer resistance, and minimal series resistance, which supports good electrochemical performance (Dongre, 2022). It also demonstrated thermal stability up to 400 °C with only 3% weight loss. The excellent cyclic stability of antimonene further demonstrated its potential as a material for energy storage applications. In both galvanostatic charging and cyclic voltammetry, antimonene dramatically increases the carbon electrode's ability to store energy. With a charging current density of 14 A/g, an energy density of 20 mWh/kg, and a power density of 4.8 kW/kg, the capacitive performance has grown to 1578 F/g. The increased electrochemical active area and electrochemical double layer on the electrode surface are the causes of the enhanced capacitance value in antimonene. Due to their relatively high surface area and active area for reactions, 2D mono-elemental materials like arsenene and antimonene can considerably improve supercapacitor efficiency. Antimonene may be an excellent anode material for sodium-ion batteries (SIBs) due to its increased surface-active sites and high theoretical capacity of 660 mAh/g. Arsenene and antimonene display high theoretical capacity as well as high specific capacity in battery applications, which can be attributed to the materials' greater surface-active sites.

High capacity and quick kinetics anode materials are needed for the creation of magnesium-ion batteries. Based on first-principles calculations, several researchers have investigated the potential of an antimonene monolayer and an antimonene/graphene heterostructure as anode materials for magnesium-ion batteries (Liang, 2021). The findings demonstrated the stability of Mg adsorbed on a monolayer of antimonene with a low diffusion barrier of 75 meV in the (424) direction, indicating a reasonably quick charge-discharge rate. The nanomaterials based on antimonene have relatively large capacities of 735 mAh/g (as $Mg_{1.67}Sb$) and 615 mAh/g (as $Mg_{1.78}SbC_{2.78}$). A monolayer of antimonene had a theoretical specific capacity that was substantially higher than many commercial batteries (735.54 mAh/g). The arsenene/graphene heterostructure is more stable and has superior electrical conductivity than an antimonene monolayer.

Due to its high theoretical volumetric capacity and comparatively low working potential, bismuthene is a viable anode material for modern rechargeable batteries (Zhou, 2019). Due to its high theoretical volumetric capacity and comparatively low working potential, bismuthene has been touted as a suitable anode material for high-energy-density potassium-ion batteries (Lei, 2018; Fan, 2021). The areal sodium storage capacity of a free-standing bismuthene/graphene composite electrode with tunable thickness is surprisingly steady and high at 12.1 mAh/cm^2, far exceeding that of most previously reported electrode materials (Zhou, 2019).

Applications of the Metallenes in Energy Storage Devices

Li^+-/Na^+-/K^+-ion batteries, supercapacitors, and Li-S batteries are among the most recent developments and research advances in the field of energy for metallenes (Liu, 2020). Metallenes are excellent candidates for energy storage devices due to their configurable band structure, configurable surface state, and high carrier mobility. The volume expansion during cycling has been demonstrated to be suppressed in metallenes including germanene, antimonene, and bismuthene with 2D mechanical anisotropy, indicating exceptional long-term stability for energy storage devices.

Applications of Metal-Organic Frameworks in Energy Storage Devices

Due to their advantages of high porosity, modifiable functional groups, and controllable chemical composition, metal-organic frameworks (MOFs) and their derivatives have received a lot of attention recently in the field of electrochemical energy storage, making them better candidate materials for the upcoming generation of advanced batteries (Zhang, 2022). Because the transition metal-ions or clusters they contain can be utilized as redox centers in MOF-based electrode materials to further improve the reaction activity, different transition metal-based MOFs have previously been regarded as improved electrode active materials and electrocatalysts.

Applications of Other 2D Materials in Energy Storage Devices

$M\text{PS}_3$ remembers the researchers' curiosity for their prospective application in the recently created study domains, such as lithium-ion batteries and electrochemical catalysts, which were inspired by the researchers' interest in 2D materials (Zhu, M. 2020). The most alluring feature of $M\text{PS}_3$ is its capacity for lithium intercalation, which makes it a suitable material for lithium battery electrodes. While the process of lithium intercalation at $x > 1.5$ is still debatable, the intercalation mechanism at low intercalation rates x of 1.5 is unquestionably explained by the movement of metal atoms from initial octahedral sites to adjacent tetrahedral sites two at a time. For instance, bulk $NiPS_3$ only manages to attain a capacity of 100 mAh/g under the same conditions, but nanosheet $NiPS_3$ produces significantly higher capacity and more stable cycling performances, with a specific capacity of 800 mAh/g after 150 cycles. Additionally used in lithium-ion batteries, sodium-ion batteries, and catalytic reactions are $M\text{PS}_3$ samples with nanosheet structures.

In electrochemical energy storage systems including supercapacitors, lithium-ion batteries, sodium-ion batteries, and other multivalent metal-ion batteries, 2D Ti-based nanostructures are used (Mao, 2021). Due to their slow lattice strain, high stability, and satisfactory cyclic stability as cathodes for other batteries (especially sodium-ion batteries), a variety of multi-functional Ti-based materials have been widely used as high-performance anode materials for metal ion batteries (e.g., Li^+, Na^+, and K^+) and supercapacitors.

FUTURE RESEARCH DIRECTIONS

The rapid development of energy applications based on 2D elemental materials is being driven by breakthroughs in fundamental physics/chemistry, synthetic capacity, and device prototypes. High capacity and quick kinetics electrode materials are needed for the implementation of rechargeable batteries and supercapacitors. Due to their layered structures with well-defined geometry, exceptional hardness, capacity to accommodate different intercalants, great mechanical strength, strong chemical stability, and vast surface area, 2D materials will continue to be investigated in energy storage applications in this century. Although there is a lot of theoretical research being done to improve the super capacitive properties of innovative 2D materials, there is not much experimental evidence to support this. For the majority of the novel 2D materials to be employed as electrodes for supercapacitors and rechargeable batteries, detailed practical evidence must yet be published. The unique 2D materials mentioned in this chapter will be the next-generation materials, applicable for a wide range of applications, with developments and optimizations in the synthesis procedures.

There are still many unresolved aspects of the two-dimensional materials' operating mechanisms when they are used as electrodes in rechargeable batteries and supercapacitors. There are still several difficulties, including a low yield of two-dimensional materials and generally weak chemical stability. Therefore, there is an urgent need for the large-scale, environmentally friendly, and inexpensive production of 2D materials with promising performance and atomic/molecular architectures. Ironically, there has been a lot of research into the electrochemical uses of two-dimensional materials, but our knowledge of their electrode durability is still incomplete. Such expertise is essential since the operational use of the materials may result in chemical or structural changes.

Batteries and supercapacitors need to have sufficiently high energy and power densities, so the lamellar distance and planar anisotropy of the 2D materials need to be adjusted to a moderate range. The assembly of two-dimensional materials monolayers into a transparent electrode film for transparent devices is another attractive paradigm. Two-dimensional materials can enable a stable solid electrolyte interphase (SEI) layer for metal-ion batteries by engineering a suitable coating layer and/or by selecting a matched electrolyte because their structure can contribute to an excessively high surface area and present some undesirable side reactions at the interfaces between electrolyte and electrode materials.

Despite the amazing qualities and benefits outlined above, metallene research is still in its infancy and confronts many challenging problems. The 2D thin layer morphology, in contrast to most bulk metals, is thermodynamically unfavorable; as a result, it is of utmost importance to adequately regulate the reaction kinetics process to encourage the 2D anisotropic growth of atoms. The development of scalable, controllable methods for the preparation of metallenes has been extremely difficult, despite the development of synthetic approaches including template-confined growth and molecular beam epitaxy. Another current research focus is investigating sophisticated modification techniques (such as thickness adjustment, oxidation, doping, crystal phase engineering, etc.) to increase energy storage and conversion efficiency.

CONCLUSION

Two-dimensional (2D) materials hold considerable potential for metal-ion battery and supercapacitor anodes because they have a large specific surface area and more open ion-transport channels. The essential features of rechargeable batteries and supercapacitors, such as capacity, charge/discharge rate, and cycling life, are determined by the electrode materials. Finding appropriate electrode materials for energy storage systems is so crucial. Electrode structures made of 2D materials must be dynamically and mechanically stable. Two-dimensional materials' metastable adsorption sites, which are created through defects and functionalization, are crucial for increasing storage capacity and lowering the hopping energy barrier. Theoretical specific capacities of rechargeable batteries from tens to thousands of mAh/g can be achieved through two-dimensional materials, however, experimental capacities can or not be achieved under specific characteristics. Therefore, studies to implement and design more two-dimensional materials will continue in this century to offer solutions that are completely reliable and can be mass-produced. The specific capacitance achieved using two-dimensional materials in supercapacitors ranges from units to thousands of F/g according to the properties of all the elements contained in the supercapacitor.

ACKNOWLEDGMENT

The author is grateful for the support of the University of Guanajuato for the development of this scientific research, and he wants to thank his wife and son for their support during the time conducting this scientific research. The author thanks the researchers who shared their articles to complete this study.

REFERENCES

AbdelHamid, A. A., Mendoza-Garcia, A., & Ying, J. Y. (2022). Advances in and Prospects of Nanomaterials' Morphological Control for Lithium Rechargeable Batteries. *Nano Energy*, *93*, 106860. doi:10.1016/j.nanoen.2021.106860

Benzidi, H., Lakhal, M., Garara, M., Abdellaoui, M., Benyoussef, A., El Kenz, A., & Mounkachi, O. (2019). Arsenene Monolayer as an Outstanding Anode Material for (Li/Na/Mg)-Ion Batteries: Density Functional Theory. *Physical Chemistry Chemical Physics*, *21*(36), 19951–19962. doi:10.1039/C9CP03230D PMID:31475997

Choi, D., Lim, S., & Han, D. (2021). Advanced Metal-Organic Frameworks for Aqueous Sodium-Ion Rechargeable Batteries. *Journal of Energy Chemistry*, *53*, 396–406. doi:10.1016/j.jechem.2020.07.024

Cui, H., Guo, Y., Ma, W., & Zhou, Z. (2020). 2D Materials for Electrochemical Energy Storage: Design, Preparation and Application. *ChemSusChem*, *13*(6), 1155–1171. doi:10.1002/cssc.201903095 PMID:31872570

Dongre, S. S., Shwetharani, R., Ravikumar, C. H., Lavanya, C., & Balakrishna, R. G. (2022). Review on 2D Arsenene and Antimonene: Emerging Materials for Energy, Electronic and Biological Applications. *Advanced Materials Interfaces*, *9*(23), 2200442. doi:10.1002/admi.202200442

Fan, F. R., Wang, R., Zhang, H., & Wu, W. (2021). Emerging Beyond-Graphene Elemental 2D Materials for Energy and Catalysis Applications. *Chemical Society Reviews*, *50*(19), 10983–11031. doi:10.1039/C9CS00821G PMID:34617521

Gao, J., & Abruña, H. D. (2014). Key Parameters Governing the Energy Density of Rechargeable Li/S Batteries. *The Journal of Physical Chemistry Letters*, *5*(5), 882–885. doi:10.1021/jz5001819 PMID:26274082

Guo, Q., Chen, N., & Qu, L. (2020). Two-Dimensional Materials of Group-IVA Boosting the Development of Energy Storage and Conversion. *Carbon Energy*, *2*(1), 54–71. doi:10.1002/cey2.34

Guo, Y., Wang, K., Hong, Y., Wu, H., & Zhang, Q. (2021). Recent Progress on Pristine Two-Dimensional Metal-Organic Frameworks as Active Components in Supercapacitors. *Dalton Transactions (Cambridge, England)*, *50*(33), 11331–11346. doi:10.1039/D1DT01729B PMID:34313288

Hao, C., Yang, B., Wen, F., Xiang, J., Li, L., Wang, W., Zeng, Z., Xu, B., Zhao, Z., Liu, Z., & Tian, Y. (2016). Flexible All-Solid-State Supercapacitors based on Liquid-Exfoliated Black-Phosphorus Nanoflakes. *Advanced Materials*, *28*(16), 3194–3201. doi:10.1002/adma.201505730 PMID:26915349

Kumar, K. S., Choudhary, N., Jung, Y., & Thomas, J. (2018). Recent Advances in Two-Dimensional Nanomaterials for Supercapacitor Electrode Applications. *ACS Energy Letters*, *3*(2), 482–495. doi:10.1021/acsenergylett.7b01169

Lei, K., Wang, C., Liu, L., Luo, Y., Mu, C., Li, F., & Chen, J. (2018). A Porous Network of Bismuth used as Anode Material for High-Energy-Density Potassium-Ion Batteries. *Angewandte Chemie International Edition*, *57*(17), 4687–4691. doi:10.1002/anie.201801389 PMID:29488300

Liang, Y.-B., Liu, K., Liu, Z., Wang, J., Liu, C.-S., & Liu, Y. (2021). A First-Principles Study of Monolayer and Heterostructure Antimonene as Potential Anode Materials for Magnesium-Ion Batteries. *Applied Surface Science*, *577*(4), 151880. doi:10.1016/j.apsusc.151880

Liu, Y., Dinh, K. N., Dai, Z., & Yan, Q. (2020). Metallenes: Recent Advances and Opportunities in Energy Storage and Conversion Applications. *ACS Materials Letters*, *2*(9), 1148–1172. doi:10.1021/acsmaterialslett.0c00280

Maksoud, M. I. A. A., Fahim, R. A., Shalan, A. E., Abd Elkodous, M., Olojede, S. O., Osman, A. I., Farrell, C., Al-Muhtaseb, A. H., Awed, A. S., Ashour, A. H., & Rooney, D. W. (2021). Advanced Materials and Technologies for Supercapacitors used in Energy Conversion and Storage: A Review. *Environmental Chemistry Letters*, *19*(1), 375–439. doi:10.100710311-020-01075-w

Mao, L., Zhao, X., Cheng, Q., Yang, G., Liao, F., Chen, L., He, P., & Chen, S. (2021). Recent Advances and Perspectives of Two-Dimensional Ti-based Electrodes for Electrochemical Energy Storage. *Sustainable Energy & Fuels*, *5*(20), 5061–5113. doi:10.1039/D1SE01203G

Mendoza-Sánchez, B., & Gogotsi, Y. (2016). Synthesis of Two-Dimensional Materials for Capacitive Energy Storage. *Advanced Materials*, *28*(9), 6104–6135. doi:10.1002/adma.201506133 PMID:27254831

Mortazavi, B., Dianat, A., Cuniberti, G., & Rabczuk, T. (2016). Application of Silicene, Germanene and Stanene for Na or Li Ion Storage: A Theoretical Investigation. *Electrochimica Acta*, *213*, 865–870. doi:10.1016/j.electacta.2016.08.027

Oh, S. M., & Hwang, S.-J. (2020). Recent Advances in Two-Dimensional Inorganic Nanosheet-based Supercapacitor Electrodes. *Journal of the Korean Ceramic Society*, *57*(2), 119–134. doi:10.100743207-020-00023-2

Peng, L., Zhu, Y., Chen, D., Ruoff, R. S., & Yu, G. (2016). Two-Dimensional Materials for Beyond-Lithium-Ion Batteries. *Advanced Energy Materials*, *6*(11), 1600025. doi:10.1002/aenm.201600025

Putungan, D. B., Lin, S.-H., & Kuo, J.-L. (2016). Metallic VS_2 Monolayer Polytypes as Potential Sodium-Ion Battery Anode via Ab Initio Random Structure Searching. *ACS Applied Materials & Interfaces*, *8*(29), 18754–18762. doi:10.1021/acsami.6b03499 PMID:27373121

Qin, H., Liang, S., Chen, L., Li, Y., Luo, Z., & Chen, S. (2020). Recent Advances in Vanadium-based Nanomaterials and Their Composites for Supercapacitors. *Sustainable Energy & Fuels*, *4*(10), 4902–4933. doi:10.1039/D0SE00897D

Rasheed, T. (2022). Mxenes as an Emerging Class of Two-Dimensional Materials for Advanced Energy Storage Devices. *Journal of Materials Chemistry. A, Materials for Energy and Sustainability*, *10*(9), 4558–4584. doi:10.1039/D1TA10083A

Shukla, V., Araujo, R. B., Jena, N. K., & Ahuja, R. (2017). The Curious Case of Two Dimensional Si_2BN: A High-Capacity Battery Anode Material. *Nano Energy*, *41*, 251–260. doi:10.1016/j.nanoen.2017.09.026

Sun, X., Chang, Y., Mu, C., Nie, A., Wang, B., Xiang, J., Zhai, K., Wen, F., & Li, Z. (2019). Two-Dimensional Layered Materials InSe Nanoflakes/Carbon Nanotubes Composite for Flexible All-Solid-State Supercapacitors. *Journal of Materials Science*, *55*(7), 2947–2957. doi:10.100710853-019-04151-x

Tang, X., Guo, X., Wu, W., & Wang, G. (2018). 2D Metal Carbides and Nitrides (MXenes) as High-Performance Electrode Materials for Lithium-based Batteries. *Advanced Energy Materials*, *8*(18), 1801897. doi:10.1002/aenm.201801897

Venkateshalu, S., Subashini, G., Bhardwaj, P., Jacob, G., Sellappan, R., Raghavan, V., Jain, S., Pandiaraj, S., Natarajan, V., Al Alwan, B. A. M., Al Mesfer, M. K. M., Alodhayb, A., Khalid, M., & Grace, A. N. (2022). Phosphorene, Antimonene, Silicene and Siloxene based Novel 2D Electrode Materials for Supercapacitors – A Brief Review. *Journal of Energy Storage*, *48*, 104027. doi:10.1016/j.est.2022.104027

Wang, T., Wang, H., Kou, Z., Liang, W., Luo, X., Verpoort, F., Zeng, Y.-J., & Zhang, H. (2020). Xenes as an Emerging 2D Monoelemental Family: Fundamental Electrochemistry and Energy Applications. *Advanced Functional Materials*, *30*(36), 2002885. doi:10.1002/adfm.202002885

Xiao, H., Wu, Z.-S., Chen, L., Zheng, S., Ren, W., Cheng, H.-M., & Bao, X. (2017). One-Step Device Fabrication of Phosphorene and Graphene Interdigital Micro-Supercapacitors with High Energy Density. *ACS Nano*, *11*(7), 7284–7292. doi:10.1021/acsnano.7b03288 PMID:28628293

Yang, B., Hao, C., Wen, F., Wang, B., Mu, C., Xiang, J., Li, L., Xu, B., Zhao, Z., Liu, Z., & Tian, Y. (2017). Flexible Black-Phosphorus Nanoflake/Carbon Nanotube Composite Paper for High-Performance All-Solid-State Supercapacitors. *ACS Applied Materials & Interfaces*, *9*(51), 44478–44484. doi:10.1021/acsami.7b13572 PMID:29192760

Zhang, S., & Pan, N. (2015). Supercapacitors Performance Evaluation. *Advanced Energy Materials*, *5*(6), 1401401. doi:10.1002/aenm.201401401

Zhang, Y., Huang, Q., Liu, J., Zhou, J.-E., Lin, X., Zeb, A., Reddy, R. C. K., & Xu, X. (2022). Recent Advances in Fe-based Metal-Organic Framework Derivatives for Battery Applications. *Sustainable Energy & Fuels*, *6*(11), 2665–2691. doi:10.1039/D2SE00319H

Zhou, J., Chen, J., Chen, M., Wang, J., Liu, X., Wei, B., Wang, Z., Li, J., Gu, L., Zhang, Q., Wang, H., & Guo, L. (2019). Few-Layer Bismuthene with Anisotropic Expansion for High-Areal-Capacity Sodium-Ion Batteries. *Advanced Materials*, *31*(12), 1807874. doi:10.1002/adma.201807874 PMID:30714223

Zhu, J. P., Xiao, G. S., & Zuo, X. X. (2020). Two-Dimensional Black Phosphorus: An Emerging Anode Material for Lithium-Ion Batteries. *Nano-Micro Letters*, *12*(1), 120. doi:10.100740820-020-00453-x PMID:34138144

Zhu, M., Kou, H., Wang, K., Wu, H., Ding, D., Zhou, G., & Ding, S. (2020). Promising Functional Two-Dimensional Lamellar Metal Thiophosphates: Synthesis Strategies, Properties and Applications. *Materials Horizons*, *7*(12), 3131–3160. doi:10.1039/D0MH00802H

ADDITIONAL READING

Anayee, M., Kurra, N., Alhabeb, M., Seredych, M., Hedhili, M. N., Emwas, A.-H., Alshareef, H. N., Anasori, B., & Gogotsi, Y. (2020). Role of Acid Mixtures Etching on the Surface Chemistry and Sodium Ion Storage in $Ti_3C_2T_x$ MXene. *Chemical Communications*, *56*(45), 6090–6093. doi:10.1039/D0CC01042A PMID:32352463

Bachinin, S., Gilemkhanova, V., Timofeeva, M., Kenzhebayeva, Y., Yankin, A., & Milichko, V. A. (2021). Metal-Organic Frameworks for Metal-Ion Batteries: Towards Scalability. *Chimica Techno Acta*, *8*(3), 20218304. doi:10.15826/chimtech.2021.8.3.04

Poyraz, A. S., Quilty, C. D., Housel, L. M., Hu, X., Bruck, A. M., Li, Y. R., Yin, J., Zhang, B., Huang, J., Wu, L., Zhu, Y., Takeuchi, E. S., Marschilok, A. C., & Takeuchi, K. J. (2019). Synthesis and Characterization of 2 × 4 Tunnel Structured Manganese Dioxides as Cathodes in Rechargeable Li, Na, and Mg Batteries. *Journal of the Electrochemical Society*, *166*(4), A670–A678. doi:10.1149/2.1341902jes

Rihm, J., Sim, E. S., Cho, S. B., & Chung, Y.-C. (2020). Enhancement of the Quantum Capacitances of Group-14 Elemental Two-Dimensional Materials by Ti-doping: A First Principles Study. *Applied Surface Science*, *530*, 147301. doi:10.1016/j.apsusc.2020.147301

Wu, C., Lu, X., Peng, L., Xu, K., Peng, X., Huang, J., Yu, G., & Xie, Y. (2013). Two-Dimensional Vanadyl Phosphate Ultrathin Nanosheets for High Energy Density and Flexible Pseudocapacitors. *Nature Communications*, *4*(1), 2431. doi:10.1038/ncomms3431 PMID:24026224

Wu, Y., Song, T.-Y., & Chen, L.-N. (2021). A Review on Recent Developments of Vanadium-based Cathode for Rechargeable Zinc-Ion Batteries. *Tungsten*, *3*(3), 289–304. doi:10.100742864-021-00091-9

Yang, G. M., Xu, Q., Fan, X., & Zheng, W. T. (2018). Quantum Capacitance of Silicene-based Electrodes from First-Principles Calculations. *The Journal of Physical Chemistry C*, *122*(4), 1903–1912. doi:10.1021/acs.jpcc.7b08955

Zang, X., Li, P., Chen, Q., Wang, K., Wei, J., Wu, D., & Zhu, H. (2014). Evaluation of Layer-by-Layer Graphene Structures as Supercapacitor Electrode Materials. *Journal of Applied Physics*, *115*(2), 024305. doi:10.1063/1.4861629

Zhang, S., Tan, H., Rui, X., & Yu, Y. (2020). Vanadium-based Materials: Next Generation Electrodes Powering the Battery Revolution? *Accounts of Chemical Research*, *53*(8), 1660–1671. doi:10.1021/acs.accounts.0c00362 PMID:32709195

KEY TERMS AND DEFINITIONS

Antimonene: It is an antimony two-dimensional allotrope having buckling honeycomb lattices as its atom arrangements.

Bismuthene: It is the bismuth allotrope that exists in two dimensions (2D) and was assumed to be a topological insulator.

Borophene: It is also referred to as a "boron sheet" and is a crystalline atomic monolayer of boron.

Germanene: It is a buckled honeycomb-shaped two-dimensional allotrope of germanium.

Phosphorene: It is the most stable allotrope of phosphorus and a two-dimensional material made of one or more layers of black phosphorus.

Plumbene: It is a lead allotrope that exists in two dimensions and resembles graphene in terms of its hexagonal honeycomb structure.

Rechargeable Battery: A kind of electrical battery that can be charged, discharged into a load, and recharged numerous times.

Silicene: It is a silicon two-dimensional allotrope that has a hexagonal honeycomb structure that resembles graphene.

Stanene: It is a topological insulator made of tin atoms organized in a single layer.

Supercapacitor: It is a high-capacity capacitor with lower voltage restrictions and a capacitance value that is significantly higher than that of conventional capacitors.

Chapter 11
Leather Industry and Its Impact on Net Zero

Mahabubur Rahman Khan
https://orcid.org/0009-0005-6627-6581
University of Lincoln, UK

Bart Vorselaars
University of Lincoln, UK

Mahfuzur Rahman
https://orcid.org/0000-0002-3386-9613
University of Lincoln, UK

Ahmed Elseragy
University of Lincoln, UK

Waqar Ahmed
University of Lincoln, UK

ABSTRACT

The leather industry is one of the oldest industries around the world and produces many different types of leather and leather goods. These products are incredibly popular among millions of people across the world for fashion and function. The leather manufacturing sector has been contributing more than $100 billion to the global markets. However, it is considered a major source of pollution with over 80% of the raw material going to waste. It produces toxic waste, environmental pollution, and greenhouse gas emissions. The main factors in greenhouse effects include energy consumption, wastewater, and solid waste. The emission of greenhouse gases (GHG) through industrial processes is a crucial problem for the environment. This chapter will provide a condensed review of relevant literature on leather manufacturing and the direction forward to achieving net zero. The importance of leather and its products, its contribution to the economic global markets, manufacturing processes, environmental pollution, net zero, and the role of mathematics for net zero have been discussed.

DOI: 10.4018/978-1-6684-6830-2.ch011

1. LEATHER AND ITS IMPORTANCE

Leather is a versatile and unique material with an assortment of complex natural fibres, skins, and hides. Small animal skin is known as skins, and skins from large animals are known as hides. Using various chemical and physical processes the hides are converted into finished leather. Raw leather is stiff and dry. After the tanning process leather becomes durable and robust.

Different types of leather have individual properties and appearances depending on the type of raw materials source, grade, environment, and treatment process. The cowhide has the largest contribution to over 65% of total skin/hide as a raw material for leather production. The history of leather dates back 7000 years (MasterClass, 2021). Around 1700s, some developed nations, mainly France, Great Britain, and Spain, had first established the world trade routes for leather manufacturing, trade of hides and its products around the globe (Liberty Leather Goods, 2023).

Leather is natural and long-lasting. It is uniquely important and very popular due to its excellent combined ability of beauty, comfort, and practicality. The main reasons for the high demand are:

- Goods made from leather are long-lasting.
- It is easy to maintain, and leather products are repairable.
- It is recyclable and can be used again and again.
- Leather is an incredibly versatile material and feels natural and comfortable.
- Leather is a good thermal insulator and keeps people warm.
- Its goods are naturally beautiful, glamorous, and stylish.

Leather is one of the oldest useful materials. Globally the leather industry creates enormous employment as well as skills for millions. It is playing an important role in the economy throughout the world.

2. IMPORTANCE OF NET-ZERO

Net zero is a term that refers to the equality in the total amount of Greenhouse Gases (GHG) that is emitted and the total amount that is removed from the atmosphere (National Grid, 2022a). Net-zero is also known as ‘climate neutrality’ and ‘carbon neutrality’ (Grantham research institute on climate change and the environment, 2021).

Net zero is the globally agreed upon target for reducing global heating by the 2nd half of this century. The IPCC has decided on the need for achieving net zero CO_2 by 2050 to stay consistent with 1.5 degrees Celsius increase in global temperature above the pre-industrial era (University of Oxford, 2022).

According to international scientists, to prevent the worst climate harm, global net human-made CO_2 emissions have to be reduced by almost 45% from the 2010 level by 2030, and 50%- 93% in 2050. This target will provide a significant path to achieve net-zero by 2050 (Sluisveld et al., 2021; University of Oxford, 2022).

World temperature is proportional to the carbon dioxide concentration of the atmosphere, which implies that the earth will remain warming for as long as CO_2 emissions stay above zero (University of Oxford, 2022). Based on current global policies, global temperatures are on the path to rise by 2.7 degrees Celsius by 2100, which could make parts of the world uninhabitable. To limit global warming, significant reductions in greenhouse gas emissions such as carbon dioxide (CO_2), methane (CH_4) are required over time. Therefore, achieving net zero is important for the world (National Grid, 2022b).

3. USES OF LEATHER

Due to the reasons mentioned in section 1. Some of the uses of leather are included in table 1.

Table 1. Applications of leather

Item	Examples	Pictures
clothing	pants, jackets, skirts, raincoats.	
footwear	leather boots, shoes, slippers	
furniture	couches, chairs, recliners, automotive interiors.	
books	binding books, albums, documents, folders	
equestrian products	saddles, horse hoof boots	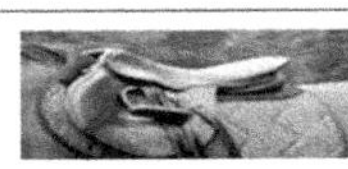
sports	footballs and baseball gloves	
bags	satchels, wallets, backpacks, purses	
cases	eyeglasses case, protective enclosure for smartphone	
holsters	arrow cuivers, gun holsters, and knife sheaths	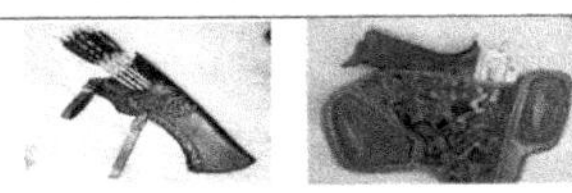
accessories	belt, gloves, wristwatch straps, bangles, and other jewellery.	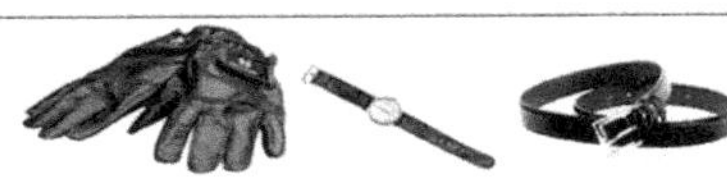
cuir Bouilli	body armour and flasks	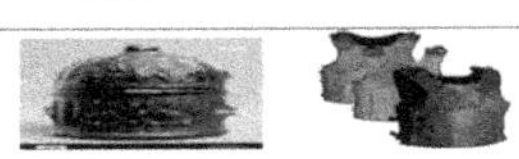

Source: Love Your Leather (2022)

Figure 1. Demand for leather products around the world
Source: Liberty Leather Goods (2022)

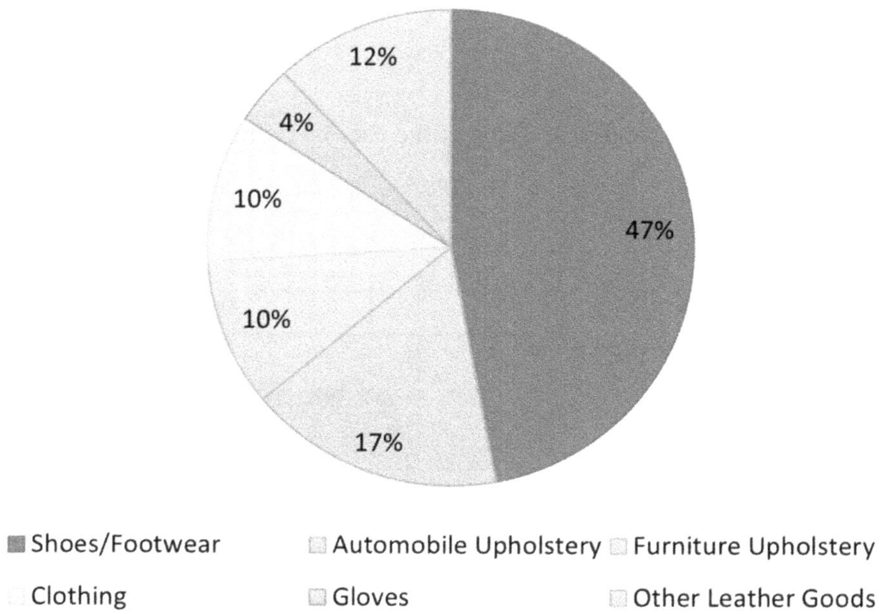

Millions of people are fond of leather and its products. Shoes/footwear is the most popular leather product all over the world, as shown in Figure 1.

4. LEATHER MARKET

4.1 Leather Producers

The global leather industry is becoming more popular with the increasing demand for leather products (BizVibe, 2022). It contributes to several countries' economic growth and employment creation. Figures 2 and 3 show the ranking of major leather producers in 2020 across the world (BizVibe, 2022).

China produces the highest amount of leather of 6.17 billion square feet every year. It accounts for 25% of world leather production. China is famous for producing belts, soles, and straps from heavy leather. Moreover, light finished leather is used for shoes, jackets, and bags. Annually China produces nearly 40% of their total leather from light bovine skin that makes China the largest manufacturer of bovine leather. Goat and sheep skins are the second biggest leather source of China's leather manufacturing industry.

Brazil is the second highest leather manufacturer producing close to 2.4 billion sq. ft of leather per year. The main market of this country's leather industry is light skin from bovine animals with over 1.8 billion square feet per year. Goats and sheep skin contribute to almost 4% in leather producing industry of Brazil.

Figure 2. Ranking order of leather producing countries in 2020
Source: BizVibe (2022)

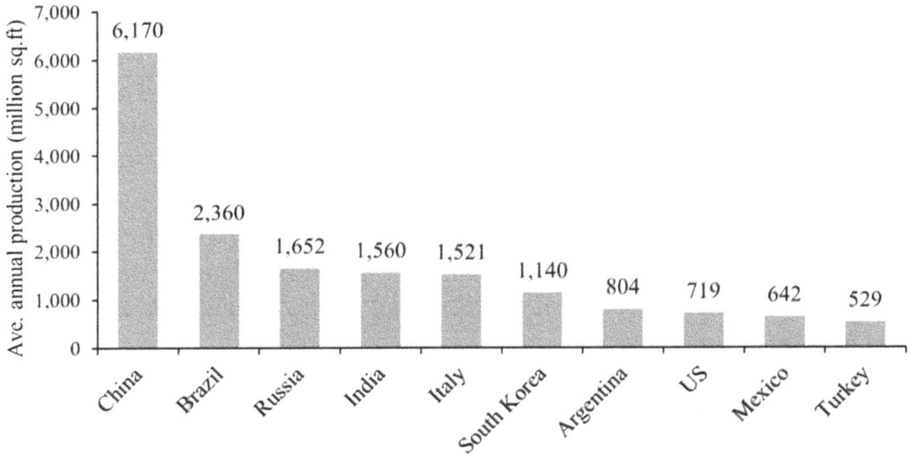

Figure 3. Top 10 largest leather producers in 2020
Source: Dalgado (2020)

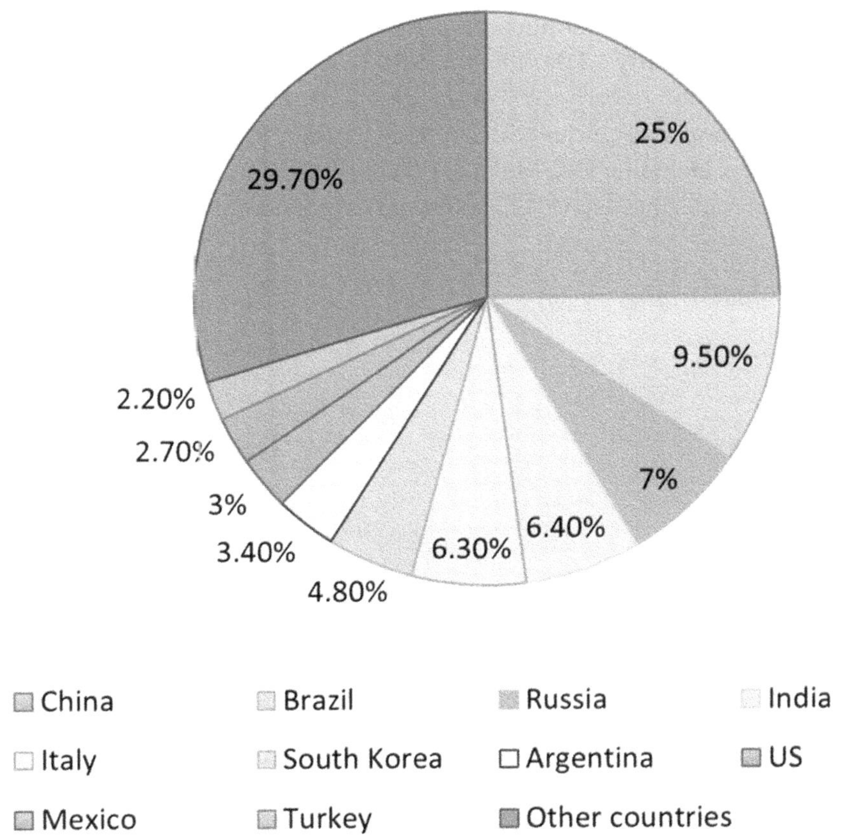

Russia ranks in third place producing over 1.6 billion sq. ft. of leather annually with bovine being the main contributor. Goats and sheep skin contributes 8% in their country's leather industry. Heavy leather represents only nearly 3% in Russian leather industry. Before 1990s Russia was the world's largest manufacturer.

India has emerged as one of the fastest-growing manufacturers of leather and ranks in fourth place on the global scale in 2020. It produces about 1.6 billion square feet of leather annually, while this nation's leather manufacturing factories mainly depend on bovine for heavy leather production. Interestingly, India makes light leather almost equally from bovine animals, goats, and sheep.

Other top leather producing nations include Italy, South Korea, USA, Argentina, Turkey, and Mexico. Italy is the largest European producer of leather. Bangladesh and Pakistan are the growing countries in leather products manufacturing. They have a vast supply of raw materials and have a huge opportunity to contribute to the global leather market.

4.2 Leather Exporters and Importers

4.2.1 Exporters

Leather is one of the oldest and most widely used materials in the world. Many countries around the world produce and export leather, but some countries are more renowned for their leather exports than others. Italy is known for producing some of the world's finest leather. It is the largest exporter of leather in the world, with exports valued close to $4.0 billion in 2019. USA is the second-largest leather exporter in the world, with exports valued nearly $2.0 billion in 2019 (BizVibe, 2022). In terms of volume, China is the largest exporter representing nearly 36% of the global market. China and South Korea are third and fourth largest leather exporter countries with exports valued at $1.0 billion and $0.5 billion in 2019 respectively. Other exporters including UK, Mexico, Poland, Pakistan, India, New Zealand, and Canada who have been making huge contributions to the leather and its products trade market as shown in Figure 4.

Figure 4. Ranking of leather exporter in 2019
Source: BizVibe (2022)

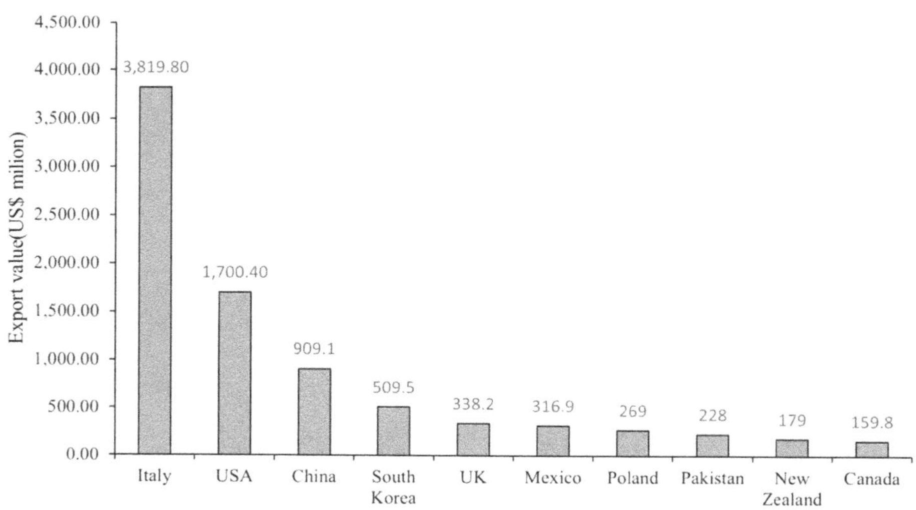

Figure 5. Top 10 leather importers in 2019
Source: BizVibe (2022)

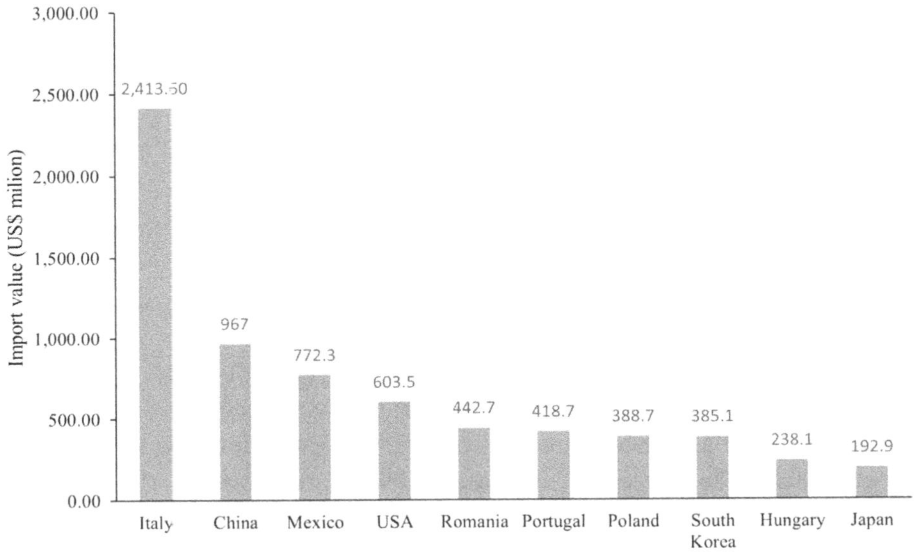

These top leather export countries are major players in the global leather industry, and they contribute significantly to the world economy. The demand for leather products is expected to continue growing, and these countries are likely to remain at the forefront of the industry for the foreseeable future.

4.2.2 Importers

Figures 4 and 5 show, Italy was the largest leather importer and exporter in the world in 2019. This country has dominated the leather market for decades due to its leather quality and increasing global demand for high quality fashion and accessories. The leather products have been popularised by Europe-based leading brands such as Gucci, Gianni Versace, Gucci, and Louis Vuitton, these brands fueling the dream and desire for exclusive and expensive products.

China is the second largest leather importer nation in the world. This nation has been making a large contribution to the fashion accessories, footwear, and furniture manufacturing sectors. Other top importer nations are USA, Mexico, Romania, South Korea, Portugal, Poland, Japan and Hungary, they have been playing a very important roles in different sectors of leather and its products. Some developing and emerging nations like Bangladesh and Pakistan, are also benefiting from booming leather manufacturing industries and its market (BizVibe, 2022).

5. TYPES OF LEATHER

There are several types of leather depending on various factors. The notable factors are:

- Animals' type, size, age, and breed

- Physical area and weather where the animals lived
- The portion of the skin/hide from where the leather was cut.
- Layer position of the skin/hide used (full grain, top grain, split)
- Leather processing and tanning qualities.

In general, there are four main types of leather (LaFalce, 2019).

- Top Grain Leather
- Full Grain Leather
- Corrected Grain/Genuine Leather
- Bonded Leather

Leather grain refers to the composition and appearance of the hide. The grain is the most outer layer of the animal hides. The fibres of this layer are the densest and tight. The corium layer is mainly comprised of collagen fibres. These collagen fibres are looser and more open compared to the grain surface. The junction of the grain and corium layers is where tight, dense fibres mix into the looser collagen fibres of the corium. The flesh layer of the hide consists of muscle and fatty tissues.

5.1 Full Grain Leather

Full grain quality leather contains all grain and is obtained from the top layer of the skin/hide. It has no surface splitting or alterations; hence it keeps its inherent hardness, and imperfections. This type of leather is the highest quality leather and is highly popular for saddlebacks. Working with this leather requires skill and high-quality manufacturing capabilities.

5.2 Top Grain Leather

This is the 2nd highest level grade leather and is commonly used in high-end leather goods. To obtain top grain leather, the upper layer of the skin from stained hides is split and has a divided layer with inherent imperfections. The thinness and smoothness of this leather help the manufacturer because it is more flexible and workable. It is used commonly for making handbags, jackets, and other high-end leather products.

5.3 Corrected Grain (Bottom Cut/Split)/ Genuine Leather

This is also known as split leather or genuine leather. It is third grade leather obtained from the layers which are collected after the upper layer, this portion is split off. Manufacturers use an artificial grain to the surface to make it more sustainable and natural by spraying it with dye or staining material. The most common uses are footwear, jackets, messenger bags, handbags, accessories, and furniture.

5.4 Bonded Leather

This leather is made from a combination of real and fake leather. Bonded leather is obtained from left-overs of the skin/hide, sometimes dust and shavings are also included. These are bonded by using latex or polyurethane on top of a fibre sheet. Manufacturers make it in such a way that looks like full grain or top grain leather. Some producers market this leather as real leather. Most of the bonded leather has only 10-20 percent genuine leather. The most common uses are bookbinding, furniture, and various fashion accessories.

6. MANUFACTURING OF LEATHER

Many types of leather are produced in the leather industry (LaFalce, 2019; The Leathersellers' Company, 2020), and all types of leather manufacturing must go through the four fundamental stages (LaFalce, 2019). These are the preparatory steps, tanning steps, post-tanning, and finishing steps. Sometimes more sub-processes must be included in the leather manufacturing process.

There are some processes before transferring to the tannery because the skin/hide needs to be cured. Though curing is a vital step for leather collection, it is not considered as a tanning process.

Figure 6. Stages and different process steps in leather manufacturing (LaFalce, 2019; The Leathersellers' Company, 2020)

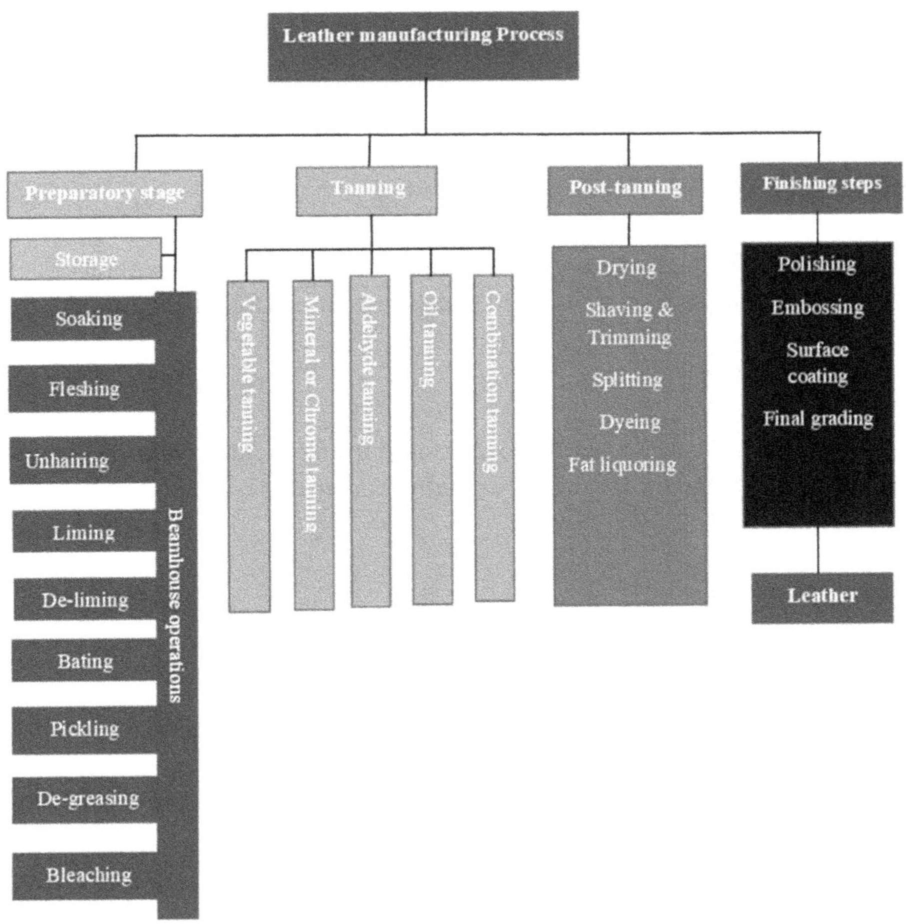

In general, when an animal is slain its skin must be separated quickly before its body temperature drops. After this, the skin or hide is cured with sodium chloride(salt) for removing the water from the skin. This skin is put in the shade to completely dry. Moreover, for further processes, this cured hide or skin is transported to the tanneries.

Figure 6 represents a flow chart of the leather manufacturing-process. Hides are going through wide range of process for producing finished leather. For different types of leather, the tanning process is different.

The various steps used in the manufacturing process are described in subsequent sections.

6.1 Preparatory Stage

In this stage, cured leather is prepared for the tanning process. This process mainly removes unwanted components from raw skin. There are different types of procedures in this step, but all these processes need not be applied for all types of skin, it depends on the type and quality of the skin and desired products. This stage has two sections, one is storage, and another is beamhouse operations. Beamhouse operations include soaking, fleshing, unhairing, liming, de-liming, bating, pickling, de-greasing, bleaching. Below we discuss the different processes of this stage. Below we discuss the different processes of this stage.

6.1.1 Storage

After the animal is slaughtered, the skin is preserved by drying, salting, or freezing so that it may make it to the tannery.

6.1.2 Soaking Process

At the soaking stage, cured skin or hide is soaked in water for several hours to several days. Lost moisture during salting is restored and debris, blood, dirt, and excess animal fats are removed.

6.1.3 Fleshing Process

In the fleshing process, subcutaneous material is removed from the flesh side. To remove the flesh, fat, and muscle the pelt is put through a machine. Normally this process is applied after soaking, slaughter, or liming.

6.1.4 Unhairing Process

The unhairing process is used for removing hair from skin/hide by using blades and rollers or using other mechanical instruments.

6.1.5 Liming Process

During the Liming process, the fibres are loosened, and the skin absorbs several tanning chemicals. Generally, wool and hair type keratinous materials are removed by using sodium sulphide and hydrated lime. Fats become hydrolysed as the pH increases. Skin fibres absorb water causing the skin structure to be swollen.

6.1.6 De-Liming Process

In the de-liming process, the hide or skin is cleaned using a mixture of ammonium sulphate or ammonium chloride with water. This removes any impurities, and water to reduce swelling.

6.1.7 Bating Process

Bating process occurs after the de-liming process which cleans the grains of the skin or hide, and the pelt becomes silky and smooth.

6.1.8 Pickling Process

To prevent decomposition, in this pickling process, soaking and cleaning of the rawhide are done in salts or acids. Tanning agents such as aldehydes and chromium penetrated are employed. Pickling agents help to preserve skins or hide for a few months, and the pH level is reduced by soaking the skin or hide in sulfuric acid.

6.1.9 De-Greasing Process

The de-greasing process eliminates natural fatty acids or excess grease from the hide or skin using water-based solutions or solvents.

6.1.10 Bleaching Process

The bleaching process is used for making hide colourless by using various chemicals to add aimed colour at other stages.

6.2 Tanning Stage

Tanning converts the raw skin/hide into a non-decomposable and stable material that is known as leather. This step permanently changes the protein structure of the skin into sturdy and durable leather.

There are several types of tanning used in leather manufacturing. These include vegetable tanning, chrome or mineral tanning, aldehyde tanning, oil tanning and combination tanning.

6.2.1 Vegetable Tanning Process

The vegetable tanning process is environmentally friendly compared to other tanning processes. Vegetable tanning uses a polyphenol astringent chemical that occurs naturally, this chemical is known as tannin. It is normally found in tree parts such as bark, branches, and leaves. Tannin produces deep brown shades, beige, red, and yellow giving the leather a unique texture and gorgeous colour. This process is eco-friendly, but it is expensive, time-consuming, and laborious. This process has two variants – a slow process normally taking one to several months because it needs multiple treatments, and the manual but quick process taking only 36 to 48 hours. Vegetable tanning produces highly durable leather.

6.2.2 Chrome or Mineral Tanning Process

The chrome or mineral tanning process is the most popular process because it is faster, less labour intensive, and affordable. It can be automated, and it takes a day at most. Chromium (III) sulphate is an effective and efficient tanning agent. Chrome tanned leather is also known as wet blue leather because of its bluish colour.

In this process chemicals and acids are used creating a negative environmental impact. This is a concern in the whole world but in developing countries such as Bangladesh, Pakistan, and India.

6.2.3 Aldehyde Process

This aldehyde tanning process uses oxazolidine or glutaraldehyde compounds which produce wet white leather due to its pale cream colour. This type of leather is soft, water absorbent, and can wash by machine. It is, consequently, suitable for using in chamois.

6.2.4 Oil tanning Process

In oil tanning a mixture of emulsified oils and aldehyde chemicals are used to produce exceptionally soft and flexible leather.

6.2.5 Combination Tanning Process

Combination tanning is also known as semi veg-tanned or veg re-tanned, this process is the combination of veg-tanned and chrome tanned. This process is used depending on the requirements of the leather type. Normally, one veg tan first and then chrome tan, or vice versa.

6.3 Post Tanning Stage

Post-tanning converts the tanned leather into a marketable product. The choice of chemicals used in this process depends on the desired colour and texture.

6.3.1 Drying Process

Drying step is used for removing water from tanned leather by using two rolling cylinders.

6.3.2 Shaving and Trimming Process

The shaving and trimming process creates a uniform level of semi-finished leather by using a shaving machine.

6.3.3 Splitting Process

In this step, splitting machine slices the material (thick leather) into the required horizontal layers. Moreover, for making high end consumer goods sometimes this process is also applied after liming.

6.3.4 Dyeing Process

Except vegetable-tanned leather dyeing process is very important for all kinds of leather because in this stage leather is coloured often using water soluble dyes. Normally, acid dyes, anionic dyes, metal complex compounds and basic dyes are used in this stage. It differs remarkably from the surface layer where dye is used just on the upper layer.

6.3.5 Fat Liquor Process

The fat liquor stage uses oil, fat, or wax to make leather flexible and soft.

6.4 Finishing Process

The final stage is where final touches are added after he post tanning stage to produce the desired leather in terms of texture, thickness, colour, and surface patterns.

6.4.1 Polishing Process

Polishing uses a velvet wheel that scours the leather to make a shiny surface.

6.4.2 Embossing Process

Embossing produces a three-dimensional print by using roller presses or heated hydraulics.

6.4.3 Surface Coating Process

The Surface coating process adds the required colour and desired designs to the leather. As per the customers' requirement, resins, pigments, and dyes are added in layers to the surface using a variety of techniques such as spraying, roller-coating, curtain-coating, or hand coating.

6.4.4 Final Grading Process

Finally, based on leather quality, pattern, colour, thickness, flexibility, and softness the manufacturers grade the leather, then dispatch it to the customers.

7. CHALLENGES AND SOLUTIONS

The leather industry has become an essential part of human living due to its incredible products. It produces a large amount of leather and its goods for markets across the world. The industry is a more than $100 billion global market manufacturing sector with many nations producing leather and its products that boosts their national economic sectors.

Starting from a thousand years ago the leather industry creates an enormous demand for its products among people all over the world. Nevertheless, it has numerous challenges.

Challenges in the leather industry and accompanying solution can be split in several categories and solutions (Brooke, 2020; Kazi, 2022; Sanskritiias, 2020).

7.1 Environmental

Problems

- Turning skin into finished leather uses a huge amount of dangerous and costly chemicals, including mineral salts, formaldehyde, and coal-tar derivatives.
- The leather manufacturing process generates a huge amount of waste, more than 80% of the raw materials are converted into waste materials which pose a threat to the environment.
- The leather industry is heavily dependent on fossil fuels and non-renewable energy.

Solutions

- The manufacturing process should be optimized to reduce chemical usage.
- There is a need to increase investment in waste management systems in the leather processing industry to mitigate the problems of pollution. The establishment of waste utilization/treatment industries is recommended.
- In this world, natural resources are limited, so circular economics practice can reduce the consumption of natural resources and can save the environment from this pollution.
- The leather industry is responsible for greenhouse gas emissions due to over-reliance on fossil fuels and non-renewable energy. Hence, energy efficiency and the use of renewable energy can be a good solution for green leather industry.

7.2 Work

Problems

- There is a lack of an adequate skilled workforce and financial support.
- Working place of most of the tanneries is very poor environmentally.
- The leather industry environment is highly affected by leather tanning and other processing.

Solutions

- Workers should be given proper training.
- To develop the tannery environment, it is urgent to take initiative for proper management of waste. All tanneries should follow the Basic Health Compliance Guidelines. Overall, governments and leather related associations should take proper initiative to develop the work environment.

7.3 Facilities and Materials

Problems

- Machinery for leather processing and making products are not available and is expensive.
- Incapacity in raw material accumulation
- Poor livestock management and poor-quality raw material supply because of ante-mortem and post-mortem handling of hides and skins
- Low utilization of industry capacity
- Poor linkages among different organizations involved with hides and skins.

Solutions

- Existing capacity should be fully utilized to move towards increased production of leather through rehabilitation and technology upgradation.
- Farmers should be given opportunities to participate in relevant training programs and learn about methods of increasing animal productivity.
- There should be a long-term plan for setting up a central slaughterhouse. There should be training in proper curing, flaking and storage techniques and widespread distribution of equipment and literature in local languages.
- To improve the collection, curing and handling of hides, price-based incentives and a quality grading system should be implemented by tanners.
- Those who make skins/hides must provide manuals for curing, preservation, and storage procedures as per environmental and market demands.

7.4 Market Forces and Policies

Problems

- Strong competition - small and medium-sized tanneries are dominated by large tanneries and brokers.
- Companies are not interested to provide value to the product with good branding and differentiator.
- Prices, processes, and facts are manipulated by the agents.
- Lack of or poor policies for the specific development of the sector.
- Cost of waste management, non-tariff barriers, quality specifications and compliance with various standards hampers export growth of the leather industry.

Solutions

- To achieve sustainable development and net zero leather industry, government and leather industry decision makers should formulate long-term fair policies and provide adequate financial support, especially for small and medium tanneries.

8. IMPACT OF THE LEATHER INDUSTRY AND PROGRESS TOWARDS NET ZERO

Over the last three decades, the industrial sector has been largely blamed for global climate degradation. Climate change has been defined as the most formidable challenge facing the contemporary world. Achieving net-zero by 2050 is the main goal for governments and industries to combat world climate change (Fankhauser et al., 2022). The leather industry is one of the world's most GHG (CO_2, N_2O, CH_4, fluorocarbons, etc.) emitting sectors. Leather manufacturing industry has a huge contribution to the global economy. But this industry is a major source of pollution with 80% of the raw material going to waste. It produces toxic waste, environmental pollution, and greenhouse gas emissions. The main factors in greenhouse effects include energy consumption, wastewater, and solid waste.

8.1 Types of Waste and Sources

In this industrialised modern world, resources entering the human environment are transformed, processed, used, and disposed in the form of solid, liquid and gaseous wastes (Understanding global change, 2022).

Industrial waste is usually produced during product manufacturing, agricultural production, or natural resource extraction. It is a process that converts raw materials into products that are used for various purposes.

There are three categories of waste, these are liquid, solid, and gaseous (Understanding global change, 2022). All types of wastes are emitted by the leather industry in the form of:

- Solid waste
- Liquid waste, and
- Gaseous waste

8.1.1 Solid Waste From Leather Industry

The most common waste is solid waste. In 2011, more than 9 billion tons of industrial solid waste was generated worldwide (Vignesh et al., 2021). The global leather industry produces more than 4 million tons of solid waste each year (Koppiahraj et al., 2019). Tannery waste is characterized by a high amount of dissolved, suspended inorganic and organic solids creating a rise to high oxygen demand, potentially chromium metal ions and toxic metal salts. The main solid wastes in the leather industry are dusted curing salt, skin trimmings, meat waste, keratin waste, chrome shaving and buffing solid waste. When processing raw skins, adhered dusted salt, which is contaminated with hair, blood, dirt, and a particular type of bacteria is removed and dumped. Trimming is the cuttings of the edges of raw hide/skins. Fleshing is the fleshy material of the calcareous skins produced during the fleshing operation. Chrome wet shaving is generated after the chrome tanning process when skins are shaved for proper thickness.

These proteins and other chemicals, which are present in chemically treated proteins, if not used properly, will create dangerous pollution problems for the environment. About 850 kg of solid waste is generated from 1000 kg of raw hide in the leather processing. Only 150 kg of raw hide/skin is converted into the finished leather (Koppiahraj et al., 2019). Tannery produces large quantities of solid waste as follows:

Fleshing- 50 to 60%; chrome shaving, buffing dust, and chrome split- 35 to 40%; skin trimming- 5 to 7%; and hair- 2 to 5%. Solid wastes generated from the leather manufacturing process include beam house- 80%; tanning-19%; completion- 1% (Kanagaraj et al., 2006; Koppiahraj et al., 2019).

8.1.2 Liquid Waste From Leather Industry

Liquid waste discharged from different sectors can be classified into different types, like domestic wastewater, sewage wastewater, agricultural wastewater, storm run-off wastewater, and industrial wastewater. In recent times, more attention has been focused on water pollution due to rapid industrialization and its adverse health effects. According to the WHO, about 80% of diseases are caused by waterborne. The most common water pollution diseases are diarrhea, giardiasis, typhoid, cholera, hepatitis, jaundice and cancer (Ahmed et al., 2021).

The leather industry produces huge amount of wastewater, mainly, the tanning process is the most significant contributor to waste in the leather industry. Water is the most essential input of the leather manufacturing process. For cleaning the hides/skin, water is the main carrier of the chemicals. The same quantity of water is drained out as wastewater after the completion of the manufacturing process as the amount it was used. Normally, 50 liter/kg water is recommended for tanneries process but in most of the cases manufacturing process needs more water, even sometimes it could be more than 150 liter/kg. Leather industrial wastewater is extremely polluted regarding biochemical oxygen demand (BOD), chemical oxygen demand (COD), organic nitrogen, suspended solids, conductivity, sulphate, and chromium (National Grid, 2022b; Sivaram, 2019b). The volume of all these pollutants in the water is very large. Sludge is another pollutant of wastewater. Wastewater from tannery process formations of pollution of different pH values. All these chemicals in wastewater are very dangerous for the environment. Tannery's wastewater is badly contaminated, and this contamination is many times higher than the limits that are set by Environmental Quality Standards.

8.1.3 Gaseous Waste From Leather Industry

The leather industry produces a variety of gaseous wastes that are harmful for the environment and human health while also adding to air pollution. Some of the notable gaseous wastes from the leather industry are discussing below (Sivaram, 2019a):

- Hydrogen sulfide (H_2S): H_2S is produced during the decomposition of skins/hides. This gas is responsible for eye irritation, respiratory problems, and other health related issues.
- Volatile Organic Compounds (VOCs): These compounds are emitted from different stages of leather manufacturing, mainly from tanning, dyeing, and finishing stages. VOCs are also cause of various health problems as do hydrogen sulfide.
- Ammonia (NH_3): NH_3 is produced in various stages of leather manufacturing, like deliming, bating, and pickling. This gas is dangerously harmful for human health.

Figure 7. Inputs and outputs associated with the leather manufacturing process

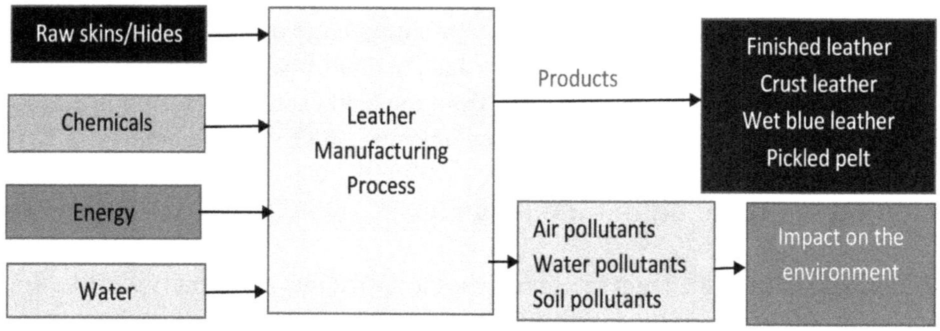

- Carbon monoxide (CO): This gas is emitted from various stages, mainly when fossil fuels are burning to generate energy for leather manufacturing. CO cause of many health problems, like headaches, nausea, and dizziness.
- Sulfur dioxide (SO_2): SO_2 is a colorless gas that is produced during the burning of fossil fuels (such as coal or oil) in the leather producing industry. Sulfur dioxide is a life-threatening gas, it causes of difficulty of breathing, damages lung function, aggravate existing heart disease, and creates so many others health-related problems.

8.2 Greenhouse Gases (GHG) Emission From Different Sources of the Leather Industry

Leather production is destructive to deforestation and biodiversity. The leather manufacturing process produces large amounts of greenhouse gases that contribute to global warming.

According to a case study published in Energy Procedia, the carbon footprint of bovine leather ranges between 65 to 150 kg CO_2 per square meter. Considering that there are more than 2 billion square meters of leather produced each year, that is a devastatingly huge economic footprint.

One square meter of hide produces 16500 litres of wastewater full of chromium, sulphates and pathogens (Somody, 2021). In addition, for every 1000 kg of raw hide processed for leather, only 255 kg of the raw material is converted into usable leather including leather splits. The rest, 745 kg is generated as solid waste in leather processing. See figure 7 for inputs and outputs in the leather manufacturing process.

However, there are several factors which contribute to greenhouse emissions in leather manufacturing process, some factors are given bellow:

8.2.1 Transport

For example, CO_2 emissions from raw material and leather goods transport are shown in Figure 8.

Figure 8. The average amount of CO_2 (in grams) emission from transportation per metric ton of freight and km
Source: United Nations (2017)

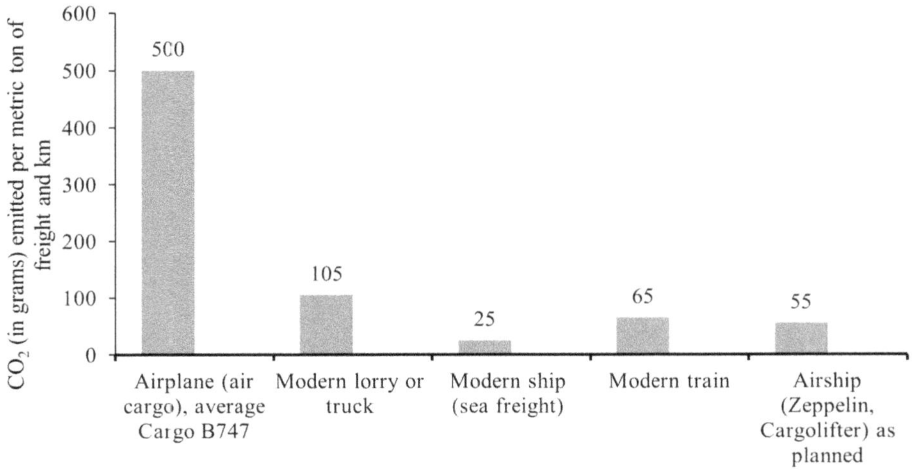

8.2.2 Energy and CO_2 Emissions

The CO_2 emissions associated with a daily input of 10 tons of wet salted hides. For making hot water liquid fuel emitted 3kg CO_2 per litre of combusted fuel, and each ton wet salted hide requires 230 litres of liquid fuel, refer to table 2.

The leather industry needs huge amounts of energy, especially in the leather manufacturing process. This tremendously contributes to the production of greenhouse gases. The projected global energy demand in this sector is shown in Figure 9.

8.2.3 Thermal Energy

Most of the required thermal energy is needed to heat the water in the tannery. This energy is usually supplied by a central boiler with fuel, which significantly impacts CO_2 emissions. Figure 10 shows the amount of emitted CO_2 per GJ of energy for various fuels.

Table 2. CO_2 emissions for a day input of 10 tons of wet salted hides in a tannery (Understanding global change, 2022)

Liquid Fuel Emission (kg CO_2/Litre)	Liquid Fuel (Litre/Ton of Raw Hide)	CO_2 Emission		
		kg CO_2/Ton of Raw Hide	kg CO_2/Day	kg CO_2/Year
3	230	690	6900	2518000

Figure 9. Projected global energy demand in the leather industry (compared to 2019)
Source: Teske et al. (2022)

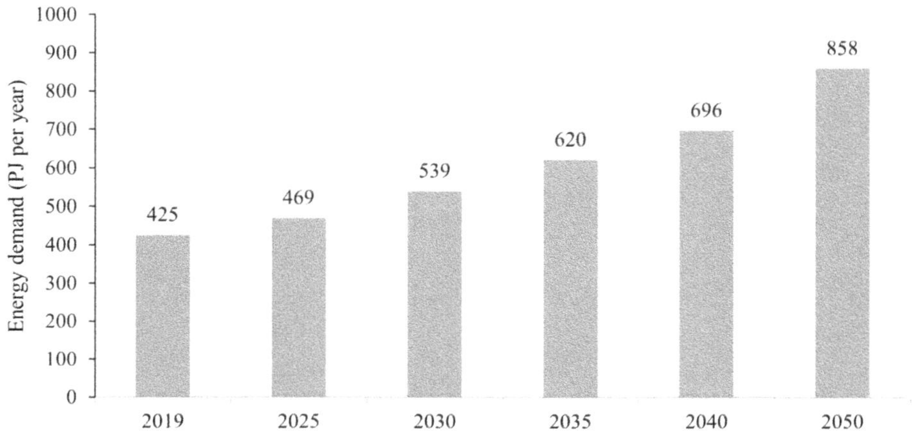

Figure 10. Amount of CO_2 emitted per GJ of energy for various fuels
Source: The European Standard EN 16887:2017 (United Nations, 2017)

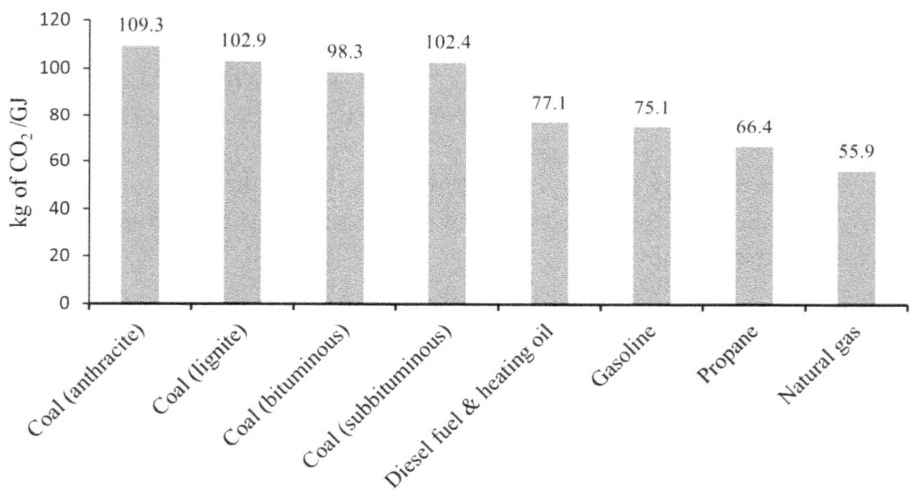

8.2.4 Solid Waste

In the leather industry a lot of solid waste is produced and most of the solid waste is generated from tanneries. Disposing of this waste generates CO_2 emissions. Figure 11 shows how much solid waste is produced from the various leather manufacturing processes and Figure 12 shows how much they contribute to emitting CO_2.

Figure 11. Quantity and types of solid wastes generated from tanneries for 1 ton of raw hides
Source: Koppiahraj et al. (2019)

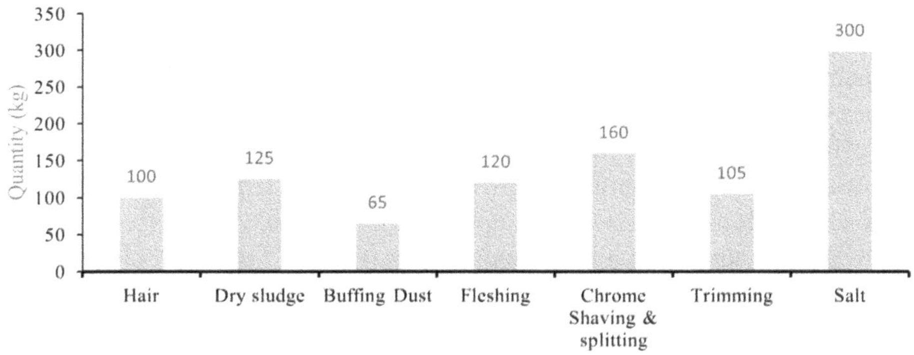

Figure 12. CO_2 emissions from solid wastes in the leather industry are shown
Source: The European Standard EN 16887:2017 (United Nations, 2017)

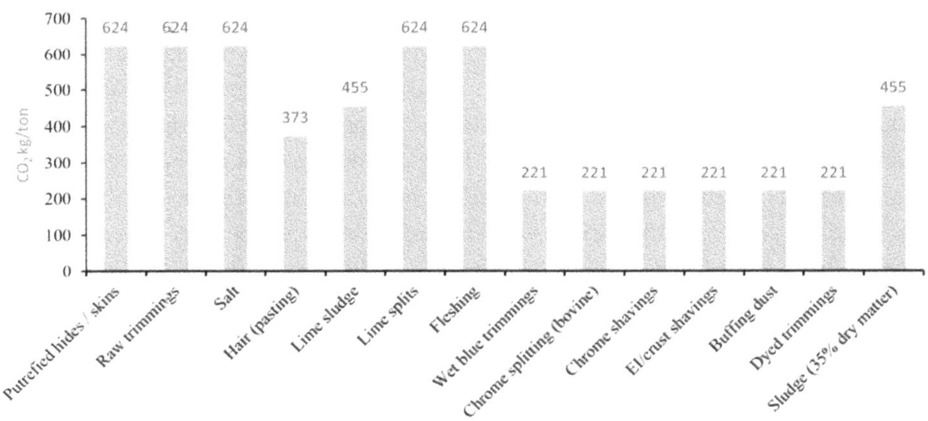

8.2.5 Wastewater Treatment

In the treatment of cleaning wastewater, huge amounts of CO_2 are emitted into the environment. In wastewater there are mainly two types of organic elements. The content of the organic element in the wastewater is expressed as biochemical oxygen demand (BOD) and chemical oxygen demand (COD). The content of carbon in the organic matter of wastewater is expressed as total organic carbon (TOC). The ratio between TOC to COD, and TOC to BOD are approximately 1: 3, and 1: 1.4 respectively. The accepted value for COD before aerobic biological treatment is 60 kg COD per ton of raw hide, 25 kg BOD per ton of raw hide, average TOC is 70kg CO_2 per ton of raw hide, and the total CO_2 emissions (direct and indirect) is 81 kg CO_2 per ton of raw hide (United Nations, 2017).

8. MATHEMATICAL MODELLING FOR NET ZERO

Of the growing warning of global heating, there is growing concern about environmental degradation, mainly arising from carbon dioxide (CO_2) outflow (Su et al., 2022; Zhou et al., 2021). Achieving net-zero by 2050 is the main goal for governments and industries to combat world climate change (Fankhauser et al., 2022). Since global warming is proportional to atmospheric greenhouse gas (GHG) concentration (CO_2, N_2O, CH_4, fluorocarbons, etc.), the planet has a limited ability to reduce greenhouse gases emission. This indicates the world will continue to warm if its emissions remain above zero. A wave of planet stimulus for industrialization arrives with the supportive effect of increased use of fossil fuels, energy, natural resources, and production of a huge amount of waste. (e.g., oil, coal, and natural gas) (Joshua & Bekun, 2020), which has been placed at a crossroads of economic development and climate change. Compared to 2010 levels, entire global carbon outflows in the industry sector are required to decrease by 50%- 93% in 2050 to fulfil the aim of the 1.5°C Paris climate objective (Sluisveld et al., 2021). All these calculations are determined by mathematical/ computational/ theoretical modelling. In the world, the climate is one of the most unpredictable objects.

Danish scientist Niels Bohr said, "it is very difficult to predict - especially the future." For making predictions a mathematician must have deep knowledge of the systems and needs the capability to apply this to the real world. A mathematician calculates mean values of all aspects and analyses various variances then identifies whether the weather has changed and find out the reasons why it has changed. Mathematical evidence helps the decision makers to making more informed decisions. The mathematical model encourages the International Panel on Climate Change (IPCC) and individuals to prevent the destructive economic, social, and environmental effects of climate change (Richardson, 2022).

In the present time, net zero is the most important phenomenon of the decision maker for the environmental challenge. For solving this crisis, decision makers depend on the scientific approach. The scientific approach tries to apply mathematical models and the potentialities of the latest computers to solve complex and unstructured real problems that confront managers. The scientific approach relates to management science. Management science has been known by a variety of other names such as operation research, systems analysis, cost-effectiveness analysis, cost-benefit analysis, resources management analysis, energy system analysis and even all-over real-world problems. In management science, mathematical model, especially linear programming model, is one of the best developed and most useful branches. Linear programming models play a vital role in the optimal use of resources in a world of limited resources. A linear programming model consists of a set of constraints imposed by the characteristics of the problem. These constraints consider economic, technological, organizational, marketing, environmental, and many other situations. In a broad sense, the mathematical programming model is a mathematical representation of planning and programming of the best possible distribution of limited resources.

Furthermore, a cautiously developed mathematical model can provide an exact visible depiction of an abstract concept. We can also benefit from using simple regression models for mathematical optimisation. A mathematical model can consider a reckonable amount of time and can identify the factors which lead the maximum and minimum changes in the environment. For example, the optimisation targets of the United Kingdom are to 'reduce emissions by 78% by 2035 compared to 1990 levels' (GOV.UK, 2021) and 'decarbonise all sectors of the UK economy to meet our net zero targets by 2050' (GOV.UK, 2022). It is urgent to linking the mathematics with the climate changes factors such that general people easily could understand the effect of climate change.

Recently, Net Zero Centre King's College London (2022), focused on four priority areas for net zero: i) materials for net zero; ii) sustainable manufacturing; iii) data and optimisation; and iv) simulation and modelling. The latter implies mathematical modelling can play a significant role to achieve net zero in this globe.

CONCLUSION

Leather manufacturing is one of the most significant and oldest industries in the world. Its products are very useful among millions of people around the world. It makes a major impact on the global economy and everyday its contribution to the economic sector is rapidly increasing. However, the leather industry is a major source of pollution with over 80% of the raw material going to waste. It produces toxic waste, environmental pollution, and greenhouse gas emissions. The main factors in greenhouse effects include energy consumption, wastewater, and solid waste. Leather processing produces more than 200 times more waste than total final products. Growing demand and limited resources are creating complex and challenging problems. Urgent and collective actions are needed to achieve net-zero by 2050 (Fankhauser et al., 2022) by governments, regulators, researchers, and industrial leaders to avoid further global warming and severe climate change. Improvements in industrial processes, applications of renewable energy instead of fossil-fuel, energy efficiency, fuel conversion, reduction of natural resources consumptions and reduction of waste produced from the manufacturing process are essential concerns for achieving net zero. Therefore, mathematical modelling plays a vital role in reducing all greenhouse gases emission from all relevant sectors through the optimisation process.

REFERENCES

Ahmed, J., Abhijeet, T., & Goyal, A. (2021). *Industrial Wastewater and Its Toxic Effects. Biological Treatment of Industrial Wastewater*, 1-14.

Bhavya. (2019, November-December). Leather processing, its effects on environment and alternatives of Chrome tanning. *International Journal of Advanced Research in Engineering and Technology*, *10*(6), 69–79.

BizVibe. (2022). *Global Leather Industry Factsheet 2020: Top 10 Largest Leather Producing Countries, Largest Exporters & Importers.* Retrieved October 25, 2022, from https://blog.bizvibe.com/blog/top-10-largest-leather-producing-countries

Brooke. (2020). *The leather industry: problems, solutions and roadblocks to environmental viability*, Retrieved on November 10, 2022, from https://www.techstyler.fashion/leather-industry-environmental-viability/

Dalgado. (2020). *All you need to know about leather grades.* Retrieved October 10, 2022, from https://dalgado.de/en/the-journal/all-about-leather/all-you-need-to-know-about-leather-grades

Fankhauser, S., Smith, S. M., Allen, M., Axelsson, K., Hale, T., Hepburn, C., Kendall, J. M., Khosla, R., Lezaun, J., Mitchell-Larson, E., Obersteiner, M., Rajamani, L., Rickaby, R., Seddon, N., & Wetzer, T. (2022). The meaning of net zero and how to get it right. *Nature Climate Change*, *12*(1), 15–21. doi:10.103841558-021-01245-w

GOV.UK. (2021). *UK enshrines new target in law to slash emissions by 78% by 2035*. Retrieved October 05, 2022, from https://www.gov.uk/government/news/uk-enshrines-new-target-in-law-to-slash-emissions-by-78-by-2035

GOV.UK. (2022). *Net Zero Strategy: Build Back Greener*. Retrieved October 05, 2022, from https://www.gov.uk/government/publications/net-zero-strategy

Grantham research institute on climate change and the environment. (2021). *Why is 'net zero' so important in the fight against climate change?* Retrieved October 05, 2022, from https://www.lse.ac.uk/granthaminstitute/explainers/why-is-net-zero-so-important-in-the-fight-against-climate-change/

Joshua, U., & Bekun, F. V. (2020). The path to achieving environmental sustainability in South Africa: The role of coal consumption, economic expansion, pollutant emission, and total natural resources rent. *Environmental Science and Pollution Research International*, *27*(9), 2735–2743. doi:10.100711356-019-07546-0 PMID:31919826

Kanagaraj, J., Velappan, K. C., Chandra Babu, N. K., & Sadulla, S. (2006). Solid wastes generation in the leather industry and its utilization for cleaner environment-A review. *Journal of Scientific and Industrial Research*, *65*(July), 541–548.

Kazi. (2022). *Leather Industry of Bangladesh: Challenges and Opportunities*, retrieved on March 9, 2022, from https://businessinspection.com.bd/leather-industry-of-bangladesh/

Koppiahraj, K., Bathrinath, S., Saravanasankar, S. (2019). Leather Waste Management Scenario in Developed and Developing Nations. *International Journal of Engineering and Advanced Technology, 9*(1).

LaFalce. (2019). *Leather insights: a comprehensive guide for buying the best leather*. Retrieved October 15, 2022, from https://octaneseating.com/blog/leather-buying-guide/

Liberty Leather Goods. (2022). *The Leather Industry – An Overview of Fascinating Facts.* Retrieved October 05, 2022, from https://www.libertyleathergoods.com/leather-industry/

Liberty Leather Goods. (2023). *The Incredible History of Leather*. Retrieved July 05, 2022, from https://www.libertyleathergoods.com/history-of-leather/

Love Your Leather. (2022). *13 Different Uses for Leather*. Retrieved October 23, 2022, from https://www.loveyourleather.ca/leather-blog/13-different-uses-for-leather/

MasterClass. (2021). *Fabric Guide: Learn About Leather*. From https://www.masterclass.com/articles/learn-about-leather

National Grid. (2022a). *What is net zero?* Retrieved November 10, 2022, from https://www.nationalgrid.com/stories/energy-explained/why-net-zero-so-important

National Grid. (2022b). *Why is net zero so important?* Retrieved November 10, 2022, from https://www.nationalgrid.com/stories/energy-explained/why-net-zero-so-important

Richardson. (2022). *Mathematics and climate change: What role do you think mathematics can play in guiding policy makers and in helping public understanding?* Retrieved November 25, 2022, from https://newn.cam.ac.uk/wp-content/uploads/2022/07/Maths3-Richardson_Emily.pdf

Sanskritiias. (2020). *Leather industry: problems & solutions*. Retrieved on November 10, 2022, from https://www.sanskritiias.com/current-affairs/leather-industry-problems-solutions

Sean, G. (2014). *India: The Toxic Price of Leather*. Retrieved November 10, 2022, from https://pulitzercenter.org/stories/india-toxic-price-leather

Sivaram, N. M. (2019a). Energy from Toxic Organic Waste for Heat and Power Generation. In Toxic Waste from Leather Industries (pp. 55-67). Woodhead Publishing.

Sivaram, N.M. (2019b). *Toxic Waste from Leather Industries, Energy from Toxic Organic Waste for Heat and Power Generation*. Science Direct.

Sluisveld, Mariesse, Boer, Daioglou, Hof, & Vuuren. (2021). A race to zero - Assessing the position of heavy industry in a global net-zero CO_2 emissions context. *Energy and Climate Change*, *2*, 100051.

Somody, F. (2021). *Why does the leather industry need to change?* Retrieved November 25, 2022, from https://www.explore-leap.com/post/why-the-leather-industry-needs-to- change.

Su, C.-W., Pang, L.-D., Tao, R., Shao, X., & Umar, M. (2022). Renewable energy and technological innovation: Which one is the winner in promoting net-zero emissions? *Technological Forecasting and Social Change*, *182*, 121798. doi:10.1016/j.techfore.2022.121798

Teske, S., Niklas, S., & Talwar, S. (2022). Decarbonisation Pathways for Industries. In S. Teske (Ed.), *Achieving the Paris Climate Agreement Goals: Part 2: Science-based Target Setting for the Finance industry — Net-Zero Sectoral 1.5þC Pathways for Real Economy Sectors* (pp. 81–129). Springer Nature.

The Leathersellers' Company. (2020). *Leather manufacturing process*. Retrieved October 05, 2022, from https://leathersellers.co.uk/leather-industry/leather-manufacturing-process/

Understanding global change. (2022). *Pollutants and waste*. Retrieved November 10, 2022, from https://ugc.berkeley.edu/background-content/pollutants-and-waste/

United Nations, Industrial Development Organization. (2017). *Leather carbon footprint, Review of the European Standard EN 16887*. Author.

University of Oxford. (2022). *What is net zero?* Retrieved July 05, 2022, from https://netzeroclimate.org/what-is-net-zero/

Vignesh, Rajadesingu, Suriyaprakash, & Kantha. (2021). Challenges, issues, and problems with zero-waste tools. In Concepts of Advanced Zero Waste Tools (pp. 69-90). Elsevier.

Zhou, Z. F., Rahat, B., & Rubbaniy, G. (2021). Carbon neutrality target for leading exporting countries: On the role of economic complexity index and renewable energy electricity. *Journal of Environmental Management*, *299*, 113558. doi:10.1016/j.jenvman.2021.113558 PMID:34425500

Chapter 12
Epoxy-Based Composite Materials for Innovative Biomedical Applications

A. B. Hemavathi
JSS Science and Technology University, India

A. S. Zarena
Teresian College, India

M. P. Chandresh
JSS Science and Technology University, India

R. S. Jagadish
JSS Academy of Technical Education, Noida, India

ABSTRACT

Epoxies are a versatile class of resins, well-known for its excellent combination of properties like light weight, chemical resistance, fatigue strength, custom-made impact resistance, and can be tailored to be biocompatible. Although ceramic and metallic composites are widely accepted for the development of biomaterials, their heaviness, non-resorbability, and necessity of second surgical operation to remove the implant after healing induces extra pain for the patients; hence, they were gradually replaced by polymer composites. Epoxies can be self-cross-linked or cross-linked with external curatives at stoichiometric ratio to achieve desired properties and purity. Advances in epoxy-based biomedical composites have been focused on the design of implants with concerns on comfort, biocompatibility, precise property matching, light weight, and closely mimicking natural structures in performance. Epoxy-based materials are also having footprints in drug delivery and tissue engineering. The chapter highlights epoxy usage in the biomedical field for various advanced applications.

DOI: 10.4018/978-1-6684-6830-2.ch012

1. INTRODUCTION

The biomedical applications of polymers in the human body range from head to toe. Research on polymer composites application in biomedical field is one of the fastest growing areas of science and it is highly interdisciplinary, where the traditional science concepts are connected to the modern engineering aspects to solve human health related issues. Biodegradable polymers can be formulated to stay in the body as long as they serve their intended function and then they dissolve without the need of a second surgical intervention, hence they are suitable to be used as temporary implants/supports. Ligament augmentation and orthopedic fixation were the primary applications of biodegradable polymers. Synthetic, hydrolytically degrading polymers are preferred for many applications as temporary implants, scaffolds and drug release/delivery system, because their degradation rate can be tuned to suit different implantation conditions (Song et al., 2018; Sowmya et al., 2021).

A biomedical composite can be defined as two or more physically and/or chemically distinct biocompatible phases, suitably arranged or distributed with an interface separating them. It exhibits characteristic properties that are not shown by the constituents in isolation. The significance of composite material is that there are numerous matrix materials and as many reinforcement/filler types that can be combined in numerous ways to produce the desired/intended properties. One can design a composite material that meets or sometimes exceeds the performance requirements (Hemavathi et al., 2021; Gowda et al., 2022). Composite materials can have one or more dispersed noncontinuous phases which usually give superior properties than the matrix in most of the cases. The transfer and distribution of stresses from the matrix to the fibers or particles occur through the interface separating them. The area at the interface and the strength of the interfacial bond greatly affect the final composite properties and its durability. The matrix phase being the bulk phase accepts the applied load over a large surface area and transfers it to the stiffer reinforcement phase resulting in minimum bulk strain on deformation or may totally resist the deformation. The matrix phase serves several important functions like it protect and envelopes the reinforcement phase either completely or partially. Holds the fibers or particles in place to maintain the preferred orientation of fibers and gives surface finish to the composite product. An ideal matrix efficiently transfers the applied load to the reinforcement and redistributes the stress. When used with brittle fibers like glass, the matrix helps to increase fracture toughness because it is typically of a lower stiffness material and can tolerate greater elongation and shear forces than the reinforcement. The matrix also determines the durability of the composite by resisting chemical, hygroscopic and thermal stresses and protects the fragile reinforcement from these stresses (Peters, 1998; Hemavathi et al., 2022).

The matrix selection/formulation greatly decides the characteristics of a biomedical composite. Ceramic matrices are used for their superior compressive properties and bioinert nature but they suffer from poor fracture toughness. Metal matrices are mainly preferred for structural and high temperature applications. However, both of these matrices are heavier compared to polymer. Biobased feedstock such as vegetable oils can be easily epoxidized with organic peracids or H_2O_2 and epoxidized precursors thus obtained can be used as an inexpensive, renewable material for polymerization. But they tend to possess lower strength triggering the need for incorporation of additives or fillers to improve the desired properties (Mustapha et al., 2019). Epoxy-amine oligomers derived from renewable feedstock (spearmint oil) with synergistic antifungal functionalities for potential use in biomedical applications was reported by O'Brien et al. (2019). Adsorbable bioglass fibers made from calcium phosphate have promising applications in implant design along with epoxy. Collagen fibers are used in tissue engineering of skin and ligament. Carbon fiber is used extensively to make high-strength lightweight composites in prosthetic structural components, where the fatigue resistance of carbon-fiber composites is an added advantage. For biomedical applications light weight, impact resistant and biocompatible components are preferred, but it is still challenging to design a composite that has all the desired set of properties. FDA approval and additional validation of each components biocompatibility has direct implications on design flexibility of polymer composites for intended applications (Hemavathi et al., 2002; Hemavathi & Siddaramaiah, 2018).

Epoxies which respond uniquely to applied stimulus/environmental conditions gained special attention as functional biomaterials. Response to physical stimuli like mechanical stress, electricity, temperature changes or light irradiation and combinations of these inputs can be used to trigger a drug release by external stimuli. Epoxies can also respond to internal chemical and biochemical triggers like pH, drugs, metabolites, bioactive agents, antigens or enzymes and so autonomously respond to a physiological status (Patel and Purohit, 2018). Due to the carbon based chemistry, polymers closely mimic biological tissue than inorganic materials like ceramics and metals. Polymer based biomaterials are increasingly used to replace those living materials which have ceased to serve their intended function/purpose. The kind of tissue reaction to an implant depends not only on the type of surrounding tissue (e.g. muscle, bone, blood, skin) but also on many parameters like the degree of compatibility, stresses experienced, strains taken, elasticity, flexibility and relative motion. All the necessary requirements can be inculcated in epoxy either during the synthesis, formulation and/or curing to get the desired functionality for the end use applications.

2. EPOXY: A VERSATILE MATRIX SYSTEM

Ever since their discovery in the late 1930s, epoxy resin is a significant member of thermoset family. Epoxy resins are mixtures of several components formulated in exact proportions to assure proper polymerization. The term epoxy resin is applied to both prepolymers and to cured resins; the prepolymer is characterized by a three-membered ring known as the epoxy or oxirane or ethoxyline group. In the cured resin, all epoxy reactive groups might have reacted or not depending on curing conditions. Epoxy can form an infusible and insoluble stable network structure upon curing due to the formation of covalent cross-links. They are generally amorphous and possess various desirable traits such as high tensile strength, modulus, good thermal and chemical resistance and dimensional stability. Epoxies are further tunable in physical, chemical, mechanical and biological properties in a wide range from synthesis to curing so as to match the requirements of specific applications. Commercial epoxy resins may contain aliphatic, cycloaliphatic or aromatic backbone. The capability of the epoxy ring to react with a variety of substrates adds versatility to the resin. In order to facilitate processing and modify cured resin properties, other additives may be included in the composition like functional fillers, reinforcements, solvents, diluents, plasticizers and accelerators. Hydrogen bonds are the most representative structural features of epoxy polymers. Each step of the cure reaction involving epoxy ring opening forms one OH group, whose concentration in a cured polymer is high enough to give polar nature to the resin. Epoxy resins are highly reactive due to the presence of strained epoxy ring structures and can readily react with many nucleophilic and electrophilic reagents. Therefore, a wide variety of organic compounds having active hydrogen atoms can be used as curatives which include amines (both aliphatic/aromatic and primary/secondary), phenols, carboxylic acids, thiols, anhydrides and so on (Ashcroft, 1993; Siddaramaiah and Jagadeesh, 1996).

An anticipated performance property of an epoxy resin is obtained by cross-linking with suitable curing agent so as to form a stable three-dimensional network. The choice of curing agent depends on the application, properties required and on the handling characteristics such as viscosity, pot life and gel time. Curing agents are either catalytic or co-reactive. Catalytic curing agents function as initiators for epoxy resin homopolymerization or as supplemental curing agents with polyamines or as accelerators for anhydride cured systems. On the other hand, the co-reactive curing agents act as a comonomer in the polymerization process. A wide variety of reagents such as amines (primary and secondary), isocyanates, carboxylic acids, mercaptans and acid anhydrides are used as co-reactive curing agents. Aliphatic polyamines are considered to be fast reacting curing agents and are used for room temperature curing (Jin et al., 2015). In general, hydroxyl groups accelerate the rate of curing with amines, among which polyfunctional alcohols are the best accelerators.

Epoxies can be cured at ambient or increased temperature via thermal curing or radiation curing (electron-beam (EB), ultraviolet (UV), microwave). The mechanism of curing differs in these methods, thermal curing happens through a step polymerization mechanism whereas radiation curing occurs by chain polymerization involving initiation, propagation and termination steps. The mechanical features of the cured matrix also differ in both cases. The cure times can be reduced to the range of few minutes from hours through the use of microwave curing (Verma et al., 2020). Reduced cure time, low-temperature cure, extended material shelf life and greater design flexibility are some of the advantages of electron beam curing over conventional thermal curing. Ambient temperature curing provides a lower T_g, greater impact resistance, higher flexibility and greater thermal shock resistance. There are also advanced shape memory epoxy polymers, which can be freely deformed and return to their original shape upon a stimulus like pH, temperature, light or magnetic field. They find applications in drug delivery devices, sutures, vascular stents, clot removal devices, aneurysm or ductus arteriosus occlusion devices and orthodontic therapy (Maitz, 2015). It is possible to mix epoxy resins having the proper refractive index to produce a matrix with superior optical properties.

Epoxy resins are most attractive thermosetting matrices still for advanced fiber reinforced composites as it adheres well with variety of fibers and also because of their excellent chemical stability, mechanical properties, ease of processing and low toxicity. In spite of these desirable properties, the major limitations are their low toughness and poor crack resistance. Hence, epoxy materials are toughened for many end-use applications. The epoxy resins become brittle mostly due to the self-polymerization that occurs during the curing process, which is inherited from its formation of a highly crosslinked network. High performance tough epoxy with excellent adhesive strength is in demand for many biomedical applications. Hence they are toughened by blending with different flexible polymers like rubbers, hyperbranched polyesters and hyperbranched epoxy resins etc. A balanced effect between the embrittlement and toughness, which are due to postcuring reactions and phase separation phenomena respectively, should be controlled to ensure optimum properties (De and Karak, 2013). Cizravi and Subramaniam (1999) reported toughening of amino propoxylate-cured bisphenol A diglycidyl ether (DGEBA) based epoxy matrices with epoxidized natural rubber.

Epoxy resin chemistry remains the most versatile of all thermoset chemistries. This versatility rests not only on the chemical building-blocks, reactivity, flexibility, rigidity, high durability, high specific strength and crosslink density, but also on the adaptability of epoxies to a wide variety of reactive processing from photocurable coatings to high-pressure lamination. The wide choice of precursor and curatives availability, numerous structural modifications and customized curing method of epoxy offer cured material with the properties of strong resistance to chemicals and water, minimum shrinkage and tunable mechanical strength suitable for biomedical applications.

3. EPOXY FOR BIOMEDICAL APPLICATIONS

Bloch in 1958 first reported a case of retention of an implant in place safely for more than 10 years in human body by epoxy resin (Bloch, 1958). Epoxy composites are highly sought materials for the manufacture of medical devices, device functional coatings, drug-eluting devices, orthopedic, dental implants and for tissue engineering applications. Mechanical behavior of epoxy can be tuned to suit both soft and hard tissue restorative and regenerative applications. The flexibility of synthesis practices and the tunability of functionalization strategies provide additional choice as an attractive and suitable material for biomedical applications. Figure 1 depicts varied applications of epoxy composites in biomedical field. Green epoxy production from naturally available resources like epoxidized vegetable oils, glycerol polyglycidyl ether and sorbitol polyglycidyl ether is a potential field to be explored further to meet sustainability.

Epoxy resin synthesis based on rosin of pine trees has been reported by a set of researchers for preparing BPA free epoxy composites for possible applications in biomedical field (Wang et al., 2008; Mantzaridis et al., 2013; Brocas et al., 2014). Additional merit of making composite is that by controlling the volume fractions and local/global arrangement of the reinforcement phase, the properties and design of an implant can be varied and tailored to suit the mechanical and physiological conditions of the host tissue. Other reasons for the use of polymer composite based biomaterials are absence of cor-

Figure 1. Applications of epoxy composites in biomedical field

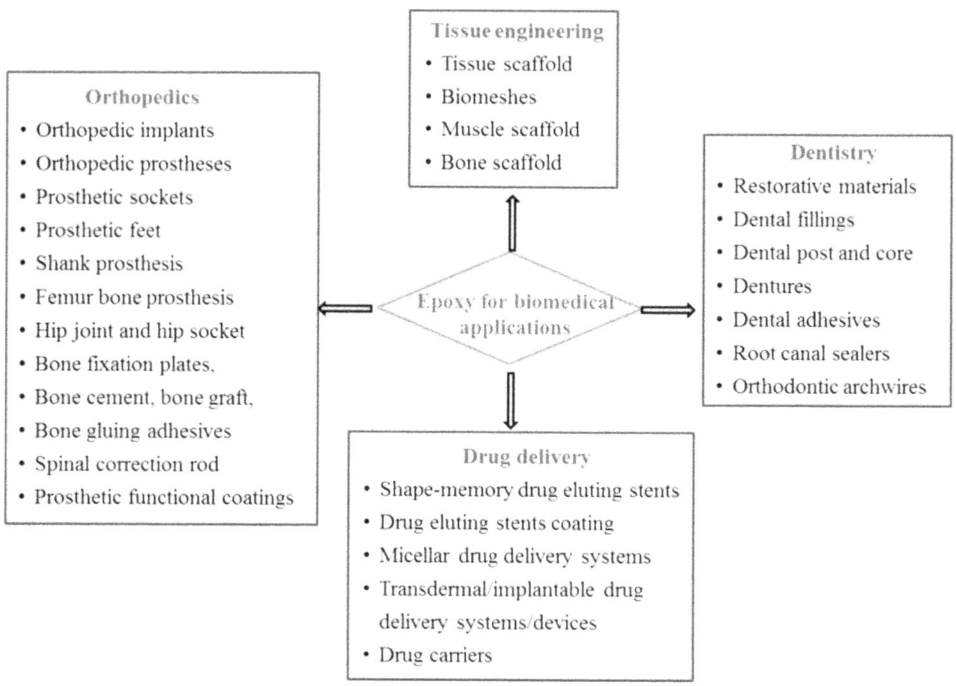

rosion, fatigue failure, metal ions release, patient discomfort and allergic skin reactions. Low fracture toughness of ceramic materials makes them a worst choice for load bearing applications. In all such cases polymer composites are highly preferred. The following sections highlights epoxy applications in varied biomedical field.

3.1 Epoxy for Dental Applications

Dental composites have high concerning requirements like esthetics, toughness, mode of polymerization besides the biocompatibility. Chosen material must support high load, shear forces, compressive forces, forces of thermal expansion and shrinkage. Dental restorative materials can be either made of ceramic, polymer or metal based composites. In recent times polymer composites are widely preferred due to their inherent ability to adapt to manipulative procedures to arrive at the optimum desired properties. Epoxy based composite materials are becoming increasingly popular for use in restorative dentistry over the past 30 years since they offer highly aesthetic restoration and are comparatively easy to use (Vitale et al., 2014). Epoxy resins have physical properties that make them perfect choice for dental and orthopedic applications as cements and adhesives. However, it has been observed that some grades of epoxy resins harden too slowly for clinical use when mixed with conventional curing agents like amines and polyamides. A promising epoxy - diepisulfide - polyamide system has been reported which hardens in 5 to 20 minutes at body temperature (Hadjinikolaou et al., 1987). There is also a need for development of low shrinkage and visible-light curable dental composites to replace dimethacrylate polymers (Tilbrook et al., 2000). Shrinkage which occurs when the composites are cured can compromise the integrity of the tooth/composite interface and is a contributing factor for secondary caries in patients. Cycloaliphatic epoxy resin formulated with polyol extenders can be considered as potential candidates to address this need. Epoxy-polyols have significant advantages over dimethacrylates as they exhibit higher strength, lower shrinkage, no oxygen inhibition layer, equivalent hardness and acceptable glass transition temperatures. Vitale et al. (2014) reported non-cytotoxic and non-mutagenic epoxy (UV30) based dental fillings which cure at acceptable time frame and to an adequate depth using visible light source. Many of the commercially available restorative composites for dental applications contain 10 - 60 wt% of an organic matrix. Among which photopolymerizable cycloaliphatic epoxy monomers are popular as they exhibit very little volume shrinkage upon irradiation than methacrylic resins, thereby lowers residual stresses.

Early diagnosis and proper treatment of the tooth decays are important to retain original strength. In some cases, teeth are severely damaged due to trauma, decay, cavities or even as a result of previous bulky restorations. They are unable to maintain the final restoration without some additional support, in such cases, dental post and core build up following endodontic treatment could be of great help for holding the future restoration (Davoudi et al., 2021). As a result there is a rising demand for tooth colored posts in recent years, conventional metallic posts are gradually replaced by various non-metallic posts. Amongst them, epoxy or methacrylate resin posts reinforced with glass or quartz fibers, epoxy resin posts reinforced with carbon fibers are notable in the field of dentistry. Ideal properties of post material include desired range of flexural modulus, compressive strength, modulus of elasticity and thermal expansion similar to that of dentin. It should also esthetically resemble and bond efficiently to the dentin. Due to mismatch of elastic modulus of the dentin and the post material, stress concentration may occur at dentin leading to potential root cracking followed by fracture, which is very common problem with metallic and ceramic post. Thus, polymer composite (epoxy resin or bis-GMA based) posts having mechanical properties similar to that of natural dentin are highly preferred. They also help in homogenous stress distribution, which greatly reduces the failure of the post due to teeth fracture (Freedman, 2001; Davoudi et al., 2021). Carbon fiber reinforced epoxy is also used as potential material for fixed partial dentures.

It was pointed out that for dental posts and cores, a material with an elastic modulus closely matched to dentin (18 GPa) is highly preferred for better distribution of stresses across the bonding interfaces of the dentin/dowel/core system to the tooth. Tey and Lui (2014) reported preparation of dowel posts from glass fiber-reinforced epoxy composites with similar moduli which can reinforce the weakened tooth minimizing probability of root fracture. Habitually dental posts made of stainless steel, titanium, Au-Pt or Ni-Cr alloys were used based on the hypothesis that the post should be rigid enough to resist failure. Even in such circumstances failures have been reported due to corrosion of posts, loss of retention, bending or fracture of posts, core fracture and root fracture. In recent years this old basic belief itself has been strongly questioned and it has been found that the modulus mismatch between the dentine and the post should be low enough so as to minimize the failure of restorations and occurrence of root fractures (Alharbi et al., 2014).

Newer dental posts made of zirconia, glass fiber reinforced polyester and carbon fiber reinforced epoxy composites were upcoming in recent times. These posts were claimed to possess adequate rigidity, corrosion and fatigue resistance compared to conventional materials. Ramakrishna et al. (2000) designed and developed a functionally graded dental post using epoxy/braided CF composite. It has a high stiffness in the coronal region and gradually reduces to a value comparable to the dentin stiffness at the apical end. The graded stiffness post usage decreases the chances of dental post loosening from the dentin by eliminating stress concentration in the dentin and reduces post-dentin interfacial shear stresses in addition it also overcomes the chances of root fracture.

Dental adhesives are used to bond composite resins to dentin and enamel. Their chemical formulation determines their adhesive performance in clinic. The primary aim of dental adhesive is to retain composite fillings or cements. In addition it should withstand mechanical forces and in particular shrinkage stress from the lining composite, a good adhesive should also be able to prevent leakage along the restoration margins. Clinically, failure of restorations occurs more often due to inadequate sealing and subsequent discoloration of the cavity margins than due to loss of retention. Epoxy resin based formulations are generally used to bond orthodontic attachments to enamel surfaces. The success and durability of endodontically treated teeth depends on effective sealing of root canal system. The aim of endodontic obturation of a root canal system is to fill empty spaces, resist coronal leakage, promote hermetic sealing and prevent infection from infiltrating the tissues surrounding the root of a tooth (Munchow and Bottino, 2017). The bonding strength of endodontic sealers to intraradicular dentin also decides fracture resis-

Table 1. Commercially available epoxy based endodontic sealers

Product Name	Manufacturer	Country
AH Plus	Dentsply De Trey GmbH, Konstanz	Germany
Adseal	Meta Biomed, Cheongju	Korea
Acroseal	Septodont, Saint-Maur-des-fossés	France
BioRoot RCS	Septodont, Saint-Maur-des-fossés	France
Endo CPM Sealer	EGEO S.R.L., Buenos Aires,	Argentina
Grey and Neo MTA Plus	NuSmile Avalon Biomed, Houston	USA
MM seal	Micro-Mega, Besancon	France

tance of treated teeth. Root canal sealers should strongly adhere to dentine to provide clinical longevity of treated tooth and should minimize the chances of endodontic failure.

Epoxy based root canal sealer AH Plus was studied by Mittal et al. (2017). They observed improved fracture resistance, good wettability and infiltration with dentin. Numerous studies have also reported that the bond strength of epoxy based AH Plus sealer is significantly higher than calcium hydroxide, zinc oxide eugenol and glass ionomer based sealers. AH plus, a biocompatible resin sealer was found to seal the canal better and has elastic modulus similar to that of dentin and exhibited a higher bond strength compared to silicate-based MTA Fillapex sealer (Yavari et al., 2017) and is conformed to the standards proposed by American Dental Association (ADA). Other epoxy based sealers like AD Seal, MM seal and Radic Sealer showed shorter setting time than the AH Plus. They also offer excellent seal strength once set, insolubility against tissue fluids, dimensional stability, good adhesion with canal walls, adequate radiopacity and biocompatibility (Lee et al., 2017). Guttapercha combined with AH Plus sealer still remain the most widely used material for such applications. Table 1 lists commercially available epoxy based endodontic sealers (Komabayashi et al., 2020).

The archwire is an integral part of orthodontic appliance, esthetic archwires are in high demand to go with brackets. Until 1930s, the only orthodontic wires available were made of gold. Archwires were designed to move teeth with light continuous forces. Tooth toned archwires coated with epoxy reported to enhance patient comfort, reduce treatment duration, has superior wear resistance and color stability. Epoxy coated archwires are available under the trade name Filaflex (American Orthodontics), have stainless steel core of high tensile strength and durable tooth colored coating. Which blends exceptionally well with plastic or ceramic brackets and does not stain or discolor and moreover they resist chipping and cracking. Appropriate design of epoxy based archwire adds versatility to orthodontic treatment (Singh, 2016). Orthodontic treatment which supports esthetic smile is need of the day; transparent orthodontic archwires were developed along this need. Orthodontic appliances made of transparent materials are highly sought by the patients in recent times. Mikulewicz et al. (2017) systematically reviewed flexible, high strength and chemical resistant, transparent orthodontic wires for clinical applications. Huang et al. (2003) developed wires with properties comparable to those of NiTi archwires using epoxy resins. Formability, biocompatibility, springback, stiffness, resilience and low friction are desirable characteristics of an archwire for optimum mechanical performance during the treatment. In general, epoxy based composites are well accepted as alternative to metal and ceramic based dental restorative materials.

3.2 Epoxy for Orthopedic Applications

Polymer composites have found wide usage in orthopedic applications particularly as bone cement, bone graft, bone pin, bone fixation plates, hip joint and hip socket replacement etc. Significant increase in peel strength, impact properties and fracture toughness has been achieved without much loss in tensile and thermal properties. The host responses to these materials were extremely diverse. Metals or ceramics were often chosen for hard tissue applications. Since the fiber reinforced polymers exhibit synchronized high strength and low elastic modulus, they are proposed for several orthopedic applications (Ghalme et al., 2016). Epoxy based nanocomposites which are perceived to be impact resistant were used for such applications. Figure 2 represents basic requirements of epoxy composites for orthopedic applications. In a noticeable work by Wetzel et al. (2003), epoxy matrix have been filled with nano and microscale particles of alumina (Al_2O_3) (13 nm) and calcium silicate ($CaSiO_3$) (4 - 15 µm) for wear resistant applications.

Figure 2. Basic requirements of epoxy composites for orthopedic applications

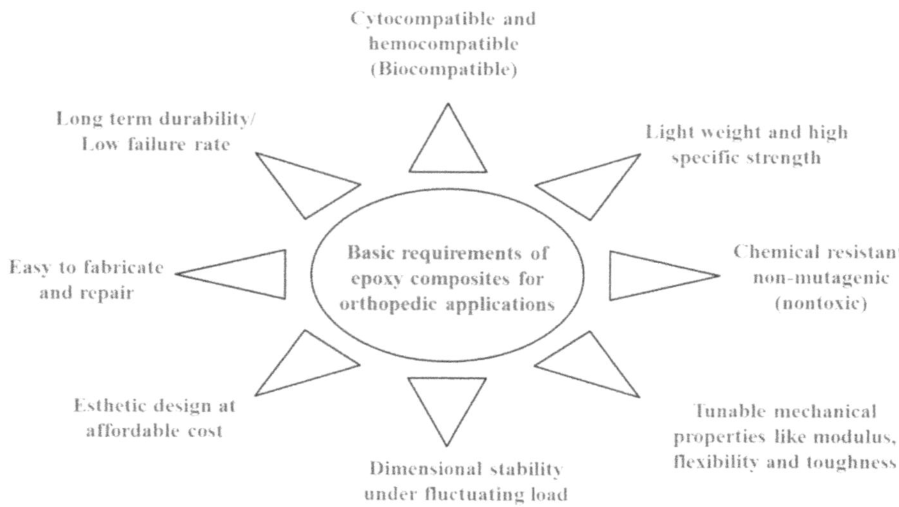

The proper selection of polymer for wear and environment conditions includes evaluation of wear pairs (contact material nature), wear configurations (like rotary, sliding), lubricity, humidity and dryness. Epoxy nanocomposites loaded with multiwalled carbon nanotubes were used in making orthopedic prosthetic sockets (Arun and Kanagaraj, 2016). This light weight material tends to increase the comfort level of transfemoral amputees by decreasing the cost of socket. The epoxy/CF/flax hybrid composites were used for orthopedic long bone fracture plates as an alternate to metal plates (Bagheri et al., 2013). These epoxy based plates were closer to human cortical bone compared to clinically used metal plates. Hybridized carbon nanofibers containing calcium phosphate nanoparticles play an important role in improving the interfacial adhesion of epoxy resin for bone repair (Gao et al., 2016). In another study by Hirwani et al. (2022), epoxy composites filled with hyaluronic acid and ultra-high molecular weight polyethylene fillers, biotribological properties were evaluated against CoCrMo alloy for hip prosthetic implant application.

Orthopedic surgeons have been conventionally using metallic bone plates made of stainless steel and titanium alloys for the fixation of humerus bone fractures. These metallic prostheses can generally cause problems like corrosion, magnetism effect, metal incompatibility, anode-cathode reactions and osteopenia including osteoporosis and can tend to delay fracture healing. Due to insufficient bone growth, refracture after the removal of prostheses were more common. It was also found that the difference in elasticity of bone and metallic implant may cause loosening of the implant (Rony et al., 2018). Thus, research on alternative implant materials have been undertaken in the past decade. Natural fiber reinforced polymer composite materials which are less rigid than metals may be good alternative because of properties closer to bone mechanical properties. It was found that they help to avoid stress shielding and increase bone remodeling.

Biodegradable thermoset composites with efficient mechanical properties were also used as a material for prosthetic sockets. A bioactive banana pseudo stem fiber reinforced epoxy composite was used to replace transtibial prosthetic sockets (Odusote et al., 2016). Sisal fibers have been used by certain researchers to reinforce epoxy matrices to produce composites for orthopedic implants (Dinesh et al.,

2014). Pineapple leaf fiber loaded thermoset polyester and epoxy resin (Odusote and Oyewo, 2016) are other examples. A transfemoral prosthetic device serves as an artificial limb replacement when the knee joint has been removed. But the individual still has a portion of the thigh bone or femur intact and consists of a socket, rotator, knee joint, pylon and foot (Otto et al., 2010). Epoxy/carbon fiber composite found to stimulate bone formation over the implant with a much higher degree of osseointegration when compared to titanium based implants (Petersen, 2014). In addition, the bone structure growth over the epoxy/carbon fibre composite was found to be porous, implying a better nutrient and oxygen accessibility compared to the titanium alloy which lacks such features. Another interesting point is that the polarity of cell membrane composed of proteins, lipids and carbohydrates which comes in direct contact with an implant is more closely matched to that of the epoxy composite as compared to that of metal alloys.

Epoxy and epoxy/polyester hybrid enriched with CaO and TiO_2 nanoparticles serve as an ideal coating material for titanium implants to enhance their biocompatibility as well as osteoinductivity (Hou et al., 2016). Self-healing properties of chitosan-reinforced epoxy composite coating (Atay et al., 2013) is worth mentioning here. Researchers in this field have aimed at developing composite bone plate for fracture repair using flax/carbon fiber with epoxy backbone (Bagheri et al., 2013, 2014, 2015). The composite was designed such that the epoxy/flax composite would form core with thin sheets of the epoxy/carbon fiber system attached to the outer surfaces of the core. The proposed composite did not show any negative consequence on osteogenesis when compared with conventional metallic fracture plates. Epoxy/CF based tubing has been used to replace stainless steel in artificial arms. Satin-weave CF fabric in epoxy prepreg has been used to make the shank of the *Endolite* transfemoral prosthesis. The *Flex-Foot* unit uses a long epoxy/CF composite beam that stores flexural energy along the whole length of the prosthesis rather than just the foot unit and is customized to the individual patient characteristics and activity level. Nonresorbable epoxy/CF bone plates with sufficient strength and fatigue properties are available from manufacturers like *Orthodesign Ltd.*, (Krishnakumar and Senthilvelan, 2021).

Biocompatible bone gluing adhesives are in demand as it is very easy and rapid to apply instead of using screws, nails and pins to connect fractured bone. The usage of glue helps in optimal load transfer from one fracture side to the other as continuous contact is made over the whole surface and not only at spotted contacts as with pins or nails. Moreover, in cases of fractures involving joints of uneven articular surface, unevenness can be well compensated by adhesives. Another potential advantage is stress-free fracture repair can be done by elastic epoxy glue instead of rigid and stiff metallic implants (Boker et al., 2019). Bone adhesives used in trauma and orthopedic surgery tend to resorb and degrade to non-toxic products after fulfilling their function in living organism. Degradation rate and healing rate should match to ensure mechanical stability, however such studies are still in evolving stages. When a bone is severely damaged due to accidents or disease like bone cancer and cannot be reset, amputation or bone graft is usually suggested by physician as a primary option. A stiff, lightweight and biocompatible plate material for the fixation of such fractured bones was developed by Chandramohan and Marimuthu (2011) using bio-epoxy resin of 3554A grade and hardner 3554B. The bone plate material can be externally coated with hydroxyapatite and calcium phosphate and can be used for both external and internal fixation of fractured bones.

The conventional method of bone fracture treatment involves use of heavy plaster cast. Disadvantage of this method is that, the joints immediately below and above the fracture gets immobilized by the plaster cast and can lead to temporary muscle atrophy and joint stiffness. Other alternative is to fix the fracture internally by one or more metal devices such as an intramedullary nail, wires or plates. But it requires two painful operations, one for fitting and one for removal of the fixation device. In the case of rigid fixation (usually metal devices), induced osteopenia/loss of bone strength may develop beneath the plate due to stress protection and this can result in refracture. Alternatively, if bone/plate system is less rigid, fatigue fracture of the implant may occur due to the excessive stresses experienced. As a solution, many studies have been attempted to explore the potential use of proper stiffness implants which are very much matching to bone elastic modulus. Epoxy based desired modulus/stiffness material with good fatigue resistance may be capable of reducing the refracture, at the same time allowing sufficient movement to promote proliferation of callus and early amalgamation of the fracture. Bradley et al. (1980) developed multi-axial carbon fibre reinforced epoxy composite plates as a low modulus yet sufficient strength material for internal fixation of fractures in the lower and upper limbs. The plates exhibited superior fatigue and flexural strength in comparison with stainless steel, at the same time having only one third the stiffness of steel. Also reported similar elastic properties as bone to prevent stress shielding and helps in homogeneous load transfer. Bone plates may be grouped as fully resorbable, partially resorbable and non-resorbable depending on the material used. Epoxy/carbon fiber and epoxy/glass fiber composites are examples of non-resorbables.

Schmitt-Thomas et al. (1998) reported development of spinal correction rod using biocompatible epoxy/unidirectional and braided carbon fibers to overcome the metal alloy related problems like interference with the diagnostic techniques and corrosion. The implanted stem or shaft, mimic the skeleton and provides structural support and helps to transmit the applied force. The epoxy reinforced by woven/braided fabric of carbon fiber is commonly used for such applications. In another study by Simoes and Marques (2001), composite stem preparation using epoxy resin/braided hybrid carbon fiber and glass fiber preforms is reported. Epoxy/carbon fiber laminated composites are used in the prosthetic feet manufacturing to give the user a higher degree of mobility with a more natural feel compared to wooden feet. In some designs, the foot unit consists of forefoot and heel components, so that the foot provides energy storing and cushioning effect. They are designed to store energy during stance and release energy as body weight progresses forward, thus helping to move the body and to achieve smooth ambulation.

Epoxy based composites are widely used for making customized and cost effective prostheses for accident victims. Orthopedic prostheses are used to totally or partially replace a segment of a deficient or missing limb. Their main purpose is to provide the possibility of restoring functional capabilities to individuals with amputations as they become extensions of the victim's body. Similarly, orthoses are externally used devices that modify the functional and structural characteristics of the skeletal and neuromuscular system. Chandramohan et al. (2014) reported fabrication of natural fiber based composite plate material with bio epoxy resin suitable for both internal and external fixation of fractured bone. Interfaces and sockets are important elements used in limb orthoses and prostheses, as they helps in proper fit and contact between the devices and the affected stump or limb. Arun and Kanagaraj (2016) reported high performance prosthetic sockets made from epoxy resins/glass fabric/multi-walled carbon nanotubes. In another study, Gouda et al. (2014) investigated use of epoxy/hybrid natural fiber composite as orthopedic implants for femur bone prosthesis. Odusote et al. (2016) used banana pseudo stem fibre reinforced epoxy composite for making transtibial prosthetic socket, which can be a potential replacement for glass fiber based composite in prosthetic socket application. Epoxy composite reinforced with ramie fiber was prepared by Irawan et al. (2011) for prosthesis socket application. It displayed highest flexural and tensile strengths when compared to polyester/ramie and polyester/fiberglass composites. An important constituent of the lower limb prosthesis is shank prosthesis. Irawan et al. (2016) tested mechanical properties of epoxy composites reinforced with rattan and fiberglass for endoskeletal shank prosthesis application. The test results showed that the prepared prosthesis has good strength and can be used by patients to walk appropriately. In another recent study by Hadi et al. (2022), fabrication of antimicrobial prosthetic socket from renewable, biobased epoxy reinforced with seashell nanoparticles and flax fibres for prosthetic socket application was reported.

Untreated knee anterior cruciate ligament (ACL) injuries often produce significant disability in athletes due to severe articular deterioration and may further increase functional instability. Clinically, these patients are suggested for an ACL reconstruction. Glutaraldehyde-fixed bioprostheses was initially used for such treatments. But this resulted in fibrosis, thrombosis, tissue stiffness and calcification problems. To address these deficiencies, Sung et al. (1994) developed epoxy based compound which can also fix biological tissues in heart valve and small-diameter vascular grafts. The development of precisely designed, light weight carbon fiber based sports prostheses allow lower-limb amputees to actively participate in competitive sports. The simultaneous presence of high strength and low elastic modulus is of particular importance in case of direct skeletal attachment (osseointegration) of artificial limbs. The ease of control and comfort of an artificial limb critically depends on quality of the interface between prosthesis and stump. Stiffness mismatch between host tissue and implant can cause severe stress shielding. In order to minimize such effects, the development and application of biocompatible materials is essential. Fiber reinforced polymer composites are able to accomplish both low elastic modulus as well as high strength, in an efficient manner. Furthermore, fatigue and corrosion resistance characteristics can be greatly improved. Sensory feedback system and functional coatings of metallic implants are further anticipated addition in new generation orthopedic devices.

3.3 Epoxy for Tissue Engineering Applications

Tissues in the human body may be damaged by injury, disease or trauma and can be treated by various methods like autograft (tissue is transplanted from one location to another within the same patient), allograft (tissue is transplanted from one individual to another) and more recently, by tissue engineering technique, which regenerate the damaged tissues using porous 3D scaffolds that provide the right environment for the growth of tissues and organs (O'Brien, 2011). Tissue engineering is another discipline within the biomedical territory where epoxy composites have found much usage. Artificial skin is an important option for the treatment of severe injuries or wounds of the skin. It also opens new possibilities for curing coetaneous ulcers, burns and congenital anomalies. The most important part of artificial skin is the tissue scaffold which helps in rapid regeneration and proliferation of tissues. However, available skin tissue scaffolds suffer a range of demerits such as poor strength and toxicity to the host tissue. Polymeric scaffolds as the extracellular matrix (ECM) are gaining stimulus in recent times due to their close resemblance with natural organic tissue components. Epoxy resins have strong footage in tissue engineering as they possess desired properties like cross linked networks to provide high rigidity, easy tailor-made processing, good thermal and mechanical properties. It is not possible to grow the cells in an artificial environment without the use of suitable scaffold. Important characteristics of a scaffold material include biocompatibility, biodegradability, sufficient mechanical strength and optimized pore architecture (O'Brien, 2011). Primary step in the construction of a tissue scaffold is decellularization for providing adequate integration of the xenomaterial into the patient's tissue without causing immunogenic effect. To address this problem epoxy cross linked decellularized bovine pericardium (DBP) biomeshes were tested and found improved angiogenesis and vascularisation (Elagin et al., 2020). The response of tissues to the external implant is largely controlled by the nature and texture of material surface. The surface texturing or surface treatment of implants is sometimes done to promote cellular activity. Textured surfaces show a greater area to improve bone bonding through osseointegration mechanisms. Furthermore, textured surfaces help the ingrowth of the tissues. Polymers used in scaffolds should be biodegradable so that as cells generate their own extracellular matrices, patient's own tissue will completely replace the polymeric material with time, but in case of polymeric heart valves, the polymeric material should be wear-resistant and nonbiodegradable so it will remain stable and not degrade with time. These degraded materials should not disturb or induce the opposite response in the host. These materials should be sterilizable and should not decompose or emit toxic gases during the sterilization process. In a study by Barua et al. (2015), a cytocompatibile, antimicrobial hyperbranched epoxy-nanoclay composites were developed by embedding silver nanoparticles and was found fit to be used as scaffolds in vivo. A potential, non-toxic tissue-scaffold nanocomposite was prepared by Barua et al. (2014) using hyperbranched epoxy with *Homalomena aromatica* rhizome oil-modified bentonite and organically modified montmorillonite clay. The adherence and proliferation of cells were optimal as per the findings reported. Glycerol-based hyperbranched epoxy with desirable properties was also reported by the same author for proliferation of dermocytes in vivo with no toxic effect on the major organs.

Fibrous scaffold-like structures are of interest not only in the field of soft-tissue engineering but also for specific applications of advanced bone tissue engineering. Traditionally, defective bones are managed by well-established treatment methods such as autografting, allografting, the application of vascularized grafts and bone marrow replacement. Present approaches are focusing on use of biocompatible, osteoconductive, osteoinductive and mechanically compatible scaffolding, as it potentially integrate with native tissue whilst stimulating contiguous bone formation. Scaffolding material may further act to carry implanted bone cells and other agents like growth promoters. Some scaffolds are capable of drug delivery and able to locally release growth factors and antibiotics, which assures improved bone ingrowth in combination with faster wound healing. A novel, porous, 3D bio mimic collagen/nano-hydroxyapatite scaffold cross-linked by epoxy based compound was successfully fabricated by Liu et al. (2020) for bone tissue engineering application. The prepared scaffold presented excellent mechanical strength, cytocompatibility and histocompatibility for bone tissue regeneration due to highly interconnected morphology and the low level of residual crosslinking agent.

Muscles reconstruction is a longstanding problem in medical science. However, with the progress of polymeric scaffolds, repairing of damaged or lost tissues has become possible.

Extremely compatible and infection resistant polymeric scaffolds are desired for tissue engineering applications to avoid post-surgical complications and thorough recovery of damaged tissues. CuO-nanofibrillar cellulose/glycerol based hyperbranched epoxy nanocomposite as a biodegradable, infection resistant and implantable muscle scaffold was developed by Barua et al. (2015). The prepared material could support growth and proliferation of L6 muscle cells without causing any impairment to cellular morphology. Further, in vivo and in vitro toxicity assessments could prove the profound biocompatibility of the material. The biodegradability of the prepared material overrules the need of repeated surgery for any implantable biomaterial. Implantable tissue scaffolds are used even for reconstruction of damaged organs. Skeletal muscles are the controlled strained muscle tissues which cover maximum part of the body. Due to burn, trauma or some congenital anomalies, these muscles may get damage. Repairing of damaged muscles is a serious problem in health discipline due to several post-operative cosmetic disorders. However, it is possible to culture muscle tissues in vitro to foster cell growth by use of three dimensional polymeric scaffolds. Successive implantation of such scaffolds within host body offers neat restoration of muscle cells (myocytes) without inducing any cosmetic defects.

The reconstruction of complex bone defects in the maxillofacial region is challenging in the field of tissue engineering. Tools such as 3D computer aided design and rapid prototyping can be exploited to fabricate custom made bone scaffolds. In a study by Scalera et al. (2014), hydroxyapatite filled UV curable epoxy suspensions were used for stereolithography of bone tissues. Carbon microlattice fabricated by 3D printing of an epoxy architecture using stereolithography followed by carbonization showed to be promising aspirant for tissue engineering experiments. Wherein murine MC3T3-E1 cells adhered and survived on the carbon microlattices with high viability. The microlacttice exhibited elastic modulus around 2.28 MPa which is within the modulus range of human tissue (Islam et al., 2020). Integration of carbon (CNTs, carbon fiber, carbon dots and graphene oxide) with epoxy matrix to form composites has been considered as a powerful strategy to upgrade scaffold properties and enhance tissue regeneration. To obtain carbon design-controlled porous structures or 3D lattice scaffolds, prototyping methods like 3D printing, stereolithography, selective laser sintering and fused deposition modelling can be adapted. In that case, materials derive their properties through structural organization rather than their intrinsic composition and tend to show qualitatively different mechanical behavior (Dogan et al. 2020). A major concern for surgical success in the long term is contact area of the implant with the tissue. The biocompatibility and surface contact of implants or scaffold materials is vital for any type of tissues. Epoxy based hydrophobic surface coated endoprosthetic implants and scaffolds was developed and used for tissue engineering applications by Geckeler et al. (2003). Thus tissue engineering is a pioneering technology to restore and reconstruct the damaged tissues.

3.4 Epoxy for Drug Delivery Applications

Many instances require preoperative and postoperative administration of drugs like antibiotics and analgesics in a metered dosage. Scaffolds can also be used as drug delivery media to minimize the systemic dosage. Some specific biomedical applications such as active substance delivery require appropriate use of conductive polymers. In dry powder and aerosol drug delivery, improper drug immobilization and subsequent incorrect drug dosage can happen due to unplanned drug administration (Bae and Park, 2020). Polymer-coated metallic drug eluting stents have transformed the treatment of obstructive coronary disease and is an ideal choice to prevent allergies caused by bare metallic stents. Drug-eluting coronary stent is used to promote blood flow to heart, to prevent plaque buildup and to relieve chest pain. They may also lower the chances of heart attack. A cyto and hemocompatible shape-memory drug eluting stent made from chitosan-based strips crosslinked by an epoxy compound is reported by Chen et al. (2009). This method of drug entrapment was able to embed lipophilic drugs, preventing accumulation and the wastage resulting from the initial release. The ability of the stent to memorize its permanent shape and mechanical strength largely depended on the degree of epoxy crosslinking. Amine cured epoxy compounds can be used to incorporate the drugs by embedding, coating and extrusion. Drugs can be incorporated in the bulk or beads of water-insoluble monomers during polymerization itself. They may serve as matrices for the sustained release of long acting oral drugs.

The ability to selectively target/kill cancerous cells while leaving healthy cells unaffected is a key objective in cancer therapeutics. A novel glutathione-responsive hydrogel synthesized through epoxy-amine click chemistry using a diepoxy crosslinker and PEG-diamine served as a platform for drug immobilization particularly for cancer therapy (Cengiz, 2020). Design of multi-responsive biodegradable micellar drug delivery systems for targeted cancer chemotherapy is another challenge. Sun et al. (2019) have demonstrated a kind of reduction and pH dual-responsive cationic polymer using epoxy-amine polymerization, which could self-assemble into micelles in an aqueous environment. These micelles exhibit specific tumor target ability and maximal drug-release controllability inside tumor cells upon changes in chemical and physical environments but presents outstanding stability under physiological conditions. These biodegradable micelles could be the efficient double-responsive nanocarriers to improve the drug delivery and enhance the antitumor efficacy. Epoxy is also used in transdermal drug delivery systems/devices (Akhtar et al., 2020). Functionalized amphiphilic epoxies play an important role in the advancement of controlled therapeutics. The release may happen at cyclic dosage, constant dosing over long periods or tunable release of the drug. Epoxy readily form nanoparticles covering hydrophobic as well as hydrophilic drugs. Epoxies are frequently used as carriers to stabilize and deliver lipophilic or sensitive drugs to their targeted sites. Most drugs require an appropriate formulation for their successful application in pharmaceuticals. The need for formulating may be drug sensitivity/instability, aqueous insolubility, fast clearance or unspecific distribution in human body. Functional polymer nanotubes with active epoxy groups were produced by Chen et al. (2012) using surface-initiated photopolymerization pattern transfer method, which can be used as the scaffold to design precise drug delivery carriers.

Developing a suitable drug delivery system is challenging in pharmaceutical industry and can emerge as novel therapeutics in upcoming years. Iron oxide magnetic nanoparticles modified with epoxy functional groups for improved drug binding was reported by Maltas et al. (2013). Wherein immobilized human serum albumin is used as drug carrier along with epoxy modified iron nanoparticles for ketoconazole drug delivery. Limited aqueous solubility is the major concern in development of new drugs formulation. Material innovation has led to the generation of biocompatible, biodegradable, stimuli-responsive and targeted drug delivery systems. To enhance the dissolution and solubility of practically insoluble drugs polymer nanocomposites are used. A promising strategy to enhance the proficiency of nanoparticle based drug delivery is to hybridize materials which help to overcome the biological barriers for drug delivery.

The traditional drug delivery methods are not advisable when predefined drug pharmacokinetics, low toxicity levels and time specific release are required for therapeutic treatment. Systemic administrations such as oral, inhalation, topical and injection are unable to lower the destructive side effects on the healthy tissues. New developments in drug delivery devices have reformed the course of therapeutic treatment in the contemporary period. These devices offer diverse strategies to overcome the restrictions of systemic administration that cannot provide high drug potency at the specific location to infected body tissues. Implantable and transdermal drug delivery devices can provide more drug efficacy based on wirelessly controlled microchips, microvalves, micropumps and magnetic robots with reliable actuating mechanism that can dispense drugs precisely to the required site with minimum side effects. The implantable devices are mostly in the form of microrobots. The electrical connections are covered with UV-cured epoxy in such devices. The targeted drug delivery technique can also efficiently release pharmaceutical agents to deep intricate pathological sites that are not directly accessible with systemic administration. With suitable design of formulation, epoxy resins can exhibit outstanding performance in biomedical applications.

3.5 Epoxy for Other Biomedical and Health-Related Applications

Epoxies have been used in many medical devices for sealing and bonding applications. A medical-grade, low-viscosity, clear epoxy adhesive has found application in fabrication of access ports which are implanted beneath the skin of patients requiring multiple injections (Han et al., 2021). By allowing access for subsequent treatments, the access ports boost patient comfort level and help to reduce complications. In amalgamation with catheter assembly, access ports are used to transport drugs to a specific part of the body. They can be used for arterial, venous, peritoneal or interspinal access. The access port assembly is done by potting the titanium or stainless steel access port with epoxy based adhesives. Epoxy resins are also used in pacemakers and ultrasound catheters electronic packages. A silver-filled epoxy has been used for bonding critical components in pacemaker circuits. Certain medical-grade epoxy adhesives can be sterilized by autoclave and chemical methods, so they are also used in medical devices that need sterilization prior to its use. Another application of epoxy in medicine is for instant fixing of end-caps to the main tube of a blood filter. Epoxies are used in surgical tapes and as bioadhesives for tissue bonding and hemostasis (Fakhar et al., 2012, Zaokari et al., 2020). The balance between strong adhesive strength and mechanical properties is critical in preparing and designing of a bio-based hydrogel adhesive for wound closure. Mussel-inspired biocompatible epoxy bioadhesive with enhanced interfacial interactions for surgical wound and tissue repair was reported by Du et al. (2021). Biomedical patches play important role in tissue regeneration and wound closure. A multifunctional biomedical patch based on hyperbranched epoxy loaded with curcumin and MXene was reported by Zou et al. (2021). The developed hybrid patch exhibited multi-substrate adhesion, shape-morphing capability, self-healing ability, strong UV-shielding, effective antibacterial activity, excellent oxidation resistance and good cytocompatibility. Snail shell-derived hydroxyapatite reinforced epoxy bio-composites for orthopaedic adhesive application was studied by Oladele et al. (2022).

Additional area where epoxy composites have found use is in the construction of platforms and electrodes for biosensing applications (Smith and Lamprou, 2014). An epoxy/graphite biocomposite containing tyrosinase enzyme was used for the effective biosensing of phenolic compounds by Wang et al. (1994). The enzyme retained its bioactivity on confinement in the epoxy resin environment. This rigid and renewable bioprobe offers convenient quantification of phenolic substrates. Amperometric biosensor based on three-dimensional hydrogel epoxy networks with high specificity was reported by Heller et al. (1993). Epoxy composites containing functionalized carbon nanotubes and nanocrystalline diamond powders were also used for the design of biosensors (Khabashesku et al., 2005). In another study by Singh et al. (2020), lead-free, zinc oxide-graphene-epoxy electroactive nanocomposite thin films was fabricated for use as secondary sensors or electron transport layers for biomedical wearable devices application.

Although the main aim of restorative and regenerative medicinal practice is to restore the functional and structural integrity of the damaged tissue, particular attention must be paid to the durability of the material and its related aesthetic feature. This is also true with assisting medical devices and equipment. Many of the medical devices have plastic-on-plastic or plastic-on-metal sliding parts (equipment elements, implants, sliding covers). The performance of such equipment and devices designed for clinical use may be strongly impacted by the durability of the plastics used for manufacture. Many of the times epoxy is the best possible chemical resistant material to sustain harsh strerilizable environment. One area where epoxy finds much use is in the development of piezo-composites for ultrasonic systems in medical imaging (Cannata et al., 2006; Ritter et al., 2002). Epoxy coated titania nanolayers have been used to produce positive replicas with accurate surface topographies to improve the reproducibility of cell cultural assays (Schuler et al., 2009). In some innovative applications, epoxy is typically not present as bulk materials but they are used as coatings on biomedical devices or as micro and nanospheres, dendrimers, micelles, solid colloids, capsules, nanogels or core-shell particles for targeted drug delivery. Coatings may be crosslinked coatings, non-structured homogeneous coatings, polymer brushes or layer-by-layer deposited films. Nickels et al. (2010) reported functionalization of iron oxide nanoparticles with epoxy amine linker, which can be used to label variety of imaging or targeting agents for biomedical in vivo diagnostics or therapies.

Considering epoxy composites light weight, chemical resistance and superior mechanical properties, they are also used as structural components in medical imaging devices. Ceramics and metal alloys are radio opaque and in some cases they cause undesirable artefacts in X-ray radiography. In the case of polymer composites the radio transparency can be tuned by adding contrast medium to the polymer. Moreover the polymer composites are fully compatible with modern diagnostic methods such as magnetic resonance imaging and computed tomography as they are non-magnetic. Implantable medical devices have experienced a rapid growth ever since 1950s with the innovation of pacemakers. Epoxy casting is used as nonhermetic encapsulation for implantable electronic devices (Boeser et al., 2016). Main packaging strategy is to protect the implanted electronics from the body environment. One of the main features of human intracorporeal implants is the biocompatibility while preserving the implants functionality for a well-defined period of time. The durability of xenograft heart valves is restricted due to calcification of the tissue, which leads to stiffening, tearing and rupture of the cusp. The relative rigidity of the stent is supposed to enhance primary prosthetic failure. To reduce immunogenicity and degradation, the porcine tissue is crosslinked with glutaraldehyde. The researchers believe that glutaraldehyde crosslinking is a major contributing factor for this failure. To reduce the occurrence of calcification, alternative crosslinking methods were investigated. Porcine aortic heart valves are often considered for heart valve replacement in elderly patients, Wachem et al. (2000) used epoxy crosslinked porcine aortic valve tissue. Epoxy crosslinked collagen materials had shown a large reduction in calcification compared to glutaraldehyde crosslinked materials. Hence bioprostheses derived from collagenous tissues can be fixed by epoxy and subsequently sterilized before they can be implanted in humans. Epoxies can thus serve multiple roles in biomedical applications.

CONCLUSION

Medical practice today utilizes a great number of implants and devices. Biomaterials in the form of implants (dental implants, bone plates, joint replacements, sutures, ligaments, heart valves, vascular grafts, intraocular lenses etc.) and medical devices (pacemakers, artificial hearts, biosensors, blood tubes etc.) are widely used to replace and/or restore the function of degenerated or disturbed tissues or organs, to improve function, to correct anomalies, to assist in healing and thus improve the overall quality of patient life. Over the centuries, advancements in biomaterials, surgical techniques and sterilization methods have solved the challenges of medical field in many ways. In all these contributions epoxy and its composites have a strong footage. The properties of epoxy composites can be customized to the required biomedical application based on type of epoxy resin, formulation, curing agent, curing methodology and processing technique. Due to which they are used in dental implants, bone replacement, prosthetic sockets, hard tissue replacements, heart valves, tissue scaffolds, drug delivery and in medical imaging devices. Epoxy usage can overcome relatively high corrosion rate pitting and crevice corrosion of stainless steel in such applications. Ceramics though are known for their corrosion resistance and good biocompatibility they suffer from drawbacks like brittleness, high density, low fracture strength and are difficult to fabricate. Epoxy based composites with tunable physicochemical properties are becoming most sought materials due to their light weight, low cost, high strength, body tissue matching stiffness, simple and easy manufacturing principles. With ever increasing demand for custom-made materials in biomedical applications, definitely epoxy can be an ideal material of choice.

REFERENCES

Akhtar, N., Singh, V., Yusuf, M., & Khan, R. A. (2020). Non-invasive drug delivery technology: development and current status of transdermal drug delivery devices, techniques and biomedical applications. *Biomedical Engineering/Biomedizinische Technik, 65*(3), 243-272. doi:10.1515/bmt-2019-0019

Alharbi, F., Nathanson, D., Morgano, S. M., & Baba, N. Z. (2014). Fracture resistance and failure mode of fatigued endodontically treated teeth restored with fiber-reinforced resin posts and metallic posts in vitro. *Dental Traumatology, 30*(4), 317–325. doi:10.1111/edt.12095 PMID:25039331

Arun, S., & Kanagaraj, S. (2016). Performance enhancement of epoxy based sandwich composites using multiwalled carbon nanotubes for the application of sockets in trans-femoral amputees. *Journal of the Mechanical Behavior of Biomedical Materials, 59*, 1–10. doi:10.1016/j.jmbbm.2015.12.013 PMID:26736176

Ashcroft, W. R. (1993). Curing agents for epoxy resins. In B. Ellis (Ed.), *Chemistry and technology of epoxy resins* (pp. 37–71). Springer. doi:10.1007/978-94-011-2932-9_2

Atay, H. Y., Dogan, L. E., & Celik, E. (2013). Investigations of self-healing property of chitosan-reinforced epoxy dye composite coatings. *Journal of Materials, 2013*, 1–7. doi:10.1155/2013/613717

Bae, Y. H., & Park, K. (2020). Advanced drug delivery 2020 and beyond: Perspectives on the future. *Advanced Drug Delivery Reviews, 158*, 4–16. doi:10.1016/j.addr.2020.06.018 PMID:32592727

Bagheri, Z. S., El Sawi, I., Bougherara, H., & Zdero, R. (2014). Biomechanical fatigue analysis of an advanced new carbon fiber/flax/epoxy plate for bone fracture repair using conventional fatigue tests and thermography. *Journal of the Mechanical Behavior of Biomedical Materials, 35*, 27–38. doi:10.1016/j.jmbbm.2014.03.008 PMID:24727574

Bagheri, Z. S., El Sawi, I., Schemitsch, E. H., Zdero, R., & Bougherara, H. (2013). Biomechanical properties of an advanced new carbon/flax/epoxy composite material for bone plate applications. *Journal of the Mechanical Behavior of Biomedical Materials, 20*, 398–406. doi:10.1016/j.jmbbm.2012.12.013 PMID:23499250

Bagheri, Z. S., Giles, E., El Sawi, I., Amleh, A., Schemitsch, E. H., Zdero, R., & Bougherara, H. (2015). Osteogenesis and cytotoxicity of a new carbon fiber/flax/epoxy composite material for bone fracture plate applications. *Materials Science and Engineering C, 46*, 435–442. doi:10.1016/j.msec.2014.10.042 PMID:25492008

Barua, S., Chattopadhyay, P., Aidew, L., Buragohain, A. K., & Karak, N. (2015). Infection-resistant hyperbranched epoxy nanocomposite as a scaffold for skin tissue regeneration. *Polymer International, 64*(2), 303–311. doi:10.1002/pi.4790

Barua, S., Dutta, N., Karmakar, S., Chattopadhyay, P., Aidew, L., Buragohain, A. K., & Karak, N. (2014). Biocompatible high performance hyperbranched epoxy/clay nanocomposite as an implantable material. *Biomedical Materials (Bristol, England), 9*(2), 025006. doi:10.1088/1748-6041/9/2/025006 PMID:24495981

Bloch, B. (1958). Interim success with the first use of epoxy resins. *The Journal of Bone and Joint Surgery. British Volume*, *40*, 804. doi:10.1302/0301-620X.40B4.804 PMID:13610994

Boeser, F., Lang, B. G., Stieglitz, T., & Plachta, D. T. (2016, August). Epoxy casting used as nonhermetic encapsulation technique for implantable electronic devices. In *38th Annual International Conference of the IEEE Engineering in Medicine and Biology Society (EMBC)* (pp. 1938-1941). IEEE. 10.1109/EMBC.2016.7591102

Boker, K. O., Richter, K., Jackle, K., Taheri, S., Grunwald, I., Borcherding, K., Byern, J. V., Hartwig, A., Wildemann, B., Schilling, A. F., & Lehmann, W. (2019). Current state of bone adhesives - Necessities and hurdles. *Materials (Basel)*, *12*(23), 3975. doi:10.3390/ma12233975 PMID:31801225

Bradley, J. S., Hastings, G. W., & Johnson-Nurse, C. (1980). Carbon fiber reinforced epoxy as a high strength, low modulus material for internal fixation plates. *Biomaterials*, *1*(1), 38–40. doi:10.1016/0142-9612(80)90057-5 PMID:7470551

Brocas, A. L., Llevot, A., Mantzaridis, C., Cendejas, G., Auvergne, R., Caillol, S., Carlotti, S., & Cramail, H. (2014). Epoxidized rosin acids as co-precursors for epoxy resins. *Designed Monomers and Polymers*, *17*(4), 301–310. doi:10.1080/15685551.2013.840504

Cannata, J. M., Williams, J. A., Zhou, Q., Ritter, T. A., & Shung, K. K. (2006). Development of a 35-MHz piezo-composite ultrasound array for medical imaging. *IEEE Transactions on Ultrasonics, Ferroelectrics, and Frequency Control*, *53*(1), 224–236. doi:10.1109/TUFFC.2006.1588408 PMID:16471449

Cengiz, N. (2020). Glutathione-responsive multifunctionalizable hydrogels *via* amine-epoxy "click" chemistry. *European Polymer Journal*, *123*, 109441. doi:10.1016/j.eurpolymj.2019.109441

Chandramohan, D., Bharanichandar, J., Karthikeyan, P., Vijayan, R., & Murali, B. (2014). Progress of biomaterials in the field of orthopaedics. *American Journal of Applied Sciences*, *11*(4), 623–630. doi:10.3844/ajassp.2014.623.630

Chandramohan, D., & Marimuthu, K. (2011). Natural fiber bone plates as substitutes for orthopaedic alloy plates. *Journal of Biotechnology. Bioinformatics and Bioengineering*, *1*(4), 405–413.

Chen, C. J., Jin, Q., Liu, G. Y., Li, D. D., Wang, J. L., & Ji, J. (2012). Reversibly light-responsive micelles constructed *via* a simple modification of hyperbranched polymers with chromophores. *Polymer*, *53*(17), 3695–3703. doi:10.1016/j.polymer.2012.06.024

Chen, M. C., Chang, Y., Liu, C. T., Lai, W. Y., Peng, S. F., Hung, Y. W., Tsai, H. W., & Sung, H. W. (2009). The characteristics and *in vivo* suppression of neointimal formation with sirolimus-eluting polymeric stents. *Biomaterials*, *30*(1), 79–88. doi:10.1016/j.biomaterials.2008.09.006 PMID:18838155

Cizravi, J. C., & Subramaniam, K. (1999). Thermal and mechanical properties of epoxidized natural rubber modified epoxy matrices. *Polymer International*, *48*(9), 889–895. doi:10.1002/(SICI)1097-0126(199909)48:9<889::AID-PI227>3.0.CO;2-L

Davoudi, A., Shadmehr, E., Savoj, M., Akhavan-Khaleghi, M., Tabesh, M., Zakizade, M., & Ebadian, B. (2021). Laser treatment of fiber post and final push-out bond strength: A systematic review and meta-analysis. *Photobiomodulation, Photomedicine, and Laser Surgery*, *39*(5), 321–333. doi:10.1089/photob.2020.4809 PMID:33275497

De, B., & Karak, N. (2013). Novel high performance tough hyperbranched epoxy by an A 2+ B 3 polycondensation reaction. *Journal of Materials Chemistry. A, Materials for Energy and Sustainability*, *1*(2), 348–353. doi:10.1039/C2TA00011C

Dinesh, K. R., Jagadish, S. P., & Thimmanagouda, A. (2014). Characterization and analysis of wear study on sisal fibre reinforcement epoxy composite materials used as orthopaedic implant. *International Journal of Advances in Engineering and Technology*, *6*(6), 2745–2757.

Dogan, E., Bhusal, A., Cecen, B., & Miri, A. K. (2020). 3D Printing metamaterials towards tissue engineering. *Applied Materials Today*, *20*, 100752. doi:10.1016/j.apmt.2020.100752 PMID:32856000

Du, D., Chen, X., Shi, C., Zhang, Z., Shi, D., Kaneko, D., Kaneko, T., & Hua, Z. (2021). Mussel-inspired epoxy bioadhesive with enhanced interfacial interactions for wound repair. *Acta Biomaterialia*, *136*, 223–232. doi:10.1016/j.actbio.2021.09.054 PMID:34610475

Elagin, V., Kuznetsova, D., Grebenik, E., Zolotov, D. A., Istranov, L., Zharikova, T., Istranova, E., Polozova, A., Reunov, D., Kurkov, A., Shekhter, A., Gafarova, E. R., Asadchikov, V., Borisov, S. M., Dmitriev, R. I., Zagaynova, E., & Timashev, P. (2020). Multiparametric optical bioimaging reveals the fate of epoxy crosslinked biomeshes in the mouse subcutaneous implantation model. *Frontiers in Bioengineering and Biotechnology*, *8*, 107. doi:10.3389/fbioe.2020.00107 PMID:32140465

Fakhar, A., Aabaadiaan, M., Keivani Hafshejani, M., & Langari, A. (2012). Use of reactive oligomer to improve fracture resistance of epoxy used in medical applications and GRP pipelines. *World Applied Sciences Journal*, *20*(2), 259–263. doi:10.5829/idosi.wasj.2012.20.02.2487

Freedman, G. A. (2001). Esthetic post-and-core treatment. *Dental Clinics of North America*, *45*(1), 103–116. doi:10.1016/S0011-8532(22)00470-0 PMID:11210689

Gao, X., Lan, J., Jia, X., Cai, Q., & Yang, X. (2016). Improving interfacial adhesion with epoxy matrix using hybridized carbon nanofibers containing calcium phosphate nanoparticles for bone repairing. *Materials Science and Engineering C*, *61*, 174–179. doi:10.1016/j.msec.2015.12.033 PMID:26838838

Geckeler, K., Wacker, R., Martini, F., Hack, A., & Aicher, W. (2003). Enhanced biocompatibility for SAOS-2 osteosarcoma cells by surface coating with hydrophobic epoxy resins. *Cellular Physiology and Biochemistry*, *13*(3), 155–164. doi:10.1159/000071866 PMID:12876386

Ghalme, S. G., Mankar, A., & Bhalerao, Y. (2016). Biomaterials in hip joint replacement. *International Journal of Materials Science and Engineering*, *4*(2), 113–125. doi:10.17706/ijmse.2016.4.2.113-125

Gouda, T. A., Jagadish, S. P., Dinesh, K. R., Gouda, H., & Prashanth, N. (2014). Characterization and investigation of mechanical properties of hybrid natural fiber polymer composite materials used as orthopaedic implants for femur bone prosthesis. *IOSR Journal of Mechanical and Civil Engineering*, *11*(4), 40–52. doi:10.9790/1684-11424052

Gowda, M. S., Hemavathi, A. B., Srinivas, S., Santhosh, G., & Siddaramaiah, H. (2022). Physico-mechanical, thermal, and tribological studies of polytetrafluoroethylene-filled polyoxymethylene/silicone composites. *Journal of Thermoplastic Composite Materials*, *35*(6), 846–859. doi:10.1177/0892705720925117

Hadi, A. N., Mohammed, M. R., & Al-Bakri, A. F. (2022). Numerical and experimental assessment of mechanical properties of biobased epoxy reinforced by flax fibres and seashell nanoparticles for prosthetic socket. *Advances in Materials Science and Engineering*, *2022*, 1–11. doi:10.1155/2022/9024498

Hadjinikolaou, I. N., Bell, J. P., & Spangberg, L. (1987). Fast curing epoxy and episulfide resins for biomedical applications. *MRS Proceedings*, *110*, 451. 10.1557/PROC-110-451

Han, L., Zhang, J., Deng, X., Kong, X., Yang, C., Peng, L., Kou, C., Zou, K., Lv, L., Li, C., Wang, S., & Wei, G. (2021). Totally implantable venous access ports: A prospective randomized study comparing subclavian and internal jugular vein punctures in children. *Journal of Pediatric Surgery*, *56*(2), 317–323. doi:10.1016/j.jpedsurg.2020.04.021 PMID:32467037

Heller, A., Maidan, R., & Wang, D. (1993). Amperometric biosensors based on three-dimensional hydrogel-forming epoxy networks. *Sensors and Actuators. B, Chemical*, *13*(1-3), 180–183. doi:10.1016/0925-4005(93)85356-F

Hemavathi, A. B., Anupama, G., Vani, K. V., Asha, P., Vijayalakshmi, N. S., & Kumar, K. R. (2002). Physico-chemical and mechanical properties evaluation of polyolefin films and composites. *Popular Plastics and Packaging*, *47*(3), 65–68.

Hemavathi, A. B., Babu, A., Murthy, P. K., & Zarena, A. S. (2022). Sustainable nanocomposite films of improved barrier properties for food packaging applications. *Materials Today: Proceedings*, *65*, 364–366. doi:10.1016/j.matpr.2022.06.373

Hemavathi, A. B., Monisha, C. A., Akhila, P., & Bhat, P. M. (2021). Environmentally Benign Food Packaging Material based on Polylactic acid - A Brief Review. *International Research Journal of Engineering and Technology*, *8*(8), 362–368.

Hemavathi, A. B., & Siddaramaiah, H. (2018). Food Packaging: Polymers as Packaging Materials in Food Supply Chains. In M. Mishra (Ed.), *Encyclopedia of Polymer Applications*. CRC Press.

Hirwani, J. K., Nishimura, R., Shinmori, H., Morita, T., Sawae, Y., & Sinha, S. K. (2022). Epoxy (SU-8) polymer composites with ultra-high molecular weight polyethylene and hyaluronic acid fillers for hip prosthetic implant application. *Tribology International*, *167*, 107399. doi:10.1016/j.triboint.2021.107399

Hou, N. Y., Zhu, J., Zhang, H., & Perinpanayagam, H. (2016). Epoxy resin-based ultrafine dry powder coatings for implants. *Journal of Applied Polymer Science*, *133*(37), 43960. doi:10.1002/app.43960

Huang, Z. M., Gopal, R., Fujihara, K., Ramakrishna, S., Loh, P. L., Foong, W. C., Ganesh, V. K., & Chew, C. L. (2003). Fabrication of a new composite orthodontic archwire and validation by a bridging micromechanics model. *Biomaterials*, *24*(17), 2941–2953. doi:10.1016/S0142-9612(03)00093-0 PMID:12742734

Irawan, A. P., Daywin, F. J., Fanando, F., & Agustino, T. (2016). Mechanical characteristics of rattan reinforced fiberglass and epoxy composites for shank prosthesis application. *IACSIT International Journal of Engineering and Technology*, *8*(3), 1543–1550.

Irawan, A. P., Soemardi, T. P., Widjajalaksmi, K., & Reksoprodjo, A. H. (2011). Tensile and flexural strength of ramie fiber reinforced epoxy composites for socket prosthesis application. *International Journal of Mechanical and Materials Engineering*, *6*(1), 46–50.

Islam, M., Lantada, A. D., Gómez, M. R., Mager, D., & Korvink, J. G. (2020). Microarchitectured carbon structures as innovative tissue-engineering scaffolds. *Advanced Engineering Materials*, *22*(6), 2000083. doi:10.1002/adem.202000083

Jin, F. L., Li, X., & Park, S. J. (2015). Synthesis and application of epoxy resins: A review. *Journal of Industrial and Engineering Chemistry*, *29*, 1–11. doi:10.1016/j.jiec.2015.03.026

Khabashesku, V. N., Margrave, J. L., & Barrera, E. V. (2005). Functionalized carbon nanotubes and nanodiamonds for engineering and biomedical applications. *Diamond and Related Materials*, *14*(3-7), 859–866. doi:10.1016/j.diamond.2004.11.006

Komabayashi, T., Colmenar, D., Cvach, N., Bhat, A., Primus, C., & Imai, Y. (2020). Comprehensive review of current endodontic sealers. *Dental Materials Journal*, *39*(5), 703–720. doi:10.4012/dmj.2019-288 PMID:32213767

Krishnakumar, S., & Senthilvelan, T. (2021). Polymer composites in dentistry and orthopedic applications-a review. *Materials Today: Proceedings*, *46*, 9707–9713. doi:10.1016/j.matpr.2020.08.463

Lee, J. K., Kwak, S. W., Ha, J. H., Lee, W., & Kim, H. C. (2017). Physicochemical properties of epoxy resin-based and bioceramic-based root canal sealers. *Bioinorganic Chemistry and Applications*, *2582849*, 1–8. Advance online publication. doi:10.1155/2017/2582849 PMID:28210204

Liu, Y., Gu, J., & Fan, D. (2020). Fabrication of high-strength and porous hybrid scaffolds based on nano-hydroxyapatite and human-like collagen for bone tissue regeneration. *Polymers*, *12*(1), 61. doi:10.3390/polym12010061 PMID:31906327

Maitz, M. F. (2015). Applications of synthetic polymers in clinical medicine. *Biosurface and Biotribology*, *1*(3), 161–176. doi:10.1016/j.bsbt.2015.08.002

Maltas, E., Ozmen, M., Yildirimer, B., Kucukkolbasi, S., & Yildiz, S. (2013). Interaction between ketoconazole and human serum albumin on epoxy modified magnetic nanoparticles for drug delivery. *Journal of Nanoscience and Nanotechnology*, *13*(10), 6522–6528. doi:10.1166/jnn.2013.7742 PMID:24245109

Mantzaridis, C., Brocas, A. L., Llevot, A., Cendejas, G., Auvergne, R., Caillol, S., Carlotti, S., & Cramail, H. (2013). Rosin acid oligomers as precursors of DGEBA-free epoxy resins. *Green Chemistry*, *15*(11), 3091–3098. doi:10.1039/c3gc41004h

Mikulewicz, M., Gronostajski, Z., Wielgus, A., & Chojnacka, K. (2017). Transparent orthodontic archwires: A systematic literature review. *Archives of Civil and Mechanical Engineering*, *17*(3), 651–657. doi:10.1016/j.acme.2016.12.012

Mittal, A., Dadu, S., Garg, P., Yendrembam, B., Abraham, A., & Singh, K. (2017). Comparative evaluation of fracture resistance of endodontically treated teeth with epoxy resin-based sealers AH plus and mineral trioxide aggregate fillapex: An in vitro study. *Indian Journal of Dental Sciences*, *9*(1), 8–11. doi:10.4103/IJDS.IJDS_83_16

Munchow, E. A., & Bottino, M. C. (2017). Recent advances in adhesive bonding: The role of biomolecules, nanocompounds, and bonding strategies in enhancing resin bonding to dental substrates. *Current Oral Health Reports*, *4*(3), 215–227. doi:10.100740496-017-0146-y PMID:29177123

Mustapha, R., Rahmat, A. R., Abdul Majid, R., & Mustapha, S. N. H. (2019). Vegetable oil-based epoxy resins and their composites with bio-based hardener: A short review. *Polymer-Plastics Technology and Materials*, *58*(12), 1311–1326. doi:10.1080/25740881.2018.1563119

Nickels, M., Xie, J., Cobb, J., Gore, J. C., & Pham, W. (2010). Functionalization of iron oxide nanoparticles with a versatile epoxy amine linker. *Journal of Materials Chemistry*, *20*(23), 4776–4780. doi:10.1039/c0jm00808g PMID:20668643

O'Brien, D. M., Vallieres, C., Alexander, C., Howdle, S. M., Stockman, R. A., & Avery, S. V. (2019). Epoxy-amine oligomers from terpenes with applications in synergistic antifungal treatments. *Journal of Materials Chemistry. B, Materials for Biology and Medicine*, *7*(34), 5222–5229. doi:10.1039/C9TB00878K PMID:31369021

O'Brien, F. J. (2011). Biomaterials & scaffolds for tissue engineering. *Materials Today*, *14*(3), 88–95. doi:10.1016/S1369-7021(11)70058-X

Odusote, J. K., & Oyewo, A. T. (2016). Mechanical properties of pineapple leaf fiber reinforced polymer composites for application as a prosthetic socket. *Journal of Engineering Technology*, *7*(1), 125–139.

Odusote, J. K., Oyewo, A. T., Adebisi, J. A., & Akande, K. A. (2016). Mechanical properties of banana pseudo stem fibre reinforced epoxy composite as a replacement for transtibial prosthetic socket. *The Journal of the Association of Professional Engineers of Trinidad and Tobago*, *44*(2), 4–10.

Oladele, I. O., Onuh, L., Taiwo, A. S., Borisade, S., Agbeboh, N. I., & Lephuthing, S. S. (2022). Mechanical, wear and thermal conductivity characteristics of snail shell-derived hydroxyapatite reinforced epoxy bio-composites for adhesive biomaterials applications. *International Journal of Sustainable Engineering*, *15*(1), 125–137. doi:10.1080/19397038.2022.2084174

Otto, J. K., Hajaj, B., Abovitz, R., Mistry, A., Nortman, S., & Brown, S. B. (2010). *U.S. Patent No. 7842092*. Washington, DC: U.S. Patent and Trademark Office.

Patel, K. K., & Purohit, R. (2018). Future prospects of shape memory polymer nano-composite and epoxy based shape memory polymer-A review. *Materials Today: Proceedings*, *5*(9), 20193–20200. doi:10.1016/j.matpr.2018.06.389

Peters, S. T. (1998). Introduction, composite basics and road map. In S. T. Peters (Ed.), *Handbook of Composites* (pp. 1–20). Springer US. doi:10.1007/978-1-4615-6389-1_1

Petersen, R. C. (2014). Titanium implant osseointegration problems with alternate solutions using epoxy/carbon-fiber-reinforced composite. *Metals*, *4*(4), 549–569. doi:10.3390/met4040549 PMID:25635227

Ramakrishna, S., Fujihara, H., Yoshida, E., & Hamada, H. (2000). Design and development of braided carbon fiber reinforced polymer composites for dental posts and bone plates. In *29th FRP symposium.* Japan Society for Materials Science.

Ritter, T. A., Shrout, T. R., Tutwiler, R., & Shung, K. K. (2002). A 30-MHz piezo-composite ultrasound array for medical imaging applications. *IEEE Transactions on Ultrasonics, Ferroelectrics, and Frequency Control*, *49*(2), 217–230. doi:10.1109/58.985706 PMID:11885679

Rony, L., Lancigu, R., & Hubert, L. (2018). Intraosseous metal implants in orthopedics: A review. *Morphologie*, *102*(339), 231–242. doi:10.1016/j.morpho.2018.09.003 PMID:30348489

Scalera, F., Corcione, C. E., Montagna, F., Sannino, A., & Maffezzoli, A. (2014). Development and characterization of UV curable epoxy/hydroxyapatite suspensions for stereolithography applied to bone tissue engineering. *Ceramics International*, *40*(10), 15455–15462. doi:10.1016/j.ceramint.2014.06.117

Schmitt-Thomas, K. G., Yang, Z. G., & Hiermer, T. (1998). Development of torsion loading facility used for evaluating shear performance of polymeric composite implant rods. *Polymer Testing*, *17*(2), 117–130. doi:10.1016/S0142-9418(97)00037-8

Schuler, M., Kunzler, T. P., de Wild, M., Sprecher, C. M., Trentin, D., Brunette, D. M., Textor, M., & Tosatti, S. G. P. (2009). Fabrication of TiO_2-coated epoxy replicas with identical dual-type surface topographies used in cell culture assays. *Journal of Biomedical Materials Research. Part A*, *88*(1), 12–22. doi:10.1002/jbm.a.31720 PMID:18257057

Siddaramaiah, & Jagadeesh, K. S. (1996). Properties of multi-functional epoxy resins cured with different amines. *Journal of Polymer Materials, 13*(3), 223-231.

Simoes, J. A. O., & Marques, A. T. (2001). Determination of stiffness properties of braided composites for the design of a hip prosthesis. *Composites. Part A, Applied Science and Manufacturing*, *32*(5), 655–662. doi:10.1016/S1359-835X(00)00157-3

Singh, D. P. (2016). Esthetic archwires in orthodontics: A review. *Journal of Oral Hygiene & Health*, *4*(1), 194. doi:10.4172/2332-0702.1000194

Singh, M., Kumar, S., Zoghi, S., Cervantes, Y., Sarkar, D., Ahmed, S., Chowdhury, S., & Banerjee, S. (2020). Fabrication and characterization of flexible three-phase ZnO-Graphene-Epoxy electro-active thin-film nanocomposites: Towards applications in wearable biomedical devices. *Journal of Composites Science*, *4*(3), 88. doi:10.3390/jcs4030088

Smith, J. R., & Lamprou, D. A. (2014). Polymer coatings for biomedical applications: A review. *Transactions of the IMF*, *92*(1), 9-19. https://doi.org/10.1179/0020296713Z.000000000157

Song, R., Murphy, M., Li, C., Ting, K., Soo, C., & Zheng, Z. (2018). Current development of biodegradable polymeric materials for biomedical applications. *Drug Design, Development and Therapy*, *12*, 3117–3145. doi:10.2147/DDDT.S165440 PMID:30288019

Sowmya, B., Hemavathi, A. B., & Panda, P. K. (2021). Poly(ε-caprolactone) based electrospun nanofeatured substrate for tissue engineering applications: A review. *Progress in Biomaterials*, *10*(2), 91–117. doi:10.100740204-021-00157-4 PMID:34075571

Sun, G., Liu, J., Wang, X., Li, M., Cui, X., Zhang, L., Wu, D., & Tang, P. (2019). Fabrication of dual-sensitive poly (β-hydroxyl amine) micelles for controlled drug delivery. *European Polymer Journal, 114*, 338–345. doi:10.1016/j.eurpolymj.2019.02.048

Sung, H. W., Tu, R., Shen, S. H., Witzel, T. H., Lin, D., Hata, C., Kingsbury, C.J., Noishiki, Y., Tomizawa, Y., & Quijano, R. C. (1994). A newly developed porcine heart valve bioprosthesis fixed with an epoxy compound. An experimental evaluation. *ASAIO Journal, 40*(2), 192-198.

Tey, K. C., & Lui, J. L. (2014). The effect of glass fiber-reinforced epoxy resin dowel diameter on the fracture resistance of endodontically treated teeth. *Journal of Prosthodontics, 23*(7), 572–581. doi:10.1111/jopr.12146 PMID:24750324

Tilbrook, D. A., Clarke, R. L., Howle, N. E., & Braden, M. (2000). Photocurable epoxy-polyol matrices for use in dental composites I. *Biomaterials, 21*(17), 1743–1753. doi:10.1016/S0142-9612(00)00059-4 PMID:10905456

Verma, N., Kumar, R., Zafar, S., & Pathak, H. (2020). Vacuum-assisted microwave curing of epoxy/carbon fiber composite: An attempt for defect reduction in processing. *Manufacturing Letters, 24*, 127–131. doi:10.1016/j.mfglet.2020.04.010

Vitale, A., Sangermano, M., Bongiovanni, R., Burtscher, P., & Moszner, N. (2014). Visible light curable restorative composites for dental applications based on epoxy monomer. *Materials (Basel), 7*(1), 554–562. doi:10.3390/ma7010554 PMID:28788474

Wachem, P. B., Brouwer, L. A., Zeeman, R., Dijkstra, P. J., Feijen, J., Hendriks, M., Cahalan, P. T., & van Luyn, M. J. A. (2000). *In vivo* behavior of epoxy-crosslinked porcine heart valve cusps and walls. *Journal of Biomedical Materials Research, 53*(1), 18–27. doi:10.1002/(SICI)1097-4636(2000)53:1<18::AID-JBM3>3.0.CO;2-J PMID:10634948

Wang, H., Liu, B., Liu, X., Zhang, J., & Xian, M. (2008). Synthesis of biobased epoxy and curing agents using rosin and the study of cure reactions. *Green Chemistry, 10*(11), 1190–1196. doi:10.1039/b803295e

Wang, J., Fang, L., & Lopez, D. (1994). Amperometric biosensor for phenols based on a tyrosinase-graphite-epoxy biocomposite. *Analyst (London), 119*(3), 455–458. doi:10.1039/AN9941900455 PMID:8192231

Wetzel, B., Haupert, F., & Zhang, M. Q. (2003). Epoxy nanocomposites with high mechanical and tribological performance. *Composites Science and Technology, 63*(14), 2055–2067. doi:10.1016/S0266-3538(03)00115-5

Yavari, H., Ghasemi, N., Divband, B., Rezaei, Y., Jabbari, G., & Payahoo, S. (2017). The effect of photodynamic therapy and polymer solution containing nano-particles of Ag/ZnO on push-out bond strength of the sealers AH-Plus and MTA Fillapex. *Journal of Clinical and Experimental Dentistry, 9*(9), e1109–e1114. doi:10.4317/jced.54069 PMID:29075413

Zaokari, Y., Persaud, A., & Ibrahim, A. (2020). Biomaterials for adhesion in orthopedic applications: A review. *Engineered Regeneration, 1*, 51–63. doi:10.1016/j.engreg.2020.07.002

Zou, Y., Jin, X., Zhang, X., Kong, X., Zhang, Q., Xie, X., Liu, C., Ke, L., Liu, W. G., & Wang, W. (2021). A multifunctional biomedical patch based on hyperbranched epoxy polymer and MXene. *Science China. Technological Sciences, 64*(12), 2744–2754. doi:10.100711431-021-1843-3

Chapter 13
The Impact of Risk Factors on the Prevalence of Cardiovascular Disease

Abiola Basirat Bakare
University of Lincoln, UK

Mubeen Ghafoor
College of Science, University of Lincoln, UK

Manuela Mura
University of Lincoln, UK

Waqar Ahmed
University of Lincoln, UK

ABSTRACT

The growing rate of cardiovascular disease all over the globe is so alarming that urgent measures need to be put in place. Approximately 18 million deaths are recorded every year globally due to cardiovascular disease. People living with high blood pressure, diabetes, high cholesterol level, or living poor are at high risk of developing cardiovascular disease. Other factors include age, gender, family history, and ethnic group. Understanding the risks of cardiovascular disease can help individuals, organizations, and governments promote a healthy lifestyle. Cleveland heart disease dataset was collected from online and open source for the purpose of this study. The dataset includes 297 medical records of patients between the ages of 29-77. The dataset was processed and analyzed in Microsoft Excel using frequency counts and percentages. The risk factors are important risk factors of cardiovascular disease for the participant of the study. Discussion on the use of biomaterials in fabricating tissue-engineered nanofibers for the treatment of cardiovascular diseases is also presented.

DOI: 10.4018/978-1-6684-6830-2.ch013

INTRODUCTION

Cardiovascular Disease (CVD) is a term used for diseases associated with or affecting the heart or blood vessels. It occurs due to blockages in the arteries from the build-up of cholesterol in the body (National Health Service, 2022). The four major types of cardiovascular disease are Coronary Heart Disease (CHD), Stroke, Peripheral Arterial Disease (PAD) and Aortic Disease. Coronary heart disease occurs when blood supply to the heart muscles is restricted or blocked due to the build-up of fatty substances in the major blood vessels that supply the heart with blood called coronary arteries. Stroke occurs because of inadequate supply of blood to the brain. Peripheral arterial disease occurs when the arteries that supply blood to the limbs especially the legs are blocked. Aortic disease occurs when there is problem such as weakness or bulginess with the largest blood vessel in the body called aorta (NHS Inform, 2023). Causes of cardiovascular disease include smoking, high stress, excess alcohol, high blood pressure, high blood cholesterol, obesity, and inactivity (British Heart Foundation, 2019).

Despite the fatality of these diseases, people still live unhealthy life by exposing themselves to the risks. This increases the mortality and morbidity rate every year. According to WHO (2022), approximately 18 million deaths are recorded every year all over the world due to cardiovascular disease. British Heart Foundation (BHF, 2022) showed that over half a billion people are living with cardiovascular disease globally, with a 90% increase every year. They mentioned that over 7 million people (with males being the higher population of 4 million) are suffering from cardiovascular disease in the UK. This number will continue to increase due to age and population growth.

Figure 1. Formation of atherosclerosis and blood clot in the arteries
Source: Cardiovascular Consultants P.C. (2021)

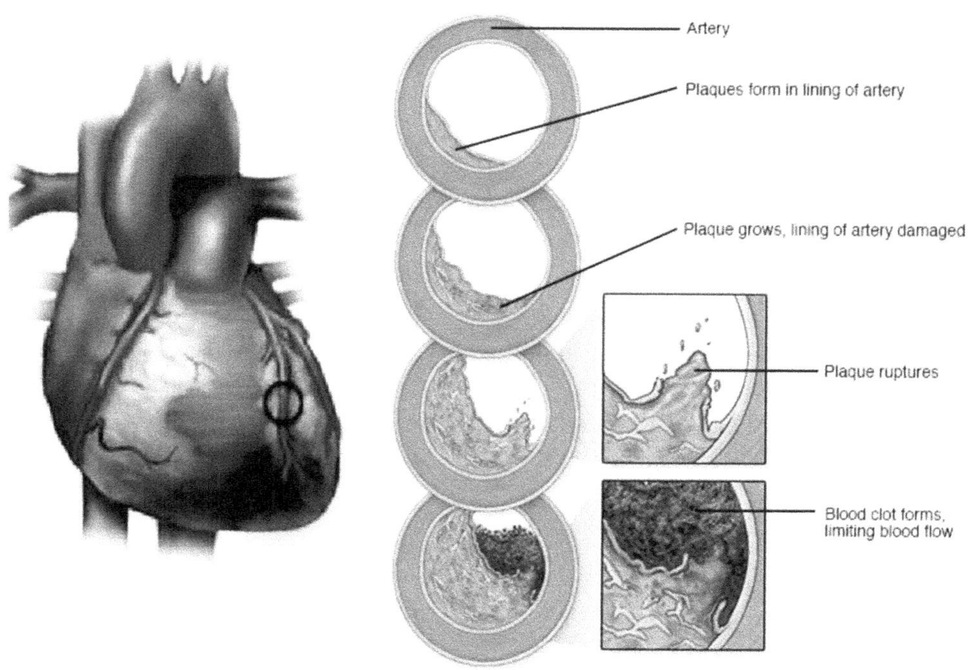

Figure 2. Formation of atherosclerosis leading to thrombosis.
Source: Fertilitypedia (2023)

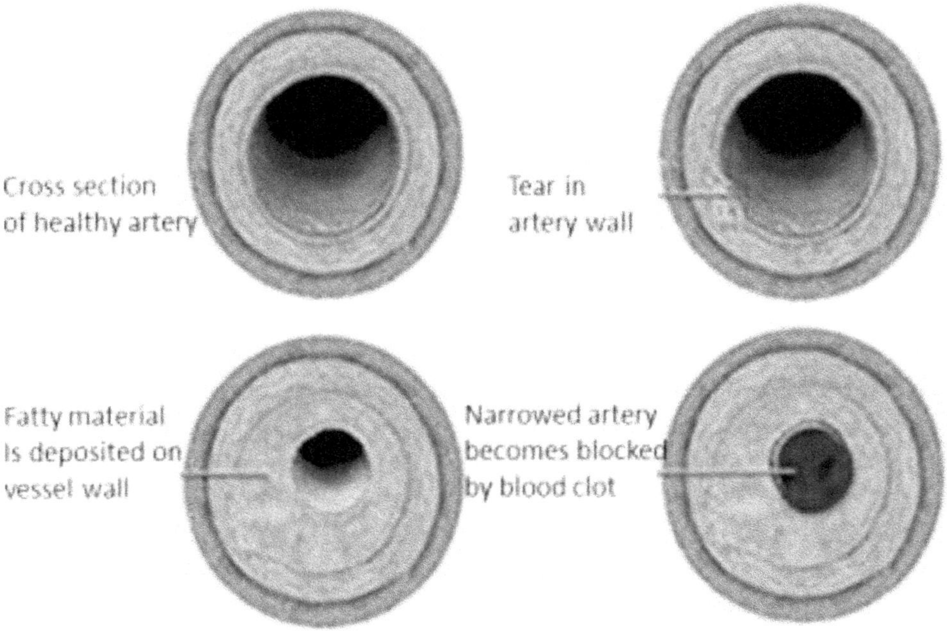

The population of people living with cardiovascular disease is double that of total number of people living with cancer and Alzheimer disease put together. They mentioned age and lifestyle change as some of the factors causing the increase in this population. The BHF fact sheet also stated that someone gets admitted every 5 minutes in UK hospitals and someone dies every 3 minutes in the United Kingdom due to cardiovascular diseases. This is equivalent to 160,000 deaths per year and 460 deaths per day. According to Centre for Disease Control and Prevention (Centers for Disease Control and Prevention, 2022), cardiovascular disease is the leading cause of death in the United States and have a record of one person per half a minute death from cardiovascular disease. Approximately 700,000 people are living with heart attack in the United State which is equivalent to one out of every five deaths recorded. Coronary artery disease is the most common of all the types of cardiovascular disease in the United States with over 20 million people aged 20 years and above and which has claimed 382, 820 lives only in the year 2020. Of these deaths recorded, 20% of them were adults less than 65 years of age. The CDC facts sheets also revealed that 20% of the 805,000 heart attacks are silent where damage is done before the person is aware of it. The United States spends over $200 billion on cardiovascular disease each year (from 2017) due to cost of health care services, medicines and lost productivity due to death. WHO gives support to all nations over the world to prevent, manage and monitor cardiovascular disease. It has provided a package called HEARTS technical package which will help ministries of health in all countries to manage cardiovascular disease health care setting (WHO, 2020).

Despite the efforts by government and organizations, the rate of cardiovascular disease increases yearly and there is need for adequate intervention to reduce its risks. Analysing the data on cardiovascular disease in order to give account of its most important risk factors can help reduce the prevalence.

Risk factors are those characteristics or behaviour that increases that chances of someone developing a disease. In this context, risk factors are health conditions being suffered from, lifestyle being practised, age and family history that increases the risk of developing cardiovascular disease (CDC, 2019). There are two categories of the risk factors of cardiovascular disease. Modifiable and non-modifiable risk factors (University of California San Francisco Health, 2022). Modifiable risk factors are those ones that can be controlled by mere lifestyle changes (e.g diabetes, smoking, obesity, inactivity, high blood pressure, alcohol, nutrition, high cholesterol level, depression) while non-modifiable are genetic and cannot be controlled (e.g age, gender, ethnicity, and family history). Risk factors of cardiovascular disease as used in this study are age, sex, chest pain, resting blood pressure, cholesterol level, fasting blood sugar, maximum heart rate achieved, resting electrocardiographic results, Exercise induced angina, Old peak, Slope of peak exercise ST segment, number of vessels coloured by fluoroscopy and thalassemia. High blood pressure is the force at which the heart pumps the blood flowing through the blood vessels is consistently too high (American Heart Association, 2023). The normal blood pressure is one measured between 90-120mmHg/60-80mmHg (NHS, 2019). The numerator measurement represents systolic pressure which is the force the heart beats to pumps blood through the body and the denominator represent the diastolic pressure which is the pressure that occur in the arteries when the heart rests between beats (CDC, 2021). Any measurement above 120mmHg for systolic and 80mmHg for diastolic is high blood pressure. Diabetes due to high of glucose/sugar in the blood (NIDDK, 2016). Fasting blood sugar level of ≤99mg/dL (or 5.5 mmol/L) is said to be normal; 100-125mg/dL (or 5.6-6.9mmol/L) is at prediabetes stage while any measurement higher than 125mg/dL (7mmol/L) is said to be at diabetes stage (Mayo clinic, 2023). Cholesterol is a fatty substance in the blood that helps in producing hormones, cell membranes and vitamin D in the body (Martin, 2023) but it becomes of concern once it's high in the blood. A healthy total cholesterol level is lower than 200mg/dL (5.17mmol/L); 200-239mg/dL (or 5.17–6.18mmol/L) is borderline high level and that which greater or equal to 240mg/dL (or 6.21mmol/L) is high.

Obesity which is also known as overweight is another risk factor of cardiovascular disease and it is the accumulation of calories from fatty and sugary foods and drinks in the body (NHS, 2023). Obesity is measured using Body Mass Index (BMI). A BMI measurement of above 25 is said to be overweight (NHS, 2023). Chest pain is also referred to as Angina. Angina is caused by reduced blood flow to the heart muscles. Although it is not life threatening but it gives sign that one is at risk of cardiovascular disease (BHF, 2022). Anginas include typical angina, atypical angina, nonanginal and asymptomatic. Typical angina is a squeezing, tightening heaviness in the centre of the chest just behind the breast bones and it can spread to the neck, arms, back and jaw. It is caused by exertion or emotional stress, and it gets relieved within few minutes of resting or taking medication (UMass Memorial Health, 2015). Atypical angina is the type that does not meet the criteria of typical angina. It is a burning, stabbing or indigestion feeling of pain in the back (DeVon et al, 2020) or any of the mentioned location and neither begin by exertion nor get relieved by rest within few minutes (Marateb & Goudarzi, 2015). Nonanginal chest pain is the one that occur feeling of pain is over 30 minutes or less than 5seconds. It may begin by just lifting of arms or bending down and it gets relieved within moments of lying down (Constant, 1990). Asymptomatic angina is a silent pain and can be observed on Electrocardiogram (ECG).

WHO (2022) has mentioned that cardiovascular disease is on the increase due to inadequate preventive measures even though they are preventable. Age has been popular among research as being a determining factor for the development of cardiovascular disease. Adults ≥65 years are more at risk of developing cardiovascular disease than the younger people (National Institute on Aging, 2018) while NHS (2022) reported that it is common in people aged 50 years and above. Also, several studies have shown that these factors are associated with a high risk of developing cardiovascular disease. The most important risks factors of cardiovascular disease are high blood pressure, diabetes, high LDL cholesterol, smoking, overweight, poor diets and physical inactivity (CDC, 2022). The reduction of heart rate by drugs is helpful in the several clinical conditions in patients living with myocardial infarction or congestive heart failure (Palatini & Julius, 2004). This means that the control of high blood pressure can reduce the risk of developing cardiovascular disease.

The control and management of the risk factors will help in preventing cardiovascular disease. Living a healthy lifestyle is another way to prevent cardiovascular disease (WHO, 2021). British Heart Foundation ascertained that high blood pressure is the cause of about half of the heart attacks and strokes. High blood pressure is a major contributor to the development of cardiovascular diseases (Fuchs &Whelton, 2020). Wu et al (2015) carried out a study to find out the relationship between high blood pressure, cardiovascular disease and mortality rate to find out the optimal blood pressure for survival for adults of age 65 and above. About 77,000 male and female adults were observed for a period of four (4) years. The data collected included the age, sex, marital status, education level, smoking status, alcohol consumption, regularity of exercise. Also, the blood pressure of the participants was measured using medical equipment called oscillometric sphygmomanometer. Although the study could not establish the optimal blood pressure rate for adults, but it ascertained that blood pressure is an important risk factor for the development of cardiovascular diseases. It revealed that stage 2-3 hypertension (i.e the systolic blood pressure ≥160mmHg over diastolic blood pressure ≥100mm Hg) had greatest effect on mortality rate. It also revealed that stage 2-3 hypertension had significant effects on mortality rate amongs women but otherwise in men.

According to BHF (2022), people living with diabetes have a double risk of developing stroke. Despite several reports identifying these factors, cardiovascular disease and the number of people living with these risk factors like diabetes and high blood pressure are increasing. There are still controversies in research over the most popular age group at risk of cardiovascular disease. Hence, it is of paramount importance to act quickly to tackle this problem, and this has geared the idea behind this study.

The aim of the study is to analyse the risk factors of cardiovascular disease in order to find out the important ones that effects the development of cardiovascular disease. The objective of the study is to analyse the risk factors of cardiovascular disease to find out the important ones based on age.

METHODS

The re-processed Cleveland heart disease dataset from University of California UCI (Irvine) was pulled online with no bias. The same result will be obtained no matter the location and time of data collection. The dataset contains 297 health records. It comprised of 14 columns with 13 features and the last column indicating the presence or absence of heart disease. The dataset which originally contains 303 records was already processed, 6 rows with missing values were removed and the dataset was filtered to 137 to have only rows with presence of heart disease. The age range used are 26-35, 36-45, 46-55, 56-65, 66-75. The youngest is 29 years and the oldest is 77 years old. The researcher decides to leave out the last range which is 76-85 because it is negligible and not significant since there is only one (1) record of heart disease patient that falls in this range. Hence, the analysis in this study was based on the 136 people living with cardiovascular disease. The analysis was carried out using frequency counts and percentages; and all these was done on Microsoft excel.

RESULTS AND DISCUSSION

The following table shows the number of people with cardiovascular disease according to age range (as used in the study) and sex.

The effects of age on the number of people suffering from heart disease where investigated and the results are plotted in Figure 1.

Table 1 and Figure 3 show the analysis of age of cardiovascular disease patients. Out of the 136 cases of cardiovascular disease, 32(23.53%) of them are patients between the ages of 46-55 and 74 (54.41%) are between the ages of 56-65. The highest number and percentage of cardiovascular disease cases are those between the age of 56-65. The number of patients with cardiovascular disease aged 46-55 is also significant. The trend in the chart also shows that the risk of developing cardiovascular disease in the patients is directly proportional to their age (i.e one is at risk of developing cardiovascular disease as one grow). Findings from the study revealed that cardiovascular disease is common to the older participants

Table 1. Number of cardiovascular disease cases according to age and sex

Age Group	Female	Male	Total
26-35	0	2	2
36-45	1	12	13
46-55	3	29	32
56-65	20	54	74
66-75	1	14	15
Total	25	111	136

Figure 3. Chart of cardiovascular disease cases based on age group

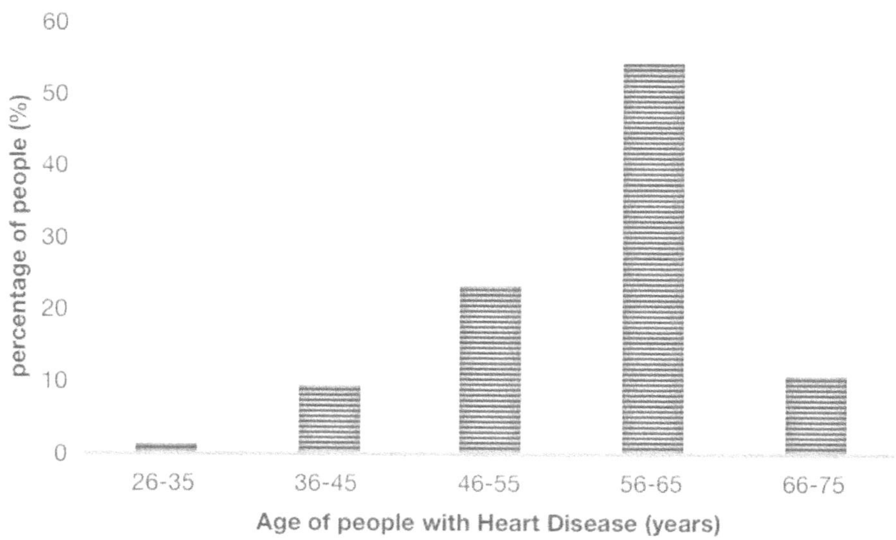

of the study. This may be because of the normal changes that occur in the heart and blood vessels as one grows older (NIA, 2018).

Therefore, participants between the age of 46-65 are the most affected by cardiovascular disease. This result disagrees with NIA (2018) who reported that adults from age 65 and above are most at risk of cardiovascular disease while it agrees with the report of NHS (2022) which says that the risk of cardiovascular disease increases in adult of 50 years and above. It also agrees with the work of Sazlina, et al (2020) who reported in their study that older people are more at risk of cardiovascular disease.

Gender has also been found to influence the number of people suffering from heart disease.

This according to Louisiana Heart and Vascular Institute (2023), is because the hearts and blood vessels of men are bigger than that of the women, and cholesterol tends to build up and form in the major arteries but build up in smaller blood vessels in women. They also mentioned that in most cases, men carry their excess fat in their bellies while women carry theirs around the abdomen and pelvises. Carrying fats around the middle body is a risk for developing heart disease (Healthdirect, 2021). High blood pressure may arise and blood flow to the heart may be impeded when anyone is going through chronic stress. People, either male or female going through depression, anxiety, stress, and even PTSD for a long period of time may develop increased heart rate and high blood pressure, high level of cortisol and the flow of blood to their heart may be reduced (CDC, 2020). These physiologic effects can results in the accummulation of calcium in the arteries, development of metabolic disease, and heart disease (Jacob et al, 2007; Bremner et al, 2018; Jurek & Neumann, 2018) .

Louisiana Heart and Vascular Institute (2023) also stated that gender differences lie in hormones, and the manner of dealing with stress. They mentioned that although both men and women experience chronic stress, women internalize their symptoms, while men externalize their own by becoming aggressive. Women are believed to be protected by estrogen and progesterone (which enhance the well-being of blood vessels) until they reach menopause (Louisiana Heart and Vascular Institute, 2023). This assertion agrees the findings from this study. The age group with the highest number of female cardiovascular disease patients is the 56-65 age group and it is coincidentally the agree group after the menopause age. The menopausal phase usually starts around age 45-55 (NIA, 2021; WHO, 2022 and NHS, 2022).

The effect of gender on development of Cardiovascular disease was investigated and the results shown in Figure 4.

Table 1 and Figure 4 show the analysis of male and female living with cardiovascular disease. There were 111 male participants and 25 female participants in the study. The table reveals that the number(**n**) and percentage(**p**) of male (**n** = 111, **p** = 81.62%) is greater that the female (**n** =25, **p**=18.38%). Out of the 136 patients, 25(18.38%) of them were females and 111(81.62%) were male patients. Out of the 25(18.38%) females, 20 (14.71%) of them are between the ages of 56-65. Also, out of the 111(81.62%) male patients 29(21.32%) are aged 46-55 and 54 (39.71%) out of the 111(81.62%) male patients are also between the ages of 56-65. These mean that more men in this study developed cardiovascular disease than the women.

The findings that men are more at risk of cardiovascular disease agrees with the work of Bots (2017). This according to WebMD (2022), men are more exposed to stress and anger that can increase blood pressure and stress hormone which in turn reduces the flow of blood to the heart. Although the rate of developing cardiovascular disease is higher in men than in women in most age groups, but there is no difference in their actual lifetime risk (Albrektsen et al, 2016).

The types of Anginas suffered is an indicator of the heart disease severity. Table 2 and Figure 5 show the types of Angina suffered by various age groups.

Figure 4. Chart showing female and male living with heart disease according to age

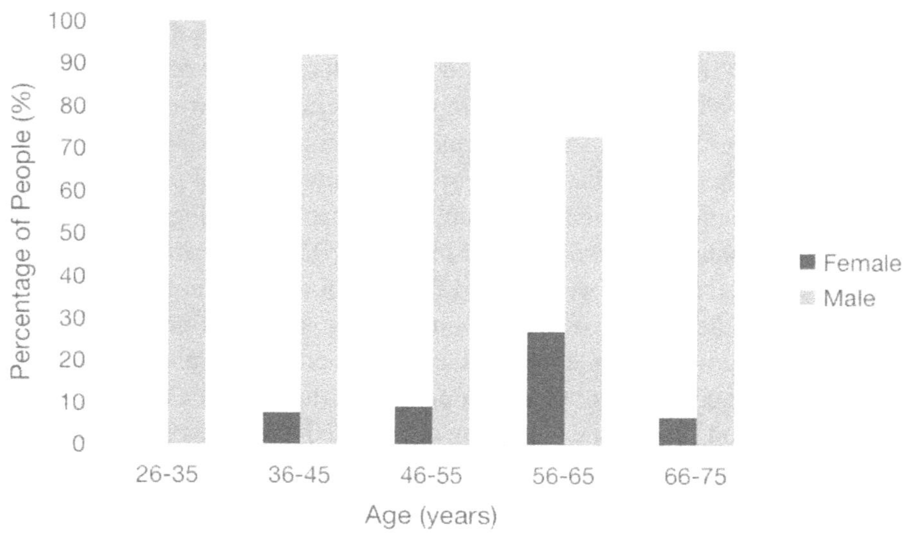

Table 2. Types of chest pain based on age group

Age	Typical Angina	Atypical Angina	Non-Angina	Asymptomatic	Total
26-35	0	0	0	2	2
36-45	2	0	0	11	13
46-55	0	2	5	25	32
56-65	5	6	9	54	74
66-75	0	1	4	10	15
Total	7	9	18	102	136

Table 2 and Figure 5 represent the analysis of chest pain based on age group. The analysis reveals that asymptomatic chest pain is the most common across age group. This type of chest pain is significant across age groups especially at age 26-35. It shows that asymptomatic chest pain contributed to the development of cardiovascular disease for the patients across all age groups.

Asymptomatic chest pain contributes mostly to development of cardiovascular disease. This may be as a result of its 'no pain' nature. Someone with asymptomatic chest pain and having cardiovascular disease will not know until the damage has been done.

High blood pressure (Fuchs, 2020; Wu, 2015) and high cholesterol level (Mortensen, 2020 and Nelson, 2013) have been found to contribute to the development of cardiovascular disease. The effects of age on the total cholesterol level were investigated and the data plotted in Figure 6.

Figure 6 represents the analysis of cholesterol level based on age of cardiovascular disease patients. The chart shows a positive correlation which simply means that cholesterol level of the participants increases as their age increases, and it is also an important risk factor of cardiovascular disease.

Figure 5. Chest pain according to age

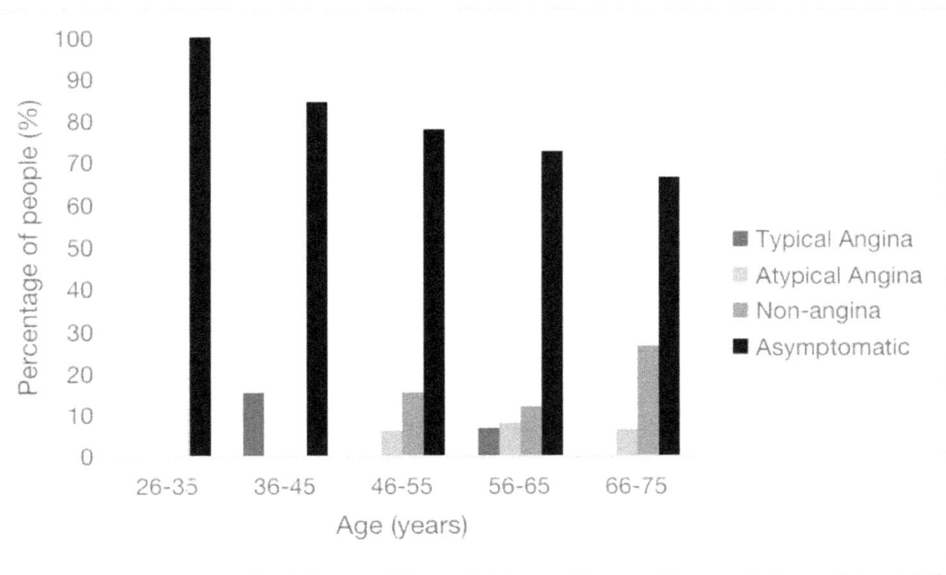

Figure 6. Analysis of cholesterol level based on age

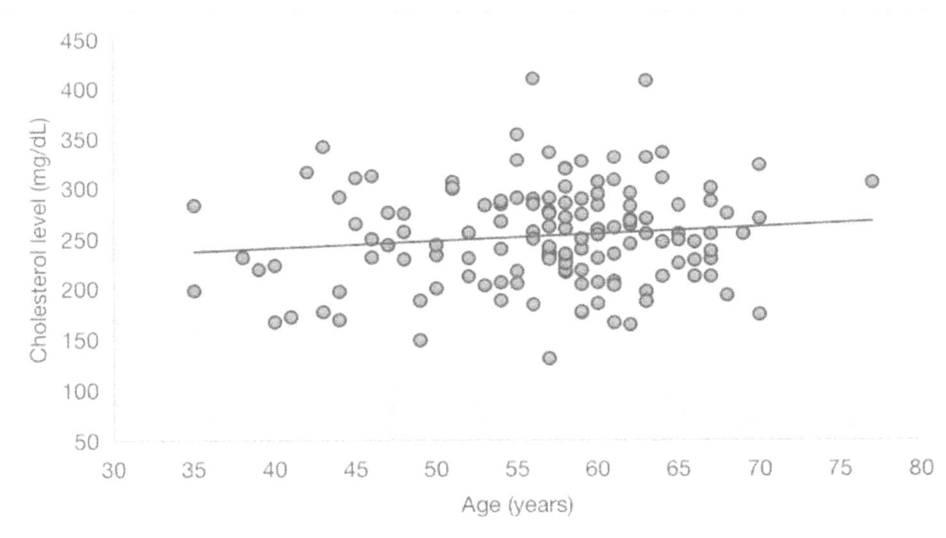

It is commonly known that blood pressure contributes to the development if cardiovascular disease. The effects of age on the blood pressure were investigated and the data plotted in Figure 7.

Figure 7 is the scatter diagram showing the correlation between age and resting blood pressure. The chart shows that there is positive correlation between the age and resting blood pressure. This simply means that high blood pressure increases as the age increases and is an important risk factor of cardiovascular disease as the participants grow with age. This is due to the high pressure at which the blood in

Figure 7. Resting blood pressure according to age

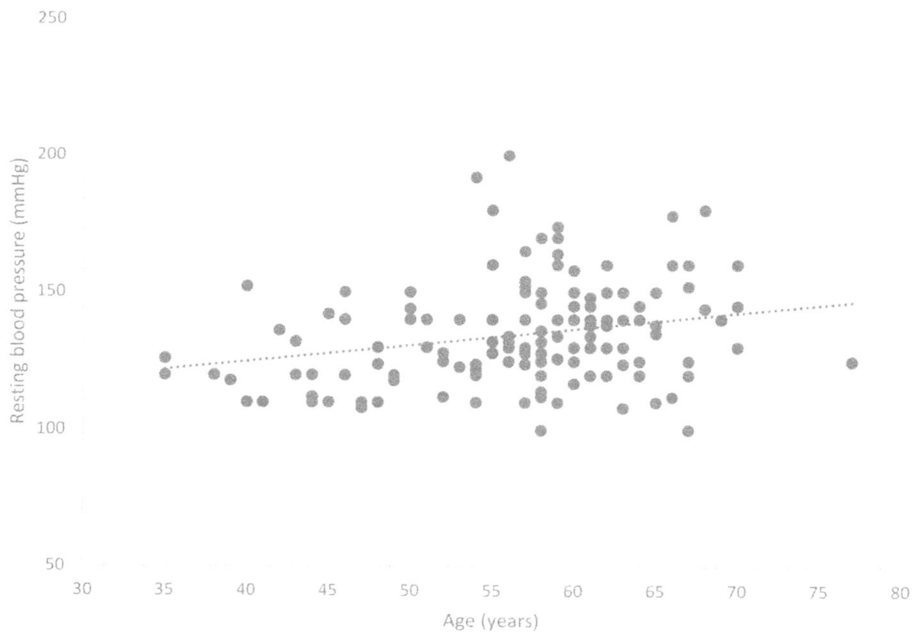

arteries and blood vessels is being pumped to the heart (CDC, 2019). The build-up of fatty substances called atherosclerosis on the walls of arteries and blood vessels reduce the flow of blood to the heart and thereby causing cardiovascular disease.

Hence, preventative measures such as controlled diet and regular exercise become more important with increasing age.

Cholesterol levels also impact the development of cardiovascular disease. The serum cholesterol, also known as total cholesterol, comprised of the three types of cholesterol. The Low-Density Lipoprotein (LDL), High-Density Lipoprotein (HDL), and Triglycerides. LDL is the bad cholesterol that builds up (plague) in the arteries and cause health problems. HDL is also referred to as good cholesterol because it eats up all bad cholesterol and transport it back to the liver. Triglycerides are the type of fats gives the body energy. High LDL and high triglycerides increase the risk developing cardiovascular disease (CDC, 2022). Normal total cholesterol in adult is less than 200mg/dL. Anyone having cholesterol level above 200mg/dL is considered at risk of cardiovascular disease. Another way of measuring cholesterol level is the ratio of total cholesterol to the HDL i.e total cholesterol divided by the HDL. The healthy cholesterol ratio should be below 5, with 3.5 as the ideal level (Healthline Editorial Team, 2021).

High blood sugar is known to be an important risk factor in heart disease. It commonly causes diabetes. The effects of age on the development of high fasting blood sugar were investigated and results shown in Table 3 and Figure 8.

Figure 8 is a chart showing the analysis of fasting blood sugar of cardiovascular disease patients based on age groups. Out of the 136 cardiovascular disease patient, 20 (14.71%) of them has high blood sugar level and 116 (85.29%) did not. This shows that diabetes does not have effect on the development of cardiovascular disease in the patients. The trend of the graph shows that although there is no

Table 3. Table showing Fasting blood sugar based on age group

Age Group	False	True	Total
26-35	2	0	3
36-45	12	1	13
46-55	30	2	32
56-65	61	13	74
66-75	11	4	15
Total	116	20	136

Figure 8. Analysis of fasting blood sugar based on age

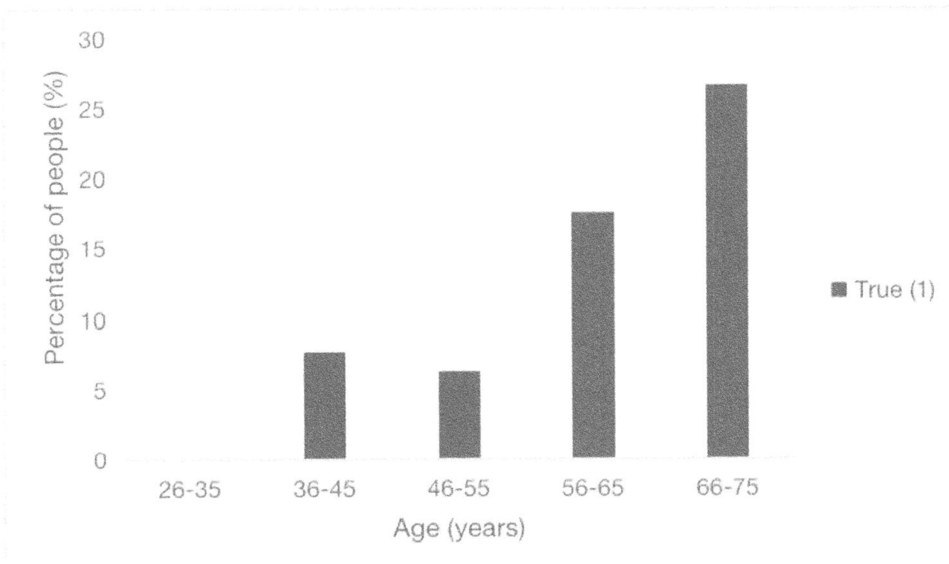

relationship between diabetes and cardiovascular disease but as one grows with age, the risk of having diabetes increases.

Maximum heart rate achieved is known to also be an important risk factor in heart disease. It is the number of heart beats per minute at maximum stress. The effects of age on the maximum heart rate achieved in cardiovascular disease patients was investigated and results shown in Figure 9.

Figure 9 reveals a negative correlation between maximum heart rate achieved and age of cardiovascular disease patients. It also shows that the number of heart beats per minute at maximum exercise stress decreases as one grows with age. The reasons behind this could be physical or psychological. Physically, as we age our muscles loses their size and their strength due to reduction in number and size of muscle fibre (Better Health Channel, 2021). This can lead to fatigue and inactivity. It is believed to be psychological because everyone develops a kind of super strength to act during stress or danger. The researcher believes that this kind of strength could also be developed and used anytime.

Figure 9. Maximum heart rate achieved based on age

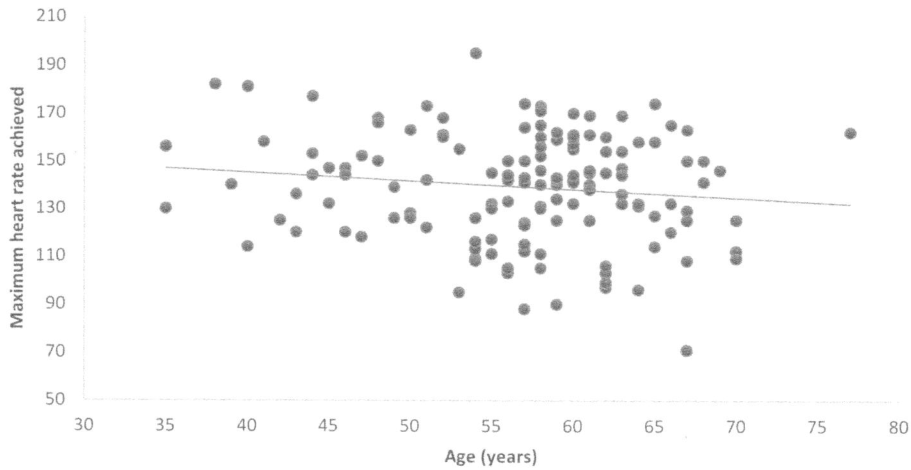

The maximum heart rate achieved decreases as age increases in this study. This shows that maximum heart rate achieved is an important factor of cardiovascular disease in the patients. The data showed that the lower the number of heart beats at maximum exercise stress of the patients, the higher their risk of developing cardiovascular disease (Das et al, 2019).

DISCUSSION

Some risk factors of crdiovascular disease in cleveland heart disease patients have been discussed and analysed in previous chapters. Meanwhile, its pertinent to also discuss some simple and affordable solutions to the care of cardiovascular disease. These factors can eventually lead to damaged tissues in heart and blood vessels like coronary and aortic valve. . Several cases of damaged tissue may require organ transplants. Although some cardiovascular disease can be controlled and treated by changing one's lifestyle like stopping smoking, eating healthily and frequently exercising the body but not all damaged tissues can be repaired. Over the years, heart transplants, coronary artery bypass graft, coronary angioplasty, aortic valve implantation and some other medical procedures have been the solution to the problems of cardiovascular disease patients. However, there is short fall of available donors that will satisfy the long list of awaiting patients of organ transplants. Complications like thrombosis and atherosclerosis also arise after the procedures of vascular graft. This has been a problem over the years, and this leads to the deaths of many patients. Balloon aortic valvuloplasty (BAV) was used to treat stenotic aortic valve through a percutaneous approach (Cribier et al, 1986) but this waned because the immediate hemodynamic and symptomatic gains were not durable and within one year, almost all the patients developed severe an dsymptomatic restenosis (Bernard et al, 1992). "According to NHLBI registry, BAV was associated with high rates of acute complications including procedural mortality up to 3%, and major hemodynamic or arrhythmic complications up to 25% and local vascular complications (20% of patients required a transfusion and 5% operative vascular repair)" (Abdelaziz & Roberts, 2020). To develop a universal cure for heart disease and reduce mortality, alternative strategies that are simple and easy procedures are required. Such cure may include the use of biomaterials in the fabrication of biodegradable and biocompatible nanofibers to support, enhance or replace damaged tissues or organs. Natural biomaterial (collagen, gelatin, silk, cellulose, chitosan e.t.c) and synthetic biomaterials (polyurethane(PU), polycaprolactone(PCL), polyvinyl acid (PVA)) have been used by tissue engineers to fabricate vascular or heart scaffolds. This is a promising area that needs serious exploits and several researchers (Huang et al, 2021; Li et al, 2022; Gossla et al, 2016; Liu et al, 2020) have given their contributions in this regard. Sequel to this study, the researcher plans to collect more datasets with larger size and ones that includes other variables like smoking, diets, alcohol intake, body mass index e.t.c Further analysis that will be a representative for all will be carried out on the datasets. The researcher also plans to carry out a study on fabrication of biodegradable and biocompatible nano fibers that mimics nature vessels or organs.

CONCLUSION

This study concludes that the participants from age 45 years and above are at higher risk of developing cardiovascular disease than the younger ones. More men in this study have developed cardiovascular disease than the women across all the age groups and they also developed cardiovascular disease at earlier age than the women.

The study also concludes that asymptomatic chest pain contributed mostly to the development of cardiovascular disease in the participants than other types of chest pain.

High blood pressure of the participants increases as their age increases and it also contributed to the development of cardiovascular disease. Cholesterol is another important risk factor of cardiovascular disease as concluded in this study. Cholesterol level of the participants increased as their age increased.

Although in this study, the rate of diabetes increased with age but diabetes does not have effects on development of cardiovascular disease in the participants.

Lastly, the number of heart beats achieved at maximum exercise stress by the participants reduced as age increased. Hence, the number of heart beats achieved at maximum exercise contributed to the development of cardiovascular disease in the participants.

ACKNOWLEDGMENT

The data set was that of Cleveland Clinic Foundation and it was collected from Janosi A., Steinbrunn, W., Pfisterer, M. &Detrano, R. Cleveland heart disease dataset. In Cherngs (2020). Heart Disease Cleveland UCI. https://www.kaggle.com/datasets/cherngs/heart-disease-cleveland-uci.

REFERENCES

Abdelaziz, H. K., & Roberts, D. H. (2020). Advances in transcatheter aortic valve implantation. In *Ahmed W. et al (2020). Advances in Medical and Surgical Engineering* (pp. 103–119). Elsevier. Academic Press. doi:10.1016/B978-0-12-819712-7.00008-5

Albrektsen, G., Heuch, I., Løchen, M. L., Thelle, D. S., Wilsgaard, T., Njølstad, I., & Bønaa, K. H. (2016). Lifelong gender gap in risk of incident myocardial infarction: The Tromsø Study. *JAMA Internal Medicine*, *176*(11), 1673–1679. doi:10.1001/jamainternmed.2016.5451 PMID:27617629

American Heart Association. (2023). *The facts about high blood pressure. High blood pressure*. Retrieved online from https://www.heart.org/en/health-topics/high-bloodpressure.

Bernard, Y., Etievent, J., & Mourand, J. L. (1986). Long-term results in percutaneous aortic valvuloplasty compared with aortic valve replacement in patients more than 75 years old. J Am Coll Cardiol 1992; 20(4): 796-801. In *Advances in Medical and Surgical Engineering* (pp. 103–119). Elsevier. Academic Press.

Better Health Channel. (2021). *Age-related changes in muscle. Ageing-Muscle Bones and Joints*. Retrieved online from Better HealthChannel. https://www.betterhealth.vic.gov.au/health/conditionsandtreatments/ageing-musclesbones-and-joints

Bots, S. H., Peters, S. A. E., & Woodward, M. (2017). Sex differences in coronary heartdisease and stroke mortality: A global assessment of the effect of ageing between 1980 and 2010. *BMJ Evidence-Based Medicine*, *2*(2), e000298. Advance online publication. doi:10.1136/bmjgh-2017-000298 PMID:28589033

British Heart Foundation. (2019). *What increases my risk of cardiovascular heartdisease? Cardiovascular heart disease*. Retrieved online from https://www.bhf.org.uk/ informationsupport/conditions/cardiovascular-heart-disease

British Heart Foundation. (2022). Retrieved online on 9th January 2022 fromhttps://www.bhf.org.uk/informationsupport/risk-factors/high-blood- pressure#Heading1

Cardiovascular Consultants, P. C. (2021). *What is atherosclerotic heart disease?* Retrieved online from https://www.cardiovascularconsultantspc.com/services- specialties/testsand-procedures/atherosclerotic-heart-disease/

Centers for Disease Control and Prevention. (2019). *Know your risk for heartdisease*. https://www.cdc.gov/heartdisease/risk_factors.htm#:~:text=Several%20health%20conditions%2C%20your%20 lifestyle,%2C%20hig%20cholesterol%2C%2 and%20smoking

Centers for Disease Control and Prevention. (2020). *Heart Disease and Mental Health disorders*. Heart Disease. retrieved online on 16th April 2020 from https://www.cdc.gov/heartdisease/mentalhealth.htm#

Centers for Disease Control and Prevention, (2022). *The nations riskfactorsand CDC's response*. Heart Disease and Stroke. National Center for Chronic Disease Prevention and Health.

Constant, J. (1990). The diagnosis of nonanginal chest pain. *Keio J Med., 39*(3), 187-92. doi:10.2302/kjm.39.187

Cribier, A., Savin, T., & Saoudi, N. (1986). Precutaneous transluminal valvuloplasty of acquired aortic stenosis in elderly patients: an alternative to valve replacement? Lancet 1986; 1(8472):63-7. In *Advances in Medical and Surgical Engineering* (pp. 103–119). Elsevier. Academic Press.

Das, R. N., Lee, Y., Devi, R. S., Sahoo, R. K., & Konai, D. (2019). Influencing factors formaximum heart rate achieved for cardiac patients. *ECCardiology Editorscolumn*, *02*, 22–26.

DeVon, H. A., Mirzaei, S., & Zegre-Hemsey, J. (2020). Typical and Atypical Symptoms of Acute Coronary Syndrome: Time to retire the Terms? *Journal of theAmerican Heart Association, 9*. https://www.ahajournals.org/doi/10.1161/JAHA.119.015539#:~:text=Atypical%20pain%20is%20frequently%20defined,heavy%2C%20tigh%2C%20or%20crushing.

Fertilitypedia. (2023). *About cardiovascular disease*. Retrieved online on 7th of March 2023 fromhttps://fertilitypedia.org/edu/symptoms/cardiovasculardisease

Fuchs, F. D., & Whelton, P. K. (2020). High blood pressure and cardiovascular disease. *Hypertension.* https://www.ahajournals.org/doi/full/10.1161/HYPERTENSIONAHA. 19.14240

Gossla, E., Tonndorf, R., Bernhardt, A., Kirsten, M., Hund, R. D., Aibibu, D., Cherif, C., & Gelinsky, M. (2016). Electrostatic flocking of chitosan fibres leads to highly porous, elastic and fully biodegradable anisotropic scaffolds. *Acta Biomaterialia, 44*, 267–276. doi:10.1016/j.actbio.2016.08.022 PMID:27544815

Healthdirect. (2021). *How to reduce visceral body fat (hidden fat).* healthdirect. Retrieved online on 17th April 2023 from https://www.healthdirect.gov.au/how-to-reduce- visceral-body-fat-hidden-fat#backToTop

Healthline Editorial Team. (2021). What's a cholesterol? Understanding the Cholesterol Ratio: What it is and Why it's Important. *Healthline.* https://www.healthline.com/health /cholesterol-ratio#:~:text=According%20to%20StatPearls%2C%20you%20should,200%20mg%2FdL%2§%20less%20

Huang, W., Huo, M., Cheng, N., & Wang, R. (2021). New forms of electrospun nanofibers applied in cardiovascular field. *Frontiers in Cardiovascular Medicine, 8.* . doi:10.3389/fcvm.2021.801077

Inform, N. H. S. (2023). *Types of cardiovascular disease.* Cardiovascular Disease. https://www.nhsinform.scot/illnesses-and-conditions/heart-and-blood- vessels/conditions/cardiovascular-disease

Jacob, L., Haro, J. M., & Koyanagi, A. (2018). Post-traumatic stress symptoms are associated with physical multimorbidity: Findings from the Adult Psychiatric Morbidity Survey 2007. *Journal of Affective Disorders, 232*, 385–392. doi:10.1016/j.jad.2018.02.063 PMID:29522958

Janosi, A., Steinbrunn, W., Pfisterer, M., & Detrano, R. (2020). *Cleveland heart disease dataset.* https://www.kaggle.com/datasets/cherngs/heart-disease-cleveland-uci

Jurek, B., & Neumann, I. D. (2018). The oxytocin receptor: From intracellular signaling to behavior. *Physiological Reviews, 98*(3), 1805–1908. doi:10.1152/physrev.00031.2017 PMID:29897293

Li, M., Wu, H., Yuan, Y., Hu, B., & Gu, N. (2022). recent fabrications and applications of cardiac patch in myorcardial infarction treatment. *VIEW, 3*(2), 20200153. doi:10.1002/VIW.20200153

Liu, H., Ahlinder, A., Yassin, M. A., Finne-Wistrand, A., & Gasser, T. C. (2020). Computational and experimental characterization of 3D-printed PCL structures toward the design of soft biological tissue scaffolds. *Materials & Design, 188*, 108488. doi:10.1016/j.matdes.2020.108488

Louisiana Heart and Vascular Institute. (2023). *Do Men Have a Higher Risk for Heart Disease?* Retrieved online on 16th April 2023 from https://www.louisianaheart.org/blog/do- men-have-a-higher-risk-for-heart-disease#:~:text=That%20said%2C%20men%20have%20larger,found%20in%20smaller%20blood%20vessels

Marateb, H. R., & Goudazi, S. (2015, March). A noninvasive method for coronary artery diagnosis using a clinically-interpretable fuzzy-based system. *Journal of Research in Medical Sciences*, *20*(3), 214–223. PMID:26109965

Martin, S. S. (2023). *Facts about cholesterol. Cholesterol in the Blood. Health.* John Hopkins Medicine. https://www.hopkinsmedicine.org/health/ conditions-and-diseases/high cholesterol/cholesterol-in-the-blood

Mayo Clinic. (2023). *Tests for type 1 and type 2 diabetes and prediabetes.Diabetes.* https://www.mayoclinic.org/diseases-conditions/diabetes/diagnosis-treatment/drc- 20371451#:~:text=A%20blood%20sugar%20level%20less,L)%20means%20you%20 have%20prediabetes

Mortensen, M. B. (2020). Elevated LDL cholesterol and increased risk of myocardial infarction and atherosclerosis cardiovascular disease in individuals aged 70-100 years: a contemporary prevention cohort. *The Lancet*. https://www.thelancet.com/article/S0140-6736(20)32233-9/fulltext

National Health Services. (2019). *What is high blood pressure?* https://www.nhs.uk/conditions/high-blood-pressure- hypertension/

National Health Services. (2022). *Cardiovascular disea*ses. NHS Inform. Retrieved online from https://www.nhsinform.scot/illnesses-and-conditions/heart-and-blood- vessels/conditions/cardiovascular-disease#about-cardiovascular-disease.

National Health Services. (2022). *Overview. Menopause*. Retrieved online on 5th May 2023 from https://www.nhs.uk/conditions/menopause/

National Health Services. (2023). *How to tell if you are living with obesity.* https://www.nhs.uk/conditions/obesity/#:~:text=Obesity%20and%20overweight%20is%20caused,and%20do%20enough%20 physical%20activity

National Institute of Diabetes and Digestive and Kidney Diseases. (2016). *What is diabetes?* Retrieved online from https://www.niddk.nih.gov/health-information/diabetes/overview/what-is-diabetes

National Institute on Aging. (2018). *The future of research on aging and the heart.* Heart Health and Aging. https://www.nia.nih.gov/health/heart-health-and-aging#:~:text=Adults%20age%2065%20and%20 older,risk%20of%20developing%20cardiovascular%20disease.

National Institute on Aging. (2021). *What is menopause?* https://www.nia.nih.gov/health/what-menopause

Nelson, R. H. (2014). *Hyperlipidemia as a risk factor for cardiovascular disease. National Centre for Biotechnology Information.* National Library of Medicine. https://www.ncbi.nlm.nih.gov/pmc/articles/PMC3572442/

Palatini, P., & Julius, S. (2004). Elevated Heart Rate: A Major Risk Factor for Cardiovascular Disease. *Clinical and Experimental Hypertension*, *26*(7-8), 7–8, 637–644. doi:10.1081/CEH-200031959 PMID:15702618

Promotion. (n.d.).https://www.cdc.gov/chronicdisease/resources/publications/factsheetsheart- diseasestroke.htm#:~:text=Leading%20risk%20factors%20for%20heart,unhealthy% 20diet%2C%20and%20 physical%20inactivity.

Sazlina, S. G., Sooryanarayana, R., Ho, B. K., Omar, M. A., Krishnapillai, A. D., & MohdTohit, N. (2020). Cardiovascular disease risk factors among older people: Data from the National Health and Morbidity Survey 2015. *PLoS ONE, 15*(10), e0240826. doi:10.1371/journal.pone.0240826

UMass Memorial Health. (2015). *Typical and Atypical Angina: What to Look For.* UMass Memorial Health. https://www.harringtonhospital.org/typical-and- atypical-angina- what-to-look-for/

University of California San Francisco Health. (2022). *Understanding your risk for heart disease.* https://www.ucsfhealth.org/education/understanding-your-risk-for-heart-disease

WebMD Editorial Contributors. (2022). *How do heart disease risk factors differ for men? Men and Heart Disease.* webMD. https://www.webmd.com/heart-disease/heart-disease- men#:~:text=Men%20are%20also%20more%20prone,raises%20risk%20for%20heart%20disease

World Health Organization. (2020). *HEARTS: technical package forcardiovascular disease management in primary health care: Risk-basedCVD management.* World Health Organisation Publications. https://www.who.int/publications/i/item/9789240001367

World Health Organization. (2021). *What are the risk factors of cardiovasculardisease?* Cardiovascular Disease. https://www.who.int/news- room/factsheets/detail/cardiovascular-diseases-

World Health Organization. (2022). *Key facts. Menopause.* Retrieved online on 5th May 2023 from https://www.who.int/news-room/fact-sheets/detail/menopause

Wu C., Hu H., Chou Y., Huang N., Chou Y., Li C. (2015). *High blood pressure and All-Cause and cardiovascular Disease Mortalities in Community-Dwelling Older Adults.* . doi:10.1097/MD.0000000000002160

KEY TERMS AND DEFINITIONS

Aortic Disease: Is a type of cardiovascular disease caused by the malfunctioning of the main artery in the body called aorta.

Cardiovascular Disease: Is the nomenclature used for any disease of the heart and blood vessels.

Coronary Artery Disease: Is a type of cardiovascular disease caused by accumulation of fats in the artery that supplies blood to the heart (coronary artery). It occurs when the flow of blood to the heart is impeded.

Fasting Blood Sugar: It is the testing of amount of sugar in the blood before eating or drinking in the morning. This is done after fasting overnight and before taking anything in the morning. It is measured in mg/dL or mmol/L.

Maximum Heart Rate Achieved: It is the highest number of heart beats per minutes an individual has at maximum stress or maximum exercise.

Peripheral Arterial Disease: This is a type of cardiovascular disease that occur when the vessels that carries blood from the heart to the limbs (especially the legs) is narrowed by build-up of fatty deposits.

Resting Blood Pressure: This is the measurement of the resting period of the heart between one heartbeat to the next. It is also known as diastolic blood pressure, and it is measured in mmHg.

Risk Factors: These are characteristics that increase the possibility of developing cardiovascular disease. These characteristics may be physical, behavioural, psychological or genetical.

Serum Cholesterol Level: This is also known as total cholesterol. It is the amount of total cholesterol in the body. Serum cholesterol contains the total amount of Low-density lipoprotein, high-density lipoprotein, and triglycerides in the body. It is measured in mg/dL or mmol/L.

Stroke: This is a type of cardiovascular disease that occur when the blood vessels that carry oxygenated blood to brain gets ruptured or blocked. This usually happens as result of fatty deposits in the vessels.

Compilation of References

Ababkov, N., Smirnov, A., Danilov, V., Popova, N., & Nikonenko, E. (2021). Structural-phase state, mechanical properties, acoustic and magnetic characteristics in the sustainable deformation localization zones of power equipment made of structural and heat resistant steels. *Metals, 11*(10), 1638.

Abdelaziz, H. K., & Roberts, D. H. (2020). Advances in transcatheter aortic valve implantation. In *Ahmed W. et al (2020). Advances in Medical and Surgical Engineering* (pp. 103–119). Elsevier. Academic Press. doi:10.1016/B978-0-12-819712-7.00008-5

AbdelHamid, A. A., Mendoza-Garcia, A., & Ying, J. Y. (2022). Advances in and Prospects of Nanomaterials' Morphological Control for Lithium Rechargeable Batteries. *Nano Energy*, *93*, 106860. doi:10.1016/j.nanoen.2021.106860

Abe, R., Higashi, M., Sayama, K., Abe, Y., & Sugihara, H. (2006). Photocatalytic activity of R_3MO_7 and $R_2Ti_2O_7$ (R = Y, Gd, La; M = Nb, Ta) for water splitting into H_2 and O_2. *The Journal of Physical Chemistry B*, *110*(5), 2219–2226. doi:10.1021/jp0552933 PMID:16471807

Abuali Galedari, S., Mousavi Khoei, S.M. (2013). Effect of pulse frequency on microstructure and surface properties of Ck45 steel treated by plasma electrolysis method. *Journal of Alloys and Compounds, 551*, 415-421.

Abuali, G. S., & Mousavi, S. M. (2013). Effect of pulse frequency on microstructure and surface properties of Ck45 steel treated by plasma electrolysis method. *Journal of Alloys and Compounds*, *551*, 415–421. doi:10.1016/j.jallcom.2012.10.176

Agrawal, P., & Agrawal, R. (2019). Glass Fiber Reinforced Polymer Composites: Manufacturing Methods, Microstructural Aspects, and Mechanical Behavior. In Reference Module in Materials Science and Materials Engineering (pp. 1-11). Elsevier. doi:10.1016/B978-0-12-803581-8.11443-4

Ahmad, M. Z., Chang, J., Ahmad, M. S., Waclawik, E. R., & Wlodarski, W. (2013). Non-aqueous synthesis of hexagonal ZnO nanopyramids: Gas sensing properties. *Sensors and Actuators. B, Chemical*, *177*, 286–294. doi:10.1016/j.snb.2012.11.013

Ahmad, M., Ahmed, E., Zhang, Y., Khalid, N. R., Xu, J., Ullah, M., & Hong, Z. (2013). Preparation of highly efficient Al-doped ZnO photocatalyst by combustion synthesis. *Current Applied Physics*, *13*(4), 697–704. doi:10.1016/j.cap.2012.11.008

Ahmed, T., Zounemat-Kermani, M., & Scholz, M. (2020). Climate change, water quality and water-related challenges: A review with focus on Pakistan. In International Journal of Environmental Research and Public Health (Vol. 17, Issue 22, pp. 1–22). MDPI AG. https://doi.org/ doi:10.3390/ijerph17228518

Ahmed, J., Abhijeet, T., & Goyal, A. (2021). *Industrial Wastewater and Its Toxic Effects. Biological Treatment of Industrial Wastewater*, 1-14.

Ahmed, S. A. (2017). Structural, optical, and magnetic properties of Mn-doped ZnO samples. *Results in Physics*, *7*, 604–610. doi:10.1016/j.rinp.2017.01.018

Akhtar, N., Singh, V., Yusuf, M., & Khan, R. A. (2020). Non-invasive drug delivery technology: development and current status of transdermal drug delivery devices, techniques and biomedical applications. *Biomedical Engineering/ Biomedizinische Technik, 65*(3), 243-272. doi:10.1515/bmt-2019-0019

Alberts, B., Johnson, A., Lewis, J., Raff, M., Roberts, K., & Walter, P. (2014). *Molecular Biology of the Cell* (6th ed.). Garland Science.

Albrektsen, G., Heuch, I., Løchen, M. L., Thelle, D. S., Wilsgaard, T., Njølstad, I., & Bønaa, K. H. (2016). Lifelong gender gap in risk of incident myocardial infarction: The Tromsø Study. *JAMA Internal Medicine*, *176*(11), 1673–1679. doi:10.1001/jamainternmed.2016.5451 PMID:27617629

Algatti, M. A., Santos, C. N., Mota, R. P., De Campos, E., Lucena, E. F., Machida, M., & Melo, F. C. (2017). Characterization of SiC and Al2O3 ceramics exposed to nitrogen ions from inverse-Z pinch plasma discharge. *Proceedings of the 16th Latin American Workshop on Plasma Physics (LAWPP),* 5–11. 10.1109/LAWPP.2017.8692198

Al-Hamdi, A. M., Rinner, U., & Sillanpää, M. (2017). Tin dioxide as a photocatalyst for water treatment: A review. In Process Safety and Environmental Protection (Vol. 107, pp. 190–205). Institution of Chemical Engineers. doi:10.1016/j.psep.2017.01.022

Alharbi, F., Nathanson, D., Morgano, S. M., & Baba, N. Z. (2014). Fracture resistance and failure mode of fatigued endodontically treated teeth restored with fiber-reinforced resin posts and metallic posts in vitro. *Dental Traumatology*, *30*(4), 317–325. doi:10.1111/edt.12095 PMID:25039331

Ali, A. M. (2020). Application of Polymers in Automotive Industry. *Polymer Science: Advanced Technologies*, *1*(1), 13–24.

Ali, M., Amrane, N., & Tit, N. (2020). Relevance of defects in ZnO nanotubes for selective adsorption of H_2S and CO_2 gas molecules: Ab-initio investigation. *Results in Physics*, *16*, 102907. Advance online publication. doi:10.1016/j.rinp.2019.102907

Allen, N. S., Mahdjoub, N., Vishnyakov, V., Kelly, P. J., & Kriek, R. J. (2018). The effect of crystalline phase (anatase, brookite and rutile) and size on the photocatalytic activity of calcined polymorphic titanium dioxide (TiO_2). *Polymer Degradation & Stability*, *150*, 31–36. doi:10.1016/j.polymdegradstab.2018.02.008

Al-Othman, A. (2013). Polymer matrix materials. In *Materials for Transportation Technology* (pp. 309–328). Elsevier.

Álvarez, P. M., Jaramillo, J., López-Piñero, F., & Plucinski, P. K. (2010). Preparation and characterization of magnetic TiO_2 nanoparticles and their utilization for the degradation of emerging pollutants in water. *Applied Catalysis B: Environmental*, *100*(1–2), 338–345. doi:10.1016/j.apcatb.2010.08.010

American Heart Association. (2023). *The facts about high blood pressure. High blood pressure*. Retrieved online from https://www.heart.org/en/health-topics/high-bloodpressure.

Anel'nik, D.Ye. Polevoy, S.N., & Yevdokimov, V.D. (1981). Remont, vosstanovleniye i ispytaniye instrumenta i tekhnologicheskoy proverki: spravochnik. *Tekhnika*, 200. Kiyev.

Anishik, V. M., Ponaryadov, V. V., & Uglov, V. V. (2011). *Difraktsionnyy analiz*.

Arshad, M., Ehtisham-Ul-Haque, S., Bilal, M., Ahmad, N., Ahmad, A., Abbas, M., Nisar, J., Khan, M. I., Nazir, A., Ghaffar, A., & Iqbal, M. (2020). Synthesis and characterization of Zn doped WO_3 nanoparticles: Photocatalytic, antifungal and antibacterial activities evaluation. *Materials Research Express*, *7*(1), 015407. Advance online publication. doi:10.1088/2053-1591/ab6380

Arshinger, I. (1982). Instrumental'nyye stali i ikh termicheskaya obrabotka: spravochnik. *Metallurgiya, 313.*

Artinger I. (1982). *Instrumental'nye stali i ih termicheskaja obrabotka*, 312.

Artinger, I. (1982). Instrumental'nyye stali i ikh khimiko-termicheskaya obrabotka. *Spravochnik*, 312.

Arun, S., & Kanagaraj, S. (2016). Performance enhancement of epoxy based sandwich composites using multiwalled carbon nanotubes for the application of sockets in trans-femoral amputees. *Journal of the Mechanical Behavior of Biomedical Materials*, *59*, 1–10. doi:10.1016/j.jmbbm.2015.12.013 PMID:26736176

Arzamasov, B.N., Bratukhin, A.G., Yeliseyev, YU.S., & Panayoti, T.A. (1999). Ionnaya khimiko-termicheskaya obrabotka splavov. *MGTU im. Baumana*, 400.

Arzamasova, B.N., Mukhina, G.G. (2001). Materialovedeniye: uchebnik dlya vuzov. N.E. Baumana, (ed.) *Izd-vo MGTU im.* 648.

Ashcroft, W. R. (1993). Curing agents for epoxy resins. In B. Ellis (Ed.), *Chemistry and technology of epoxy resins* (pp. 37–71). Springer. doi:10.1007/978-94-011-2932-9_2

Astaf'yev A.A. (1997). Nekotoryye zakonomernosti vodorodnogo okhrupchivaniya konstruktsionnykh staley. *Metallovedeniye i termicheskaya obrabotka metallov*, (2), 5 -8.

ASTM G133–95. (1995). *Standard Test Method for Linearly Reciprocating Ball-on-Flat Sliding Wear.* ASTM International.

ASTM G99–05. (2005). *Standard Test Method for Wear Testing with a Pin-on-Disk Apparatus.* ASTM International.

Atay, H. Y., Dogan, L. E., & Celik, E. (2013). Investigations of self-healing property of chitosan-reinforced epoxy dye composite coatings. *Journal of Materials*, *2013*, 1–7. doi:10.1155/2013/613717

Auffan, M., Rose, J., Proux, O., Borschneck, D., Masion, A., Chaurand, P., Hazemann, J.-L., Chaneac, C., Jolivet, J.-P., Wiesner, M. R., van Geen, A., Bottero, J.-Y., Doherty, L., & Observatory, E. (2008). Enhanced adsorption of arsenic onto maghemites nanoparticles: As (III) as a probe of the surface structure and heterogeneity. *Langmuir*, *24*(7), 3215–3222. doi:10.1021/la702998x PMID:18266393

Azarenkov, N. A., Sobol, O. V., & Pogrebnjak, A. D. (2012). Materyalovedenye neravnovesnogo sostoyania modificirovannoy poverhnosti. Sumy, Sumy State University.

Babrauskas, V., & Peacock, R. D. (2016). Heat Release Rate: The Single Most Important Variable in Fire Hazard. *Fire Technology*, *52*(4), 851–872. doi:10.100710694-015-0512-1

Badisch, E., & Mitterer, C. (2003). Abrasive wear of high speed steels: Influence of abrasive particles and primary carbides on wear resistance. *Tribology International*, *36*(10), 765–770. doi:10.1016/S0301-679X(03)00058-6

Bae, Y. H., & Park, K. (2020). Advanced drug delivery 2020 and beyond: Perspectives on the future. *Advanced Drug Delivery Reviews*, *158*, 4–16. doi:10.1016/j.addr.2020.06.018 PMID:32592727

Bagdasaryan A.A., Smirnova E., Konarski P., Misnik M. & Zawada A. (2014). The analysis of elemental composition and depth profiles of nitride nanostructures coating based on the TiHfVNbZr high-entropy alloy. *JNEP*, (6), 1–5.

Bagheri, Z. S., El Sawi, I., Bougherara, H., & Zdero, R. (2014). Biomechanical fatigue analysis of an advanced new carbon fiber/flax/epoxy plate for bone fracture repair using conventional fatigue tests and thermography. *Journal of the Mechanical Behavior of Biomedical Materials*, *35*, 27–38. doi:10.1016/j.jmbbm.2014.03.008 PMID:24727574

Bagheri, Z. S., El Sawi, I., Schemitsch, E. H., Zdero, R., & Bougherara, H. (2013). Biomechanical properties of an advanced new carbon/flax/epoxy composite material for bone plate applications. *Journal of the Mechanical Behavior of Biomedical Materials*, *20*, 398–406. doi:10.1016/j.jmbbm.2012.12.013 PMID:23499250

Bagheri, Z. S., Giles, E., El Sawi, I., Amleh, A., Schemitsch, E. H., Zdero, R., & Bougherara, H. (2015). Osteogenesis and cytotoxicity of a new carbon fiber/flax/epoxy composite material for bone fracture plate applications. *Materials Science and Engineering C*, *46*, 435–442. doi:10.1016/j.msec.2014.10.042 PMID:25492008

Baig, A., Baig, A., Rathinam, V., & Ramya, V. (n.d.). *Facile fabrication of Zn-doped SnO_2 nanoparticles for enhanced photocatalytic dye degradation performance under visible light exposure*. https://doi.org/ doi:10.1007/s42114-020-00195-9

Baig, N., Kammakakam, I., Falath, W., & Kammakakam, I. (2021). Nanomaterials: A review of synthesis methods, properties, recent progress, and challenges. *Materials Advances*, *2*(6), 1821–1871. doi:10.1039/D0MA00807A

Bakovets, V. V., Polyakov, O. V., & Dolgovesova, I. P. (1991). Plazmenno-elektroliticheskaya anodnaya obrabotka metallov. Novosibirsk, Nauka. Sib. otd.

Balanovskii, A.E. (2016). Visualization of the process of metal heating and melting in the anode zone in an arc discharge with a tungsten electrode. *High Temperature, 54*, 627-631.

Balanovskij A.E. (2016). Osnovnye voprosy teorii plazmennogo poverhnostnogo uprochneniya metallov (Obzor. Chast 2). *Uprochnyayushie tehnologii i pokrytiya, 1*, 25-34.

Balanovskij, A.E. & Pletnikov, I.A. (1999). Kompleksnaya ocenka kachestva tehnologii plazmennogo poverhnostnogo uprochneniya bandazhej lokomotivov. *Svarka i diagnostika, 3,* 45-50.

Balanovskij, A.E. (2006). Plazmennoe poverhnostnoe uprochnenie metallov. *Irkutsk: Izdatelstvo IrGTU*, 180.

Balanovskij, A.E. (2006). Rezultaty vnedreniya tehnologii plazmennogo uprochneniya na VSZhD. *Zheleznodorozhnyj transport, 4,* 28-32.

Balanovskij, A.E. (2015). Ocenka zerna austenita pri plazmennom poverhnostnom uprochnenii sredneuglerodistyh stalej. *Uprochnyayushie tehnologii i pokrytiya, 6*, 27-32.

Balanovskij, A.E. (2015). Osnovnye voprosy teorii plazmennogo poverhnostnogo uprochneniya metallov (Obzor. Chast 1). *Uprochnyayushie tehnologii i pokrytiya, 12,* 18-30.

Balanovskij, A.E. (2016). Osnovnye voprosy teorii plazmennogo poverhnostnogo uprochneniya metallov (Obzor. Chast 3). *Uprochnyayushie tehnologii i pokrytiya, 2,* 20-30.

Balanovskij, A.E. (2016). Vizualizaciya processa nagreva i plavleniya metalla v anodnoj oblasti pri dugovom razryade s neplavyashimsya elektrodom. *Teplofizika vysokih temperature, 54*, 1-8.

Balanovskij, A.E. (2016). Vozmozhnosti cifrovoj vizualizacii processa nagreva i plavleniya metalla pri dugovom razryade s neplavyashimsya elektrodom. *Svarochnoe proizvodstvo, 6,* 31-39.

Balanovskij, A.E., & Grechneva, M.V., Vu Van, G. (2015). Issledovanie struktury relsovoj stali posle plazmennogo poverhnostnogo uprochneniya. *Uprochnyayushie tehnologii i pokrytiya, 11*, 23-32.

Balanovskij, A.E., & Nesterenko, H.A. (1992). Plazmennoe ciklicheskoe uprochnenie stalej. *Svarochnoe proizvodstvo, 11,* 19-20.

Balanovskij, A.E., & Pletnikov, I.A. (1999). Plazmennoe poverhnostnoe uprochnenie specialnyh stalej. *Metallurgiya mashinostroeniya, 5,* 24-30.

Balanovskij, A.E., & Pletnikov, I.A. (1999). Kompleksnaya ocenka kachestva tehnologii plazmennogo poverhnostnogo uprochneniya bandazhej lokomotivov. Chast 2. *Svarka i diagnostika, 4,* 46-50.

Balazs, A. C., Emrick, T., & Russell, T. P. (2006). Polymer-based materials for environmental and energy applications. *Science*, *314*(5802), 1107–1110. doi:10.1126cience.1130557 PMID:17110567

Balta, Z., Bilgin Simsek, E., & Berek, D. (2019). Solvothermal synthesis of WO_3/TiO_2/carbon fiber composite photocatalysts for enhanced performance under sunlight illumination. *Photochemistry and Photobiology*, *95*(6), 1331–1338. doi:10.1111/php.13117 PMID:31074885

Banerjee, D. K. (1986). Water pollution and human health. *Water and Sanitation at Mid-Decade: Proceedings of the 12th WEDC Conference*, 47–50. 10.4066/2529-8046.100020

Bannier, E., Vincent, M., Rayon, E., Benavente, R., Salvador, M. D., & Sanchez, E. (2014). Effect of TiO2 addition on the microstructure and nanomechanical properties of Al2O3 suspension plasma sprayed coatings. *Applied Surface Science*, *316*, 141–146. doi:10.1016/j.apsusc.2014.07.168

Barabash R.I., Belobashskiy A.V., & Permyakov V.G. (1974). Tonkaya struktura i uprochneniye azotirovannykh volokon zheleza, legirovannogo khromom. Izvestiya vuzov. *Chernaya metallurgiya,* (10), 118–120.

Baranchikov V.I., Zharikov A.B., Yudina N.D., & Sadykhov A.I. (1990). Progressivnyye otvety i rezhimy rezaniya metallov, 400. *Mashinostroyeniye.*

Baranova L.V., & Demina E.L. (1986). Metallograficheskoye travleniye metallov i splavov: spravochnik, 256. *Metallurgiya.*

Baranova, L.V., & Demina, E.L. (1986). *Metallograficheskoe travlenie metallov i splavov*. Spravochnik.

Barua, S., Chattopadhyay, P., Aidew, L., Buragohain, A. K., & Karak, N. (2015). Infection-resistant hyperbranched epoxy nanocomposite as a scaffold for skin tissue regeneration. *Polymer International*, *64*(2), 303–311. doi:10.1002/pi.4790

Barua, S., Dutta, N., Karmakar, S., Chattopadhyay, P., Aidew, L., Buragohain, A. K., & Karak, N. (2014). Biocompatible high performance hyperbranched epoxy/clay nanocomposite as an implantable material. *Biomedical Materials (Bristol, England)*, *9*(2), 025006. doi:10.1088/1748-6041/9/2/025006 PMID:24495981

Bashir, I., Lone, F. A., Bhat, R. A., Mir, S. A., Dar, Z. A., & Dar, S. A. (2020). Concerns and threats of contamination on aquatic ecosystems. In Bioremediation and Biotechnology: Sustainable Approaches to Pollution Degradation (pp. 1–26). Springer International Publishing. doi:10.1007/978-3-030-35691-0_1

Bayatanova, L.B. (2014). *Formirovanie modificirovannyh sloev na poverhnosti nizkouglerodistoj stali metodom vozdejstviya elektrolitnoj plazmoj: dissertaciya na soiskanie uchenoj stepeni doktora filosofii.* Vostochno-Kazahstanskij gosudarstvennyj tehnicheskij universitet im. D.Serikbaeva.

Bayatanova, L.B., Sapataev, E.E., & Tleukenov E.O. (2012). Uprochnenie poverhnosti podshipnikovoj dorozhki burovogo instrumenta elektrolitno-plazmennoj obrabotkoj. *Materialy III Mezhdunarodnoj nauchno-prakticheskoj konferencii s elementami nauchnoj shkoly dlya molodyh uchenyh «Innovacionnye tehnologii i ekonomika v mashinostroenii.* YuTI TPU.

Bayatanova, L., Rakhadilov, B., Kurbanbekov, S., Skakov, M., & Popova, N. (2021). Fine structure of low-carbon steel after electrolytic plasma treatment. *Materials Testing*, *63*(9), 842–847. doi:10.1515/mt-2020-0119

Belkin P.N. (2010). Anodnaya elektrokhimiko-termicheskaya modifikatsiya metallov i splavov. *Elektronnaya obrabotka materialov, 46*(6), 29–41.

Belkin P.N., & Pasinkovskiy Ye.A. (1989). Termicheskaya i khimiko-termicheskaya obrabotka staley pri nagreve v rastvorakh elektrolitov. *Metallovedeniye i termicheskaya obrabotka metallov*, (5), 12 - 17.

Belkin P.N., Ganchar V.I., & Tovarkov A.K. (1986). Teploobmen mezhdu anodom i parogazovoy obolochkoy pri elektrolitnom nagreve. *Inzhenerno-fizicheskiy zhurnal. T. 51*(1), 154-155.

Belkin P.P. (1997). Anodnyy nagrev v vodnykh rastvorakh. *Vestnik Kostromskogo gosudarstvennogo pedagogicheskogo universiteta*, (4), 55-58.

Belkin P.P., & Ganchar V.I. (1988). Prokhozhdeniye toka cherez parogazovuyu obolochku pri elektrolitnom elektrolitnom nagreve. *Elektronnaya obrabotka materialov*, (5), 59-62.

Belkin, P.N. (2005). *Elektrohimiko-termicheskaya obrabotka metallov i splavov*. P. N. Belkin *Mir*.

Belkin, P.N. (2005). Stacionarnaya temperatura anoda, nagrevaemogo v vodnyh elektrolitah / P. N. Belkin, A. B. Belihov, *Inzhenerno-fizicheskij zhurnal*, *75*(6), 19-24.

Belotskiy A.V., & Permyakov V.G. (1972). O prirode vysokoprochnogo azotirovannogo sostoyaniya sloyev na legirovannom zheleze. *Zashchitnyye pokrytiya na metallakh*, (6), 83-86.

Belotskiy, A. V., Marchevskaya, E. I., & Permyakov, V. G. (1973). Osobennosti obrazovaniya azotistykh faz v sisteme Fe–V–N. Izv. AN SSSR. *Metally*, (3), 116–119.

Belver, C., Bedia, J., Gómez-Avilés, A., Peñas-Garzón, M., & Rodriguez, J. J. (2018). Semiconductor Photocatalysis for Water Purification. In *Nanoscale Materials in Water Purification* (pp. 581–651). Elsevier. doi:10.1016/B978-0-12-813926-4.00028-8

Belyy A.V., Karpenko G.D., & Myshkin N.K. (1991). *Struktura i metody formirovaniya iznosostoykikh sloyev,* 209. Mashinostroyeniye.

ben Chobba, M., Messaoud, M., Weththimuni, M. L., Bouaziz, J., Licchelli, M., de Leo, F., & Urzì, C. (2019). Preparation and characterization of photocatalytic Gd-doped TiO_2 nanoparticles for water treatment. *Environmental Science and Pollution Research, 26*(32), 32734–32745. doi:10.1007/s11356-019-04680-7

Benhebal, H., Chaib, M., Léonard, A., Tasseroul, L., & Paèz, C., Salmon, T., Poelman, D., Crine, M., Heinrichs, B., & Pirard, J.-P. (2013). Sol-gel preparation and characterisation of SnO_2 powders employed as catalyst for phenol photodegradation. *Scientia Iranica*, *20*(6), 1891–1898.

Benzidi, H., Lakhal, M., Garara, M., Abdellaoui, M., Benyoussef, A., El Kenz, A., & Mounkachi, O. (2019). Arsenene Monolayer as an Outstanding Anode Material for (Li/Na/Mg)-Ion Batteries: Density Functional Theory. *Physical Chemistry Chemical Physics*, *21*(36), 19951–19962. doi:10.1039/C9CP03230D PMID:31475997

Berdnikov, A.A., Filippov, M.A., & Studenok, E.S. (2007). Struktura zakalennyh uglerodistyh stalej posle plazmennogo poverhnostnogo nagreva. *Metallovedenie i termicheskaya obrabotka metallov*, *6,* 2-4.

Beresnev, V. M., Sobol, O. V., Grankin, S. S., Nemchenko, U. S., Novikov, V. Y., Bondar, O. V., Belovol, K. O., Maksakova, O. V., & Eskermesov, D. K. (2016). Physical and mechanical properties of (Ti–Zr–Nb)N coatings fabricated by vacuum-arc deposition. *Inorganic Materials: Applied Research*, *7*(7), 388–394. doi:10.1134/S2075113316030047

Bernard, Y., Etievent, J., & Mourand, J. L. (1986). Long-term results in percutaneous aortic valvuloplasty compared with aortic valve replacement in patients more than 75 years old. J Am Coll Cardiol 1992; 20(4): 796-801. In *Advances in Medical and Surgical Engineering* (pp. 103–119). Elsevier. Academic Press.

Better Health Channel. (2021). *Age-related changes in muscle. Ageing-Muscle Bones and Joints.* Retrieved online from Better HealthChannel. https://www.betterhealth.vic.gov.au/health/conditionsandtreatments/ageing-musclesbones-and-joints

Bhavya. (2019, November-December). Leather processing, its effects on environment and alternatives of Chrome tanning. *International Journal of Advanced Research in Engineering and Technology*, *10*(6), 69–79.

BizVibe. (2022). *Global Leather Industry Factsheet 2020: Top 10 Largest Leather Producing Countries, Largest Exporters & Importers.* Retrieved October 25, 2022, from https://blog.bizvibe.com/blog/top-10-largest-leather-producing-countries

Bloch, B. (1958). Interim success with the first use of epoxy resins. *The Journal of Bone and Joint Surgery. British Volume*, *40*, 804. doi:10.1302/0301-620X.40B4.804 PMID:13610994

Boeser, F., Lang, B. G., Stieglitz, T., & Plachta, D. T. (2016, August). Epoxy casting used as nonhermetic encapsulation technique for implantable electronic devices. In *38th Annual International Conference of the IEEE Engineering in Medicine and Biology Society (EMBC)* (pp. 1938-1941). IEEE. 10.1109/EMBC.2016.7591102

Boker, K. O., Richter, K., Jackle, K., Taheri, S., Grunwald, I., Borcherding, K., Byern, J. V., Hartwig, A., Wildemann, B., Schilling, A. F., & Lehmann, W. (2019). Current state of bone adhesives - Necessities and hurdles. *Materials (Basel)*, *12*(23), 3975. doi:10.3390/ma12233975 PMID:31801225

Bokov, D., Turki Jalil, A., Chupradit, S., Suksatan, W., Javed Ansari, M., Shewael, I. H., Valiev, G. H., & Kianfar, E. (2021). Nanomaterial by Sol-Gel Method: Synthesis and Application. *Advances in Materials Science and Engineering*, *5102014*, 1–21. Advance online publication. doi:10.1155/2021/5102014

Bots, S. H., Peters, S. A. E., & Woodward, M. (2017). Sex differences in coronary heartdisease and stroke mortality: A global assessment of the effect of ageing between 1980 and 2010. *BMJ Evidence-Based Medicine*, *2*(2), e000298. Advance online publication. doi 10.1136/bmjgh-2017-000298 PMID:28589033

Boxman R.L. & Zhitomirsky V.N. Vacuum-arc deposition devices. *Review Science Instruments* (77), 1 – 15.

Boyarskaya YU.S., Grabko D.Z., & Kats M.S. (1986). *Fizika protsessov mikroidentirovaniya*, 295. Kishinev, Shtiinitsa.

Bradley, J. S., Hastings, G. W., & Johnson-Nurse, C. (1980). Carbon fiber reinforced epoxy as a high strength, low modulus material for internal fixation plates. *Biomaterials*, *1*(1), 38–40. doi:10.1016/0142-9612(80)90057-5 PMID:7470551

Brant, P., & Keener, T. (2017). *Polymer Processing Principles and Design.* Carl Hanser Verlag GmbH & Co. KG.

Braun E.D., Bushe N.A., Chichinadze A.V., & Buyanovskiy I.A. (2001). Osnovy tribologii (treniye, iznos, smazka). *Uchebnik dlya tekhnicheskikh VUZov*, 2-ye izdaniye, 778. Tsentr «Nauka i tekhnika».

Braun, E. D., Bushe, N. A., & Buyanovsky, I. A. (2001). Fundamentals of tribology (friction, wear, lubrication). In A.V. Chichinadze (ed.) Textbock for technical schools, 2nd edition. Center "Science and technology", 778.

Briffa, J., Sinagra, E., & Blundell, R. (2020). Heavy metal pollution in the environment and their toxicological effects on humans. In Heliyon (Vol. 6, Issue 9). Elsevier Ltd. doi:10.1016/j.heliyon.2020.e04691

British Heart Foundation. (2019). *What increases my risk of cardiovascular heartdisease? Cardiovascular heart disease*. Retrieved online from https://www.bhf.org.uk/ informationsupport/conditions/cardiovascular-heart-disease

British Heart Foundation. (2022). Retrieved online on 9th January 2022 fromhttps://www.bhf.org.uk/informationsupport/risk-factors/high-blood- pressure#Heading1

Brocas, A. L., Llevot, A., Mantzaridis, C., Cendejas, G., Auvergne, R., Caillol, S., Carlotti, S., & Cramail, H. (2014). Epoxidized rosin acids as co-precursors for epoxy resins. *Designed Monomers and Polymers*, *17*(4), 301–310. doi:10.1080/15685551.2013.840504

Brooke. (2020). *The leather industry: problems, solutions and roadblocks to environmental viability*, Retrieved on November 10, 2022, from https://www.techstyler.fashion/leather-industry-environmental-viability/

Budilov V.V., Agzamov R.D., & Ramazanov K.N. (2004). *Sposob azotirovaniya izdeliy v tleyushchem razryade s effektom pologo katoda.* Patent for invention No. 2276201, IPC 7, C23C8/36, C23C8/80.

Budilov V.V., Agzamov R.D., & Ramazanov K.N. (2004). Sposobnost' vzryvnogo uprochneniya detaley. *Patent na izobreteniye Nº 2275433*, MPK 7, C21D1/09, C21D1/38. 3.

Budilov, V. V., Agzamov, R. D., & Ramazanov, K. N. (2007). Tekhnologiya ionnogo azotirovaniya v tleyushchem razryade s polym katodom. *MiTOM*, (7), 25–29.

Buitkenov, D., Rakhadilov, B., Erbolatuly, D., & Sagdoldina, Z. (2020). Influence of Heat Treatment on the Phase Composition and Microhardness of Coatings Based on Ti_3SiC_2/TiC. *Key Engineering Materials*, *839*, 137–143. doi:10.4028/www.scientific.net/KEM.839.137

Callister, W. D. Jr, & Rethwisch, D. G. (2014). *Materials science and engineering: an introduction*. John Wiley & Sons.

Callister, W. D. Jr, & Rethwisch, D. G. (2018). *Materials Science and Engineering: An Introduction* (10th ed.). John Wiley & Sons.

Cannata, J. M., Williams, J. A., Zhou, Q., Ritter, T. A., & Shung, K. K. (2006). Development of a 35-MHz piezo-composite ultrasound array for medical imaging. *IEEE Transactions on Ultrasonics, Ferroelectrics, and Frequency Control*, *53*(1), 224–236. doi:10.1109/TUFFC.2006.1588408 PMID:16471449

Cardiovascular Consultants, P. C. (2021). *What is atherosclerotic heart disease?* Retrieved online from https://www.cardiovascularconsultantspc.com/services- specialties/testsand-procedures/atherosclerotic-heart-disease/

Cecchini, R., Fabrizi, A., Cabibbo, M., Paternoster, C., Mavrin, B. N., Denisov, V. N., Novikova, N. N., & Haidopoulo, M. (2011). Mechanical, microstructural and oxidation properties of reactively sputtered thin Cr/N coatings on steel. *Thin Solid Films*, *519*(19), 6515–6521. doi:10.1016/j.tsf.2011.04.115

Cengiz, N. (2020). Glutathione-responsive multifunctionalizable hydrogels *via* amine-epoxy "click" chemistry. *European Polymer Journal*, *123*, 109441. doi:10.1016/j.eurpolymj.2019.109441

Centers for Disease Control and Prevention, (2022). *The nations riskfactorsand CDC's response.* Heart Disease and Stroke. National Center for Chronic Disease Prevention and Health.

Centers for Disease Control and Prevention. (2019). *Know your risk for heartdisease*. https://www.cdc.gov/heartdisease/risk_factors.htm#:~:text=Several%20health%20conditions%2C%20your%20lifestyle,%2C%20hig%20 cholesterol%2C%2 and%20smoking

Centers for Disease Control and Prevention. (2020). *Heart Disease and Mental Health disorders*. Heart Disease. retrieved online on 16th April 2020 from https://www.cdc.gov/heartdisease/mentalhealth.htm#

Chakhtouna, H., Benzeid, H., Zari, N., Qaiss, A., & Bouhfid, R. (2021). Recent progress on Ag/TiO_2 photocatalysts: Photocatalytic and bactericidal behaviors. *Environmental Science and Pollution Research International*, *28*(33), 44638–44666. doi:10.100711356-021-14996-y PMID:34212334

Chanda, M., & Roy, S. K. (2020). Enhancing Polymer Properties through Fillers and Reinforcements. In *Plastics Technology Handbook* (pp. 107–141). CRC Press.

Chandramohan, D., Bharanichandar, J., Karthikeyan, P., Vijayan, R., & Murali, B. (2014). Progress of biomaterials in the field of orthopaedics. *American Journal of Applied Sciences*, *11*(4), 623–630. doi:10.3844/ajassp.2014.623.630

Chandramohan, D., & Marimuthu, K. (2011). Natural fiber bone plates as substitutes for orthopaedic alloy plates. *Journal of Biotechnology. Bioinformatics and Bioengineering*, *1*(4), 405–413.

Chang, Y. C. (2015). Cadmium hydroxide and oxide nanoporous walls with high performance photocatalytic properties. *Journal of Alloys and Compounds*, *637*, 112–118. doi:10.1016/j.jallcom.2015.02.214

Chatterjee-Fisher, R., & Supova, A.B. (1990). Azotirovaniye i karbonitrirovaniye. 12. *Metallurgiya*.

Chaturvedi, S., & Methodologies, P. D.-C. (2019). undefined. (2019). Water purification using nanotechnology an emerging opportunities. *Iranjournals.Nlai.Ir*, *3*, 115–144. doi:10.22034/CHEMM.2018.143461.1069

Chaus, A. S. (2008). On the wear resistance of high-speed steels. *Journal of Friction and Wear*, *29*(1), 24–34. doi:10.3103/S1068366608010054

Chaus, A. S., & Hudakova, M. (2009). Wear resistance of high-speed steels and cutting performance of tool related to structural factors. In *17th International Conference on Wear of materials*. Las Vegas, USA. 10.1016/j.wear.2008.12.101

Chen, C. J., Jin, Q., Liu, G. Y., Li, D. D., Wang, J. L., & Ji, J. (2012). Reversibly light-responsive micelles constructed *via* a simple modification of hyperbranched polymers with chromophores. *Polymer*, *53*(17), 3695–3703. doi:10.1016/j.polymer.2012.06.024

Chen, D., Sivakumar, M., & Ray, A. K. (2000). Heterogeneous Photocatalysis in Environmental Remedia tion. In *Dev* (Vol. 3). Chem. Eng. Mineral Process.

Cheng, J., Ge, Y., Wang, B., Zhang, L., Hu, X., Hong, S., Liang, X., & Zhang, X. (2020). Microstructure and Tribocorrosion behavior of Al2O3/Al composite coatings: Role of Al2O3 addition. *Journal Thermal Spray Technology*, *29*(7), 1741–1751. doi:10.100711666-020-01062-1

Cheng, Y., Huang, W., Zhang, Y., Zhu, L., Liu, Y., Fan, X., & Cao, X. (2010). Preparation of TiO_2 hollow nanofibers by electrospining combined with sol-gel process. *CrystEngComm*, *12*(7), 2256–2260. doi:10.1039/b922564a

Chen, M. C., Chang, Y., Liu, C. T., Lai, W. Y., Peng, S. F., Hung, Y. W., Tsai, H. W., & Sung, H. W. (2009). The characteristics and *in vivo* suppression of neointimal formation with sirolimus-eluting polymeric stents. *Biomaterials*, *30*(1), 79–88. doi:10.1016/j.biomaterials.2008.09.006 PMID:18838155

Chen, X., Jing, X., Wang, J., Liu, J., Song, D., & Liu, L. (2013). Self-assembly of ZnO nanoparticles into hollow microspheres via a facile solvothermal route and their application as gas sensor. *CrystEngComm*, *15*(36), 7243–7249. doi:10.1039/c3ce40654g

Chen, Y., Tian, G., Ren, Z., Tian, C., Pan, K., Zhou, W., & Fu, H. (2011). Solvothermal synthesis, characterization, and formation mechanism of a single-layer anatase TiO_2 nanosheet with a porous structure. *European Journal of Inorganic Chemistry*, *5*(5), 754–760. doi:10.1002/ejic.201000999

Cherepivska, M. K., & Prihod'Ko, R. V. (2014). Sol-Gel Synthesized Semiconductor Oxides in Photocatalytic Degradation of Phenol. *ISRN Physical Chemistry*, *2014*, 1–7. Advance online publication. doi:10.1155/2014/724095

Chernyavskiy, K.S. (1977). Stereologiya v metallovedenii. 240. *Metallurgiya*.

Choi, D., Lim, S., & Han, D. (2021). Advanced Metal-Organic Frameworks for Aqueous Sodium-Ion Rechargeable Batteries. *Journal of Energy Chemistry*, *53*, 396–406. doi:10.1016/j.jechem.2020.07.024

Chudina O.V. (2003). Kombinirovannye metody poverhnostnogo uprochneniya stalej s primeneniem lazernogo nagreva. *Teoriya i tehnologiya*. M.: Izdatelstvo MADI (GTU).

Cizravi, J. C., & Subramaniam, K. (1999). Thermal and mechanical properties of epoxidized natural rubber modified epoxy matrices. *Polymer International*, *48*(9), 889–895. doi:10.1002/(SICI)1097-0126(199909)48:9<889::AID-PI227>3.0.CO;2-L

Colm, G. (n.d.). Simplified diagram of a transmission electron microscope. Drawing by Graham Colm, courtesy of Wikimedia Commons. https://www.ccber.ucsb.edu/ucsb-natural-history-collections-botanical-plantanatomy / transmission-electron-microscope.

Constant, J. (1990). The diagnosis of nonanginal chest pain. *Keio J Med.*, *39*(3), 187-92. doi:10.2302/kjm.39.187

Cribier, A., Savin, T., & Saoudi, N. (1986). Precutaneous transluminal valvuloplasty of acquired aortic stenosis in elderly patients: an alternative to valve replacement? Lancet 1986; 1(8472):63-7. In *Advances in Medical and Surgical Engineering* (pp. 103–119). Elsevier. Academic Press.

Cui, H., Guo, Y., Ma, W., & Zhou, Z. (2020). 2D Materials for Electrochemical Energy Storage: Design, Preparation and Application. *ChemSusChem*, *13*(6), 1155–1171. doi:10.1002/cssc.201903095 PMID:31872570

da Silva-Rêgo, L. L., de Almeida, L. A., & Gasparotto, J. (2022). Toxicological effects of mining hazard elements. In Energy Geoscience (Vol. 3, Issue 3, pp. 255–262). KeAi Communications Co. doi:10.1016/j.engeos.2022.03.003

Dalgado. (2020). *All you need to know about leather grades.* Retrieved October 10, 2022, from https://dalgado.de/en/the-journal/all-about-leather/all-you-need-to-know-about-leather-grades

Das, R. N., Lee, Y., Devi, R. S., Sahoo, R. K., & Konai, D. (2019). Influencing factors formaximum heart rate achieved for cardiac patients. *ECCardiology Editorscolumn*, *02*, 22–26.

Dave, B. C., & Lockwood, S. B. (2016). *Sol–Gel Method BT.* In B. Bhushan (Ed.), *Encyclopedia of Nanotechnology* (pp. 3748–3761). Springer Netherlands. doi:10.1007/978-94-017-9780-1_359

Davoudi, A., Shadmehr, E., Savoj, M., Akhavan-Khaleghi, M., Tabesh, M., Zakizade, M., & Ebadian, B. (2021). Laser treatment of fiber post and final push-out bond strength: A systematic review and meta-analysis. *Photobiomodulation, Photomedicine, and Laser Surgery*, *39*(5), 321–333. doi:10.1089/photob.2020.4809 PMID:33275497

de Matteis, V., Cascione, M., Brunetti, V., Toma, C. C., & Rinaldi, R. (2016). Toxicity assessment of anatase and rutile titanium dioxide nanoparticles: The role of degradation in different pH conditions and light exposure. *Toxicology In Vitro*, *37*, 201–210. doi:10.1016/j.tiv.2016.09.010 PMID:27622577

De, B., & Karak, N. (2013). Novel high performance tough hyperbranched epoxy by an A 2+ B 3 polycondensation reaction. *Journal of Materials Chemistry. A, Materials for Energy and Sustainability*, *1*(2), 348–353. doi:10.1039/C2TA00011C

Deng, W., Shuangjian, L., Guoliang, H., Xia, L., Xiaoqin, Z., Yulong, A., Huidi, Z., & Jianmin, C. (2017). Comparative study on wear behavior of plasma sprayed Al2O3 coatings sliding against different counterparts. *Ceramics International, 43*(9), 6976–6986. doi:10.1016/j.ceramint.2017.02.122

Deng, Y., Feng, C., Tang, L., Zeng, G., Chen, Z., & Zhang, M. (2018). Nanohybrid Photocatalysts for Heavy Metal Pollutant Control. In *Nanohybrid and Nanoporous Materials for Aquatic Pollution Control* (pp. 125–153). Elsevier. doi:10.1016/B978-0-12-814154-0.00005-0

Dette, C., Pérez-Osorio, M. A., Kley, C. S., Punke, P., Patrick, C. E., Jacobson, P., Giustino, F., Jung, S. J., & Kern, K. (2014). TiO_2 anatase with a bandgap in the visible region. *Nano Letters, 14*(11), 6533–6538. doi:10.1021/nl503131s PMID:25252265

DeVon, H. A., Mirzaei, S., & Zegre-Hemsey, J. (2020). Typical and Atypical Symptoms of Acute Coronary Syndrome: Time to retire the Terms? *Journal of theAmerican Heart Association, 9*. https://www.ahajournals.org/doi/10.1161/JAHA.119.015539#:~:text=Atypical%20pain%20is%20frequently%20defined,heavy%2C%20tigh%2C%20or%20crushing.

Dey, A., Singh, A., Das, D., & Iyer, P. K. (2015). Organic semiconductors: A new future of nanodevices and applications. In *Thin Film Structures in Energy Applications* (pp. 97–128). Springer International Publishing. doi:10.1007/978-3-319-14774-1_4

Dhakar, B., Namdeo, A., Chatterjee, S., & Sabiruddin, K. (2018). Heat treatment of plasma sprayed alumina-chromia composite coatings. *Surface Engineering, 34*(10), 737–746. doi:10.1080/02670844.2017.1316029

Dinesh, K. R., Jagadish, S. P., & Thimmanagouda, A. (2014). Characterization and analysis of wear study on sisal fibre reinforcement epoxy composite materials used as orthopaedic implant. *International Journal of Advances in Engineering and Technology, 6*(6), 2745–2757.

Dogan, E., Bhusal, A., Cecen, B., & Miri, A. K. (2020). 3D Printing metamaterials towards tissue engineering. *Applied Materials Today, 20*, 100752. doi:10.1016/j.apmt.2020.100752 PMID:32856000

Dongre, S. S., Shwetharani, R., Ravikumar, C. H., Lavanya, C., & Balakrishna, R. G. (2022). Review on 2D Arsenene and Antimonene: Emerging Materials for Energy, Electronic and Biological Applications. *Advanced Materials Interfaces, 9*(23), 2200442. doi:10.1002/admi.202200442

Doyan, A. (1816). The effect of dopant material to optical properties: Energy band gap Tin Oxide thin film. *Journal of Physics: Conference Series*, (1). Advance online publication. doi:10.1088/1742-6596/1816/1/012114

Du, D., Chen, X., Shi, C., Zhang, Z., Shi, D., Kaneko, D., Kaneko, T., & Hua, Z. (2021). Mussel-inspired epoxy bioadhesive with enhanced interfacial interactions for wound repair. *Acta Biomaterialia, 136*, 223–232. doi:10.1016/j.actbio.2021.09.054 PMID:34610475

Du, E. H., Marchie, V. V., Boerma, D. O., & Chechenin, N. C. (2002). Low-temperature extension of the Lehrer diagramand the iron-nitrogenp has ediagram. *Metallurgical and Materials Transactions*, (33), 34–39.

Dunleavy C.S., Golosnoy I.O., Curran J.A., Clyne T.W. (2009). Characterisation of discharge events during plasma electrolytic oxidation. *Surface & Coatings Technology, 203*, 3410-3419.

Duradzhi V. N., & Bryantsev I.V. (1979). O nekotorykh osobennostyakh issledovaniya magnitnogo polya na anodnyy i katodnyy protsessy. *Elektronnaya obrabotka materialov*, (5), 15 - 19.

Duradzhi V. N., & Parsadanyan A. S. (1988). *Nagrev metallov v elektrolitnoy plazme*. Kishinev, Shtiintsa.

Duradzhi V. N., Parsadanyan A. S. (1988). *Nagrev v elektrolite*. Kishinev: Shtinca.

Dyachenko, S. S. (1981). *Formation of austenite in carbon alloys. M.* Metallurgy.

Echegoyen, Y. (2020). Biodegradable Polymers. In Handbook of Bioplastics and Biocomposites Engineering Applications (pp. 51-76). CRC Press.

Elagin, V., Kuznetsova, D., Grebenik, E., Zolotov, D. A., Istranov, L., Zharikova, T., Istranova, E., Polozova, A., Reunov, D., Kurkov, A., Shekhter, A., Gafarova, E. R., Asadchikov, V., Borisov, S. M., Dmitriev, R. I., Zagaynova, E., & Timashev, P. (2020). Multiparametric optical bioimaging reveals the fate of epoxy crosslinked biomeshes in the mouse subcutaneous implantation model. *Frontiers in Bioengineering and Biotechnology*, *8*, 107. doi:10.3389/fbioe.2020.00107 PMID:32140465

Elayaraja, M., Punithavathy, I. K., Jothibas, M., Muthuvel, A., & Jeyakumar, S. J. (2020). Effect of rare-earth metal ion Ce^{3+} on the structural, optical and photocatalytic properties of CdO nanoparticles. *Nanotechnology for Environmental Engineering*, *5*(3), 25. Advance online publication. doi:10.100741204-020-00091-z

Endryus K., Dayson D., & Kioun S. (1971). Elektronogrammy i ikh interpretatsiya. *Mir*, 56.

Engel, L. (1986). *Rastrovaya electronnaya microskopia*. Metallurgy.

Erkan, A., Bakir, U., & Karakas, G. (2006). Photocatalytic microbial inactivation over Pd doped SnO_2 and TiO_2 thin films. *Journal of Photochemistry and Photobiology A Chemistry*, *184*(3), 313–321. doi:10.1016/j.jphotochem.2006.05.001

Erkinbekkyzy T.E., Zhurerova L.G, Daryn B. (2020). Influence of electrolyte-plasma hardening technological parameters on the structure and properties of banding steel 2. *Key Engineering Materials, 839*.

Erkinbekkyzy, T. E., Zhurerova, L. G., & Daryn, B. (2020). Influence of electrolyte-plasma hardening technological parameters on the structure and properties of banding steel 2. *Key Engineering Materials*, (839), 57–62.

Espitia, P. J. P., Soares, N. de F. F., Coimbra, J. S. dos R., de Andrade, N. J., Cruz, R. S., & Medeiros, E. A. A. (2012). Zinc Oxide Nanoparticles: Synthesis, Antimicrobial Activity and Food Packaging Applications. In Food and Bioprocess Technology (Vol. 5, Issue 5, pp. 1447–1464). doi:10.100711947-012-0797-6

Fakhar, A., Aabaadiaan, M., Keivani Hafshejani, M., & Langari, A. (2012). Use of reactive oligomer to improve fracture resistance of epoxy used in medical applications and GRP pipelines. *World Applied Sciences Journal*, *20*(2), 259–263. doi:10.5829/idosi.wasj.2012.20.02.2487

Fakirov, S., Bhattacharyya, D., & Bhattacharya, S. N. (2017). *Polymer composites: from nano- to macro-scale*. CRC Press.

Fan, F. R., Wang, R., Zhang, H., & Wu, W. (2021). Emerging Beyond-Graphene Elemental 2D Materials for Energy and Catalysis Applications. *Chemical Society Reviews*, *50*(19), 10983–11031. doi:10.1039/C9CS00821G PMID:34617521

Fang, W., Yang, Y., Yu, H., Dong, X., Wang, T., Wang, J., Liu, Z., Zhao, B., & Yang, M. (2016). One-step synthesis of flower-shaped WO_3 nanostructures for a high-sensitivity room-temperature NOx gas sensor. *RSC Advances*, *6*(108), 106880–106886. doi:10.1039/C6RA21322G

Fang, Y., Kong, X., Wang, D., Cui, S., & Liu, J. (2018). Role of dopant Ga in tuning the band gap of rutile TiO_2 from first principles. *Zhongguo Wuli Xuekan*, *56*(4), 1370–1377. doi:10.1016/j.cjph.2018.04.011

Fankhauser, S., Smith, S. M., Allen, M., Axelsson, K., Hale, T., Hepburn, C., Kendall, J. M., Khosla, R., Lezaun, J., Mitchell-Larson, E., Obersteiner, M., Rajamani, L., Rickaby, R., Seddon, N., & Wetzer, T. (2022). The meaning of net zero and how to get it right. *Nature Climate Change*, *12*(1), 15–21. doi:10.103841558-021-01245-w

Faudoa-Arzate, A., Arteaga-Durán, A., Saenz-Hernández, R. J., Botello-Zubiate, M. E., Realyvazquez-Guevara, P. R., & Matutes-Aquino, J. A. (2017). HRTEM microstructural characterization of β-WO_3 thin films deposited by reactive RF magnetron sputtering. *Materials (Basel)*, *10*(2), 200. Advance online publication. doi:10.3390/ma10020200 PMID:28772559

Fertilitypedia. (2023). *About cardiovascular disease*. Retrieved online on 7th of March 2023 fromhttps://fertilitypedia.org/edu/symptoms/cardiovasculardisease

Fillipov, M. A., Fillipenkov, A. A., & Plotnikov, G. N. (2009). *Iznosostoykiye stali dlya otlivok*.

Foletto, E. L., Battiston, S., Collazzo, G. C., Bassaco, M. M., & Mazutti, M. A. (2012). Degradation of leather dye using CeO_2-SnO_2 nanocomposite as photocatalyst under sunlight. *Water, Air, and Soil Pollution*, *223*(9), 5773–5779. doi:10.100711270-012-1313-3

Freedman, G. A. (2001). Esthetic post-and-core treatment. *Dental Clinics of North America*, *45*(1), 103–116. doi:10.1016/S0011-8532(22)00470-0 PMID:11210689

Fried, J. R. (2014). *Polymer Science and Technology*. Pearson.

Fu, Y., & Li, J. (2019). *Metal/semiconductor nanocomposites for photocatalysis: Fundamentals, structures, applications and properties*. Retrieved July 29, 2022, from https://www.mdpi.com/421488

Fuchs, F. D., & Whelton, P. K. (2020). High blood pressure and cardiovascular disease. *Hypertension*.https://www.ahajournals.org/doi/full/10.1161/HYPERTENSIONAHA. 19.14240

Gallyamov, A. A., Nenashev, M. V., Ibatullin, I. D., & Murzin, A. Y. (2018). Application detonation coatings to design a new metal cutting tool. *Transactions of the IMF*, *96*(6), 290–294. 10.1080/00202967.2018.1477267

Ganeyev, A. A., Gubal', A. R., Uskov, K. N., & Potapov, S. V. (2012). Analiticheskaya mass-spektrometriya s tleyushchim razryadom. Izvestiya Akademii Nauk. *Seriya Khimicheskaya*, (4), 1–17.

Gao, J., & Abruña, H. D. (2014). Key Parameters Governing the Energy Density of Rechargeable Li/S Batteries. *The Journal of Physical Chemistry Letters*, *5*(5), 882–885. doi:10.1021/jz5001819 PMID:26274082

Gao, W., Lu, J., Zhang, S., Zhang, X., Wang, Z., Qin, W., Wang, J., Zhou, W., Liu, H., & Sang, Y. (2019). Suppressing Photoinduced Charge Recombination via the Lorentz Force in a Photocatalytic System. *Advancement of Science*, *6*(18), 1901244. Advance online publication. doi:10.1002/advs.201901244 PMID:31559139

Gao, X., Lan, J., Jia, X., Cai, Q., & Yang, X. (2016). Improving interfacial adhesion with epoxy matrix using hybridized carbon nanofibers containing calcium phosphate nanoparticles for bone repairing. *Materials Science and Engineering C*, *61*, 174–179. doi:10.1016/j.msec.2015.12.033 PMID:26838838

Gavaleiro, A., & De Hosson, J. T. (2006). *Nanostructured Coating*. Springer-Verlag. doi:10.1007/978-0-387-48756-4

Geckeler, K., Wacker, R., Martini, F., Hack, A., & Aicher, W. (2003). Enhanced biocompatibility for SAOS-2 osteosarcoma cells by surface coating with hydrophobic epoxy resins. *Cellular Physiology and Biochemistry*, *13*(3), 155–164. doi:10.1159/000071866 PMID:12876386

Geller, Y.A. (1983). *Instrumental'nyye stali,* 527. Metallurgiya.

Geller, Yu. A., & Rakhshadt, A. G. (1989). *Materialovedeniye. Metody analiza, laboratornyye raboty i zadachi*. Metallurgiya.

George, R., Bahadur, N., Singh, N., Singh, R., Verma, A., & Shukla, A. K. (2016). Environmentally Benign TiO_2 Nanomaterials for Removal of Heavy Metal Ions with Interfering Ions Present in Tap Water. *Materials Today: Proceedings*, *3*(2), 162–166. doi:10.1016/j.matpr.2016.01.051

Gerasimov S.A., Kuksenova L.I., & Lapteva V.G. (2012). *Struktura I iznosostoykost' azotirovannykh staley i splavov*, 518. APAN.

Gerasimov, S.A., Kuksenova, L.I., & Lapteva, V.G. (2012). *Struktura i iznosostoykost' azotirovannykh staley i splavov*, 518.

Ghalme, S. G., Mankar, A., & Bhalerao, Y. (2016). Biomaterials in hip joint replacement. *International Journal of Materials Science and Engineering*, *4*(2), 113–125. doi:10.17706/ijmse.2016.4.2.113-125

Giamello, E., & Pacchioni, G. (2020). Applied vs fundamental research in heterogeneous photocatalysis: Problems and perspectives. An introduction to "physical principles of photocatalysis." *Journal of Physics Condensed Matter*, *32*(36), 360301. Advance online publication. doi:10.1088/1361-648X/ab926b PMID:32396884

Gladkikh, A. I., Malykhin, S. V., & Pugachev, A. T. (2006). *Difraktsionnyye metody analiza vnutrennikh napryazheniy, Teoriya i eksperiment*. NTU HPI.

Glagolev, A. A. (1941). Geometricheskie metody kolichestvennogo analiza agregatov pod mikroskopom. Lvov: Gosgeolizdat.

Gnanaprakasam, A., Sivakumar, V. M., Sivayogavalli, P. L., & Thirumarimurugan, M. (2015). Characterization of TiO_2 and ZnO nanoparticles and their applications in photocatalytic degradation of azodyes. *Ecotoxicology and Environmental Safety*, *121*, 121–125. doi:10.1016/j.ecoenv.2015.04.043 PMID:25937630

Gnyusov S.F., Khazanov I.O., & Sovetchenko B.F. (2008). Primeneniye effekta sverkhplastichnosti staley v instrumental'nom proizvodstve. *Tomsk: Izd-vo nauch.-tekhn. lit.*, 237.

Godet, M., Berthier, Y., Vincent, L., & Flamand, L. (1991). Hard coatings for tribological applications: A pluridisciplinary approach. *Surface and Coatings Technology*, *45*(1-3), 1–8. doi:10.1016/0257-8972(91)90199-7

Gol'dshmidt KH.Dzh. (1971). Splavy potrebleniya. *Mir*, 424.

Gol'dshteyn M.I., Grachev S.V., & Veksler YU.G. (1985). Spetsial'nyye stali, 408. Metallurgiya.

Golovin, Yu. I. (2008). Nanoindentirovaniye i mekhanicheskiye svoystva tverdykh tel v submikroob"yemakh, tonkikh pripoverkhnostnykh sloyakh i plenkakh. *FTT*, *50*(12), 2113–2140.

González-Borrero, P. P., Sato, F., Medina, A. N., Baesso, M. L., Bento, A. C., Baldissera, G., Persson, C., Niklasson, G. A., Granqvist, C. G., & Ferreira Da Silva, A. (2010). Optical band-gap determination of nanostructured WO3 film. *Applied Physics Letters*, *96*(6), 061909. Advance online publication. doi:10.1063/1.3313945

Gonzalez-Hernandez, R., Perez, W., & Rodriguez, M. J. (2012). Magnetic ordering and electronic properties of Pd-doped ZnO. *Physica Status Solidi. B, Basic Research*, *249*(1), 198–201. doi:10.1002/pssb.201147113

Gonzalez-Perez, A., Persson, K. M., & Samuelson, L. (2018). Semiconductor Eco-Materials for Water Treatment. In *Handbook of Ecomaterials* (pp. 1–27). Springer International Publishing. doi:10.1007/978-3-319-48281-1_170-1

Gorelik S.S., Skakov YU.A., & Rastorguyev L.N. (2002). Rentgenograficheskiy i elektronno-opticheskiy analiz. *MISIS*, 360.

Gorelik, S.S., Skakov, Y.A., & Rastorguev, L.N. (2002). *Rentgenograficheskij i elektronno-opticheskij analiz*. M.: MISIS.

Gossla, E., Tonndorf, R., Bernhardt, A., Kirsten, M., Hund, R. D., Aibibu, D., Cherif, C., & Gelinsky, M. (2016). Electrostatic flocking of chitosan fibres leads to highly porous, elastic and fully biodegradable anisotropic scaffolds. *Acta Biomaterialia*, *44*, 267–276. doi:10.1016/j.actbio.2016.08.022 PMID:27544815

GOST 9450-76. (n.d.). *Measurement of microhardness by intrusion of diamond tips* [Data set].

Gouda, T. A., Jagadish, S. P., Dinesh, K. R., Gouda, H., & Prashanth, N. (2014). Characterization and investigation of mechanical properties of hybrid natural fiber polymer composite materials used as orthopaedic implants for femur bone prosthesis. *IOSR Journal of Mechanical and Civil Engineering*, *11*(4), 40–52. doi:10.9790/1684-11424052

Gouldsteyn, Dj., N'yuberi, D., & Echlin, P. (1984). Rastrovaya elektronnaya mikroskopiya i rentgenovskiy mikroanaliz. *Mir, Moscow*, (2), 348.

GOV.UK. (2021). *UK enshrines new target in law to slash emissions by 78% by 2035*. Retrieved October 05, 2022, from https://www.gov.uk/government/news/uk-enshrines-new-target-in-law-to-slash-emissions-by-78-by-2035

GOV.UK. (2022). *Net Zero Strategy: Build Back Greener*. Retrieved October 05, 2022, from https://www.gov.uk/government/publications/net-zero-strategy

Gowda, M. S., Hemavathi, A. B., Srinivas, S., Santhosh, G., & Siddaramaiah, H. (2022). Physico-mechanical, thermal, and tribological studies of polytetrafluoroethylene-filled polyoxymethylene/silicone composites. *Journal of Thermoplastic Composite Materials*, *35*(6), 846–859. doi:10.1177/0892705720925117

Grantham research institute on climate change and the environment. (2021). *Why is 'net zero' so important in the fight against climate change?* Retrieved October 05, 2022, from https://www.lse.ac.uk/granthaminstitute/explainers/why-is-net-zero-so-important-in-the-fight-against-climate-change/

Grechneva, M. V. (1992). Plasma hardening of metals in liquid media. Welding, 7, 18–22.

Grechneva, M. V. (2017). Kratkij analiz sostoyaniya tehnologij plazmennogo poverhnostnogo uprochneniya stalej i splavov v RF. *International conference on modern researches in science and technology. Conference Proceedings*. Scientific public organization.

Grechneva, M.V. (1992). Plazmennoe uprochnenie metallov v zhidkih sredah. *Svarochnoe proizvodstvo, 7,* 8-12.

Grigor'yev, S. (2004). Kak povysit' nadezhnost' obnaruzheniya instrumenta. *TekhnoMIR.*, *21*(3), 53–57.

Grigorevich, V. K. (1976). Tverdost' i mikrotverdost' metallov. Nauka, 230.

Grigoriants, A.G., Safonov, A.N. (1987). *Methods of surface laser treatment*. Higher School Publ.

Grigoriev, S. N., Tabakov, V. P., & Volosova, M. A. (2011). *Tekhnologicheskiye metody povysheniya iznosostoykosti kontaktnykh ploshchadok rezhushchego instrumenta*.

Grishunin, V. A. (2014). *Increasing the fatigue life of rail steel by electron beam processing*. [Dissertation of Candidate of Technical Sciences. Siberian State Industrial University, Novokuznetsk].

Guenzel, R., Matz, W., Ivanov, Yu. F., & Rothtein, V. P. (2000). Pulsed electron-beam treatment of high-speed steel current tools: struchire-phase transformation and wear resistance. *1st International Congress on Radiation Physics, high current electronics, and modification of materials*, (3), 303-307.

Guenzel, R., Matz, W., Ivanov, Yu. F., & Rothtein, V. P. (2000). *Pulsedelectron-beamtreatmentofhigh-speedsteelcurrenttools: struchire-phase transformation and wear resistance*. 1st International Congress on Radiation Physics, high current electronics, and modification of materials, Tomsk, Russia.

Guizard, C. (1996). Chapter 7 Sol-gel chemistry and its application to porous membrane processing. In *Fundamentals of Inorganic Membrane Science and Technology* (Vol. 4, pp. 227–258). Elsevier. https://doi.org/https://doi.org/10.1016/S0927-5193(96)80010-6

Gulyaev, A. P. (1986). *Metallurgy. M.* Metallurgy.

Guo, Q., Chen, N., & Qu, L. (2020). Two-Dimensional Materials of Group-IVA Boosting the Development of Energy Storage and Conversion. *Carbon Energy*, *2*(1), 54–71. doi:10.1002/cey2.34

Guo, Y., Wang, K., Hong, Y., Wu, H., & Zhang, Q. (2021). Recent Progress on Pristine Two-Dimensional Metal-Organic Frameworks as Active Components in Supercapacitors. *Dalton Transactions (Cambridge, England)*, *50*(33), 11331–11346. doi:10.1039/D1DT01729B PMID:34313288

Guo, Z., Li, M., & Liu, J. (2008). Highly porous CdO nanowires: Preparation based on hydroxy- and carbonate-containing cadmium compound precursor nanowires, gas sensing and optical properties. *Nanotechnology*, *19*(24), 245611. Advance online publication. doi:10.1088/0957-4484/19/24/245611 PMID:21825823

Gupta, P., Tenhundfeld, G., Daigle, E. O., & Ryabkov, D. (2007). Electrolytic plasma technology: Science and engineering – an overview. *Surface and Coatings Technology*, *201*(25), 87–96. doi:10.1016/j.surfcoat.2006.11.023

Gupta, V. K., Fakhri, A., Tahami, S., & Agarwal, S. (2017). Zn doped CdO nanoparticles: Structural, morphological, optical, photocatalytic and anti-bacterial properties. *Journal of Colloid and Interface Science*, *504*, 164–170. doi:10.1016/j.jcis.2017.05.026 PMID:28550746

Guriev A. M., Ivanov S. G., Lygdenov B. D., Zemlyakov S. A., Vlasova O. A., Kosheleva E. A., & Guriev M.A. (2009). *MPK Sposob uprochneniya detaley iz konstruktsionnykh i instrumental'nykh staley.* Patent. 2345175 Russian Federation, N° 2007112368/02, (3), 9.

Guryev, A. M., Vlasova, O. A., & Lygdenov, B. D. (2007). Povysheniye prochnosti instrumental'nykh staley metodom termotsiklicheskogo borirovani. XVII Petersburg readings on the problems of strength, St. Petersburg, Russia.

Guryev, A. M., Kozlov, E. V., & Popova, N. A. (2001). Izmeneniye fazovogo sostava i mekhanizm formirovaniya struktury perekhodnoy zony pri termotsiklicheskom borirovanii ferritno-perlitnoy stali. *Izv. Vuzov. Fizika*, (2), 58–63.

Hadi, A. N., Mohammed, M. R., & Al-Bakri, A. F. (2022). Numerical and experimental assessment of mechanical properties of biobased epoxy reinforced by flax fibres and seashell nanoparticles for prosthetic socket. *Advances in Materials Science and Engineering*, *2022*, 1–11. doi:10.1155/2022/9024498

Hadjinikolaou, I. N., Bell, J. P., & Spangberg, L. (1987). Fast curing epoxy and episulfide resins for biomedical applications. *MRS Proceedings*, *110*, 451. 10.1557/PROC-110-451

Han, L., Zhang, J., Deng, X., Kong, X., Yang, C., Peng, L., Kou, C., Zou, K., Lv, L., Li, C., Wang, S., & Wei, G. (2021). Totally implantable venous access ports: A prospective randomized study comparing subclavian and internal jugular vein punctures in children. *Journal of Pediatric Surgery*, *56*(2), 317–323. doi:10.1016/j.jpedsurg.2020.04.021 PMID:32467037

Hao, C., Yang, B., Wen, F., Xiang, J., Li, L., Wang, W., Zeng, Z., Xu, B., Zhao, Z., Liu, Z., & Tian, Y. (2016). Flexible All-Solid-State Supercapacitors based on Liquid-Exfoliated Black-Phosphorus Nanoflakes. *Advanced Materials*, *28*(16), 3194–3201. doi:10.1002/adma.201505730 PMID:26915349

Hariharan, C. (2006). Photocatalytic degradation of organic contaminants in water by ZnO nanoparticles: Revisited. *Applied Catalysis A, General*, *304*(1–2), 55–61. doi:10.1016/j.apcata.2006.02.020

Harper, C. A. (2003). *Handbook of plastics, elastomers, and composites.* McGraw-Hill Professional.

Healthdirect. (2021). *How to reduce visceral body fat (hidden fat).* healthdirect. Retrieved online on 17th April 2023 from https://www.healthdirect.gov.au/how-to-reduce- visceral-body-fat-hidden-fat#backToTop

Healthline Editorial Team. (2021). What's a cholesterol? Understanding the Cholesterol Ratio: What it is and Why it's Important. *Healthline.* https://www.healthline.com/health /cholesterol-ratio#:~:text=According%20to%20 StatPearls%2C%20you%20should,200%20mg%2FdL%2§%20less%20

He, H. (2020). Metal oxide semiconductors and conductors. In *Solution Processed Metal Oxide Thin Films for Electronic Applications* (pp. 7–30). Elsevier. doi:10.1016/B978-0-12-814930-0.00002-5

Heller, A., Maidan, R., & Wang, D. (1993). Amperometric biosensors based on three-dimensional hydrogel-forming epoxy networks. *Sensors and Actuators. B, Chemical*, *13*(1-3), 180–183. doi:10.1016/0925-4005(93)85356-F

Hemavathi, A. B., Anupama, G., Vani, K. V., Asha, P., Vijayalakshmi, N. S., & Kumar, K. R. (2002). Physico-chemical and mechanical properties evaluation of polyolefin films and composites. *Popular Plastics and Packaging*, *47*(3), 65–68.

Hemavathi, A. B., Babu, A., Murthy, P. K., & Zarena, A. S. (2022). Sustainable nanocomposite films of improved barrier properties for food packaging applications. *Materials Today: Proceedings*, *65*, 364–366. doi:10.1016/j.matpr.2022.06.373

Hemavathi, A. B., Monisha, C. A., Akhila, P., & Bhat, P. M. (2021). Environmentally Benign Food Packaging Material based on Polylactic acid - A Brief Review. *International Research Journal of Engineering and Technology*, *8*(8), 362–368.

Hemavathi, A. B., & Siddaramaiah, H. (2018). Food Packaging: Polymers as Packaging Materials in Food Supply Chains. In M. Mishra (Ed.), *Encyclopedia of Polymer Applications.* CRC Press.

He, Z., & Zhou, J. (2013). Synthesis, characterization, and activity of tin oxide nanoparticles: Influence of solvothermal time on photocatalytic degradation of rhodamine B. *Mod. Res. Catal*, *2*(3), 13–18. doi:10.4236/mrc.2013.23A003

Hirwani, J. K., Nishimura, R., Shinmori, H., Morita, T., Sawae, Y., & Sinha, S. K. (2022). Epoxy (SU-8) polymer composites with ultra-high molecular weight polyethylene and hyaluronic acid fillers for hip prosthetic implant application. *Tribology International*, *167*, 107399. doi:10.1016/j.triboint.2021.107399

Holleck, H. (1988). Metastable coatings – Prediction of composition and structure. *Surface and Coatings Technology*, *36*(1-2), 151–159. doi:10.1016/0257-8972(88)90145-4

Holmberg, K., & Matthews, A. (1994). Coating's tribology: A concept, critical aspects and future directions. *Thin Solid Films*, *253*(1-2), 173–178. doi:10.1016/0040-6090(94)90315-8

Holt, J. K., Park, H. G., Wang, Y., Stadermann, M., Artyukhin, A. B., Grigoropoulos, C. P., Noy, A., & Bakajin, O. (2006). Fast mass transport through sub-2-nanometer carbon nanotubes. *Science*, *312*(5776), 1034–1037. doi:10.1126cience.1126298 PMID:16709781

Hones, P., Sanjine, R., & Levy, F. (1998). Sputter deposited chromium nitride based ternary compounds for hard coatings. *Thin Solid Films*, *332*(1-2), 240–246. doi:10.1016/S0040-6090(98)00992-4

Hou, N. Y., Zhu, J., Zhang, H., & Perinpanayagam, H. (2016). Epoxy resin-based ultrafine dry powder coatings for implants. *Journal of Applied Polymer Science*, *133*(37), 43960. doi:10.1002/app.43960

Hsieh, M. H., Tsai, M. H., Shen, W. J., & Yeh, J. W. (2013). Structure and properties of two Al-Cr-Nb-Si-Ti high-entropy nitride coatings. *Surface and Coatings Technology*, *221*, 118–123. doi:10.1016/j.surfcoat.2013.01.036

Huang, F., Yan, A., & Zhao, H. (2016). *Influences of Doping on Photocatalytic Properties of TiO2 Photocatalyst*. doi:10.5772/63234

Huang, W., Huo, M., Cheng, N., & Wang, R. (2021). New forms of electrospun nanofibers applied in cardiovascular field. *Frontiers in Cardiovascular Medicine, 8*. . doi:10.3389/fcvm.2021.801077

Huang, Z. M., Gopal, R., Fujihara, K., Ramakrishna, S., Loh, P. L., Foong, W. C., Ganesh, V. K., & Chew, C. L. (2003). Fabrication of a new composite orthodontic archwire and validation by a bridging micromechanics model. *Biomaterials*, *24*(17), 2941–2953. doi:10.1016/S0142-9612(03)00093-0 PMID:12742734

Huang, Z., Zheng, B., Zhu, S., Yao, Y., Ye, Y., Lu, W., & Chen, W. (2014). Photocatalytic activity of phthalocyanine-sensitized TiO_2-SiO_2 microparticles irradiated by visible light. *Materials Science in Semiconductor Processing*, *25*, 148–152. doi:10.1016/j.mssp.2013.10.014

Hussain, A., & Asi, M. R. (2008). Pesticides as water pollutants. In Groundwater for Sustainable Development: Problems, Perspectives and Challenges (pp. 95–101). CRC Press. https://doi.org/ doi:10.4018/978-1-5225-6111-8.ch001

Hussain, M. M., & Iqbal, M. (2019). Polymer Additives and Blends. In Advanced Polymers in Medicine (pp. 141-162). Springer.

Hu, W., & Yuan, X. (2017). Facile Hydrothermal Synthesis of SnO_2 Nanospheres as Photocatalysts. *Journal of Nanomaterials*, *2017*, 1–6. Advance online publication. doi:10.1155/2017/6976203

Ibrahim, S. D., Ismail, R., & Mazumder, M. A. J. (2020). Polymers in Aerospace Applications. In Handbook of Polymers for Aerospace Applications (pp. 3-24). Elsevier.

Imran, M. A., Miraj, F., Khan, I., & Tlili, I. (2018). MHD fractional Jeffrey's fluid flow in the presence of thermo diffusion, thermal radiation effects with first order chemical reaction and uniform heat flux. *Results in Physics*, *10*, 10–17. doi:10.1016/j.rinp.2018.04.008

Inform, N. H. S. (2023). *Types of cardiovascular disease*. Cardiovascular Disease. https://www.nhsinform.scot/illnesses-and-conditions/heart-and-blood- vessels/conditions/cardiovascular-disease

Irawan, A. P., Daywin, F. J., Fanando, F., & Agustino, T. (2016). Mechanical characteristics of rattan reinforced fiberglass and epoxy composites for shank prosthesis application. *IACSIT International Journal of Engineering and Technology*, *8*(3), 1543–1550.

Irawan, A. P., Soemardi, T. P., Widjajalaksmi, K., & Reksoprodjo, A. H. (2011). Tensile and flexural strength of ramie fiber reinforced epoxy composites for socket prosthesis application. *International Journal of Mechanical and Materials Engineering*, *6*(1), 46–50.

Islam, M., Lantada, A. D., Gómez, M. R., Mager, D., & Korvink, J. G. (2020). Microarchitectured carbon structures as innovative tissue-engineering scaffolds. *Advanced Engineering Materials*, *22*(6), 2000083. doi:10.1002/adem.202000083

Ivanov Yu.F., Kozlov E.V. (2002). Volumetric and surface hardening of structural steel morphological analysis of the structure. Physics. - 2002. No. 3. pp. 5-23.

Ivanov, Y.F. & Kozlov, E.V. (1991). Electron microscopic analysis of the martensitic phase of steel 38KHN3MFA. *Ferrous metallurgy, 8,* 38-41.

Ivanov, Y.F., Gromov, V.E., & Vorobev, S.V. (2011). Fazovyj sostav i defektnaya substruktura stali 20H13, obrabotannoj elektronnym puchkom v rezhime oplavleniya poverhnosti. *Fundamentalnye problemy sovremennogo materialovedeniya*.

Ivanov, Yu., Matz, W., Rotshtein, V., Günzel, R., & Shevchenko, N. (2002). Pulsed electronbeams melting of high-speed steel: Structural phase transformations and wear resistance. *Surface and Coatings Technology*, *150*(2-3), 188–198. doi:10.1016/S0257-8972(01)01542-0

Jacob, L., Haro, J. M., & Koyanagi, A. (2018). Post-traumatic stress symptoms are associated with physical multimorbidity: Findings from the Adult Psychiatric Morbidity Survey 2007. *Journal of Affective Disorders*, *232*, 385–392. doi:10.1016/j.jad.2018.02.063 PMID:29522958

Jain, A., Jain, A., & Narula, S. (2016). *Introduction to Polymer Chemistry and Applications*. CRC Press.

Janosi, A., Steinbrunn, W., Pfisterer, M., & Detrano, R. (2020). *Cleveland heart disease dataset*. https://www.kaggle.com/datasets/cherngs/heart-disease-cleveland-uci

Jeevanandam, J., Barhoum, A., Chan, Y. S., Dufresne, A., & Danquah, M. K. (2018). Review on nanoparticles and nanostructured materials: history, sources, toxicity and regulations. *Beilstein Journal of Nanotechnology, 9*(1), 1050–1074. doi:10.3762/bjnano.9.98

Jiang, D., Otitoju, T. A., Ouyang, Y., Shoparwe, N. F., Wang, S., Zhang, A., & Li, S. (2021). A Review on Metal Ions Modified TiO2 for Photocatalytic Degradation of Organic Pollutants. In Catalysts (Vol. 11, Issue 9). doi:10.3390/catal11091039

Jin, L., Liu, T., Wu, C., & Xiao, R. (2018). Plasticizers: Types, Properties, and Applications. In Functional Polymeric Composites (pp. 255-278). Elsevier.

Jin, X., Wang, B., Xue, W., Du. J., Wu, X., & Wu, J. (2013). Characterization of wear-resistant coatings on 304 stainless steel fabricated by 148 cathodic plasma electrolytic oxidation. *Surface & Coatings Technology, 236*, 22-28.

Jin, F. L., Li, X., & Park, S. J. (2015). Synthesis and application of epoxy resins: A review. *Journal of Industrial and Engineering Chemistry*, *29*, 1–11. doi:10.1016/j.jiec.2015.03.026

Joshua, U., & Bekun, F. V. (2020). The path to achieving environmental sustainability in South Africa: The role of coal consumption, economic expansion, pollutant emission, and total natural resources rent. *Environmental Science and Pollution Research International*, *27*(9), 2735–2743. doi:10.100711356-019-07546-0 PMID:31919826

Jurek, B., & Neumann, I. D. (2018). The oxytocin receptor: From intracellular signaling to behavior. *Physiological Reviews*, *98*(3), 1805–1908. doi:10.1152/physrev.00031.2017 PMID:29897293

Kabir, R., Saifullah, M. A. K., Ahmed, A. Z., Masum, S. M., & Molla, M. A. I. (2020). Synthesis of n-doped zno nanocomposites for sunlight photocatalytic degradation of textile dye pollutants. *Journal of Composites Science*, *4*(2), 49. Advance online publication. doi:10.3390/jcs4020049

Kajihara, K. (2013). Recent advances in sol–gel synthesis of monolithic silica and silica-based glasses. *Journal of Asian Ceramic Societies*, *1*(2), 121–133. doi:10.1016/j.jascer.2013.04.002

Kanagaraj, J., Velappan, K. C., Chandra Babu, N. K., & Sadulla, S. (2006). Solid wastes generation in the leather industry and its utilization for cleaner environment-A review. *Journal of Scientific and Industrial Research*, *65*(July), 541–548.

Kandasamy, M., Seetharaman, A., Sivasubramanian, D., Nithya, A., Jothivenkatachalam, K., Maheswari, N., Gopalan, M., Dillibabu, S., & Eftekhari, A. (2018). Ni-Doped SnO_2 Nanoparticles for Sensing and Photocatalysis. *ACS Applied Nano Materials*, *1*(10), 5823–5836. doi:10.1021/acsanm.8b01473

Kanjwal, M. A., Barakat, N. A. M., Sheikh, F. A., & Kim, H. Y. (2010). Electronic characterization and photocatalytic properties of TiO_2/CdO electrospun nanofibers. *Journal of Materials Science*, *45*(5), 1272–1279. doi:10.100710853-009-4078-3

Kannan, K., Radhika, D., Reddy, K. R., Raghu, A., Sadasivuni, K. K., Palani, G., & Gurushankar, K. (2021). $Gd^{3+}$and Y^{3+}co-doped mixed metal oxide nanohybrids for photocatalytic and antibacterial applications. *Nano Express*, *2*(1). Advance online publication. doi:10.1088/2632-959X/abdd87

Kantay N., Rakhadilov B., Kurbanbekov Sh., Yeskermessov D., Yerbolatova G. & Apsezhanova A. (2021). Influence of detonation-spraying parameters on the phase composition and tribological properties of Al2O3 coatings. *Coatings, 11*(7), 793, 1-10.

Kantay, N., Kasmamytov, N., Rakhadilov, B., Plotnikov, S., Paszkowski, M., & Kurbanbekov, S. (2020). Influence of temperature on structural-phase changes and physical properties of ceramics on the basis of aluminum oxide and silicon. *Materials Testing*, *62*(7), 716–720. doi:10.3139/120.111537

Kardonina, N. I., Yurovskikh, A. S., & Kolpakov, A. S. (2010). Prevrashcheniya v sisteme Fe – N. *MiTOM*, (10), 5–15.

Kargari, A., & Shirazi, M. M. A. (2014). Water Desalination: Solar-Assisted Membrane Distillation. Encyclopedia of Energy Engineering and Technology, 2095–2109. doi:10.1081/E-EEE2-120051388

Kar, P., Shukla, K., Jain, P., Sathiyan, G., & Gupta, R. K. (2021). Semiconductor based photocatalysts for detoxification of emerging pharmaceutical pollutants from aquatic systems: A critical review. *Nano Materials Science*, *3*(1), 25–46. doi:10.1016/j.nanoms.2020.11.001

Karunakaran, C., Sakthiraadha, S., & Gomathisankar, P. (2013). Photocatalytic and bactericidal activities of hydrothermally and sonochemically prepared Fe_2O_3-SnO_2 nanoparticles. *Materials Science in Semiconductor Processing*, *16*(3), 818–824. doi:10.1016/j.mssp.2012.12.030

Kasem Mahmoud, E., & Mohamed Ghoneim, A. (2016). Effect of polluted water on soil and plant contamination by heavy metals in El-Mahla El-Kobra, Egypt. *Solid Earth*, *7*(2), 703–711. doi:10.5194e-7-703-2016

Kashchenko, M. P., Dzhemilev, K. N., & Chashchina, V. G. (2012). The connection of various habituses with variants of orientation relations under γ-α martensitic transformation in dynamic theory. Fundamental problems of modern. *Materials Science*, (1), 50–56.

Kaya, G., Gunhan, B., Ozer, I. O., & Poyraz, H. B. (2020). Production of TiO2 coated α-Al2O3 platelets by flame spray pyrolysis and their characterization. *Ceramics International*, *46*(16), 25512–25519. doi:10.1016/j.ceramint.2020.07.022

Kazi. (2022). *Leather Industry of Bangladesh: Challenges and Opportunities*, retrieved on March 9, 2022, from https://businessinspection.com.bd/leather-industry-of-bangladesh/

Keşir, M. K., Sökmen, M., & Bıyıklıoğlu, Z. (2021). Photocatalytic efficiency of metallo phthalocyanine sensitized tio2 (Mpc/tio2) nanocomposites for Cr(vi) and antibiotic amoxicillin. *Water (Basel)*, *13*(16), 2174. Advance online publication. doi:10.3390/w13162174

Khabashesku, V. N., Margrave, J. L., & Barrera, E. V. (2005). Functionalized carbon nanotubes and nanodiamonds for engineering and biomedical applications. *Diamond and Related Materials*, *14*(3-7), 859–866. doi:10.1016/j.diamond.2004.11.006

Khan, M. E., Khan, M. M., & Cho, M. H. (2016). Fabrication of WO_3 nanorods on graphene nanosheets for improved visible light-induced photocapacitive and photocatalytic performance. *RSC Advances*, *6*(25), 20824–20833. doi:10.1039/C5RA24575C

Khan, M. M., Adil, S. F., & Al-Mayouf, A. (2015). Metal oxides as photocatalysts. *Journal of Saudi Chemical Society*, *19*(5), 462–464. doi:10.1016/j.jscs.2015.04.003

Kheyker D.M., & Zevin A.S. (1963). *Rentgenovskaya difraktometriya*. 540. M.:Fizmatgiz,

Khrushchov, M. M., & Babichev, M. A. (1970). *Abrazivnoye iznashivaniye*. Nauka.

Kidin I.N. (1978). Elektrohimiko-termicheskaya obrabotka metallov i splavov Tekst. *Metallurgiya*, 320.

Kim, H., Choi, S. R., & Yoon, S. Y. (2002). Superhard Ti-Si-N coatings by a hybrid system of arc ion plating and sputtering techniques. *Surface and Coatings Technology*, *161*(2-3), 243–248. doi:10.1016/S0257-8972(02)00499-1

Kim, S. K., & Cha, B. C. (2005). Deposition of tantalum nitride thin films by D.C. magnetron sputtering. *Thin Solid Films*, *475*(1-2), 202–207. doi:10.1016/j.tsf.2004.08.059

Kim, S. P., Choi, M. Y., & Choi, H. C. (2016). Photocatalytic activity of SnO_2 nanoparticles in methylene blue degradation. *Materials Research Bulletin*, *74*, 85–89. doi:10.1016/j.materresbull.2015.10.024

Kittel, C. (1976). *Introduction to Solid State Physics*. John Wiley & Sons.

Klaine, S. J., Alvarez, P. J. J., Batley, G. E., Fernandes, T. F., Handy, R. D., Lyon, D. Y., Mahendra, S., McLaughlin, M. J., & Lead, J. R. (2008). Nanomaterials in the environment: Behavior, fate, bioavailability, and effects. *Environmental Toxicology and Chemistry*, *27*(9), 1825–1851. doi:10.1897/08-090.1 PMID:19086204

Ko, Y., Lee, K., & Shin, D. (2013). Effect of ammonium metavanadate on surface characteristics of oxide layer formed on Mg alloy via plasma electrolytic oxidation. *Surface & Coatings Technology, 236*, 70-74.

Kolmakov, A., Klenov, D. O., Lilach, Y., Stemmer, S., & Moskovitst, M. (2005). Enhanced gas sensing by individual SnO_2 nanowires and nanobelts functionalized with Pd catalyst particles. *Nano Letters*, *5*(4), 667–673. doi:10.1021/nl050082v PMID:15826106

Komabayashi, T., Colmenar, D., Cvach, N., Bhat, A., Primus, C., & Imai, Y. (2020). Comprehensive review of current endodontic sealers. *Dental Materials Journal*, *39*(5), 703–720. doi:10.4012/dmj.2019-288 PMID:32213767

Kondrat'ev, V. V., Balanovskii, A. E., Ivanov, N. A., Ershov, V. A., & Kornyakov, M. V. (2014). Evaluation of the Effect of Modifier Composition with Nanostructured Additives on Grey Cast Iron Properties. *Metallurgist*, *58*(5-6), 377–387. doi:10.100711015-014-9919-x

Kondrat'ev V.V., Balanovskii A.E., Ivanov N.A., Ershov V.A., Kornyakov M.V. (2014). Evaluation of the Effect of Modifier Composition with Nanostructured Additives on Grey Cast Iron Properties. *Metallurgist, 58,* 377-387.

Kong, J. H., Takeda, T., Okumiya, M., Tsunekawa, Y., Yoshid, M., & Kim, S. G. (2012). The Study about Surface Modification of Steel by Water Plasma. *Proc. of 13th International Conference on Plasma Surface Engineering*, Garmisch-Partenkirchen.

Konstantinova, M. V., Balanovskiy, A. E., Gozbenko, V. E., Kargapoltsev, S. K., Karlina, A. I., Shtayger, M. G., & Kuznetsov, B. O. (2019). Application of plasma surface quenching to reduce rail side wear. *IOP Conference Series. Materials Science and Engineering*, *560*(1), 012146. doi:10.1088/1757-899X/560/1/012146

Koppiahraj, K., Bathrinath, S., Saravanasankar, S. (2019). Leather Waste Management Scenario in Developed and Developing Nations. *International Journal of Engineering and Advanced Technology, 9*(1).

Korotkov, V A. (2016). The 15th Anniversary of the Manual Plasma Quenching. Ferrous Metallurgy. *Bulletin of Scientific, Technical and Economic Information*, *10*, 62-68.

Korotkov, V.A., Baskakov, L.B., & Tolstov, I.A. (1991). Vosstanovitelnaya naplavka i uprochnenie rolikov rolgangov. *Svarochnoe proizvodstvo, 3*, 31-33.

Kozlov, E., Skakov, M., Uazyrkhanova, G., & Popova N. (2014). Redistribution of second phase particles in surface layers of deformed roll. *Advanced Materials Research, 1013*, 158-165.

Kozlov, E.V., Popova, N.A., & Koneva, N.A. (2004). Fragmented substructure formed in BCC steels during deformation. *RAS. The series is physical*, *68*(10), 1419-1428.

Kozlov, E., Skakov, M., Uazyrkhanova, G., & Popova, N. (2014). Redistribution of second phase particles in surface layers of deformed roll. *Advanced Materials Research, 1013*, 158–165. doi:10.4028/www.scientific.net/AMR.1013.158

Kremnev L.S. (1985). Zaevtektoidnyye bystrorezhushchiye stali. *Metallovedeniye i termicheskaya obrabotka metallov*, (8), 24 – 25.

Kremnev, L.S. (2008) Teoriya legirovaniya i sozdaniye na yeye osnove teplostoykikh instrumental'nykh staley i splavov optimal'nogo sostava. *Metallovedeniye i termicheskaya obrabotka metallov*, (11),18-28.

Kremnev, L.S. (2008). Teoriya legirovaniya i sozdaniye na yeye osnove teplostoykikh instrumental'nykh staley i splavov optimal'nogo sostava. *Metallovedeniye i termicheskaya obrabotka metallov*, (641), 18-28.

Kremnev, L. S. (2008). Teoriya legirovaniya i sozdaniya na yeye osnove teplostoykikh instrumental'nykh staley i splavov razlichnogo sostava. *Metallovedeniye i termicheskaya obrabotka metallov*, (11), 18–28.

Krishnakumar, S., & Senthilvelan, T. (2021). Polymer composites in dentistry and orthopedic applications-a review. *Materials Today: Proceedings*, *46*, 9707–9713. doi:10.1016/j.matpr.2020.08.463

Kuksenova, L.I., Lapteva, V.G., & Berezina, Y.V. (2004). *Struktura i iznosostoykost' azotirovannoystali*, (1), 31-34.

Kuksenova, L. I., Lapteva, V. G., & Berezina, Ye. V. (2004). Struktura i iznosostoykost' azotirovannoy stali. *MiTOM*, (1), 31–34.

Kulikov, I.S., Vashenko, S.V., & Kamenev A.Ya. (2010). *Elektrolitno-plazmennaya obrabotka materialov.* Minsk: Belarus.navuka.

Kumar, K. S., Choudhary, N., Jung, Y., & Thomas, J. (2018). Recent Advances in Two-Dimensional Nanomaterials for Supercapacitor Electrode Applications. *ACS Energy Letters*, *3*(2), 482–495. doi:10.1021/acsenergylett.7b01169

Kumar, S., Ojha, A. K., & Walkenfort, B. (2016). Cadmium oxide nanoparticles grown in situ on reduced graphene oxide for enhanced photocatalytic degradation of methylene blue dye under ultraviolet irradiation. *Journal of Photochemistry and Photobiology. B, Biology*, *159*, 111–119. doi:10.1016/j.jphotobiol.2016.03.025 PMID:27045279

Kumar, V., Singh, R. K., & Singh, D. (2015). Polymer-based nanocomposites for aerospace applications. In *Handbook of Polymer Nanocomposites for Industrial Applications* (pp. 491–508). Elsevier.

Kuo, C. Y., Jheng, H. K., & Syu, S. E. (2021). Effect of non-metal doping on the photocatalytic activity of titanium dioxide on the photodegradation of aqueous bisphenol A. *Environmental Technology*, *42*(10), 1603–1611. doi:10.1080/09593330.2019.1674930 PMID:31580775

La Cruz, A. M.-D., Martínez, D. S., & Cuéllar, E. L. (2010). Synthesis and characterization of WO_3 nanoparticles prepared by the precipitation method: Evaluation of photocatalytic activity under vis-irradiation. *Solid State Sciences*, *12*(1), 88–94. doi:10.1016/j.solidstatesciences.2009.10.010

LaFalce. (2019). *Leather insights: a comprehensive guide for buying the best leather*. Retrieved October 15, 2022, from https://octaneseating.com/blog/leather-buying-guide/

Lahtin Y.M., & Arzamasov B.N. (1985). *Himiko-termicheskaya obrabotka metallov: uchebnoe posobie dlya vuzov.* M.: Metallurgiya.

Lakhtin Y.M., & Arzamasov B.N. (1985). *Chemical and thermal treatment of metals.* Metallurgy Publ.

Lakhtin, Y.M. & Silina, N.V. (1977). Kharakter vysokoy tverdosti legirovannogo ferrita posle azotirovaniya. *MiTOM*, 23-29.

Lakhtin, Y.M. (1995). Diffuzionnyye osnovy protsessa azotirovaniya. *Metallovedeniye termicheskaya obrabotka metallov*, (7),14-17.

Lakhtin, Y.M. (2009). *Materialovedeniye i termicheskaya obrabotka metallov*, (5),448.

Lakhtin, Y.M. (2009). *Materialovedeniye i termicheskaya obrabotka metallov: uchebnik dlya vuzov. 5-ye izd., pererab,* 448. TID Az – kniga.

Lakhtin, Y.M., Kogan, YA.D., Shpis, G.Y., & Bemer, Z. (1991). Teoriya i tekhnologiya azotirovaniya. Metallurgiya, 320.

Lakhtin, Y.M., Neustroyev, G.N., & Ivanov, YU.P. (1973). Nizkotemperaturnoye tsianirovaniye instrumental'nykh staley. *Metallovedeniye i termicheskaya obrabotka metallov*, (12), 27-31.

Lakhtin, YU. M., & Arzamasov, B. N. (1985). Khimiko-termicheskaya obrabotka metallov: ucheb. posobiye dlya vuzov po spets. "Metallovedeniye, oborud. i tekhnologiya term, obrab. metallov. Metallurgiya, 256.

Lakhtin, YU.M., & Kogan, YA.D. (1982). Gazovoye azotirovaniye detaley mashin i instrumenta. Mashinostroyeniye, 60.

Lakhtin, Y. U. M. (1993). Sovremennoye sostoyaniye protsessa azotirovaniya. *Metallovedeniye i termicheskaya obrabotka metallov*, (JV27), 6–11.

Lakhtin, Y. U. M. (1995). Diffuzionnyye osnovy protsessa azotirovaniya. *MiTOM*, (7), 14–17.

Lazarenko, B.R., Duradzhi, V.N., & Bryantsev, I.V. (1979). O vliyanii povysheniya induktivnosti na kharakteristiki anodnogo i katodnogo protsessov *Elektronnaya obrabotka materialov*, (5), 8 - 13.

Lee K.M., & Polycarpou A.A. (2005). Wear of conventional pearlitic and improved bainitic rail steels. *Wear, 259*, 391-399.

Lee, G. R., Kim, H., Choi, H. S., & Lee, J. J. (2007). Superhard tantalum-nitride films formed by inductively coupled plasma-assisted sputtering. *Surface and Coatings Technology*, *201*(9-11), 5207–5210. doi:10.1016/j.surfcoat.2006.07.207

Lee, J. K., Kwak, S. W., Ha, J. H., Lee, W., & Kim, H. C. (2017). Physicochemical properties of epoxy resin-based and bioceramic-based root canal sealers. *Bioinorganic Chemistry and Applications*, *2582849*, 1–8. Advance online publication. doi:10.1155/2017/2582849 PMID:28210204

Lei, K., Wang, C., Liu, L., Luo, Y., Mu, C., Li, F., & Chen, J. (2018). A Porous Network of Bismuth used as Anode Material for High-Energy-Density Potassium-Ion Batteries. *Angewandte Chemie International Edition*, *57*(17), 4687–4691. doi:10.1002/anie.201801389 PMID:29488300

Lei, P., Wang, F., Gao, X., Ding, Y., Zhang, S., Zhao, J., Liu, S., & Yang, M. (2012). Immobilization of TiO_2 nanoparticles in polymeric substrates by chemical bonding for multi-cycle photodegradation of organic pollutants. *Journal of Hazardous Materials*, *227–228*, 185–194. doi:10.1016/j.jhazmat.2012.05.029 PMID:22658211

Lellis, B., Fávaro-Polonio, C. Z., Pamphile, J. A., & Polonio, J. C. (2019). Effects of textile dyes on health and the environment and bioremediation potential of living organisms. *Biotechnology Research and Innovation*, *3*(2), 275–290. doi:10.1016/j.biori.2019.09.001

Leng, Y. X., Sun, H., Yang, P., Chen, J. Y., Wang, J., Wan, G. J., Huang, N., Tian, X. B., Wang, L. P., & Chu, P. K. (2001). Biomedical properties of tantalum nitride films synthesized by reactive magnetron sputtering. *Thin Solid Films*, *398-399*, 471–475. doi:10.1016/S0040-6090(01)01448-1

Leshchinsky L. K., & Samotugin S. S. (2001). *Mechanical properties of plasma-hardened 5% chromium tool steel deposited by arc welding. Welding,* 25-30.

Leshinskij, L. K., Samotugin, S. S., Pirch, I. I., & Komar, V. I. (1990). Plazmennoe poverhnostnoe uprochnenie. Kiev: Tehnika.

Letifi, H., Dridi, D., Litaiem, Y., Ammar, S., Dimassi, W., & Chtourou, R. (2021). High Efficient and Cost Effective Titanium Doped Tin Dioxide Based Photocatalysts Synthesized via Co-precipitation Approach. In Catalysts (Vol. 11, Issue 7). doi:10.3390/catal11070803

Letifi, H., Litaiem, Y., Dridi, D., Ammar, S., & Chtourou, R. (2019). Enhanced Photocatalytic Activity of Vanadium-Doped SnO_2 Nanoparticles in Rhodamine B Degradation. *Advances in Condensed Matter Physics*, *2019*, 1–11. Advance online publication. doi:10.1155/2019/2157428

Lewis, P. (2004). *Polymer Science: A Comprehensive Reference*. Elsevier.

Li, Q., Mahendra, S., Lyon, D., Brunet, L., & Liga, M. (n.d.). *Antimicrobial nanomaterials for water disinfection and microbial control: potential applications and implications.* Elsevier. Retrieved July 29, 2022, from https://www.sciencedirect.com/science/article/pii/S0043135408003333?casa_token=mmxoSvxfy3AAAAAA:i05GpT1KbFBTdQCFDuXMLsQIM8Oi7_r0Pn32bDBKf-6InqxNHbBSs_68bjCZCj4inhzUNAXyzw

Li, X., Yang, K., Yu, C., Zhang, K., & Yang, S. (2019). *Broadband photocatalysis using a Z-scheme heterojunction of Au/NaYF 4: Yb, Er/WO 3· 0.33 H 2 OW 18 O 49 via a synergetic strategy of upconversion function and.* Retrieved July 29, 2022, from https://pubs.rsc.org/en/content/articlehtml/2019/qi/c9qi00823c

Li, Y., Yang, X., Wei, G., Zhang, Z., & Cheng, S. (2019). Advanced Functional Block Copolymers. In Functional Polymers in Food Science: From Technology to Biology, (pp. 41-75). Springer.

Liang, S. C., Chang, Z. C., Tsai, D. C., Lin, Y.-C., Sung, H.-S., Deng, M.-J., & Shieu, F.-S. (2011). Effects of substrate temperature on the structure and mechanical properties of (TiVCrZrHf)N coatings. *Applied Surface Science*, *257*(17), 7709–7713. doi:10.1016/j.apsusc.2011.04.014

Liang, S., Xiao, K., Mo, Y., & Huang, X. (2012). A novel ZnO nanoparticle blended polyvinylidene fluoride membrane for anti-irreversible fouling. *Journal of Membrane Science*, *394–395*, 184–192. doi:10.1016/j.memsci.2011.12.040

Liang, Y. C., & Chang, C. W. (2019). Preparation of orthorhombic WO_3 thin films and their crystal quality-dependent dye photodegradation ability. *Coatings*, *9*(2), 90. Advance online publication. doi:10.3390/coatings9020090

Liang, Y.-B., Liu, K., Liu, Z., Wang, J., Liu, C.-S., & Liu, Y. (2021). A First-Principles Study of Monolayer and Heterostructure Antimonene as Potential Anode Materials for Magnesium-Ion Batteries. *Applied Surface Science*, *577*(4), 151880. doi:10.1016/j.apsusc.151880

Liao, D. L., & Liao, B. Q. (2007). Shape, size and photocatalytic activity control of TiO_2 nanoparticles with surfactants. *Journal of Photochemistry and Photobiology A Chemistry*, *187*(2–3), 363–369. doi:10.1016/j.jphotochem.2006.11.003

Liberty Leather Goods. (2022). *The Leather Industry – An Overview of Fascinating Facts.* Retrieved October 05, 2022, from https://www.libertyleathergoods.com/leather-industry/

Liberty Leather Goods. (2023). *The Incredible History of Leather.* Retrieved July 05, 2022, from https://www.libertyleathergoods.com/history-of-leather/

Li, D., & Qu, J.LI. (2009). The progress of catalytic technologies in water purification: A review. *Journal of Environmental Sciences (China)*, *21*(6), 713–719. doi:10.1016/S1001-0742(08)62329-3 PMID:19803071

Li, H., Bian, Z., Zhu, J., Zhang, D., Li, G., Huo, Y., Li, H., & Lu, Y. (2007). Mesoporous titania spheres with tunable chamber stucture and enhanced photocatalytic activity. *Journal of the American Chemical Society*, *129*(27), 8406–8407. doi:10.1021/ja072191c PMID:17571890

Li, M., Wu, H., Yuan, Y., Hu, B., & Gu, N. (2022). recent fabrications and applications of cardiac patch in myorcardial infarction treatment. *VIEW*, *3*(2), 20200153. doi:10.1002/VIW.20200153

Lima, A. A. M., Moore, S. R., Barboza, M. S. Jr, Soares, A. M., Schleupner, M. A., Newman, R. D., Sears, C. L., Nataro, J. P., Fedorko, D. P., Wuhib, T., Schorling, J. B., & Guerrant, R. L. (2000). Persistent Diarrhea Signals a Critical Period of Increased Diarrhea Burdens and Nutritional Shortfalls: A Prospective Cohort Study among Children in Northeastern Brazil. *The Journal of Infectious Diseases*, *181*(5), 1643–1651. doi:10.1086/315423 PMID:10823764

Ling, B., Zhang, G., Liao, H., Coddet, C., & Ding, C. (2009). Friction and wear behavior of ZrO2–Al2O3 composite coatings deposited by air plasma spraying: Correlation with physical and mechanical properties. *Surface and Coatings Technology*, *203*(20, 21), 3235–3242. doi:10.1016/j.surfcoat.2009.03.056

Li, T., Cai, H., Li, C., Liu, X., & Huang, F. (2020). Rocksalt-Zincblende–Wurtzite Mixed-Phase ZnO Crystals With High Activity as Photocatalysts for Visible-Light-Driven Water Splitting. *Frontiers in Chemistry*, *8*, 351. Advance online publication. doi:10.3389/fchem.2020.00351 PMID:32411673

Liu, Z., Yu, Y., Zhu, X., Fang, J., Xu, W., Hu, X., Li, R., Yao, L., Qin, J., & Fang, Z. (2022). Semiconductor heterojunctions for photocatalytic hydrogen production and Cr(VI) Reduction: A review. In Materials Research Bulletin (Vol. 147). Elsevier Ltd. doi:10.1016/j.materresbull.2021.111636

Liu, Z.Y., Loh NH, Khor KA, & Tor S.B. Microstructure evolution during sintering of injection molded M2 high speed steel. *Mater SciEng*, -P. 46–55.

Liu, H., Ahlinder, A., Yassin, M. A., Finne-Wistrand, A., & Gasser, T. C. (2020). Computational and experimental characterization of 3D-printed PCL structures toward the design of soft biological tissue scaffolds. *Materials & Design*, *188*, 108488. doi:10.1016/j.matdes.2020.108488

Liu, X., Li, C., Han, S., Han, J., & Zhou, C. (2003). Synthesis and electronic transport studies of CdO nanoneedles. *Applied Physics Letters*, *82*(12), 1950–1952. doi:10.1063/1.1562331

Liu, Y., Dinh, K. N., Dai, Z., & Yan, Q. (2020). Metallenes: Recent Advances and Opportunities in Energy Storage and Conversion Applications. *ACS Materials Letters*, *2*(9), 1148–1172. doi:10.1021/acsmaterialslett.0c00280

Liu, Y., Gu, J., & Fan, D. (2020). Fabrication of high-strength and porous hybrid scaffolds based on nano-hydroxyapatite and human-like collagen for bone tissue regeneration. *Polymers*, *12*(1), 61. doi:10.3390/polym12010061 PMID:31906327

Lobanov, M. L., Rusakov, G. M., Redikul'tsev, A. A., Belikov, S. V., Karabanalov, M. S., Struina, E. R., & Gervas'ev, A. M. (2016). Investigation of Special Misorientations in Lath Martensite of Low Carbon Steel Using the Method of Orientation Microscopy. *The Physics of Metals and Metallography*, *117*(3), 254–259. doi:10.1134/S0031918X1603008X

Loladze T.N. (1982). Prochnost' i iznosostoykost' rezhushchego instrumenta. *Mashinostroyeniye*, 320.

Louisiana Heart and Vascular Institute. (2023). *Do Men Have a Higher Risk for Heart Disease?* Retrieved online on 16th April 2023 from https://www.louisianaheart.org/blog/do- men-have-a-higher-risk-for-heart-disease#:~:text=That%20 said%2C%20men%20have%20larger,found%20in%20smaller%20blood%20vessels

Love Your Leather. (2022). *13 Different Uses for Leather*. Retrieved October 23, 2022, from https://www.loveyourleather.ca/leather-blog/13-different-uses-for-leather/

Lu, H. B., Liao, L., Li, H., Tian, Y., Wang, D. F., Li, J. C., Fu, Q., Zhu, B. P., & Wu, Y. (2008). Fabrication of CdO nanotubes via simple thermal evaporation. *Materials Letters*, *62*(24), 3928–3930. doi:10.1016/j.matlet.2008.05.010

Luttrell, T., Halpegamage, S., Tao, J., Kramer, A., Sutter, E., & Batzill, M. (2015). Why is anatase a better photocatalyst than rutile? - Model studies on epitaxial TiO_2 films. *Scientific Reports*, *4*(1), 4043. Advance online publication. doi:10.1038rep04043 PMID:24509651

Maazouz, A., Notta-Cuvier, D., & Krawczak, P. (2013). *Polypropylene-based polymer blends and composites: processing, morphology and properties*. CRC Press.

Macak, J. M., Zlamal, M., Krysa, J., & Schmuki, P. (2007). Self-organized TiO_2 nanotube layers as highly efficient photocatalysts. *Small*, *3*(2), 300–304. doi:10.1002mll.200600426 PMID:17230591

Maitz, M. F. (2015). Applications of synthetic polymers in clinical medicine. *Biosurface and Biotribology*, *1*(3), 161–176. doi:10.1016/j.bsbt.2015.08.002

Majorov V.S. (2009). *Lazernye tehnologii obrabotki materialov: sovremennye problemy fundamentalnyh issledovanij i prikladnyh razrabotok*. M.: Fizmatlit.

Majorov, V. S. (2009). *Lazernoe uprochnenie metallov. V kn.: Lazernye tehnologii obrabotki materialov: sovremennye problemy fundamentalnyh issledovanij i prikladnyh razrabotok*. M. Fizmatlit.

Makarov A.V. (2009). Povyshenie iznosostojkosti splavov zheleza za schet sozdaniya metastabilnyh i nanokristallicheskih struktur: avtoref. …dokt.teh.nauk. *Institut mashinovedeniya Uralskogo otdeleniya RAN*. Chelyabinsk.

Maksakova, O. V., Beresnev, V. M., & Eskermesov, D. K. (2017). Effects of Cr and Si additions under the various deposition conditions on the mechanical properties of the (Zr-Ti-Nb)N coatings. *Proceedings of the 2017 IEEE 7th International Conference on Nanomaterials: Applications & Properties*, (1), 1-6.

Maksakova, O. V., Plotnikov, S. V., Eskermesov, D. K., & Yurchenko, D. (2016). The effect of Si additions and deposition conditions on the structure and properties of the (Zr-Ti-Cr-Nb-Si)N coatings fabricated by vacuum-arc deposition. *Proceedings of the 6th International Conference Nanomaterials: Applications and Properties*. NAP.

Maksoud, M. I. A. A., Fahim, R. A., Shalan, A. E., Abd Elkodous, M., Olojede, S. O., Osman, A. I., Farrell, C., Al-Muhtaseb, A. H., Awed, A. S., Ashour, A. H., & Rooney, D. W. (2021). Advanced Materials and Technologies for Supercapacitors used in Energy Conversion and Storage: A Review. *Environmental Chemistry Letters*, *19*(1), 375–439. doi:10.100710311-020-01075-w

Malekshoar, G., Pal, K., He, Q., Yu, A., & Ray, A. K. (2014). Enhanced solar photocatalytic degradation of phenol with coupled graphene-based titanium dioxide and zinc oxide. *Industrial & Engineering Chemistry Research*, *53*(49), 18824–18832. doi:10.1021/ie501673v

Malinovskaya V.A. (2006). *Raspredelenie azota i ugleroda, fazovaya struktura gradientnyh sloev i mehanicheskie svojstva stali 20H2N4A posle nitrocementacii: dis… kand. tehn*. Tomsk.

Mali, S. S., Patil, J., Kim, H., & Hong, C. K. (2018). Synthesis of SnO_2 nanofibers and nanobelts electron transporting layer for efficient perovskite solar cells. *Nanoscale*, *10*(17), 8275–8284. doi:10.1039/C8NR00695D PMID:29687121

Mallick, P. K. (2017). Introduction to Polymer Composites Manufacturing. In Fiber-Reinforced Composites (pp. 3-14). CRC Press.

Maltas, E., Ozmen, M., Yildirimer, B., Kucukkolbasi, S., & Yildiz, S. (2013). Interaction between ketoconazole and human serum albumin on epoxy modified magnetic nanoparticles for drug delivery. *Journal of Nanoscience and Nanotechnology*, *13*(10), 6522–6528. doi:10.1166/jnn.2013.7742 PMID:24245109

Mantzaridis, C., Brocas, A. L., Llevot, A., Cendejas, G., Auvergne, R., Caillol, S., Carlotti, S., & Cramail, H. (2013). Rosin acid oligomers as precursors of DGEBA-free epoxy resins. *Green Chemistry*, *15*(11), 3091–3098. doi:10.1039/c3gc41004h

Mao, L., Zhao, X., Cheng, Q., Yang, G., Liao, F., Chen, L., He, P., & Chen, S. (2021). Recent Advances and Perspectives of Two-Dimensional Ti-based Electrodes for Electrochemical Energy Storage. *Sustainable Energy & Fuels*, *5*(20), 5061–5113. doi:10.1039/D1SE01203G

Marateb, H. R., & Goudazi, S. (2015, March). A noninvasive method for coronary artery diagnosis using a clinically-interpretable fuzzy-based system. *Journal of Research in Medical Sciences*, *20*(3), 214–223. PMID:26109965

Martin, J., Melhem, A., Shchedrina, I., Duchanoy, T., Nominé, A., Henrion, G., Czerwiec, T., & Belmonte T. (2013). Effects of electrical parameters on plasma electrolytic oxidation. *Surface & Coatings Technology*, *221*.

Martin, S. S. (2023). *Facts about cholesterol. Cholesterol in the Blood. Health.* John Hopkins Medicine. https://www.hopkinsmedicine.org/health/ conditions-and-diseases/high cholesterol/cholesterol-in-the-blood

Martin, P. J., Bendavid, A., Cairney, J. M., & Hoffman, M. (2005). Nanocomposite Ti-Si-N, Zr-Si-N, Ti-Al-Si-N, Ti-Al-V-Si-N thin film coatings deposited by vacuum-arc deposition. *Surface and Coatings Technology*, *200*(7), 2228–2235. doi:10.1016/j.surfcoat.2004.06.012

MasterClass. (2021). *Fabric Guide: Learn About Leather*. From https://www.masterclass.com/articles/learn-about-leather

Matalkeh, M., Nasrallah, G. K., Shurrab, F. M., Al-Absi, E. S., Mohammed, W., Elzatahry, A., & Saoud, K. M. (2022). Visible Light Photocatalytic Activity of Ag/WO_3 Nanoparticles and its Antibacterial Activity Under Ambient Light and in The Dark. *Results in Engineering*, *13*, 100313. doi:10.1016/j.rineng.2021.100313

Maxwell's Demon: Nano Machine Of The Future Captures Great Scientist's Bold Vision -- ScienceDaily. (n.d.). Retrieved July 29, 2022, from https://www.sciencedaily.com/releases/2007/02/070201144706.htm

Mayo Clinic. (2023). *Tests for type 1 and type 2 diabetes and prediabetes.Diabetes.* https://www.mayoclinic.org/diseases-conditions/diabetes/diagnosis-treatment/drc-20371451#:~:text=A%20blood%20sugar%20level%20less,L)%20means%20you%20 have%20prediabetes

Medvedev, S.I., Nezhivlyak, A.E., Grechneva, M.V., Balanovsky, A.E., & Ivakin, V.L. (2015). Optimization of plasma hardening conditions of the side surface of rails in PUR-1 experimental equipment. *Welding International, 29,* 643-649.

Medvedovskaya L. A., & Shur N. F. (1983). Oborudovanie i tehnologiya lazernoj termicheskoj obrabotki (obzor). *Metallovedenie i term.*

Mendoza-Sánchez, B., & Gogotsi, Y. (2016). Synthesis of Two-Dimensional Materials for Capacitive Energy Storage. *Advanced Materials*, *28*(9), 6104–6135. doi:10.1002/adma.201506133 PMID:27254831

Menin, R., Ramos, E., & Magrini, M. (1981). Laser surface treatment of carbon and alloy steels. *Proc. 4 Inter. conf. of rapid quench. metals*. Sendai.

Messaadi, C., Ghrib, T., Jalali, J., Ghrib, M., Alyami, A., Gaidi, M., Manso-Silván, M., & Ezzaouia, H. (2018). Synthesis and Characterization of SnO2-TiO2 Nanocomposites Photocatalysts. *Current Nanoscience, 14*. Advance online publication. doi:10.2174/1573413714666180927110912

Mikulewicz, M., Gronostajski, Z., Wielgus, A., & Chojnacka, K. (2017). Transparent orthodontic archwires: A systematic literature review. *Archives of Civil and Mechanical Engineering, 17*(3), 651–657. doi:10.1016/j.acme.2016.12.012

Minkevich, A.N., & Suchevyanu, G. (1968). Khimiko-termicheskaya obrabotka staley v smesyakh soley, soderzhashchikh mochevinu. *Metallovedeniye i termicheskaya obrabotka metallov*, (10), 11-16.

Mirkin L.I. (1961). Spravochnik po rentgenostrukturnomu analizu polikristallov. *Gosizdat. Fiz. mat. literatury*, 863.

Mittal, A., Dadu, S., Garg, P., Yendrembam, B., Abraham, A., & Singh, K. (2017). Comparative evaluation of fracture resistance of endodontically treated teeth with epoxy resin-based sealers AH plus and mineral trioxide aggregate fillapex: An in vitro study. *Indian Journal of Dental Sciences, 9*(1), 8–11. doi:10.4103/IJDS.IJDS_83_16

Moiseev, V.F., & Grigor'ev, S.N. (2004). *Instrumental'nye materialy*, 248.

Moiseyev V.F., & Grigor'yev S.N. Instrumental'nyye materialy. *Izd. 2-ye*, 248. MGTUStankin.

Montgomery, M. A., & Elimelech, M. (2007). Water and sanitation in developing countries: Including health in the equation - Millions suffer from preventable illnesses and die every year. *Environmental Science & Technology, 41*(1), 17–24. doi:10.1021/es072435t PMID:17265923

Moon, S., & Jeong, Y. (2009). Generation mechanism of microdischarges during plasma electrolytic oxidation of Al in aqueous solutions. *Corrosion Science, 51*, 1506-1512.

Morris, D. G. (1998). Material. Science foundation, Trans Technical publication LVD.

Mortazavi, B., Dianat, A., Cuniberti, G., & Rabczuk, T. (2016). Application of Silicene, Germanene and Stanene for Na or Li Ion Storage: A Theoretical Investigation. *Electrochimica Acta, 213*, 865–870. doi:10.1016/j.electacta.2016.08.027

Mortensen, M. B. (2020). Elevated LDL cholesterol and increased risk of myocardial infarction and atherosclerosis cardiovascular disease in individuals aged 70-100 years: a contemporary prevention cohort. *The Lancet*. https://www.thelancet.com/article/S0140-6736(20)32233-9/fulltext

Munawar, T., Rehman, M. N., Nadeem, M. S., Mukhtar, F., Manzoor, S., Ashiq, M. N., & Iqbal, F. (2021). Facile synthesis of Cr-Co co-doped CdO nanowires for photocatalytic, antimicrobial, and supercapacitor applications. *Journal of Alloys and Compounds, 885*, 160885. Advance online publication. doi:10.1016/j.jallcom.2021.160885

Munchow, E. A., & Bottino, M. C. (2017). Recent advances in adhesive bonding: The role of biomolecules, nanocompounds, and bonding strategies in enhancing resin bonding to dental substrates. *Current Oral Health Reports, 4*(3), 215–227. doi:10.100740496-017-0146-y PMID:29177123

Muras, V. S. (1955). *O svyazi mezhdu temperaturoy i napryazheniyem toka pri elektrolitnom nagreve metallov. Mashinostroitel' Belorussii*. Gosizdat BSSR.

Mushtaq, K., Saeed, M., Gul, W., Munir, M., Firdous, A., Yousaf, T., Khan, K., Sarwar, H. M. R., Riaz, M. A., & Zahid, S. (2020). Synthesis and characterization of TiO_2 via sol-gel method for efficient photocatalytic degradation of antibiotic ofloxacin. *Inorganic and Nano-Metal Chemistry, 50*(7), 580–586. doi:10.1080/24701556.2020.1722695

Mustapha, R., Rahmat, A. R., Abdul Majid, R., & Mustapha, S. N. H. (2019). Vegetable oil-based epoxy resins and their composites with bio-based hardener: A short review. *Polymer-Plastics Technology and Materials, 58*(12), 1311–1326. doi:10.1080/25740881.2018.1563119

Nakata, K., & Fujishima, A. (2012). TiO2 photocatalysis: Design and applications. In Journal of Photochemistry and Photobiology C: Photochemistry Reviews (Vol. 13, Issue 3, pp. 169–189). doi:10.1016/j.jphotochemrev.2012.06.001

Nam, C. T., Yang, W.-D., & Duc, L. M. (2013). Solvothermal Synthesis of TiO_2 Photocatalysts in Ketone Solvents with Low Boiling Points. *Journal of Nanomaterials*, *627385*, 1–11. Advance online publication. doi:10.1155/2013/627385

Nassiri Koopaei, N., & Abdollahi, M. (2017). Health risks associated with the pharmaceuticals in wastewater. In DARU, Journal of Pharmaceutical Sciences (Vol. 25, Issue 1). BioMed Central Ltd. doi:10.118640199-017-0176-y

National Grid. (2022a). *What is net zero?* Retrieved November 10, 2022, from https://www.nationalgrid.com/stories/energy-explained/why-net-zero-so-important

National Grid. (2022b). *Why is net zero so important?* Retrieved November 10, 2022, from https://www.nationalgrid.com/stories/energy-explained/why-net-zero-so-important

National Health Services. (2019). *What is high blood pressure?* https://www.nhs.uk/conditions/high-blood-pressure-hypertension/

National Health Services. (2022). *Cardiovascular diseases*. NHS Inform. Retrieved online from https://www.nhsinform.scot/illnesses-and-conditions/heart-and-blood- vessels/conditions/cardiovascular-disease#about-cardiovascular-disease.

National Health Services. (2022). *Overview. Menopause*. Retrieved online on 5th May 2023 from https://www.nhs.uk/conditions/menopause/

National Health Services. (2023). *How to tell if you are living with obesity.* https://www.nhs.uk/conditions/obesity/#:~:text=Obesity%20and%20overweight%20is%20caused,and%20do%20enough%20physical%20activity

National Institute of Diabetes and Digestive and Kidney Diseases. (2016). *What is diabetes?* Retrieved online from https://www.niddk.nih.gov/health-information/diabetes/overview/what-is-diabetes

National Institute on Aging. (2018). *The future of research on aging and the heart.* Heart Health and Aging. https://www.nia.nih.gov/health/heart-health-and-aging#:~:text=Adults%20age%2065%20and%20older,risk%20of%20developing%20cardiovascular%20disease.

National Institute on Aging. (2021). *What is menopause?* https://www.nia.nih.gov/health/what-menopause

Nelson, R. H. (2014). *Hyperlipidemia as a risk factor for cardiovascular disease. National Centre for Biotechnology Information.* National Library of Medicine. https://www.ncbi.nlm.nih.gov/pmc/articles/PMC3572442/

Nesterov, D.K., & Sapozhkov, V.E. (1989). i dr. Uprochnenie relsov iz zaevtektoidnoj stali kombinirovannoj termoobrabotkoj. *MiTom, 12*, 2-5.

Neustroyev G.P., Bogdanov V.V., & Ivanov YU.P. (1972). Nizkotemperaturnoye tsianirovaniye instrumental'nykh staley. *Progressivnyye metody termicheskoy i khimiko-termicheskoy obrabotki*, 89-91. Mashinostroyeniye.

Nguyen, V. T., Astafeva, N. A., Tikhonov, A. G., Balanovskiy, A. E., & Vu, V. H. (2022). Evaluation of the Hardness and Wear Resistance of Alloyed Coatings From Fastening CuSn/CrxCy Mixture Hardened by Plasma and Laser. *Tribology in Industry*, *44*(3), 518–527. doi:10.24874/ti.1074.03.21.07

Nickels, M., Xie, J., Cobb, J., Gore, J. C., & Pham, W. (2010). Functionalization of iron oxide nanoparticles with a versatile epoxy amine linker. *Journal of Materials Chemistry*, *20*(23), 4776–4780. doi:10.1039/c0jm00808g PMID:20668643

Nie, S., & Emory, S. R. (1997). Probing Single Molecules and Single Nanoparticles by Surface-Enhanced Raman Scattering. *Science*, *275*(5303), 1102–1106. doi:10.1126cience.275.5303.1102 PMID:9027306

Noonuruk, R., & Wattanawikkam, C. (2020). Visible-light-driven photodegradation of commercial dyes by the cooperation of co-doped TiO_2 material. *Current Applied Science and Technology*, 20(1), 43–51. doi:10.14456/cast.1477.4

Nordin, M., Larsson, M., & Hogmark, S. (1998). Mechanical and tribological properties of multilayered PVD TiN/CrN, TiN/MoN, TiN/NbN and TiN/TaN coatings on cemented carbide. *Surface and Coatings Technology*, *106*(2-3), 234–241. doi:10.1016/S0257-8972(98)00544-1

Nordin, M., Larsson, M., & Hogmark, S. (1999). Mechanical and tribological properties of multilayered PVD TiN/CrN. *Wear*, *232*(2), 221–225. doi:10.1016/S0043-1648(99)00149-0

Novikov, I. I. (1986). Teoriya termicheskoy obrabotki metallov. *Metallurgiya*, 480.

O'Brien, D. M., Vallieres, C., Alexander, C., Howdle, S. M., Stockman, R. A., & Avery, S. V. (2019). Epoxy-amine oligomers from terpenes with applications in synergistic antifungal treatments. *Journal of Materials Chemistry. B, Materials for Biology and Medicine*, *7*(34), 5222–5229. doi:10.1039/C9TB00878K PMID:31369021

O'Brien, F. J. (2011). Biomaterials & scaffolds for tissue engineering. *Materials Today*, *14*(3), 88–95. doi:10.1016/S1369-7021(11)70058-X

Odusote, J. K., & Oyewo, A. T. (2016). Mechanical properties of pineapple leaf fiber reinforced polymer composites for application as a prosthetic socket. *Journal of Engineering Technology*, *7*(1), 125–139.

Odusote, J. K., Oyewo, A. T., Adebisi, J. A., & Akande, K. A. (2016). Mechanical properties of banana pseudo stem fibre reinforced epoxy composite as a replacement for transtibial prosthetic socket. *The Journal of the Association of Professional Engineers of Trinidad and Tobago*, *44*(2), 4–10.

Ohno, T., Sarukawa, K., Tokieda, K., & Matsumura, M. (2001). Morphology of a TiO_2 photocatalyst (Degussa, P-25) consisting of anatase and rutile crystalline phases. *Journal of Catalysis*, *203*(1), 82–86. doi:10.1006/jcat.2001.3316

Oh, S. M., & Hwang, S.-J. (2020). Recent Advances in Two-Dimensional Inorganic Nanosheet-based Supercapacitor Electrodes. *Journal of the Korean Ceramic Society*, *57*(2), 119–134. doi:10.100743207-020-00023-2

Okello, C., Tomasello, B., Greggio, N., Wambiji, N., & Antonellini, M. (2015). Impact of population growth and climate change on the freshwater resources of Lamu Island, Kenya. *Water (Basel)*, *7*(3), 1264–1290. doi:10.3390/w7031264

Oladele, I. O., Onuh, L., Taiwo, A. S., Borisade, S., Agbeboh, N. I., & Lephuthing, S. S. (2022). Mechanical, wear and thermal conductivity characteristics of snail shell-derived hydroxyapatite reinforced epoxy bio-composites for adhesive biomaterials applications. *International Journal of Sustainable Engineering*, *15*(1), 125–137. doi:10.1080/19397038.2022.2084174

Oliver, W. C., & Pharr, G. M. (1992). An improved technique for determining hardness and elastic modulus using load and displacement sensing indentation experiments. *Journal of Materials Research*, *7*(7), 1564–1583. doi:10.1557/JMR.1992.1564

Omran Alkhayatt, A. H., Al-Hussainy, I. A. D., & Al-Rikaby, O. A. C. (2014). *Annealing Effect on the Structural and Optical Properties of Sol-Gel Deposited Nanocrystalline CdO Thin Films* (Vol. 34). www.iiste.org

Ong, W. J., Tan, L. L., Chai, S. P., Yong, S. T., & Mohamed, A. R. (2014). Facet-dependent photocatalytic properties of TiO2-based composites for energy conversion and environmental remediation. In ChemSusChem (Vol. 7, Issue 3, pp. 690–719). doi:10.1002/cssc.201300924

Ong, C. B., Ng, L. Y., & Mohammad, A. W. (2018). A review of ZnO nanoparticles as solar photocatalysts: Synthesis, mechanisms and applications. In *Renewable and Sustainable Energy Reviews* (Vol. 81, pp. 536–551). Elsevier Ltd. doi:10.1016/j.rser.2017.08.020

Otto, J. K., Hajaj, B., Abovitz, R., Mistry, A., Nortman, S., & Brown, S. B. (2010). *U.S. Patent No. 7842092*. Washington, DC: U.S. Patent and Trademark Office.

Palatini, P., & Julius, S. (2004). Elevated Heart Rate: A Major Risk Factor for Cardiovascular Disease. *Clinical and Experimental Hypertension*, *26*(7-8), 7–8, 637–644. doi:10.1081/CEH-200031959 PMID:15702618

Palatnik, L. S., Fuks, M. Y. A., & Kosevich, V. M. (1972). *Mekhanizm obrazovaniya i substruktura kondensirovannykh plenok*. Nauka.

Pappu, A., Roy, A., & Dey, S. (2009). Applications of polymers in drug delivery. *International Journal of Green Pharmacy*, *3*(1), 9–22.

Parfenov, E. V., Yerokhin, A. L., & Mattews, A. (2007). Frequency response studies for the plasma electrolytic oxidation process. Surface and coatings technology. Science Direct. doi:10.1016/j.surfcoat.2007.04.044

Parthibavarman, M., Karthik, M., Sathishkumar, P., & Poonguzhali, R. (2018). Rapid synthesis of novel Cr-doped WO_3 nanorods: An efficient electrochemical and photocatalytic performance. *Journal of the Indian Chemical Society*, *15*(6), 1419–1430. doi:10.100713738-018-1342-y

Parvulescu, V. I., Epron, F., Garcia, H., & Granger, P. (2022). Recent Progress and Prospects in Catalytic Water Treatment. *Chemical Reviews*, *122*(3), 2981–3121. doi:10.1021/acs.chemrev.1c00527 PMID:34874709

Pastukh, I. M., & Zdybel', A. S. (2008). Kharakteristiki obrazovaniya nitridov v stalyakh pri azotirovanii v tleyushchem razryade. OTTOM - 9: materialy mezhdunarodnoy konferentsii. Khar'kov, 62-68.

Pastukh, I.M. (2006). *Teoriya i praktika bezvodorodnogo azotirovaniya v tleyushchem razryade*, 364.

Pastukh, I.M. (2006). *Teoriya i praktika bezvodorodnogo azotirovaniya v tleyushchem razryade*. Khar'kov, NNTS KHFTI, 364.

Patanjali, P., Singh, R., Kumar, A., & Chaudhary, P. (2019). Nanotechnology for water treatment: A green approach. *Green Synthesis, Characterization and Applications of Nanoparticles*, 485–512. doi:10.1016/B978-0-08-102579-6.00021-6

Patel, K. K., & Purohit, R. (2018). Future prospects of shape memory polymer nano-composite and epoxy based shape memory polymer-A review. *Materials Today: Proceedings*, *5*(9), 20193–20200. doi:10.1016/j.matpr.2018.06.389

Patent na poleznuyu model N° 5365 Sposob uprochneniya zheleznodorozhnyh koles. Data registracii v Gosudarstvennom reestre poleznyh modelej Respubliki Kazahstan.

Paulmier, T., Bell, J.M., Fredericks, P.M. (2008). Development of a novel cathodic plasma-electrolytic deposition technique. Part 2: Physico-chemical analysis of the plasma discharge. *Journal of materials processing technology*, *208*, 117-123.

Paulmier, T., Bell, J. M., & Fredericks, P. M. (2008). Development of a novel cathodic plasma-electrolytic deposition technique. Part 2: Physico-chemical analysis of the plasma discharge. *Journal of Materials Processing Technology*, (208), 117–123. doi:10.1016/j.jmatprotec.2007.12.101

Paul, S., & Choudhury, A. (2014). Investigation of the optical property and photocatalytic activity of mixed phase nanocrystalline titania. *Applied Nanoscience*, *4*(7), 839–847. doi:10.100713204-013-0264-3

Peng, L., Zhu, Y., Chen, D., Ruoff, R. S., & Yu, G. (2016). Two-Dimensional Materials for Beyond-Lithium-Ion Batteries. *Advanced Energy Materials*, *6*(11), 1600025. doi:10.1002/aenm.201600025

Permyakov V.G., Belotskiy A.V., & Barabash R.I. (1972). Struktura i svoystva diffuzionnoy zony pri azotirovanii khromistogo zheleza. Izvestiya vuzov. *Chernaya metallurgiya*, (40), 129–131.

Permyakov, V.G., Belotskiy, A.V., & Barabash, R.I. (1972). Struktura i svoystva diffuzionnoy zony pri azotirovanii khromistogo zheleza. Izvestiya vuzov. *Chernaya metallurgiya*, (4), 129–131.

Petersen, R. C. (2014). Titanium implant osseointegration problems with alternate solutions using epoxy/carbon-fiber-reinforced composite. *Metals*, *4*(4), 549–569. doi:10.3390/met4040549 PMID:25635227

Peters, S. T. (1998). Introduction, composite basics and road map. In S. T. Peters (Ed.), *Handbook of Composites* (pp. 1–20). Springer US. doi:10.1007/978-1-4615-6389-1_1

Petrova L.G. & Zyuzin D.M. (2005). Vysokotemperaturnoye azotirovaniye austenitnoy stali. *Uprochnyayushchiye tekhnologii pokrytiy*, (3), 29–36.

Petrova L.G. (2001). Vnutrenneye azotirovaniye zharoprochnykh staley i splavov *MiTOM9*, (1), 10-17.

Petrova L.G. (2001). *Vnutrenneye azotirovaniye zharoprochnykh staley i splavov*, (1), 10-17.

Petrova L.G. (2004). *Vysokotemperaturnoye azotirovaniye zharoprochnykh splavov*, (1), 18-24.

Petrova L.G. (2004). *Vysokotemperaturnoye azotirovaniye zharoprochnykh splavov*, (1),18-24.

Petrova, L. G., Aleksandrov, V. A., & Zyuzin, D. M. (2003). Reguliruyemyye protsessy azotirovaniya korrozionnostoykikh staley. *Vestnik MADI*, (1), 20–26.

Petukhov, B. V. (2001). Statisticheskaya teoriya dvizheniya dislokatsiy pri nalichii spontannykh protsessov blokirovki-deblokirovki. *FTT*, *43*(5), 813–817.

Pilloud, D., Pierson, J. F., Marques, A. P., & Cavaleiro, A. (2004). Structural changes in Zr-Si-N films vs. their silicon content. *Surface and Coatings Technology*, *180-181*(180), 352–356. doi:10.1016/j.surfcoat.2003.10.087

Plotnikov, S. V., Pogrebnjak, A. D., Yerokhina, L. N., Yeskermessov, D. K., & Erdybaeva, N. K. (2016). Study of nanostructured (Ti-Zr-Nb)N coatings' physical- mechanical properties obtained by vacuum arc evaporation. *IOP Conference Series. Materials Science and Engineering*, *110*(1), 12–31. doi:10.1088/1757-899X/110/1/012031

Pogrebnjak, A. D., & Tyurin, Yu. (2000). Preparation and investigation of the structure and properties of Al_2O_3 plasma-detonation coatings. *Techn. Phys. Lett.*, *26*(11), 960-964.

Pogrebnjak, A. D., Tyurin, Y. N., & Kobzev, A. P. (2003). Mass transfer and doping during electrolyte-plasma treatment of cast iron. *Tech. Phys. Lett.*, *29*(4), 312-315.

Pogrebnjak, A. D., Maksakova, O., Kozak, C., Koltunowicz, T. N., Grankin, S., Bondar, O., Eskermesov, D. K., Drozdenko, A., Petrov, S., & Erdybaeva, N. (2016). *Przegląd Elektrotechniczny*, (8), 180–183.

Pogrebnjak, A. D., Shpak, A. P., Azarenkov, N. A., & Beresnev, V. M. (2009). Structure and properties of hard and superhard nanocomposite coatings. *Physics Uspekhi*, *52*(1), 29–54. doi:10.3367/UFNe.0179.200901b.0035

Pogrebnjak, A. D., & Tyurin, Y. N. (2005). Modification of material properties and coating deposition using plasma jets. *Physics Uspekhi*, (48), 515–544.

Pogrebnjak, A. D., Yakushchenko, I. V., Bondar, O. V., Beresnev, V. M., Oyoshi, K., Ivasishin, O. M., Amekura, H., Takeda, Y., Opielak, M., & Kozak, C. (2016). Irradiation resistance, microstructure and mechanical properties of nanostructured (TiZrHfVNbTa)N coatings. *Journal of Alloys and Compounds*, *679*, 155–163. doi:10.1016/j.jallcom.2016.04.064

Pogrebnjak, A. D., Yakushchenko, I. V., Bondar, O. V., Sobol', O. V., Beresnev, V. M., Oyoshi, K., Amekura, H., & Takeda, Y. (2015). Influence of Implantation of Au- Ions on the Microstructure and Mechanical Properties of the Nanostructured Multielement (TiZrHfVNbTa)N Coating. *Physics of the Solid State*, *57*(8), 1559–1564. doi:10.1134/S1063783415080259

Pogrebnyak, A. D., Ponomarev, A. G., Shpak, A. P., & Kunitskiy, Yu. A. (2012). Primeneniye mikro- i nanozondov dlya analiza malorazmernykh 3D materialov, nanosistem i nanoobyektov. *UFN*, *182*(3), 287–321.

Pogrebnyak, A. D., Tyurin, Y. N., Ivanov, Y. F., Kobzev, A. P., Kulmentieva, O. P., & Ilyashenko, M. I. (2000). Obtaining and investigation of the structure and properties of plasma-detonation coatings from Al2O3. *Letter Journal Technology Physics*, (21), 53–60.

Poletika, I.M., Golkovskij, M.G., & Perovskaya, M.B. (2006). Elektronno-luchevaya zakalka poverhnostnogo sloya stali vne vakuuma. *Fizicheskaya mezomehanika, 9,* 181-184.

Polyak M.S. (1995). Tehnologiya uprochneniya. *Tehnologicheskie metody uprochneniya: spravochnik v 2 t.* Mashinostroenie.

Popilov L.YA., & Zaytseva L.P. (1963). Elektropolirovaniye i elektrotravleniye metallograficheskikh shlifov. *Metallurgiya*, 125.

Popova, N.A., Nikonenko, E.L, Tabieva, E.E., & Uazyrkhanova, G.K. (2020). Structure and phase composition of ferritic-perlitic steel surface after electrolytic plasma quenching. *Russian Physics Journal, 63*(5), 791-796.

Popova, N. A., Nikonenko, E. L., Tabieva, E. E., & Uazyrkhanova, G. K. (2020). Structure and phase composition of ferritic-perlitic steel surface after electrolytic plasma quenching. *Russian Physics Journal*, *63*(5), 791–796. doi:10.100711182-020-02099-z

Popova, N. A., Nikonenko, E. L., Tabieva, E. E., Uazyrkhanova, G. K., & Gromov, V. E. (2020). Influence of surface quenching on morphology and phase composition of ferritic-pearlitic steel. *Izvestiya Ferrous Metallurgy*, *63*(11-12), 915–921. doi:10.17073/0368-0797-2020-11-12-915-921

Poznyak, L. A. (1996). Instrumental'nyye stali, 488. Kiyev: Naukova dumka.

Poznyak, L.A., Tishayev, S.I., & Skrynchenko, Y.M. (1977). Instrumental'nyye stali: spravochnik. *Metallurgiya*, 167.

Precious Sibiya, N., Rathilal, S., & Kweinor Tetteh, E. (2021). Coagulation treatment of wastewater: Kinetics and natural coagulant evaluation. *Molecules (Basel, Switzerland)*, *26*(3), 698. Advance online publication. doi:10.3390/molecules26030698 PMID:33572703

Prokoshkin D A. (1984). *Khimiko-termicheskaya obrabotka metallov karbonitratsiya*, Metallurgiya, mashinostroyeniye, 204.

Promotion. (n.d.).https://www.cdc.gov/chronicdisease/resources/publications/factsheetsheart- diseasestroke.htm#:~:text=Leading%20risk%20factors%20for%20heart,unhealthy%20diet%2C%20and%20physical%20inactivity.

Przeździecka, E., Strąk, P., Wierzbicka, A., Adhikari, A., Lysak, A., Sybilski, P., Sajkowski, J. M., Seweryn, A., & Kozanecki, A. (2021). The Band-Gap Studies of Short-Period CdO/MgO Superlattices. *Nanoscale Research Letters*, *16*(1), 59. Advance online publication. doi:10.118611671-021-03517-y PMID:33835276

Przhenosil B. (1969). Nitrotsementatsiya. *Mashinostroyeniye*, 212.

Putungan, D. B., Lin, S.-H., & Kuo, J.-L. (2016). Metallic VS_2 Monolayer Polytypes as Potential Sodium-Ion Battery Anode via Ab Initio Random Structure Searching. *ACS Applied Materials & Interfaces*, *8*(29), 18754–18762. doi:10.1021/acsami.6b03499 PMID:27373121

Qi, K., Cheng, B., Yu, J., & Ho, W. (2017). Review on the improvement of the photocatalytic and antibacterial activities of ZnO. In *Journal of Alloys and Compounds* (Vol. 727, pp. 792–820). Elsevier Ltd. doi:10.1016/j.jallcom.2017.08.142

Qin, H., Liang, S., Chen, L., Li, Y., Luo, Z., & Chen, S. (2020). Recent Advances in Vanadium-based Nanomaterials and Their Composites for Supercapacitors. *Sustainable Energy & Fuels*, *4*(10), 4902–4933. doi:10.1039/D0SE00897D

Qiu, R., Zhang, D., Mo, Y., Song, L., Brewer, E., Huang, X., & Xiong, Y. (2008). Photocatalytic activity of polymer-modified ZnO under visible light irradiation. *Journal of Hazardous Materials*, *156*(1–3), 80–85. doi:10.1016/j.jhazmat.2007.11.114 PMID:18248886

Qu, X., & Alvarez, P. (n.d.). *Applications of nanotechnology in water and wastewater treatment.* Elsevier. Retrieved July 29, 2022, from https://www.sciencedirect.com/science/article/pii/S0043135413001772

Quan, H., Gao, Y., & Wang, W. (2020). Tungsten oxide-based visible light-driven photocatalysts: Crystal and electronic structures and strategies for photocatalytic efficiency enhancement. In Inorganic Chemistry Frontiers (Vol. 7, Issue 4, pp. 817–838). Royal Society of Chemistry. doi:10.1039/C9QI01516G

Qu, X., Brame, J., Li, Q., & Alvarez, P. J. J. (2013). Nanotechnology for a safe and sustainable water supply: Enabling integrated water treatment and reuse. *Accounts of Chemical Research*, *46*(3), 834–843. doi:10.1021/ar300029v PMID:22738389

Radi, P. A., Brito-Madurro, A. G., Madurro, J. M., & Dantas, N. O. (2006). Characterization and Properties of CdO Nanocrystals Incorporated in Polyacrylamide. *Brazilian Journal of Physics*, *36*(2A), 412–414. doi:10.1590/S0103-97332006000300048

Radwan, E. H., Saber, M. A. K., Saber, M. E. K., & Fahmy, G. H. (2017). The Impact Of Some Organic And Inorganic Pollutants On Fresh Water (Rashid, River Nile), Egypt. *Journal of Advances in Biology*, *10*(2), 2133–2145. doi:10.24297/jab.v10i2.6481

Rahadilov, B.K. (2014). *Elektrolitno-plazmennoe azotirovanie poverhnostnyh sloev bystrorezhushih stalej: dissertaciya na soiskanie uchenoj stepeni doktora filosofii* [Dissertation, Vostochno-Kazahstanskij gosudarstvennyj tehnicheskij universitet im. D.Serikbaeva].

Rahadilov, B. K., Zhurerova, L. G., Sagdoldina, Z. B., Kenesbekov, A. B., & Bayatanova, L. B. (2021). Morphological Changes in the Dislocation Structure of Structural Steel 20GL after Electrolytic-Plasma Hardening of the Surface. *Journal of Surface Investigation. X-ray, Synchrotron and Neutron Techniques*, *15*(2), 408–413. doi:10.1134/S1027451021020300

Raizada, P., Soni, V., Kumar, A., Singh, P., Parwaz Khan, A. A., Asiri, A. M., Thakur, V. K., & Nguyen, V. H. (2021). Surface defect engineering of metal oxides photocatalyst for energy application and water treatment. *Journal of Materiomics*, *7*(2), 388–418. doi:10.1016/j.jmat.2020.10.009

Rakhadilov B. K., Baizhan D. R., Uazyrkhanova G. K. & Tabieva Y.Y. (2020). *Patent for utility model No. 5365.* State Register of Utility Models of the Republic of Kazakhstan.

Rakhadilov B.K., Skakov M.K., & Karipbayeva G.S. (2011). Elektrolitno-plazmennoye azotirovaniye bystrorezhushchey stali Sbornik trudov IV mezhdunarodnoy nauchno-prakticheskoy konferentsii s ispol'zovaniyem nauchnykh shkol dlya molodykh uchenykh *Innovatsionnyye tekhnologii i ekonomika mashinostroyeniya*, 151-155. Yurginskiy tekhnologicheskiy institut. Izd-vo Tomskogo politekhnicheskogo universiteta.

Rakhadilov, B. K., Tabieva, E. E., & Zhurerova, L. G. (2019). Plasma-electrolytic Nitriding of 0.3Cr-1Mn-1Si-Fe Construction Steel. *METAL 2019 - 28th International Conference on Metallurgy and Materials, Conference Proceedings*, (pp. 1174-1179). Research Gate.

Rakhadilov, B.K., Tabiyeva, Y.Y., Uazyrkhanova, G.K., Zhurerova, L.G., & Baizhan, D. (2020). Inñuence of electrolytic-plasma surface quenching on the structure and strength properties of ferritic-pearlite class wheel steel. *Eurasian Journal of Physics and Functional Materials, 4*, 167-173.

Rakhadilov, B.K., Tabiyeva, Y.Y., Uazyrkhanova, G.K., Zhurerova, L.G., & Popova N.A. (2020). Effect of electrolyte-plasma surface hardening on structure wheel steel 2. *Bulletin of Karaganda State University, 2* (98), 68-74.

Rakhadilov, B.K., Tabiyeva, Y.Y., Uazyrkhanova, G.K., Zhurerova, L.G., Maulit, A., & Baizhan, D. (2019). Surface modification of steel mark 2 electrolytic-plasma exposure. *Eurasian Journal of Physics and Functional Materials, 3*, 355-362.

Rakhadilov, B. K., Buitkenov, D. B., Tuyakbaev, B. T., Sagdoldina, Zh. B., & Kenesbekov, A. B. (2019). Structure and properties of detonation coatings based on titanium carbosilicide. *Key Engineering Materials*, (821), 301–306.

Rakhadilov, B. K., Buranich, V. V., Satbayeva, Z. A., Kozhanova, R. S., & Pogrebnjak, A. D. (2020). The cathodic electrolytic plasma hardening of the 20Cr2Ni4A chromium-nickel steel. *Journal of Materials Research and Technology, 9*(4), 6969–6976. doi:10.1016/j.jmrt.2020.05.020

Rakhadilov, B. K., Buytkenov, D. B., Kakimzhanov, D., Kozhanova, R. S., & Bektasova, G. S. (2020). The effect of detonation spraying on the phase composition and hardness of Al2O3 coatings. *Eurasian Journal of Physics and Functional Materials, 4*(2), 160–166. doi:10.29317/ejpfm.2020040207

Rakhadilov, B. K., Sagdoldina, Zh. B., Baizhan, D. R., Zhurerova, L. G., Yeskermessov, D. K., Kalitova, A. A., & Smaiylova, M. (2021). Obtaining of hydroxyapatite coatings on a titanium substrate by detonation-gas spraying. *Eurasian Physical Technical Journal, 3*(37), 30–36. doi:10.31489/2021No3/30-36

Rakhadilov, B. K., & Skakov, M. K. (2012). Issledovaniye tribologicheskikh svoystv stali bystrorezhushchey R6M5, uprochnennoy vysokoenergeticheskoy otsenki. *Vestnik VKGTU*, (4), 43–48.

Rakhadilov, B. K., & Skakov, M. K. (2014). Vysokoeffektivnaya tekhnologiya elektrolitno-plazmennogo azotirovaniya bystrorezhushchikh staley. *Vestnik VKGTU*, (1), 50–55.

Rakhadilov, B. K., Tabiyeva, Y. Y., Uazyrkhanova, G. K., Zhurerova, L. G., Maulit, A., & Baizhan, D. (2019). Surface modification of steel mark 2 electrolytic-plasma exposure. *Eurasian Journal of Physics and Functional Materials*, (3), 355–362.

Rakhadilov, B. K., Zhurerova, L. G., & Pavlov, A. V. (2016). Electrolyte-plasma surface hardening of 65G and 20GL low-alloy steels. *Bulletin of the University of Karaganda, 4*(84), 8–13.

Rakhadilov, B. K., Zhurerova, L. G., Scheffler, M., & Khassenov, A. K. (2018). Change in high-temperature wear resistance of high-speed steel by plasma nitriding. *Bulletin of the University of Karaganda, 3*(91), 59–65.

Rakhadilov, B., Kantay, N., Sagdoldina, Z., Erbolatuly, D., Bektasova, G., & Paszkowski, M. (2021). Experimental investigations of Al2O3 - and ZrO2 -based coatings deposited by detonation spraying. *Materials Research Express, 8*(5), 1–12. doi:10.1088/2053-1591/abfbb7

Rakhshtadt A.G., Bernshteyn L.M., Kurdyumov G.V., & Mes'kin V.S. (2005). Metallovedeniye i termicheskaya obrabotka stali i chuguna: spravochnik v 3 t.. *Intermet Inzhiniring,(*T.2), 528.

Ramakrishna, S., Fujihara, H., Yoshida, E., & Hamada, H. (2000). Design and development of braided carbon fiber reinforced polymer composites for dental posts and bone plates. In *29th FRP symposium.* Japan Society for Materials Science.

Ramalingam, S., & Zheng, L. (1995). Film-substrate interface stresses and their role in the tribological performance on surface coatings. *Tribology International*, *28*(3), 145–161. doi:10.1016/0301-679X(95)98963-E

Ramazanov K. N., & Vafin R. K. (2011). *Razrabotka khorosho ionno-stal'nogo azotirovaniya instrumental'noy KH12 v skreshchennykh elektricheskikh i magnitnykh polyakh*, (41), 101-104.

Ramazanov K.N., Budilov V.V., & Vafin R.K. (2010). Azotirovaniye stalirezhushchey R6M5 v tleyushchem razryade s nalozheniyem magnitnogo polya. *Uprochnyayushchiye bystrodeystvuyushchiye tekhnologii i pokrytiya*, (5), 39-42.

Ramazanov, K. N., & Vafin, R. K. (2011). Razrabotka khorosho ionno-stal'nogo azotirovaniya instrumental'noy KH12 v skreshchennykh elektricheskikh i magnitnykh polyakh. *Vestnik UGATU*, (41), 101–104.

Rameshkumar, C., Anderson, A., Ananth, D., Hinduja Mohan, T., & Subalakshmi, R. (2020). An investigation of SnO_2 nanofilm for solar cell application by spin coating technique. *Materials Today: Proceedings*, *45*, 6042–6045. doi:10.1016/j.matpr.2020.09.732

Randall, N. (2002). Overview of mechanical testing standards. *CSM Instruments Applications Bulletin*, (18), 3.

Rane, A. V., Kanny, K., Abitha, V. K., & Thomas, S. (2018). Chapter 5 - Methods for Synthesis of Nanoparticles and Fabrication of Nanocomposites. In *Micro and Nano Technologies* (pp. 121–139). Woodhead Publishing. https://doi.org/https://doi.org/10.1016/B978-0-08-101975-7.00005-1

Rane, Y. N., Shende, D. A., Raghuwanshi, M. G., Koli, R. R., Gosavi, S. R., & Deshpande, N. G. (2019). Visible-light assisted CdO nanowires photocatalyst for toxic dye degradation studies. *Optik (Stuttgart)*, *179*, 535–544. doi:10.1016/j.ijleo.2018.10.215

Rasheed, T. (2022). Mxenes as an Emerging Class of Two-Dimensional Materials for Advanced Energy Storage Devices. *Journal of Materials Chemistry. A, Materials for Energy and Sustainability*, *10*(9), 4558–4584. doi:10.1039/D1TA10083A

Rathi, B. S., & Kumar, P. S. (2021). Application of adsorption process for effective removal of emerging contaminants from water and wastewater. *Environmental Pollution*, *280*, 116995. Advance online publication. doi:10.1016/j.envpol.2021.116995 PMID:33789220

Ray, S. S. (2017). Polymer Blends and Composites. In Polymer Science: A Comprehensive Reference (pp. 317-357). Elsevier.

Reghunath, S., Pinheiro, D., & KR, S. D. (2021). A review of hierarchical nanostructures of TiO2: Advances and applications. In *Applied Surface Science Advances* (Vol. 3). Elsevier B.V. doi:10.1016/j.apsadv.2021.100063

Ren, G., Han, H., Wang, Y., Liu, S., Zhao, J., Meng, X., & Li, Z. (2021). Recent advances of photocatalytic application in water treatment: A review. In Nanomaterials (Vol. 11, Issue 7). MDPI AG. doi:10.3390/nano11071804

Rezaei, M., & Habibi-Yangjeh, A. (2013). Simple and large scale refluxing method for preparation of Ce-doped ZnO nanostructures as highly efficient photocatalyst. *Applied Surface Science*, *265*, 591–596. doi:10.1016/j.apsusc.2012.11.053

Richard, R. (2007). *Track Demonstration of Laser-Treated Rail to Reduce Friction and WearTransportation Technology Center / R.* Richard. Inc.

Richardson. (2022). *Mathematics and climate change: What role do you think mathematics can play in guiding policy makers and in helping public understanding?* Retrieved November 25, 2022, from https://newn.cam.ac.uk/wp-content/uploads/2022/07/Maths3-Richardson_Emily.pdf

Riffat, R., & Husnain, T. (2022). *Fundamentals of Wastewater Treatment and Engineering*. doi:10.1201/9781003134374

Ritter, T. A., Shrout, T. R., Tutwiler, R., & Shung, K. K. (2002). A 30-MHz piezo-composite ultrasound array for medical imaging applications. *IEEE Transactions on Ultrasonics, Ferroelectrics, and Frequency Control*, *49*(2), 217–230. doi:10.1109/58.985706 PMID:11885679

Rizzo, A., Signore, M. A., Penza, M., Tagliente, M. A., De Riccardis, F., & Serra, E. (2006). RF sputtering deposition of alternate TiN/ZrN multilayer hard coatings. *Thin Solid Films*, *515*(2), 500–504. doi:10.1016/j.tsf.2005.12.279

Rodríguez, C., Tapia, C., Leiva-Aravena, E., & Leiva, E. (2020). Graphene oxide–zno nanocomposites for removal of aluminum and copper ions from acid mine drainage wastewater. *International Journal of Environmental Research and Public Health*, *17*(18), 1–18. doi:10.3390/ijerph17186911 PMID:32967362

Rony, L., Lancigu, R., & Hubert, L. (2018). Intraosseous metal implants in orthopedics: A review. *Morphologie*, *102*(339), 231–242. doi:10.1016/j.morpho.2018.09.003 PMID:30348489

Rowell, R. M. (2006). *Handbook of Wood Chemistry and Wood Composites*. CRC Press.

Rusli, N. I., & Chee, W. S. (2019). *Handbook of Polymer Applications in Medicine and Medical Devices*. Elsevier.

Ruzaidi, C. M., Maulud, A. S., & Azhari, C. H. (2013). Feasibility of using biomass-derived polymer in composite materials: A review. *International Journal of Automotive and Mechanical Engineering*, *8*(1), 1354–1366.

Sablev, L. P., Andreyev, A. A., & Kunchenko, V. V. (2000). Plazmennoye azotirovaniye rezhushchego instrumenta iz bystrorezhushchey stali, 133-137. Trudy simpoziuma OTTOM, Khar'kov.

Sadovsky, V. D., & Malyshev, K. A. (1984). *Phase and structural transformations during heating of steel*. Metallurgizdat.

Saidin, N. U., Choo, T. F., Kok, K. Y., Yusof, M. R., & Ng, I. K. (2017). Fabrication and characterization of ZnO nanofibers by electrospinning. *Materials Science Forum, 888 MSF*, 309–313. https://doi.org/ doi:10.4028/www.scientific.net/MSF.888.309

Sakthivel, S., Neppolian, B., Shankar, M. V., Arabindoo, B., Palanichamy, M., & Murugesan, V. (2003). Solar photocatalytic degradation of azo dye: comparison of photocatalytic efficiency of ZnO and TiO2. In Solar Energy Materials & Solar Cells (Vol. 77). Academic Press.

Sal'nikova S.S., & Rudman V.A. (1987). Opyt primeneniya ionnogo azotirovaniya v mashinostroyenii. *LDNTP, 20.*

Saltykov S.A. (1976). Stereometricheskaya metallografiya. *Metallurgiya.*

Samotugin, S.S., Muratov, V.A, & Kovalchuk, A.V. (1997). Plazmennoe uprochnenie instrumentov kolcevoj formy. *Metallovedenie i termicheskaya obrabotka metallov*.

Sampaio, M. J., Lima, M. J., Baptista, D. L., Silva, A. M. T., Silva, C. G., & Faria, J. L. (2017). Ag-loaded ZnO materials for photocatalytic water treatment. *Chemical Engineering Journal*, *318*, 95–102. doi:10.1016/j.cej.2016.05.105

Samsonov G.V. (1969). Nitridy. *Scientific Opinion*, 380.

Samuel, O., Othman, M. H. D., Kamaludin, R., Sinsamphanh, O., Abdullah, H., Puteh, M. H., & Kurniawan, T. A. (2022). WO3–based photocatalysts: A review on synthesis, performance enhancement and photocatalytic memory for environmental applications. In Ceramics International (Vol. 48, Issue 5, pp. 5845–5875). Elsevier Ltd. doi:10.1016/j.ceramint.2021.11.158

Sanskritiias. (2020). *Leather industry: problems & solutions*. Retrieved on November 10, 2022, from https://www.sanskritiias.com/current-affairs/leather-industry-problems-solutions

Saxena, V. K. (2005). *Impact of Pollution due to Tanneries on Groundwater Regime*. https://www.researchgate.net/publication/216087091

Sazlina, S. G., Sooryanarayana, R., Ho, B. K., Omar, M. A., Krishnapillai, A. D., & MohdTohit, N. (2020). Cardiovascular disease risk factors among older people: Data from the National Health and Morbidity Survey 2015. *PLoS ONE, 15*(10), e0240826. doi:10.1371/journal.pone.0240826

Scalera, F., Corcione, C. E., Montagna, F., Sannino, A., & Maffezzoli, A. (2014). Development and characterization of UV curable epoxy/hydroxyapatite suspensions for stereolithography applied to bone tissue engineering. *Ceramics International*, *40*(10), 15455–15462. doi:10.1016/j.ceramint.2014.06.117

Schastlivtsev V.M., Mirzayev D.A., & Yakovleva I.L. (1994). Struktura termicheski obrabotannoy stali. *Metallurgiya*, 288.

Schmitt-Thomas, K. G., Yang, Z. G., & Hiermer, T. (1998). Development of torsion loading facility used for evaluating shear performance of polymeric composite implant rods. *Polymer Testing*, *17*(2), 117–130. doi:10.1016/S0142-9418(97)00037-8

Schuler, M., Kunzler, T. P., de Wild, M., Sprecher, C. M., Trentin, D., Brunette, D. M., Textor, M., & Tosatti, S. G. P. (2009). Fabrication of TiO_2-coated epoxy replicas with identical dual-type surface topographies used in cell culture assays. *Journal of Biomedical Materials Research. Part A*, *88*(1), 12–22. doi:10.1002/jbm.a.31720 PMID:18257057

Sean, G. (2014). *India: The Toxic Price of Leather*. Retrieved November 10, 2022, from https://pulitzercenter.org/stories/india-toxic-price-leather

Seriya. (1975). Tekhnologicheskiy protsess poverkhnostnogo uprochneniya rezhushchego instrumenta. *Informatsionnyy listok*, 10-09.

Serpone, N., & Emeline, A. V. (2012). Semiconductor Photocatalysis — Past, Present, and Future Outlook. *The Journal of Physical Chemistry Letters*, *3*(5), 673–677. doi:10.1021/jz300071j PMID:26286164

Shannon, M. A., Bohn, P. W., Elimelech, M., Georgiadis, J. G., Marĩas, B. J., & Mayes, A. M. (2008). Science and technology for water purification in the coming decades. *Nature, 452*(7185), 301–310. doi:10.1038/nature06599

Sharma, N. (n.d.). *Nanotechnology in Stem Cell Research by Neha Sharma, Keshav Mahavidyalaya :* SSRN. Retrieved July 29, 2022, from https://papers.ssrn.com/sol3/papers.cfm?abstract_id=3958890

Shestopalova L.P. (2011). *Issledovaniye khimicheskogo sostava i sostava nitrid-oksidnoy zony oksiazotirovannogo sloya legirovannykh konstruktsionnykh staley,* (54), 78-81.

Shestopalova, L. P. (2011). Issledovaniye khimicheskogo sostava i sostava nitrid-oksidnoy zony oksiazotirovannogo sloya legirovannykh konstruktsionnykh staley. *Vestnik KHNADU*, (54), 78–81.

Shin, C. S., Kim, Y. W., Gall, D., Greene, J. E., & Petrov, I. (2002). Phase composition and microstructure of polycrystalline and epitaxial TaNx layers grown on oxidized Si(001) and MgO(001) by reactive magnetron sputter deposition. *Thin Solid Films*, *402*(1-2), 172–182. doi:10.1016/S0040-6090(01)01618-2

Shoja, S., Alm, O., Norgren, S., Andrén, H. O., & Halvarsson, M. (2021). Calculated and experimental Schmid factors for chip flow deformation of textured CVD α-alumina coatings. *Surface and Coatings Technology*, *412*, 126991. doi:10.1016/j.surfcoat.2021.126991

Shukla, V., Araujo, R. B., Jena, N. K., & Ahuja, R. (2017). The Curious Case of Two Dimensional Si_2BN: A High-Capacity Battery Anode Material. *Nano Energy*, *41*, 251–260. doi:10.1016/j.nanoen.2017.09.026

Shulayev, V. M., Anreyev, A. A., & Rudenko, V. P. (2006). Moderniacia vacuumno-dugovih ustanovok dlya sinteza pokriti i azotirovania metodom ionnoy implantacii i osajdenia. *FIP*, *4*(3-4), 136–142.

Shypylenko, A., Pshyk, A. V., Grzeskowiak, B., Medjanik, K., Peplinska, B., Oyoshi, K., Pogrebnjak, A., Jurga, S., & Coy, E. (2016). Effect of ion implantation on the physical and mechanical properties of Ti-Si-N multifunctional coatings for biomedical applications. *Materials & Design*, *110*, 821–829. doi:10.1016/j.matdes.2016.08.050

Siddaramaiah, & Jagadeesh, K. S. (1996). Properties of multi-functional epoxy resins cured with different amines. *Journal of Polymer Materials, 13*(3), 223-231.

Simoes, J. A. O., & Marques, A. T. (2001). Determination of stiffness properties of braided composites for the design of a hip prosthesis. *Composites. Part A, Applied Science and Manufacturing*, *32*(5), 655–662. doi:10.1016/S1359-835X(00)00157-3

Singh, D. P. (2016). Esthetic archwires in orthodontics: A review. *Journal of Oral Hygiene & Health*, *4*(1), 194. doi:10.4172/2332-0702.1000194

Singh, M., Kumar, S., Zoghi, S., Cervantes, Y., Sarkar, D., Ahmed, S., Chowdhury, S., & Banerjee, S. (2020). Fabrication and characterization of flexible three-phase ZnO-Graphene-Epoxy electro-active thin-film nanocomposites: Towards applications in wearable biomedical devices. *Journal of Composites Science*, *4*(3), 88. doi:10.3390/jcs4030088

Sisson, R. D. (2016). *The Plastics Compendium.* Carl Hanser Verlag GmbH & Co KG.

Sivaram, N. M. (2019a). Energy from Toxic Organic Waste for Heat and Power Generation. In Toxic Waste from Leather Industries (pp. 55-67). Woodhead Publishing.

Sivaram, N.M. (2019b). *Toxic Waste from Leather Industries, Energy from Toxic Organic Waste for Heat and Power Generation.* Science Direct.

Skakov M.K., Rakhadilov B.K., & Rakhadilov M.K. (2014). Wear-resistance of nitrided W-Mo-high speed steel in abrasive wear conditions. *Key engineering materials*, (594 – 595).

Skakov M.K., Rakhadilov B.K., Karipbayeva G.S., & Manapbayeva A.B. (2014). *Structural-phase state of R6M5 high-speed steel after thermal processing*, (48), 53-59

Skakov, M. K., & Rakhadilov, B. K. (2012). Change of structure and wear-resistance of P6M5 steel for processing in electrolyte plasma. *32th All-Polish Tribology conference Autumnal school of Tribology 2012.* Wroclaw University of Technology Institute of Machine Design and Operation, Wroclaw, Poland.

Skakov, M., & Rakhadilov, B. (2012). Change of structure and wear-resistance of P6M5 steel for processing in electrolyte plasma. *32th All-Polish Tribology conference Autumnal school of Tribology 2012.* Wroclaw University of Technology Institute of Machine Design and Operation.

Skakov, M., Bayatanova, L., & Sheffler, M. (2012). The drilling tool electrolyte-plasma material hardening technology planning and research. *The 7th International Forum on Strategic Technology*. Tomsk.

Skakov, M., Kurbanbekov, S., Tabieva, Y., & Zamanbekuly, E. (2013). Nitriding and carbonitriding influence on stainless steels surface layers changes. *Applied Mechanics and Materials, 379*, 105-109.

Skakov, M., Rakhadilov, B., & Scheffler, M. (2013). Wear-resistance increase of high speed steel surface layer. 5 World Tribological Congress. *Tribology of materials. –Wroclaw*, (3), 491.

Skakov, M.K., & Rakhadilov, B.K. (2013b). Modifikatsiya poverkhnosti obnaruzheniya stali instrumentom iz bystrorezhushchey al'ternativno-plazmennoy obrabotki. *Sbornik tezisov V konferentsii Sovremennyye metody v teoreticheskoy i eksperimental'noy elektrokhimii i IV konferentsii Elektrokhimicheskiye i elektrolitno-plazmennyye metody obrabotki metallov*. Ples: Izd-vo Institut khimii rastvorov im. G.A. Krestova RAN – Ivanovo.

Skakov, M.K., Bayatanova, L.B., & Scheffler, M. (2012). *Research of Surface Treatment on the Microstructure, Microhardness and Wear Resistance of 18CrNi3MoA-Sh Steel*. Tribologia.

Skakov, M.K., Bayatanova, L.B., & Scheffler, M. (2013). Changes of Structural-Phase Condition In *18CrNi3MoA-Sh Steel After Elektrolyte-Plasma Processing*. Advanced Materials Research.

Skakov, M. K., & Rakhadilov, B. K. (2013a). *Ustroystvo dlya nagrevaniya detaley v elektrolite: Patent na poleznuyu model' Respubliki Kazakhstan: MPK6 S21D 1/44 - N° 912*. Zayavl, Byul.

Skakov, M. K., Rakhadilov, B. K., Batyrbekov, E. G., & Manapbayeva, A. B. (2014). Povysheniye iznosostoykosti bystrorezhushchikh staleyel ektrolitno-plazmennym azotirovaniyem. *Vestnik KazNU*, (48), 44–52.

Skakov, M. K., Rakhadilov, B. K., Batyrbekov, E. G., Sheffler, M., Manapbayeva, A. B., Ayapbergenova, G. T., & Karipbayeva, G. S. (2014). Izmeneniye rezhimov elektrolitno-plazmennogo azotirovaniya na strukturno-fazovoye sostoyaniye i iznosostoykost' stali R6M5. *Vestnik KazNTU, 103*(3), 65–71.

Skakov, M. K., Rakhadilov, B. K., & Karipbaeva, G. S. (2013). Specifics of microstructure and phase composition of high-speed steel R6M5. *Applied Mechanics and Materials, 404*, 20–24. doi:10.4028/www.scientific.net/AMM.404.20

Skakov, M. K., Rakhadilov, B. K., Karipbayeva, G. S., & Manapbayeva, A. B. (2014). Strukturno-fazovoye sostoyaniye bystrorezhushchey stali R6M5 posle termicheskoy obrabotki. *Vestnik KazNU, 48*(1), 53–59.

Skakov, M. K., Rakhadilov, B. K., & Rakhadilov, M. K. (2014). Wear-resistance of nitrided W-Mo-high speed steel in abrasive wear conditions. *Key Engineering Materials*, 1117–1121.

Skakov, M. K., Rakhadilov, B. K., & Sheffler, M. (2013). Modification of structure and properties of steel R6M5 at electrolyteplasma treatment. *Advanced Materials Research*, (601), 64–68.

Skakov, M. K., Uazyrkhanova, G. K., & Popova, N. A. (2013). Electron microscopic analysis of 30CrMnSiA steel surface layers after hot deformation. *Applied Mechanics and Materials, 395-396*(396), 336–340. doi:10.4028/www.scientific.net/AMM.395-396.336

Skakov, M., Kurbanbekov, S., Tabieva, Y., & Zamanbekuly, E. (2013). Nitriding and carbonitriding influence on stainless steels surface layers changes. *Applied Mechanics and Materials, 379*, 105–109. doi:10.4028/www.scientific.net/AMM.379.105

Skripkin, A.A., Necvetaev, V.A., Sherbakov, V.E., & Minenko, N.Y. (1992). Poluchenie teplostojkih sloev na stali 20 s ispolzovaniem plazmennogo nagreva. *Svarochnoe proizvodstvo, 11*, 15-17.

Slovetskiy D.I., Terent'yev S.D., & Plekhanov V.G. (1986). Mekhanizm plazmenno-elektrolitnogo nagreva. *Teplofizika vysokoy temperatury*, (2), 353-363.

Slovetskiy, D. I. (1991). Mekhanizmy neravnovesnykh plazmokhimicheskikh reaktsiy, Novosibirsk, Nauka. Khimiya plazmy. T.3. Nizkotemperaturnaya plazma, (3), 94-100.

Sluisveld, Mariesse, Boer, Daioglou, Hof, & Vuuren. (2021). A race to zero - Assessing the position of heavy industry in a global net-zero CO_2 emissions context. *Energy and Climate Change*, *2*, 100051.

Smith, J. R., & Lamprou, D. A. (2014). Polymer coatings for biomedical applications: A review. *Transactions of the IMF*, *92*(1), 9-19. https://doi.org/10.1179/0020296713Z.000000000157

Solntseva Yu.P. (1988). Metallovedeniye i tekhnologiya metallov. *Metallurgiya*, 491.

Somasundaram, G., Rajan, J., Sangaiya, P., & Dilip, R. (2019). Hydrothermal synthesis of CdO nanoparticles for photocatalytic and antimicrobial activities. *Results in Materials, 4*, 100044. https://doi.org/https://doi.org/10.1016/j.rinma.2019.100044

Somody, F. (2021). *Why does the leather industry need to change?* Retrieved November 25, 2022, from https://www.explore-leap.com/post/why-the-leather-industry-needs-to- change.

Song, H., Li, Y., Lou, Z., Xiao, M., Hu, L., Ye, Z., & Zhu, L. (2015). Synthesis of Fe-doped WO3 nanostructures with high visible-light-driven photocatalytic activities. *Applied Catalysis B: Environmental*, *166–167*, 112–120. doi:10.1016/j.apcatb.2014.11.020

Song, J., Wang, X., Yan, J., Yu, J., Sun, G., & Ding, B. (2017). Soft Zr-doped TiO_2 Nanofibrous Membranes with Enhanced Photocatalytic Activity for Water Purification. *Scientific Reports*, *7*(1), 1636. Advance online publication. doi:10.103841598-017-01969-w PMID:28487571

Song, R., Murphy, M., Li, C., Ting, K., Soo, C., & Zheng, Z. (2018). Current development of biodegradable polymeric materials for biomedical applications. *Drug Design, Development and Therapy*, *12*, 3117–3145. doi:10.2147/DDDT.S165440 PMID:30288019

Song, X. C., Yang, E., Liu, G., Zhang, Y., Liu, Z. S., Chen, H. F., & Wang, Y. (2010). Preparation and photocatalytic activity of Mo-doped WO_3 nanowires. *Journal of Nanoparticle Research*, *12*(8), 2813–2819. doi:10.100711051-010-9859-8

Sowmya, B., Hemavathi, A. B., & Panda, P. K. (2021). Poly(ε-caprolactone) based electrospun nano-featured substrate for tissue engineering applications: A review. *Progress in Biomaterials*, *10*(2), 91–117. doi:10.100740204-021-00157-4 PMID:34075571

Srinivasan, M. (2018). Filled Polymer Composites. In Polymeric Composite Materials (pp. 319-377). Wiley.

Stalinskij, D.V., & Rudyuk, A.S. (2007). Sferidizaciya karbidnoj fazy v zaevtektoidnoj stali i ee vliyanie na svojstva relsov. *Metallurgicheskaya i gornorudnaya promyshlennost, 2*, 48-54.

Su, C.-W., Pang, L.-D., Tao, R., Shao, X., & Umar, M. (2022). Renewable energy and technological innovation: Which one is the winner in promoting net-zero emissions? *Technological Forecasting and Social Change*, *182*, 121798. doi:10.1016/j.techfore.2022.121798

Suhomlin, G.D. (1974). Elektronno-difrakcionnoe issledovanie orientacionnyh sootnoshenij ferrit-cementit v perlite. *FMM*, *38*(4), 878-880.

Suminov I.V., Belkin P.N., & Epel'fel'd A.V. (2011). Plazmenno-elektrolitnoye modifitsirovaniye poverkhnosti metallov i splavov. *Tekhnosfera*, (T.1), 464.

Suminov I.V., Epel'fel'd A.V., Lyudin V.B., Krit B.L., & Borisov A.M. (2005). Mikrodugovoye oksidirovaniye, teoriya tekhnologii, oborudovaniye. *EKOMET*, 368.

Suminov, I.V. (2005). Mikrodugovoye oksidirovaniye teoriya, tekhnologiya oborudovaniye. *EKOMET*, 368.

Sung, H. W., Tu, R., Shen, S. H., Witzel, T. H., Lin, D., Hata, C., Kingsbury, C.J., Noishiki, Y., Tomizawa, Y., & Quijano, R. C. (1994). A newly developed porcine heart valve bioprosthesis fixed with an epoxy compound. An experimental evaluation. *ASAIO Journal, 40*(2), 192-198.

Sun, G., Liu, J., Wang, X., Li, M., Cui, X., Zhang, L., Wu, D., & Tang, P. (2019). Fabrication of dual-sensitive poly (β-hydroxyl amine) micelles for controlled drug delivery. *European Polymer Journal*, *114*, 338–345. doi:10.1016/j.eurpolymj.2019.02.048

Sun, X., Chang, Y., Mu, C., Nie, A., Wang, B., Xiang, J., Zhai, K., Wen, F., & Li, Z. (2019). Two-Dimensional Layered Materials InSe Nanoflakes/Carbon Nanotubes Composite for Flexible All-Solid-State Supercapacitors. *Journal of Materials Science*, *55*(7), 2947–2957. doi:10.100710853-019-04151-x

Sun, X., Wei, D., Liu, J., & Du, Y. (2022). Free radical-enhanced formation of toxic byproduct benzoyl benzoquinone during the combined UV-chlorine treatment on BP-1. *Chemical Engineering Journal*, *433*, 134344. Advance online publication. doi:10.1016/j.cej.2021.134344

Suresh, S. (2013). Semiconductor Nanomaterials, Methods and Applications: A Review. *Nanoscience and Nanotechnology*, *3*(3), 62–74. doi:10.5923/j.nn.20130303.06

Szkoda, M., Siuzdak, K., & Lisowska-Oleksiak, A. (2016). Non-metal doped TiO_2 nanotube arrays for high efficiency photocatalytic decomposition of organic species in water. *Physica E, Low-Dimensional Systems and Nanostructures*, *84*, 141–145. doi:10.1016/j.physe.2016.06.004

Tabieva, E.E., Popova, N.A., Nikonenko, E.L., Uazyrkhanova, G.K., & Gromov, V.E. (2020). Influence of surface quenching on morphology and phase composition of ferritic-pearlitic steel. *Izvestiya Ferrous Metallurgy, 63*(11-12), 915–921.

Tabieva, E.E., Rahadilov, B.K., Uazyrhanova, G.K., & Bajzhan, D. (2020). Struktura i mehanicheskie svojstva kolesnoj stali do i posle elektrolitno-plazmennoj poverhnostnoj zakalki. *Vestnik VKTU, 3*.

Tabieva, E.E., Uazyrhanova, G.K., Popova, N.A., & Nikonenko E.L. (2020). Vliyanie poverhnostnoj zakalki na izmenenie strukturno-fazovogo sostoyaniya st2. *Fundamentalnye problemy sovremennogo materialovedeniya, 2*, 169-175.

Tabieva, E. E., Zhurerova, L. G., & Baizhan, D. (2020). Influence of Electrolyte-Plasma Hardening Technological Parameters on the Structure and Properties of Banding Steel 2. *Key Engineering Materials*, *839*, 57–62. doi:10.4028/www.scientific.net/KEM.839.57

Tabiyeva, E., Rahadilov, B., Uazyrkhanova, G., Satbaeva, Z., & Kengesbekov A. (2019). Influence of electrolytic-plasma surface hardening on the surface properties of bandage steel. *TRIBOLOGIA, 4*, 105-111.

Tabiyeva, Y. Y., Popova, N. A., & Nikonenko, E. L. (2020). Influence of electrolytic-plasma treatment on the morphological components of the matrix of steels of ferrite-perlite class. *International Scientific and Technical Conference PRFN-2020*. Tomsk.

Tabiyeva, Y. Y., Rakhadilov, B. K., Uazyrkhanova, G. K., Zhurerova, L. G., Sagdoldina, Zh. B., & Kurbanbekov, S. K. (2019). *Research of change in tribological properties of steel grade two after electrolytic-plasma exposure*. 11th International Scientific Conference "Chaos and Structures in Nonlinear Systems, Karaganda..

Tang, X., Guo, X., Wu, W., & Wang, G. (2018). 2D Metal Carbides and Nitrides (MXenes) as High-Performance Electrode Materials for Lithium-based Batteries. *Advanced Energy Materials*, *8*(18), 1801897. doi:10.1002/aenm.201801897

Teske, S., Niklas, S., & Talwar, S. (2022). Decarbonisation Pathways for Industries. In S. Teske (Ed.), *Achieving the Paris Climate Agreement Goals: Part 2: Science-based Target Setting for the Finance industry — Net-Zero Sectoral 1.5þC Pathways for Real Economy Sectors* (pp. 81–129). Springer Nature.

Tey, K. C., & Lui, J. L. (2014). The effect of glass fiber-reinforced epoxy resin dowel diameter on the fracture resistance of endodontically treated teeth. *Journal of Prosthodontics*, *23*(7), 572–581. doi:10.1111/jopr.12146 PMID:24750324

The Leathersellers' Company. (2020). *Leather manufacturing process*. Retrieved October 05, 2022, from https://leathersellers.co.uk/leather-industry/leather-manufacturing-process/

Theron, J., & Walker, J. (2008). *Nanotechnology and water treatment: applications and emerging opportunities*. Taylor & Francis. doi:10.1080/10408410701710442

Thi Quyen, V., Jitae, K., Thi Huong, P., Thi Thu Ha, L., My Thanh, D., Minh Viet, N., & Quang Thang, P. (2021). Copper doped titanium dioxide as a low-cost visible light photocatalyst for water splitting. *Solar Energy*, *218*, 150–156. doi:10.1016/j.solener.2021.02.036

Thirumalaikumarasamy, D., Shanmugam, K., & Balasubramanian, V. (2014). Corrosion performance of atmospheric plasma sprayed alumina coatings on AZ31B magnesium alloy under immersion environment. *Journal of Asian Ceramic Societies*, *2*(4), 403–415. doi:10.1016/j.jascer.2014.08.006

Thornton, J. A. (1982). *Deposition technologies for films and coating*. Noyes Publications.

Tian, L., Yang, X., Lu, P., Williams, I. D., Wang, C., Ou, S., Liang, C., & Wu, M. (2008). Hollow single-crystal spinel nanocubes: The case of zinc cobalt oxide grown by a unique Kirkendall effect. *Inorganic Chemistry*, *47*(13), 5522–5524. doi:10.1021/ic702457b PMID:18507455

Tilbrook, D. A., Clarke, R. L., Howle, N. E., & Braden, M. (2000). Photocurable epoxy-polyol matrices for use in dental composites I. *Biomaterials*, *21*(17), 1743–1753. doi:10.1016/S0142-9612(00)00059-4 PMID:10905456

Tiwari, A., Krishna, N. V., Giribabu, L., & Pal, U. (2018). Hierarchical Porous TiO_2 Embedded Unsymmetrical Zinc-Phthalocyanine Sensitizer for Visible-Light-Induced Photocatalytic H2 Production. *The Journal of Physical Chemistry C*, *122*(1), 495–502. doi:10.1021/acs.jpcc.7b09759

Tlili, I., & Alkanhal, T. A. (2019). Nanotechnology for water purification: Electrospun nanofibrous membrane in water and wastewater treatment. *Journal of Water Reuse and Desalination*, *9*(3), 232–248. doi:10.2166/wrd.2019.057

Toloman, D., Popa, A., Stefan, M., Silipas, T. D., Suciu, R. C., Barbu-Tudoran, L., & Pana, O. (2020). Enhanced photocatalytic activity of Co doped SnO_2 nanoparticles by controlling the oxygen vacancy states. *Optical Materials*, *110*, 110472. Advance online publication. doi:10.1016/j.optmat.2020.110472

Tomilov A.P. (1968). Elektrokhimiya pishchevykh soyedineniy. Izd-vo *Khimiya*, 115.

Torres-Martínez, J. A., Mora, A., Mahlknecht, J., Kaown, D., & Barceló, D. (2021). Determining nitrate and sulfate pollution sources and transformations in a coastal aquifer impacted by seawater intrusion—A multi-isotopic approach combined with self-organizing maps and a Bayesian mixing model. *Journal of Hazardous Materials*, *417*, 126103. Advance online publication. doi:10.1016/j.jhazmat.2021.126103 PMID:34229392

Troyannikov, A.Y. (2021). Sravnitelnyj analiz vysokoenergeticheskih metodov poverhnostnogo uprochneniya stali. *Molodoj uchenyj, 19,* (361).

Troyan, V. I., Pushkin, M. A., Borman, V. D., & Tronin, V. N. (2008). *Fizicheskie osnovi metodov issledovania nanostruktur i poverhnosti tverdogo tela*. MIFI.

Tsai, D. C., Huang, Y. L., Lin, S. R., Liang, S.-C., & Shieu, F.-S. (2010). Effect of nitrogen flow ratios on the structure and mechanical properties of (TiVCrZrY)N coatings prepared by reactive magnetron sputtering. *Applied Surface Science*, *257*(4), 1361–1367. doi:10.1016/j.apsusc.2010.08.078

Types of Drinking Water Contaminants | US EPA. (n.d.). Retrieved July 29, 2022, from https://www.epa.gov/ccl/types-drinking-water-contaminants

Tyuftyaev, A. S. (2002). *Zakonomernosti strukturoobrazovaniya pri plazmennoj obrabotke stali 60G i promyshlennoe osvoenie tehnologii poverhnostnogo uprochneniya grebnej zheleznodorozhnyh kolesnyh par*. FGUP CNIIchermet im. I.P.Bardina. Moskva.

Tyuftyaev, A.C. (2002). *Zakonomernosti strukturoobrazovaniya pri plazmennoj obrabotke stali 60 G i promyshlennoe osvoenie tehnologii poverhnostnogo uprochneniya grebnej zheleznodorozhnyh kolesnyh par*. OIVTRAN.

Tyuirin, Y.N. & Pogrebnjak, A.D. (2001). *Electric Heating Using a Liquid Electrode*, 142-144. Surf. And Coat. Tech.

Tyurin, Y. (1999). *Technology and equipment electrolytic-plasma quenching*. 3rd Intern. conf. on quenching and control of distortion. Prague.

Tyurin, Y. (2000). Features and advantages of surface modification of metallic pieces using electrolytic plasma thermal cycling hardening. *Plasma Surface Eng*, 1230-1233.

Tyurin, Y.I. (1999). *Osobennosti elektrolitno-plazmennoj zakalki. 3-ya mezhdunarodnaya konferenciya. Vzaimodejstvie izluchenij s tverdym telom*. Minsk.

Uazyrkhanova, G. K., Rakhadilov, B. K., Tabieva, E. E., & Baizhan, D. (2020). The influence of electrolyte-plasma surface hardening on the strength characteristics of grade two steel. *VI International Scientific and Practical Conference "Science and education in the modern world: challenges of the XXI century"*. Nur-Sultan.

Uddin, M. T., Nicolas, Y., Olivier, C., Toupance, T., Servant, L., Müller, M. M., Kleebe, H. J., Ziegler, J., & Jaegermann, W. (2012). Nanostructured SnO_2-ZnO heterojunction photocatalysts showing enhanced photocatalytic activity for the degradation of organic dyes. *Inorganic Chemistry*, *51*(14), 7764–7773. doi:10.1021/ic300794j PMID:22734686

Uglov, V. V., Chereda, N. N., & Anishin, V. M. (2007). *Metodi analiza elementnogo sostava poverhnostnih sloev. Learning guide*. Belarusian State University.

Ulianitsky, V. Y., Batraev, I. S., Shtertser, A. A., Dudina, D. V., Bulina, N. V., & Smurov, I. (2018). Detonation spraying behaviour of refractory metals: Case studies for moand Ta-based powders. *Advanced Powder Technology*, *29*(8), 1859–1864. doi:10.1016/j.apt.2018.04.023

Ulianitsky, V., Shtertser, A., Zlobin, S., & Smurov, I. (2011). Computer-Controlled Detonation Spraying: From Process Fundamentals Toward Advanced Applications. *Journal of Thermal Spray Technology*, *20*(4), 791–801. doi:10.100711666-011-9649-6

Uma, H. B., Ananda, S., & Nandaprakash, M. B. (2019). High efficient photocatalytic treatment of textile dye and antibacterial activity via electrochemically synthesized Ni-doped ZnO nano photocatalysts. *Chemical Data Collections*, *24*, 100301. Advance online publication. doi:10.1016/j.cdc.2019.100301

Umanskiy, Ya. S., Skakov, Yu. A., Ivanov, A. N., & Rastorguyev, L. N. (1982). *Kristallografiya, rentgenografiya i elektronnaya mikroskopiya*. Metallurgiya.

UMass Memorial Health. (2015). *Typical and Atypical Angina: What to Look For.* UMass Memorial Health. https://www.harringtonhospital.org/typical-and- atypical-angina- what-to-look-for/

Understanding global change. (2022). *Pollutants and waste.* Retrieved November 10, 2022, from https://ugc.berkeley.edu/background-content/pollutants-and-waste/

United Nations, Industrial Development Organization. (2017). *Leather carbon footprint, Review of the European Standard EN 16887.* Author.

University of California San Francisco Health. (2022). *Understanding your risk for heart disease.* https://www.ucsfhealth.org/education/understanding-your-risk-for-heart-disease

University of Oxford. (2022). *What is net zero?* Retrieved July 05, 2022, from https://netzeroclimate.org/what-is-net-zero/

Usmanov, K. B., & Yakunin, G. I. (1984). Vliyaniye vneshney sredy na iznos i stoykost' obnaruzheniya instrumentov. Tashkent, 158.

Ustinovshchikov, Y. U. I. (1988). *Vydeleniye vtoroy fazy v tverdykh rastvorakh.* Nauka, 172.

Utari, A. W., & Herdiansyah, H. (2020). Using filtration as a technology to remove pollutants in domestic wastewater. *IOP Conference Series. Materials Science and Engineering, 725*(1), 012025. Advance online publication. doi:10.1088/1757-899X/725/1/012025

Utevskij, L.M. (1973). *Difrakcionnaya elektronnaya mikroskopiya v metallovedenii.* Metallurgiya.

Utevskiy L.M. (1973). Difraktsionnaya elektronnaya mikroskopiya v metallosoderzhanii. *Metallurgiya*, 584.

Van Gyui, V. (2015). Parovodyanaya plazmennaya cementaciya v tverdoj faze na ustanovke MULTIPLAZ 3500. *Innovacii v nauke, 51* (1), 95-102.

Van Gyui, V. (2015). Plazmennaya cementaciya uglerodistyh stalej s ispolzovaniem past v tverdoj faze. *Vestnik nauki i obrazovaniya Severo-Zapada Rossii, 1*, 205-211.

Van Viet, P., Thi, C. M., & Van Hieu, L. (2016). The High Photocatalytic Activity of SnO_2 Nanoparticles Synthesized by Hydrothermal Method. *Journal of Nanomaterials, 4231046*, 1–8. Advance online publication. doi:10.1155/2016/4231046

Vanin, V. S. (1967). Nagrev metallov v elektrolite. *Elektrotermiya*, (55), 18–19.

Vargas Hernández, J., Coste, S., García Murillo, A., Carrillo Romo, F., & Kassiba, A. (2017). Effects of metal doping (Cu, Ag, Eu) on the electronic and optical behavior of nanostructured TiO_2. *Journal of Alloys and Compounds, 710*, 355–363. doi:10.1016/j.jallcom.2017.03.275

Vasanthan, N., & Wang, Y. (2013). Polymer nanocomposites for energy storage. *Energy Storage Materials, 1*(2), 82–102.

Venkateshalu, S., Subashini, G., Bhardwaj, P., Jacob, G., Sellappan, R., Raghavan, V., Jain, S., Pandiaraj, S., Natarajan, V., Al Alwan, B. A. M., Al Mesfer, M. K. M., Alodhayb, A., Khalid, M., & Grace, A. N. (2022). Phosphorene, Antimonene, Silicene and Siloxene based Novel 2D Electrode Materials for Supercapacitors – A Brief Review. *Journal of Energy Storage, 48*, 104027. doi:10.1016/j.est.2022.104027

Veprek S. (2013). Recent search of new superhard materials: Go nano. *Journal Vacuum Science Technology A*(31), 1-33.

Verma, N., Kumar, R., Zafar, S., & Pathak, H. (2020). Vacuum-assisted microwave curing of epoxy/carbon fiber composite: An attempt for defect reduction in processing. *Manufacturing Letters, 24*, 127–131. doi:10.1016/j.mfglet.2020.04.010

Vershinin D.S., Smolyakova M.Yu., Manokhin S.S., Druchinina O.A. & Akhmadeyev Yu.H. (2010). Issledovaniye tribologicheskikh svoystv azotirovannogo titanovogo splava VT16. Zavodskaya laboratoriya. *Diagnostika materialov, 76*(12), 45-49.

Veynberga F. (1973). Pribory i metody fizicheskogo metallovedeniya, (1), 427.

Vignesh, Rajadesingu, Suriyaprakash, & Kantha. (2021). Challenges, issues, and problems with zero-waste tools. In Concepts of Advanced Zero Waste Tools (pp. 69-90). Elsevier.

Vikesland, P. J., & Wigginton, K. R. (2010). Nanomaterial enabled biosensors for pathogen monitoring - A review. *Environmental Science & Technology*, *44*(10), 3656–3669. doi:10.1021/es903704z PMID:20405829

Vitale, A., Sangermano, M., Bongiovanni, R., Burtscher, P., & Moszner, N. (2014). Visible light curable restorative composites for dental applications based on epoxy monomer. *Materials (Basel)*, *7*(1), 554–562. doi:10.3390/ma7010554 PMID:28788474

Voevodin, A. A., Shtansky, D. V., Levashov, E. A., & Moore, J. J. (2004). *Nanostructured Thin Films and Nanodispersion Strengherred Coatings*. Kluger Academic. doi:10.1007/1-4020-2222-0

Vorob'yeva, G. A., & Skladnova, Ye. Ye. (2003). Instrumental'nyye stal'nyye materialy: Instrumental'nyye i splavy. *SPb.*, (1), 100.

Vorob'yeva, G.A., & Skladnova, Y.Y. (2003). *Instrumental'nyye materialy: instrumental'nyye stali i splavy,* 100.

Vu, T. H. T., Au, H. T., Tran, L. T., Nguyen, T. M. T., Tran, T. T. T., Pham, M. T., Do, M. H., & Nguyen, D. L. (2014). Synthesis of titanium dioxide nanotubes via one-step dynamic hydrothermal process. *Journal of Materials Science*, *49*(16), 5617–5625. doi:10.100710853-014-8274-4

Vyacheslav, F.B. & Szczerek, M. (2018). The influence of surface layer quality on the performance characteristics of machine parts under machining. *Tribologia, 271*(1), 23-29.

Wachem, P. B., Brouwer, L. A., Zeeman, R., Dijkstra, P. J., Feijen, J., Hendriks, M., Cahalan, P. T., & van Luyn, M. J. A. (2000). *In vivo* behavior of epoxy-crosslinked porcine heart valve cusps and walls. *Journal of Biomedical Materials Research*, *53*(1), 18–27. doi:10.1002/(SICI)1097-4636(2000)53:1<18::AID-JBM3>3.0.CO;2-J PMID:10634948

Wagner Ferreira Sabará, E., Pereira, V., Luiz Molisani, A., & Mary Ealias, A. (2017). A review on the classification, characterisation, synthesis of nanoparticles and their application. *IOP Conference Series. Materials Science and Engineering*, *263*(3), 032019. doi:10.1088/1757-899X/263/3/032019

Wagner, C. D. (1995). Handbook of XPS. Physical Electronics.

Wang, B., Jin, X., Xue, W., Wu, Z., Du, J., & Wu. J. (2013). High temperature tribological behaviors of plasma electrolytic borocarburized Q235 low-carbon steel. *Surface & Coatings Technology, 232*, 142-149.

Wang, Y., Sun, C., Zhao, X., Cui, B., Zeng, Z., Wang, A., Liu, G., & Cui, H. (2016). The Application of Nano-TiO2 Photo Semiconductors in Agriculture. In Nanoscale Research Letters (Vol. 11, Issue 1). Springer New York LLC. doi:10.118611671-016-1721-1

Wang, C., & Cao, L. (2011). Preparation, spectral characteristics and photocatalytic activity of Eu^{3+}-doped WO_3 nanoparticles. *Journal of Rare Earths*, *29*(8), 727–731. doi:10.1016/S1002-0721(10)60531-5

Wang, G., & Coppens, M. O. (2010). Rational design of hierarchically structured porous catalysts for autothermal reforming of methane. *Chemical Engineering Science*, *65*(7), 2344–2351. doi:10.1016/j.ces.2009.09.079

Wang, H., Liu, B., Liu, X., Zhang, J., & Xian, M. (2008). Synthesis of biobased epoxy and curing agents using rosin and the study of cure reactions. *Green Chemistry*, *10*(11), 1190–1196. doi:10.1039/b803295e

Wang, J. L., & Xu, L. J. (2012). Advanced oxidation processes for wastewater treatment: Formation of hydroxyl radical and application. *Critical Reviews in Environmental Science and Technology*, *42*(3), 251–325. doi:10.1080/10643389.2010.507698

Wang, J., Fang, L., & Lopez, D. (1994). Amperometric biosensor for phenols based on a tyrosinase-graphite-epoxy biocomposite. *Analyst (London)*, *119*(3), 455–458. doi:10.1039/AN9941900455 PMID:8192231

Wang, L., Hu, H., Xu, J., Zhu, S., Ding, A., & Deng, C. (2019). WO_3 nanocubes: Hydrothermal synthesis, growth mechanism, and photocatalytic performance. *Journal of Materials Research*, *34*(17), 2955–2963. doi:10.1557/jmr.2019.189

Wang, L., Pyzalla, A., Stadlbauer, W., & Werner, E. A. (2003). Microstructure features on rolling surfaces of railway rails subjected. *Materials Science and Engineering A*, *359*(1-2, A359), 31–43. doi:10.1016/S0921-5093(03)00327-7

Wang, T., Wang, H., Kou, Z., Liang, W., Luo, X., Verpoort, F., Zeng, Y.-J., & Zhang, H. (2020). Xenes as an Emerging 2D Monoelemental Family: Fundamental Electrochemistry and Energy Applications. *Advanced Functional Materials*, *30*(36), 2002885. doi:10.1002/adfm.202002885

Wang, Y., Yang, C., Liu, Y., Fan, Y., Dang, F., Qiu, Y., Zhou, H., Wang, W., & Liu, Y. (2021). Solvothermal Synthesis of ZnO Nanoparticles for Photocatalytic Degradation of Methyl Orange and p-Nitrophenol. *Water (Basel)*, *13*(22), 3224. Advance online publication. doi:10.3390/w13223224

Wang, Z. L. (2004). Zinc oxide nanostructures: Growth, properties and applications. *Journal of Physics Condensed Matter*, *16*(25), R829–R858. Advance online publication. doi:10.1088/0953-8984/16/25/R01

WebMD Editorial Contributors. (2022). *How do heart disease risk factors differ for men? Men and Heart Disease.* webMD. https://www.webmd.com/heart-disease/heart-disease- men#:~:text=Men%20are%20also%20more%20prone,raises%20risk%20for%20heart%20disease

Weinberg, F. (1973). *Pribory i metody fizicheskogo metallovedeniya, 1.*

Wetzel, B., Haupert, F., & Zhang, M. Q. (2003). Epoxy nanocomposites with high mechanical and tribological performance. *Composites Science and Technology*, *63*(14), 2055–2067. doi:10.1016/S0266-3538(03)00115-5

World Health Organization. (2020). *HEARTS: technical package forcardiovascular disease management in primary health care: Risk-basedCVD management.* World Health Organisation Publications. https://www.who.int/publications/i/item/9789240001367

World Health Organization. (2021). *What are the risk factors of cardiovasculardisease?* Cardiovascular Disease. https://www.who.int/news- room/factsheets/detail/cardiovascular-diseases-

World Health Organization. (2022). *Key facts. Menopause.* Retrieved online on 5th May 2023 from https://www.who.int/news-room/fact-sheets/detail/menopause

Wu C., Hu H., Chou Y., Huang N., Chou Y., Li C. (2015). *High blood pressure and All-Cause and cardiovascular Disease Mortalities in Community-Dwelling Older Adults.* . doi:10.1097/MD.0000000000002160

Wu, Y., Liu, Z., Li, Y., Chen, J., Zhu, X., & Na, P. (2019). WS 2 nanodots-modified TiO_2 nanotubes to enhance visible-light photocatalytic activity. *Materials Letters*, *240*, 47–50. doi:10.1016/j.matlet.2018.12.056

Xiao, H., Wu, Z.-S., Chen, L., Zheng, S., Ren, W., Cheng, H.-M., & Bao, X. (2017). One-Step Device Fabrication of Phosphorene and Graphene Interdigital Micro-Supercapacitors with High Energy Density. *ACS Nano*, *11*(7), 7284–7292. doi:10.1021/acsnano.7b03288 PMID:28628293

Xu, Y., Xu, Y., & Qian, J. (2019). Flame Retardants: Mechanisms, Current Trends, and Future Directions. In Advances in Flame Retardant Polymeric Materials (pp. 3-30). Elsevier.

Xu, F., Shen, Y., Sun, L., Zeng, H., & Lu, Y. (2011). Enhanced photocatalytic activity of hierarchical ZnO nanoplate-nanowire architecture as environmentally safe and facilely recyclable photocatalyst. *Nanoscale*, *3*(12), 5020–5025. doi:10.1039/c1nr11033k PMID:22057641

Yakuphanoglu, F. (2010). Nanocluster n-CdO thin film by sol-gel for solar cell applications. *Applied Surface Science*, *257*(5), 1413–1419. doi:10.1016/j.apsusc.2010.08.045

Yamamoto, K., Sato, T., Takahara, K., & Hanaguri, K. (2003). Properties of (Ti,Cr,Al)N coatings with high Al content deposited by new plasma enhanced arc-cathode. *Surface and Coatings Technology*, *174-175*(174), 620–626. doi:10.1016/S0257-8972(03)00580-2

Yang, H. G., Sun, C. H., Qiao, S. Z., Zou, J., Liu, G., Smith, S. C., Cheng, H. M., & Lu, G. Q. (2008). Anatase TiO2 single crystals with a large percentage of reactive facets. *Nature*, *453*(7195), 638–641. doi:10.1038/nature06964

Yang, B., Hao, C., Wen, F., Wang, B., Mu, C., Xiang, J., Li, L., Xu, B., Zhao, Z., Liu, Z., & Tian, Y. (2017). Flexible Black-Phosphorus Nanoflake/Carbon Nanotube Composite Paper for High-Performance All-Solid-State Supercapacitors. *ACS Applied Materials & Interfaces*, *9*(51), 44478–44484. doi:10.1021/acsami.7b13572 PMID:29192760

Yang, K., & Xing, B. (2010). Adsorption of organic compounds by carbon nanomaterials in aqueous phase: Polanyi theory and its application. *Chemical Reviews*, *110*(10), 5989–6008. doi:10.1021/cr100059s PMID:20518459

Yao, G., An, X., Lei, H., Fu, Y., & Wu, W. (2013). Electronic and Optical Properties of Rocksalt CdO: A first-Principles Density-Functional Theory Study. In *Modeling and Numerical Simulation of Material Science* (Issue 3). http://www.SciRP.org/journal/mnsms

Yao, P., Liu, H., Wang, D., Chen, J., Li, G., & An, T. (2018). Enhanced visible-light photocatalytic activity to volatile organic compounds degradation and deactivation resistance mechanism of titania confined inside a metal-organic framework. *Journal of Colloid and Interface Science*, *522*, 174–182. doi:10.1016/j.jcis.2018.03.075 PMID:29601959

Yaqoob, A. A., Parveen, T., Umar, K., Nasir, M., & Ibrahim, M. (n.d.). Role of nanomaterials in the treatment of wastewater: A review. *Mdpi.Com*. doi:10.3390/w12020495

Yasnorodskiy I.Z. (1971). Elektrolitnyy nagrev metallov. Elektrokhimicheskaya i elektromekhanicheskaya obrabotka metallov. *Mashinostroyeniye*, 117-168.

Yavari, H., Ghasemi, N., Divband, B., Rezaei, Y., Jabbari, G., & Payahoo, S. (2017). The effect of photodynamic therapy and polymer solution containing nano-particles of Ag/ZnO on push-out bond strength of the sealers AH-Plus and MTA Fillapex. *Journal of Clinical and Experimental Dentistry*, *9*(9), e1109–e1114. doi:10.4317/jced.54069 PMID:29075413

Yelyutina O.P. (1979). Prakticheskiye voprosy ispytaniy metallov. *Metallurgiya*, 280.

Yerokhin, A.L., Nie, X., Leyland, A., Matthews, A., & Dowey S.J. (1999). Plasma electrolysis for surface engineering. *Surface and Coatings Technology*, *122*, 73-93.

Yerokhin, A. L., Nie, X., Leyland, A., Matthews, A., & Dowey, S. J. (1999). Plasma electrolysis for surface engineering. *Surface and Coatings Technology*, *122*(2-3), 73–93. doi:10.1016/S0257-8972(99)00441-7

Younesi, Y., Bahrololoom, M. E., & Fooladfar, H. (2010). Development of Wear Resistant NFSS-HA Novel Biocomposites and Study of their Tribological Properties for Orthopedic Applications. *Journal of the Mechanical Behavior of Biomedical Materials, 3*(3), 178–188. doi:10.1016/j.jmbbm.2009.08.003 PMID:20129417

Yudoyono, G., Ichzan, N., Zharvan, V., Daniyati, R., Santoso, H., Indarto, B., Pramono, Y. H., Zainuri, M., & Darminto. (2016). Effect of calcination temperature on the photocatalytic activity of TiO_2 powders prepared by co-precipitation of $TiCl_3$. *AIP Conference Proceedings, 1725*(1), 20099. doi:10.1063/1.4945553

Yu, J., & Yu, X. (2008). Hydrothermal Synthesis and Photocatalytic Activity of Zinc Oxide Hollow Spheres. *Environmental Science & Technology, 42*(13), 4902–4907. doi:10.1021/es800036n PMID:18678024

Yushenko, K.A., Borisov, Y.S., & Tyurin, Y.N. (1994). *Teoriya i praktika impulsno-plazmennogo modificirovaniya poverhnosti detalej mashin i instrumenta.* Svarochnoe proizvodstvo. Kiev.

Zaides S.A., Van Huy, V., & Van, D. (2020). Developing installation to increase cylindrical part surface hardness. *Proceedings of Irkutsk State Technical University, 24*(2), 262-274.

Zaokari, Y., Persaud, A., & Ibrahim, A. (2020). Biomaterials for adhesion in orthopedic applications: A review. *Engineered Regeneration, 1*, 51–63. doi:10.1016/j.engreg.2020.07.002

Zargar, R. A., Chackrabarti, S., Arora, M., & Hafiz, A. K. (2016). Synthesis, characterization and interpretation of screen-printed nanocrystalline CdO thick film for optoelectronic applications. *International Nano Letters, 6*(2), 99–104. doi:10.100740089-015-0172-5

Zekić, E., Vuković, Ž., Halkijević, I., Zelić, E., Hidrokon doo, M., & Ivan Halkijević, A. (2018). Primjena nanotehnologije u pročišćavanju otpadnih voda. *Casopis-Gradjevinar.Hr, 4*, 315–323. doi:10.14256/JCE.2165.2017

Zhang P., Nie X., & Northwood D.O. (2009). Influence of coating thickness on the galvanic corrosion properties of Mg oxide in an engine coolant. *Surface & Coatings Technology, 203*, 3271-3277.

Zhang W. (2002). Kinetics of Ferrite to Austenite Phase Transformation during Welding of 1005 Steel. *Scripta Materialia, 46*. 753-757.

Zhang W., Elmer J.W., & Debroy T. (2005). Integrated modeling of thermal cycles, austenite formation, grain growth and decomposition in the heat affected zone of carbon steel. *Science and Technology of Welding and Joining, 10* (5), 574-582.

Zhang, W. (2002). Integrated modeling of thermal cycles, austenite formation, grain growth and decomposition in the heat affected zone of carbon steel. *Science and Technology of Welding and Joining, 5*, 574-582.

Zhang, W. (2002). Modeling and Real Time Mapping of Phases during GTA Welding of 1005 Steel. *Materials Science and Engineering A, 333*, 320-335.

Zhang, G., Ren, J., Liu, B., Tian, M., Zhou, H., & Zhao, J. (2018). In situ hydrothermal preparation and photocatalytic desulfurization performance of metallophthalocyanine sensitized SnO_2. *Inorganica Chimica Acta, 471*, 782–787. doi:10.1016/j.ica.2017.12.025

Zhang, J., Zhou, P., Liu, J., & Yu, J. (2014). New understanding of the difference of photocatalytic activity among anatase, rutile and brookite TiO_2. *Physical Chemistry Chemical Physics, 16*(38), 20382–20386. doi:10.1039/C4CP02201G PMID:25144471

Zhang, P., Nie, X., & Northwood, D. O. (2009). Influence of coating thickness on the galvanic corrosion properties of Mg oxide in an engine coolant. *Surface and Coatings Technology, 203*(20-21), 3271–3277. doi:10.1016/j.surfcoat.2009.04.012

Zhang, Q., Liu, P., Miao, C., Chen, Z., Lawrence Wu, C. M., & Shek, C. H. (2015). Formation of orthorhombic SnO_2 originated from lattice distortion by Mn-doped tetragonal SnO2. *RSC Advances*, *5*(49), 39285–39290. doi:10.1039/C5RA04946F

Zhang, S. H., Cho, T. Y., Yoon, J. H., Fang, W., Song, K. O., Li, M. X., Joo, Y. K., & Lee, C. G. (2008). Characterization of microstructure and surface properties of hybrid coatings of WC–CoCr prepared by laser heat treatment and high velocity oxygen fuel spraying. *Materials Characterization*, *59*(10), 1412–1418. doi:10.1016/j.matchar.2008.01.003

Zhang, S., & Pan, N. (2015). Supercapacitors Performance Evaluation. *Advanced Energy Materials*, *5*(6), 1401401. doi:10.1002/aenm.201401401

Zhang, T. (2020). *Heterogeneous Catalytic Process for Wastewater Treatment*. Advanced Oxidation Processes - Applications, Trends, and Prospects. doi:10.5772/intechopen.90393

Zhang, Y., Huang, Q., Liu, J., Zhou, J.-E., Lin, X., Zeb, A., Reddy, R. C. K., & Xu, X. (2022). Recent Advances in Fe-based Metal-Organic Framework Derivatives for Battery Applications. *Sustainable Energy & Fuels*, *6*(11), 2665–2691. doi:10.1039/D2SE00319H

Zheng, J. Y., Song, G., Hong, J., Van, T. K., Pawar, A. U., Kim, D. Y., Kim, C. W., Haider, Z., & Kang, Y. S. (2014). Facile fabrication of WO_3 nanoplates thin films with dominant crystal facet of (002) for water splitting. *Crystal Growth & Design*, *14*(11), 6057–6066. doi:10.1021/cg5012154

Zhou, J., Chen, J., Chen, M., Wang, J., Liu, X., Wei, B., Wang, Z., Li, J., Gu, L., Zhang, Q., Wang, H., & Guo, L. (2019). Few-Layer Bismuthene with Anisotropic Expansion for High-Areal-Capacity Sodium-Ion Batteries. *Advanced Materials*, *31*(12), 1807874. doi:10.1002/adma.201807874 PMID:30714223

Zhou, Z. F., Rahat, B., & Rubbaniy, G. (2021). Carbon neutrality target for leading exporting countries: On the role of economic complexity index and renewable energy electricity. *Journal of Environmental Management*, *299*, 113558. doi:10.1016/j.jenvman.2021.113558 PMID:34425500

Zhu, D., & Zhou, Q. (2019). Action and mechanism of semiconductor photocatalysis on degradation of organic pollutants in water treatment: A review. In Environmental Nanotechnology, Monitoring and Management (Vol. 12). Elsevier B.V. https://doi.org/ doi:10.1016/j.enmm.2019.100255

Zhu, C., Zhang, W., Wang, L., Ning, Z., Feng, Y., Feng, K., Liao, J., Yang, Y., Liu, N., & Yang, J. (2020). Effect of thermal cycles on structure and deuterium permeation of Al2O3 coating prepared by MOD method. *Fusion Engineering and Design*, *159*, 111750. doi:10.1016/j.fusengdes.2020.111750

Zhu, J. P., Xiao, G. S., & Zuo, X. X. (2020). Two-Dimensional Black Phosphorus: An Emerging Anode Material for Lithium-Ion Batteries. *Nano-Micro Letters*, *12*(1), 120. doi:10.100740820-020-00453-x PMID:34138144

Zhu, M., Kou, H., Wang, K., Wu, H., Ding, D., Zhou, G., & Ding, S. (2020). Promising Functional Two-Dimensional Lamellar Metal Thiophosphates: Synthesis Strategies, Properties and Applications. *Materials Horizons*, *7*(12), 3131–3160. doi:10.1039/D0MH00802H

Zou, Y., Jin, X., Zhang, X., Kong, X., Zhang, Q., Xie, X., Liu, C., Ke, L., Liu, W. G., & Wang, W. (2021). A multifunctional biomedical patch based on hyperbranched epoxy polymer and MXene. *Science China. Technological Sciences*, *64*(12), 2744–2754. doi:10.100711431-021-1843-3

Zubtsov M.E., & Korsakov V.D. (1971). Stoykost' shtampov. *Mashinostroenie*, 200.

About the Contributors

Zarina Aringozhina is a Master of technical sciences in the specialty "6M072300-Technical Physics", Deputy Dean Faculty of Basic Engineering Training, D. Serikbayev East Kazakhstan Technical University. Qualifications: - 6M072300-Technical Physics MSc - D. Serikbayev East Kazakhstan Technical University 2015; - 6B04101 - Economy BSc - Sarsen Amanzholov East Kazakhstan University 2013; - 050716- Instrumentation BSc - D. Serikbayev East Kazakhstan Technical University 2011. Internships: - MSc - Wroclaw Polytechnic University, Poland 2015; - Program of Bolashak – University of Lincoln, United Kingdom 2022-23. Zarina Aringozhina is actively engaged in research work, is a highly qualified scientist in the field of materials science, condensed matter physics, analysis of patterns of formation of nanocrystalline coatings and films, studying the correlation of their structural-phase state and properties, modification of the surface of metals and alloys and technology for obtaining new materials.

M. P. Chandresh is working as Teaching Faculty in the Department of Polymer Science and Technology, SJCE, JSSSTU, Mysuru. Secured Bachelor of Engineering, Master of Technology and PhD degrees in the same domain. Has more than 11 years of teaching and 10 years of research experience. Has two patents and four papers in International Journals. Main area of interest includes fracture safety enhancement and replacement of carcinogenic oils using liquid rubbers. 'Passionate in teaching – learning process'. 'Engaged in inter-disciplinary studies for holistic life experience & quality system establishment'. 'Freelance counselor for: i) overall personality development based on Universal Human Values & ii) technical consultations in the domain of polymer technology based on 'DiLiPo' approach'.

Ahmed Elseragy holds a PhD from the University of Nottingham. Ahmed is the Director of Enterprise at the School of Engineering at the University of Lincoln since 2019. Ahmed is also an Associate Professor of Engineering Management and Programme leader for MSc Engineering Management. Between March 2021 and February 2022 Ahmed was Interim Head of School of Engineering. Ahmed is HEA Senior Fellow (AdvanceHE) since 2020. Prior to joining Lincoln Ahmed held the position of Professor of Sustainable Buildings and Architectural Engineering and the Head of School of Architectural Engineering and Environmental Design (AEED), College of Engineering, at the Arab Academy for Science and Technology and Maritime Transportation (AASTMT), Alexandria, Egypt. He has held various leadership and management roles where he was the Deputy Director of the International Research and Projects Centre at AASTMT, with the responsibility of coordinating partnerships with many higher education institutions in over 20 different countries (2016-2018). Dr Elseragy has worked as Internationalisation Consultant for the Higher Education Academy (HEA) currently part of AdvanceHE in the UK to develop and implement national higher education reform and internationalisation strategies for different countries in the MENA region. Ahmed has worked on many projects, that have included building university strategic partnerships in different countries and enhancing international institution cooperation in research, teaching and learning, and enterprise. He was a consultant for the Minister of Trade and Industry in Egypt advising on skills and professional development. Ahmed was the Dean of Internationalisation and Global Engagement between 2009-2012 at the Arab Academy for Science and Technology and Maritime Transportation, London office with the responsibility of the development and implementation of the institution's internationalisation strategy at home and abroad. His role included liaising with international universities, NGOs, and international governmental bodies including UN International Maritime Organisation IMO.

Mubeen Ghafoor is Senior Lecturer in Machine Learning at the School of Computer Science (SoCS), University of Lincoln, UK. Earlier he was Post-Doctoral Research Fellow in Data Science at the Dept. of Computer Science and Creative Technologies at the University of the West of England (UWE), UK. From 2015 to 2019, he worked as Assistant Professor at the Dept. of computer science, COMSATS University Islamabad (CUI). His areas of interest include Computer Vision Applications, Machine Learning, Data Analytics, Biometrics, Medical Imaging and Light-Field Data Compression.

A. B. Hemavathi is working as an Associate Professor, Department of Polymer Science and Technology, SJCE, JSS STU, Mysuru. Recipient of CSIR-GATE-SRF fellowship and SERB (TARE) research grant. Published more than 40 research papers in national and international peer reviewed journals and conference proceedings. Has more than 15 years of research and teaching experience.

R. S. Jagadish is a Professor & Dean Academics, Department of Chemistry, JSS Academy of Technical Education, Noida. He carried out his research in CFTRI and obtained his PhD in Polymer Science from University of Mysore and published more than 60 research papers in repute peer reviewed journals.

Dauir Kakimzhanov is a PhD student of East Kazakhstan Technical University named after D. Serikbayev (Kazakhstan).

Nurgamit Kantay is actively engaged in research work, is a highly qualified scientist in the field of materials science, condensed matter physics, analysis of patterns of formation of nanocrystalline coatings and films, studying the correlation of their structural-phase state and properties, modification of the surface of metals and alloys and technology for obtaining new materials. He is proficient in direct methods of materials and coatings research, as well as analytical research methods using TEM/SEM-EDX, AFM and XRD analyses.

Lizzie Mampane is a PhD researcher at the University of the Witwatersrand in the Chemistry department working on the synthesis of thin film membrane for the desalination of sea water. Lizzie has worked on different project including green nanotechnology and bio-renewable energy. She filed a patent for her masters work.

Sivuyisiwe Mapukata recently completed her PhD having specialized in nanotechnology and environmental remediation research. With a niche for applied chemistry, catalysis and materials fabrication, her research experience is mainly based on water purification and gas sensing. She is currently working on heterogenous photocatalysis.

Manuela Mura is a Senior Lecturer in College of Science. In 2010 Manuela received a doctorate from the King's College London, where she wrote a PhD thesis on "Theoretical characterization of STM images of assemblies of flat organic molecules on metal surfaces" under the supervision of Prof Lev Kantorovitch. The work included collaboration with experimentalists from the universities of Nottingham, Oxford, Aarus in Denmark and CAE Saclay in France. Manuela is an expert in a variety of computer simulation techniques in Condensed Matter Physics such as quantum mechanics ab-initio tools to reproduce self-assembled supra-molecular assemblies on metallic and semiconducting surfaces and classical molecular dynamics tools to study the interaction of antimicrobial peptides with modelled lipid bilayer membrane. Qualifications: PhD in Computational Physics - King's College London 2010; MSc in Physics - University of Cagliari 2004. Awards: Springer PhD Thesis Prize - Springer 2010; Tadion-Rideal Prize for Molecular Science - King's College London 2010.

Athule Ngqalakwezi has 5+ working R&D on different catalysis projects. She graduated from the University of the Witwatersrand with an MSc in Chemical engineering.

Mahfuzur Rahman is an Associate Professor at the University of Lincoln, UK. He is also an Associate Editor of the International Journal of Public Sociology and Sociotherapy (IJPSS) and the Journal of Management and Organisation. His current academic and research interests are in the area of Sustainability and Innovation with a focus on Emerging Markets. Investigating the potential environment friendly manufacturing, such as, adoption of textile enzymes, for a sustainable future motivates his most recent research works. Dr Rahman is an Academic Adviser to the Commonwealth Scholarship Commission, UK. As an adviser, Dr Rahman is representing UK Higher Education to Commonwealth countries. He is also playing a critical role in the selection of Commonwealth Scholar and Fellows. His research projects (as Principal Investigator/Co-Investigator) and Scholarships (as Supervisor) are funded by the Commonwealth Commission, Government of Bangladesh, British Council, Government of Ghana and UNIDO. He is an active team member of the University of Lincoln UNESCO Chair on Responsible Foresight for Sustainable Development. Dr Rahman has published in internationally refereed journals, presented his research at a range of international conferences, and has contributed to a number of research projects. He has received a number of awards including University of Lincoln Merit Award 2018, Individual Merit Award 2019, Two best paper awards, University of Lincoln Team Award 2018 and Technology and Management for Development Award 2008 by the University of Oxford. Dr Rahman has supervised 5 doctoral candidates towards successful completion at the University of Lincoln. Currently he is supervising six PhD students and one Post-Doctoral Fellow funded by Commonwealth Commission and University of Lincoln.

Bauyrzhan Rakhadilov is PhD, Associate Professor in the specialty "6D072300-Technical Physics", Director and founder of the PlasmaScience LLP, Republic of Kazakhstan. Bauyrzhan Rakhadilov is actively engaged in research work, is a highly qualified scientist in the field of materials science, condensed matter physics, analysis of patterns of formation of nanocrystalline coatings and films, studying the correlation of their structural-phase state and properties, modification of the surface of metals and alloys and technology for obtaining new materials. He is proficient in direct methods of materials and coatings research, as well as analytical research methods using TEM/SEM-EDX, AFM and XRD analyses. He is one of the best specialists in the field of surface modification of materials and tribology. It also has its own scientific school and followers, is an active promoter of science in Kazakhstan.

Yerkezhan Tabiyeva is a PhD in the specialty "6D072300-Technical Physics", Head of Department of Physics, Faculty of Basic Engineering Training, D. Serikbayev East Kazakhstan Technical University Qualifications • 6B072300 - Technical Physics PhD – D. Serikbayev East Kazakhstan Technical University 2021 • 6M060400 - Physics MSc - D. Serikbayev East Kazakhstan Technical University 2013 • 5B050110 - Physics BSc - S. Amanzholov East Kazakhstan University 2010 Internships • MSc - Wroclaw Polytechnic University, Poland 2011 • PhD - Wroclaw Polytechnic University, Poland 2019 • Program of Bolashak – University of Lincoln, United Kingdom 2022-2023 Yerkezhan Tabiyeva conducts active research work, is a highly qualified scientist in the field of materials science, condensed matter physics, analysis of the patterns of formation of hardening phases during high-temperature modifications of the metal surface, studying the relationship between the structural-phase state and properties. She is the leader and one of the developers of the educational program in the field of "Technical Physics".

Zhanerke Toleukhanova is a PhD student in the specialty "6D072300-Technical Physics" D. Serikbayev East Kazakhstan Technical University. Qualifications: - 6M072300-Technical Physics MSc - D. Serikbayev East Kazakhstan Technical University 2018; - 5B072300-Technical Physics BSc - D. Serikbayev East Kazakhstan Technical University 2014. Internships: - MSc - National Nuclear Center2014; - PhD - University of Lincoln, United Kingdom 2022. Zhanerke Toleukhanova is a 3rd year doctoral student. At the moment I have been working as a medical physicist in the multidisciplinary regional center of oncology and surgery as a medical physicist since 2010.

Gulzhaz Uazyrkhanova is PhD, Associate Professor in the specialty "6D072300-Technical Physics", Director of the Department of Research Activities, D. Serikbayev East Kazakhstan Technical University. Gulzhaz Uazyrkhanova is engaged in research work, is a highly qualified scientist in the field of materials science, condensed matter physics, analysis of the patterns of formation of hardening phases during deformation of metals and alloys, on the problems of structural-phase transformations of welded joint zones and deformation.

Rafael Vargas-Bernal received a Bachelor's degree in Communications and Electronics Engineering from the University of Guanajuato in 1995, and the degrees of Master of Science and Doctorate in Science with Specialty in Electronics from the National Institute of Astrophysics, Optics and Electronics (INAOE) in 1997 and 2000, respectively. He is a professor-researcher at the University of Guanajuato. He has edited two books. He has authored 17 articles in journals, 46 chapters in books, and about 150 conference articles. He is a member of the National System of Researchers (SNI-Mexico). He regularly serves as a reviewer of scientific articles in RSC, Elsevier, Wiley, Hindawi, MDPI, Taylor and Francis as well as Standards in Semiconductor Equipment and Materials International (SEMI). His research interests include two-dimensional materials, nanomaterials, space materials, composite materials, gas sensors, and biosensors.

Bart Vorselaars is a senior lecturer at the School of Mathematics and Physics, University of Lincoln, UK. His research is in the field of computational physics, soft matter and nano-materials. He employs computational, simulational, mathematical and theoretical techniques and has an interest in self-assembly, mechanical and dynamical properties, as well as algorithmic developments, machine learning, societal, health and environmental issues.

Didar Yeskermessov is a PhD in the specialty "6D072300-Technical Physics", Acting Associate Professor of the Department of Physics, Faculty of Basic Engineering Training, D. Serikbayev East Kazakhstan Technical University Qualifications: - 6B072300-Technical Physics PhD – D. Serikbayev East Kazakhstan Technical University 2018; - 6M072300-Technical Physics MSc - D. Serikbayev East Kazakhstan Technical University 2014; - 5B072300-Technical Physics BSc - D. Serikbayev East Kazakhstan Technical University 2012. Internships: - MSc - Wroclaw Polytechnic University, Poland 2013; - PhD - Gdansk Polytechnic University, Poland 2016; - PhD - Wroclaw Polytechnic University, Poland 2016; - Program of Bolashak – University of Lincoln, United Kingdom 2022-23. Didar Yeskermessov is actively engaged in research work, is a highly qualified scientist in the field of materials science, condensed matter physics, analysis of patterns of formation of nanocrystalline coatings and films, studying the correlation of their structural-phase state and properties, modification of the surface of metals and alloys and technology for obtaining new materials. He is proficient in direct methods of materials and coatings research, as well as analytical research methods using TEM/SEM-EDX, AFM and XRD analyses.

Index

A

B

C

D

E

F

G

H

I

L

M

N

O

P

Q

R

S

T

V

W

X

www.ingramcontent.com/pod-product-compliance
Lightning Source LLC
LaVergne TN
LVHW062108300926
849161LV00010B/2405